THE REPTILE EAR

Iguana iguana. Drawing by Anne Cox

THE REPTILE EAR

Its Structure and Function

By Ernest Glen Wever

Princeton University Press

Published by Princeton University Press, Princeton, New Jersey
In the United Kingdom:
Princeton University Press, Guildford, Surrey

Library of Congress Cataloging in Publication Data will be
found on the last printed page of this book

Publication of this book has been aided by a grant from
the Beltone Institute for Hearing Research, Chicago, Illinois

This book has been composed in Linotype Times Roman

Clothbound editions of
Princeton University Press books are printed on
acid-free paper, and binding materials are
chosen for strength and durability

Printed in the United States of America by
Princeton University Press, Princeton, New Jersey

CONTENTS

PART IV: CONCLUDING OBSERVATIONS

PREFACE

This treatment of the ear and hearing in reptiles had its beginnings more than two decades ago in a brief series of measurements, in terms of the electrical potentials of the cochlea, of auditory sensitivity in a few species of turtles, snakes, and crocodilians, and was continued shortly thereafter in similar measurements on several species of lizards. The results of these early observations, when comparisons were made among the different animals, revealed well-defined species characteristics and evident relationships also to systematic status. It thus appeared that the variations found in the ear's actions are dependent upon features of morphology and physiology that themselves should warrant further study.

At that time the structure of the reptilian ear was known only in a general way, and its modes of operation in response to sounds were poorly understood. This situation reflected the relatively primitive condition then prevailing in histological procedures for the preparation of ear tissues for microscopic study. Accordingly, in setting up a program of research in this area a dual effort was required: a systematic determination of hearing capabilities among existing reptiles over a wide range of species and the development of suitable preparatory methods for the histological study of their ears.

Work on histological procedures for the ear had already been under way in our laboratories over a number of years, with some progress. With the demands of this new program the efforts in this direction were intensified, and in a few more years our methods began to produce fully satisfactory results. The significant improvements concern primarily the preservation of the tectorial tissues, which with proper treatment lose the indefinite, almost formless character commonly seen and present an orderly, meaningful structure, fully predictable in form and relations in all specimens of a given species. In this development the critical roles taken by tectorial mechanisms were brought to light, and also revealed were the wide variations in these structures and their modes of operation among the different reptilian groups. The determination of these tectorial relationships becomes a central theme in the following pages.

The book begins with some suggestions about the sense of hearing and its biological values in reptiles, and then considers the methods of treatment to be pursued. Included is an indication of the probable origins and relationships among the living reptiles.

Then follows a presentation of the general anatomy of the reptilian ear and a consideration of the problems of sound transmission in these ears. This treatment for the reptiles as a whole is presented at this stage to avoid repetition in the topical sections to follow.

Part II is concerned with the lizards and is by far the largest section of the book. It begins with the problem of classification within this suborder, presenting some of the difficulties encountered in the systematic handling of the group, and indicates the arrangement to be followed in the treatment of the ear. Then comes a chapter on sound conduction in which three general forms of middle ear mechanisms are recognized, along with a number of divergent types. Thereafter are 13 chapters in which are included the 16 lizard families on which observations were made.

Part III contains five chapters of varying length that deal with the remaining reptiles from snakes to crocodilians.

Part IV is a concluding chapter made up of two sections. The first of these presents a view of the evolution of the vertebrate ear that is largely derived from the observations having to do with tectorial structures and their roles in the stimulation process in reptiles, along with corresponding observations (though of more limited scope) on the other classes of vertebrates.

The last section points out significant areas left untreated, or only summarily dealt with, and that deserve further research effort.

In each chapter or section the treatment varies according to the degree of diversity of the group and the particular problems encountered. The lizards are given the most detailed attention because their ears exhibit the greatest diversity of all the reptiles, and indeed of all the vertebrates. For the most part each lizard family is accorded a chapter to itself. In general, the treatment of these families begins with a detailed description of the ear structures in one representative species, and then goes on to other species that show departures from this general type. The choice of type species is more or less arbitrary, and often is determined by the material available.

The anatomical treatment is frequently interspersed with considerations of special problems in the reception of sound and its transmission through the cochlea, and with accounts of new experiments carried out to clarify these problems.

This morphological treatment is followed by a presentation of sensitivity functions for the species examined, and an effort is made to relate the ear's performance to its peculiarities of structure. In this relation, special attention is given to the degree of differentiation of the basilar membrane and auditory papilla and the distribution of the hair cells along the cochlea.

Many of the results have appeared previously in journal articles, but more than half are presented here for the first time.

Illustrations are used freely, as the most effective way of clarifying anatomical relationships. At the beginning of many of the chapters, to convey to the reader a touch of acquaintance with the animals concerned, are brush drawings, largely based on living specimens, executed by Anne Cox.

The remaining drawings are prepared by the author and are mainly based on serial sections, with the major structures traced in outline from photomicrographs and details filled in from examination under the microscope. A number of the drawings represent reconstruction models based on the sections.

In the study of anatomical structures the matter of orientation is a constant concern. The reptilian ear presents particular difficulties in this respect because its parts vary somewhat in location in the head among the different groups, and this location is usually oblique, not conforming to conventional planes. To simplify the matter, the reader is usually taken from general orienting views to successively more detailed presentations without change in the angle of view, or if a change is necessary the new angle is made clear.

In the process of sectioning, the block of tissue was oriented so as to compensate in large degree for the obliquity of the inner ear's position, to provide sections that for the most part cut across the basilar membrane at right angles to its long axis. The orientation required for such cross-sectional views of the basilar membrane and its sensory elements is one that will be designated as the frontal plane, but which always will depart from this plane by a moderate amount.

In further simplification of the problem of orientation, all representations are of the right ear.

The references are given at the end of the text. These are cited by author and year, and further identified by a letter when more than one have appeared in the same year.

The appendices include a number of symbols and definitions, and formulas are given for some of the solutions used in the histological processing.

ACKNOWLEDGMENTS

The research here described was made possible by the cooperative efforts of many friends and colleagues, to whom I am much indebted. I first acknowledge the contributions of the entire staff of the Auditory Research Laboratories of Princeton University. Particular mention is due to Jerry Palin and Joseph M. Pylka, who designed, constructed, and maintained the electronic equipment, to Wilmer C. Ames, who provided many forms of laboratory assistance, to Anne Cox, Jeanne Shelton, Rochelle Margolis, and Ewald Pauming, who carried out the exacting work of histological preparation, to Uta Runyan, who provided extraordinary secretarial assistance, and to a number of animal caretakers.

Several present and former colleagues have taken part in the experiments themselves, and their names appear in the list of references as joint authors. Especially to be mentioned in this relation are the contributions of Carl Gans, Yehudah L. Werner, Jack A. Vernon, and William F. Strother.

Most of the specimens used in the study were obtained from commercial sources, but for a number urgently needed to fill gaps in the series I have had the assistance of specialists and collectors in the field. I express my thanks to the following: to Robert L. Bezy for specimens of *Xantusia riversiana*, to H. Robert Bustard for several species of Australian geckos, to Max K. Hecht for specimens of the sea snake *Pelamis platurus*, to Donna J. Howell for a number of species collected in the West Indies, to Brian Johnstone for specimens of the giant Australian skink *Trachydosaurus*, to Arnold G. Kluge for a specimen of the pygopodid *Lialis burtonis*, to Iseli Kraus for specimens of *Anolis*, to Clive D. Jorgensen for specimens of *Crotaphytus wislizenii* and tape recordings of their vocalizations, to Joseph M. Pylka for several reptiles collected from the New Jersey area, to James A. Simmons for lizards collected in Jamaica, to A. H. Whitaker (Department of Scientific and Industrial Research, New Zealand) for specimens of *Hoplodactylus pacificus*, and to John Visser for specimens of *Typhlops* and *Acontias* from South Africa. An important acquisition was a specimen of the large iguanid *Cyclura stejnegeri*, authorized by Hon. Pedro Negron Ramos, Departamento de Recursos Naturales, Puerto Rico, and captured by the naturalist for the area, Thomas Wiewandt. I am especially indebted to R. G. Northcutt for generously making available a specimen of *Sphenodon punctatus* supplied to him by the New Zealand Government, and to Carl Gans for additional preserved material of this species. Further major contributions to the series of specimens were made by Yehudah L.

Werner, who sent a number of geckos from Israel, and by Carl Gans, who supplied nearly all the amphisbaenians, three species of primitive snakes, and the *Crotalus viridis* specimens.

The identification of species presents a difficult problem, and in this regard I am pleased to acknowledge the continued assistance over several years of C. J. McCoy, Curator of Amphibians and Reptiles of the Carnegie Museum of Natural History at Pittsburgh. For most of the specimens studied, the bodies, after fixation and removal of the head for histological processing, were sent to Dr. McCoy for species confirmation, usually along with one or more complete specimens from the same batch of animals. These specimens remain in the Carnegie Museum for further reference.

The identification of a few specimens have required the assistance of specialists; thus A. E. Greer has named the *Leiolopisma virens* skinks, A. Schwartz the *Leiocephalus barahonensis* hybrid, and Robert L. Bezy the *Lepidophyma* specimens. Carl Gans has identified the amphisbaenians and several snake specimens, and Yehudah L. Werner a number of geckos. Joseph M. Pylka has been of particular assistance in making many preliminary identifications and several final ones for species from the Atlantic seaboard area.

Finally I express my special gratitude to Carl Gans and to Yehudah L. Werner, who have critically examined several chapters of the manuscript. Also from these two I have had the pleasure and benefit of many days of cooperative work in the laboratory, and of much good advice generously extended.

The research has been made possible by grants, extending over many years, from the National Institutes of Health (NINCDS), Public Health Service.

The publication of this book has been assisted by a grant from the Beltone Institute for Hearing Research.

PART I. INTRODUCTION

1. THE SENSE OF HEARING
IN REPTILES

The sense of hearing in its advanced forms, as it occurs in ourselves and our near relatives among the mammals, is clearly one of the major instruments through which an animal is informed about the outside world. The ear is a distance receptor of wide latitude, capable of conveying a wealth of detail quickly and with precision. Our own use of sounds in speech and their elaboration in the art of music testify to the wide scope of the auditory sense. Many other animals, especially those with highly developed group relations such as baboons and wolves, have a great variety of vocalizations that provide for an interchange of information and serve to maintain the social order.

One of the most sophisticated uses of hearing is found in the bats and porpoises, which produce sounds and utilize their echoes to orient themselves in the environment and to locate, identify, and pursue their prey in a manner corresponding in effectiveness with other animals' use of vision. Another remarkable performance is that of the owl, which from a perch in a tree can locate a mouse scurrying among leaves and grass on the ground below, and can swoop down and seize it, all in total darkness, with the aid of hearing alone.

In its more primitive forms this sense has more limited usefulness, yet its biological significance is profound. Its primary value in animals with the simpler ears is as an alerting sense, to warn of new events and impending danger. A second application is as a means of finding and recognizing mates, and in carrying out other activities in the breeding process and in the care of the young.

The alerting function alone might have been sufficient to bring the auditory sense into being and to give it its differentiation and development among the early vertebrates.

A careful study of the ear and its characteristics throughout the animal series indicates that this sense had a number of distinct origins— about ten among the invertebrates and five or six among the vertebrates (Wever, 1974d). Though these appearances of the auditory sense can be considered distinct they were not altogether independent, inasmuch as in every instance the ear was derived from another mechanoreceptor of simpler form.

If we could reveal all stages of the development of ears in early animals, their several origins might ultimately be traced to a sense of touch: to a simple organ at the skin surface reporting a local deformation; but more immediately they seem to have been derived in most of the arthropods from a kinesthetic organ—a detector of relative motion between parts of the body—and in the vertebrates from the labyrinthine system —a set of detectors of head orientation and motion.

As will be discussed in detail in Chapter 25, the ear of vertebrates appeared in primitive forms in the fishes and amphibians, and then in the reptiles took on a more advanced configuration from which it was further elaborated along separate lineages in the birds and mammals. This book is concerned with the various developments of this receptor in the living reptiles and its utilization, especially among the lizards, of a variety of mechanisms and principles of action.

This study of the reptilian ear requires first a general understanding of the reptiles themselves and their position in the vertebrate series. Here we come to the problem of animal classification.

CLASSIFICATION OF ANIMALS

The first attempts at a classification of animals were made when the species were regarded as fixed and independent, and the purpose was simply to group together those creatures with obvious similarities and life habits. Later, after the principles of evolutionary development came to be understood and accepted, this grouping acquired a deeper significance, one of representing actual genetic relationships and of indicating, in a general way at least, the historical sequences of the changes passed through in the evolutionary process.

In theory, a system of classification ought to take into account all the characteristics, both structural and functional, of the animals concerned. In practice, however, most of the systems that have been proposed have been based upon a very few features, and these have almost exclusively been structural, and more particularly have included the parts of the skeleton that are best preserved as fossils.

Because the different classifications of a particular group are often based on different features, there are many variations among them as to what animals belong together and what are their true historical relationships. Indeed, the evidence often seems contradictory, and various indications have to be weighed one against another. Any system, therefore, is to be regarded as a selection among numerous possibilities made in the light of the evidence available at the time. These systems are never final, but are subject to almost continuous revision as new evidence comes to light and old evidence is reevaluated.

At the present time this field of systematics, though still largely based on structural relations, is being extended by the inclusion of functional relations of various kinds. Comparative studies of physiological processes and patterns of behavior have added a significant amount of new information.

It is clear that sensory function is an important aspect of adaptation and survival, embracing as it does the avenues by which organisms are informed about conditions and events in the outside world. A notable use of sensory information in animal phylogeny was made by Walls (1942a, b) in his treatment of the eye.

An early suggestion for the use of the ear and auditory function in the systematics of reptiles was offered by Shute and Bellairs (1953), and a further application was made by Miller (1966a, 1968), who made extensive measurements of the form and size of the cochlear duct in lizards and snakes, and discussed these features in relation to what was then known about auditory sensitivity in these animals.

The systematic arrangement of the major groups or classes of vertebrates is firmly established and is not expected to be changed, though the relationships among the protochordates, the earlier members of the chordate phylum, are uncertain and require further study. We feel sure about the derivation of the reptiles from early amphibians, and of these from certain of the fishes, and also of the development from the reptiles of the two highest branches of the vertebrates, the birds and mammals.

CLASSIFICATION OF THE REPTILES

The classification of the reptiles has passed through many stages. The systems that have been developed for them, like the ones derived for other orders of vertebrates, have been based largely upon skeletal features and especially the skull. Indeed, for the reptiles the use of these features is practically dictated by the consideration that the species now living are only a small remnant of the diverse and populous forms that once lived and in a true sense ruled the earth. We must look to the fossil record for evidence about them and their relationships, and the existing reptiles can best be understood by fitting them into the total picture. It is a great achievement of paleontology to have developed classification systems that allow the whole reptilian array to be viewed in a meaningful way. Among current accounts there is good agreement on the general plan, though there are many differences of detail.

In Romer's system (1933, 1956, 1959), which is one of the most widely accepted, the class Reptilia is divided into six subclasses, and within these are 15 or 16 orders, containing some of the most remarkable animals ever known, from the flying reptiles to the gigantic herbivorous dinosaurs.

Only four of these orders of reptiles have living representatives; all the others, after having reached the peak of their ascendency about 200 million years ago, declined and finally disappeared. Remaining are the Testudines (turtles), Crocodilia (crocodiles, alligators, and gavials), Squamata (lizards, snakes, and amphisbaenians), and Rhynchocephalia (now represented by a single species, *Sphenodon punctatus*). Of these, the turtles and squamates have been highly successful and have continued up to modern times to maintain themselves and to adapt to the many changes of climate and habitat. The crocodilians have persisted also, but with much less diversification, and because of pressures by man in recent years are seriously endangered. The rhynchocephalians, reduced to a single species living on a few islands off the coast of New Zealand, are maintaining only a tenuous existence. The development of these existing reptiles and their relations to one another are indicated in Fig. 1-1.

As the figure shows, the earliest reptiles were the cotylosaurs. These arose by gradual and largely unknown stages from advanced amphibians during the late Carboniferous period, reached their zenith in the Permian, then disappeared toward the end of the Triassic. Out of these stem reptiles came all the other groups, including the few represented here. The figure depicts the independent lines of development of the turtles directly from the cotylosaurs, of the rhynchocephalians and squamates by different paths out of the eosuchians, and the crocodilians from the

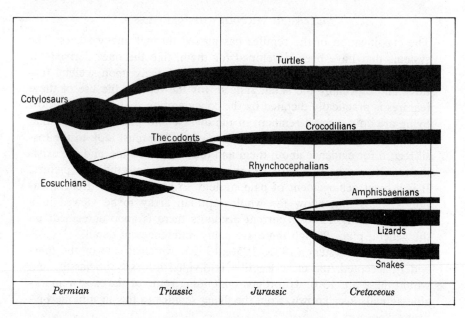

FIG. 1-1. Developmental history of the living reptiles. The width of the ribbons represents roughly the varying numbers of known species. Based largely on Broom, 1924, with extinct lines omitted.

eosuchians also by way of (or parallel with) the thecodonts. The latest of these general groups to appear were the squamates, with the lizards (or their immediate forerunners) constituting the parent line, and the snakes and amphisbaenians differentiating from these sometime late in the Jurassic. The similarities of ear structure and function among these forms that will be revealed in the subsequent treatment reflect their common origins, and the differences have arisen during the many millions of years during which they have followed separate lines of descent.

THE PRESENT STUDY

Ideally, an examination of the reptilian ear would include three approaches—morphological, physiological, and behavioral. The morphological method is clearly the basic one, revealing the form of the auditory mechanism and the interrelations of its parts. Physiological methods then make it possible to study the working of the mechanism both as a whole and as a series of interrelated processes. Finally, behavioral methods show the range of operation of the system, its sensitivities, and the discriminations that it makes possible. This last method also reveals the manner in which the sense is utilized in the animal's adjustments to the outside world.

In the study of hearing in higher animals, the birds and mammals, and a few of the lower ones such as the fishes, all three types of approaches have been used to advantage. For the reptiles, however, the behavioral methods have had only limited application. General observations of behavior under natural conditions often give indications of hearing. Some animals, such as the geckos and crocodiles, are active in producing sounds, and the males appear to use calls to proclaim their territorial rights and to repel other males. The presence of such behavior is a good indication of hearing ability, but the quality of this ability is difficult to determine by field observations alone.

A few attempts have been made to observe startle movements in reptiles such as lizards and crocodiles in response to sudden sounds, but these have met with only slight success. For the most part these animals fail to exhibit spontaneous responses to sound stimuli, even very intense ones.

Other efforts have been made to employ conditioned response methods to establish some type of positive reaction to an acoustic signal, but most of these efforts have been unavailing. A few experiments with turtles have been successful, but this technique applied to other reptiles has thus far yielded doubtful or negative results.

Because of the paucity of behavioral information, the present treatment depends almost entirely on results obtained by anatomical and electrophysiological methods. Nearly all these results have been obtained by the writer and his associates by the use of special methods developed for application to the reptilian group.

The anatomy has been studied in considerable detail by coarse and fine dissection and by microscopic examination of serial sections prepared by new and improved procedures.

The performances of these ears have been observed chiefly in terms of the electrical potentials of the cochlea. These potentials reflect the activities of the sensory cells in response to sounds, and provide information on the operation of the peripheral portion of the auditory system as far as the cochlea.

Auditory nerve responses, which represent the activities of the auditory system one or more steps farther along the sensory stream, have so far been studied only to a limited extent, and in a very few species.

The electrophysiological observations provide a basis for comparisons of the ear's performances in the various reptilian species. The differences revealed at this stage can be expected to remain, at least in a general way, as the sensory information is relayed to the higher centers of the nervous system and becomes effective in the determination of behavior.

In the treatment to follow, the lizards are given the most extensive consideration both because they exhibit many features that may be regarded as primitive or basic in nature, and because they present wide variations of ear structure and function. The significant variations are in the cochlea, and concern the mechanisms by which vibrations set up in the cochlear fluids by sounds are able to stimulate the sensory cells and thereby excite the fibers of the auditory nerve. These cochleas display two general types of processes by which stimulation is achieved, and these two types occur in different forms and combinations, often varying in different regions of a single cochlea.

Such diversity is not found elsewhere among the reptiles, or indeed anywhere else along the vertebrate line. In contrast, the inner ear structures and processes of other reptiles, and also of the birds and mammals, are relatively uniform, and represent only minor variations on a single stimulation system. These relatively stereotyped forms of ears can be reasonably well represented by a limited sample, whereas the lizard group requires wider treatment. Nearly all the existing lizard families are covered here, though sometimes less extensively than might be wished because of difficulties in obtaining specimens.

Thus far, in the present study, 247 species of reptiles have been investigated, including 186 species of lizards belonging to 16 families, 19 species of snakes, 14 species of amphisbaenians, 24 species of turtles, 3 species of crocodilians, and the single existing species of rhynchocephalians.

2. METHODOLOGICAL APPROACHES

As mentioned in the introduction, the ear of reptiles has been investigated chiefly by two methods, anatomical and electrophysiological. These methods will now be described in detail.

ANATOMICAL PROCEDURES: PREPARATORY

The first method ordinarily used in the study of the ear in an unfamiliar species is a general examination of superficial features, usually with the aid of a medium power dissection microscope. This examination is followed by a careful dissection to reveal the deep structures. The animal may be under anesthesia or freshly killed (preferably by an overdose of anesthetic), so that the blood vessels remained filled and are easily recognizable. Animals that have died for some reason may of course be utilized, but the postmortem deterioration of tissues often gives difficulty. Specimens killed by the perfusion of fixative are especially useful, because the tissues are hardened somewhat and better withstand the manipulations. At times it is desirable to employ latex injection of blood vessels, preferably with different colors for arteries and veins.

Some investigators have developed great skill in certain types of dissection techniques. The early anatomists, in times before the development of serial sectioning, relied altogether on such methods. Retzius is renowned for the unusual delicacy and precision of his dissections of the labyrinth, and for the exquisite illustrations that he and his lithographers achieved. These methods as applied to the mammalian ear have recently been "rediscovered," and under the new name of "surface preparation" are employed extensively for the demonstration of such anatomical features as the arrangements of the hair cells and their relations to supporting structures. This technique has proved especially useful in determinations of the degree of damage produced by overstimulation, ototoxic drugs, and other agencies.

Miller (1966, 1968) developed a technique for exposing the cochlear duct of reptiles and removing it with only limited defacing of the specimen, which made it possible to study the inner ears of preserved specimens. With the generous cooperation of a number of museums and their directors he was able to study a great variety of preserved reptilian species, including many seldom encountered in the field and difficult to collect.

Some of the specimens used in the present experiments were so rare and difficult to obtain that they could not be spared for a dissection type of study. These specimens were examined only superficially, to the extent possible without damage to essential structures, then were used for cochlear potential measurements, and finally were prepared for microscopic examination. For such specimens a thorough understanding of the structural relations of many parts of the ear is difficult to obtain in a simple examination of the serial sections. In a number of instances reconstructions based on the sections were employed to provide a clearer picture.

Histological Techniques

The microscopic preparation of the ear is a distinct and special branch of histology. The conventional procedures developed for other tissues fail to give satisfactory results, and procedures have been developed particularly for this purpose. The ear is an unusually difficult object with a number of physical properties that present special problems. It combines some of the most fragile materials with others that are dense and rigid. Thus Reissner's membrane in most animals is only two cell layers thick, and one of these layers is so attenuated that it is often difficult to identify with ordinary methods of examination. The supporting framework for the hair cells along the basilar membrane is made up of long, thin elements with delicate connections, and the form is easily disarranged or damaged. The tectorial tissues of the ear, which play a fundamental role in the process of hair-cell stimulation, are particularly tenuous and by ordinary treatments are so shrunken and distorted as to be hardly recognizable. Further, in many species the inner ear structures are enclosed in a capsule of bone that is the densest and most brittle of the entire skeleton.

Because of the great variation in physical properties, the ear specimen must be carefully protected from mechanical stresses of all kinds.

Because the ear contains many widely different tissues, there is no single histological treatment or staining procedure that is fully satisfactory for all parts. A choice must be made according to the dominant research interests. Usually this choice is a compromise that will provide a reasonably good representation of the majority of elements.

In the present study the focus has been the hair cell and the means by which acoustic vibrations bring about its stimulation. Attention is directed to the manner of suspension of the hair cells in the auditory papilla and the relations of these cells to the tectorial membrane and other structures involved in the stimulation process. As it turns out, these inner ear structures are among the most delicate parts of the ear, and when conditions are such as to preserve them well and exhibit their relations in a reliable manner the other tissues are presented in an acceptable form also.

The principal steps in the histological procedure followed in this study will be discussed in sequence.

FIXATION

The most critical stage of histological preparation is the initial one of fixation. After proper fixation the tissues in subsequent handling can withstand a number of minor departures from best practice—though of course such departures (usually accidental) are to be avoided as far as possible. If the fixation is inadequate, there is nothing that can be done later to bring the material to an acceptable level.

If fixation fails it is best to discard the material at once without further waste of time, unless the specimen is irreplaceable and there is at least a faint hope that something can be learned from it.

In the fixation process, the living tissues—the cells with their enclosing membranes and contents—are quickly killed, changed to a relatively stable and more solid state, and their contained enzymes are rendered inactive. At the same time the abundant non-living substances, which are cell products of a more or less permanent character, are rendered stable also. This set of alterations in cells and cellular materials must be made rapidly, within a very few minutes of the withdrawal of oxygen, for in oxygen lack the delicate balance within the cell is upset and some of the enzymes begin to digest the cellular substance and eventually the extracellular material as well.

The maximum permissible delay from the moment that respiration ceases until fixation is complete is not accurately known, and it probably varies in different animals and with circumstances such as the state of health, recent activities affecting physiological conditions, and temperature.

A few experiments have been carried out to study the time course of this deterioration process. Wersäll, Kimura, and Lindquist (1965), working on guinea pigs, found no differences between controls in which there was no delay and experimental animals in which the fixation was delayed five minutes after death, but observed swelling and other tissue changes when this delay was 15 minutes.

These observations set an outside limit on the period after death during which tissues are not seriously altered. The true limit may well be less, because the procedure in the above experiment was somewhat crude, in that the animals were killed by a violent method (removal of the heart) and the control specimens may have suffered deterioration from this cause. It has been shown that the form of death has a considerable effect on the rate of decline of the cochlear potentials. In experiments on guinea pigs and cats, Wever, Bray, and Lawrence (1941) found that violent forms of death, caused by such procedures as pithing the medulla or clamping the heart, caused initial declines of cochlear potentials that were two or three times as rapid as those caused by death resulting from an overdose of curare. Thus the controls in the experiment of Wersäll *et al.* may have suffered some degree of deterioration from this cause, and such deterioration could have obscured minor changes in the tissues occurring in times less than 15 minutes.

From my own experience I have the impression that any delay of fixation over four or five minutes is deleterious, and it is most desirable to get the perfusion under way within three minutes or less from the time of opening the chest.

The early signs of postmortem changes in the ear tissues are usually seen in the nuclei of the hair cells, first as changes in staining properties and then as changes of form and size. More serious effects are the appearance of vacuoles within the hair cells, the swelling of these cells, and the extrusion of globules from their surfaces.

At more advanced stages of deterioration, the effects extend to the supporting cells and include gross alterations in the form and position of membranes and component structures. Finally there is a complete disintegration of cochlear structure. These advanced stages are not of particular concern here, because they represent a condition only encountered when perfusion fails.

Complete perfusion failure has occurred on several occasions, in a few instances as a result of equipment breakdown, but more often from a blocking of the cannula or a rupture of the aorta. Such failures are obvious, and the specimen is discarded or used for other purposes. The cannula blockings and aorta ruptures have occurred most often in the very small animals in which the vessels are of fine caliber and correspondingly small cannulas must be used.

Choice of Fixative.   There is little in the way of established theory to guide the choice of a fixative for a particular purpose. Progress in this area has been almost entirely empirical. Hundreds of fixative formulas have been designed (Gray, 1954). Some of these work well for particular purposes, but few are even moderately satisfactory for the ear.

The fixative used in the present study is a variation on one employed by Maximow (1909), containing mercuric chloride, potassium dichromate, sodium sulfate, and formaldehyde, and made up as shown in Appendix B.

The Perfusion Procedure.   With the animal under anesthesia, the fixative is applied through the circulatory system so as to reach the tissues in the region of the ear as promptly as possible. The common practice of killing the animal, dissecting out the ear tissues, and immersing them in fixative gives relatively crude results, even when this procedure is carried out rapidly. After immersion the fixative can reach the inner ear only by diffusion through surrounding tissues. This is a slow process, proceeding at a rate of about a millimeter per hour in soft tissues, and even more slowly in the presence of bony and cartilaginous materials as present about the inner ear. A large block of tissue might well require several hours for the fluids to penetrate to the deeper layers in suitable concentration.

An intermediate method is to expose the inner ear in the anesthetized animal, puncture the round window, extract the stapes from the oval window, and instill a fixative through the cochlea under pressure. This method is superior to simple immersion but inferior to a proper procedure of perfusion. It is better, of course, than a poor perfusion in which the fixative is not brought to the ear tissues promptly, in suitable concentration, and for a sufficient period of time.

For special purposes the regular (and successful) perfusion has sometimes been followed by this instillation procedure. This has usually been done to study the effects of a combination of fixatives, such as Maximow's solution and an osmic tetroxide mixture. Most cells and tissues are well preserved by this double fixation, but the more delicate elements and especially the tectorial networks are not improved over primary fixation, and often are damaged.

It should be pointed out that all inner ear fixation involves diffusion at the final stage, because of the absence of blood vessels in the organ of Corti itself. In perfusion through the circulatory system the fixative is brought to the vessels of the otic capsule and to the capillaries adjacent to the cochlear duct, but the final passage through the endolymph to the hair cells is by diffusion. These diffusion paths are short, however, and with sufficient perfusion time a suitable concentration of fixative at the hair cells can be established.

The procedure begins with a flushing of the blood vessels by injection of physiological saline solution, of a composition suitable for the particular animals. The formula for reptiles is given in Appendix B.

The saline solution is made up fresh, or if made up in quantity it should be sterilized with heat or stored in a refrigerator to prevent the growth of molds. Just before use a small amount of amyl nitrite is added to the saline as a vasodilator. It is convenient to use the 0.3 ml ampoules of this material prepared for the use of cardiac patients; the ampoule is broken beneath the surface of the saline solution to avoid escape of the volatile material into the air; these fumes can have highly disagreeable effects on anyone breathing them.

The flushing of the blood vessels is continued for only a short time, until the larger part of the blood is removed. The object is to remove enough of the blood to prevent coagulation in the vessels, but not to take so long as to permit changes in the ear tissues before the fixative reaches them.

The perfusion apparatus, shown in Fig. 2-1, consists of two flasks, one for the saline solution and the other for the fixative, connected through a 3-way stopcock to a pump, and connected through another 3-way stopcock to the cannula that is tied into the blood vessel. This first 3-way stopcock permits a rapid changeover from flushing solution to fixative. The second stopcock connects either to the main line or to a waste tube. The main line contains first a bubble catcher and then a tube with a side branch. The side branch leads to a second waste tube, and the main line goes to the cannula.

The figure shows a heating jacket around the saline flask; this jacket is used for endothermic animals to keep the saline solution at body temperature. For reptiles this heating arrangement is not used, and the solutions are at room temperature.

A flowmeter, not indicated in the figure, could be inserted into the line when desired. It was not used routinely, but only for test purposes.

The pump is of the tubing type with pulsating flow. Adjustments can be made of pulse rate and pressure, which with the flow resistance in the system and especially in the animal determine the flow rate. This flow rate varied somewhat, being greater in the larger animals. In a medium-sized lizard (18 grams body weight) it was about 12 ml per minute.

The cannula was a hypodermic needle with the tip cut off straight and the edges carefully rounded. A set of these cannulas contained all the numbered sizes from 20 to 32 gauge. For a given animal a size is chosen according to the caliber of the blood vessel to be entered.

The detailed procedure is as follows. After checking that the entire apparatus is thoroughly clean, the upper stopcock is turned so that the saline flask is open to the central tube, and the two flasks are filled with an appropriate quantity of their solutions, which for a medium-sized lizard is about 150 ml of saline solution and 300 ml of fixative. The lower stopcock is set for connection between the upper one and the main line, and the second side branch is clamped closed.

The saline solution is run through the main line until the bubble catcher is completely filled (it must be inverted to achieve this), and all other tubes, including the side branch, are filled. The clamp at the end of the side branch is opened briefly until fluid escapes. Special care is necessary to exclude all air bubbles, and the use of transparent tubing aids in checking this condition. The tube between the two stop-

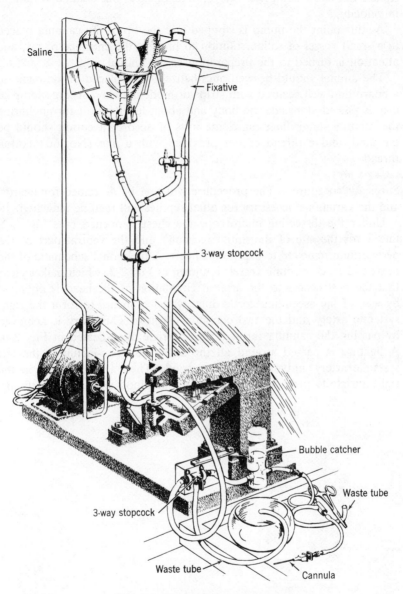

Fig. 2-1. The perfusion apparatus.

cocks is inserted into the jaws of the pump mechanism, and adjustments are made to provide a pulsating stream out of the cannula. This stream should not show complete interruptions of flow, but definite variations in velocity.

At this point the pump is stopped and the tip of the cannula placed in a small vessel of saline solution to prevent the entrance of air, and attention is turned to the preparation of the animal.

The animal should be well anesthetized. It is placed on its back on a board and well secured with strips of modeling clay. A large lump of clay is placed alongside the body and about level with it for anchoring the cannula shank later on. Some tufts of absorbent cotton should be on hand, and a thread carrier provided with a fine (No. 60) cotton thread.

Surgical Procedure. The procedure will first be described for lizards, and the variations necessary for other reptiles will then be indicated.

Under the dissecting microscope the chest is opened (see Fig. 2-2, line 2 for the site of the initial incision) and the ventral part of the pericardium removed to expose the heart. A functional schematic of the heart and its connecting vessels is shown in Fig. 2-3, which makes clear that the best access to the arterial circulation of the head is obtained by way of the ascending aortic duct, which branches to form the right systemic artery and the two common carotids. This entry is achieved by passing the cannula through the ventricle as indicated in Fig. 2-4. A ligature is passed around all the aortic vessels—including the left systemic artery and the pulmonary trunk. Then a slit is made in the right auricle to provide an escape path for blood and perfusion fluids.

FIG. 2-2. Surgical approaches in lizards. At 1 is the incision for exposure of the round window, and at 2 is the incision for exposure of the heart.

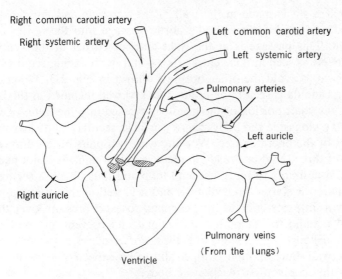

FIG. 2-3. The heart and its vessels;
shown for *Sceloporus magister*.

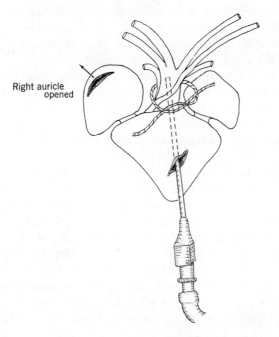

FIG. 2-4. The perfusion cannula in the heart.

Next a slit is made in the wall of the ventricle and the cannula inserted and guided through the aortic passage into the aortic duct as shown. This passage is at the right anterior corner of the ventricle— to the left as seen in this ventral view—and its threading requires a certain care to avoid the other openings shown in Fig. 2-3. When the tip of the cannula is seen to be in the duct and well included in the ligature loop, the shaft portion of the needle is pressed into the lump of modeling clay alongside the body where it is held securely, thus leaving both hands of the operator free. Working with two pairs of fine forceps, the loop of thread that is already in place is pulled tight. It is not necessary or even desirable to form a second loop and thus obtain a secure knot. The pump is started immediately, and it is well at this point to place a finger a little posterior to the heart and to exert pressure over the vertebral column to immobilize the animal and prevent any motion that might dislodge the cannula or, even worse, cause it to penetrate the aortic wall. Such motion is especially to be expected at the next stage when the fixative first reaches the body.

The pumping of saline is continued until the fluid escaping from the right auricle just begins to become pale; then the upper stopcock is turned to start the flow of fixative. The perfusion of fixative is usually continued for about one hour, with replenishment of the supply in its flask as needed. For the smaller animals, in which cannulas of small size must be used, with correspondingly slow rates of flow, it is often desirable to use even longer times, up to two hours.

The perfusion times mentioned are much longer than commonly employed, but the necessity of such extended perfusion has been proved by the experience of many years. The procedure results in the building up of a high concentration of fixative in the tissues peripheral to the ear, and thereby leads to an effective diffusion into the ear tissues themselves.

The second waste tube is not used for medium and large animals, but has an important application for small ones. For the small animals the quantity of solution in the connecting tubes between the lower stopcock and the cannula is more than sufficient for flushing the circulatory system, and even if switching is done almost at the beginning of the procedure the time taken for this fluid to pass through may be excessive. After the flow out of the right auricle shows that the flushing is sufficient, the clamp is opened on this waste tube and fluid allowed to escape until the fixative reaches the junction with the cannula branch. Sometimes the pump is speeded up momentarily to reduce the time for this outflow.

As already indicated, the flow rate varies with pump adjustments and with the size of the animal. The proper regulation is determined by experience. A useful guide is the pulsation of the aortic vessels, which should be watched carefully at the beginning and checked again after fixation of the blood vessel walls has increased the flow resistance of the circulatory system. The aortic vessels should show clear pulsations in synchrony with the pump, but should not swell very noticeably beyond their normal size. If they do so the pressure is excessive and they are in danger of rupturing.

A further guide to proper pressure regulation is observation of the general condition of the animal as the perfusion proceeds. The tissues should color promptly, taking a yellow hue from the potassium dichromate component, but they ought not to show marked edema. A slight expansion of legs and body is expected, especially in a lean and somewhat dehydrated animal; but a very noticeable bloating signifies that the pressure has been excessive.

When the perfusion is complete the cannula is removed and the head is severed well back of the ear openings. Except in very large specimens, those over 5 cm in head width, there is only moderate trimming and both ears are kept in a single block. The skin is removed except that surrounding the ear openings. The tissues of the floor of the mouth, including the tongue, are removed, but usually the mandible is retained. Except in very small specimens, the nose is cut off. The right half of the remaining forepart of the skull, including the eye, is then removed. This is done to make the specimen asymmetrical, and thus to assist the orientation after embedding in celloidin has obscured many surface features, and also to give a sure identification of right and left sides in the final sections.

For large animals—an extreme example is the sea turtle—for which the united ears form a block too large for convenient handling, the specimen is split along the midline and the two pieces processed separately. Then it is best to insert a large-headed stainless steel pin in some tissues away from the ear to indicate how the specimen is to be mounted and sectioned.

A thread is tied around the specimen with a tag bearing an identifying number; if the head has been split, the tag carries an indication of right or left. In our practice the body is preserved also, tagged with the same number, and later is deposited in a museum with which an arrangement has been made for confirmation of species identity. For many species in which the chief identifying characters are on the head it is desirable to have one specimen in each lot, obtained from the same source, which retains the head.

At this point the perfusion apparatus is thoroughly cleaned by passing through it about a liter of tap water, and by the separate washing of the terminal tubing with its bubble catcher and cannula. It is essential that no trace of fixing fluid be left to contaminate the preliminary flushing solution, for its entrance into the circulatory system would produce clotting.

A pulsating pressure is used in the perfusion because it is much more effective than simple direct pressure. The terminal arteries supplying the ear region are particularly small, and in such small vessels a blockage can occur through the clumping of blood corpuscles before these have been cleaned out. A pulsating pressure breaks up these clumps, whereas a steady pressure tends to consolidate them. It is no accident that the heart is a pulsating device.

Speed is a factor in the success of the procedure. It is gained by careful arrangement of the equipment and by practice.

In the procedure as described the vessels to the posterior part of the body are not tied off. In large animals it is desirable to limit the fixation to the head region because of the pressure loss into the large vessels of the body, but this is unnecessary in small reptiles. As noted above, pressure is exerted by a finger in the region posterior to the heart at the beginning of perfusion primarily to immobilize the animal, but this pressure is applied in such a way as also to compress the vessels to the posterior part of the body and is maintained long enough to favor the head region in the early stages of perfusion. Then this pressure is removed and the whole body allowed to receive the fixing fluids. This fixation of the entire animal has been done mainly as a matter of convenience, but it has been of benefit also to workers in other areas of reptile morphology by providing well fixed tissues and organs over the entire body.

Fixation in Other Reptiles.   In turtles and snakes the procedure departs from that just described in that the cannula is not introduced through the heart but into one of the arteries in the neck region. With these animals a perfusion through the heart is usually ineffective, except in some of the smaller specimens. In turtles the heart is located about the middle of the shell or a little forward, and in snakes it lies about a third of the distance from head to tail. Thus the distance from heart to head is considerable, and the arteries carrying the blood over this long course are relatively small.

For turtles, the neck region is dissected and a portion of the carotid artery close to the head selected for insertion of the injection needle. Also the jugular vein is dissected free and a ligature placed around it loosely. Two ligatures are passed around the artery, one anterior and the other posterior to the site chosen for insertion of the cannula. When these ligatures are in place, a clamp is applied to the artery ahead of the anterior ligature, and when the artery has swollen a little the posterior ligature is pulled tight, leaving the segment of artery distended. With fine scissors an angular nick is made through one wall of the artery, the cannula is inserted to a point beyond the anterior ligature, and this ligature is tightened around it. The clamp is then removed. The loop around the vein is pulled out and the vein is cut, whereupon the perfusion pump is started. From here on the procedure is as described above.

With this procedure the fluids are directed to the head, but through anastomoses they can find their way back through other vessels to the body, so that the entire animal is fixed. At times this backward flow has been restricted by passing a strong cord around the neck posterior to the region where the cannulation is made and pulling this cord as tight as possible. Good results have been obtained both with and without this neck constriction.

In snakes the procedure is similar to that in turtles except that the left carotid artery is used (many species lack the right carotid), and the site of the cannulation is located somewhat farther from the head where this artery maintains a suitable size.

Crocodilians and amphisbaenians are perfused through the heart in the same manner as lizards.

For all animals, the fixation process is continued after the perfusion by immersion of the specimen in 40-50 times its volume of Maximow's solution for 7-10 days, followed by 10 per cent formol for the same period, in each case with daily changes of solution.

Signs of Good Fixation.   There are a number of indications of the degree of success of the fixation process, and from these a decision can be made as to whether it is worth while to continue the long process of histological treatment.

Some of these signs appear early in the procedure. The flow should be free, with a noticeable amount of blood escaping from the opening in the right auricle during the first few seconds. The fluid should begin to clear promptly, within half a minute or so, though in the smallest animals, in which a cannula of very small bore (No. 28 and smaller) must be used, this time may be doubled. Within seconds after the fixative begins to flow the animal commonly shows a sharp muscular spasm. Later, after pressure on the abdominal vessels is removed, there are tremors of hind legs and tail. Still later in the perfusion the skin and other exposed tissues turn yellow, there is exudation through the incisions, and usually a drop or two of fluid escapes from the nostrils. The body finally becomes strongly rigid.

When the head is removed, the cut surfaces should be strongly colored and should exude the fixing fluid. All muscles should be well colored and rigid. The pterygomandibular muscle is a particularly good indicator; it is usually the last tissue to be thoroughly fixed. Finally, the cut end of the spinal cord should be yellow and its substance firm; if it remains white and soft it is likely that there has been poor penetration to the ear also.

If any of these signs of good fixation are negative, a decision must be made regarding further action. If the specimen is particularly needed it may be carried forward, but with faint expectations of fully satisfactory results. In our laboratory all perfusions are given a rating as soon as completed, and almost never has a procedure rated anything below "excellent" yielded a series of sections of first quality. Fortunately the failures have been few. They have mostly occurred in lizards of 1 gram weight and below and in slender snakes, in which the blood vessels are so small that they have not always been successfully cannulated. In these animals the success has been around 50 per cent, rising to 70 per cent when several animals of the same species have been tried in succession and cumulative practice has exerted its beneficent effects.

The further preparation, from decalcification to final mounting of the sections, will be described somewhat briefly. Some of the details of staining are given in Appendix B, but mainly these must be learned by experience.

DECALCIFICATION

The specimen is decalcified in a 0.5% solution of nitric acid in 10% formol. Nitric acid is used in preference to other acids because all its salts are soluble, and it is effective at a low concentration with minimal effects on cell structure. The action is facilitated by continuous agitation provided by a shaking machine set at a low speed. The time required is 10 days for an average-sized lizard, up to 40 days for a large turtle, with daily changes of solution.

At the end of the decalcification the specimen is exposed to a 5% solution of sodium sulfate for 2 days, with one change of solution.

DEHYDRATION

The specimen is carried through 13 steps of increasing concentration of ethyl alcohol, beginning with 10% and increasing by 10% daily until 90% is reached, then through 95% (2 days) to 100% (2 days). It is essential that the alcohol used for the final two days be actually anhydrous, as traces of water will retard the infiltration of celloidin in the next stage. Contamination of the absolute alcohol by water can be determined by testing a sample with a lump of anhydrous copper sulfate, which turns from white to blue if water is present.

Celloidin Embedding

The specimen is transferred for 1 day to a 25:75% mixture of anhydrous ether and anhydrous alcohol, then for 1 day to a 50:50 mixture, which is the celloidin solvent. Infiltration then proceeds through four steps of increasing concentration of celloidin, beginning with 4% and going through 8 and 12 to 16% over a period of 12 weeks, all with the container sealed. At this point the specimen is oriented in the container according to the plane of sectioning to be used. Slow evaporation of the solvent is permitted and controlled so that at the end of about 4 weeks a firm consistency is obtained.

The specimen is carefully separated from the container and removed with a large block of celloidin around it, and exposed to chloroform vapor for four days or longer until well hardened. The block is then trimmed to a convenient size and placed in 80% alcohol for 2 weeks or more for final hardening. The 80% alcohol should be changed daily for the first three days so as to remove the last traces of chloroform. The block may remain in the alcohol for months.

Sectioning

The hardened block of tissue is finally trimmed with view to its orientation for sectioning, and mounted on a fiber block. It is best to harden the tissue block further by keeping it in 80% alcohol for a week or two before sectioning.

For sectioning, the block is secured in the holder of a sliding microtome, and sections cut serially, usually at 20 μ. Every section in the region of the ear is saved.

Planes of Sectioning. The usual plane of sectioning was transverse to the long axis of the basilar membrane. This is not one of the normal planes of the head, because in reptiles this membrane is positioned obliquely. In lizards, as shown in Fig. 2-5, the longitudinal axis of this membrane is rotated counterclockwise from the vertical by 24°. Accordingly, the nose was tipped downward 24° as shown in Fig. 2-5c, and sections cut from above downward.

In amphisbaenians the nose was tipped downward 20°, and in snakes 24°. In turtles it was tipped upward 15°. In crocodiles it was tipped downward 25°; this gave a nearly transverse section of the basilar membrane near the posterodorsal end, but because this basilar membrane makes a sharp bend near its midregion the section becomes oblique for its distal portion. The effects of this obliquity are discussed in Chapter 24.

The plane of sectioning just described corresponds approximately to the horizontal plane commonly used: the plane parallel to the ground

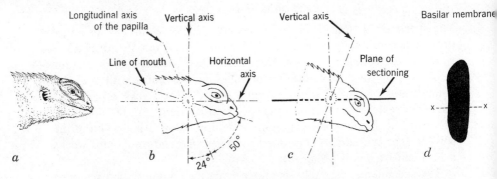

FIG. 2-5. Planes of sectioning in lizard ears. *a*, the head of *Sceloporus magister*; *b*, the principal axes; *c*, head position for sectioning transverse to the basilar membrane; *d*, the basilar membrane in an enlarged view. From Wever, 1965a.

when the reptile is in its usual resting position. This plane is often referred to as the frontal plane by transfer from human anatomy. For convenience the designations horizontal or frontal will be used hereafter for this plane despite the fact that the orientation is modified for the different reptiles as just indicated.

A second plane often used is the transverse plane, which cuts through the vertical axis perpendicular to the plane of Fig. 2-5. If unmodified this plane passes obliquely through the long axis of the basilar membrane. Usually the orientation was adjusted so that the section passed longitudinally through this membrane; in lizards this was done by rotating the nose upward 66°. Sometimes it was possible to obtain sections in the plane of the basilar membrane surface, across the long axes of the hair cells. Such sections are especially useful in displaying the patterning of the hair cells.

Of particular value is a sagittal plane, which extends in a series from right to left and in the median position divides the head into two equal halves. Sections in this plane show the location of the cochlea in the otic capsule and its relations to supporting structures, as in Fig. 7-2.

STAINING

A number of staining procedures were used in connection with the present investigation, but three in particular are worthy of special mention. These are our HAO stain, Pollak's trichrome stain, and Held's phosphomolybdic acid-hematoxylin stain. (See Appendix B for the formulas.)

The HAO stain was developed for the differential staining of hair-cell nuclei to determine sizes of hair-cell populations. It consists of successive treatments with hematoxylin, azocarmine, and orange G and has been outstanding for its purpose. In addition it gives satisfactory rendering of most other tissues in and around the ear. This stain colors the nucleus of the hair cell rather densely while leaving the cytoplasm lightly colored, and thus makes it possible to recognize the nuclei in the 20 μ sections where they may appear at two to four levels.

Pollak's trichrome stain is useful in the study of various structures and especially the connective tissues. It was used for the general study of specimens that were not intended to be utilized for hair-cell counting, because it stains the cytoplasm rather heavily. Mallory's triple stain was used in a few instances as an alternative to Pollak's.

Held's phosphomolybdic acid-hematoxylin stain was used by him to reveal the supporting elements of the mammalian organ of Corti, and was employed in certain specimens of the present study for a similar purpose, to bring out the special elements of the supporting structure in the auditory papilla of lizards. It was discovered, after a number of trials, that Held's stain gives a satisfactory result only if the solution is adjusted to a pH within a range of 7.5-8.1. It then reveals the pillars running between the hair cells as described below.

MOUNTING

After the staining, the sections are cleared in oil (a mixture of cedarwood and bergamot oils, 3:1) then oriented and mounted, one section to a slide. The slides are arranged in order for each of the three subseries separately, and then these three are interleaved to obtain the final continuous series.

ANATOMICAL PROCEDURES: DESCRIPTION

The study of ear structure in a particular species is best carried out by the use of dissections to provide a general orientation, and then a filling in of details from serial sections. When dissection material is unavailable it is difficult to visualize the forms and relations of structures, especially the parts of the middle ear. At times in this study it was necessary to use reconstructions based on the sections to provide an objective model. This was done, for example, to study the middle ear of *Rhinophis drummondhayi*, for which only a single specimen was available.

The plan of study of the anatomy of the ear emphasized those features relating most directly to the ear's performance in the reception, transmission, and sensory representation of sounds. Consideration of the middle ear centered upon the effectiveness of this mechanism in the initial reception of sound waves and their conduction inward to the cochlea. The examination of the inner ear included first the mechanical actions by which acoustic vibrations in the cochlear fluids bring about a stimulation of the hair cells, and then considered the processes by which the particular characteristics of sounds might be discriminated and represented in the excitation of auditory nerve processes. As will be brought out, the reptiles display a variety of mechanisms, involving the tectorial membrane and related structures, by which stimulation of the hair cells is brought about. The inner ear structures are of concern in relation to the range of frequencies to which the ear responds and the sensitivity of its operations within this range. Of particular moment are the conditions in the cochlea that have to do with differential action: the degree to which sounds of different frequencies are responded to selectively, and how their differences can be represented first in activity at the sensory level and then in neural response.

Accordingly, in this study the conditions within the cochlea are examined quantitatively, with measurements of variations along the cochlea in such features as width and thickness of the basilar membrane and the size of the auditory papilla, and a determination of the numbers and distribution of the hair cells along the papilla.

The numbers of hair cells were determined for each species by counting in successive sections throughout the cochlea. For this purpose a magnification of 1200 times was used, with fine focusing to cover the depth of the section. This depth was ordinarily 20 μ, which usually presented from two to three levels of hair cells depending on their size, but in some series the sections were thicker, up to 30 μ. Actually, the nuclei of the hair cells were counted, as these are most readily recognizable.

No allowance has been made for the possibility that nuclei might be split in the sectioning process so as to appear in two successive sections. A careful scrutiny of the sectioned material has led to the conviction that such splitting is rare. The nuclei appear to be relatively firm bodies that tend to move aside when encountered by the knife edge, and so remain entire. More likely than an overcount due to splitting, it seems, is the missing of nuclei, perhaps caused by a failure to discriminate two of these at different depths in the same optical line.

In several specimens a check on the count of nuclei was made by counting the ciliary tufts, and this procedure yielded closely similar results.

One species, *Gekko gecko*, that is included in the present experiment was also studied by Miller (1973a) by the use of surface preparations examined with the scanning electron microscope. By this method he obtained counts around 2000 for the hair cells of his specimens, whereas the section counting method as just described shows a total of about 1600 cells for this species (Wever, 1965a). Assuming that Miller's method is the more accurate, as seems justifiable, this comparison would appear to indicate an undercounting of 20% for the section counting method.

An alternative explanation of this discrepancy is that the specimens in Miller's study and mine, although identified as of the same species, are actually different in the size of their hair-cell populations. This suggestion arises partly because of a doubt that so large a proportion as 20% of the nuclei could have been missed in the section counting method, but primarily because of another feature in which the two groups of animals are clearly at variance. Miller described and pictured along the surface of the auditory papilla a mid-axial hiatus, a relatively wide gap separating the longitudinal rows of hair cells having tectorial connections from other rows containing cells with connections to the sallets. No such hiatus appears in any of the specimens of *Gekko gecko* in the present study. It is possible that the two groups of animals represent different geographical variants or races within this widely distributed species.

A measure of the accuracy of the counting process, and at the same time an indication of the uniformity of arrangement of the hair cells along the papilla, is obtained by correlating the counts obtained from alternate sections. In a specimen of *Gekko gecko* one ear gave a correlation between odd and even counts of 0.93, and the other ear gave a correlation of 0.92; in a specimen of *Hemitheconyx caudicinctus* these measures were 0.96 and 0.97 for the two ears; and for a specimen of *Zonosaurus madagascariensis* they were 0.80 for both ears; for all these coefficients the p values were <0.0001, indicating highest reliability. The differences in these correlations for the three species probably represent actual variations in the regularity of patterning of the hair cells.

All these correlations are high enough to show that the counting procedure used, whether it contains a bias or not, is sufficiently reliable for the intercomparison of species.

ELECTROPHYSIOLOGICAL METHODS

Several types of electrical potentials may be recorded from the ear, including three kinds of direct-current potentials, the alternating cochlear potentials, and the auditory nerve potentials (Wever, 1966). Most of these potentials are altered by stimulation of the ear by sounds, and the observation of their variations provides a certain amount of information on the performance of the ear. Only two of these, however—the alternating potentials of the cochlea and the auditory nerve potentials—appear to be indicative of the ear's sensitivity. The cochlear potentials have had the most widespread use for this purpose, and will be given the principal attention here.

THE COCHLEAR POTENTIALS AND THEIR SIGNIFICANCE

The cochlear potentials are invariably present in every functioning ear; they have been recorded as a product of acoustic stimulation in all classes of vertebrates from fishes to mammals, and their counterpart is found in invertebrate ears as well. When these potentials are absent, the ear is deaf by any other test that may be applied.

Origin in Hair-cell Activity.   More specifically, the cochlear potentials may be correlated with the presence and level of activity of the cochlear hair cells. A number of hereditary anomalies are known among animals in which the hair cells are absent—in albinotic cats and a strain of Dalmatian dogs, for example—and these animals are congenitally deaf. In other hereditary strains, including several kinds of mice, the hair cells are greatly reduced in numbers, and the hearing is correspondingly defective. Of particular interest in this relation is a strain of mice known as the Shakers in which the hair cells are present at birth but degenerate after about three weeks. Both cochlear potentials and behavioral responses to sounds are present in the younger animals, but these decline and disappear as the degeneration proceeds (Grüneberg, Hallpike, and Ledoux, 1940; Wever, 1965b).

Further evidence on the close association between cochlear potentials and hair cells comes from experiments on injury to the inner ear by a number of drugs such as streptomycin and injury by overstimulation with sounds. The ototoxic drugs have a predilection for the hair cells, and after their administration a stage of deterioration may be found in which the cochlear potentials are reduced and the hair cells are atrophied, though other tissues retain a normal or nearly normal appearance. In experiments on overstimulation with sounds in guinea pigs a correlation of 0.91 was obtained between hair-cell loss and the reduction in cochlear potentials (Alexander and Githler, 1951, 1952).

Stimulus Relations.   The cochlear potentials show further relations that reveal their close implication in the sensory processes set up by sounds. These relations have been extensively treated elsewhere (Wever and Lawrence, 1954; Wever, 1959), and will be only briefly mentioned here.

Frequency Relations. — The cochlear potentials follow exactly the frequencies of the stimulating sounds within the limits of the ear's tonal range.

Wave Form. — For pure tones these potentials reproduce the wave form with great fidelity at low and moderate intensity levels, but show overloading at high levels. For complex tones the reproduction is inexact because of the presence of frequency and phase distortion, which is characteristic of any acoustic system with range and temporal limitations.

Linearity. — For stimuli of low and moderate intensities these potentials show a linear or nearly linear variation in magnitude as a function of sound pressure, and then become nonlinear beyond some limit that varies with species and also with tonal frequency. In auditorily proficient animals such as guinea pigs and cats the range of linearity for tones of middle frequencies may be as great as 75 db.

Overloading. — As the stimulus is raised to high levels, the response in all ears shows first a gradual and then an increasingly rapid departure from linearity, attains a maximum, and then rapidly declines. In the region of the maximum, and increasingly so beyond it, the ear may be gravely endangered. There are large species and individual variations in susceptibility to injury by sounds applied at these high levels, and there are wide variations also in the progressive character of the injuries and in the time course and degree of recovery from them. Overloading of the ear is attended not only by a reduction in the output of potentials but also by wave-form distortion, which may be regarded as a conversion of vibratory energy into additional frequences that typically are simple multiples of the stimulating frequency.

Stability of Response. — For a steady stimulus of moderate magnitude, producing a response well below the level of overloading, the response is highly stable as long as the physiological condition of the animal is unchanged and the middle ear muscles (if present) remain inactive. It has been possible to maintain a response to a moderately loud tone of 1000 Hz in an anesthetized cat over a period of 85 hours with variations of no more than 4 db (Rahm, Strother, and Gulick, 1958). This was the maximum period over which the physiological state of the animal could be kept stable under the conditions of the experiment. Incidental to other observations, and with intermittent stimulation, a corresponding stability has been observed in lizards over periods of up to four days (Werner and Wever, 1972; Werner, 1976).

Relation to Behavioral Evidences of Hearing. Among the birds and mammals there is abundant evidence of a close relation between cochlear potential measurements and hearing capability as determined by behavioral tests. Such functions as shown in Fig. 2-6 indicate for a number of tones along the frequency scale the sound pressure required for a standard response. This response for the cochlear potentials is an

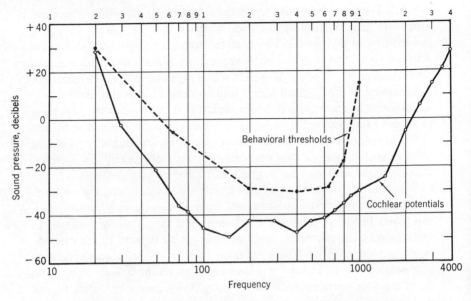

FIG. 2-6. Behavioral and cochlear potential functions for the turtle *Chrysemys scripta*. The behavioral data are from Patterson, 1966, and represent the mean thresholds for four animals. The cochlear potential data represent the 0.1 μv level for one ear.

arbitrary reading of 0.1 μv, and for the behavioral function is a certain required performance; for the turtle represented in this figure it was a retraction of the head.

The relations between these two measures of auditory sensitivity were first observed for five species: marmosets, rhesus monkeys, cats, guinea pigs, and pigeons, which were the first animals on which both cochlear potential functions and behavioral tests of sensitivity were carried out with sufficient precision to permit a meaningful comparison (Wever, 1959).

Close but not perfect correspondences were found between the sensitivity functions obtained by the two methods. The two types of curves exhibit a general resemblance, in that they cover about the same span along the frequency scale, present some middle region of best sensitivity, and show increasingly poorer sensitivity on either side of this region, and especially at the high-frequency end.

In three respects the two types of functions were found to differ: in the rate of change in sensitivity with frequency at the lower end of the frequency scale, in the prominence of the region of maximum sensitivity, and in the upper frequency limit attained. These differences are readily accounted for by the involvement of the auditory nervous system in the behavioral performance.

Low-tone Sensitivity. — The sensitivity to low tones as shown by the cochlear potentials improves nearly proportionally as the frequency rises, but as determined behaviorally it improves more rapidly, about in proportion to the square of the frequency.

The improvement in sensitivity in the low frequencies as seen in the cochlear potentials can be accounted for, at least in part, by a progressive narrowing of the response pattern along the basilar membrane and a concentration of the available acoustic energy upon fewer hair cells toward the basal end of the cochlea. The behavioral sensitivity function also reflects these changes, and in addition is affected by neural factors: an increasing density of innervation in this basal region of the cochlea, providing greater summation of nerve fibers; increasing excitability of nerve fibers to pulsating currents (i.e., to the cochlear potentials) as the frequency of these currents is increased; and the rising rate of the synchronized nerve responses themselves (cf. Wever, 1949).

Maximum Regions. — The behavioral functions all show a well-defined region of maximum sensitivity, whereas the cochlear potential functions are generally flatter, with only a broad region of best response within which irregularities are common. It can be presumed, as Békésy (1967) strongly contended, that higher order processes in the auditory nervous system, which he conceived under the general rubric of lateral inhibition, carry out the function of sharpening up the relatively minor variations of response along the basilar membrane.

The Upper Limit. — In general, in the five species examined, the behavioral acuity functions show a rapid loss of sensitivity for tones a little above the maximum region, and then quickly reach a limit beyond which no responses can be elicited at sound intensities ordinarily available. The cochlear potential functions are not so severely limited, and usually can be extended well beyond the behavioral curves.

These results show that, as the frequency rises, the mechanical impedances of the ear become increasingly restrictive as manifested in cochlear and neural activities alike, but that in the upper frequency range additional limitations are encountered in neural processes that affect the behavioral function. As sound frequencies increase, the nerve fibers reach a limit to their synchronous firing rate and begin to skip waves, so that their effectiveness in the representation of sound intensity is impaired. Above the middle range, as frequency increases further the pattern of action on the basilar membrane is narrowing with great rapidity. Also it is likely that in these mammals (as is known to be the case in man) there is a sharp reduction in the density of innervation at the basal end of the cochlea, where the high tones have their main effect. Accordingly, the number of nerve fibers responsive to the stimulating tone rapidly declines as frequency rises.

For all these reasons the acuity for the uppermost tones decreases and comes to a practical end while the cochlear potential function continues.

A further review of the relations between cochlear potentials and auditory sensitivity as determined by behavioral tests was carried out by Raslear in 1974; by this time 16 species had been assessed by both methods in a manner that could be considered reliable enough for this purpose. His survey included 13 mammals, one bird, one reptile, and one fish.

The results of this extended study are in good agreement with the ones already discussed. Raslear calculated correlation coefficients by the Pearson product-moments method between these two determinations of sensitivity for all 16 species, and obtained values over a range from 0.40 for the monkey *Macaca mulatta* to 0.89 for the opossum *Didelphys virginiana*. (A small value of 0.05 reported by Raslear for the marmoset *Hapale jacchus* is clearly in error; a recalculation gives 0.47 with a probability figure $p = 0.03$. Also new data for the turtle *Chrysemys scripta* gives a coefficient of 0.86 and $p = <0.0001$, well above Raslear's figure of 0.45.)

The mean of these coefficients for the 16 species (corrected as just indicated) is 0.68. From these results it may be concluded, as Raslear did, that the cochlear potential curve for a species provides a good approximation to the form of its behavioral acuity curve.

The prediction about behavioral acuity that may be made on the basis of cochlear potential observations is greatly improved if account is taken of the consistent differences that have been pointed out, and which seem properly attributable to the action of the auditory nervous system.

PREDICTION PROBLEM IN REPTILES

The above evidence is chiefly concerned with the mammals, and the conclusions are obviously more directly applicable to species within this class. Yet as we turn to the reptiles the problem of prediction from cochlear potential data to hearing in its broader sense seems even more simple. There are two reasons for this expected simplification of the problem, one based on morphological and the other on neural considerations.

Structural Arrangement. In the mammals the cochlea is greatly extended. Though compactly arranged in a coil of one and a half to four turns, the bony walls between turns probably insulate the parts sufficiently that as an electrical conductor this structure can be considered as having almost its full tortuous length. Therefore an electrode on the round window at the basal end of the structure, where it was located in all the mammalian experiments reported above, will represent the potentials arising along the various cochlear regions in a biased manner: with the actions of hair cells in the basal areas more faithfully represented and those of hair cells located apically suffering an attenuation according to their distance away.

The magnitude of this attenuation along the cochlea was determined in cats by Wever and Lawrence (1952, 1954) by applying alternating potentials at one end of the cochlea and recording them with electrodes applied at various places along its length. In one series of measurements the potential was applied at the round window and the pickup electrode was inserted in a minute hole drilled at the apex. The average loss over the whole length of the cochlea was 23.7 db when the potential was applied through concentric electrodes on the round window and 19.0 db when it was applied through two separate basal electrodes, one on the round window and the other at the edge of the stapes. In cats the length of the basilar membrane as measured in 6 ears was found to vary from 19.4 to 25.4 mm, with a mean of 22.5 mm (Freedman, 1947). The shortest fluid pathway along the cochlea is somewhat less, about 17.4 mm on the average. Thus the attenuation along the cochlea, averaged for the two methods of potential injection, comes to 1.2 db per mm. This attenuation does not vary significantly with frequency, and evidently is due to the ohmic resistance of fluids and tissues and to leakage loss along the path.

In the reptiles the cochlea is short in comparison with that of mammals, but its cross section is not much less. Accordingly, it seems reasonable as a first approximation to use the same attenuation figure for reptilian cochleas that was worked out for the cat, at least until direct measurements can be made. Then the conduction from one end of the cochlea to the other can be estimated as suffering an attenuation of about 6 db in crocodiles, whose cochleas are about 5 mm long, about 2.5 db in certain lizards such as the varanids and some of the geckos in which cochlear lengths of 2 mm or so are found, and as only 1 db or less in others.

However, even these small biases are not to be expected in the recording of cochlear potentials in these animals. The round window in reptiles is located not at one end of the cochlea as in the higher forms, but on one side, in a direction about at right angles to the long axis of the cochlea, and at a distance of a millimeter or two away. Therefore an electrode at the round window will record from all regions of the cochlea about equally well.

It should also be borne in mind that although in a few reptiles like the geckos there are evidences of a fair amount of differentiation along the cochlea, the majority exhibit a high degree of uniformity, so that the responses to sounds spread widely with relatively little spatial specificity.

Neural Factors.   Our knowledge about the auditory nervous system of reptiles is limited, but there is good agreement that this system is much simpler than that of mammals. Many of the primary processes conceived for the cochlear nerve fibers of higher animals are no doubt operative, but some of the more complex interactions at cochlear and higher levels may be lacking.

It seems likely, on both structural and neural grounds, that in reptiles the activities of the hair cells are reflected in the neural channels and operate at higher neural levels in the determination of behavior with much less modification than in the higher animals. The cochlear potentials in these forms therefore may be expected to provide an even better indication of hearing ability than they do in the mammals.

It is unfortunate that there is little direct evidence with which to test this speculation. For the reptiles we have only the behavioral experiments of Patterson on hearing in turtles to compare with cochlear potential measurements. These experiments, to be described in further detail in a later chapter, provide the behavioral function for hearing in the turtle *Chrysemys scripta* shown by the broken curve of Fig. 2-6, which is to be compared with the cochlear potential function for this same species represented by the solid curve of this figure. The two curves have much the same form, showing a rapid increase in sensitivity at the extreme low-frequency end of the range, a region of maximum sensitivity in the low tones, and a rapid loss as the high tones are approached. The regions of maximum in the two functions show a good degree of correspondence, and might show a better one if additional tones around 100 Hz had been used in the behavioral tests. The decline in the high frequencies occurs earlier and is more abrupt in the behavioral function than in the cochlear potential function, as is generally the case in other animals.

These two functions correlate closely, with a coefficient of 0.86 and a probability figure $p = <0.0001$. In Raslear's summary only the raccoon and opossum showed closer correlations.

PROCEDURES FOR COCHLEAR POTENTIAL MEASUREMENTS

The determinations of auditory sensitivity in terms of cochlear potentials were carried out on the animals under deep anesthesia, and required certain surgical procedures for the placing of recording electrodes.

Anesthesia.   Most anesthetics in common use produce inactivity of general body muscles while respiratory movements and the heart beat are continued at acceptable levels. Usually these functions are depressed below waking levels, but not to a degree that is serious in an animal at a minimum of energy expenditure. At times, when a particularly deep stage of anesthesia is required, it is advisable to employ artificial respiration.

Choice of Anesthetic. — A great many anesthetic drugs are available, most of which are derivatives of barbituric acid. These vary in many properties: in the time required for induction, in the duration of the anesthetic state after a single dose, in the course of recovery, and in the margin of safety between an anesthetic dose and a dose that is lethal. A choice of anesthetic must take account of these properties in relation to the purposes of the experiment, such as the time required for the experimental procedures, whether the animal is expendable or not, and whether the experiment is "chronic" or "terminal," i.e., whether the animal is expected to revive and be available for further study or whether the experiments are completed in one session or a limited series of sessions.

In most of the experiments to be described below the experiments were "terminal," and at their close the animal was prepared for histological study of the ear structures.

The experimental conditions of the present experiments called for an anesthetic with fairly rapid induction and then a steady maintenance of physiological condition for several hours thereafter. For all reptiles except the sea turtles the drug ethyl carbamate, commonly known as Urethane, was found to fill these requirements more satisfactorily than any other. It was used in a 20% solution in reptilian physiological saline, and injected intraperitoneally in a dosage usually of 0.01 ml per gram of body weight for the smaller reptiles up to about 30 grams weight, and in a smaller dosage, of 0.007 ml per gram, in the larger animals. There are species differences, and some animals required as much as 0.012 ml per gram. In the rarer animals an initial dose was usually given at a rate of 0.007 ml per gram, and then supplementary amounts added as required.

Administration Methods. — For nearly all the animals a transparent plastic bag was used to confine the animal and to weigh it, after which the bag was folded about the animal so as to immobilize it, and then the anesthetic dose was injected through the bag. This method eliminates the need of an assistant to hold the animal and protects the operator from being bitten.

The anesthetic effect of ethyl carbamate is long-lasting. Though most of the experiments were of relatively short duration, requiring only a few hours for completion, a few extended through several sessions over periods up to two to four days. Some of the reptiles could be maintained for this time without further anesthetic dosage, and with only the intra-peritoneal injection from time to time of small amounts of physiological saline to provide a proper body fluid level. In such experiments it was usually found that results obtained on the second or third day were in close agreement with those of the first day, indicating a satisfactory maintenance of physiological condition.

The venomous reptiles required more cautious handling. For the liz-ard *Heloderma horridum* and some of the moderately venomous snakes, a squeeze box was used to immobilize the animal. When the dangerously venomous rattlesnakes were studied, the assistance of a colleague with long experience and special equipment for the handling of these reptiles was welcomed—and the proper antivenom kit was kept at hand.

It is important to mention that success in the use of ethyl carbamate as an anesthetic depends upon the obtaining of a suitable grade of this chemical. For reasons that remain obscure, the products of different manufacturers, and even of a single one, are subject to wide variations, and only by trial is it possible to determine whether a given lot of mate-rial is satisfactory for anesthetic use. The best policy is to try out several of these products in succession until one is found suitable, and then to purchase a good stock with the same lot number.

For sea turtles a gas method of anesthesia was used, employing a mix-ture of Halothane and oxygen in a respirator that had been modified by the addition of an apneustic plateau device. This device makes it possi-ble to fill the lungs quickly and to hold them inflated for a time, then to deflate the lungs and begin another cycle. The duration of the cycle was varied from 5 to 10 minutes for different animals, the longer times being used for the larger specimens. This procedure reproduces the nor-mal respiratory rhythm of a diving animal.

Temperature Regulation.   It is well known that the general physiologi-cal activities of reptiles are a function of body temperature, and this condition holds true for both cochlear and auditory nerve potentials. The relations between these potentials and temperature were worked out and discussed by Werner (1968, 1972, 1976) and Campbell (1967a, 1969).

As best shown in the experiments of Werner, the ear displays its great-est sensitivity, as measured by the cochlear potentials, when the body temperature is held at the "preferred" level. This level varies with spe-cies, and is determined by measurements of body temperature made on the animals in their natural environment when they appear to be normally active or, perhaps better, by measurements made on animals in an artificial environment in which a wide temperature gradient is maintained and they are free to choose a position along the gradient.

At the preferred or optimum temperatures, as best shown in Werner's experiments, the sensitivity function is generally regular in form and its middle portion is unimodal, passing smoothly through its region of best sensitivity. When the temperature is decreased below the optimum the responses to high tones (those above the region of best sensitivity) suffer a marked decline, while those to the low tones are less affected and sometimes even show an increase. The amount of this effect varies greatly with species and conditions of adaptation, but may reach 20-40 db or even more on occasion. When the temperature is raised a few degrees above the optimum there is also a decrease in sensitivity, usually of moderate amount and involving all tones.

The effects on the potentials of substantial elevations in temperature beyond the optimum have not been thoroughly explored, and there are difficulties in doing so: the animal often becomes hyperactive and moves its head relative to the sound tube or otherwise disrupts the testing arrangement, or the ear suffers irreversible damage (Gans and Wever, 1974; see below, Fig. 21-65).

The changes of sensitivity with body temperature are independent of the sound conductive mechanism of the ear, and continue when the ossicular chain is broken and the ear is stimulated either with aerial sounds as before or with mechanical vibrations applied to the columella (Werner, 1972, 1976). The changes with temperature therefore are located in the cochlea. Though it has been suggested that such mechanical effects within the cochlea as changes in the viscosity of the endolymph might be implicated, such changes almost certainly must be minor in character, and the primary alterations can be attributed to the hair cells, and more specifically to the processes in these cells through which the cochlear potentials are generated.

In the present experiments the body temperature was maintained at a constant level by draping an electric heating blanket over the animal, though not in direct contact with it, and varying the heating current by means of a control device set for the desired temperature and monitored with a thermoelectric probe inserted deep in the rectum. This system held the temperature within 1° C of the desired value. Direct current was used for the heating blanket, because alternating current would have introduced undesirable electrical noise in the vicinity of the animal.

In nearly all the experiments the cochlear potential measurements were made at a standard body temperature of 24-25° C. This temperature lies toward the lower end of the range of temperatures that can be regarded as normal and as permitting the usual activities of the majority of the animals used in this study (see Brattstrom, 1965).

This choice of a single temperature level for all species, in spite of greatly varying temperature preferences among them, was made after careful consideration of a number of factors both practical and theoretical.

The practical considerations are many. Though the existence of temperature preferences among reptiles and their relevance to numerous aspects of their behavior is well established, the nature of these preferences has been studied for very few species. The results available for even these species are often of uncertain merit for a number of reasons. In the first place, several methods for the determination of temperature preferences have been used, and their results vary. Even more important, the effects of thermal preadaptation are not well understood; most likely there are seasonal variations in temperature relations, probably these relations change in a diurnal cycle, and certainly they are subject to considerable variations with immediately preceding thermal exposures (see Werner, 1976, for references).

With a limited number of lizard species and subspecies (6 iguanids and 14 gekkonids), Werner worked out the relations between body temperature and cochlear potential sensitivity with attention to the more recent temperature conditions. A determination of all seasonal and short-term aspects of temperature adaptation, however, presents a formidable problem.

To have determined optimal temperatures for all species of the present study would have required a number of specimens of each species and long periods of testing. For many species of this study the number of specimens was severely limited; for many of these only three to five specimens were obtainable, and for several species only one or two.

Further practical considerations relate to technical aspects of the recording of cochlear potentials and account for the choice of a relatively low temperature within the acceptable range. A low body temperature gives a more quiet animal, with comparatively little muscle activity and thus a lower level of physiological noise. Then the tests can be carried out with a minimum output from the ear and reduced risk of overstimulation damage. Also the ear's responses are more stable at the lower levels. Further, the higher temperatures produce a copious secretion of mucous from the pharyngeal membranes, and this mucous, by acting as an insulating film over the round window membrane, reduces the potentials conducted to the recording electrode. Observations on auditory sensitivity made with a single standard temperature on a variety of reptiles will provide a basis for comparison among them that though perhaps not the best possible should still prove useful. Within particular groups with similar ecological relations and temperature adaptations these comparisons may be made with a good deal of confidence. Between widely differing groups, such as desert reptiles and others living in more temperate regions, the comparisons will be more tentative, and allowance must be made for the less favorable conditions under which the hearing of the desert species is measured.

Recording Sites.   The potentials set up in the hair cells are conducted through the endolymph bounding these cells, and produce electrical fields that extend into the perilymph spaces and surrounding tissues. These potentials are most conveniently recorded with an electrode on the round window membrane, or alternatively with an electrode inserted into some part of the perilymphatic system. They may readily be recorded from the endolymph spaces also, of course, but such recording requires a penetration of the cochlear duct itself and runs the risk of injury to the delicate structures within; therefore this procedure has not been used in the present series of experiments.

The round window position was employed in all the species in which such a window is present, because it is favorably located as an opening in the exoccipital wall of the cochlear duct and is easily approached by surgical procedures that cause no serious modifications of the acoustic mechanism. Its thin membranous covering provides good electrical contact with the perilymph of the scala tympani.

Surgical Approach to the Round Window — In most of the lizards the exposure of the round window is relatively simple and involves only an incision through the floor of the mouth and pharynx. The procedure will be described for the right ear in the species *Sceloporus magister*.

The anesthetized animal is placed on its back on an adjustable stand, and an incision is made first through the skin along a line about halfway between the midline and the right side of the head (shown by line 1 in Fig. 2-2). After removal of the skin and a thin layer of superficial fascia (including the posterior intermandibular muscle), the arches of the hyoid apparatus will come into view as shown in *a* of Fig. 2-7. These arches are paired but are shown in the drawing only on one side (plus the second ceratobranchial immediately to the left of the midline).

The hyoid arches vary in form and location with species, but in all iguanid lizards are much as illustrated. In the midline the single hypohyal extends from the body of the hyoid anteriorly and the second ceratobranchial extends posteriorly. The first ceratobranchial runs obliquely in a lateral and posterior direction from a socket joint on the hyoid body. The ceratohyal arises also from the hyoid body, runs anterolaterally for a short distance, then turns sharply and extends posterolat-

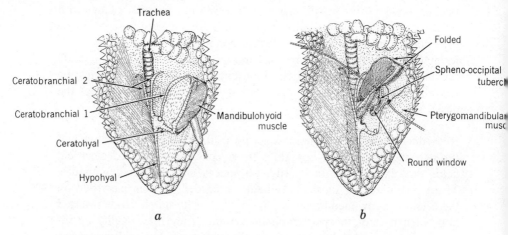

FIG. 2-7. Exposure of the round window in lizards. The head of *Sceloporus magister* is shown in a ventral view, *a*, after removal of ventral skin and superficial fascia, and *b*, after displacement of a part of the mandibulohyoid muscle and the floor of the mouth and pharynx.

erally almost parallel to the first ceratobranchial. The ends of these last two processes are close together and deeply embedded in muscles ventral to the paroccipital region. Included between the mandible and all three processes, and closely attached to them, are the three layers of the mandibulohyoid muscle.

The anterolateral edge of a portion of the mandibulohyoid, indicated in *a* of Fig. 2-7 by dark shading, is the specific site of the approach for exposure of the round window. As suggested by the elevator instrument depicted in the drawing, this muscle portion is undermined, proceeding deeply along the edge of the pterygomandibular muscle. In this dissection the ceratohyal process is divided at the knee and its lateral portion is lifted away, and also the posterolateral edge of the muscle is severed. The whole triangular mass of tissue is then raised and folded back as indicated in *b* of this figure. The floor of the mouth and pharynx is included in this tissue displacement, and after retraction of adjacent tissues there is a free exposure of the spheno-occipital tubercle and its surrounding area.

A deep recess lies dorsal to the ledge formed by the spheno-occipital tubercle, and contains the round window along with the internal carotid artery and jugular vein as shown in the two sketches of Fig. 2-8. Within the recess the internal carotid branches, sending off a large stapedial artery and then continuing its anterior course as a much smaller vessel.

The oval window, containing the footplate of the columella, is in the same area and may be exposed in the same operation. It is not seen from

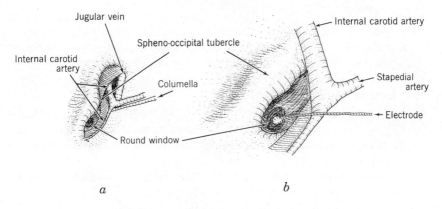

FIG. 2-8. Two views of the round window area.

the angle of view of these drawings, as it lies more dorsal and is obscured by the internal carotid and the interfenestral crest beyond. Its exposure involves turning the animal onto its left side and further retraction of adjoining tissues.

Other species require much the same treatment as just described for *Sceloporus*, though more extensive retraction is often needed when the pterygomandibular muscles are highly developed. More difficulty is experienced in some of the larger lizards, such as *Varanus*, in which the pterygomandibulars are so much enlarged as almost to meet at the midline. Then it may be necessary to remove a portion of the muscle mass. This removal is accompanied by bleeding that must be controlled by thermocoagulation and sometimes by the use of ligatures.

It should be noted that the use of electrocoagulation with high-frequency currents must be scrupulously avoided in the vicinity of the ear. Such currents may take unexpected pathways through the cochlea and produce serious damage to its sensory contents. A simple thermal cautery, heated by direct or slow alternating currents, applied for brief times and kept away from the round window, is entirely safe.

The drawings indicate an exposure that for purposes of orientation is made much more extensive than the operation requires. In actual practice the invasion of tissue is held to a minimum. Indeed, the procedure ordinarily is carried out with only the incision shown in Fig. 2-2 above and no removal of tissue except in those instances in which a bulging pterygomandibular must be reduced.

Procedures When Round Window Is Absent. — The round window is absent in chameleons, anniellids, snakes, turtles, amphisbaenians, and *Sphenodon*, and in these reptiles a minute opening was made in some part of the enclosure of the perilymphatic space and a needle, insulated except at the tip, was inserted so as to make contact with the perilymph. The needle was inserted so as to plug the opening that had been made, and a loss of fluid was carefully avoided. Tests showed that any of the perilymphatic ducts or cavities will serve for the recording of potentials, but these potentials are stronger for locations immediately adjacent to the cochlea.

In chameleons the hole was drilled close to the oval window, usually slightly anterior and medial to it. The recording needle thus entered the large extension of the scala vestibuli often called the vestibular or peri-lymphatic cistern.

In amphisbaenids and anniellids the hole for the electrode was drilled in the roof of one of the saccules, usually the left one. The saccular organs are particularly large in these animals, and bulge dorsally and laterally in the extreme posterior region of the skull.

In snakes the hole was drilled in the dorsolateral part of the prootic bone, likewise giving access to the perilymphatic cistern.

In turtles a dorsal or dorsolateral approach was made to one of the semicircular canals, usually the posterior canal. In many species the superior adductor mandibulae muscle was simply retracted laterally, exposing the floor of the temporal fossa. This floor was cleaned sufficiently to expose the suture between supraoccipital and opisthotic bones, and a minute hole was drilled along this suture. In some instances it was necessary to remove a portion of the muscle mass, and then it was usually helpful to use one or two ligatures to control bleeding. Careful application of the thermocautery on the cut muscle surfaces was of assistance also. When the drilling was deep enough that a little fluid could be seen to well up in the hole, the electrode needle, of the same size as the drill, was quickly inserted.

Electrodes. — The round window electrode consisted of a fine silver wire that had been given a balled tip by holding it momentarily in a flame. The diameter of the ball varied from 0.2 to 0.5 mm, and always was small relative to the size of the round window.

The needle electrodes were of stainless steel, especially sharpened and coated with insulating varnish except at the tip. In the large turtles these needles were further protected by enclosure, up to the tip, with thin, close-fitting plastic tubing.

Two reference electrodes were used, both located in inactive tissues and one of them grounded, as required to produce a balanced input to the preamplifier. One electrode was located relatively close to the ear, often at the edge of the incision, and the other was farther away, usually beneath the skin in the shoulder region. These electrodes also were stainless steel needles insulated except near the tip.

The electrodes led to a preamplifier that provided a voltage gain of

1000 (or in later experiments 10,000) times, and presented the amplified signal to a wave analyzer that was used as a selective voltmeter. This analyzer was set to provide a pass band of 3 cycles per second.

The balanced input to the preamplifier gave a considerable rejection of extraneous noise and the narrow pass band of the wave analyzer attenuated this noise further, so that the limiting condition for signal detection was essentially the physiological noise level. This level varies with frequency and with the state of the animal (especially the level of anesthesia). Under most conditions, with deep anesthesia and over the main range of frequencies, it was possible in these experiments to record responses of the ear down to 0.04 µv, and occasionally as low as 0.03 µv. In the low frequencies the physiological noise level increased, mainly because of muscle action, especially the heart beat, so that below 600 Hz it became difficult to record signals smaller than 0.1 µv. In some animals, and nearly always when the anesthesia was light, the level of the noise rose to as much as 0.2 µv in the region below 200 Hz. Then the responses were measured at a higher level, perhaps 0.4 µv. The practice, as a general rule, was to set the wave analyzer at each stimulus frequency and to note the existing noise level, and then to adjust the sound stimulus so as to produce a response at least twice the voltage of the noise.

THE STIMULI AND THEIR CALIBRATION

The acoustic signals were pure tones produced by applying a sinusoidal current either to a loudspeaker for aerial stimulation or to a mechanical vibrator for vibratory stimulation. Aerial sounds were used for all the animals for the routine tests, and vibrations were used for certain special purposes, such as the study of the transmission characteristics of the middle ear.

The sinusoidal voltages were derived from the generator circuit of the wave analyzer, more recently with the aid of a tone synthesizer, and were monitored with an electronic counter. The frequency was maintained within 1 Hz of that specified.

The use of the generator circuit of the wave analyzer is convenient because the analyzer is automatically in tune with the stimulating frequency, and thus with the fundamental frequency component of the animal's output. Although in lizards there is often a second harmonic component in the output, this component ordinarily was not measured. At the low levels of stimulation normally employed, this harmonic and all others are 20 db or more below the fundamental magnitude.

For aerial stimulation, the sinusoidal signal was passed through an attenuator network to a power amplifier, and then through a second attenuator network to the loudspeaker. For vibratory stimulation the signal was passed through an attenuator to a voltage amplifier and then into a crystal vibrator. Block diagrams of these circuits are given in Fig. 2-9.

Aerial Stimulation.   The loudspeaker was of a wide-range type (Western Electric Type 555) no longer manufactured. It was capable of delivering large sound outputs, of the order of 1000 dynes per sq cm or more (except at a few specific frequencies) over a range from 100 to 15,000 Hz and with low distortion. This speaker is designed to operate into a horn, but here was provided with a plastic tube 8 mm in inside diameter and 80 cm long, whose terminal end connected to a sound cannula that ordinarily was sealed over the animal's external ear opening, as shown in Fig. 2-10.

This sealing of the ear is critical, and must be carried out so as to achieve an airtight connection. Variations of the order of 2-4 db are common, and larger amounts up to 10 db may occasionally result from the absence of a seal or the accidental loss of one, as may be caused by head movements in an incompletely anesthetized animal. The effects of sealing are indicated in Fig. 2-11, in which curves are presented for two different specimens of *Crotaphytus collaris* in each of which observations were made with the sound tube sealed around the meatal opening (solid line) and with its end close to the skin surface but not sealed (broken line).

Several short cannulas were provided to give a wide choice of openings, so that for any species one could be chosen to fit over the auditory meatus. Sometimes, as in the spiny lizards, it was necessary to trim away the protruding scales about the meatal orifice so that the cannula would fit satisfactorily. The cannula was placed almost in contact with the skin surface, leaving only a narrow gap. Then a strand of cotton wool that had been soaked in petroleum jelly was passed around the end of the cannula, and gently pressed into contact with the skin surface. The narrow gap that had been left between cannula and skin was thereby sealed. This arrangement avoids any possibility that mechanical vibrations will be transmitted from the cannula to the tissues of the head. A standard series of frequencies was used that included 19 or 20 tones over the range from 100 to 10,000 Hz, and sometimes was extended upward by adding 15,000 and 20,000 Hz, and in species with good low-tone sensitivity (notably the snakes and turtles) was extended downward by including the tones 30, 40, 50, 70, and 80 Hz. A tone of 60 Hz was avoided because it is the power-line frequency and was sometimes present as a spurious signal in the wave analyzer circuit. Also usually avoided was 300 Hz, a prominent harmonic of the line frequency, and

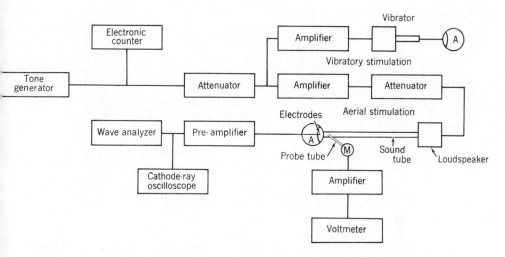

Fig. 2-9. Stimulating and measuring circuits. The animal is shown at *A* in two locations, with stimulation with a vibrator at the upper right and with aerial stimulation to the right of center. *M* represents the calibrated microphone.

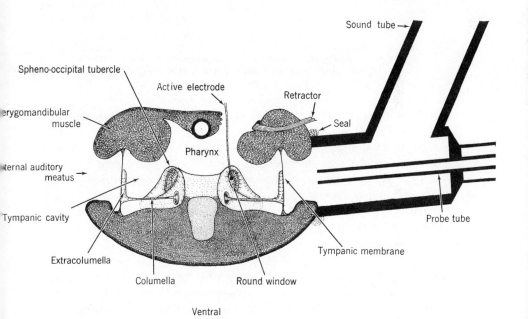

Fig. 2-10. Aerial stimulation: the sound tube applied to the ear of the lizard, *Sceloporus magister*.

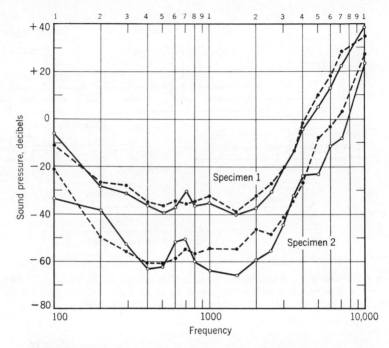

F<small>IG</small>. 2-11. Effect of sealing the sound tube over the ear, shown for two specimens of *Crotaphytus collaris*. Solid lines represent measurements with good sealing; broken lines represent measurements with the sound tube close over the meatus but unsealed. Tests carried out in collaboration with Y. L. Werner.

also often present at an undesirably high level in the measuring circuits. The tone 290 Hz was used in its place in many of the tests.

The aerial stimulus was monitored by means of a condenser microphone and probe tube incorporated into the sound tube near the animal's ear as Fig. 2-10 shows. The probe tube ended concentrically with the sound tube at the ear opening, and the system was calibrated by the use in place of the animal of a "dummy ear," which consisted of a short rubber tube enclosing a volume of air about equal to that of the average lizard's outer ear.

Vibratory Stimulation. The mechanical vibrator was of our own design and consisted of a Rochelle salt crystal made up of laminar plates with the piezoelectric axes chosen so that, when the plates were cemented together, the crystal produced longitudinal variations in response to alternating voltage inputs. The crystal assembly was housed in a heavy metal case, with one end cemented to the interior of the case. To the other end of the crystal was cemented a plastic rod that ran out through a hole in the case. This rod carried near its end a small accelerometer, and ended in a screw socket by which various stimulating tips could be attached. Ordinarily a tip was used that terminated in a steel needle with a slightly blunted end that could be applied to various parts of the receptive mechanism of the ear as required.

The metal case served as a reaction mass for the vibrating crystal and also as an electrostatic shield. The case was hermetically sealed by the use of rubber dam and silicone grease to prevent a loss or gain of water of crystallization by the Rochelle salt. By the use of this moisture barrier, crystals of this type have been maintained in good condition for many years.

This vibratory stimulus was calibrated by means of a "Fotonic sensor," an instrument using fiber optics to measure small amplitudes of displacement. The calibration was carried out by attaching to the plunger of the vibrator a special tip carrying a reflecting plane surface, and adjusting the distance from the sensor to a suitable value. For a given input to the vibrator, the amplitude of alternating motion was determined from a meter reading. A simultaneous reading from the accelerometer near the plunger's tip showed no reduction of vibratory amplitude when the needle tip was applied to some part of the animal, such as the tympanic membrane or the surface of the skull. This is the case because the mechanical impedance of this type of crystal vibrator is exceedingly great. Accordingly, the calibration figures obtained for the vibrator when operating unloaded, as described, could be used without correction for its performance when in contact with auditory and cranial structures.

The use of a vibrator as indicated for driving parts of the ear or other structures depends upon the elasticity of the structures to produce a faithful following of the motions. The application of the vibrator tip therefore requires a certain care. The vibrator was held in a heavy manipulator, and by the use of an operating microscope the needle tip was lined up with the structure to be driven, with the point just out of contact. Power was applied to the vibrator at a frequency that experience had shown to be appropriate and in an amount that was expected to produce an easily detected response. Then the crystal was slowly advanced with the manipulator until a first indication of the ear's response was observed. Usually this initial response was irregular, because the pressure against the structure was slight and its elastic reaction was insufficient to give a complete following of the alternating movements. A slight additional advance of the plunger then produced a steady response. Further advance of the plunger usually caused no change in the response, until suddenly the response began to diminish, which was a sign that the pressure had become excessive and that a further advance was likely to be disruptive. There was always a comfortable margin between the point at which the response first became regular and the extreme position at which the response began to decline. A sketch of the vibrator applied to the stump of the cut columella is shown in Fig. 2-12.

Oscilloscopic Monitoring.   The output from the animal was always displayed on a cathode-ray oscilloscope. Such a display provides a continuous indication of the physiological noise level and ordinarily includes the electrocardiogram. From this information a judgment can be made of the state of the animal, including the level of anesthesia.

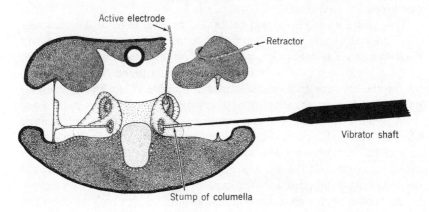

FIG. 2-12. Vibratory stimulation: the mechanical vibrator applied to the stump of the columella.

OBSERVATIONAL PROCEDURES AND PROBLEMS

SENSITIVITY DETERMINATIONS

With the sound tube in place and the electrodes suitably located, tests of sensitivity were carried out for a number of tones throughout the animal's frequency range. These tests ordinarily began with a tone of middle frequency, which experience with this or a related species had indicated to lie in a region of good sensitivity. The tone was presented first at a low level of intensity and then increased in steps of 10 db until a response was observed on the meter of the wave analyzer. Thereafter the intensity was varied in 1 db steps until the desired output was obtained. This procedure was then repeated for all the other tones in the standard series, sometimes working upward in frequency and sometimes working downward, and usually varying the order for different runs on the same animal and for different animals of the same species. Experience has shown, however, that the order of testing is unimportant as long as stimuli of extreme intensities are avoided.

The most hazardous stimuli are the high tones, to which most of the reptiles are particularly insensitive, so that high sound intensities were required for an observable response. Therefore it was the usual practice, at least until experience with a species had revealed its characteristics in some degree, to leave the testing of the uppermost part of the range until the very end. With these high-frequency tones also care was taken to demand the minimum of output from the ear, often an output as low as 0.04 μv.

The results of these tests, plotted as the sound pressure required to produce a standard cochlear potential of 0.1 μv, give a sensitivity function of the form shown in Fig. 2-13 for the lizard *Urosaurus ornatus*. In this figure the sound pressure is indicated in two equivalent ways, in pressure units (dynes per square centimeter) and in decibels relative to a zero level of 1 dyne per sq cm. In nearly all later figures only the second manner of representation is used, for it is the more convenient.

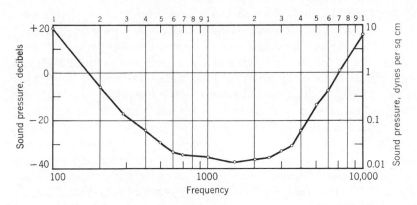

FIG. 2-13. A sensitivity function for a specimen of the lizard *Urosaurus ornatus*. The stimulus intensity is represented on right and left ordinate scales in two equivalent ways, and at various frequencies indicates the levels required to produce a standard response of 0.1 μv.

The choice of 0.1 μv as a standard level is arbitrary but realistic: it is a response voltage that is obtainable from the great majority of reptilian ears for a wide range of tones applied at safe intensities, and within a range that does not overload the ear. At times it was necessary to elicit stronger responses, as when noise was present in amounts obscuring the smaller signals. On the other hand, when exploring the limits of high-tone sensitivity it was expedient to work at the minimum possible response levels. In these instances the curves are drawn to indicate the expected sound pressures for the standard response, obtained from the actual observations by a calculation that assumes a linear relation between sound pressure and response magnitude. This assumption is a valid one at low response levels, as will be shown.

It is well to bear in mind that the standard curves for the cochlear potentials, like the one in Fig. 2-13, do not represent thresholds of sensitivity. This caution is necessary because some experimenters have spoken of "thresholds" for these potentials, evidently having been misled by the limitations of their measuring equipment into considering the lowest signals seen as representing true limits of the ear's activity. There is no reason to believe that a lower limit exists, for every improvement in equipment and physiological conditions that has reduced the noise level has permitted measurements at progressively lower levels. In a series of experiments on cats in which a special effort was made to reduce the noise level, it was possible to record potentials as low as 0.014 μv (Wever, Rahm, and Strother, 1959), and correspondingly low levels have been recorded in reptilian ears in the course of the present study. Evidence obtained by McGill (1959) on cats whose behavioral thresholds were first determined by a training method and whose ears were then used for cochlear potential measurements indicated that these potentials were usually lost in the noise background when tones at true threshold levels were presented. Measurements made at higher levels made possible an extrapolation to determine the level of the cochlear potentials for threshold stimulation, and this level was found to be 0.034 to 0.0023 μv, varying as a function of frequency. For the turtle, according to the data of Fig. 2-6, the thresholds correspond to cochlear potential readings varying from 2.6 to 0.00033 μv over the whole range and averaging 0.0024 μv, and with the values in the more sensitive region between 200 and 600 Hz averaging 0.00066 μv. Werner (1972) obtained similar results from data reported earlier.

From this comparison it appears that over the best range of this turtle's ear, from 100 to 800 Hz, a tone that elicits a round window potential of 0.1 μv should be easily audible, and indeed might present a comfortable level of loudness.

These relationships and those available for the cat on the basis of McGill's observations, together with a consideration of the distribution of sensitivity levels encountered among the reptiles, may serve as a basis for a tenuous evaluation of the quality of hearing revealed in the cochlear potential tests. Hazardous though such an evaluation may be if taken in absolute terms, it serves a useful purpose in making comparisons among individuals and species. With the necessary reservations in mind, the descriptions of the sensitivity results to follow will include characterizations of the level of hearing performance in such terms as "poor" for ears that require sound pressures around +10 to 0 db for the standard response, "fair" to "good" for those showing regions around −10 to −20 db, and "very good" to "excellent" for those with best regions at −30 to −40 db and beyond. As will be seen, the great majority of ears in which there are no obvious abnormalities are in the "good" to "excellent" category, and a few excel with levels up to −60 or rarely to −80 db.

Intensity Functions.   To determine the relations between stimulation level and cochlear output, the sound pressure was varied for particular tones, usually in steps of 5 db, and the resulting potential noted. Usually the readings began at a level about twice that of the background noise and continued at increasingly higher sound pressures until signs of overloading of the ear appeared.

The first sign of the ear's overloading is a failure of the response to increase in proportion to the increase in sound pressure. Thus for the fainter stimuli an increase of 5 db will raise the output to a value nearly double that just obtained (strictly the increase should be 1.78 times), although noise and instabilities may sometimes obscure the picture. When the increase is noticeably less than this, the ear has entered into its range of nonlinearity. A still further increase of stimulus intensity may produce no rise in the response at all, or may even produce a decline. At these levels of stimulation the ear is endangered, and may suffer temporary or permanent damage if the stimulation is prolonged. Some ears withstand this overstimulation more successfully than others, and some show better recovery after an impairment has been sustained.

In most of the intensity function determinations, the level was raised only to the point where definite nonlinearity was exhibited, and in the great majority of these ears the sensitivity remained unimpaired. However, because there was always a risk of injury in species that had not been thoroughly studied in this respect, the determination of intensity functions was carried out in only a few of the specimens, usually in one or two of a given species, while the others were tested only for sensitivity. This precaution was used so that the majority of specimens would retain a normal picture in subsequent histological examination.

An example of an intensity function that was held within moderate limits of stimulation for the ear under study is shown as *a* of Fig. 2-14. Another function that was carried a little beyond the maximum is shown as *b* in this figure.

It is evident in Fig. 2-14 that at low levels of response the cochlear potential closely approximates a linear function of sound pressure. Departures from linearity are often seen as an upward curl at the lower

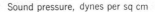

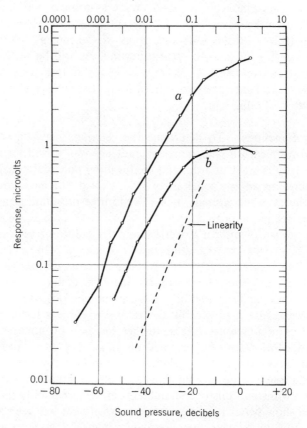

FIG. 2-14. Intensity functions for individual reptile specimens; *a*, the turtle *Geoemyda pulcherrima manni* and *b*, the lizard *Iguana iguana*. A line is added to show perfect linearity.

end of the curve, as in *a* of this figure, and are due to the presence of noise. Often in plotting such data the noise level (the meter indication in the absence of a stimulus) is first subtracted from the readings. This has not been done in Fig. 2-14; if it were the curve would be more nearly linear.

When efforts are made to reduce the noise level, which often may be done by special adjustments of the measuring apparatus and the use of particularly deep anesthesia, the intensity function can be followed to very low levels, sometimes entering the region that McGill's observations on the cat indicated as containing the auditory thresholds for this ear. It is significant that the cochlear potential function maintains its linear form, at least approximately, as far as it can be extended in this manner. It seems likely that this function continues all the way to zero and lacks a threshold. Only at the next step of action of the auditory response system, in the excitation of the auditory nerve fibers, does a true threshold appear.

This testing soon discloses that an intensity function can have more than one maximum. If the stimulus intensity is raised further after a decline from the first maximum is seen, the response curve often reverses again to resume its upward course, and then attains a second maximum higher than the first. Even a third maximum can sometimes be obtained, and vary rarely a fourth. An example of a multi-maximum function is presented in Fig. 2-15 for the gecko *Ptyodactylus hasselquistii guttatus*. In this series of measurements three maximums were passed through and a fourth was approached; this fourth maximum could not be clearly determined because at the last measurement the response began a rapid decline, indicating injury to the ear. Histological examinations made after experiments of this kind show widespread cochlear injury as a breaking of the connections between the tectorial membrane and the ciliary tufts of the hair cells. In one such specimen every tectorial connection was broken along the cochlea.

The changes in form of the intensity functions at high levels represent complex processes and are only partially understood. These changes are believed to arise in the activities of multiple groups of hair cells, and a maximum appears as the hair cells in one area of the cochlea are reaching their limit of action. Then when the stimulus is further increased other hair cells with lower inherent sensitivity enter the action and produce a secondary rise in the curve (Wever and Lawrence, 1949a).

Maximum Response Functions. A further procedure consists of a determination of maximums for a series of tones throughout the animal's frequency range. This procedure is difficult because the dangers of damage to the ear are multiplied, and it can be carried out successfully only in species that are resistant to such damage.

When primary maximums are determined for a number of tones throughout the frequency range, with care to present the stronger stimuli for only the briefest times necessary to obtain readings, a picture is obtained like that of Fig. 2-16, which shows the maximum values as a function of frequency. This function is for *Gekko gecko*, a species that withstands the procedure particularly well. These observations show that the largest maximums are in the low frequencies, which are also the ones for which this ear is most sensitive. Above 500 Hz, the maximums decline at an decreasing rate.

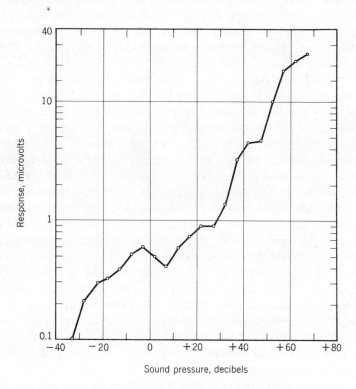

FIG. 2-15. An extended intensity function for a specimen of the gecko *Ptyodactylus hasselquistii guttatus.*

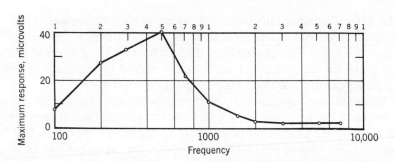

FIG. 2-16. A maximum response curve for a specimen of the lizard *Gekko gecko.* From Wever, Vernon, Peterson, and Crowley, 1963.

Auditory Nerve Potentials in Relation to Hearing

It has long been known that potentials arise in the auditory nerve and in higher acoustic channels up to the temporal cortex when the ear is stimulated with sounds, and many attempts have been made to utilize these nerve actions for at least a general indication of hearing capabilities. The majority of these experiments have been carried out in birds and mammals, beginning with the early efforts by Beck in 1890 and Danilewsky in the year following, who connected an electrode on the brain of an animal to a sensitive galvanometer and obtained deflections when stimulating with loud sounds.

In later experiments of this kind the electrode was located on the auditory nerve or inserted in the cochlear nucleus or in higher ganglia of the acoustic system. The object of these procedures was to record from a large number of active units, and ideally to obtain a representative sample of the total response.

These procedures have encountered some serious difficulties. An electrode located on a nerve bundle or inserted into a ganglion can be brought in close contact with only a limited number of active elements. To be sure, the electric fields of these elements spread in the tissues and may be recorded at a distance, but this spreading varies greatly with local conditions, so that the results are highly variable. Moreover, the form of summation of the potentials as reported by an electrode is not likely to represent in any precise way the effects of these units at higher levels of the nervous system. Although this procedure has produced qualitative results of great value, it so far has not proved very serviceable for intercomparisons of species.

A more refined method, much more extensively employed in recent times, is the use of a microelectrode that ideally records from a single nerve fiber or cell body. Its small size makes it possible to locate it on or in an active unit and thus to indicate the changing potential at its tip in relation to an inactive electrode elsewhere.

The drawback of this procedure is that only a quantum of information is obtained at a given site, and a picture of the whole response must be built up by placements near many units over a period of time. Each unit is tested by presenting a large number of tones to ascertain the characteristic frequency of the unit, which is the frequency showing the lowest threshold, and to discover also the variations of sensitivity for higher and lower tones.

Kiang and his associates (1966), in their observations on cats, pointed out that plotting the composite function representing many points for a number of animals produced a form resembling the behavioral audibility curve for this species. This relation suggested the use of the single unit method, preferably based on a large number of measurements on a single specimen, as a way of obtaining audiometric information.

Konishi (1969, 1970) reported a close relation between the composite neural unit function and behavioral evidences of hearing in two species of songbirds, the canary and the starling. He also noted that the frequency of appearance of neural units bears a relation to the ear's sensitivity: units are most common with characteristic frequencies in the region of the scale where the sensitivity is greatest as measured by behavioral tests.

A number of observations on reptiles has been made by the composite neural unit method. Suga and Campbell (1967) reported results on the lizard *Coleonyx variegatus*. Weisbach and Schwartzkopff (1967) worked on *Caiman crocodilus*. The most extensive observations were made by G. A. Manley (1970a, b), who tested the caiman, a turtle, and three species of lizards.

The results obtained by Suga and Campbell on *Coleonyx variegatus* are presented in Fig. 2-17. In this experiment the auditory region of the medulla (probably the olivary cortex and the dorsal nucleus magnocellularis) was explored with a glass pipette electrode during stimulation with tone pulses over a wide range of frequencies. The responses of six units are shown in the figure, and the probable course of the auditory

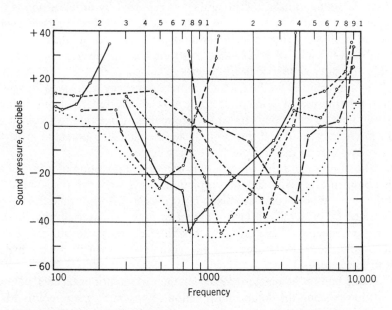

FIG. 2-17. A sensitivity function derived from thresholds observed in six neural units in the lizard *Coleonyx variegatus*. From Suga and Campbell, 1967.

threshold is suggested by the dotted line, drawn to follow the peaks of sensitivity.

Many further studies need to be carried out in this area, with a view toward correlations with cochlear potential measurements, and also with behavioral evidences of hearing if these are possible to obtain.

DATA PROCESSING METHODS

The statistical treatment of audiometric data presents some special problems. Because the ear in its operation covers a very wide range of acoustic intensities, often of the order of 10 to 1000 million-fold from the faintest tone found to be effective to the strongest that is likely to be applied, it is customary to calibrate the instruments and make the readings in logarithmic terms, usually in decibels with reference to some arbitrary standard. In the present studies the intensity is expressed in decibels relative to 1 dyne per sq cm (defined in Appendix A).

The decibel readings may be used as such in statistical manipulations or may be converted first into absolute values, such as sound pressures expressed as dynes per sq cm, then carried through the mathematical processing, and the result finally reconverted to decibels. These two procedures may yield greatly different results.

The absolute conversion method is often defended as the more logical, in that it deals with real physical quantities, but it has some practical drawbacks. The ear is in some sense a logarithmic converter, and in its functioning it compresses the intensity range in a manner that at least approximates the decibel scale. The variations from the standard manner of operation seem most meaningful when considered in these terms.

Also any given type of performance, as for example the sensitivity displayed by a group of animals of a given species in response to a presented tone, appears usually in a scattered distribution, with most values varying moderately from a modal value and a small number of measures showing large decrements in performance. When the number of measurements is large and clustering is apparent, so that the anomalous cases are readily identified, the experimenter is justified in excluding these latter cases and dealing only with the main group. In many situations, however, the measurements are few, and a procedure must be used that provides the most probable measure of this small sample.

An example will bring out the differences between these two procedures in an actual situation. Sensitivity was measured in a group of 10 lizards of the species *Trachydosaurus r. rugosus*, and for the tone 500 Hz, which falls in the more sensitive region of frequency for this ear. The readings obtained in the eliciting of a standard cochlear potential of 0.1 μv for this tone were as follows: −57, −55, −49, −45, −43, −42,

−39, −36, −22, and −18 db. There is a moderate decline from the best animal with a sensitivity of −57 db to the eighth one with −36 db, and then a sharp drop for the last two animals. The absolute conversion method when used for these data yields a mean of −31.7 db; and omission of the two poorest animals and a similar treatment of the remaining eight gives a mean of −43.4 db. The simple decibel averaging of the group of 10 gives a mean of −40.6 db, which is in good agreement with the absolute conversion figure for the reduced group.

The decibel averaging method clearly is less affected than the other by a widely divergent measurement, such as might have some accidental cause. It is employed throughout the present study.

STATISTICAL TREATMENTS

For the determination of correlations, the Pearson product-moments procedure was used, which yields a correlation coefficient designated by the symbol r, whose values vary between $+1$ (perfect correlation) through 0 (absence of relationship) to -1 (perfect inverse relationship). The reliability of this correlation is assessed by a further procedure that yields the probability figure p, which indicates the probability that the correlation obtained is due to errors of sampling or other chance conditions. A value of p of no more than 0.05 is taken to indicate that the correlation is meaningful; one of 0.01 shows a good level of significance, and one of 0.001 or less that it is highly significant. Further discussion and methods of calculation will be found in standard texts on statistics.

In the graphic representation of hair-cell density along the basilar membrane (first appearing in Fig. 7-26 below) a moderate smoothing of the curves is obtained by a rolling average method: the points shown represent averages of observations in three successive sections.

3. GENERAL ANATOMY OF THE
REPTILIAN EAR

In this chapter the anatomy of the ear of reptiles is considered in a general way to serve as the basis for an analysis of the processes of sound reception in these animals, and as a background for more detailed inquiries in the systematic sections to follow.

Anatomical studies of the ear of reptiles have been comparatively few; research in this area has lagged far behind the corresponding treatment of the mammalian ear. Early investigators, such as Hasse (1871, 1873), Retzius (1880, 1884), and Kuhn (1882), made notable advances in providing good general descriptions of the reptilian ear and its relations to adjoining structures.

These authors were principally concerned with the membranous labyrinth and cochlea, though Retzius included in his historical reviews what was known up to his time about the middle ear structures. The peripheral mechanisms were more specifically treated for certain European lizards by Leydig (1872) and then more extensively for lizards and *Sphenodon* by Versluys (1898, 1904).

These accounts not only laid a foundation for reptilian ear anatomy but also developed a working terminology. For the most part these writers took over the terms already in use for the mammalian ear, often with assumptions of homology that now may seem questionable.

In recent times there has appeared a new wave of interest in the anatomy of the reptilian ear, no doubt stimulated by the appreciation of evolutionary relationships between the reptilian ear and the ears of birds and mammals, as well as by new developments in our understanding of function in these more primitive animals.

Among many significant contributions are those of Shute and Bellairs (1953), Baird (1960, 1970, 1974), Hamilton (1960, 1964), Miller (1966, 1968, 1973, 1974), and Mulroy (1968). The present writer, in association with Jack A. Vernon, began the examination of the reptilian ear in 1956 with a series of studies on turtles, snakes, and crocodilians, in which the results of both anatomical and electrophysiological investigations were applied to the problems of auditory function (Wever and Vernon, 1956, 1957, 1960). In these and continuing investigations attention was directed especially to the processes of sound reception and the cochlear stimulation process (Wever, 1965a, 1967a, b, 1971c).

Baird in three general reviews has dealt extensively with the anatomy of the ear in representative reptile species, and his work should be consulted for many details, especially concerning the non-auditory labyrinth, which in the present treatment is given only limited attention. Particularly thorough are his descriptions of the forms and spatial orientations of labyrinthine and cochlear structures.

The present treatment of reptilian ear anatomy and function includes as extensive a range of specimens as could be obtained in living, healthy condition, and the anatomical study of these specimens, after histological preparation by methods developed over a number of years, is carried as far as the resolution of the light microscope will permit. When the integrity of tissues is maintained by suitable histological treatment, the serviceability of the light microscope is extended beyond the usual limits, and in the present material many new features have come to attention, especially concerning the tectorial structures and their role in the stimulation process.

The light microscope obviously has limits even with the best material, and further investigation with transmission and scanning electron microscope methods is needed. Preliminary studies of this kind are already beginning to appear, with reports by Mulroy (1968), Miller (1973a, b, 1974a, b) and brief notices by Baird (1967, 1969, 1970b). These studies have substantiated some of the new observations with the light microscope, and have added details on intracellular structure. Further investigations of the finer anatomy may be expected, especially if improvements can be made in the fixation and preparatory procedures for electron microscopy.

Principal Divisions of the Ear. In mammals three general divisions of the ear are recognized: the outer ear, which receives sound waves from the external environment; the middle ear, which converts the wave pressures into vibratory movements of a chain of ossicles; and the inner ear, in which the movements continue as fluid oscillations and finally are applied to the cochlear hair cells. In reptiles the same three divisions are commonly accepted, though in most species the outer ear is poorly developed and in some is absent altogether.

Outer Ear

In a number of lizards, such as the geckos, there is an external auditory meatus leading inward to the tympanic membrane, which is thus well protected from injury. In most of the geckos and in a few other species a meatal closure muscle is present which affords further protection. A corresponding feature is found also in crocodilians, in which the shallow tympanic cavity is covered by earlids that can be opened and closed by muscles under reflex control.

In many lizards, such as iguanids and agamids, the tympanic membrane is superficial or lies within a millimeter or two of the surface, so that an outer ear hardly exists. Sometimes there is a slight depression of the skin in this region, or the anterior edge of the tympanic membrane is bounded by a little fold of skin or by a row of enlarged scales, perhaps providing some slight protection for the membrane. In turtles, snakes, amphisbaenians, and *Sphenodon* an external ear is entirely absent.

MIDDLE EAR

Typically the middle ear of reptiles consists of a tympanic membrane and an ossicular chain of two elements, the columella and extracolumella, suspended in an air space (the tympanic cavity), and leading inward to the oval window of the cochlea. These structures and their relations to the inner ear are represented in a semischematic fashion in Fig. 3-1.

There are wide variations in the form and size of the tympanic cavity.

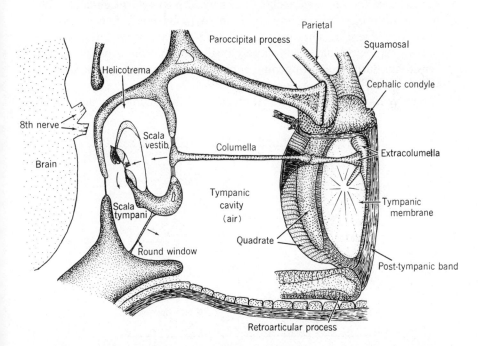

FIG. 3-1. General relations of middle and inner ear structures in the lizard *Sceloporus magister*. The right ear is viewed from behind and from the left, so that the inner surface of the tympanic membrane is exposed.

In turtles this cavity is large and extends far backward as a blind pouch in the squamosal bone. In crocodilians it is even more extensive, with a complex of passages, including a connection between right and left sides. In most lizards this cavity is hardly distinguishable from the pharyngeal space, though in others, such as the chameleons, the two cavities are separated by a membrane in which there is a small hole usually considered as corresponding to the Eustachian tube of higher forms. In snakes the tympanic cavity is reduced to a narrow fissure. In *Sphenodon* it is filled with adipose and other tissues. It is absent in amphisbaenians.

The tympanic membrane of lizards is held in a frame largely formed by the quadrate bone, supplemented along the posterior side by soft tissues. The posterior support is provided by skin and muscles, together with a strong band of connective tissue that runs along the forward edge of the depressor mandibulae muscle and attaches to the squamosal above and the retroarticular process below. Though many of these attachment structures are involved in jaw movements and other movements of the head, the action of the tympanic membrane in sound reception appears to remain relatively uniform. Fig. 3-2 shows sensitivity curves for a specimen of the fan-toed gecko *Ptyodactylus hasselquistii guttatus* with the mouth open (broken line) and closed (solid line). Except around 1000 Hz, where an improvement of 6 db appears when the mouth is opened, the variations seem entirely negligible. A similar uniformity in sensitivity during mouth opening is observed in snakes, as described in Chapter 20.

The columella is the proximal element of the middle ear apparatus.

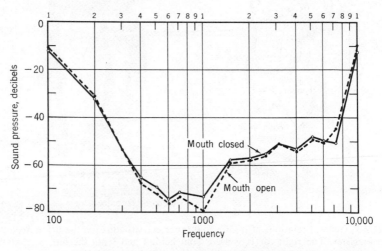

Fɪɢ. 3-2. The effects of opening the mouth on sound conduction in the ear of the lizard *Ptyodactylus hasselquistii guttatus*.

It is typically a slender rod with its inner end greatly expanded to form a footplate in the oval window. The main portion is osseous, with cartilage covering the two ends, or sometimes with a good part of the outer shaft consisting of cartilage where it connects with the extracolumella.

The extracolumella is entirely cartilaginous and in typical lizards consists of a shaft portion and a number of processes.

Typically four processes radiate over the surface of the tympanic membrane, stiffening this surface and transmitting its vibratory movements. In some species this number is reduced to two or three. An additional process, known as the internal process, is found in iguanids and related species. It arises from the shaft portion of the extracolumella or from the region of the junction with the columella, and extends roughly at right angles to the shaft to form an attachment on the surface of the quadrate bone. It serves as a support for the columellar system, permitting in-and-out or angular movements but reducing lateral deflections. Other lizard species lack this process, and lateral support for the columellar mechanism is provided by a fold of mucous membrane.

In most reptiles the columella has a form much as described, but the extracolumella varies greatly and may be absent. The extracolumella is lacking in certain amphisbaenians (*Bipes, Blanus*) and also in snakes (although some believe a remnant may be present). In turtles it is greatly simplified and consists only of a shaft portion and a widely expanded external plate that forms the inner layer of the tympanic membrane. In crocodilians it forms a complex structure on the inner surface of the tympanic membrane.

The connection between columella and extracolumella is sometimes a more or less flexible joint (as in crocodilians and a few lizards) but more often is a simple union, often strengthened by an overlay of dense connective tissue.

A middle ear muscle is present in gekkonid and pygopodid lizards, in crocodilians, and in *Bipes biporus* among the amphisbaenians. Though no direct evidence is available on its function, by analogy with higher animals it seems likely that this muscle in its contractions applies tension and friction to the columellar system and reduces sound transmission to the inner ear. The frequent observation of transient reductions in cochlear potentials, especially for intense sounds, in lightly anesthetized geckos lends support to this assumption.

The service of the middle ear mechanism in sound transmission is treated quantitatively for typical species of lizards in Chapter 6.

INNER EAR

The labyrinthine system of vertebrates is a complex of six to eight sensory receptors, most of which serve equilibrial functions, but this system as a whole is commonly referred to as "the ear" or "the auditory labyrinth." Similarly the bony capsule in which these organs are contained is known as the otic or auditory capsule, and many other parts have names implying an auditory function. The reason for this auditory bias is historical: the early anatomists who developed the terminology were aware of only one function, that of hearing.

The labyrinth of reptiles is located toward the posterior end of the skull, lateral to the brain, and is enclosed in a capsule formed mainly by prootic and opisthotic bones, with some contribution from the exoccipital, often fused with the opisthotic.

The labyrinth as a whole consists of an inner portion, the membranous or endolymphatic labyrinth, made up of a series of interconnected tubes and sacs within which the sense organs are contained, and an outer portion, the perilymphatic labyrinth, consisting of supporting tissues along with a series of fluid-filled spaces.

The Membranous Labyrinth. Fig. 3-3 shows the membranous labyrinth of the lizard *Iguana tuberculata*, which Retzius considered to present the most elegant form of all the reptiles that he examined. This system of interconnected ducts and sacs is often considered in two divisions, superior and inferior. The superior division includes the three semicircular canals, the utricle, and the endolymphatic duct and sac. The inferior division includes the saccule and the cochlear duct, with the latter containing two more or less distinct portions, the cochlea (in a narrow sense) and the lagena.

The Superior Division. — The three semicircular canals embrace the remaining structures. Each canal is a narrow duct that begins and ends in the utricle, and has a bulbous expansion, or ampulla, at one end. The anterior and posterior canals unite to form the common crus that extends downward, first connecting with the end of the lateral canal and then giving off two branches that are considered to be parts of the utricle.

The utricle consists of a main part, which is a broad tube running from the common crus to a junction with the paired ampullae of anterior and lateral semicircular canals, and a sharply bent portion that is confluent with the common crus and extends posteroventrally as the posterior sinus. These relations are more readily seen in Fig. 3-4, where the utricle is represented in isolation except for the superposition of the saccule in outline. The expanded anterior end of the utricle adjacent to the paired ampullae is designated as the utricular recess, and contains the macular endorgan. A connection with the saccule is made through a narrow utriculosaccular duct, which extends laterally from the utricle at a point near its sharp flexure.

The endolymphatic duct arises from the medial wall of the saccule

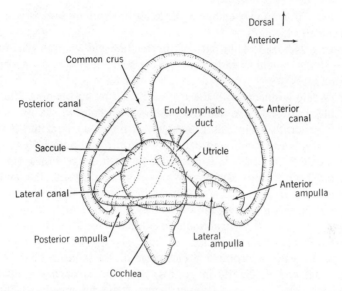

FIG. 3.3. Form of the membranous labyrinth in the lizard *Iguana tuberculata*. The right ear is seen in a lateral view. After Retzius, 1884.

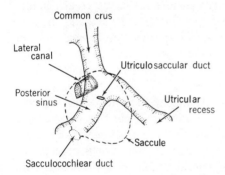

FIG. 3-4. The utricle as in the preceding figure, with an outline of the saccule superimposed.

(as shown in Fig. 3-3), swings beneath the main part of the utricle and runs dorsally alongside the common crus, finally expanding as the endolymphatic sac (only partially indicated in this figure). This sac extends into the brain cavity and ends in close relation to the choroid plexus of the fourth ventricle. The sac contains calcareous granules, sometimes in considerable quantities.

The Inferior Division. — The saccule in lizards is rather highly placed, but is nevertheless included in the inferior division. Its position lateral to the utricle as shown in Fig. 3-3 is characteristic of lizards and snakes, but in amphisbaenians it is anterior, and in other reptiles is more ventrally located.

Covering a considerable part of the medial and ventral wall of the saccule is the macular endorgan, which bears a relatively enormous statolithic mass.

The saccule connects with the cochlear duct by an opening (the sacculocochlear duct) whose size and position vary somewhat with species, but more generally in lizards (as shown in Fig. 3-4) is at the posteroventral end of the saccule and extends ventrally and anteriorly.

The cochlear duct of lizards is a chamber with the general form of an inverted pyramid, and contains two sensory endorgans, the auditory papilla and the lagenar macula. The auditory papilla occupies the more dorsal and posterior part of the cavity, and the lagenar macula spreads over the anterior and ventral walls and often covers the lower part of the posteroventral wall as well.

In species with a prominent elevation of the anterior part of the limbic plate, and especially in geckos, this duct is partially separated into lagenar and cochlear compartments (as seen in Fig. 3-15). In snakes there is a constriction of the middle portion of the wall of the cochlear duct itself (well depicted by Miller, 1968), that makes this separation more obvious. In *Sphenodon* the separation is nearly complete.

The relative size of cochlear and lagenar areas varies greatly among the reptiles. In general the cochlear portion of the duct and the area occupied by the papillar structures are larger than the lagenar portion, and this is especially true in crocodilians. In a few lizards and nearly all the snakes the reverse is true: the lagena occupies the major portion of the cochlear duct (cf. Evans, 1936).

The Labyrinthine Endorgans. — Among the reptiles there are eight labyrinthine endorgans, contained in the different parts of the membranous labyrinth as just described. These include three macular organs: one each in the utricle, saccule, and lagena; three crista organs: one each in the anterior, posterior, and lateral ampullae of the semicircular canals; and two others commonly called papillae: one in the cochlea named the auditory papilla and another in the wall of the utricle, or close by in the common crus, named the papilla neglecta.

Of these organs only one, the auditory papilla or cochlear organ, is definitely known to serve the sense of hearing. All the others most likely are equilibratory, though for the lagenar macula and the papilla neglecta the evidence concerning function is scanty. The principal basis for excluding from an auditory function all but the cochlear organ is indirect but nevertheless telling: it is that the auditory papilla is located in a path of vibratory fluid flow and is provided with accessory structures that facilitate a stimulation of its hair cells by fluid vibrations. No other labyrinth organ is so situated.

The equilibrial organs vary in function according to their mechanical construction. The macular organs consist of a plate of hair cells whose ciliary tufts are attached to a fine network or layer of tectorial tissue. Overlaying this network is a mass of dense crystalline material in a thin tissue matrix usually characterized as gelatinous (Fig. 3-5). Therefore these organs respond most readily to changes of head orientation and sudden displacements. The crista organs consist of a mound of hair cells whose cilia extend into an elevated body, the cupula, formed of tectorial material (Fig. 3-6), and this cupula extends as a valve-like barrier across one region of a continuous fluid loop. This loop is formed by a semicircular canal together with the cavity of the utricle. When the head is rotated in the plane of the loop, the inertia of the contained fluid imposes pressure on the cupula, causing it to bend, and in this bending the hair cells at the base of the cupula are stimulated.

The papilla neglecta has been little studied, and its function is not well understood. The structure varies; in some forms (as in amphisbaenids) this organ contains a mass of statolithic crystals and is properly regarded as a macula; in others, including most of the lizards, it lacks such loading and more nearly resembles a crista (Figs. 3-7, 3-8).

The lagenar macula deserves special consideration in this relation because it has often been regarded as a possible auditory receptor. This role for it had no doubt been suggested by its presence in the cochlear duct in close physical relation to the auditory papilla. Some further evidence was offered by Weston (1939) and Hamilton (1963), who observed that the lagenar and cochlear nerve fibers run to the same area of the brain, from which they inferred that these organs have a common auditory function.

The lagenar hair cells, however, lie on the stationary wall of the cochlear duct, remote from the path of vibratory fluid flow that passes through the basilar membrane. Though some small amount of fluid pressure can be presumed to extend to these hair cells, the pressure will be exerted upon them from all sides, and not in such a way as to produce a displacement of these cells or of the foundation on which they lie. Sound pressure as such is not stimulating, as extensive experiments have shown (Wever and Lawrence, 1954), but only such pressure that sets up motion in the hair cells, and more specifically one that produces a relative motion between cell body and ciliary tuft.

Under the conditions existing for the lagenar hair cells, the level of such relative motion must be far below threshold limits for all acoustic stimuli within the normal physiological range. To be sure, at extraordinary levels of sound exposure these lagenar hair cells, and many other

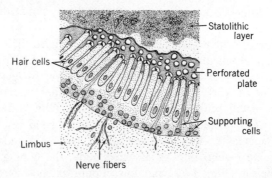

FIG. 3-5. A portion of the saccular macula in a specimen of *Iguana iguana*. The tectorial layer is a perforated plate in the openings of which the ciliary tufts of the hair cells protrude. Scale 375×.

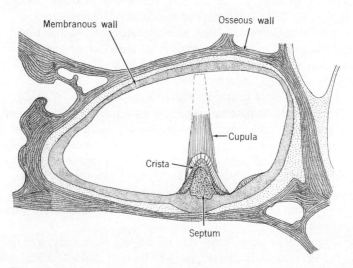

FIG. 3-6. A crista organ in the lizard *Iguana iguana*. The cupula in this specimen extended only partway to the roof of the cavity, no doubt because of shrinkage in the preparatory process. Its probable form is indicated by the broken outline.

sensory cells within the labyrinth and elsewhere, might be set in activity, but the result most likely would not be hearing but postural and other disturbances. Such disturbances in man we know to be the result of severe overstimulation by sounds.

The similarity of brain pathways for lagenar and cochlear fibers evidently represents a continuation of their close peripheral relationships, and does not necessarily indicate any community of function.

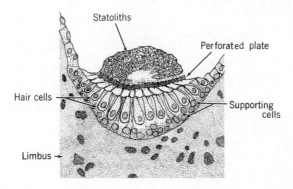

FIG. 3-7. The papilla neglecta in the amphisbaenian *Diplometopon zarudnyi*. The tectorial layer is a perforated plate, and is surmounted by a dense mass of statolithic crystals. Scale 400×.

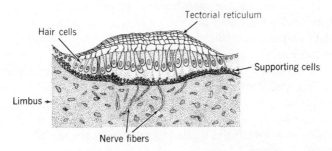

FIG. 3-8. The papilla neglecta in the
lizard *Iguana iguana*. Scale 250×.

The innervation of the labyrinthine sense organs was well worked out by Retzius (1884) and varies only slightly among the reptilian groups. The eighth nerve forms two principal branches, designated as anterior and posterior. The anterior branch sends out a small twig to the crista of the anterior ampulla and another to that of the lateral ampulla, and supplies several bundles of fibers to the utricular macula. In some forms at least, as in turtles and crocodilians, this branch also contributes a small nerve to the saccular macula. The posterior branch has a more extensive role. It contributes a large trunk to the saccular macula, and an even larger one runs as the cochlear nerve, which divides its fibers into a main radiation to the auditory papilla and a smaller group of bundles to the lagenar macula. This posterior branch also sends out a long twig that runs to the posterior crista, and this twig on its way gives off a thin strand to the papilla neglecta. This innervation pattern is shown in highly schematized form in Fig. 3-9.

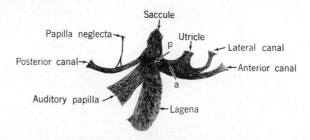

Saccule

Papilla neglecta

Utricle

p

Posterior canal

Lateral canal

Anterior canal

a

Auditory papilla

Lagena

FIG. 3-9. Divisions of the eighth nerve of *Iguana tuberculata*; *a*, anterior branch; *p*, posterior branch. The labels indicate the structure served. Based on Retzius, 1884.

The Perilymphatic Labyrinth. The perilymphatic labyrinth is a complex series of fluid-filled spaces and tissue networks contained in the otic capsule, with a portion extending outside. It surrounds the epithelial structures constituting the membranous labyrinth and maintains these structures in position. The supporting tissues are of two kinds, delicate strands of connective tissue that serve as anchoring threads or trabeculae but more often have the form of delicate open networks, and relatively thick masses of the same or similar material known as limbic plates. The trabeculae and tissue networks are primarily suspensory, and the limbic plates provide more rigid support in critical places. The most prominent of these plates is the cochlear limbus, which contains an opening over which the basilar membrane is suspended.

The more compact form of perilymphatic tissue seen in the limbic plates was often referred to by early anatomists as "cartilage" or "spindle cartilage," but histologically it bears no close relationship to this form of connective tissue. It is different in detailed structure and in staining qualities, and evidently is peculiar to the ear. Miller, Kasahara, and Mulroy (1967) carried out a number of histochemical and ultrastructural tests of limbic tissue from several species of lizards and concluded that this material is distinctive and bears no close relation to cartilage.

A special fluid, the perilymph, fills all available spaces, including the meshes of the reticular tissue. These spaces form a connected series of recesses and passages that constitute the perilymphatic labyrinth in a strict sense. This labyrinth deserves our special attention because it provides the route over which fluid vibrations set up by sounds are channeled through the ear and bring about a stimulation of the cochlear hair cells.

The perilymphatic system varies in form in the different species. The most extensive comparative study of the system was made by de Burlet (1934), who presented a number of pictorial representations for the various vertebrate groups, one of which for the lizard *Hemidactylus* is reproduced in modified form in Fig. 3-10.

In his conception of the perilymphatic system de Burlet made the assumption that only the open fluid pathways are traversed by fluid vibrations, and the reticular tissues are impervious to these vibrations. This view obfuscates the situation, for a fluid-filled network, in which the tissue strands are thin and flexible and have about the same density as the fluid itself, will be practically as transparent to sound as a clear fluid. The acoustic boundaries are only the solid walls of bone and dense limbic tissue, and the specific paths of vibratory fluid flow will be determined by these and by the areas of input and relief of acoustic pressure.

A more complete conception of the perilymphatic system and its relations to endolymphatic structures can be obtained from a series of cross sections, shown for *Iguana iguana* in Figs. 3-11 to 3-13.

The perilymphatic cistern, seen in Figs. 3-11 and 3-12, is a relatively large recess located medial to the stapedial footplate. This space as it extends anteriorly along the cochlear duct is known as the scala vestibuli. In the region just medial to the basilar membrane is the scala tympani. This scala in Fig. 3-11 appears only as a shallow trough in the

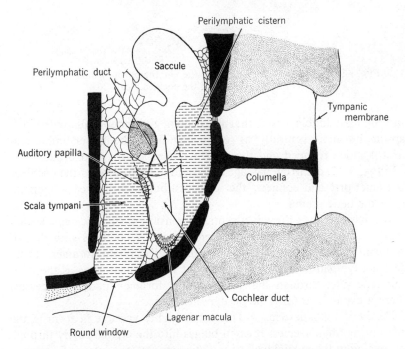

FIG. 3-10. The perilymphatic labyrinth of the lizard *Hemidactylus*, after de Burlet, 1934a.

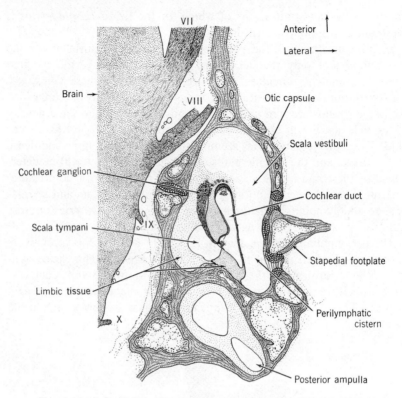

FIG. 3-11. A frontal section through the ear region of *Iguana iguana* at a level near the middle of the cochlea. Scale 20×.

mass of light limbic tissue that runs along the inner wall of the otic capsule, but more ventrally, as seen in Fig. 3-12, the lighter limbic material is much reduced and the scala tympani is expanded broadly.

In Fig. 3-12, a narrow passage appears that loops around the cochlear duct anteriorly and connects the scala vestibuli with the scala tympani; this is the helicotrema.

At this same level there is a ventral extension of the otic capsule that forms the recessus scalae tympani, and its cavity connects with that of the main part of the capsule through the perilymphatic foramen. From the scala tympani a short tube, often not well defined, runs as the perilymphatic duct through the perilymphatic foramen into the recessus, where it expands to form the perilymphatic sac.

The Perilymphatic Sac. — This sac is highly variable in form in the different reptilian species. It often bulges into the cranial cavity through an opening in the medial wall of the otic capsule (the jugular foramen), and comes in contact with an elaborated portion of the arachnoid mem-

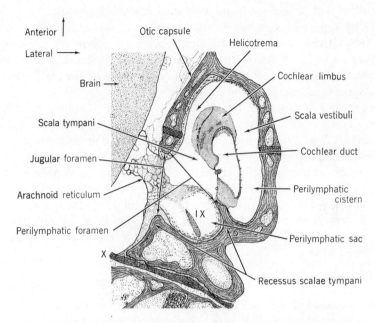

FIG. 3-12. A frontal section through the ear region of the same specimen as in the preceding figure, at a more ventral level. Scale 20×.

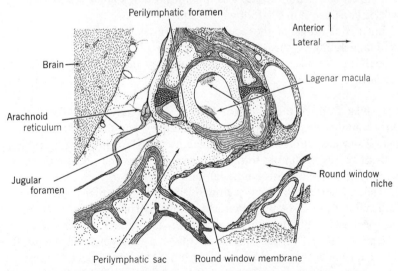

FIG. 3-13. A frontal section through the ear region of the same specimen as in the preceding figure, at a still more ventral level. Scale 20×.

brane, which in *Iguana iguana* forms a complex network whose meshes are filled with a homogeneous dark-staining material (indicated in Fig. 3-12 by stippling). Sometimes also a portion of this arachnoid network extends laterally and posteriorly, often seeming to follow the glossopharyngeal nerve in its course through the perilymphatic sac, and forms a membrane that walls off the main part of the sac from the scala tympani (as appears in Fig. 3-12).

A further extension of the perilymphatic system is still to be considered. Toward the ventral end of the otic capsule, as shown for *Iguana iguana* in Fig. 3-13, a wide opening appears in the lateral wall. This opening is the round window niche, and bulging into it is a portion of the epithelial lining of the middle ear cavity, which is thinned over its inner, free aspect to form the deeper layer of the round window membrane. The cavity of the niche, like the middle ear space, is air-filled. As will be noted presently, the round window membrane in many species constitutes a region of pressure relief in the response of the ear to sound waves.

THE REPTILIAN COCHLEAR DUCT

Many features of the cochlear duct have already been noted in the general discussion of the labyrinth. The central location of this structure between the stapedial footplate in the oval window and the pathway leading to the round window is significant for sound stimulation and will be considered in the following chapter.

Further features of the cochlear duct now to be examined are the epithelial lining of this duct, Reissner's membrane, the tectorial membrane, the basilar membrane, the auditory papilla, and the lagenar macula.

Epithelial Cells of the Cochlear Duct.   As already noted, the cochlear duct contains two sensory areas, the auditory papilla and lagenar macula. Elsewhere along its inner surface its walls are lined with epithelial cells of relatively simple form. The sensory areas may be considered to be specializations of this epithelial lining.

The epithelial cells vary regionally in form, and their distribution is indicated in Fig. 3-14 by the circled numbers. The part of the epithelium just posterior to the lagenar organ, shown as region 1, consists in *Iguana iguana* of entirely simple squamous cells in a uniform arrangement. The squamous cells sometimes are a little more randomly placed in the area of transition between regions 1 and 2. Then, farther posteriorly as the limbic mound is approached, these covering cells become orderly once more and gradually assume a cuboid or cylindrical form. This layer runs over the peak of the mound, usually changing to a taller cylindrical form

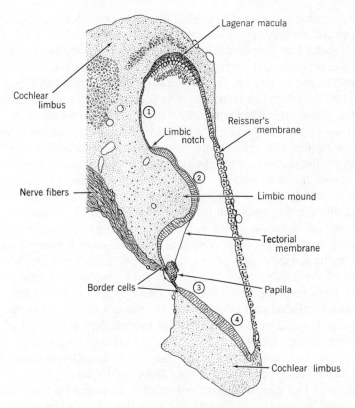

FIG. 3-14. The cochlear duct of *Iguana iguana* in frontal section, at a
level near the middle of the cochlea. Scale 75×.

(region 2). These cells reach their maximum height along the slope of
the limbus, and then become nearly cubic once more as they approach
the auditory papilla.

At the papilla this line of cells ceases and then resumes on the other
side. However, those cells immediately adjacent to the papilla, in this
and nearly all other species, differ from the other cells in appearance:
they do not stain as densely as the other cells, but appear clear and
translucent; these are designated as border cells.

Posterior to the papilla, in region 3, the limbus is covered by a relatively thick layer of columnar cells with randomly placed nuclei. In the
specimen illustrated this is considered to be pseudostratified epithelium;
more ventrally these cells become less closely packed, more orderly, and
can be followed clearly through the whole layer.

In this species much the same cell structure continues in region 4,
but in some other species the cells over this posterior end of the limbus
are specialized and show evidence of secretory function.

The epithelial lining of the cochlear duct was examined in some detail in the gecko *Eublepharis macularius*, and is shown in Fig. 3-15. This lining is formed by low squamous or cuboidal epithelium anteriorly and along the anterolateral surface of the limbic lip (regions 1 and 2). Toward the end of this lip, on its lateral side, the cells enlarge and become cubic and then low columnar. The columnar cells continue around the edge of the lip, and then rapidly flatten and continue along the undersurface of the lip (region 3) as particularly thin squamous cells. Over the part of the limbic plate that approaches the auditory papilla (region 4) the epithelium rapidly thickens, so that a moderately heavy layer covers the area anterior to the papilla. Adjacent to the papilla (anteromedial to it) are two rows of border cells, and three or four rows of these cells appear on its posterolateral side. Thereafter, in region 5, the epithelium is pseudostratified, consisting of columnar cells of increasing height. This form continues to the edge of Reissner's membrane in the dorsal region of the cochlea, but more ventrally, toward the end of the papilla (as shown in Fig. 3-15, region 6), it changes to a lower form made up of goblet cells. These cells have broad free surfaces and a rounded part (the bowl of the goblet) in which the nucleus is contained, and a long slender process that extends to the bed of tissue below. These stem processes often take relatively straight courses, but sometimes are curved or slanted, and end in a moderate expansion. Surrounding these stem processes are large fluid spaces.

The tissue bed over which the goblet cells lie is not the usual limbic plate found at more dorsal levels, for here the limbic plate is invaded by a body of tissue of much less density and with different staining properties. Many clear spaces appear around elongated cells that mostly run parallel to the free surface, and posteriorly the spaces increase until finally the tissue becomes a reticulum containing scattered nuclei. A small blood vessel (of about 30 μ diameter) extends throughout this body of tissue. It seems likely that the goblet cells have a secretory function, probably through interaction with the special tissue below.

Reissner's Membrane.   In most ears, as shown for *Iguana iguana* in Fig. 3-16, Reissner's membrane consists of two layers, of which the one on the scala vestibuli side is often so thin as to appear under the light microscope in cross section as only a fine line except in places where it is elevated by the nuclei of its constituent cells. This layer may be noticeably thicker, however, toward the two ends of the membrane.

The second layer, lining the cochlear duct, consists usually of cuboidal or short columnar cells, sometimes in an orderly array, especially near the middle of the membrane, but more often with variations in cell size and shape and in the locations of the nuclei. Accordingly, the

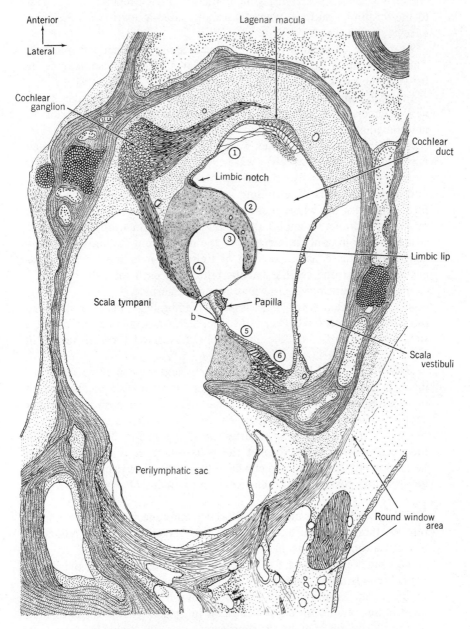

Anterior

Lateral

Lagenar macula

Cochlear
ganglion

Limbic notch

①

②

③

④

Scala tympani

b

Papilla

⑤

⑥

Perilymphatic sac

Cochlear
duct

Limbic lip

Scala
vestibuli

Round window
area

FIG. 3-15. The cochlear duct of *Eublepharis macularius* in frontal section.
b, border cells. Scale 50×.

cochlear surface of Reissner's membrane is often irregular, in contrast to the vestibular surface, which is usually smooth apart from the presence of capillaries.

A number of capillaries may be seen along Reissner's membrane. These sometimes appear on the vestibular surface and sometimes lie between the two cell layers.

In some reptile species, notably among the xantusiid lizards, Reissner's membrane is relatively thick, especially toward its anterior or anteromedial edge.

In the crocodilians Reissner's membrane becomes greatly thickened at its medial edge, especially toward the dorsal end of the cochlear duct, and then more dorsally still this portion of the membrane is thrown into numerous folds. The folding increases toward the dorsal end, and finally includes the whole membrane as shown in Fig. 3-17. This form of the membrane resembles that found regularly in birds, where this structure is commonly referred to as the tegmentum vasculosum. Characteristic of this convoluted form of Reissner's membrane is the presence of at least one capillary at the bottom of each fold, though occasional capillaries appear elsewhere along the surface.

The crocodilian form of Reissner's membrane appears to take an intermediate position between the simple form seen in most reptiles, such as the lizards, and the elaborately convoluted form occurring in birds.

It is likely that Reissner's membrane, and especially its cochlear layer, is the seat of secretory activity for the formation of endolymph or of essential components of this fluid. The presence of a prominent capillary network supports this idea. Further evidence is the presence of signs of active secretion. In turtles, as shown in Fig. 3-18, bubbles of fluid can sometimes be seen at the ends of the cells of the cochlear layer. Sometimes also these cells have little masses of deep-staining material in this position that seem to be the remains of exudates.

The Tectorial Membrane.   A tectorial membrane is present in all the reptiles, though in two lizard families, the Scincidae and Cordylidae, it is vestigial and plays no part in hair-cell stimulation.

This membrane appears to arise from a particular area of the limbic epithelium. In most lizards this origin is most likely the region of the limbic notch, for often the membrane shows a root process and attachment in this region. In many species, as in *Iguana iguana* as represented in Fig. 3-14, this membrane runs close over the limbic surface, well beyond the peak, and then leaves the surface and extends freely to the auditory papilla. At the papilla this membrane connects with the ciliary

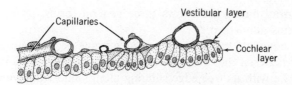

Fig. 3-16. A portion of Reissner's membrane in the lizard
Iguana iguana. Scale 300×.

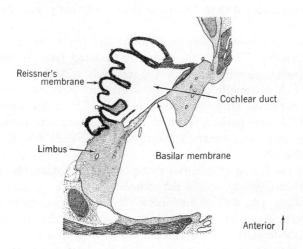

Fig. 3-17. A section through the ear region of *Caiman crocodilus*
showing the convoluted form of Reissner's membrane.

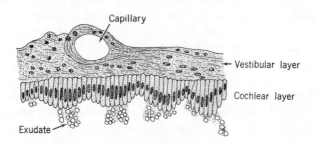

Fig. 3-18. A portion of Reissner's membrane in the turtle
Chelonia mydas. Scale 250×.

tufts of at least some of the hair cells either directly or through intermediary structures.

The tectorial membrane shows three general forms, apart from the vestigial forms alluded to.

(a) The complete type of tectorial membrane occurs as a thin sheet of reticular material that arises from the limbic epithelium and extends outward to lie with its free edge running along almost the whole of the auditory papilla. A portion of this structure is shown for the lizard *Gekko gecko* in Fig. 3-19. As in the majority of lizards, this membrane for the most part is exceedingly thin, but in this family (as also in pygopodids) it has a prominent thickening in a strip just off the limbus and running parallel with it, known as the spindle body from its form in cross section.

(b) Another common form of tectorial membrane, occurring in the iguanids, agamids, and others, is the abbreviated type. It also arises from the limbic epithelium, where it is relatively wide, but after leaving this surface it is rapidly reduced in width and ends at the auditory papilla as a narrow strip. As shown in Fig. 3-20, which represents a specimen of *Iguana iguana*, this terminal strip attaches to a small plate of tectorial tissue that covers a few rows of hair cells near the middle of the papilla.

(c) The third type of tectorial membrane, called dendritic, is found within the lizard family only in the chameleons, but is the standard type in crocodilians (as well as in birds). This membrane in one of the chameleons is shown in Fig. 3-21. It arises on the limbic epithelial surface as a moderately narrow ribbon, which first forms a thicker and more compact bundle in which numerous fascicles can be recognized, and as it leaves the limbic surface and approaches the auditory papilla it subdivides further into separate threads, each of which attaches to the end of a ciliary tuft. Sometimes, as shown here, a portion of the tectorial ribbon separates and runs to the side of the papilla. In crocodilians this type of membrane is thicker and wider, extending all along the cochlea and running out from the limbic epithelium as a laminar structure that grows progressively thinner as it gives off fibrous strands to the hair cells.

The tectorial membrane consists of a reticular material of a peculiar type that stains in a distinctive manner. The same staining characteristics are found in tectorial plates, finger processes, sallets, and culmens (to be described in the next chapter), and hence all these structures appear to be tectorial derivatives, probably produced by the same group of cells during embryonic growth.

In its most common form, the tectorial membrane consists of a curtain of extremely thin, weblike material. Usually the strands of the net-

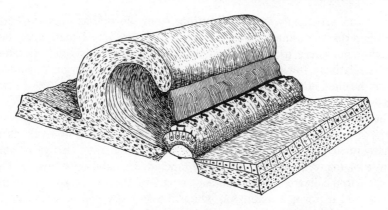

FIG. 3-19. A portion of the inner ear of the lizard *Gekko gecko*, with the complete form of tectorial membrane. From Wever, 1967b.

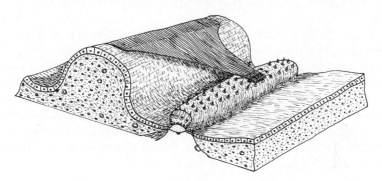

FIG. 3-20. The inner ear of *Iguana iguana* showing almost the whole of the auditory papilla. At the lower edge of the tectorial membrane is the short tectorial plate. From Wever, 1967b.

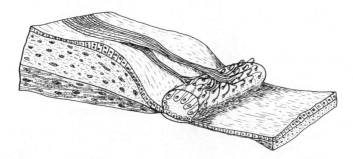

FIG. 3-21. The inner ear of a chameleon, *Chamaeleo c. calcarifer*, showing the dendritic form of tectorial membrane. From Wever, 1967b.

work run principally in one direction, from the limbus outward, with fewer strands running randomly crosswise. The outer margin of this membrane where it approaches the auditory papilla is often stiffened by a relatively thick rod running through it, or this margin is bent to form a rolled edge.

The Basilar Membrane. The basilar membrane of reptiles varies greatly with species in size and shape. Some membranes are short ovals as in many turtles and some of the snakes, most are moderately extended, and those of many geckos are long and tapered. Usually, as shown in Fig. 3-22 for the lizard *Iguana iguana*, a part of the membrane is greatly thickened to form a fundus on which the auditory papilla is borne. In snakes the fundus is lacking, and in turtles it is particularly thin.

In most reptiles with long, tapered basilar membranes the variation in width is only moderate, of the order of two- or threefold, but in a few the variation is as much as sixfold, equaling that of advanced mammals.

Under the light microscope the basilar membrane has a fibrous appearance. This structure has been little studied with the electron microscope, but the evidence available indicates the presence of microfibrils, usually without the cross banding characteristic of collagen (Mulroy, 1968a). The structure is regarded as similar to that of the limbus, and this whole supporting mechanism is thought to have developed in the embryo from mesenchyme and not from ectoderm (Shute and Bellairs, 1953; Mulroy, 1968a; Baird, 1970a). The fundus differs from the remaining tissue of basilar membrane and limbus in that it is relatively clear and homogeneous.

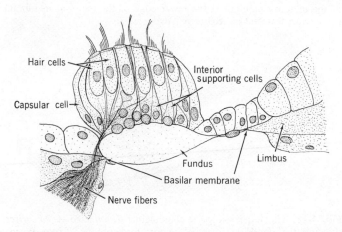

FIG. 3-22. The basilar membrane and auditory papilla of *Iguana iguana* in cross section. Scale 500×.

The Auditory Papilla.   As shown in Fig. 3-22 for *Iguana iguana*, the auditory papilla consists of a sensory portion, the organ of Corti, resting on the thickened fundus segment of the basilar membrane. The sensory structure is made up of two types of cells, hair cells and supporting cells.

Most of the supporting cells have their cell bodies resting on the fundus, others are at a higher level and send base processes to the fundus or to other cells. These cells form columns that extend upward, some to constitute the walls of the papillar capsule and others to produce an internal palisade and roof structure by which the hair cells are suspended. A special study of the supporting cells and their columnar processes was made in a number of specimens by the use of Held's phosphomolybdic acid-hematoxylin stain, which brings out these cells conspicuously. In the relatively simple papilla of *Urosaurus ornatus* the supporting structure has the form shown in Fig. 3-23. The cells at the edges of the papilla produce thin, flat sheets (seen as fine columns in cross section) that run over the outside of the structure to its top surface as the papillar capsule. The interior cells have broad bases on the fundus from which they send up slender columns that expand near the roof of the capsule into conical wedges between the upper ends of the hair cells. Figure 3-24 presents a more elaborate papillar framework as seen in a specimen of *Cnemidophorus tessellatus aethiops* in which for some unknown reason the hair cells were lacking.

It is notable that, in contrast to the condition seen in the vestibular sense organs, in which the hair cells are closely surrounded by the supporting cells, the hairs cells of the auditory papilla are largely free along their lower portions except for the presence of nerve endings.

Nerve fibers could often be seen coming from the cochlear tract through the limbus, losing their medullary sheaths at the edge of the fundus, then passing between the border cells and the fundus to enter the papillar capsule. In this passage upward along the edge of the limbic lip the naked nerve fibers come together in a series of tight bundles, evidently able to penetrate the barrier only at the junctions between border cells and fundus and between adjacent cells of the capsular wall. This area of penetration is known as the habenula perforata. After entering the papillar capsule, the fibers run obliquely to the lower ends of the hair cells.

Occasionally a nerve ending could be seen at the base of a hair cell, but details of this innervation are beyond the resolution of the light microscope.

Hair Cells and Their Cilia. — The hair cells are relatively large cells of globular form with the free end of each cell bearing a ciliary tuft that consists of a number of stereocilia in a pattern of graduated lengths, and

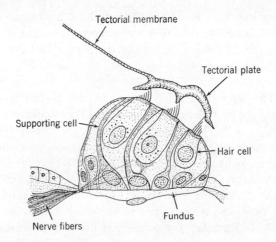

FIG. 3-23. The auditory papilla of the lizard *Urosaurus ornatus*, stained by Held's method.

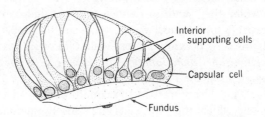

FIG. 3-24. The papillar framework in a specimen of the lizard *Cnemidophorus tessellatus aethiops* that lacked hair cells.

a single kinocilium at one side adjacent to the longest stereocilia. The ciliary tuft in a section that cuts across the array, as in Figs. 3-22 or 3-23, is usually seen as containing 6 or 7 rows of cilia, though this number varies with species and also between individual tufts of a single specimen. The cilia are arranged in a characteristic pattern, most readily seen in an end-on view, or when the section passes through the tuft parallel to the free surface.

Such an axial view of the hair tuft is presented in Fig. 3-25 for a specimen of the alligator lizard *Gerrhonotus multicarinatus*. The ciliary pattern varies a little even for adjoining hair cells, but most generally in this species there are 9 or 10 rows in an area of roughly oval form, with a notch on one side that contains the kinocilium.

The stereocilia, as shown, occur in slightly curving rows, and are spaced in these rows in a regular pattern so that crossing rows are formed. In the example given, there are 10 rows in each direction, with the number of cilia in a row reduced at the edges, giving a total in this instance of 76 stereocilia.

The stereocilia appear to be interconnected in these rows, confirming earlier reports of lightly staining "intervening material" between the cilia in the more common cross-sectional views (Wever, 1965a).

That the cilia are interconnected within the tufts, and that these interconnections have sufficient strength to maintain the tuft in a particular form and integrity is further evidenced by observations on specimens that have been subjected to mechanical stresses. Especially revealing are experiments in which damage has been produced in a somewhat controlled manner by overstimulation with sounds.

After such overstimulation the most commonly observed damage is a break between the ciliary tuft and the tectorial membrane or other tectorial structure to which it normally attaches. This separation often involves the fine fibrous extension of the tectorial membrane or tectorial plate that connects to the kinocilium (or to the kinocilium plus a few of the longer stereocilia) or else the break occurs in the tip ends of these cilia themselves.

At times, however, the break is between the ciliary tufts and their hair cells, and then the tectorial membrane usually retracts a little and remains above the papilla with the ciliary tufts clinging to it. Such a condition is represented in Fig. 3-26 for a specimen of the gecko *Hemitheconyx caudicinctus*. In such instances the ciliary tufts that remain attached to the tectorial membrane all maintain their normal form.

This maintenance of form by the ciliary tuft in the presence of damaging stimulation would not be expected if the individual cilia were independent or were in only light contact with one another. There must be strong surface cohesion, or possibly a cementing substance between them, to withstand the intense vibratory forces.

In these instances in which the cilia are detached from the hair cell, the break point is clearly at a level close to the hair-cell surface, where each cilium presents a sharp constriction. This form is shown for a single cilium in part *d* of Fig. 4-18. When examined with the light microscope the cilium is often invisible at this point. Miller (1973a) measured the diameter of the stereocilia of *Gekko gecko* as 0.24 μ at their bases, increasing to 0.5 μ or more at higher levels.

It is of course more convenient to examine the cilia and their spatial patterns with the aid of the scanning electron microscope, as Mulroy (1968a), Miller (1973a, b), and others have done with a number of reptilian species. The pattern characteristic of the lizard *Gekko gecko* as observed by Miller (1973a) is represented in Fig. 3-27. The stereocilia form a regular array, with the single kinocilium at one side. There is a 'V' of 5 stereocilia adjacent to the kinocilium and partially enclos-

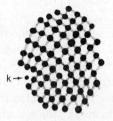

FIG. 3-25. The ciliary pattern in a hair cell of the lizard *Gerrhonotus multicarinatus*. *k*, kinocilium. Scale 3000×.

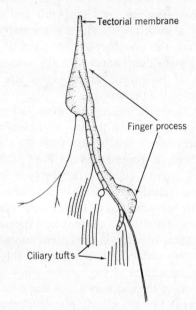

FIG. 3-26. Ciliary breaks in a specimen of *Hemitheconyx caudicinctus*. The ciliary tufts are detached from their hair cells but remain connected to the finger process of the tectorial membrane.

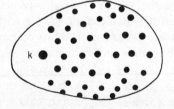

FIG. 3-27. The ciliary pattern in *Gekko gecko*. *k*, kinocilium. After Miller, 1973a. Scale 7320×.

ing it, and five other 'V's of 7 stereocilia each that are farther away and spaced so that the rows are precisely aligned. The heights of the cilia are greatest in the 'V' adjacent to the kinocilium, and progressively less in the more distant rows, varying in this species from 7 to 2.6 μ in height according to Miller's measurements. The kinocilium is usually a little taller than the five tallest stereocilia, and appears often to be embraced by these. In this, as in other gecko species and in some others, the kinocilium ends in a large bulb. The number of stereocilia varies somewhat, because partial 'V's may occur; Miller found the average number in *Gekko gecko* to be 42. In other species studied the number of stereocilia varies from about 32 in *Coleonyx variegatus* and *Eublepharis macularius* to 100 in *Ameiva ameiva* (Miller, 1973b).

The Lagenar Macula.   Still to be described is the macular organ of the lagena, which in *Iguana iguana* extends over nearly three-fifths of the length of the cochlear duct. A cross section near the middle of this extent has been shown in Fig. 3-15, and further details are given in Fig. 3-28. This endorgan has the same basic structure as the other maculae in utricle and saccule: a plate of sensory cells is overlaid by a tectorial network, and over this is a statolithic mass.

The sensory plate consists of tall columnar hair cells held in a framework of supporting cells whose cell bodies have a deep position, mostly resting on the limbus, but some standing a little higher with their base processes extended to the limbus. All send up relatively thin columnar processes that run between the hair cells and expand to clasp these cells at their upper ends. Each hair cell bears a tuft of cilia consisting of a single kinocilium and several stereocilia of graduated length, so that the end of the tuft has a tapered form.

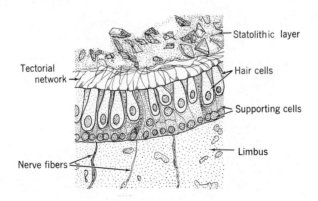

FIG. 3-28. A portion of the lagenar macula in
Iguana iguana. Scale 500×.

The sensory arrangement in this and other macular organs is closely similar to that of the auditory papilla, but the ciliary tufts of the macular organs are more slender and appear to contain fewer and generally finer stereocilia. The similarity continues in that here also the tips of the ciliary tufts are embedded in tectorial tissues. The form of these tectorial tissues varies with species; and it varies also in some degree among the different macular organs, tending to be more elaborate in the saccular maculae, simplest in the utricular maculae, and of intermediate character in the lagenar maculae. In the lizard *Iguana iguana* the tectorial tissues of the lagenar macula have the form of a loose network to the meshes of which the ciliary tufts make their connections.

Overlaying this tectorial reticulum and closely connected with its upper surface is a statolithic layer consisting of numerous crystals usually considered to be a form of calcium carbonate, scattered in random fashion within a matrix commonly characterized as gelatinous but doubtless of very complex composition.

Contrary to a frequently expressed opinion, there is never a direct contact between the statolithic mass and the sensory cells themselves; a tectorial network always intervenes. This rule holds for all types of labyrinthine endorgans as well as the auditory receptors throughout the vertebrate series (Wever, 1973e).

4. SOUND TRANSMISSION TO

THE COCHLEA AND

THE STIMULATION PROCESS

In the reptiles the sound stimulation process reaches a high level of mechanical efficiency, because here for the first time in the evolutionary series a basilar membrane lies athwart the vibratory fluid pathway through the cochlea and serves for the sensing of acoustic vibrations. When this flexible membrane is caused to move by displacements of the fluid particles, the sensory structures borne by the membrane are involved also, with a stimulation of their sensory cells.

THE VIBRATORY CIRCUIT

The vibratory fluid flow through the cochlea in response to the action of sounds is accomplished in the reptiles in two ways, because there are two methods by which the cochlear fluid can be mobilized. One method, employed in most lizards and all the crocodilians, and the more familiar because it is continued among the birds and mammals, involves the round window as a path of pressure relief. When at the positive phase of the action of a sound wave the stapedial footplate is caused to move inward in the oval window and exerts pressure on the cochlear fluid, this fluid can undergo displacement because of the presence of the round window in one wall of the fluid chamber. The thin and flexible membrane covering this window permits an expansion by bulging into the air cavity of the middle ear space. This mode of fluid mobilization is represented in Fig. 4-1. The pressure discharge for the positive phase of a sound wave, in which there is an inward thrust of the columella, follows a course through the cochlea from columellar footplate to round window membrane as indicated by the arrows. A model of this system is presented in Fig. 4-2. Here the plunger corresponds to the ossicular chain, the sensing membrane represents the basilar membrane with its sensory structures, and the terminal membrane corresponds to the round window membrane.

The bypass as shown in this figure represents the heliocotrema, and deserves special notice. It is generally supposed that this bypass has the

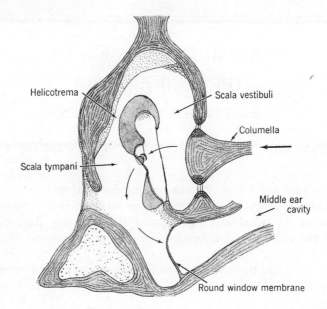

FIG. 4-1. A frontal section through the inner ear of the lizard *Iguana iguana*. Arrows indicate the path of fluid flow in response to the positive phase of a sound wave.

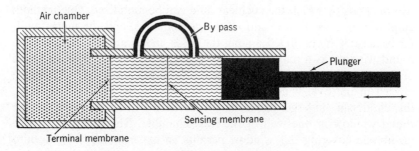

FIG. 4-2. A model of the round window system of cochlear mobilization. After Wever, 1969c.

basic function of equalizing the fluid pressures in scala vestibuli and scala tympani, preventing any accumulation of fluid on one side of the basilar membrane that would distend and perhaps disrupt it. This bypass obviously reduces the ear's sensitivity, but the amount of this loss is unknown. Clearly the effect varies with frequency. At very low frequencies the impedance of the bypass is slight compared with that of the basilar membrane. Accordingly a large proportion of the fluid movement is through the bypass, and the ear's sensitivity is impaired.

The helicotrema therefore imposes a lower limit on the ear's frequency range.

At higher frequencies the impedance of the bypass increases, because the friction along the walls of this constricted channel increases with the velocity of the wave motion. At the same time the impedance through the basilar membrane decreases, especially in the frequency range in which the moving structures exhibit some degree of mechanical tuning. Therefore the effect of the bypass most likely becomes negligible within the principal part of the ear's working range.

THE REENTRANT FLUID CIRCUIT

The second method of fluid mobilization, found in turtles, snakes, amphisbaenians, *Sphenodon*, and a few of the lizards, employs a reentrant fluid circuit. In these animals the round window is absent or (as in certain lizards of the genus *Phyrnosoma*) its outer surface is covered with fluid and rendered relatively immobile. These ears contain a fluid passage that leads from the inner boundary of the cochlea by a circuitous path outward to the lateral face of the columellar footplate, as represented for the turtle in Fig. 4-3. Therefore an inward movement of the footplate produces a fluid displacement that involves not only the cochlear pathway but the complete circuit back to the footplate, as shown by the arrows in the figure. A reversal of the footplate movement of course reverses this action. Sounds therefore set up a churning motion in this circular body of fluid, with the basilar membrane standing across the path in the cochlear portion of the circuit. A model of this arrangement is shown in Fig. 4-4.

The system serves its purpose well at low vibratory frequencies, but a burden is imposed on the cochlear action at high frequencies because of the considerable mass of fluid that must be moved and the friction encountered within the fluid itself and along its confining walls. The impedance imposed on the motion of this fluid column increases very rapidly with the velocity of the motion, which itself increases as the square of the sound frequency.

THE HAIR-CELL STIMULATION PROCESS

A hair cell is stimulated through a relative motion between its ciliary tuft and the body of the cell. More specifically, this stimulation appears to involve the kinocilium in relation to some portion of the cell body, perhaps its cuticular plate (Flock and Wersäll, 1962; Wersäll, 1967). According to prevailing theory, the cell normally is electrically polarized, and this polarization is reduced, producing a stimulating effect, when the single kinocilium in the hair tuft is displaced in a direction

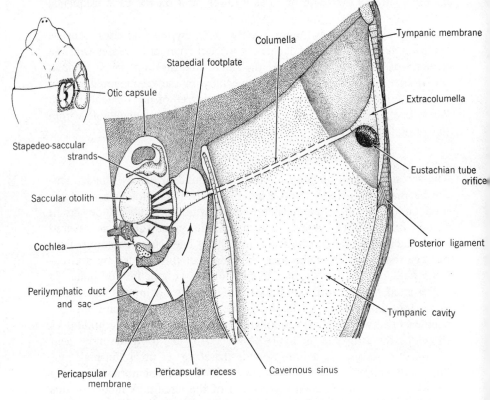

FIG. 4-3. The reentrant fluid circuit in a turtle. Arrows indicate the fluid flow produced by an inward thrust of the stapedial footplate. After Wever and Vernon, 1956b.

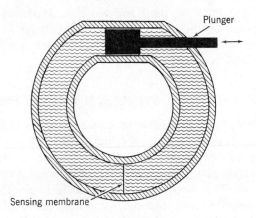

FIG. 4-4. A model of the reentrant fluid system of cochlear mobilization. From Wever, 1969c.

away from the group of stereocilia in the tuft (Fig. 4-5). A contrary motion increases the existing polarization, and has an inhibitory effect. Consequently an alternating motion of the kinocilium will produce one stimulation per cycle, and cells with a common ciliary orientation will be stimulated in phase.

According to this hypothesis, hair cells on the basilar membrane will be stimulated when their cell bodies undergo displacement while at the same time their ciliary tufts are somehow restrained. Such restraint takes many forms.

METHODS AND MECHANISMS FOR CILIARY RESTRAINT

The most common form of ciliary restraint is that imposed directly or indirectly by a tectorial membrane.

Direct Tectorial Connections.   Immediate attachments of the ciliary tuft of a hair cell to a tectorial membrane may sometimes be found, though they are not very common. In certain instances this connection seems to be made only by the elongated kinocilium itself, but more often a few of the longer stereocilia run close to the kinociiium, probably with at least tenuous attachments to it, and appear to share the connection.

A direct attachment to the rolled edge of the tectorial membrane in the gecko *Hemidactylus turcicus* is shown in Fig. 4-6.

This direct form of tectorial attachment occurs only in a few species and in certain regions of the papilla, more often in geckos toward the dorsal end.

Simple Fiber Connections.   Sometimes thin fibers make the connections between tectorial membrane and ciliary tuft, as shown on the left in Fig. 4-7.

Fiber connections occur in all the crocodilians, in which, as mentioned above, the tectorial membrane breaks up into strands that run outward along the hair-cell layer, with one strand attaching to every ciliary tuft. The arrangement in chameleons has already been seen in Fig. 3-21.

Connections by Fibrous Strands.   A second form of fibrous connection is by relatively thick strands that extend over the hair cells, connecting usually to the ciliary tufts of all the hair cells of a transverse row. Such a strand is shown for the lizard *Uma notata* in Fig. 4-8.

Connections by Finger Processes.   A more elaborate type of connection was first observed in *Gekko gecko,* and is characteristic of the gecko

Resting stage · Stimulation · Inhibition

FIG. 4-5. The kinocilium in hair-cell stimulation.

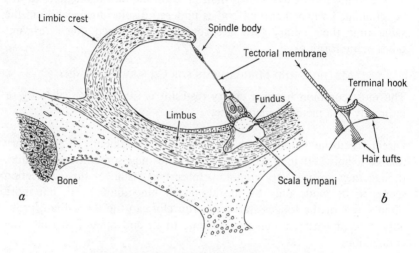

FIG. 4-6. Direct tectorial connections in the lizard *Hemidactylus turcicus*. From Wever, 1967a.

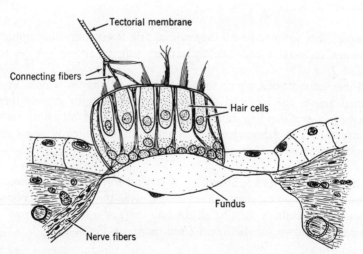

FIG. 4-7. Simple fiber connections for three cells in the auditory papilla of *Iguana iguana*, and free-standing cilia for three others. From Wever, 1967b.

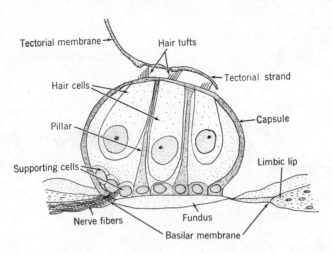

FIG. 4-8. Ciliary connections by a fibrous strand in the lizard *Uma notata*. From Wever, 1965a.

ear. In this group of lizards, described in detail in a later chapter, the form of ciliary connection varies regionally. Over the dorsal segment of the papilla the connections to the ciliary tufts are made by the tectorial membrane directly or by fine fibers extending from its edge. Throughout the ventral segment there is a division between medial and lateral hair cells along the papilla: the medial ones have tectorial connections and the lateral ones are served by sallets, as will be described presently.

The connections to the more medial hair cells are made by a series of processes that arise from a prominent thickening that runs like an embedded rod along the edge of the tectorial membrane. These processes extend like little fingers, one to each transverse row of hair cells, each making connections to the hair cells in its own row. Usually the fingers exhibit prominent swellings equal in number to the hair cells in the row, with a spinous process on each swelling to which the tip of the ciliary tuft connects. This structure in *Gekko gecko* is pictured in Fig. 4-9.

Further Intermediary Structures; The Tectorial Plate.   In many lizards and all the turtles, snakes, and amphisbaenians, and also in *Sphenodon*, a special intermediary structure is present to make the ciliary connections. In lizards this structure takes a number of forms from a simple fibrous network to a heavy plate. In all instances the longest cilia of the hair tufts are connected to the undersurface of the intermediary structure, and this structure is attached at some point to the tectorial membrane.

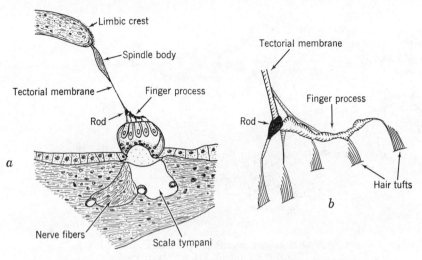

FIG. 4-9. A finger process in *Gekko gecko*. From Wever, 1967a.

These intermediary structures in lizards form a continuous series as indicated in Fig. 4-10, varying from a simple network of fine fibers through coarser networks to thick, substantial plates. These plates take a variety of forms, among which the one shown in Fig. 4-11 for the monitor lizard, *Varanus griseus*, is one of the most bizarre.

In general the tectorial plate is restrained by an attachment by way of the tectorial membrane to the medial limbus. Usually this connection is direct, but in many of the iguanids (as described in detail in Chapter 7) the tectorial membrane takes a wide looping course, and the direct restraining effect seems to be limited.

In turtles, snakes, amphisbaenians, and also in *Sphenodon*, the intermediary structure is highly standardized, and always consists of a tectorial plate with deep pits on the underside (the side facing the hair cells), and the ciliary tufts can usually be seen to extend into these pits to attach either to their side walls or their deepest recesses. Figure 4-12 shows the form of this plate in the turtle *Deirochelys reticularia*.

INERTIAL AND INERTIA-LIKE RESTRAINT

Another form of restraint of the ciliary tufts is produced through the action of inertial forces or other forces operating in a similar manner.

Inertial Bodies; The Sallets. It has long been known that the macular organs of all the vertebrates include a statolithic mass whose inertia in the presence of gravity or a sudden displacement is instrumental in their stimulation. This mass is effective in the stimulation process because

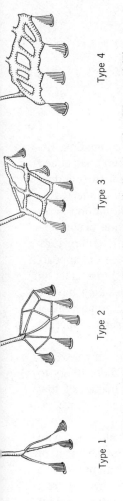

Type 1   Type 2   Type 3   Type 4

Fig. 4-10. Four types of intermediary structures in lizards. After Wever, 1967b.

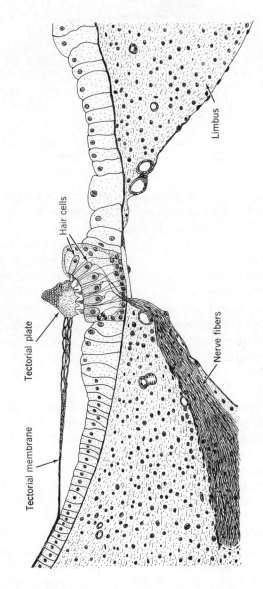

Fig. 4-11. The auditory papilla of *Varanus griseus* in cross section, showing an unusual form of tectorial plate.

Tectorial membrane

Tectorial plate

Hair cells

Limbus

Nerve fibers

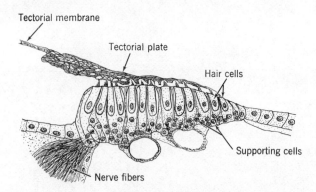

Fig. 4-12. The tectorial plate in a turtle,
Deirochelys reticularia.

the specific gravity of the statoliths exceeds that of the surrounding tissues and fluids, so that an applied force will have a differential effect, producing a smaller displacement of this mass than of surrounding structures. Because the cilia of the macular hair cells are connected to the statolithic mass through a tectorial network, a relative motion is set up between these cilia and their cell bodies, and stimulation occurs.

In the inner ears of many lizard species, resting on the ciliary tufts of the hair cells are small bodies known as sallets that have the appearance of inertial masses. Though these bodies do not contain calcium carbonate particles or other obvious weighting materials, they have a compact structure that perhaps gives them a density exceeding that of the fluid in which they lie. If this is the case they can act as inertial bodies as the macular statoliths do.

These bodies were first discovered in the lizard *Gekko gecko*, where they have the form indicated in Fig. 4-13. The term "sallet" refers to their resemblance, in this species, to an ancient French helmet of that name. In other species the form varies widely, as well as the size, as shown in Fig. 4-14.

Among the geckos, as already indicated, the sallets occur along with the tectorial membrane system of direct restraint, and some form of combination of these two systems occurs in several other families. The geckos and pygopodids are peculiar in that both these systems are present side by side, as Fig. 4-15 indicates, over the whole ventral segment of the auditory papilla.

Commonly the sallets form a long chain, with connecting processes consisting of strands of tectorial tissue giving an appearance much like a string of beads as indicated in Fig. 4-16. For the most part there is one sallet for each transverse row of hair cells, though in some species

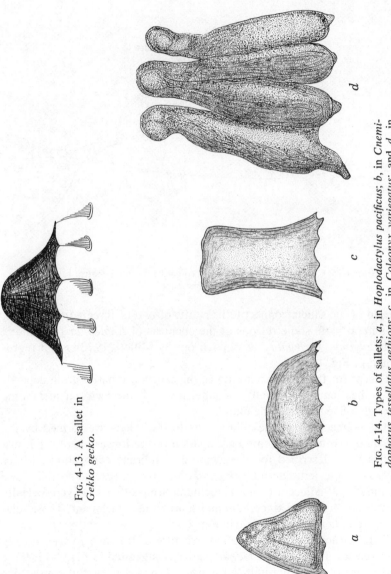

Fig. 4-13. A sallet in
Gekko gecko.

Fig. 4-14. Types of sallets; *a*, in *Hoplodactylus pacificus; b,* in *Cnemidophorus tessellatus aethiops; c,* in *Coleonyx variegatus;* and *d,* in
Gonatodes sp. Scale 1500×.

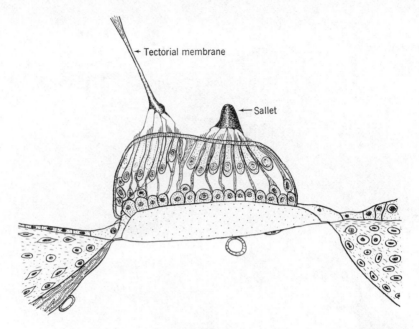

FIG. 4-15. Ciliary connections in *Gekko gecko* over the ventral segment of the auditory papilla.

certain of the sallets connect to hair tufts of two or three adjacent rows. This happens in some regions of the cochleas of *Eublepharis macularis* and *Coleonyx variegatus*, as pointed out by Miller (1973b) and represented in Fig. 4-17.

Except for their interconnections, the sallets ordinarily are independent bodies, but occasionally a connection with the tectorial membrane through a fine fiber may be found.

The connections of the ciliary tufts to the sallets are formed by the kinocilium, often with some participation of the longest stereocilia, just as has been described for the tectorial membrane connections. Miller (1973a) indicated that in *Gekko gecko* the five longest stereocilia, those forming a 'V' adjacent to the kinocilium, seem often to have especially close relations to the kinocilium and share in its attachment. These relations are indicated in parts *a-c* in Fig. 4-18.

Sallet Structure. — The internal structure of the sallets varies greatly with species. In some, like *Gekko gecko* as suggested in Fig. 4-13, these bodies are of relatively uniform density throughout, or the consistency varies in only moderate degree. Some minor variations are indicated in Fig. 4-14 as striations along the outer surfaces.

Other sallets are less unitary, and appear to be composed of several parts closely joined, as in *d* of Fig. 4-14. In the skinks the structure is

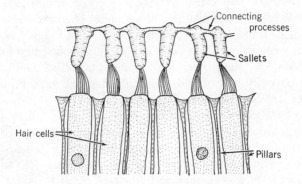

Fig. 4-16. A portion of the chain of sallets in *Gekko gecko* in a longitudinal section, showing the hair-cell connections. Scale 1250×.

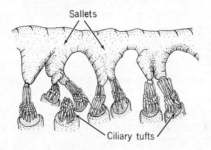

Fig. 4-17. Sallets from a region of the auditory papilla of *Eublepharis macularius* showing partial twinning. Drawing based upon a photomicrograph by Miller, 1973b.

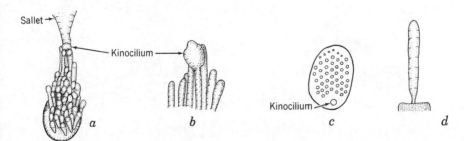

Fig. 4-18. Cilia and their connections in *Gekko gecko*. *a*, a ciliary tuft and its attachment to a sallet; *b*, the longer stereocilia surrounding the kinocilium; *c*, the hair-cell surface with its ciliary pattern; *d*, a single stereocilium in a lateral view. Drawings based upon photomicrographs by Miller, 1973a.

still less compact, and often appears as a loose assemblage of tissue strands, with each strand made up of a number of coarse fibers in a thin matrix. These strands appear to correspond at least roughly to the number of transverse rows of hair cells, and are mainly oriented in a "vertical" direction, i.e., along extensions of the longitudinal axes of the hair cells. The fibers in these strands, though somewhat wayward and intertwined, are largely oriented in the same direction.

The most elaborate sallet structure is found in the sallets of gerrhosaurids and cordylids. Fig. 4-19 shows sketches of a sallet of *Gerrhosaurus v. validus* in cross section, at *a* with the focus on the more peripheral portion of the structure and at *b* with the focus near the center. Coarse fibers form a strengthening framework along the sides and over the top of the mound, though the center remains free of them. Many of the peripheral fibers extend into the basal projections to which the ciliary tufts are attached.

A more extensive study was made of the sallets in *Cordylus warreni depressus*, a species for which a series of sections was prepared that passed longitudinally through the line of sallets. In this species, as appears to be true for cordylids in general, the sallets are so closely connected as to form a continuous band of tissue with only moderate expansions at each transverse row of hair cells to mark the constituent elements.

The interconnections between sallets are made by fibers that arise in dense networks of tectorial tissue near the base of the structure, the "basal knots" in Fig. 4-20. These fibers, indicated as longitudinal fibers in this figure, appear to extend only as far as the basal knot of the sallet immediately adjoining either dorsally or ventrally. Two other kinds of fibers arise from the basal knot, some running peripherally over the peak of the sallet (arcade fibers), and others that extend out into the basal projections of the sallet (skirting fibers). Figure 4-21 gives a transverse

FIG. 4-19. Two views of a sallet of *Gerrhosaurus v. validus*, representing at *a* a section through the periphery and at *b* one through the middle.

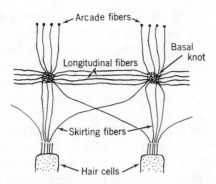

FIG. 4-20. A schematic representation of the basal knots and their outgoing fibers in two adjacent sallets based on sagittal sections in a specimen of *Cordylus warreni depressus*. A dot at the end of a fiber indicates that here the fiber turns into a plane perpendicular to the drawing.

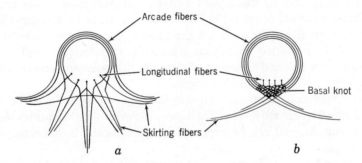

FIG. 4-21. Two sectional representations of a sallet in *Cordylus warreni depressus*; *a*, cut across near the periphery, and *b*, cut through the middle. The view is at right angles to that of the preceding figure. Again the dots at the ends of fibers indicate a course out of the plane of the drawing.

view of these fibers to show more clearly their relations to the sallet structure.

As already suggested, if the sallets have a density appreciably greater than that of the endolymph in which they lie, they can provide inertial restraint on the ciliary tufts of the hair cells. If this is not the case, then we must look for other ways in which these bodies can produce the restraint necessary for hair-cell stimulation.

Sallets and Radiation Forces.   Acoustic theory relating to the radiation of sound in air by a loudspeaker diaphragm driven by its voice coil will be called upon in the present situation to account for the behavior of a hair cell surmounted by a sallet and made to vibrate in a fluid medium. The loudspeaker diaphragm may be considered as a vibrating piston that is radiating sound energy into the medium. Accordingly, the system incurs an energy loss. Conceptually this loss can be regarded in an alternative manner as the result of the imposition of a counterforce (reaction force) acting on the vibrating piston (Kinsler and Frey, 1962; Wever, 1971c).

If we consider the sallet as taking the place of the piston in this situation, and the basilar membrane and hair cell correspond to the voice coil of the loudspeaker, then a reaction force can be considered to act upon the sallet so as to restrain its movements under the influence of a stimulating sound. This effect is shown schematically in Fig. 4-22.

A further consideration is necessary here. Both sallet and hair cell are radiating vibratory energy into the medium in this situation, and both are subject to restraint by reaction forces. If these forces were equal for the two bodies they would behave alike, no relative motion would arise, and no stimulation would occur. But these forces are not equal: the surface of the sallet is much the larger, and the reactive force acting upon it will be so much the greater that the force on the hair cell can be neglected altogether. Relatively, therefore, the sallet is restrained, the hair cell undergoes the greater displacement, and the ciliary tuft is bent so that stimulation results.

The radiation reactance that has just been described has the same kind of effect as an inertial reactance; it varies with frequency in the same manner, and can be expressed mathematically in the same form.

Sallets and Viscous Forces.   Endolymph has an appreciable viscosity, and the vibratory movements of the cochlear structures encounter a viscous force. This force is especially great for the sallets with their broad surfaces, and therefore viscous forces operate along with the reaction forces to restrain the sallet movements.

We have identified three types of possible restraining forces for cochlear structures, operating most effectively on the sallets with their considerable surface and bulk. If these bodies have a density above that of the fluid in which they lie, then all three types of force will operate jointly in restraining these structures and producing a relative motion between them and their hair cells. If no density difference exists between the sallets and the surrounding fluid then reaction force and viscosity remain to produce the necessary stimulating effect.

The Culmen.   The culmen is a tectorial body with a general resemblance to a sallet but relatively of enormous size, found in four lizard families, the Scincidae, Cordylidae, Xantusiidae, and Gerrhosauri-

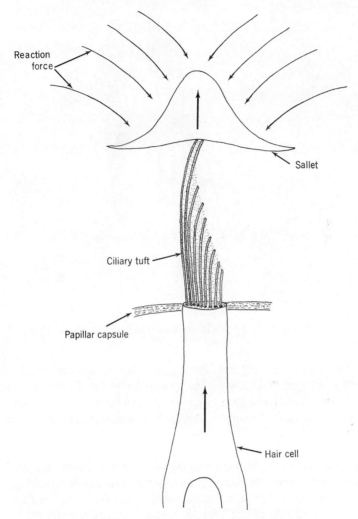

FIG. 4-22. The reaction force generated by a sallet when
vibrating in a fluid. From Wever, 1971c.

dae. As Fig. 4-23 shows, this body consists of two parts, a base portion
of compact form that sends out a number of jutting processes over the
papilla and an upper portion of light material containing many openings large and small, giving it a frothy appearance.

In skinks and cordylids this body stands free, and evidently imposes
restraint on the large group of hair cells over which it lies through the
operation of inertia and inertia-like principles such as govern the action
of sallets.

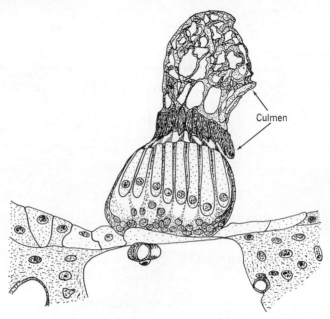

FIG. 4-23. A culmen in the skink *Chalcides c. striatus.*
From Wever, 1970b.

In xantusiids and gerrhosaurids this structure has much the same form, but its upper end is attached to a tectorial membrane. The action in these ears, therefore, must combine the inertial type of restraint with the direct mechanical restraint imposed by the tectorial membrane attachment.

Free-standing Cilia.   In a great many lizard species, including iguanids, agamids, and others, the tectorial membrane makes connections through a tectorial plate with only a limited number of hair cells, sometimes located in the middle of the cochlea and sometimes toward the ventral end. The remaining hair cells, making up the large majority of the population, are free-standing. The problem arises as to how these hair cells are stimulated.

It was first noted that the hair cells outside the area covered by the tectorial plate bear cilia that are particularly long, often four to six times as long as the ones beneath the plate. Figure 4-7 shows a degree of variation of these cilia in one of the iguanid lizards. Also it was found in some species that a delicately staining amorphous material is distributed along these outlying hair cells, lightly embedding their ciliary tufts and extending to the tectorial plate and sometimes also to the limbic region at both ends of the papilla. This embedding material may link these structural elements and impose a degree of restraint on the ciliary tufts. Figure 4-24 represents this embedding material as it appears in *Iguana iguana*.

Other species of iguanids and agamids, however, seem to lack this embedding material, and to present hair cells with no immediately obvious means of ciliary restraint.

For the stimulation of these completely free-standing hair cells we can invoke the reaction force hypothesis or the viscous force hypothesis, or both. The ciliary tuft itself may be considered as a piston, though an exceedingly small one, that is being driven by the cell body, and that encounters viscous forces from the surrounding fluid and radiates acoustic energy into this fluid. The reaction and viscous forces generated in this manner will be of small magnitude, yet may be sufficient to stimulate the cell when the amplitude of basilar membrane movement is great.

That a large difference exists in the level of stimulation of hair cells beneath the tectorial plate and others outside it was evidenced by experiments on specimens of the iguanid *Crotaphytus collaris* in which sounds of extreme intensity were presented. In some of these animals whose cochleas were examined histologically the hair cells beneath the tectorial plate showed early stages of overstimulation injury, whereas those outside this plate remained entirely normal.

STIMULATION BY RELATIVE DISPLACEMENTS

A theory of hair-cell stimulation developed long ago by ter Kuile (1900) with particular relation to the mammalian cochlea is still to be considered. Ter Kuile postulated that the movement of the cochlear fluids un-

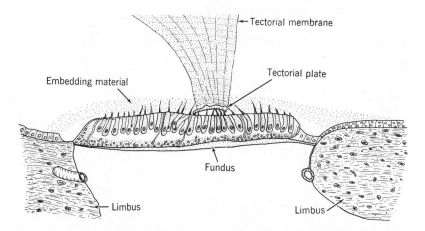

FIG. 4-24. The auditory papilla of *Iguana iguana* in a longitudinal view. A tangle of fine material is represented by stippling. From Wever, 1967a.

der the influence of sound causes the basilar membrane and tectorial membrane to execute movements about separate axes of rotation, as illustrated in Fig. 4-25. The basilar membrane, attached along both edges, moves up and down with the greatest amplitude at its midregion. The tectorial membrane, anchored only along one edge, swings about this line of attachment with its outer portion free. The tips of the ciliary tufts, embedded in the undersurface of the tectorial membrane, must follow its angular path of motion. The bodies of the hair cells, on the other hand, must conform to the up-and-down movements of the basilar membrane. Because these modes of motion are different, a relative displacement is set up between cilia and cell bodies, and the result is a stimulation of the cells.

Ter Kuile emphasized the relative displacement between the tectorial membrane and the hair-cell surface, but others have pointed to a derivative of these displacements, namely, the shearing forces set up between them. Békésy (1960, pp. 485-500; 703-710) suggested that these shearing forces might serve to match the mechanical impedances of cochlear fluids and cellular structures, and that the system operates as a mechanical amplifier to enhance the sensitivity.

This relative displacement hypothesis applies with some plausibility to the relatively thick and rigid tectorial membrane of mammals, but seems less applicable to this membrane in most of the reptiles, in which

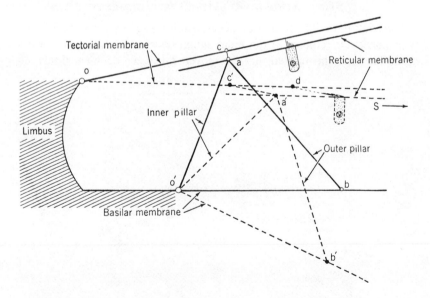

FIG. 4-25. A representation of ter Kuile's theory of the action of the tectorial membrane. The spatial relations at one instant are shown by the solid lines, and those an instant later by the broken lines. From Wever, 1971c.

it is thin and delicate, and whose manner of anchorage appears too tenuous and remote to determine a mode of movement along the lines of ter Kuile's hypothesis. Only in a few gekkonid species, the diplodactylines and *Thecadactylus rapicaudus*, has the tectorial membrane been found to present a noticeable thickening along its midportion, as shown in Fig. 4-26 and in Fig. 14-37 below. In these species it is possible that ter Kuile's hypothesis may apply.

INVERSE TECTORIAL SYSTEMS

The discussion so far has dealt with systems in which the body of the hair cells is set in motion by the basilar membrane while its ciliary tuft is restrained by one of several means. Of course the contrary process is equally effective: the ciliary tufts may be made to move while the cell body remains at rest. Indeed this manner of stimulation is the standard one for fishes and amphibians. In the amphibians, for example, the hair cells rest on a solid base, and their ciliary tufts are attached to tectorial tissues that are suspended in the cochlear fluid and set in motion by the vibratory stream. This "inverse" mode of action is so called only because it is less common in the higher types of ears and therefore less familiar.

An inverse mode of action is utilized in addition to the usual one in the ears of turtles, and also in birds. In these ears the auditory papilla

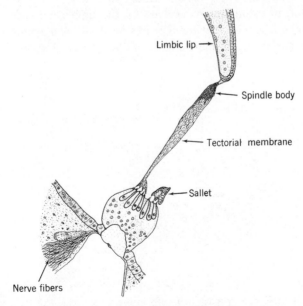

FIG. 4-26. The thickened tectorial membrane of the gekkonine *Thecadactylus rapicaudus*. Scale 200×.

extends well beyond the basilar membrane, with hair cells located on the medial surface of the limbus and in some turtle species on the lateral surface as well. The arrangement in the ear of the chicken turtle, *Deirochelys reticularia*, is shown in Fig. 4-27. These limbic hair cells are well innervated.

These cells, resting on a solid foundation, are not directly involved in the basilar membrane movements. It is more than likely, however, that they are indirectly involved, at least when the stimulation is strong. The ciliary tufts of these limbic cells are attached to strands of the tectorial membrane, and through this membrane they can receive vibrations immediately from the fluid stream, and also, and no doubt much more effectively, they can receive vibrations from the basilar membrane through the common connections with the ciliary tufts of the papillar hair cells. This stimulation process is represented schematically in Fig. 4-28. In this figure are pictured a limbic hair cell and one on the basilar membrane, each suspended between supporting cells, with a tectorial strand connecting the two ciliary tufts. At *a* these cells are shown in the absence of sound, and at *b* during stimulation when the basilar membrane undergoes a downward deflection. The tilting of the system on the right caused by the basilar membrane movement produces a displacement away from the limbic cell, and because of the tectorial linkage each kinocilium is pulled away from its stereocilia. This happens because the two ciliary tufts have contrary orientations as shown.

In general, the stimulation received by the limbic hair cells can be expected to be less than that for the cells on the basilar membrane, because the tectorial membrane along its midportion (as Fig. 4-27 shows)

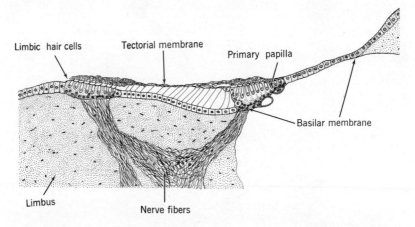

Limbic hair cells   Tectorial membrane   Primary papilla   Basilar membrane   Limbus   Nerve fibers

FIG. 4-27. Limbic and primary hair cells in the dorsal region of the cochlea in the turtle *Deirochelys reticularia*, with tectorial linkages between the ciliary tufts of the two groups. Scale 125×.

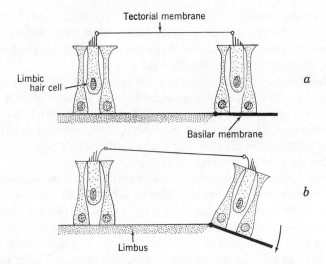

FIG. 4-28. A schema representing the manner of stimulation of the limbic hair cells. A resting stage is shown at *a*, and the effect of a basilar membrane movement caused by a sound is shown at *b*.

is connected by numerous oblique fibers to the epithelial surface of the limbus, and these anchoring strands will have a damping effect on the transmitted motion. The tectorial membrane can be expected to stretch a little, with a reduction of the transmitted motion. Nevertheless, when the stimulus is strong, the limbic hair cells should be brought into action.

As will be shown later, in most turtles a significant proportion of the hair cells is located on the limbus, and in a few species these outlying cells equal or even outnumber the others. Because, as just shown, the ciliary orientation is contrary in direction in the two groups of hair cells, the stimulation of the limbic cells at high levels will have an additive effect, producing an extension of the ear's dynamic range.

Limbic hair cells have been observed only rarely in reptiles other than turtles. However, other ears probably exhibit the inverse form of hair-cell stimulation also in some degree. For this inverse type of stimulation to operate for cells on the basilar membrane, two conditions must be met. The cochlea must display a significant measure of frequency discrimination: a given tone sets some particular area of the basilar membrane in motion while other areas are much less affected. Second, some means must be provided for the interconnection of the ciliary tufts of the more favored cells and the others. When the favored cells are set in vigorous action, the motion is transmitted secondarily along the interconnections to the other cells, which then join in the response.

Four possible means for this interconnection of ciliary tufts have been observed. The most common form of interconnection of areas over the basilar membrane is the tectorial membrane itself and its extensions in the form of fiber networks, strands, and plates.

Another form of interconnection is the intervening substance already described for certain iguanids and agamids, and considered as a means of extending the restraining influence of a tectorial membrane or of transferring such an effect from the limbus. This material could serve also as a route for the transfer of motion to the ciliary tufts of cells remote from the primary response area.

Another means of interconnection between areas of the basilar membrane is a tissue strand extending along an array of hair cells, sometimes over short distances but occasionally over a large part of the papilla.

A further manner of interconnection is found regularly along the rows of sallets in the geckos and others, as already mentioned. These sallets occur usually in well-defined rows, and those of each row are interconnected as shown in Fig. 4-16 for geckos and in Fig. 4-20 for cordylids. Ordinarily the effect of this connecting band is to stimulate the line of hair cells in concerted action, but over more remote regions, when there is an effective degree of differentiation, the action can be much more complex. There are numerous possibilities of facilitative and inhibitory effects in the action of tones through these cochlear interconnections.

It must be emphasized that the interconnection of areas of the basilar membrane as here considered will extend the region of hair-cell action in significant degree only when the stimulation is strong. At ordinary levels this effect will be slight, and whatever frequency differentiation exists for the cochlea will not be seriously impaired.

COMBINED STIMULATION SYSTEMS

In most lizard ears the stimulation of the hair cells involves two or more methods of ciliary restraint. This condition has been noted in the geckos and pygopodids, in which tectorial and sallet systems exist side by side, and also in lacertids (see Chapter 13), in which tectorial membrane connections are present over the middle portion of the dorsal papilla, with sallets serving a few cells at either end of this papilla and all the cells of the ventral papilla. In a number of families, such as the iguanids and agamids, there are tectorial connections only for the hair cells in one region of the papilla, usually the middle or ventral portion, and the remaining cells have free-standing ciliary tufts. Other combinations of stimulating mechanisms will be encountered later on; thus at least five forms of ciliary restraint can be identified in the teiids. The advantages of this multiplicity of stimulation systems will now be considered.

Sensitivity.   Clearly there are two ways of producing a relative motion between the body of a hair cell and its ciliary tuft: to move these two parts in different directions or to move one and hold the other stationary. The first of these methods is suggested in ter Kuile's theory of hair-cell stimulation in mammalian ears, but seems of doubtful application in lizards except possibly for certain of the geckos with thickened tectorial membranes. The second method is most commonly found, as in the tectorial system when the cell body is caused to vibrate and the ciliary tuft is restrained. More rarely, as observed for the limbic hair cells of turtles, the cell body remains stationary and the ciliary tuft is deflected by movements of the tectorial membrane.

The inertial and inertia-like systems of ciliary restraint would seem to be less effective than the tectorial system because in these indirect systems the restraint is never absolute: the sallet is free to vibrate in some degree along with the hair cells, though with a smaller amplitude.

It seems obvious that the sensitivity must be still less for those hair cells with free-standing cilia, with perhaps two degrees of effectiveness, one for those cells whose ciliary tufts are partially immobilized by intervening substances, and the lowest grade for those lacking such contacts and therefore truly free.

Evidence in support of these expected sensitivity variations among hair cells will be presented below in a description of experiments on the effects of overstimulation of lizard ears by sounds.

Frequency Relations.   Tectorial restraint, because it is achieved by direct mechanical coupling, is not expected to vary appreciably as a function of frequency, but restraint by sallets should vary greatly. Insofar as sallet restraint depends upon a mass reactance, it must increase as the square of the frequency. To the extent that this restraint also depends upon reaction force and viscosity it will also increase in the same manner as the frequency is raised. This increase in restraint will tend to sustain the response of the ear in the higher frequencies, and in some degree will compensate for other conditions, mainly in the operation of the middle ear mechanism, that produce a decline as frequency rises.

Specificity.   For purposes of tonal differentiation a high degree of specificity in the stimulation of the hair cells is desirable. This specificity is most effectively provided by tectorial systems in which restraint is imposed on individual hair cells by separate fiber connections. Such systems are found in a few of the lizards, especially in the dorsal segment of the auditory papilla of geckos, in the chameleons, and in most highly developed form in the crocodiles. More often fiber networks or tectorial plates connect to large numbers of ciliary tufts, so that the hair cells are stimulated in groups and specificity is of a low order.

Still less specificity is expected for the sallet and culmen systems. A given sallet usually serves a whole transverse row of hair cells, or the cells of two or three adjacent rows, and often these bodies are linked longitudinally by interconnections. The culmens of the scincomorphs connect with a large group of hair cells at the ventral end of the cochlea so that all these cells appear to operate largely in concerted fashion.

Summation.   The systems just mentioned that tend to bring in large numbers of hair cells in a general response, though working against a specificity of action, have the great advantage of increasing the sensitivity. Parts moving together do not generate the friction that they would if they moved separately. A summation of their effects adds to the excitation of the nerve fibers, and unison in the discharge of neural elements gives a more effective signal at higher levels.

It is apparent that the principles of specificity and summation are alternatives. A single stimulation system must represent some form of compromise between the highest sensitivity, which is achieved by broad summation, and the best discrimination of tonal frequency, which requires individual action of elements and minimum summation. The dilemma can be avoided by the use of different systems in separate regions of the auditory papilla, a stratagem that appears to be best realized in the geckos, in which the dorsal segment of the papilla seems best suited to specific responses and the ventral segment is more adapted to generalized activity. Other reptiles exhibit various degrees of compromise between the different alternatives. The variety of stimulating arrangements as found especially among the lizards thus appears to be of positive value in the hearing process.

Dynamic Range.   An important characteristic of an ear is its dynamic range: the variation in intensities between the least perceptible and the maximum that the structures can withstand without damage. This range in some ears is surprisingly great; in man it is of the order of 10 million times (140 db) in terms of vibratory amplitude. Little information is available concerning this range in other animals, but there is reason to believe that it is less in the lower vertebrates. In some of the geckos, such as *Teratoscincus scincus*, the range is particularly narrow, at least when these animals are anesthetized and probably the extracolumellar muscle is rendered inactive, for it is difficult to carry out routine tests of sensitivity in these specimens without risk of damage to the cochlea, even when working at lowest possible response levels.

Intense sounds injure the ear because of their large amplitudes, which cause displacements in some of the responsive structures that exceed their elastic limits. Disruption caused by sounds is most serious in tissues of complex structure, containing elements of greatly varying densities and elasticities, for some parts move relative to others and the normal connections are broken. Anchor points are lost, cell junctions are torn away, and layers are permanently separated. The most serious effects of sounds occur within the organ of Corti and in tectorial connections to the ciliary tufts, for these disruptions affect the stimulation process at the hair cell and also the capability of the cell to convey the effect of its stimulation to its nerve fibers.

Certain mechanisms in the ear reduce the amplitude of motion of cochlear structures and provide a degree of protection against overstimulation injury. The middle ear is such a mechanism. The ossicular chain serves as a reducing lever, so that the amplitude of motion presented to the inner ear is less than that at the tympanic membrane. This action of the middle ear has been studied most extensively in mammals, where the reduction is found to be of the order of two- to threefold. In reptiles it has been measured in one of the iguanid lizards, *Crotaphytus collaris*, where it is about twofold (Wever and Werner, 1970).

This amplitude reduction is advantageous both because it decreases the disruptive effect of loud sounds and because it operates, along with other mechanisms in the conductive system, to provide a better match between the impedance of the aerial medium and the impedance of the inner ear, and consequently produces a larger energy transfer.

A second mechanism for the increase of force in the process of hair-cell stimulation, with a concomitant reduction of amplitude, occurs in the tectorial system. The amount of this mechanical effect varies greatly in different species and to some degree in different regions of a given papilla. This amplitude reduction depends upon the angle at which the tectorial membrane or its fibrous extension meets the ciliary tuft. When the angle is close to the vertical as for many geckos (*a* in Fig. 4-9), there is little or no amplitude reduction; the ciliary tuft is subject to a pulling force directly proportional to the basilar membrane displacement. In this arrangement (and perhaps more generally) there is reason to believe that the tectorial membrane is normally in a state of stretch, and the vibratory movement of basilar membrane and hair cells produces an alternating increase and decrease in this stretching. The evidence for this view (shown below in Fig. 4-32) is that after its connection to the hair cell is broken the tectorial membrane retracts to about two-thirds to one-half of its normal width.

In many species, such as *Varanus*, the angle between tectorial membrane and ciliary tuft more nearly approaches 90° (Fig. 4-11), and then there is a great reduction in the relative amplitude; the ciliary tuft is displaced only slightly in a lateral direction as the hair cell moves up and down, and the force is amplified accordingly. When, as is more often the case, the angle is intermediate between these two extremes, there is a moderate reduction of ampltiude and increase of force.

A curious arrangement is found in many of the iguanids in which the tectorial membrane takes a wide looping course from its attachment on the limbus to its connection by way of the tectorial plate to the hair cells. This arrangement probably provides little or no amplitude reduction.

The vertical or gecko type of connection is especially subject to disruption by overstimulation; the iguanid type has not shown such mechanical injury.

OVERSTIMULATION EFFECTS

The effect of overstimulation with sounds was studied in three species of lizards (7 specimens of each) selected as representing different types of papillar structure. These species were *Crotaphytus collaris*, an iguanid with a tectorial membrane and tectorial plate toward the middle of the cochlea and hair cells with free-standing ciliary tufts elsewhere; *Coleonyx variegatus*, a gecko with a combination of tectorial and sallet systems; and *Trachydosaurus r. rugosus*, a skink with the sallet and culmen system. Also some earlier observations were made on four specimens of the gecko *Ptyodactylus hassselquistii guttatus* that are worthy of mention.

The stimulating tones were 500, 1000, 3000, and 4000 Hz, presented through a tube sealed over the meatus of the right ear at intensities up to 1000 dynes per sq cm and sustained for periods up to 5 minutes and sometimes longer. Immediately following the stimulation the effects were observed as reductions in cochlear potential sensitivity, and then after histological processing the ears were examined for changes in structure.

This series of experiments was not extensive enough to explore this problem thoroughly, but gives some general information about susceptibility to acoustic trauma in these three types of ears and indicates the nature of the resulting injury.

Crotaphytus collaris. Great variability was found among the collared lizard specimens in susceptibility to overstimulation. In several specimens only slight effects were seen. An animal exhibiting striking losses is represented in Fig. 4-29. The right ear of this animal was stimulated with a tone of 1000 Hz at a level of 10 dynes per sq cm for a period of 5 minutes. The resulting loss of sensitivity amounted to 64 db at 1000

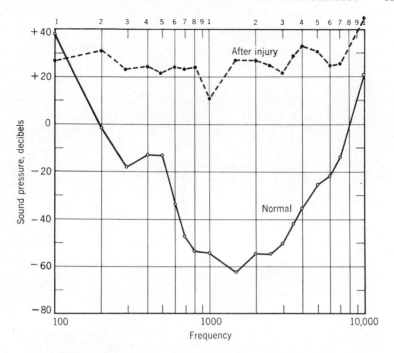

FIG. 4-29. Sensitivity functions for a specimen of *Crotaphytus collaris* before and after overstimulation with a tone of 1000 Hz.

Hz, and averaged 63.7 db over the range from 500 to 7000 Hz. A histological examination of this animal showed significant changes in the region of the tectorial plate, but seemingly normal conditions elsewhere. The changes consisted of extensive vacuolization of the nuclei of hair cells and of supporting cells, as shown in Fig. 4-30. Numerous small vacuoles were present in the hair-cell nuclei, especially in the three middle cells of the row of five occurring at this level of the papilla. The changes in the supporting cells involved the entire transverse rows, and consisted usually of the appearance of a single vacuole that occupied almost the whole basal portion of the nucleus. Exudate was often found also around the upper ends of the hair cells.

Ptyodactylus hasselquistii guttatus. In a specimen of the fan-toed gecko in which stimulation was carried to high levels, up to 300 dynes per sq cm for the tone 1000 Hz, a loss of sensitivity amounting to 58 db was noted. The histological picture showed a loss of all tectorial connections throughout the cochlea, but a retention of the sallets seemingly in normal relations to their hair cells.

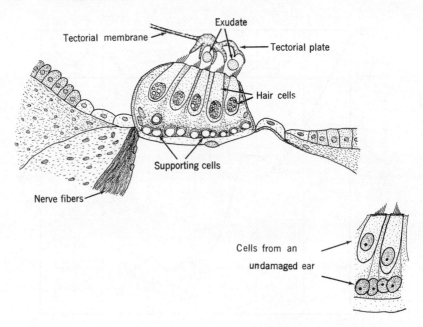

Exudate

Tectorial membrane

Tectorial plate

Hair cells

Supporting cells

Nerve fibers

Cells from an
undamaged ear

FIG. 4-30. Cellular changes in *Crotaphytus collaris* resulting from overstimulation with sound. Vacuoles are present in the nuclei of hair cells and supporting cells, and exudate appears above the hair cells. For comparison, the insert at the lower right of the figure shows a few cells from a corresponding region of the cochlea in a specimen that was not exposed to overstimulation.

Coleonyx variegatus. Specimens of the banded gecko all showed serious effects after overstimulation, both in sensitivity loss and in cochlear damage. An example of especially severe changes is given in Fig. 4-31, with the solid curve showing the normal sensitivity function and the broken curve the function after stimulating with 1000 Hz at a level of 10 dynes per sq cm for 5 minutes and then with 100 dynes per sq cm for 7 minutes. As shown, the sensitivity after the exposure suffered a general leveling around +15 to +20 db over the whole test range, with a loss of about 90 db in the region of best sensitivity between 700 and 2500 Hz.

Histologically this ear showed much the same picture as just indicated for *Ptyodactylus hasselquistii guttatus*: retention of the sallets in place and a detachment of the tectorial membrane over nearly the whole of the cochlea. As Fig. 4-32 shows, the detachment is at the ciliary tufts, and these tufts are left in position on their hair cells. The tectorial membrane is greatly retracted and shrunken to a fraction of its normal width. The spindle body appears to be involved in this shrinkage also, but to a much smaller extent.

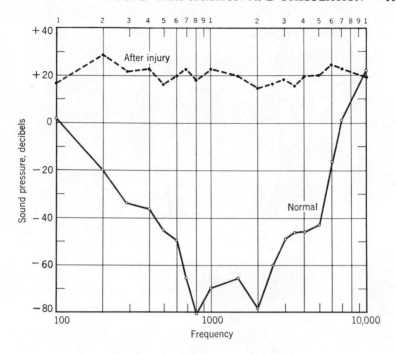

F<small>IG</small>. 4-31. Sensitivity functions for a specimen of *Coleonyx variegatus* before and after overstimulation with 1000 Hz.

At the extreme dorsal end of the cochlea, including the first two sections in which the tectorial membrane was present, this membrane was of normal appearance and made its usual connections to the hair cells.

A second specimen of *Coleonyx variegatus* was exposed to a tone of 1000 Hz at 10 dynes per sq cm for 12 minutes and then at 30 dynes per sq cm for 1.5 minutes, and suffered a loss of 16-29 db over the range from 600 to 2000 Hz, with generally smaller losses elsewhere, as shown in Fig. 4-33. In this ear the structural damage was much as described for the preceding specimen over the middle and ventral regions of the cochlea, but was much less in the dorsal region. Over the first seventh of the papilla, the condition seemed normal, with the tectorial membrane attached to the ciliary tufts in the usual way.

Trachydosaurus r. rugosus. Observations on the bob-tailed skink *Trachydosaurus* showed only moderate effects of overstimulation in a few animals and little or no effect in others. One of the more susceptible ears is represented in Fig. 4-34, with curves showing the sensitivity before and after exposure to a tone of 1000 Hz at a level of 10 dynes per

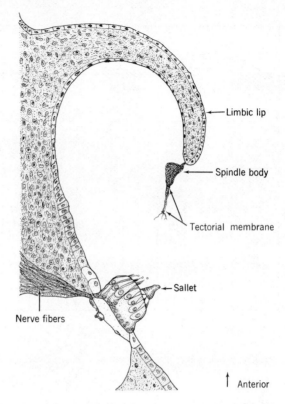

FIG. 4-32. The effects of overstimulation in *Coleonyx variegatus*. The tectorial membrane is pulled away from the ciliary tufts and strongly retracted.

sq cm for 2 minutes and then 30 dynes per sq cm for 2 minutes. A loss was suffered for all tones, varying from 6 to 32 db, with the largest effects around 1500-2000 Hz.

A histological study of this ear, as well as of six others similarly stimulated, gave no indications of inner ear damage.

It may be added that none of the lizards exposed to overstimulation showed any evidence of impairment of the middle ear structures. It was found also that, in general, when the stimulated ear gave clear evidence of injury the other ear did likewise, sometimes to about the same degree but more often to a lesser extent.

Some Concluding Observations. Though this overstimulation experiment includes only a small number of species and few specimens in each, its results suggest a number of significant points.

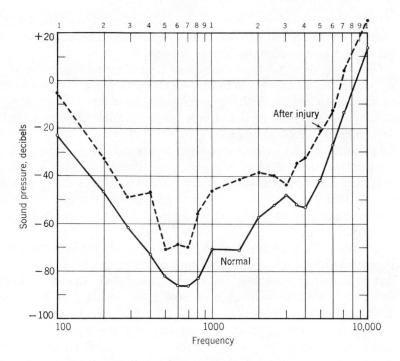

FIG. 4-33. Overstimulation effects on sensitivity in a second specimen of *Coleonyx variegatus.*

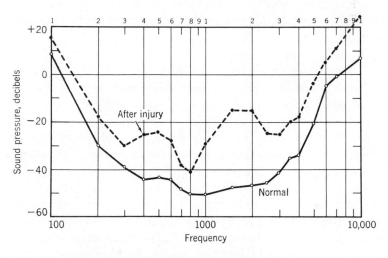

FIG. 4-34. Overstimulation effects on sensitivity in a specimen of *Trachydosaurus r. rugosus.*

The tectorial membrane system with connections to the ciliary tufts by direct contact or through fine fibers appears to be particularly fragile and subject to disruption by intense sounds.

The arrangement in which the tectorial membrane connects by way of a tectorial plate appears to be more resistant to injury, though certain specimens of *Crotaphytus collaris* exhibited grave losses of sensitivity after an exposure that produced only limited cellular changes without obvious mechanical disruptions.

The sallet system of ciliary restraint appears to be highly resistant to injury. No signs of sallet disruption were found in geckos in which all the tectorial connections were broken. Also the skinks with only sallet and culmen forms of restraint showed only slight effects on sensitivity and no signs of mechanical disruption after exposure to sounds that produced extreme damage in other species.

It is not to be supposed that all the injury produced by sounds becomes evident in histological examination. The mechanical changes described are unmistakable, but other, more subtle alterations must have been present as well. This assumption is supported by two lines of evidence.

1. There is always an immediate recovery following a period of overstimulation. During the first few minutes after a considerable loss is suffered, the potential produced by a test tone gradually increases. The amount of this recovery and its time course have not been thoroughly studied, but a few observations have shown that it may reach 20 db or so. Evidently this recovery represents some reversible process in the bioelectric activity. Such a process is hardly expected to be histologically detectable.

2. In the tests with *Trachydosaurus* and for a few of the other specimens as well there were moderate losses of sensitivity for which no clear structural basis could be found. These losses probably arose from intracellular changes that were not microscopically evident with the techniques employed.

The advantages of the presence in the same ear of two or more forms of stimulation systems are now obvious. A system with great delicacy of operation may be combined with one that responds only at high levels of sound action, and the dynamic range is thereby extended. Further, if the ear is likely to be exposed to sounds of extreme intensities, which will produce temporary or even permanent loss of function in some of the responsive elements, it is clearly an advantage to have other elements that continue in operation, so that complete deafness does not occur.

CONTRALATERAL EFFECTS

In the overstimulation experiments it was usually found that both ears suffered damage even though the sounds were presented to only one ear through a tube sealed over the meatal opening. This contralateral effect is the result of the conduction of sounds across the head. Such conduction is prominent in animals with small heads, with the ears separated by short distances. It has been noted especially in the small geckos, in which it is often difficult to carry out routine tests of sensitivity without producing a certain amount of injury, later revealed in both ears on histological examination.

Measurements of the contralateral transmission of sounds were carried out in specimens of *Crotaphytus collaris* by stimulating one ear and recording cochlear potentials from electrodes on both round windows. Figure 4-35 shows the results of one of these experiments. The solid curve represents the responses of the ear to which the sounds were presented, and the dashed curve shows the responses of the other ear. It is evident that at many frequencies the contrary ear receives the stim-

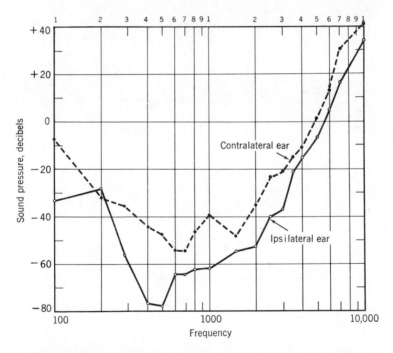

FIG. 4-35. Sound conduction across the head in *Crotaphytus collaris*. Tones were applied to the right ear and responses recorded from electrodes on both round windows.

ulus at an intensity approaching or even exceeding that at the source, and the greatest attenuation across the head in this instance amounted to only 34 db (at 400 Hz). The results of experiments of this kind vary widely, evidently because of resonances in the tissues and cavities of the head.

PART II. THE LIZARDS
ORDER SQUAMATA;
SUBORDER LACERTILIA

5. THE CLASSIFICATION OF
THE LIZARDS

Many proposals have been made for the orderly arrangement of the lizards. According to Camp (1923), there were more than a score of these proposals prior to 1864, when Cope presented his first system of lizard classification, a system that became a turning point in this field. Until then the systems had been based largely upon external appearances, including such features as body form and size, scale patterns, limbs, and the tongue. Cope for the first time placed emphasis upon skeletal structures. Twice in later years he revised and extended his system, and his final scheme, published after his death, included 9 main divisions (suborders or superfamilies) that embraced 24 families. It is a tribute to his keen perception that of these 24 families 18 are still generally accepted as valid, and 3 others appear in a few of the existing systems.

In the years following its original formulation, a number of revisions of Cope's system were proposed, in addition to his own two revisions. These revisions consisted mainly of regroupings of the families and the erection of higher categories, all with the aim of obtaining more meaningful assemblies and sequences of evolutionary import. The considerable number of these attempts, and the cool reception accorded most of them, give evidence of the complexity of existing lizards and the variety of their points of resemblance.

Camp's Classification. A highly significant event was the appearance of Camp's monograph on "Classification of the lizards" in 1923. His main departure from previous methods was an extensive utilization of soft tissues: lungs, palate, hemipenes, and musculature, and particularly the muscles of body and throat, along with other features previously considered. Altogether he listed 34 characters that he tried to take into account in the determination of familial relationships. The system that emerged had two major divisions, the Ascalabota and the Autarchoglossa, the first with three subdivisions (called sections) and the second with two. The major divisions were based primarily upon the presence or absence of an abdominal muscle, the rectus superficialis, which Camp considered to have a fundamental involvement in crawling or creeping movements. In the Autarchoglossa this muscle is invariably present and in the Ascalabota it is absent except in a few agamids. It is further of interest that many families among the Ascalabota have developed arboreal or semiaquatic practices, though the Autarchoglossa never have. On the other hand, many members of the latter group have lost their limbs and have adopted burrowing habits.

Camp's system of classification is presented in Table 5-I except for the omission of subfamilies and extinct families.

TABLE 5-I

CAMP'S SYSTEM OF LIZARD CLASSIFICATION (1923)
(subfamilies and extinct families omitted)

SUBORDER SAURIA

Division Ascalabota	
Section Gekkota	Superfamily Lacertoidea
Family Gekkonidae	Family Gerrhosauridae
Family Uroplatidae	Family Lacertidae
Section Iguania	Family Teiidae
Family Iguanidae	Superfamily Amphisbaenoidea
Family Agamidae	Family Amphisbaenidae
Section Rhiptoglossa	Section Anguimorpha
Family Chamaeleontidae	Subsection Platynota
Division Autarchoglossa	Superfamily Varanoidea
Section Scincomorpha	Family Varanidae
Superfamily Xantusioidea	Subsection Diploglossa
Family Xantusiidae	Superfamily Pygopodoidea
Superfamily Scincoidea	Family Pygopodidae
Family Scincidae	Superfamily Anguioidea
Family Anelytropsidae	Family Helodermatidae
Family Feyliniidae	Family Anguidae
Family Dibamidae	Family Xenosauridae
	Family Anniellidae
	Superfamily Zonuroidea
	Family Zonuridae

Romer's Revisions. During the half-century since its presentation, Camp's system has had widespread acceptance, with only a few suggestions for amendment. Romer, in the second edition of his *Vertebrate paleontology* in 1945 abandoned the rather cursory treatment of the Sauria in his first edition (1933) and adopted Camp's system with only minor changes. Romer left out the two great divisions and many of the superfamilies while retaining most of Camp's sections and subsections, which he mainly referred to as infraorders. The Amphisbaenidae were separated from the Scincomorpha but located adjacent to these in an infraorder Annulata. Camp's section Anguimorpha, which consisted of two subsections Platynota and Diploglossa, became the infraorders Platynota and Anguimorpha containing the same families except for the addition of Shinisauridae to the Platynota.

Further Developments. A searching examination of lizard relationships was made by McDowell and Bogert in 1954, directed particularly at the anguimorphs and carried out in the course of a special investigation of the species *Lanthanotus borneensis*, the rare earless monitor of Borneo.

They followed Camp and Romer (1945) in a general way, but made some important changes. The Pygopodidae were removed from the infraorder Anguimorpha and placed in the Gekkota. The Cordylidae also were removed from this infraorder and transferred to the Lacertoidea. Three changes concerned the Platynota. Added to this group were the Helodermatidae, which had been in the Anguimorpha (Camp's Diploglossa), and also the Lanthanotidae, recognized earlier by Gadow (1901) but not by Camp and Romer. The Shinisauridae, which Romer had placed in this infraorder, were changed to the second line of anguimorphs and made a subfamily of the Xenosauridae.

Romer in *Osteology of the reptiles* (1956) adopted many of the changes suggested by McDowell and Bogert. Here and again in the third edition of his *Vertebrate paleontology* (1966) may be found the most detailed treatment of reptile classification, which has been widely accepted and regarded as a standard. This system is summarized for existing forms in Table 5-II.

TABLE 5-II
ROMER'S SYSTEM OF LIZARD CLASSIFICATION (1956)
(several subfamilies and all extinct families omitted)

SUBORDER LACERTILIA

Infraorder Iguania
    Family Iguanidae
    Family Agamidae
    Family Chameleontidae
Infraorder Nyctisauria (Gekkota)
    Family Gekkonidae
    Family Pygopodidae
Infraorder Leptoglossa
    (Scincomorpha)
    Family Xantusiidae
    Family Teiidae
    Family Scincidae
    Family Lacertidae
    Family Cordylidae
        Subfamily Gerrhosaurinae

        Subfamily Cordylinae
        (Zonurinae)
    Family Dibamidae (?)
Infraorder Diploglossa
    Superfamily Anguoidea
    Family Anguidae
    Family Anniellidae
    Family Xenosauridae
    Superfamily Varanoidea
    (Platynota)
    Family Helodermatidae
    Family Varanidae
    Family Lanthanotidae
Infraorder Annulata (Amphisbaenia)
    Family Amphisbaenidae

Several features may be noted. The Chameleontidae no longer are given independent status, but are placed with the agamids, along with the iguanids, in the infraorder Iguania. The Gekkonidae and Pygopodidae are together in one infraorder, now called Nyctisauria, and the uroplatids are reduced to a genus among the Gekkonidae. The Cordylidae now embrace two groups formerly separated; following the suggestion of McDowell and Bogert this family is made to include the gerrhosaurids, which Romer earlier placed among the scincomorphs, and the zonurids, which he had included in the anguimorphs. The Amphisbaenidae are retained as an infraorder, now designated as Annulata.

The Persisting Problems.   In the classification systems from Cope to Romer, a certain few points of difference have appeared repeatedly. The following eight debatable points are singled out as the most common and vexatious, and will be commented on briefly.

1. Chamaeleonidae: Camp accorded them a section of their own among the Ascalabota, and Romer at first followed this practice. Later Romer included them among the Iguania with the suggestion that they are derived from agamids. Actually they do not fit well anywhere.

2. Pygopodidae: Camp placed them in the subsection Diploglossa adjacent to the Anguoidea. McDowell and Bogert (1954) moved them into the Nyctisauria (Gekkota) along with the Gekkonidae, and with this Romer (1956) and many others have agreed.

3. Xantusiidae: Camp had this family in the section Scincomorpha adjacent to the Scincoidea, and Romer (1945, 1956) agreed. McDowell and Bogert (1954) predicted that "they will be referred to the Gekkota." Underwood (1957) somewhat hesitatingly accorded them an infraorder of their own adjacent to the Gekkota. More recently Edmund (1969) definitely moved them into the Nyctisauria (Gekkota) along with Gekkonidae and Pygopodidae. This family appears to have several features that make its relationships equivocal between two distant locations. Bezy (1972) in a review of the evidence found the relations to other families still obscure, but in his chromosome studies obtained the greatest similarities with the teiids.

4. Helodermatidae: Camp had this group in the superfamily Anguoidea, adjacent to the Anguidae. McDowell and Bogert (1954) moved them into the superfamily Varanoidea (Platynota), close to the varanids, to which Romer (1956) agreed.

5. Anelytropsidae, Feyliniidae, and Dibamidae: Camp placed these three close to the skinks in the superfamily Scincoidea. Romer (1956) denied family status to the first two and raised doubts about the third, regarding all as degenerated skink-like forms.

6. Zonuridae, Gerrhosauridae, and Cordylidae: Camp placed the Zonuridae in a superfamily of its own in the section Anguimorpha close to the Anguoidea, and located the Gerrhosauridae a long distance away in the section Scincomorpha near the Lacertidae. McDowell, Romer, and others regard these two as subfamilies in the family Cordylidae, close to the Lacertidae among the Leptoglossa (Scincomorpha).

7. Amphisbaenidae: Camp located these animals among the Scincomorpha, adjacent to the Lacertidae and Teiidae. Romer (1956) placed them in a separate infraorder of Annulata. Several recent opinions favor their removal from the Lacertilia and the creation of a new suborder or order of Amphisbaenia (Gans, 1967a, 1974; Edmund, 1969; Carroll, 1969).

8. Lacertidae and Teiidae: Camp placed these two families in the section Scincomorpha, in the company of the Gerrhosauridae. Romer (1956) followed much the same plan, as has practically everyone else. Miller (1966a), however, gave them a separate grouping on the basis of his study of the cochlear duct.

These questions concerning the lines of kinship among the lizards need to be examined further after consideration of the structure and function of the ear. As Miller pointed out, this organ, and especially its cochlear portion, appears to be highly conservative in the evolutionary process, and the nature of its relationships among the lizards ought to throw light on the systematic problems.

THE PRESENT ARRANGEMENT

For the present treatment of the lizard ear a scheme is adopted, as presented in Table 5-III, that bears a close resemblance to the classification system of Romer (1956) but departs from it in certain respects, especially in the sequence and grouping of families. The changes are introduced to facilitate the descriptive treatment that follows by bringing closer together those species exhibiting similar types of ear structure. This scheme is not intended as a new proposal for lizard classification, though the lines of relationship here indicated for the ear may well be taken into account along with other structural and functional characters in the further development of classification systems.

The departures from Romer's system are four: (1) the moving of the infraorder Diploglossa, with its two superfamilies Anguoidea and Varanoidea, to a position adjacent to the Iguania, (2) the removal of two families, Teiidae and Lacertidae, from Romer's infraorder Leptoglossa (here called the Scincomorpha after Camp), and their insertion in the middle of the series between the Varanoidea and the Gekkota, (3) the recognition of two separate families, the Gerrhosauridae and

TABLE 5-III

LIZARD FAMILIES AND THEIR GROUPINGS
(as followed in this study)

Family	Superfamily	Group	Middle Ear Type	Existing Species (approx.)	Number of Species Studied
1. Iguanidae	I		I †	700	47
2. Agamidae	Iguania		I †	300	13
3. Chamaeleonidae			d	80	9
4. Anguidae	II	A	S †	60	7
5. Anniellidae	Anguoidea	Tectorial	d	2	1
6. Xenosauridae		System	d	3	1
7. Varanidae	III		I	28	3
8. Helodermatidae	Varanoidea		I	2	1
9. Lanthanotidae*			d	1	0
10. Teiidae	IV	B	I †	200	7
11. Lacertidae	Lacertoidea	Combined	I	150	11
12. Gekkonidae	V	Tectorial and	G	650	46
13. Pygopodidae	Gekkota	Sallet System	G	13	1
14. Gerrhosauridae			I	25	6
15. Xantusiidae			I, S	15	6
16. Feyliniidae*	VI		d	4	0
17. Dibamidae*	Scincomorpha	C	d	3	0
18. Scincidae		Sallet System	S †	600	17
19. Cordylidae			I	50	10
		Totals		2,878	186

* no specimens available

† some species divergent

I—iguanid type, G—gekkonid type, S—scincid type, d—divergent or degenerated

Cordylidae, instead of Romer's inclusive family Cordylidae, and (4) the removal of the Annulata from the lacertilian category altogether and the recognition of a suborder of Amphisbaenia.

These changes hardly affect Romer's assembly of families into higher categories (his infraorders or superfamilies; here called "superfamilies"), except for the sequestration of Teiidae and Lacertidae. These two are now placed under the Lacertoidea, and correspond to Camp's superfamily of that name except for the omission of the Gerrhosauridae.

Table 5-III further presents a higher grouping based upon the modes of ciliary restraint utilized in the different lizard species in the process of hair-cell stimulation. As the preceding chapter has brought out, three basic systems are employed in this restraint.

1. Free-standing Ciliary Tufts.   The simplest of these restraints is provided by the free-standing ciliary tufts, operating by principles of reaction force and viscosity in the absence of any specialized structure, and doubtless representing the most primitive stage of the restraining process.

Sometimes these ciliary tufts are in contact with a delicate embedding material that provides additional restraint by connecting the ciliary tufts with one another or with adjacent tissues. It seems likely that this embedding material is an unorganized form of tectorial tissue, and represents the earliest stage of development of specific restraining mechanisms.

2. Tectorial Membranes and Plates.   The most general type of ciliary restraint is provided by a tectorial membrane. This membrane may form a direct attachment to the ciliary tuft or, much more commonly, may make its connections through intermediate structures. These structures vary widely from simple fibers through fiber networks to complex tectorial plates.

3. Sallets and Culmens.   The third restraining mechanism consists of a compact mass of tectorial tissue, the sallet or culmen, that makes an intimate contact with the ciliary tufts of a number of hair cells and provides restraint through the operation of inertia or inertia-like principles. In two families the restraint afforded by the culmen is augmented by its connection to a tectorial membrane.

GROUPINGS OF THE FAMILIES

As shown in Table 5-III, the lizard families may be placed in three general groups according to their use of different combinations of these forms of ciliary restraint.

Group A.   The first group includes nine families in which ciliary restraint is provided by tectorial membrane connections for at least some of the hair cells, and the hair cells not served in this manner are left with free-standing ciliary tufts. Typically the tectorial membrane makes its connections through a tectorial plate.

Group B.   A second group contains four families with combined tectorial and sallet systems of ciliary restraint. An unusual feature found in the Gekkota is the presence of these two systems side by side over the ventral segment of the cochlea, with tectorial connections for the more medial rows of hair cells and sallets serving the cells of the lateral rows.

Group C.   The scincomorphs form a third group with practically all the hair cells having sallet or culmen connections to their ciliary tufts. In the Gerrhosauridae and Xantusiidae there is a functional tectorial membrane that attaches to the culmen and adds to its restraining effect. In the Scincidae and Cordylidae a tectorial membrane is present only in vestigial form.

Some exceptions to these general groupings may be found among the species examined. The most notable of these is the presence in a single species of Agamidae (*Leiolepis belliana*) of a line of sallets beyond the tectorial plate that serves the hair cells which in all others of this group have free-standing ciliary tufts. Also in both B and C groups there are a few hair cells, most often found at the extreme ends of the auditory papilla and in transitional zones, whose ciliary tufts lack any specific connections. These hair cells, if functional at all, must operate like the free-standing elements of Group A.

Living lizard families for which representative specimens were unobtainable are Dibamidae, Feyliniidae, and Lanthanotidae. These are located in the table largely on the basis of Miller's observations (1966a and b) on the form of the cochlear duct. He reported this structure in *Feylinia currori* to be "undoubtedly scincid," in agreement with Romer (1956), who considered *Feylinia* (along with *Anelytropsis*) as skink derivatives. Miller's observations on *Lanthanotus borneensis* showed close similarities with *Heloderma* and *Varanus*, as earlier reported by McDowell and Bogert (1954).

The relationships indicated in Table 5-III primarily reflect the characteristics of the inner ear, as will be further brought out in the treatment to follow. The forms of middle ear mechanism, indicated in the fourth column of the table, do not closely follow the variations of inner ear structure. As treated in the next chapter, the lizard middle ear takes three principal forms, together with divergencies that seem to represent degenerations from these three. In six families all members present these divergent forms of conductive mechanism, and in five others such variations appear sporadically in a few species.

6. SOUND CONDUCTION IN
THE LIZARD EAR

In its usual course, hearing begins with the transmission of sound waves through the air and their reception by way of an open passage, the external auditory meatus, at the end of which they impinge on the tympanic membrane. Here the acoustic pressures set up motions of the tympanic membrane and attached middle ear ossicles, and these motions are conducted inward to produce corresponding motions of the cochlear fluids within which the auditory hair cells are contained. These steps in the sound-conductive process will now be considered in turn.

THE EXTERNAL EAR IN LIZARDS

The outer ear of lizards was treated in considerable detail by Versluys (1898), who summarized his own observations along with the more extensive ones of Boulenger. The structures vary widely, from the completely superficial or only slightly depressed tympanic membranes generally found in Iguanidae and Agamidae, in which no outer ear exists in any strict sense, to the more or less deep-lying tympanic membrane and meatal cavity with its opening to the surface as found in Gekkonidae, Helodermatidae, nearly all Pygopodidae, and many Anguidae and Scincidae. Between these extremes are a number of forms in which the tympanic membrane is moderately depressed but no true meatal cavity exists, as in Varanidae, in which there is a deep notch with the tympanic membrane forming its anterolateral surface, and the Cordylidae, in which the tympanic membrane lies in an opening bordered by spinous processes.

In a few lizard families, including Anniellidae, Chamaeleonidae, Dibamidae, Lanthanotidae, and Xenosauridae, there is no ear opening and the tympanic membrane is lacking. This condition is found also in certain species within other families in which normality of structure is the rule: in some of the Agamidae, Iguanidae, and Scincidae, in one species of Anguidae, and in *Aprasia* species among the Pygopodidae.

No outstanding difference is found between the superficial and deep-lying forms of the tympanic membrane, except in a few, such as certain of the agamids, in which the superficial membrane is thick and stiff, or is covered with scales. When the tympanic membrane is thus modified so as to lose its normal flexibility, the reflection of sounds from its surface becomes greater than usual, and poor sensitivity is the result. Still more serious impairment of function is to be expected when the tympanic membrane is absent and its role in sound reception is taken by unmodified skin. The effects of these deviations, usually accompanied by other variations in the structure of the middle ear, will be assessed in such species as became available in the course of this study.

THE MEATAL CLOSURE MUSCLES

In some of the lizards with a definite meatus, a muscle is present in the meatal wall by which the external opening can be closed. Such a meatal closure muscle was discovered by Versluys (1898) in four species of geckos: in *Gekko gecko, Thecadactylus rapicaudus, Pachydactylus bibronii*, and *Tarentola annularis*. He failed to find the muscle in other geckos that he examined, but suggested that he may have overlooked it. Iordansky (1968), evidently unaware of the early observations of Versluys, reported the closure muscle in *Gekko gecko* and *Teratoscincus scincus*. He also described a muscle in *Ophisaurus apodus* that he considered to be a meatal dilatator.

Versluys described and pictured the closure muscle in *Gekko gecko* as arising from the posterior fascia in the temporal region dorsal to the ear opening, then coursing ventrally behind this opening to an insertion in the skin below and in front of the opening. He indicated that when the muscle contracts it pulls the skin at the posterior edge of the meatal wall in a forward and upward direction, thus closing the aperture. Paulsen and Mares (1971) described this muscle in *Gekko gecko* and discussed its form of insertion. Contrary to Versluys, they indicated the insertion as on the fascia of the external adductor mandibulae muscle rather than on the skin.

A further investigation of the meatal closure muscle in Gekkonidae included observations on 30 species, with representatives of all four subfamilies (Wever, 1973d). This study showed that the muscle takes two distinct forms, one in Eublepharinae and Gekkoninae that follows the description of Versluys, running along the posterior and ventral sides of the auditory meatus and, because of its form, designated as the L-type of muscle, and another found in the Diplodactylinae, in which the muscle encircles the meatus, and hence is designated as the loop type. A loop type of muscle was discovered also in *Lialis burtonis*, the only member of the Pygopodidae examined (Wever, 1974c).

Additional gekkonid species have been added more recently, so that the total number explored for this structural feature now amounts to 39. This group includes 4 species of Eublepharinae, all of which possess the L-type of closure muscle, 23 species of Gekkoninae, of which 20 display this type of muscle, and 6 species of Diplodactylinae, all of which are considered to have a loop-type of closure muscle, though in one of these, *Lucasium damaeum*, the form is divergent. In three of the Gekkoninae and all six of the sphaerodactyline species studied, a meatal closure muscle is lacking. The three gekkonines in which no meatal closure muscle was found are *Phelsuma dubia*, *Phelsuma madagascariensis*, and *Lygodactylus picturatus*, all of which are of diurnal habit in contrast to the others, which are nocturnal.

The L-type Closure Muscle.   Structural details concerning the L-type of meatal closure muscle were worked out in specimens of *Gekko gecko*, and are represented in Fig. 6-1. The form follows closely the

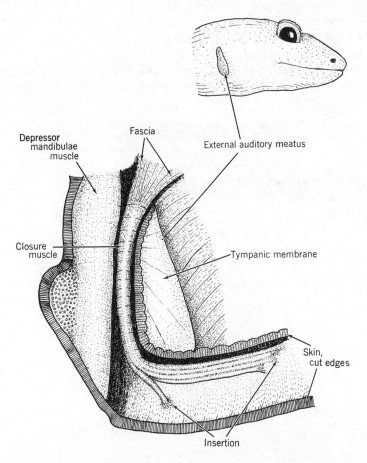

FIG. 6-1. The L-type of meatal closure muscle, from a dissected specimen of *Gekko gecko*. A strip of skin was removed along the posterior and ventral edges of the meatus; the cut surfaces are indicated by hatching. From Wever, 1973d.

early description given by Versluys. The muscle runs close along the posterior edge of the meatal wall, and over most of this course it forms a compact bundle of nearly circular cross section. Toward the dorsal end the fibers fan out and attach to the fascia above the meatal opening. Ventral to the meatus the bundle makes a sharp bend and runs forward, breaking up into a number of fascicles as it does so. These fascicles terminate in insertions both on the fascia of the external adductor mandibulae muscle and on the skin. For the most part, those fascicles turning earlier and running close along the lower border of the meatus insert on the muscle fascia, and others turning later, and sometimes less abruptly, insert on the deep skin layer. Often, as the figure shows, a large fascicle runs obliquely downward and forward and attaches to the skin just below the meatal opening.

A cross section along the midpart of the muscle of *Gekko gecko* is shown in Fig. 6-2. The relations to the meatus and to other muscles of the area are indicated.

The meatal closure muscle varies greatly in size in the different species. It is particularly large in *Eublepharis macularius* and *Hemithe-*

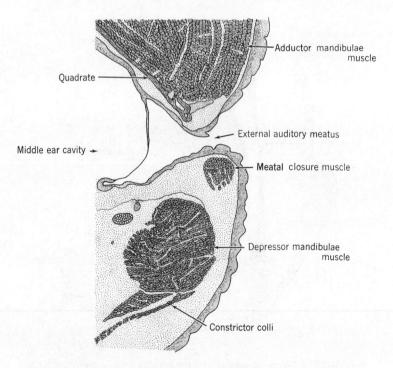

FIG. 6-2. A section through the auditory meatus of *Gekko gecko*, with the meatal closure muscle in the posterior wall. Scale 7.5×.

conyx caudicinctus, in which more than 330 fibers were seen. In the two other eublepharine species (of smaller body size) this muscle was much smaller: in a specimen of *Coleonyx brevis* there were about 35 fibers, and in three specimens of *Coleonyx variegatus* the number varied from 23 to 90.

In a specimen of *Gekko gecko* the muscle contained about 250 fibers. Several other gekkonines showed between 50 and 100 fibers. A few had extremely small muscles. In specimens of *Gehyra variegata* and *Hemidactylus mabouia* the number was around 20, and in *Cyrtodactylus kotschyi orientalis* it was even smaller. Among five *Cyrtodactylus* specimens studied in some detail, the number of fibers varied from 2 to 11, and in one of these there were 3 fibers on the right side and none on the left.

These wide variations in the size of the meatal closure muscle no doubt are related to body size, but almost certainly are species dependent also. With only small numbers of specimens available for any one species, these two conditions could not be suitably separated.

The Loop-type of Meatal Closure Muscle. As already mentioned, a distinctive form of meatal closure muscle was found in the diplodactylines. As shown in Fig. 6-3 for *Hoplodactylus pacificus* this muscle arises in fascia anteroventral to the meatal opening and runs in the deep skin

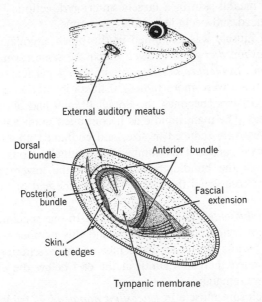

FIG. 6-3. The loop type of meatal closure muscle, from a specimen of *Hoplodactylus pacificus*. The skin was removed in an area around the meatus. From Wever, 1973a.

layer along the ventral and posterior borders of the meatus to a point near the upper posterior edge, where it sends off a small dorsal branch. This branch extends a short distance to the dorsolateral border of the adductor mandibulae muscle, and then suddenly ends. The main bundle runs around the dorsal edge of the meatal opening, but as it approaches the anterodorsal corner it is reduced in size and continues as a smaller bundle along the anterior side of the meatus. In this anterior region the bundle gradually loses its compact form as the fibers spread out in the skin tissues and become intermingled with fascial strands that form a loose bundle extending anteroventrally over the region where the meatal closure muscle originates. The upper and posterior portion of this fascial extension has connections with the sheath of the adductor mandibulae muscle and with other tissues of the region, and its anteroventral end makes connections in the region of origin of the meatal closure muscle. Thus this fascial extension supplements the few remaining muscle fibers in completing the loop around the meatal opening, as the figure indicates.

The appearance of this muscle when cut across near the middle of the loop is indicated in Fig. 6-4.

A contraction of this muscle evidently has two effects, one a constriction of the opening and the other, because of the strong fascial attachment at the anteroventral border, a pulling of the dorsoposterior portion of the meatal skin in a largely anterior direction. The opening is thereby reduced and occluded at the same time.

The closure muscle was examined in the other diplodactylines, and in all but *Lucasium damaeum* showed much the same form and action as just described. In *Oedura ocellata* this muscle is particularly large; in one specimen there were more than 220 fibers in the bundle posterior to the meatus. In this specimen the dorsal branch had about half this number of fibers. The main bundle as it continued in an anterior direction was progressively reduced in size, and contained about 100 fibers at its anterodorsal corner. This number fell gradually along the anteroventral course of the bundle, reaching 70 at the lower edge of the meatus where it completed the loop by joining the end of the posterior bundle.

In *Diplodactylus elderi* the muscle is small. In one specimen the posterior bundle contained about 10 fibers, and this number was reduced to half as the bundle ran around the meatus to its anterior side. The reduction continued thereafter, until at the end below the ear opening only 2 or 3 fibers remained.

The meatal closure muscle in *Lucasium damaeum* is evidently of the loop type, but takes an abortive form, with large variations from one specimen to another. In a specimen in which this muscle was rather well

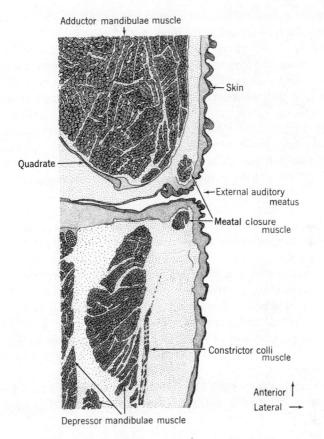

FIG. 6-4. A section through the meatal region of a specimen of
Hoplodactylus pacificus. Scale 20×.

developed there were 70-80 fibers in the posterior bundle. This bundle
then ran around the dorsal side of the meatus with a rapid reduction
in its fibers, until only about 20 remained in the bundle that evidently
corresponded to the usual anterior bundle. However, this bundle did
not reach the anterior side of the meatus, but turned and ran ventrally
just anterior to the main posterior bundle and ended as a group of 4-8
fibers. Thus a loop was formed by this muscle, but the loop did not
enclose the meatus: it remained posterior to it.

In another specimen only a partial loop was formed; the posterior
bundle took much the usual course, then continued in an anterior direction only a short distance above the dorsal edge of the meatal wall and
ended there. In three other specimens the anterior bundle ran a little
farther, but was reduced in size to 2-4 fibers and likewise failed by a

wide margin to reach the lower edge of the meatus to join the posterior bundle and thereby complete the loop.

At the time this study was made, living specimens of *Lucasium damaeum* were no longer available for use in determining whether this muscle has any closure effect. If it does, the effect may be expected to be a forward displacement of the skin similar to the action produced by the L-type of muscle, rather than a constriction of the opening.

It was suggested by Versluys that the meatal closure muscle is derived from the constrictor colli. There is ample evidence for this view, in that at least a tenuous connection persists in many species. An example is given in Fig. 6-5 for the species *Thecadactylus rapicaudus*.

Function of the Closure Muscle. Closure of the meatus is clearly protective, but it is not altogether clear whether this protection is against mechanical damage only—against the entrance of foreign objects such as twigs, grass, or sand—or includes protection against excessive sounds as well. Observations made on specimens of *Gekko gecko* and *Eublepharis macularius* showed active responses producing meatal closure when the skin was touched at the side of the head or when a moderate puff of air was directed at the ear region. No responses were observed, however, on the presentation of sounds of considerable intensity within the range of 300 to 1500 Hz, which is the range of greatest sensitivity in these animals.

Paulsen and Mares (1971) reported that *Gekko gecko* closes the meatus when it assumes a threatening posture toward an intruder, a

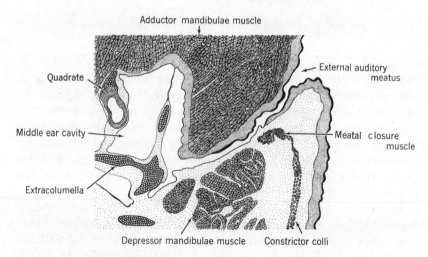

FIG. 6-5. A section through the meatal closure muscle of *Thecadactylus rapicaudus*, showing its connection with the constrictor colli.

posture in which it opens the mouth widely and sometimes produces vocal sounds.

Kästle (1964) suggested that hearing in geckos mainly serves a function of communication between members of a particular species. Thus males produce their cries to proclaim their territory rights and warn off other males. Meatal closure then may serve to protect the animal's ears against its own sounds, which in some species can be extremely loud. A corollary of this hypothesis is that mute species or those with infrequent or faint vocalizations would have no need for the closure muscle. The Sphaerodactylinae and the day geckos among the Gekkoninae, which lack a meatal closure muscle, are usually considered to be mute, though Weber and Werner have reported the day geckos as producing an audible cheep when first caught in the wild.

On the other hand, there are mute or generally silent species in which a closure muscle is present. Thus *Gehyra variegata* has not been observed to produce sounds (beyond a squeak when caught in the hand), and the same is reported for *Oedura ocellata* and *Diplodactylus elderi* (Bustard, 1965, 1970, and pers. comm. 1972), yet all these possess a meatal closure muscle. It is evident that the possession of a loud voice might assist the differentiation of a meatal closure muscle, but that other conditions could bring about this development also.

The Meatal Closure Muscle in *Lialis burtonis*.   In the single species of the Pygopodidae available for study a meatal closure muscle is present in the form of a completely closed loop, as pictured in Fig. 6-6. This

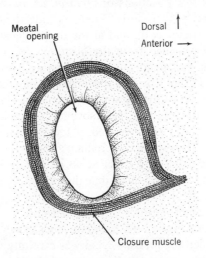

FIG. 6-6. The meatal closure muscle in *Lialis burtonis*.
From Wever, 1974c.

muscle forms a bundle of 55-70 fibers running in the skin close to the edge of the meatal opening on the posterior side and somewhat farther away anterior to the meatus. Ventrally the posterior bundle turns and runs in an anterior direction close along the edge of the meatus, and extends far enough anteriorly to join the end of the anterior bundle, thus completing the loop.

This muscle evidently operates simply as a constrictor, narrowing the meatal aperture in a regular manner.

Meatal Muscles in Anguidae. As mentioned above, Iordansky reported the discovery of a meatal muscle in *Ophisaurus apodus*, which he considered to be a meatal dilatator. This species was further examined in serial sections as a part of the present study, along with others among the Anguidae, but the results must be considered preliminary because of a shortage of material that could be devoted to this purpose. A further study is needed in which living specimens can be examined to determine what changes occur in the size and form of the meatal opening as a result of muscle action.

In *Ophisaurus apodus* a meatal muscle was found, as Iordansky reported. As he described it, this muscle is an anterior part of the constrictor colli that splits into two portions, a dorsal portion that enters the dorsal fascia of the neck and a posterior portion that inserts in the skin of the roof of the external auditory meatus. From these connections Iordansky concluded that contraction of the anterior fibers of the muscle would dilate the external auditory meatus.

My own studies of serial sections of *Ophisaurus apodus* leave some doubt about Iordansky's hypothesis. There is indeed an anterior enlargement of the constrictor colli, and at the deeper levels the terminal part of the muscle is detached from the posterior portion of the constrictor to produce an independent strand, seen as a small bundle of very minute fibers in Fig. 6-7. In the specimens examined, however, no attachment of the muscle to the meatal roof could be discovered. This strand was found to run only in a dorsoventral direction, and never passed over or under the meatus. Indeed, it failed to enter the superficial wall of the meatus anywhere, but remained deep-lying. A contraction of these fibers, if it has any effect at all on the meatus, would be moderately constrictive. It is possible that this strand of the constrictor colli acts only as a part of the whole muscle in regulating the skin of the neck and has no effect on the size of the meatus.

Much the same conditions were found in *Ophisaurus ventralis* and *Barisia gadovii*, in which the meatal region was carefully studied also in serial sections. In *Ophisaurus ventralis* the muscle is present as a bundle of 33-40 fibers in the skin posterior to the meatus, and a few fibers ex-

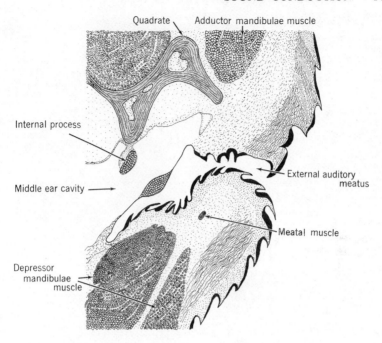

FIG. 6-7. The meatal muscle in *Ophisaurus apodus*.

tend a short distance dorsal to the meatus. Below the meatus a connection with the constrictor colli is maintained.

In *Barisia gadovii* there is an enlargement of the anterior end of the constrictor colli, which ventrally becomes separated and appears as an independent muscle of about 25 fibers posterior to the meatal opening, as shown in Fig. 6-8. This muscle extends well dorsal to the meatus, where it dwindles and disappears. Ventrally the bundle ends before the floor of the meatus is reached.

In *Gerrhonotus multicarinatus* no sign of a meatal muscle was found. The constrictor colli extends anteriorly well beyond the end of the depressor mandibulae muscle, and its tip curls around medially as a sharp hook. No part of this muscle is separated off to come in close relation with the meatal skin.

In the three anguid species in which a meatal muscle was found—*Ophisaurus apodus*, *Ophisaurus ventralis*, and *Barisia gadovii*—the muscle shows much the same form and course. It lies deep in the skin posterior to the meatus, extends from a point a little dorsal to the meatal opening, and nowhere encloses any part of the opening. To determine the action of such a muscle will require direct observations on living specimens. From the anatomical situation, it seems difficult to imagine

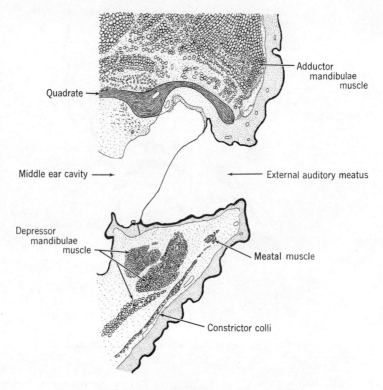

Quadrate

Adductor
mandibulae
muscle

Middle ear cavity ⟶   ⟵ External auditory meatus

Depressor
mandibulae
muscle

Meatal muscle

Constrictor colli

FIG. 6-8. The meatal muscle in *Barisia gadovii*.

any dilator action, as the fibers run dorsoventrally and lack firm attachments to skeletal structures. A constrictor or folding effect on the skin
of this region is more likely, but its extent appears to be severely limited.

Possible evolutionary relationships among the three forms of definitely
functional meatal closure muscles were suggested earlier (Wever,
1973d). The purse-string type of muscle in *Lialis burtonis* appears to
be the most elementary form of closure muscle, with its nearly uniform
bundle completely enclosing the meatal opening. Contraction of the fibers of this muscle simply constricts the opening.

The closure muscle found in the diplodactylines can be considered a
modified constrictor type. The muscle forms a loop around the meatus,
but its bundle is not uniform over this course, undergoing considerable
reduction along the anterior side of the meatus. This reduction is so
great in some species as to leave very few fibers to complete the loop
on the anteroventral side, though a band of fascia is present here to
assist the connection of the ends of the muscle and make the constrictor

action effective. Because the posterior portion of the muscle is considerably larger and therefore stronger, the contraction probably has a double effect, that of constriction and also of shifting the meatal skin. Thus the closing of the meatus is partly a reduction of its aperture and partly its occlusion by the shifted skin.

It is apparent that the diplodactyline type of closure muscle could arise from the purse-string type by a partial degeneration of the anterodorsal portion of the muscle loop. In the species *Lucasium damaeum* the muscle is incomplete, and fails to enclose the meatus; this muscle, if functional, can only operate by shifting the meatal skin in the manner of the L-type of muscle.

The L-type of muscle, as found in eublepharines and gekkonines, could arise from the loop type by a further degeneration in which only the posterior and ventral part of the loop remains.

The simpler muscles observed in the anguids can be accounted for as resulting from still further degenerations, in which only the posterior portion of the primitive loop is left, and no effective enclosure of the meatus occurs. Such a muscle can be regarded as vestigial and perhaps nonfunctional or practically so.

Finally, the complete absence of the meatal closure muscle in sphaerodactylines and certain of the gekkonines can be regarded as the final step in the degeneration process.

THE LIZARD MIDDLE EAR

The anatomy of the middle ear of lizards was first described in detail by Versluys in 1898. He examined 30 species belonging to 12 lizard families, and for these gave extensive descriptions of the general structure, including the tympanic cavity and its enclosure in the skull, the form and suspension of the tympanic membrane, and the ossicular system and its connections. No comparable treatment has appeared since that time. A certain amount of attention has been given to deviations from the usual middle ear structures in articles by Toerien (1948, 1963) and Earle (1961). A number of more recent experiments have made use of the cochlear potentials to assess the normal performance of the ears of representative species, and in some of these experiments the conductive structures have been manipulated, especially by interrupting the continuity of the ossicular chain, to bring out the functional characteristics of the various parts (Wever and Peterson, 1963; Wever, Vernon, Peterson, and Crowley, 1963; Wever, Peterson, Crowley, and Vernon, 1964; Crowley, 1964; Wever and Hepp-Reymond, 1967). These observations were made mainly with aerial sounds, but sometimes were supplemented by the use of mechanical vibrations, and the performance then was related to the anatomy of the sound conductive system. A more systematic series of experiments along these lines provides the basis for the present review of our knowledge of the lizard middle ear and its part in the determination of hearing in these animals (Wever and Werner, 1970; Werner and Wever, 1972, Wever, 1973a, b).

The middle ear of lizards takes three principal forms, apart from a number of variants that probably represent degenerations. These three forms are somewhat arbitrarily designated as the iguanid type, the gekkonid type, and the scincid type.

In all three types, an ossicular chain of two elements, the columella and extracolumella, forms a conductive path from the tympanic membrane to the oval window of the cochlea. The columella is a rod consisting largely of bone whose expanded end, the stapedial footplate, is inserted in the oval window of the cochlea, where it is securely held by an annular ligament. The footplate is usually broadly flared, but in a few instances is a rounded knob only a little larger than the shaft itself. Its edges, and usually the inner surface, are covered with cartilage. The rod or shaft portion is ordinarily slender and moderately flexible, but in a few species is thick and sturdy. This shaft usually goes over into cartilage at its lateral end where it joins the extracolumella.

The extracolumella consists of a shaft portion and a number of processes, and is composed of cartilage. Versluys identified four processes that extend from the lateral end of the shaft and radiate over the inner surface of the tympanic membrane. These are the pars superior, pars inferior, anterior process, and posterior process, collectively referred to by Versluys as the insertion piece. In some species the anterior and posterior processes are small and poorly defined, or may be absent. The pars superior and pars inferior usually form an elongated bar that extends across the tympanic membrane from its posterodorsal edge to its midpoint, stiffening the membrane and transmitting its vibratory movements.

The junction between columella and extracolumella varies greatly in form. In some species there is a joint with at least a slight amount of mobility. In others there is a discontinuity, most often consisting of a change in the type of tissue, sometimes seen as the intrusion of dense connective tissue between the cartilaginous ends of the two elements. More often this connective tissue surrounds the cartilaginous elements at the junction and fuses them so as to leave no possibility of joint action. Still more often the columella changes to cartilage and then this cartilage grades into that of the extracolumella without interruption.

A strong ligament, the extracolumellar ligament, is inserted on the pars superior or on the posterior process and runs to the intercalary cartilage adjacent to the cephalic condyle of the quadrate bone, or sometimes runs to the condyle itself or to the paroccipital process of the exoccipital. Versluys regarded this ligament as continuing outward through the fibrous layer of the tympanic membrane all the way to the tip of the pars inferior, but in many species there are two distinct ligaments: one from pars superior to intercalary and another running along the pars inferior. Here the term *extracolumellar ligament* is used for the first of these and the term *intratympanic ligament* for the second (Wever and Werner, 1970).

In the iguanid type of middle ear there is an additional process of the extracolumella that Versluys called the internal process. Typically it arises from the extracolumellar shaft close to its connection to the columella and runs dorsally and anteriorly to the inner rim of the quadrate, spreading out fanwise as it does so, and attaching to the surface of the quadrate or more often inserting in a narrow channel along this surface. The majority of lizard species have this type of middle ear (see Table 5-III).

In the gekkonid type of middle ear the internal process is lacking. The columella and shaft portion of the extracolumella form a simple rod without lateral support apart from its enclosure in a fold of mucous membrane. The extracolumella inserts into the dorsal portion of the tympanic membrane and the pars inferior extends anteroventrally to support the central area of the membrane.

A tympanic muscle is present in the gekkonid ear, as Versluys first observed. This muscle runs from the tip of the pars superior and fans out to attach to the ceratohyal process. Versluys thought that this muscle in its contractions produced a reduction in tension of the tympanic membrane and named it the laxator tympani. This action is unlikely; there is good reason to believe that the muscle exerts tension on the membrane and on the ossicular system as a whole. At any rate, it seems more suitable to employ a noncommittal term and call it the extracolumellar muscle (Wever and Werner, 1970). This type of middle ear is found in two lizard families, the Gekkonidae and Pygopodidae.

The third type of middle ear is the scincid type, in which both the internal process and the extracolumellar muscle are lacking. This form of structure occurs in Scincidae, Anguidae, and *Lepidophyma* species among the Xantusiidae.

Divergent types of middle ear occur in all species of Anniellidae, Chamaeleonidae, Xenosauridae, Dibamidae, Feyliniidae, and Lanthanotidae, but no specimens were available for the last three families and no further consideration will be given to them. It is likely that the conductive structures in the first three families represent degenerated forms of the iguanid type; these will be described later. In addition to these, several species among the Agamidae, Iguanidae, Pygopodidae, and Scincidae exhibit divergent types of structure also, and several of these were available for study.

IGUANID TYPE OF MIDDLE EAR

The form of middle ear identified as the iguanid type will first be examined in the collared lizard *Crotaphytus collaris*, following closely the account given by Wever and Werner (1970).

Middle Ear of *Crotaphytus collaris*.   A sketch of the head of this lizard appears in Fig. 6-9. There is a simple external ear opening of oval form, which in an adult of about 95 mm snout-to-vent length measured 2.4 mm wide by 5.5 mm high, and as shown is inclined at an angle of about 35° from the vertical and faces posterolaterally. The surrounding skin is somewhat loose, and the opening varies a little with the position of the head. The tympanic membrane lies about 1 mm below the surface anteriorly and a little deeper posteriorly. This membrane is semitransparent, and the pars inferior on its inner surface can be seen extending from the dorsoposterior edge obliquely forward and downward, ending near the middle of the membrane and imparting to the membrane a slight outward bulge.

After removal of the surrounding skin the tympanic membrane and its supporting framework can more clearly be seen. In Fig. 6-10 much of the membrane itself has been torn away to expose the supporting structures more fully. The frame is of roughly quadrilateral form, with a number of skeletal elements and soft tissues entering into its composition. The quadrate bone provides the anterior and most of the dorsal support, the retroarticular process of the mandible forms the short ventral side, and skin and fascia constitute the posterior border. This post-tympanic support includes a strong ligamentous band extending along the border of the depressor mandibulae muscle, anchored between the cephalic condyle above and the retroarticular process below.

In the figure the ossicular elements are exposed by removal of the central area of the tympanic membrane and of the midportion of the post-tympanic band. Three of the processes of the extracolumella are clearly in view, but only the tip of the fourth may be seen behind the pars superior. The pars inferior is joined at a slight angle to the pars

FIG. 6-9. Head of *Crotaphytus collaris*.
From Wever and Werner, 1970.

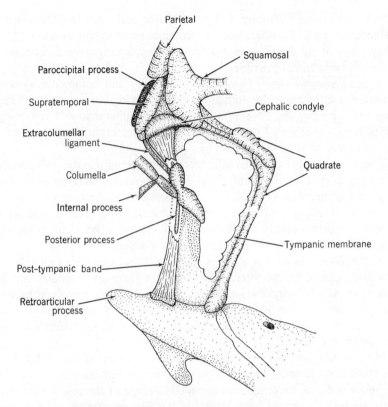

FIG. 6-10. The tympanic frame and tympanic membrane of *Cro-taphytus collaris*, oriented as in Fig. 6-9. From Wever and Werner, 1970.

superior and extends downward and anteriorly. The slender posterior process points directly downward.

The pars superior runs upward in the fibrous layer of the tympanic membrane and from this process the extracolumellar ligament extends to an anchorage on the intercalary cartilage, a part of which is shown in the figure between the cephalic condyle and the supratemporal bone.

The shaft of the extracolumella is shown running inward from its junction with the pars inferior, and at its medial end where it joins the columella it gives off the internal process (only a portion of which is indicated in the figure).

The axis of rotation of the extracolumellar system runs from the cephalic condyle through the pars superior and the proximal end of the posterior process. This axis can be located by moving the tip of the pars inferior in and out, which causes the larger part of the tympanic membrane to move with it. This part of the membrane, immediately adjacent to the pars superior and pars inferior and extending for a short distance ventrally, is stiffer than the remainder and moves as a unit. The other parts of the membrane are thinner and undergo the flexion that makes the movement possible.

A more general view of the tympanic cavity and columellar system is obtained by placing the animal on its back with the nose toward the observer, opening the floor of the mouth and pharynx, and removing the posterior portion of the pterygomandibular muscle. Such an exposure is represented in Fig. 6-11.

There is no definite demarcation of the tympanic cavity: this cavity is simply a part of the pharyngeal space, bounded anteriorly by the mass of the pterygomandibular muscle, laterally by the tympanic membrane, skin area, and quadrate bone, and medially by bones of the otic-occipital region—chiefly the basioccipital. There is no clear posterior boundary, and no Eustachian tube can be defined.

As the figure shows, the shaft of the columella comes into view from a position dorsal to the interfenestral crest (where its stapedial end is hidden in the oval window), and runs posterolaterally, continued by the shaft portion of the extracolumella. The extracolumellar shaft approaches the tympanic membrane and sends its terminal processes over the membrane surface.

Further details of the columellar system are shown in Fig. 6-12. The orientation is the same as in the previous figure, but the essential parts are more isolated. Indicated are the ends of three of the four bones that unite to form a secure anchorage for the dorsoposterior edge of the quadrate: the parietal, squamosal, and the paroccipital process of the exoccipital; the supratemporal, which also enters into this complex, is not visible from this side. The intercalary cartilage assists in the union of these bones, and provides an anchorage for the extracolumellar ligament. The cephalic condyle is an outgrowth of the quadrate, and serves for the dorsal attachment of the post-tympanic band.

This figure also shows the internal process and its attachment along the quadrate. The columella is represented over its entire course with its footplate in the oval window.

Middle Ear of *Sceloporus magister*.   A similarly detailed study was made of a second iguanid species, the desert spiny lizard *Sceloporus magister*. The middle ear structure for this species is represented in the following five figures, and shows only minor points of variation from what has been seen in *Crotaphytus collaris*. The external ear opening of *Sceloporus*, shown in Fig. 6-13, is a short oval and is bordered on its anterior side by a row of 5-7 guard scales. The end of the pars inferior

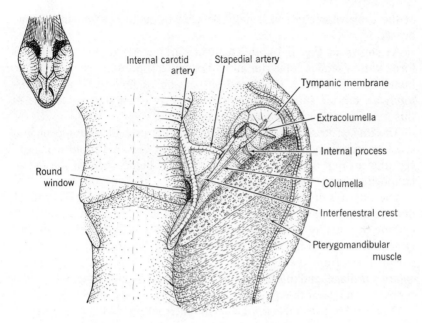

FIG. 6-11. The tympanic cavity and its contents in *Crotaphytus collaris*.
From Wever and Werner, 1970.

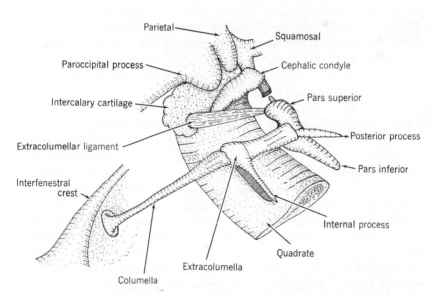

FIG. 6-12. Details of the ossicular mechanism in *Crotaphytus collaris*.
From Wever and Werner, 1970.

of the extracolumella can usually be seen through the tympanic membrane.

As shown in Fig. 6-14, the tympanic frame has a sturdy shell-like form, with a semicircular rim and a heavy deep portion. The same four bones are involved in the dorsal suspension of this frame as in *Crotaphytus*, but the supratemporal is reduced to a thin slip, not visible in this figure and often not readily recognizable.

The four processes of the extracolumella are present, and appear in a cruciate arrangement in this figure and also in Figs. 6-15 and 6-16, except that in Fig. 6-15 the posterior process is hidden behind the extracolumellar shaft.

The internal process is shown in Fig. 6-15 arising from the end of the extracolumellar shaft near its junction with the columella and extending in a nearly dorsal direction, spreading out fanwise with its rim lying in a groove in the surface of the quadrate. The extracolumellar ligament arises from the end of the pars superior, runs along the intercalary cartilage and partially encloses it, and finally attaches to the cephalic condyle of the quadrate.

Figure 6-16 gives a clear view of the internal process and its relation to the quadrate.

A reconstruction of the *Sceloporus* middle ear, based on serial sections, is given in Fig. 6-17. This form of representation is particularly useful for observing the degree of mobility of the system and inferring its responsiveness to sound pressures. Here may be noted the location of the two principal extracolumellar processes along the dorsoposterior surface of the tympanic membrane, with the long shaft portions of extracolumella and columella leading to the stapedial footplate in the oval window, braced laterally by the internal process.

The Function of the Internal Process.   It has already been suggested that the internal process of the extracolumella has a supportive function for the columellar shaft. A series of experiments was concerned with this function, and consisted of first measuring the sensitivity to a range of tones with the columellar system in its normal condition, and then repeating the measurements after the internal process was freed of its connection with the quadrate. The results of one of these experiments on a specimen of *Crotaphytus collaris* are shown in Fig. 6-18, where the solid curve represents the normal condition and the broken curve the performance after sectioning the internal process. The differences between the two sensitivity curves are small, and at some points along the frequency scale represent improvements and at other points indicate losses. In general, the tendency is for the responses to the low tones to be improved a little, and for those to the high tones to be

a   *b*

FIG. 6-13. The head of *Sceloporus magister* and an enlarged view of the tympanic membrane. From Wever, 1973b.

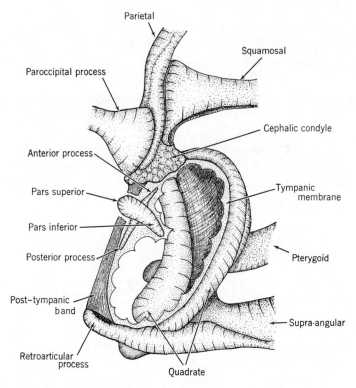

FIG. 6-14. The tympanic frame of *Sceloporus magister*. The view is from the right as in the preceding figure. Much of the tympanic membrane is removed to disclose the deeper structures. From Wever, 1973b.

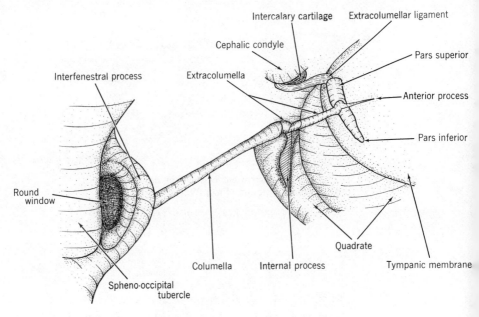

FIG. 6-15. The columellar apparatus of *Sceloporus magister*, seen from an anterior, medial, and ventral position. Scale 15×. From Wever, 1973b.

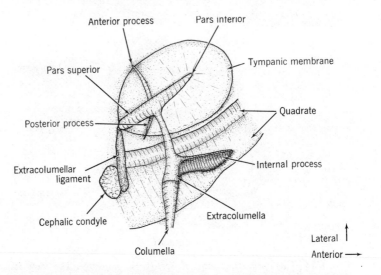

FIG. 6-16. The tympanic membrane and extracolumella of *Sceloporus magister*, seen from a ventral and medial position. Scale 15×. From Wever, 1973b.

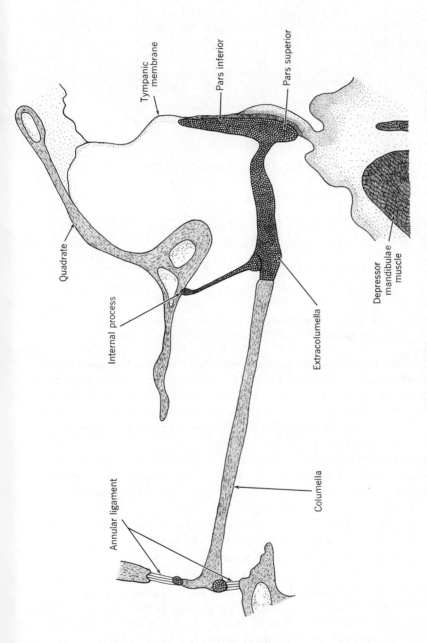

Fig. 6-17. Reconstruction of the middle ear of *Sceloporus magister*, based on serial sections. From Wever, 1973b.

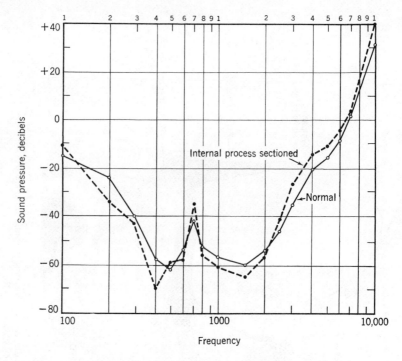

FIG. 6-18. Effects on sensitivity of sectioning the internal process in *Crotaphytus collaris*. From Wever and Werner, 1970.

slightly impaired. Because the observed effects are small, it seems that the internal process contributes little to the general effectiveness of the ossicular mechanism in this ear, and its function most likely is simply protective, to give lateral bracing that guards in some degree against injury to the mechanism during vigorous activity such as pursuit of prey, fighting, or the like.

In other species in which the internal process is less flexible, or consists of a substantial mound-like elevation as in *Callisaurus draconoides*, the effects of sectioning this process are more significant. Figure 6-19 gives a sectional view of the middle ear of this species, and Fig. 6-20 presents the structure in a more complete form, based upon a dissection. The effects on sensitivity of breaking the connection between extracolumella and quadrate are presented in Fig. 6-21. There is little change in the low frequencies, but beyond 1000 Hz there are losses up to 14 db and averaging 11 db over the high-frequency range. Such losses can be explained as a result of a reduction in the stiffness of the responsive mechanism. Changes of similar amount in the medium high frequencies, and extending in slight degree to the low tones as well, were observed in *Uma notata* as Fig. 6-22 shows.

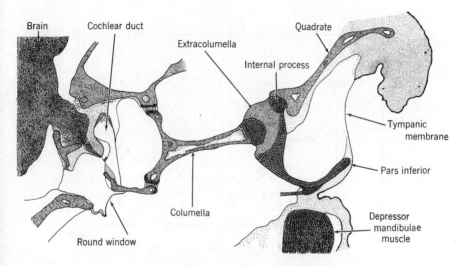

FIG. 6-19. The middle ear of *Callisaurus draconoides*.
Scale 25×. From Wever, 1973b.

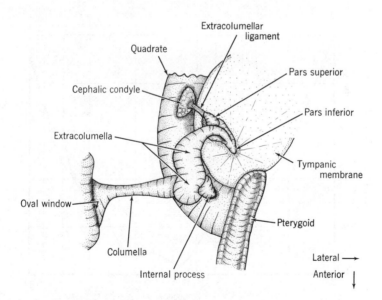

FIG. 6-20. The ossicular mechanism in *Callisaurus draconoides* in a
ventrolateral view. The specimen is rotated about 180° from its position in the preceding figure. The intercalary lies on the cephalic condyle and is covered by the end of the ligament. Scale 15×.

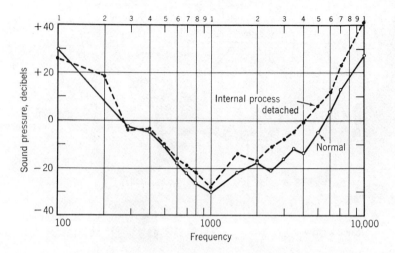

FIG. 6-21. Effects on sensitivity of detaching the internal process from the quadrate in *Callisaurus draconoides*.

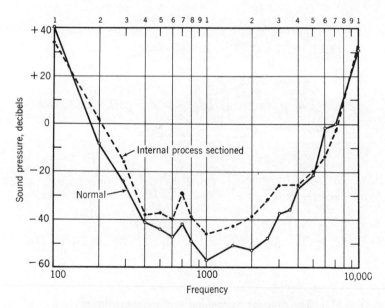

FIG. 6-22. Effects on sensitivity of sectioning the internal process in *Uma notata*.

The Transformer Function of the Middle Ear

Extensive investigations on the middle ear of mammals, especially on the cat (Wever and Lawrence, 1954; Békésy, 1960), have established that the function of the middle ear is to provide a mechanical transformer to match the relatively high mechanical impedance of the inner ear to the low impedance of the aerial medium from which sounds are most commonly received. Abundant evidence shows that this same function is served by the middle ear of reptiles.

In a specimen of *Crotaphytus collaris* sensitivity measurements in terms of cochlear potentials as recorded from the round window were first obtained under normal conditions, as shown by the solid curve of Fig. 6-23. Here the best sensitivity is in the region of 200 to 2500 Hz, reaching −40 db at two points along this range. The columella was then severed at a point close to the oval window. This operation was done with particular care to avoid any other disturbance of the preparation, and especially of the sound delivery system, which consisted of a tube sealed over the external auditory meatus. Also after the initial sectioning of the columella a small piece of bone was removed from

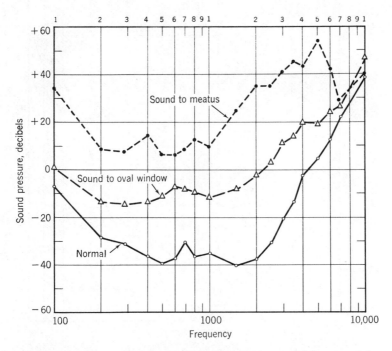

Fig. 6-23. Effects of sectioning the columella on sensitivity to aerial sounds in a specimen of *Crotaphytus collaris*. From Wever and Werner, 1970.

the peripheral part to prevent any reestablishment of contact between the severed ends.

The results of stimulating the ear in the usual way, through a sound tube over the meatus, are shown in the upper curve of Fig. 6-23. The sensitivity is greatly reduced, except for tones at the extreme upper end of the range, with a mean loss up to 4000 Hz of 52 db and a maximum reduction at 2000 Hz of 73 db.

A second animal treated similarly gave the curves of Fig. 6-24, and three others produced results in close agreement. These results show clearly that cutting the columella causes a great reduction in the acoustic vibrations reaching the inner ear. The effect of this procedure is complicated, however, and warrants further consideration.

Cutting the columella produces two principal changes; one is the interruption of the conductive path from tympanic membrane to cochlear fluid, with loss of the transformer action of the middle ear, and the other is an interference effect now established between two sound

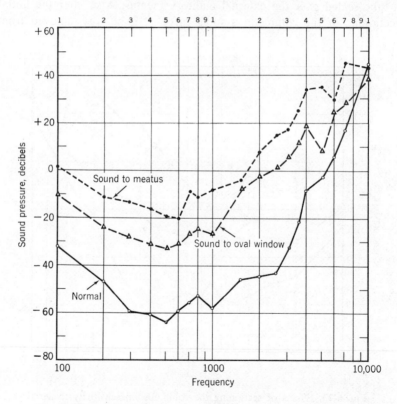

FIG. 6-24. Results similar to the preceding, in a second specimen of *Crotaphytus collaris*. From Wever and Werner, 1970.

paths, one by way of the oval window and the other through the round window. Ordinarily the round window pathway may be ignored, for on account of the transformer action of the tympanic membrane and ossicles the oval window pathway is so much the more efficient that the other is negligible by comparison. When the transformer action is lost, however, these two pathways are of comparable effectiveness and interaction between them becomes a significant matter. When sound pressures are exerted simultaneously at the two cochlear windows they act on the basilar membrane in opposite directions, and tend to cancel one another's effects.

The degree of cancellation that occurs under the conditions of the experiment, with the sound stimulus applied as usual at the external auditory meatus, depends on such factors as the relative lengths of the two sound paths and the impedances along these paths. Also involved is the wavelength of the tones, for this wavelength, along with path and impedance differences, determines the form of interaction of the two pathways. Under certain conditions the two pathways could cancel one another when the effective phases are equal; and under other conditions they might reinforce one another—when the effective phases are opposite.

There is still another complication caused by the presence of the disconnected portion of the columella adherent to the tympanic membrane. This loaded membrane acts as a reflecting surface, and the sounds applied to it only partially pass through as aerial waves into the tympanic cavity. Of course this obstruction effect is present normally, but is small relative to the transmitted sound when the usual transformer action is in force.

To separate these effects of columella cutting—the transformer loss as distinct from interaction of pathways and membrane loading effects —the mode of stimulation was changed so that the sound waves were permitted access to only one cochlear pathway, the one through the oval window. A small sound tube was sealed over this window, and the sounds applied through it. Reflections at the tympanic membrane surface and interfering stimuli through the round window were thus prevented.

Stimulation by this method gave the results shown by the long-dashed curves of Figs. 6-23 and 6-24. The loss of sensitivity is less and represents simply the elimination of the transformer action of the middle ear. This loss is still substantial, however, and for the range from 300 to 4000 Hz its average is 35 db for each of these two specimens. Other animals gave closely comparable results.

A further analysis of the action of the middle ear involved the use of vibratory stimuli. Fig. 6-25 shows results obtained by stimulating with a vibrating needle applied in two ways, by bringing the point of the needle in contact with the tympanic membrane just over the middle portion of the pars inferior (short-dashed curve), and then, after cutting the columella, by driving its remaining proximal end with the needle tip inserted into its small marrow cavity so as to produce movements in line with its longitudinal axis (long-dashed curve).

Also in this animal, before the columella was cut, a normal aerial curve was obtained in the usual way, with results indicated by the solid curve. Similar results for another animal are presented in Fig. 6-26.

The manner of plotting the sound intensity in these curves should be noted. For the vibratory functions the intensity is expressed in decibels relative to a amplitude of 1 mμ. Also for the aerial curves in these figures the intensity is in somewhat comparable terms, in decibels relative to a mean amplitude of motion of the air particles of 1 mμ as measured at the entrance to the auditory meatus.

There is a good deal of similarity among these curves. In all the best sensitivity is in the medium low and middle frequencies. The minimum stimulation required for the standard response of 0.1 μv is at 500 Hz for two of the curves and at 700 Hz for the other, and a secondary minimum lies between 1500 and 3000 Hz for all three. Again, for all curves there is a rapid decline in sensitivity in the low frequencies and an even more rapid one for the high frequencies. From these results we gain insight into the action of the middle ear and the important part played by the cochlea in the determination of sensitivity.

Let us first examine the two conditions of stimulation. For aerial sounds under usual conditions the tympanic membrane is the receptive surface, and to a considerable degree it provides a summation of the forces exerted by the air particles striking its surface. The resultant force, after a further mechanical transformation (if a lever action is present), is conveyed to the stapedial footplate and thereby impressed upon the fluid of the cochlea.

The mechanical actuation of the columellar system is simpler. The vibrating needle applied to the pars inferior causes this element to move in the direction of the applied force and with the applied amplitude. The remaining parts follow this motion to a degree depending on their coupling with the pars inferior. Evidently this coupling is close along the path through columella and footplate to the cochlear fluid.

The situation is still simpler when the vibrator is applied to the stump of the columella after the more peripheral parts have been removed. The columella and footplate follow the movements of the vibrating needle and transmit them directly to the cochlear fluid. In mechanical driving of the pars inferior or of the columellar shaft as described, the mechanical impedances of the cochlear structures do not play any part.

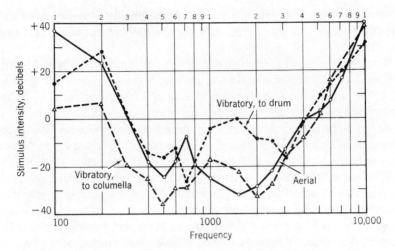

FIG. 6-25. Vibratory sensitivity curves in a specimen of *Crotaphytus collaris*, obtained under two conditions, together with an aerial curve. For the two vibratory functions the sound intensity is in decibels relative to a zero amplitude of 1 mμ. For the aerial function the intensity is in decibels relative to an air particle amplitude of 1 mμ at the entrance to the external auditory meatus. From Wever and Werner, 1970.

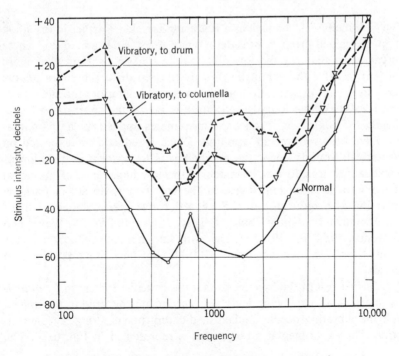

FIG. 6-26. Results similar to the preceding, in a second specimen of *Crotaphytus collaris*.

These impedances are small relative to the inherent impedance of the driving crystal, and the driven structures follow the impressed motions directly.

Because the general form of the sensitivity function persists for the different conditions of stimulation, two general conclusions are drawn. (1) The middle ear operates over the middle frequency range so as to compensate in considerable degree for impedance variations, giving at least an approximation to uniform stapedial displacements for given sound amplitudes over a wide frequency range. This same conclusion was reached by Guinan and Peake (1967) for the action of the cat's middle ear on the basis of extensive measurements of ossicular motions in response to aerial sounds. The middle ear provides a very effective transfer of acoustic energy from air to cochlea.

(2) The cochlea plays a major role in the determination of the ear's sensitivity. Evidently there is a certain freedom of motion of the basilar membrane and its related structures, so that these can conform to a pattern of motion determined largely by their own mechanical characteristics and do not merely reflect the actions imposed by the stapedial footplate. These characteristics include the mass and stiffness of the moving structures along the cochlea, together with such features as the numbers and distribution of the hair cells and the character of the tectorial stimulating mechanisms for these hair cells.

The Columellar Mechanism as a Reducing Lever.   Further information is to be gained from a consideration of Figs. 6-25 and 6-26. These curves show that for the low and middle tones it is somewhat more efficient to drive the columella directly than to drive the entire middle ear system with the vibrating needle on the pars inferior: for most tones up to 2500 Hz the long-dashed curve is below the short-dashed one. Clearly for this range of tones the movements applied to the pars inferior have been reduced in amplitude. Such a reduction is expected on theoretical grounds, for if the middle ear is to perform its function as a mechanical transformer in matching the low impedance of the aerial medium to the high impedance of the ear it must do so, in part, by operating as a reducing lever. Such action has been demonstrated for the mammalian ossicular chain (Békésy, 1960, pp. 95-126; Wever and Lawrence, 1954, pp. 69-114). Evidently an amplitude reduction occurs in this lizard ear for tones up to 2500 Hz, but fails above this point in the scale.

A further test of the hypothesis that the ossicular system in *Crotaphytus* operates as a reducing lever was carried out by driving the system with the vibrating needle applied at different points along the pars inferior. The experimental arrangement is represented in Fig. 6-27. The

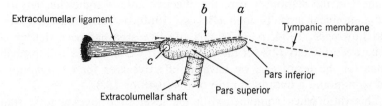

FIG. 6-27. The lever action of the middle ear in *Crotaphytus collaris*.
From Wever and Werner, 1970.

tympanic membrane is indicated in cross section by the broken line, and only a portion of the extracolumellar shaft is shown. The pars inferior was driven at two points, *a*, at its tip, and *b*, near its middle close to the junction with the pars superior. The axis of rotation of the system is indicated at *c*. The vibratory frequency was 500 Hz, and the driving amplitude at each position was adjusted to produce a cochlear potential of 0.1 µv. The difference between the two inputs to the vibrator was significant, and indicated that on the average the required driving amplitude at *a* was 2.14 times that at *b*.

From these results and measurements of the physical dimensions, the rotational axis at *c* could be located. The distances *c-a* and *c-b* are the two arms of a lever of the second class (a reducing lever) whose ratio is 2.14 and whose difference in length *a-b* was 1.27 mm. From these data the length of the arm *a-c* was determined as 2.38 mm—a distance consistent with visual observation of the movement of the system under static displacement in response to a force impressed at *a*. The location of point *b* was chosen as roughly over the place of attachment of the extracolumellar shaft, as the figure shows.

Thus it appears that the lever action in this animal produces an amplitude reduction of approximately twofold. This reduction is accompanied by a corresponding increase in force, and such an increase is desirable as a means of matching the ear's impedance to that of the aerial medium.

Further Impedance Matching. The impedance matching by ossicular lever action just indicated is limited in amount, and further matching is needed. No doubt the principal amount in this ear, as found in the ears of mammals, is provided by the areal ratio between the effective surface of the tympanic membrane and the surface of the stapedial footplate. In one of the *Crotaphytus* specimens the tympanic surface was measured as 9.7 sq mm and the stapedial surface was 0.48 sq mm.

Because the edges of the tympanic membrane are fixed, and a gradient of motion must exist between the free middle portion of the membrane and the stationary rim, the effective area is somewhat less than the anatomical area. In man (Békésy, 1960, p. 102) and cat (Wever and Lawrence, 1954, p. 113) this effective area has been determined as between 60 and 72% of the total area. If we take 66% as a probable value for the lizard ear, we have an effective area for *Crotaphytus* of 6.4 sq mm and a hydraulic ratio of 6.4/0.48 or 13.3.

This ratio when combined with the lever ratio gives a total transformer ratio of 2 × 13.3 or 26.6 times. The impedance transformation provided by the mechanism is the square of this amount, which is 710. This means that the impedance of the air, which is 41.5 mechanical ohms per sq cm, will appear as 710 × 41.5 or 29,465 mechanical ohms per sq cm as the sounds are presented to the cochlea.

We do not know what the impedance of the lizard cochlea is, but there are some rough indications of the magnitude of this impedance in human ears. A number of attempts have been made to determine the impedance in human ears by measuring the reflection of sounds applied at the external auditory meatus, and though this method is technically difficult it has yielded impedance values varying from 1.6 megohms at 100 Hz to 121,000 ohms at 1600 Hz (see the discussion in Wever and Lawrence, 1954, pp. 57 ff., 385 ff.). Békésy (1960, pp. 172 ff.) made observations on the volume displacements of the cochlear fluids produced by sounds, and from these data calculated the impedance as 1.9 megohms at 160 Hz, and falling progressively to 200,000 ohms around 2000-4000 Hz. The inner ear of the lizard, being much smaller in dimensions and mass, probably has an impedance much below that of man, and the figure given above of about 30,000 mechanical ohms per sq cm may be of the proper order of magnitude.

If this calculated impedance matching for the lizard ear is considered inadequate to account for the high degree of sensitivity found, then the impedance matching problem will need to be explored further. A line of experimentation that is promising in this relation has been carried out by Khanna and Tonndorf (1972) in which the movements of the cat's tympanic membrane in response to sounds have been measured by holography. Their observations support the hypothesis that the local flexions of the tympanic membrane itself provide a form of lever action that may add to the other actions mentioned. Helmholtz conceived that a curved membrane, operating on a catenary principle, can amplify the force impressed upon its center to produce a relatively large force along the plane of the membrane. If this catenary action holds for the lizard tympanic membrane, then a still greater transformer action is possible and a more effective impedance matching may be produced, if indeed more is needed than that already indicated.

THE GEKKO TYPE OF MIDDLE EAR

The type of middle ear found among the geckos will be discussed in two species, *Gekko gecko* and *Eublepharis macularius*, following the account presented earlier by Werner and Wever (1972).

Middle Ear of *Gekko gecko.* The head of *Gekko gecko* is pictured in Fig. 6-28, with the external auditory opening appearing as an oval whose long axis is inclined about 20° from the vertical. The end of the pars inferior is just visible through the tympanic membrane, and careful scrutiny of a living specimen shows that the membrane is slightly bulged outward at this point.

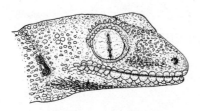

FIG. 6-28. The head of *Gekko gecko.* From Werner and Wever, 1972.

The supporting framework for the tympanic membrane is represented in Fig. 6-29. The principal support is provided by the quadrate, to whose curved rim the anterior edge of the membrane is attached. The remaining edges of the membrane are less firmly anchored to skin and fascia, which themselves are attached to bony and cartilaginous elements. The dorsal elements of the framework are the ends of three bones, the squamosal, parietal, and paroccipital process of the exoccipital, which fuse to the end of the quadrate with particular aid from the cephalic condyle and the intercalary cartilage. The quadrate is firmly seated on the end of the mandible, with bracing by the pterygoid. The retroarticular process of the mandible is directly involved in the anchorage of the ventral side of the tympanic membrane. Posterior support is provided by the ceratohyal process, a rod of cartilage that extends dorsally and expands in spatulate form to attach to the paroccipital process. A small ligament, the hyoarticular, runs from the ceratohyal process to the tip of the retroarticular process.

The tympanic membrane and the columellar mechanism are portrayed in Fig. 6-30 with the angle of view, which is from below, anterior, and lateral, indicated by the inset drawing. The inner surface of the tympanic membrane is seen, with the insertion piece on its dorsal half. In this inset the inner end of the columella is hidden beneath the interfenestral crest, and the round window niche also is out of view ventral to this crest.

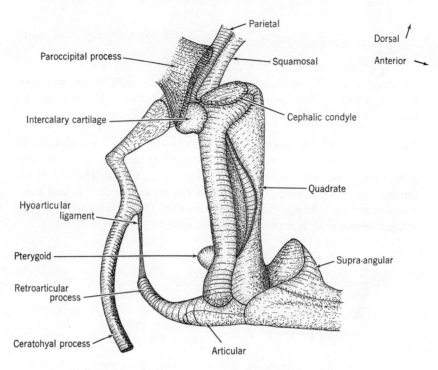

Dorsal ↗
Anterior →

FIG. 6-29. The framework of the tympanic membrane in *Gekko gecko*.
From Werner and Wever, 1972.

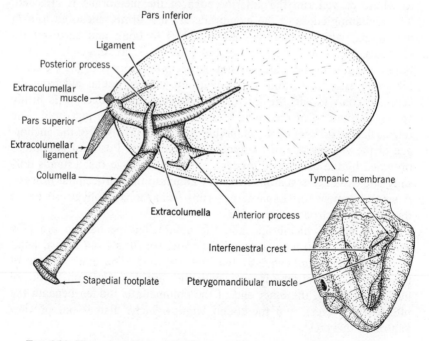

FIG. 6-30. The columellar mechanism in *Gekko gecko*, oriented as shown by
the inset drawing. From Werner and Wever, 1972.

The main part of this figure shows the middle ear mechanism in isolation. The usual four processes of the extracolumella are present. The pars superior lies near the dorsal edge of the membrane, and runs into the pars inferior that extends in rapidly tapering form to the center of the membrane. A shaft portion runs off from the junction of pars superior and pars inferior, and at the same time the anterior process, of rather complex form, extends over the anterior-superior quadrant of the membrane. The posterior process branches off from the shaft of the extracolumella as shown.

From the pars superior a thick ligamentary band (the extracolumellar ligament) runs anterosuperiorly, and opposite its insertion is a thin ligament that runs over the inner surface of the membrane. Also attached to the pars superior is the conical tip of the extracolumellar muscle, a structure peculiar to the gekkonids and pygopodids. This small muscle is difficult to locate by dissection, but may be exposed by removing a part of the tympanic rim.

Middle Ear of *Eublepharis macularius*.   The middle ear structures of *Eublepharis macularius* differ from those of *Gekko gecko* in only minor respects except for their orientation in the head. The tympanic membrane in this species has its longer axis nearly horizontal, and the extracolumellar processes appear as if rotated about a quarter circle in comparison with other species. Thus the pars inferior extends nearly anteriorly, the pars superior is posterior, the anterior process is nearly dorsal, and the posterior process is posteroventral. These processes extend into the fibrous layer of the tympanic membrane much as described for *Gekko gecko*. The columella and extracolumella are firmly joined, as in all the geckos.

The extracolumellar muscle is relatively large. It arises from an extension of the ceratohyal process and inserts on the end of the pars superior. The extracolumellar ligament also inserts on the pars superior, and is anchored on the intercalary cartilage. Here is a variation from *Gekko gecko*, in which this ligament mainly attaches to the paroccipital process.

This variation in *Eublepharis macularius* may be seen in Fig. 6-31, which is a cross-sectional view of relevant portions of this ear with the ligament and muscle attachments.

Functional Observations.   The contributions of the middle ear mechanism to sensitivity are indicated in Fig. 6-32 for a specimen of *Gekko gecko* as the difference between the two curves. In this figure the solid curve represents sensitivity under normal conditions, and the broken curve the sensitivity after the columella was cut close to the oval win-

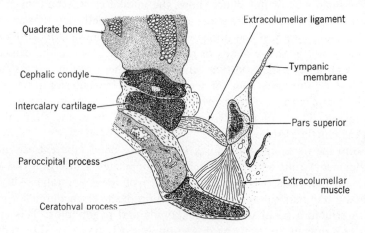

FIG. 6-31. The extracolumellar muscle and ligament of *Eublepharis macularius*. From Werner and Wever, 1972.

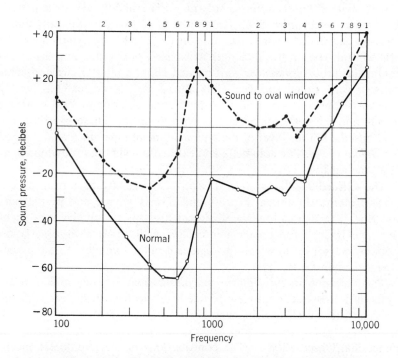

FIG. 6-32. Auditory sensitivity in *Gekko gecko* for aerial sounds under normal conditions and after cutting the columella and presenting the sounds to the oval window. From Werner and Wever, 1972.

dow and the sound tube sealed over this window. The differences between these two functions, representing the transformer effect of the
middle ear, amount to 30 db on the average for the entire range. The
greatest effect is at 700 Hz, where it amounts to 70 db, but it is considerable for all the low tones, averaging 41 db for the range of 400 to
3000 Hz. The transformer effect declines in the upper frequencies, and
is relatively small beyond 4000 Hz.

A second animal examined in the same manner gave the results of
Fig. 6-33. In this ear the average loss from columella sectioning was
28.2 db, averaged over the entire range of the measurements. The maximum effect was 59 db at 600 Hz, and the mean for the low tones up to
1500 Hz was 41.5 db. There is good agreement between these two
specimens, and clearly the middle ear performs its function well in
this species, with its greatest effectiveness in the medium low frequencies.

Further indications of the ear's operation were obtained by the use
of vibratory stimulation. Figure 6-34 presents results for the same animal whose aerial curves are given in Fig. 6-32. Here the observations
were made with two ways of applying the mechanical stimuli, with the
tip of the vibrating needle located on the tympanic membrane just over
the end of the pars inferior and driving the ossicular system as a whole
(solid line), and then, after the columella was cut close to the footplate,
by applying the needle to this cut end and thereby driving the footplate
(broken line).

The two functions shown in Fig. 6-34 are similar, in that both show
the best sensitivity in the low frequencies, around 500 Hz, and a secondary maximum in the medium high frequencies from 2000 to 4000 Hz.
It is notable that these maximums are in the same regions as observed
for aerial stimulation.

Results on a second animal with vibratory stimulation are presented
in Fig. 6-35. Again the greatest sensitivity is found in the medium low
tones, and a secondary region of good responsiveness occurs in the high
tones.

Observations similar to the above were made on specimens of *Eublepharis macularius*. With aerial stimulation one animal gave the results shown in Fig. 6-36, where the solid line represents the normal
condition and the broken line represents the performance with the columella sectioned and the sound tube sealed over the oval window. The
difference between these two curves, which shows the effectiveness of the
middle ear transformer action, amounts to 35.2 db on the average over
the range up to 3500 Hz, and is most marked between 400 and 1000
Hz where it averages 49.3 db. Over the entire range the average is
29.6 db.

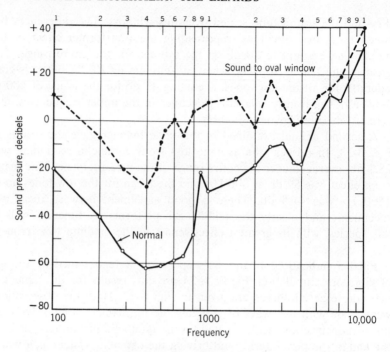

FIG. 6-33. Sensitivity functions in a second specimen of *Gekko gecko* under the same conditions as in the preceding figure. From Werner and Wever, 1972.

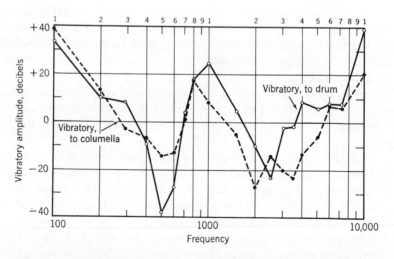

FIG. 6-34. Sensitivity functions in a specimen of *Gekko gecko* with mechanical stimuli applied in two ways. The curves show the amplitude in decibels relative to 1 mμ required for a response of 0.1 μv. From Werner and Wever, 1972.

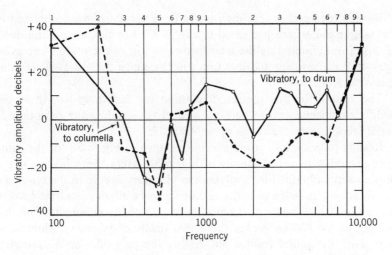

FIG. 6-35. Sensitivity in a second specimen of *Gekko gecko* under the same conditions as in the preceding figure. From Werner and Wever, 1972.

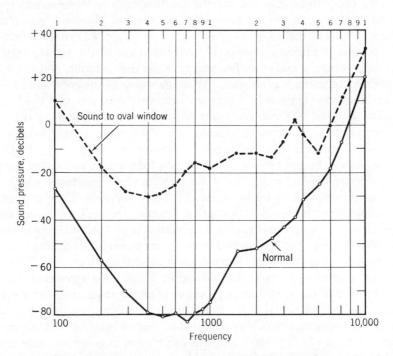

FIG. 6-36. Sensitivity to aerial stimulation in a specimen of *Eublepharis macularius*. From Werner and Wever, 1972.

Results for another specimen of *Eublepharis macularius*, obtained in the same manner, are presented in Fig. 6-37. For this animal the over-all loss from eliminating the transformer action of the middle ear averages 45.7 db, and for the low tones in the range 290 to 1000 Hz this loss is 56.3 db.

The results for this species, like those for *Gekko gecko*, show the middle ear mechanism to be highly effective in aiding the reception of aerial stimuli over the range up to 4000 Hz.

Results for vibratory stimulation are shown in Fig. 6-38 for the same animal used with aerial sounds in Fig. 6-36. Here the solid-line curve shows results obtained by applying the vibratory needle to the tympanic membrane at its bulge over the end of the pars inferior, and the broken curve shows the results of driving the stump of the columella. For this animal, as for *Gekko gecko*, these two forms of vibratory stimulation give generally similar results, but driving the pars inferior is somewhat more effective in the medium low frequencies, and driving the columella somewhat better in the high frequencies.

In an additional specimen of *Eublepharis macularius* this analysis of middle ear performance was carried still further. The vibrator was applied in successive tests at four different positions along the ossicular chain: (1) over the tip of the pars inferior as before, (2) over the middle portion of the pars inferior, (3) over the head of the extracolumellar shaft where it makes its junction with the pars inferior, and finally, after the columella was severed, (4) to the stump of the columella. The four functions obtained in this series were similar except that the first, and to some degree the second, showed greater sensitivity than the others at the primary and secondary peaks, with differences around 6-12 db. Outside these peaks the variations were never more than 4 db. Figure 6-39 shows (by the broken line) the first function of this series, with the vibrator applied at the tip of the pars inferior. This figure also includes an aerial curve for the same ear. The comparison between these functions must be made with caution, because the ordinate scales are different, and sound pressure in its production of fluid motions in the cochlea involves the mechanical impedances of middle and inner ears in ways more critical than direct mechanical driving does.

The results for the two gekkonid species are in good agreement. Loss of the middle ear transformer produces a severe impairment of sensitivity in both, which signifies that these two species similarly benefit from the impedance matching provided by this mechanism. In both species the most effective service of this mechanism is in the medium low frequencies, around 500-800 in *Gekko gecko* and around 500-1000 in *Eublepharis macularius*. The gain from this transformation is similar in amount, reaching maximum values of 59 and 70 db in the two *Gekko*

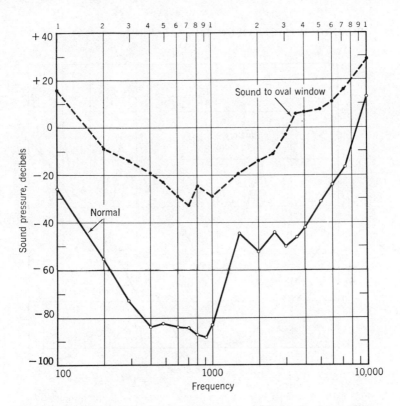

FIG. 6-37. Aerial sensitivity in a second specimen of *Eublepharis macularius.*
From Werner and Wever, 1972.

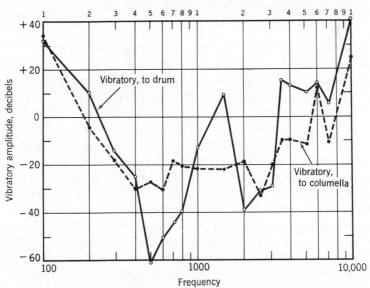

FIG. 6-38. Vibratory sensitivity in a specimen of *Eublepharis macularius* (the same ear shown in Fig. 6-36 for aerial stimulation). From Werner and Wever, 1972.

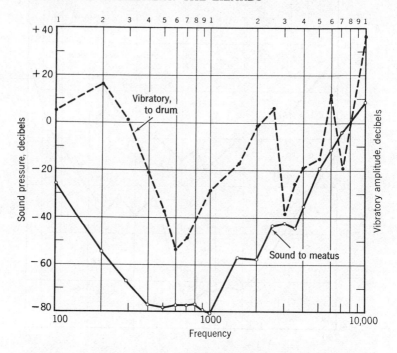

FIG. 6-39. A comparison of aerial and vibratory stimulation in *Eublepharis macularius*. The solid curve refers to the left-hand ordinate and the broken curve to the right-hand ordinate, on the same decibel scale. From Werner and Wever, 1972.

specimens and 57 and 66 db in the two *Eublepharis* specimens. For the entire range of the tests, the average effectiveness is 28.2 and 29.6 db for the two *Gekko* specimens and 29.6 and 45.7 for the two *Eublepharis* specimens. This comparison is perhaps more meaningful if the range is curtailed, for it appears that the transformer action fails to operate for the extreme high tones. If only the range up to 4000 Hz is considered, the values become 35.0 and 35.2 for the two *Gekko* specimens and 35.2 and 49.6 for the two *Eublepharis* specimens.

The consideration of the variations between these two gekkonid species made earlier (Werner and Wever, 1972) led to the conclusion that the middle ear mechanisms of these two species differ in effectiveness to a significant degree, accounting for a large part of the normal sensitivity difference between them, which amounts to about 20 db in the specimens reported here and to a somewhat larger amount, around 27 db, in a larger group. However, this transformer action does not explain all the difference, and evidently the cochleas themselves are responsible for an important share.

As brought out in the discussion of *Crotaphytus collaris*, the transformer action of the middle ear is considered as involving two steps, a kind of hydraulic action and an ossicular lever action.

The hydraulic action involves a summation of the forces exerted by the air particles over the surface of the tympanic membrane and a transmission of this total force to the smaller area of the stapedial footplate. In a specimen of *Gekko gecko* the tympanic membrane had an area of 48 sq mm and the columellar footplate an area of 0.8 sq mm, for a ratio (uncorrected) of 60 times. In a specimen of *Eublepharis macularius* the area of the tympanic membrane was measured as 28.6 sq mm and that of the columellar footplate was 0.84 sq mm, yielding an uncorrected ratio of 34 times. A correction of these values to take account of the fact that the edges of the tympanic membrane are restrained, by the use of an effective area of 66% of the total as was done for *Crotaphytus*, gives final values for the hydraulic ratio of 39.6 for *Gekko gecko* and 22.4 for *Eublepharis macularius*.

A comparison of the sensitivity curves for these gekkonid species (a) when the ear is normal and is stimulated with aerial sounds and (b) when the columella is interrupted and the stimulation is with mechanical vibrations reveals some striking similarities despite the great differences in these forms of stimulation. This comparison may be made between the solid-line curves of Figs. 6-32 and 6-34 for *Gekko gecko*, and the solid-line curves of Figs. 6-36 and 6-38 for *Eublepharis macularius*, or more conveniently for a specimen of this second species in Fig. 6-39. This comparison shows a similar form for aerial and vibratory functions in all instances, except that the vibratory curves undergo much more abrupt alterations in their courses. Both functions exhibit primary and secondary regions of best sensitivity, and at nearly the same positions along the frequency scale. In all there is a rapid decline of sensitivity at the upper end of the frequency range.

The persistence of these features despite such drastic alterations as cutting the columella points to the strong involvement of the inner ear, as was inferred from the similar relations found for *Crotaphytus collaris*.

SKINK TYPE OF MIDDLE EAR

The middle ear in the skinks is the simplest of the three normal types found among lizards; there is no internal process and no extracolumellar muscle. This conductive mechanism is considered here in much the same way as has been done for the other types, by examining the details of structure in representative species and assessing the effectiveness of the sound-conductive system in terms of the changes in sensitivity caused by interrupting the ossicular chain.

The middle ear structure of skinks was examined most thoroughly in specimens of *Mabuya brevicollis* and *Mabuya homalocephala*. Several other species were studied as well, and showed only minor variations from these. The variations were mainly in the accessory processes of the extracolumella and in relations to the hyoid apparatus, and are considered as having little or no functional significance.

A view of the head of *Mabuya homalocephala* is presented in Fig. 6-40. In this species the external ear opening is nearly round (it appears oval in this view because it is directed posterolaterally and therefore is seen obliquely) and has a diameter of about 1 mm. A short meatus leads inward to a depth of about 1 mm and then expands almost suddenly in a posterior direction to a diameter of about 3 mm to accommodate the tympanic membrane. The pars inferior extends along the posterodorsal radius of this membrane, but its tip is not usually visible through the small aperture, though it may be disclosed by pushing back the skin.

The location of the tympanic membrane and its surroundings in the head are shown in Fig. 6-41 in a dissected specimen of *Mabuya brevicollis*. Shining through the semitransparent membrane may be seen the whole of the pars inferior and a part of the pars superior. The tip of the pars inferior, along with the tympanic ligament on its external surface, produces a slight outward bulge of the middle of the membrane.

The structures that form the supporting frame of the tympanic membrane are shown in this drawing, and in further detail in Fig. 6-42. The principal support of the membrane is provided by the quadrate, whose shell-like rim describes a semicircle and contains the membrane on its dorsal, anterior, and ventral sides. Posterior support is afforded by the post-tympanic band, a strong ribbon of ligament extending from the cephalic condyle above to the retroarticular process below. Additional support comes from membranous tissue that extends posteriorly from the post-tympanic band and embeds the slender ends of the ceratohyal and ceratobranchial processes.

The quadrate is anchored to the skull by a number of bony processes: by the pterygoid anteroventrally and by a fusion of four bones dorsally that includes the squamosal, parietal, supratemporal, and paroccipital process of the exoccipital. The prootic lends rigidity in a medial direction.

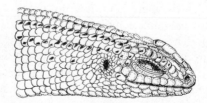

FIG. 6-40. The head of *Mabuya homalocephala*.

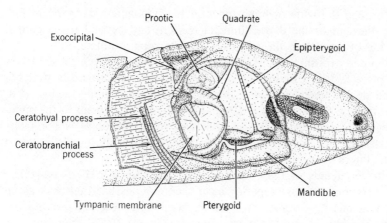

FIG. 6-41. A dissection of the head of *Mabuya brevicollis* to show the tympanic structures. From Wever, 1973a.

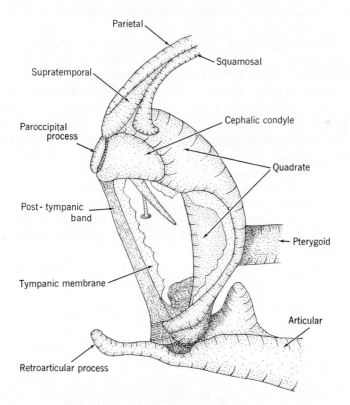

FIG. 6-42. The tympanic frame and its attachments in *Mabuya brevicollis*. Much of the tympanic membrane is removed to reveal the supporting structures. Scale 8×. From Wever, 1973a.

Through the wide opening made in the tympanic membrane in Fig. 6-42 the end of the extracolumella is seen, and beyond it the medial portion of the columella with its widely flanging footplate.

A broad exposure of the middle ear region from the ventral side gives the picture shown in Fig. 6-43. The tympanic membrane is well exposed on the inner side, with the ossicular chain running medial to it. The round window lies in a depression anterior and dorsal to the spheno-occipital tubercle. Between oval and round windows is the interfenestral crest, largely hidden in this view by the anterior continuation of the internal carotid artery.

Further details of the ossicular chain are shown in Fig. 6-44. The pars inferior and pars superior form a continuous bar that runs along the tympanic membrane, with the intratympanic ligament inserted between the fibrous and the more superficial layers of this membrane. A short posterior process is sometimes present, running off from the pars superior as shown. In one specimen of *Mabuya brevicollis* this process was present on one side but not on the other. An anterior process was not found.

From the end of the pars superior an extracolumellar ligament runs to an attachment on the cephalic condyle. Versluys described the intratympanic and extracolumellar ligaments as continuous, but in the species examined there were two separate ligaments as represented here. The relations vary somewhat with species, but the form shown is the most common.

The extracolumellar ligament encloses a small intercalary cartilage (not visible without dissecting the ligament) that assists in the connection to the cephalic condyle.

The extracolumella has a short shaft, which is attached to the columella without any joint, and presents only the discontinuity of a transition from cartilage to bone. The columella is thin and delicate, and rather flexible in its midportion. Its medial end flares suddenly to form the footplate.

A fold of mucous membrane adheres to the extracolumellar shaft, and perhaps gives lateral support to the ossicular chain.

PERFORMANCE OF THE SCINCID MIDDLE EAR

For the skinks, the same experimental procedures were used as for other types of middle ears, with measurements of sensitivity made for aerial sounds at the meatus first with the ear in its normal condition, then after cutting the columella, and finally by applying the sounds through a tube sealed over the oval window. In some of the small skinks used in these experiments, such as *Eumeces skiltonianus*, this last procedure required special care to avoid any contact with the stump of the columella in the

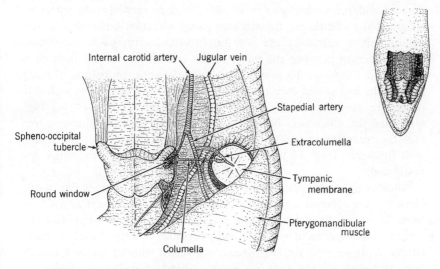

FIG. 6-43. The right middle ear cavity and its contents in *Mabuya brevicollis*. The view is ventral and anterior. From Wever, 1973a.

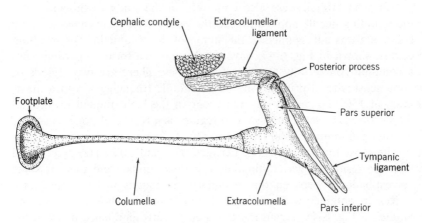

FIG. 6-44. The conductive mechanism in *Mabuya brevicollis*, oriented as in the preceding figure. From Wever, 1973a.

placing of the sound cannula at the oval window, and extra pains were required to obtain a tight seal. Bone wax was often used to complete the sealing.

Results on a specimen of *Mabuya carinata* are shown in Fig. 6-45. The solid curve indicates the normal sensitivity to aerial sounds, and the long-dashed curve the sensitivity after the columella was severed, with the sounds still applied at the external auditory meatus. As explained earlier, the differences between these two functions represent the effect of a loss of the middle ear transformer along with two other effects: the obstruction of sound by the now nonfunctioning tympanic membrane and interaction between the paths to the cochlea by way of both oval and round windows. To avoid these complications, as before, a small sound tube was sealed over the oval window. The observations with this procedure gave the results shown by the short-dashed curve of Fig. 6-45, and the differences between this curve and the normal one represent the transformer action of the middle ear. These differences vary with frequency from 14 to 48 db, and average 30.2 over the range investigated.

Another specimen of this same species gave the results of Fig. 6-46. Here the transformer action of the middle ear is indicated as varying from 25 to 52 db and averaging 40.2 db. Contrary to what was observed in the species of iguanids and gekkonids described earlier, these effects continue into the high-frequency range without marked reduction, and even for the very low tones the reduction is only moderate. Evidently the transformer action in these ears operates very broadly over the frequency scale.

Vibratory stimuli were also employed in the study of this ear, with the vibrating needle applied first to the pars inferior and then, after the columella was cut, applied to the stump of the columella. The resulting curves, shown in Fig. 6-47, have much the same form, departing from a common course only at a few frequencies. There are two regions of good sensitivity, a primary one in the middle frequencies with a maximum at 1500 Hz, and a secondary one in the high frequencies around 4000-7000 Hz. Between these two regions is a region of poor sensitivity with its minimum at 3500 Hz.

The most favorable region, between 400 and 2000 Hz, agrees well with the results of aerial stimulation in this animal, and there is some correspondence between the secondary maximums in these functions.

Results obtained with mechanical stimulation are shown for a second animal in Fig. 6-48. Again there is a close correspondence between the functions for stimulation on the pars inferior (solid line) and on the medial end of the columella (broken line). Such differences as exist are mostly in favor of columellar stimulation, though a region from 1500 to 4000 Hz shows the contrary relation. Comparison of these results with those obtained on the same ear by air conduction (Fig. 6-46) shows good agreement in the primary region of sensitivity, and fair correspondence elsewhere.

Observations made on another species, *Mabuya macularia*, are represented in Fig. 6-49. The loss of sensitivity after elimination of the middle ear transformer was strikingly great, reaching a maximum of

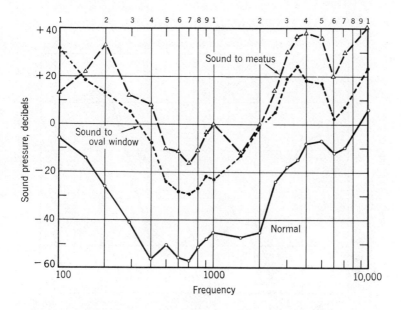

FIG. 6-45. Sensitivity functions for a specimen of *Mabuya carinata* for aerial stimulation under normal conditions and after cutting the columella, with the sound applied through a tube over the meatus or over the oval window. From Wever, 1973a.

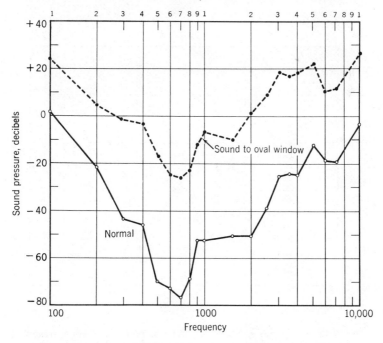

FIG. 6-46. Aerial sensitivity functions for a second specimen of *Mabuya carinata* under normal conditions and after cutting the columella and applying the sounds to the oval window. From Wever, 1973a.

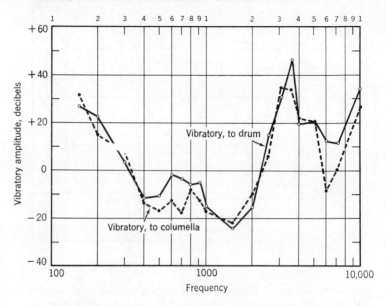

FIG. 6-47. Sensitivity curves for vibratory stimulation in *Mabuya carinata*, applied in two ways. From Wever, 1973a.

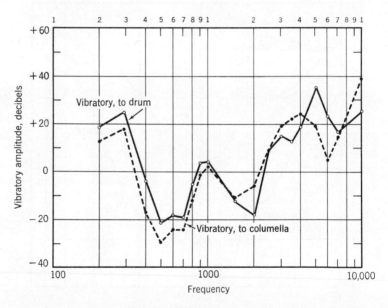

FIG. 6-48. Results for a second specimen of *Mabuya carinata*, treated as in the preceding figure. From Wever, 1973a.

62 db at 700 Hz, and evident even for the extreme frequencies. For the whole range this loss averaged 40.6 db.

Results on a third species, *Eumeces gilberti*, are shown in Fig. 6-50. The measurements indicate a transformer action in this ear varying in effectiveness from 27 to 62 db over the frequency range and averaging 42.2 db.

Another specimen of *Eumeces gilberti* was studied with vibratory stimulation, and gave the results shown in Fig. 6-51. The solid curve represents the application of the vibrating needle to the pars inferior, and the broken curve its application to the stump of the columella. The two curves are similar, but everywhere the stimulation of the medial end of the columella shows greater sensitivity, though at one point, at 500 Hz, the difference is slight.

The fourth species of skink examined was *Eumeces skiltonianus*. Results for one of the specimens with aerial stimulation are presented in Fig. 6-52. In general there is a great difference between these functions attributable to the transformer effect, though at 100 Hz the effect is negligible. The maximum difference appears at 800 Hz, where it amounts to 52 db. The mean over the range shown is 36.1 db, and for the more sensitive range between 300 and 5000 the average is 40.4 db. Thus in this species the middle ear serves its purpose with great effectiveness, especially throughout the middle range.

Another animal of this species was used in measurements with vibratory stimuli, with results given in Fig. 6-53. In this ear, as was generally true for the others, the sensitivity is greater for direct driving of the medial end of the columella (broken line) than for the application of this stimulus to the pars inferior (solid line).

For the middle ear of skinks, as for other types of middle ear, we can accept the hypothesis that the transformer action involves at least two mechanisms, a lever action of the ossicular chain and a hydraulic action involving the areal ratio between tympanic membrane and stapedial footplate. No experiments have been carried out in this group of lizards, however, to obtain direct evidence of the lever action.

In one of the specimens of *Mabuya carinata* the area of the tympanic membrane was measured as 29.2 sq mm and that of the columellar footplate was 1.51 sq mm. If the effective area of the tympanic membrane is taken as two-thirds the anatomical area, this area becomes 19.5 sq mm, and the ratio to the stapedial area becomes 19.5/1.51 or 12.8. This ratio compares favorably with the one of 13.3 observed in *Crotaphytus collaris* but is much less than the values of 39.6 and 22.4 obtained in *Gekko gecko* and *Eublepharis macularius*.

The results just presented for four species of skinks are consistent with what was found for the other two types of middle ear in iguanids and

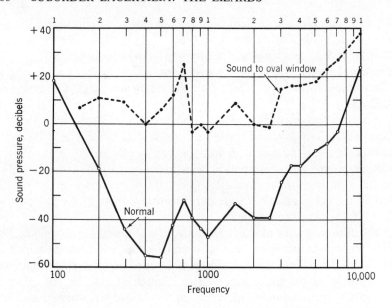

FIG. 6-49. Aerial sensitivity in a specimen of *Mabuya macularia* under normal conditions and, after columella sectioning, with the sounds applied through a tube sealed over the oval window. From Wever, 1973a.

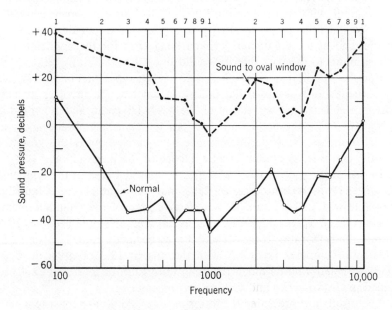

FIG. 6-50. Aerial sensitivity in a specimen of *Eumeces gilberti*, observed in the same manner as for the preceding species. From Wever, 1973a.

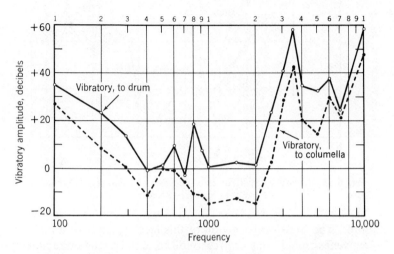

FIG. 6-51. Vibratory sensitivity in *Eumeces gilberti* with two forms of stimulation as indicated. From Wever, 1973a.

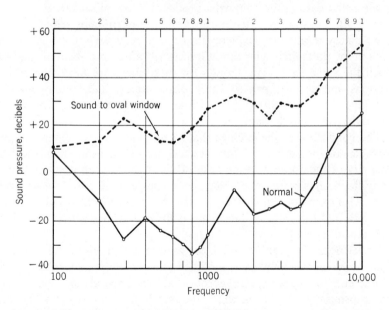

FIG. 6-52. Sensitivity to aerial sounds in a specimen of *Eumeces skiltonianus*. From Wever, 1973a.

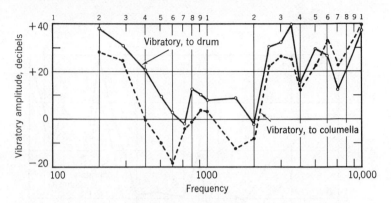

FIG. 6-53. Vibratory sensitivity in a specimen of *Eumeces skiltonianus*.
From Wever, 1973a.

gekkonids. The transformer action of this mechanism is of a high order of effectiveness, improving the reception of aerial sounds in amounts of the order of 40 db over most of the frequency range, and attaining levels of 60 db or more in certain favored regions of frequency.

It is observed in these results, as in those obtained for other lizard species, that cutting the columella produces a loss in sensitivity without greatly altering the form of the function: the maximums remain in the same places along the frequency scale and only undergo a moderate amount of flattening. The vibratory curves after this sectioning show much the same form also, though for these there are more abrupt deflections and sharper peaks. These observations favor the hypothesis that inner ear mechanisms play a major role in the determination of sensitivity, as already suggested.

The conditions in the cochlea that seem most likely to determine the character of the ear's sensitivity are the size of the fluid chambers, the elasticity of membranes in the path of vibratory fluid movement (especially the basilar membrane), and the masses of fluid and tissues in this path. Also of major significance are the numbers and distribution of the hair cells and the tectorial mechanisms through whose action these hair cells are stimulated.

Under normal conditions the basic pattern of response determined by these inner ear mechanisms is modified and generally augmented by the middle ear. In its role in the reception of aerial sounds and their effective transmission inward, in which the matching of impedances between aerial and cochlear media is primary, this mechanism adds numerous mechanical resonances, which enhance the transmission for some frequencies and reduce it for others. Evidently in most species the extreme frequencies, those above 5000 Hz and below 100 Hz, are greatly discriminated against.

The above experiments have dealt with the normal performance of the middle ear mechanism in lizards, but still to be considered are a number of species in which this mechanism takes different forms, and in general appears to represent standard types that have undergone various degrees of degeneration. A study of these divergent species and comparisons with normal structures brings out many of the more significant features of the conductive system.

DIVERGENT TYPES OF MIDDLE EAR IN LIZARDS

Divergent forms of middle ear structure are found in at least eleven families of lizards. They occur in all members of the Anniellidae, Chamaeleonidae, and Xenosauridae, as well as in the rare Dibamidae and Lanthanotidae. They appear also in certain species among the Agamidae, Iguanidae, Pygopodidae, and Scincidae, and more rarely in Anguidae and Teiidae.

For the most part the deviations from normal consist of a loss of the tympanic membrane. Versluys (1898) discussed this occurrence and listed the species in which it was then observed. Mertens (1971) brought this listing up to date.

The absence of the tympanic membrane is often accompanied by other deviations, such as reductions in the extracolumella by loss of one or more of its processes, and sometimes includes further variations in ossicular structure. More rarely there are still more serious modifications, including loss of the round window in Anniellidae and Chamaeleonidae.

A special effort was made to obtain representatives of these divergent species, and it was possible to examine specimens belonging to 22 such species. The relations between the ear's sensitivity and the forms taken by the middle ear structures will be considered for 11 of these in this chapter, roughly in order of the seriousness of the deviation from normal.

The remaining eleven species include one species each of Anniellidae and Zenosauridae and nine species of Chamaeleonidae. Because the middle ear anomalies are common to all members of these three families, their treatment will be taken up later as a part of the topical discussion of lizard families. As will be shown, the problems of sound conduction for these species are much the same as for the species now to be examined.

Acontias plumbeus (Scincidae). — This is a burrowing skink of elongate body form, without legs, and with no sign of an external ear: the scalation continues over the side of the head without interruption. Versluys named nine genera of skinks, including 31 species, in which this condition was found, but this is the only one obtained for the present study.

A columella of unusually sturdy structure runs outward from the oval window to end in a thick mass of connective tissue just beneath the skin. Its large footplate covers almost the whole of the lateral area of the otic capsule. The columellar shaft is at first osseous and then is continued as cartilage. This cartilage, probably representing the extracolumella, extends laterally as a flattened rod and then swings anteriorly, flaring to a spatulate form as it passes around the posterolateral edge of the quadrate. Though this element runs close to the quadrate it has no firm connection with it.

The skin at the side of the head serves as the sound-receptive surface, and conveys the vibratory movements through the thick subdermal tissues to the terminal cartilage. The effectiveness of this mode of sound conduction is indicated in Fig. 6-54. The sensitivity is best in the region of 400 to 1000 Hz, and reaches a maximum of −8 db at 800 Hz. This is a fair degree of sensitivity in an animal lacking a tympanic membrane and external ear opening.

Cophosaurus texanus (Iguanidae). — In this species there is no tympanic membrane. The skin continues over the ear region at the side of the head with no noticeable change, as Fig. 6-55 shows. The depressor mandibulae muscle extends far forward, but does not bar access of the extracolumellar mechanism to the surface.

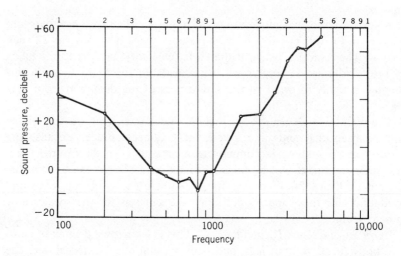

Fig. 6-54. Sensitivity to aerial sounds in a specimen of *Acontias plumbeus*.

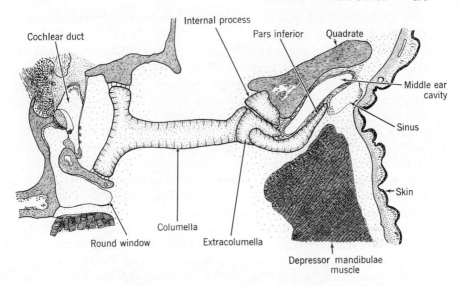

FIG. 6-55. The middle ear of *Cophosaurus texanus* as reconstructed
from serial sections. From Wever, 1973b.

The columella is thick and sturdy, and its footplate covers the entire
lateral wall of the otic capsule. Its shaft connects with a compact body
that constitutes the extracolumella, and this element bears two processes,
a platelike internal process that connects broadly with the quadrate, and
a pars inferior that extends laterally. The pars inferior passes above the
upper end of the depressor mandibulae muscle, and its lateral portion
lies alongside an anterolateral recess of the middle ear cavity (filled
with air), with its tip between this recess and a sinus adjacent to the
skin layer.

The joint between the body of the extracolumella and its internal
process appears to be a gliding surface with great mobility. A joint also
appears between the body of the extracolumella and its inferior process.
The air cavity should impart considerable mobility to the skin layer in
its vicinity, and the pars inferior is well located to take up the movements produced by sounds impinging on the skin surface.

Sensitivity curves for two specimens of *Cophosaurus texanus* are presented in Fig. 6-56. These curves show a rather high degree of sensitivity within a narrow region of frequency from 500-800 Hz, and for one
of the animals there is a sharp peak at 700 Hz reaching −48 db. The
sensitivity declines for higher and lower tones at the moderate rate of
about 20 db per octave.

It is suggested that this ear attains relatively good sensitivity by mechanical tuning, so that within a limited frequency range it performs

well in spite of the handicap of utilizing a relatively stiff skin layer as the sound-receptive surface.

Phrynosoma platyrhinos (Iguanidae). — In the desert horned lizard the form of the skin in the temporal region varies greatly from one individual to another. This skin is always thick and stiff, but sometimes is rough and scaly much like the skin of the surrounding area, so that an ear region is hardly discriminable; in other specimens this skin is smooth and well marked off from the outlying area. In a few individuals an ear region is easily recognizable on one side but not on the other. In the specimen represented in Fig. 6-57 the tympanic area was slightly differentiated in that the skin here was smoother and with finer scalation than elsewhere. Much as in the preceding species, the tympanic area of skin lies close over a large air-filled recess of the middle ear cavity, which gives it mobility. The columella has a long, slender shaft, with a good-sized stapedial footplate that covers most of the lateral wall of the otic capsule. Its lateral end articulates with an internal process in the form of a thick disk lying on the quadrate, and also with an extracolumella that sends out a tapering pars inferior. The pars inferior runs laterally to a position between the skin layer and the large air cavity, and accompanying it is a ligamentous sheet, which is probably an extension of the extracolumellar ligament. This tympanic ligament no doubt assists in conveying the vibratory movements of the skin layer to the extracolumella.

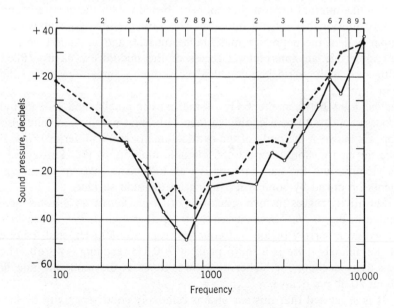

FIG. 6-56. Auditory sensitivity to aerial sounds in two specimens of *Cophosaurus texanus*. From Wever, 1973b.

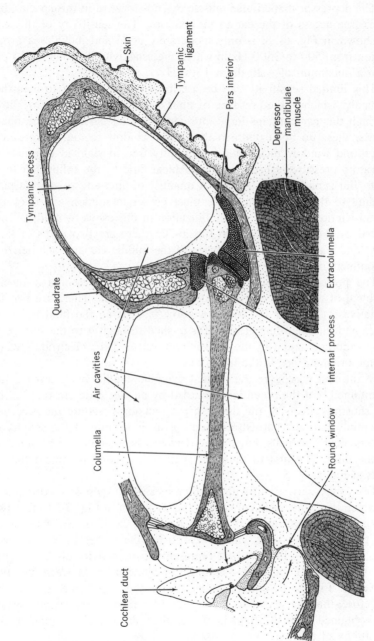

Fig. 6-57. The middle ear of *Phrynosoma platyrhinos*. A reconstruction from serial sections. From Wever 1973b.

The depressor mandibulae muscle extends forward in this species, but gives free access of the ear to the exterior. The sensitivity of this ear, as shown in Fig. 6-58, is only moderately good and includes a broad region from 500 to 3500 Hz in which the sensitivity is around −15 db, with a maximum of −26 db at 1500 Hz.

This limited sensitivity can be ascribed in part at least to a further anomaly, which is the presence of fluid adjacent to the round window. Though this ear contains large interconnected air cavities as the figure shows, these do not extend to the round window region. Accordingly the round window membrane is not fully free to yield in response to vibratory pressures exerted on the cochlear fluid by the columellar footplate, but is partially restrained. A quantity of fluid and tissue in a path leading to the nearest air space must be set in motion ahead of the round window. Probably the mobilization in this ear is twofold, as suggested by the double arrow in Fig. 6-57: there are two paths of fluid flow, one leading into the air space of the middle ear and another that is confined to the fluid.

The presence of fluid over the round window presented a special problem of recording cochlear potentials in this ear. There are two alternatives: (1) after the round window is exposed the fluid is left in place and the recording electrode is pushed through it to the surface of the membrane, and (2) the fluid is removed and the electrode located on the dried surface in the usual way.

In the first procedure the normal dynamic conditions in this ear are maintained, but the potentials recorded by an electrode are reduced by the shorting effects of the fluid. In the second procedure the recorded potentials can be regarded, as usual, as an indication of the activity of the cochlear hair cells, but this activity has been enhanced by removal of the fluid. The first of these alternatives was used to obtain the results of Fig. 6-58.

Tests were carried out with another species, *Phrynosoma coronatum*, in which this same problem exists, and as shown in Fig. 6-59 these two procedures yielded results that differ by about 28 db on the average for the range 100-4000 Hz, and by 11 db for the higher tones. The true sensitivity curve—the one best representing the activity of this ear under its normal conditions—should run somewhere between the two curves of Fig. 6-59, and probably nearer the upper one. It appears, therefore, that the sensitivity shown for *Phrynosoma platyrhinos*, which was obtained with the fluid in place, gives an underestimation of the sensitivity of this ear.

Phrynosoma coronatum (Iguanidae). — Ten specimens of the coast horned lizard were examined and, in all, the tympanic membranes were smooth and well differentiated from the surrounding skin. This mem-

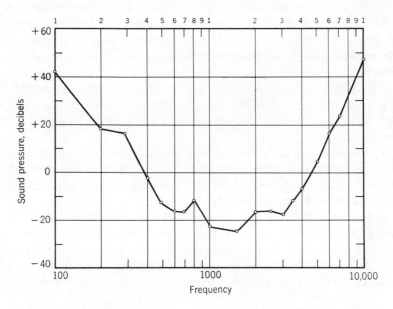

FIG. 6-58. Sensitivity to aerial sounds in a specimen of *Phrynosoma platyrhinos*, measured with fluid over the round window. From Wever, 1973b.

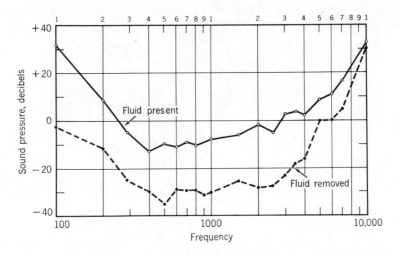

FIG. 6-59. Auditory sensitivity in a specimen of *Phrynosoma coronatum*, measured under two conditions, with the fluid over the round window left in place, and with this fluid removed. From Wever, 1973b.

brane is of oval form, and in the specimens studied measured about 1.3 × 3.3 mm, with its main axis inclined about 30° from the vertical. In most other respects the structure resembles that of *Phrynosoma platyrhinos*, though a definite posterior process of the extracolumella is present.

The sensitivity curves for this species have just been presented in Fig. 6-59 for the two conditions of "fluid present" and "fluid removed," and the true sensitivity function can be considered as falling between these two curves.

Holbrookia maculata (Iguanidae). — In *Holbrookia maculata* the skin continues unmodified over the ear region, and the depressor mandibulae muscle extends anteriorly all the way to the quadrate, excluding the ear from access to the exterior.

As shown in Fig. 6-60 the columella is short and sturdy, with a relatively large footplate. Its shaft connects with a flattened cartilage that represents the body of the extracolumella. This body sends out a tiny process (labeled "accessory process") that probably represents the pars superior and pars inferior in miniature. A little ligament runs along

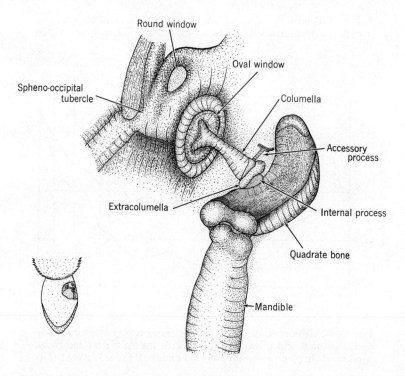

FIG. 6-60. The middle ear of *Holbrookia maculata*. Drawn from a dissected specimen. From Wever, 1973b.

these elements and ends on the cephalic condyle of the quadrate, just as in most normal ears. Between the extracolumellar plate and the quadrate is a wedge of cartilage that constitutes the internal process. It is securely attached to the quadrate, but is only in loose contact with the extracolumellar plate and with the lateral end of the columella on its anterior side. This appears to be a gliding contact, for no ligamentary attachments were found.

In the dissection for Fig. 6-60 the depressor mandibulae muscle was removed to expose the other structures more fully, but in Fig. 6-61 this muscle is seen with its upper end extending into the concavity of the quadrate, in close contact with the skin along its lateral side, and with the extracolumellar plate resting in a shallow depression on its medial side. There is contact also between the "accessory process" and this muscle, though this is probably of little significance. Medial to the main mass of the depressor mandibulae muscle are portions of the middle ear cavity that give mobility to the skin and muscle layers when acted upon by sound pressures. The vibratory motions are transmitted from the muscle to the extracolumella, and thereafter through the columella in the usual way.

In Fig. 6-62 is shown a sensitivity curve for a specimen of *Holbrookia maculata* obtained by presenting aerial sounds through a tube sealed over an area of skin at the side of the head. This curve shows poor sensitivity, reaching a level of only −9 db in the most favorable region around 800 Hz, and running around 0 db elsewhere in the range between 300 and 2000 Hz. This performance is to be expected from the absence of a tympanic membrane and a utilization of skin and the depressor mandibulae muscle as receptive and transmitting tissues.

Anguis fragilis (Anguidae). — In a few lizard species the auditory structures show variations according to the regions in which the specimens are found, and this appears to be true of *Anguis fragilis*. From a review of the early literature on this species Versluys (1898) found such geographic variations: specimens from Hungary and Russia often have a small external ear opening, whereas those from Europe do not. The specimens of the present study, obtained from dealers without information as to origin, all lacked a meatal opening, though one ear in a single specimen showed a minute pore that could be traced through serial sections all the way from the skin surface to the middle ear cavity. However, this pore appeared too small and too much obstructed to serve as a path of sound conduction.

In all the specimens examined, and as shown in Fig. 6-63, the skin was continuous over the ear region, and three masses of the depressor mandibulae muscle were found to extend forward to obstruct the path between skin and ear. There is a small but well-formed columella that

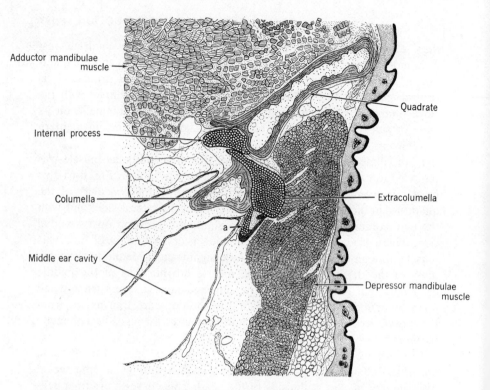

Adductor mandibulae muscle

Internal process

Columella

Middle ear cavity

Quadrate

Extracolumella

a

Depressor mandibulae muscle

FIG. 6-61. The middle ear of *Holbrookia maculata* seen at one level in frontal section, showing the relations between the ossicular elements and peripheral tissues. *a*, accessory process.

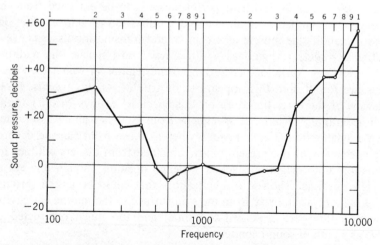

FIG. 6-62. Auditory sensitivity in a specimen of *Holbrookia maculata*. The sounds were applied through a tube sealed to the skin at the side of the head. From Wever, 1973b.

connects closely to a simple extracolumella that lies in loose tissue adjacent to the innermost mass of the depressor mandibulae muscle as shown. Medial to this region is a part of the middle ear cavity, which can be expected to give some degree of mobility to the more peripheral tissues. Another cavity was found closer to the skin layer that is considered to be a part of the external ear cavity that has become isolated. It is almost certainly filled with fluid, as no means of ventilating it could be discovered, and therefore it can give no benefit to tissue mobilization. The extracolumella sends off a short pars inferior and an even shorter pars superior, but there are no other processes. An internal process is absent as in all anguids.

Sounds acting on the skin surface may be expected to be transmitted through this tissue and the loose material beneath to the extracolumella, largely involving the muscle masses in this area. The material moves relative to the remaining parts of the head by compressing the air in the middle ear cavity, and thereby sets up a vibratory motion of the columella.

The sensitivity of the ear of a specimen of *Anguis fragilis* is represented in Fig. 6-64. This sensitivity is poor, with responses over the range from 300-2000 Hz around −5 db, and the best peak only reaching −15 db at 400 Hz.

Draco volans (Agamidae). — In *Draco volans* an area of skin in the temporal region is readily distinguished from the remaining skin, for it lacks the usual scales and is only lightly tuberculated. This area is of oval form, and in one specimen measured 0.9 × 1.4 mm, with the long axis obliquely placed.

As shown in Fig. 6-65, the depressor mandibulae muscle extends forward but stops short of the ear structures, leaving the extracolumella free to make contact with the skin. The columella has a broad and noticeably asymmetrical footplate that constitutes the whole lateral wall of the inner ear. It connects with the extracolumella, which sends a short internal process to the quadrate, and then extends at an angle to a broad expansion on the skin. This expansion consists of a prominent pars inferior and a short, broad process that probably corresponds to the pars superior.

A relatively large air space, formed as a recess of the tympanic cavity, lies anterior to the columella along the skin area and extends into the concavity of the quadrate, but probably provides only limited mobilization for the region of skin over the pars inferior.

The sensitivity of this animal is represented in Fig. 6-66, and must be regarded as poor. Over a range from 500 to 3000 Hz the curve runs around −5 db with a maximum point of −6 db.

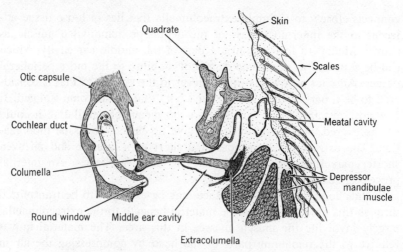

FIG. 6-63. The middle ear of *Anguis fragilis*. Scale 17×.
From Wever, 1973b.

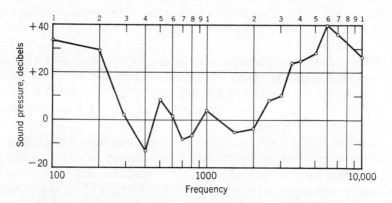

FIG. 6-64. Auditory sensitivity in a specimen of *Anguis fragilis*, obtained by presenting aerial sounds through a tube sealed over the skin at the side of the head. From Wever, 1973b.

Phrynocephalus maculatus (Agamidae). — In *Phrynocephalus maculatus* the tympanic membrane is lacking, and the auditory area is covered by unmodified skin. As Fig. 6-67 shows, the depressor mandibulae muscle extends far forward, ahead of the greater part of the quadrate, and intervenes between the ear and the skin surface.

There is a long, slender columella connecting with an extracolumella that consists of a small body portion with two unusually long processes. One of these, the internal process, runs to the quadrate, and the other, the pars inferior, takes a nearly semicircular course to end in soft tissue

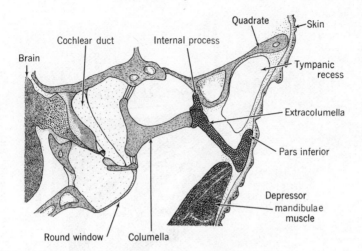

FIG. 6-65. The middle ear of *Draco volans*, reconstructed from serial sections. Scale, 25×. From Wever, 1973b.

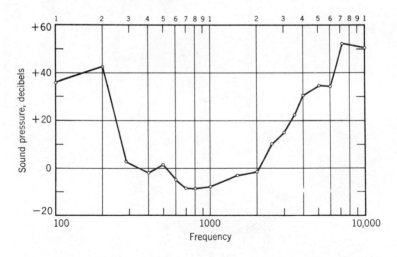

FIG. 6-66. Auditory sensitivity in a specimen of *Draco volans*, measured with aerial sounds applied through a tube sealed to the skin over the ear region. From Wever, 1973b.

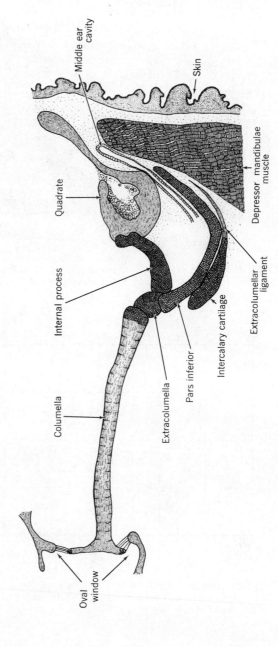

FIG. 6-67. The middle ear of *Phrynocephalus maculatus*. A reconstruction from serial sections. Scale 25×. From Wever, 1973b.

medial to the mass of the depressor mandibulae muscle. Between this process and the quadrate is a flat air space, which is an extension of the middle ear cavity. Adjacent to the inner part of the pars inferior is a relatively large cartilage, identified as the intercalary, from which a slender ligament runs along the outer half of the pars inferior. Not shown is a continuation of the intercalary to its termination between the quadrate and the paroccipital process.

In this ear the skin and muscle mass is mobilized by the air cavity beneath it, and the movements produced by sounds are transmitted to the pars inferior. The effectiveness of this mode of sound reception is indicated in Fig. 6-68, where curves for three specimens are given. The sensitivity is best over a range of 300 to 2000 Hz, with peaks around 500-700 Hz, varying among the three ears. The degree of sensitivity is fair.

Phrynocephalus sp. (Agamidae). — An additional specimen of *Phrynocephalus*, which could not be identified as to species, showed a closely corresponding structure and a generally similar sensitivity function. The maximum sensitivity was at 400 Hz, but was poorer than that shown in Fig. 6-68.

Ceratophora stoddarti (Agamidae). — In *Ceratophora stoddarti* the tympanic membrane is absent and the temporal area is covered by unmodified skin. The depressor mandibulae muscle extends far beyond the quadrate, completely blocking the ear from the exterior, as Fig. 6-69 shows. The columella is well-developed, with a large footplate occupying the lateral wall of the otic capsule. It connects through a definite articulation with an extracolumella of unusual form. The body of the extracolumella is small and sends out three long processes. One of these extends posteriorly and enters a deep depression in the quadrate; it is probably the internal process. Another process (supposedly the pars superior) runs anteriorly, where it branches and sends out one process to make a surface contact on the quadrate and another process over the surface of the cephalic condyle to attach to the paroccipital process. A third process, indicated as the pars inferior, extends into a depression in the belly of the depressor mandibulae muscle and is clearly the receiver of vibratory stimuli.

The operation of this ear does not seem to be aided by an air space that would give mobility to the peripheral tissues. The skin and muscle mass therefore can only be set in motion by compressing the deeper lying tissues. The sensitivity therefore is poor, as shown in Fig. 6-70. Here are shown the curves obtained in two specimens, one with sensitivity that only reaches −2 db at one point in the range and another that goes to −13 db, both in the region of 300 Hz. This relatively low-

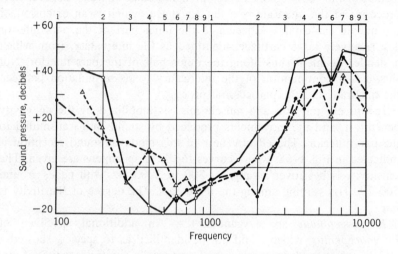

FIG. 6-68. Auditory sensitivity in three specimens of *Phrynocephalus maculatus*, stimulated with aerial sounds through a tube sealed to the skin at the side of the head. From Wever, 1973b.

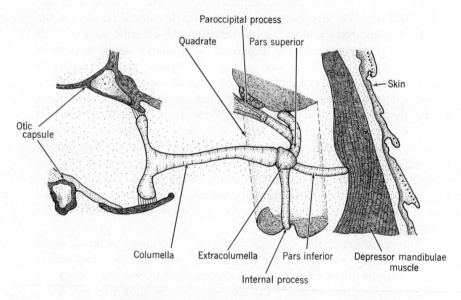

FIG. 6-69. The middle ear of *Ceratophora stoddarti*, reconstructed from serial sections. Scale 25×. From Wever, 1973b.

frequency response is probably determined by the large mass of skin and muscle involved.

Ceratophora tennenti (Agamidae). — Two specimens of *Ceratophora tennenti* were studied. The structure of the ear corresponds closely to that shown for *Ceratophora stoddarti* in Fig. 6-69. The sensitivity function is similar also, as shown for one of the animals in Fig. 6-71. The maximum sensitivity was in the region of 500-1000 Hz, and amounted to −3 db, which must be considered poor.

Common Characteristics of Divergent Middle Ears.   The preceding lizard species with divergent forms of middle ear structure show three general conditions that appear to be responsible for impairments in the reception of sounds and their transmission inward to the cochlea. These are the absence of the tympanic membrane and substitution of the stiffer skin as a receptive surface, the intrusion of the depressor mandibulae muscle as a barrier between the ear and the periphery, and in two species an encumbrance of the round window. There are also changes in the conductive mechanism, especially in the form of the extracolumella and its processes, but these changes do not appear to be of primary importance.

After the tympanic membrane is lost, there is no longer any significant purpose to be served by the four processes that form the normal insertion piece of the extracolumella in contact with the tympanic surface. The accessory processes (anterior and posterior processes) are commonly reduced or lacking. The pars inferior and pars superior usually remain, but these may be reduced (as they are in *Holbrookia* and *Anguis*) or are much modified (as in *Ceratophora* and *Phrynocephalus*).

Certain modifications, especially of the pars inferior, appear to have a positive function of compensating in some measure for the isolation of the ear caused by the anterior growth of the depressor mandibulae muscle. Thus the pars inferior may move anteriorly to pass around the muscle (as in *Cophosaurus*), may extend along the inner border of the muscle (as in *Phrynocephalus* and *Anguis*), or may make a direct connection to it (*Ceratophora*).

The pars superior may be reduced (as in *Holbrookia* and *Draco*) or may be absent (as in *Phrynocephalus*). When present it sends off an extracolumellar ligament that is anchored to the skull in a variety of ways. It may attach to the intercalary cartilage (as in *Holbrookia* and *Cophosaurus*), may go directly to the cephalic condyle (as in *Phrynosoma* and *Draco*), to the end of the paroccipital process (in *Ceratophora*), or is inserted between paroccipital process and cephalic condyle (*Anguis*). In *Phrynocephalus* the extracolumellar ligament extends

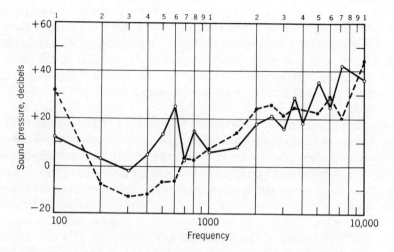

FIG. 6-70. Auditory sensitivity in two specimens of *Ceratophora stoddarti*, stimulated with aerial sounds through a tube applied to the skin at the side of the head. From Wever, 1973b.

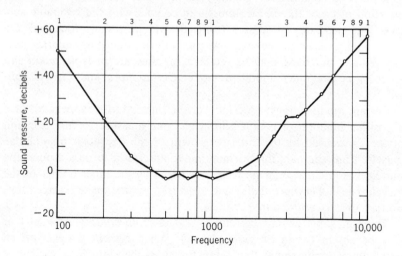

FIG. 6-71. Auditory sensitivity in a specimen of *Ceratophora tennenti*, stimulated with aerial sounds through a tube sealed to the skin over the ear region.

from the intercalary cartilage but has no specific attachments; it dwindles to a thin strand embedded in loose tissue and is evidently vestigial.

A pars inferior is present in all these species, though in *Holbrookia* it is reduced to a tiny remnant. Except in this species the process performs its usual function of receiving the peripheral vibrations and transmitting them to the inner structures.

The effectiveness of this initial reception by the pars inferior varies according to its location on a tympanic membrane (as in normal species and *Phrynosoma coronatum*), on skin (as in *Phrynosoma platyrhinos* and *Draco*), on muscle (as in *Ceratophora*), or in other strategic locations. In *Cophosaurus* the pars inferior is located in connective tissue adjacent to an air cavity, and clearly is involved in the movement of skin and other tissue into this air space. In *Anguis* and *Phrynocephalus* the pars inferior lies in connective tissue alongside the depressor mandibulae muscle, with an air space medial to it, so that the entire mass of skin, muscle, and connective tissue can undergo inward displacements. In *Holbrookia* the transmission is direct from the muscle mass to columella and extracolumella.

In contrast to the varied development of the extracolumella, the columella retains an essentially normal form. It is sometimes enlarged, acquiring a heavy shaft (in *Holbrookia* and *Cophosaurus*), or a particularly broad footplate, as in several species, but it maintains its usual ossified condition and remains fully mobile in the oval window.

An internal process is present in all the species examined except *Anguis fragilis*, though in several others it is greatly reduced. Ordinarily this process makes a connection between the extracolumella and the quadrate, but in *Phrynosoma* the connection is between columella and quadrate. There is never a direct connection of columella to quadrate.

In normal middle ears of the iguanid type, in which the internal process is a thin, flexible vane, it carries out its functions as a lateral support for the conductive mechanism without imposing any serious loading or restraint on the vibratory mechanism, as experimental evidence has shown. When the internal process is short or is a compact body, as found in *Callisaurus draconoides*, its presence may modify the transmission by adding rigidity to the conductive system. Such effects are probably present in most of these divergent ears, and especially in *Cophosaurus texanus*, the *Phrynosoma* species, and *Holbrookia maculata*, in which the connection made with the quadrate appears to be especially strong.

In the lizards under discussion, the round window is always present. In *Anguis* this window is small, but its size is entirely adequate for its purpose. In the two *Phrynosoma* species a round window is present but is bounded by fluid, so that its movements are strongly damped. The results obtained on *Phrynosoma coronatum*, shown in Fig. 6-59 before and after removal of this fluid, provide some basis for judging the seriousness of this damping effect, though a part of the observed change must be ascribed to the elimination of a shorting effect when the fluid was removed.

Hearing in Divergent Ears. If the tympanic membrane were merely thickened or replaced by skin, and all other features of the conductive system were retained, we should expect only the changes in sensitivity resulting from an increased reflection of sound waves from the hardened surface. This membrane ordinarily, in the low and middle ranges of frequency, reflects only a moderate fraction of the vibratory energy incident upon it, because it is thin and flexible, and moves with the air particles. A thick and heavy membrane, on the contrary, has an acoustic impedance much greater than that of the aerial medium, and reflects a large part of the sound energy. But if the transformer function of the middle ear is maintained the inner ear impedance is made to appear much smaller than it actually is, and a significant fraction of the incident energy is transmitted to the sensory sites in the cochlea. Without the transformer action the solidity of the external surface would appear even greater because the high impedance of the cochlear fluids and other structures would be conveyed outward to the surface. Then little effective hearing would be possible. It is clear, therefore, that even in these divergent ears the usual functions of the conductive mechanism are performed even though imperfectly.

Certain experiments were carried out to demonstrate that this is the case. In a specimen of *Holbrookia maculata* a sensitivity curve was first obtained by stimulating with aerial sounds applied through a tube sealed over an area of skin at the side of the head, with results shown by the solid line of Fig. 6-72. The columella was then severed, and a small piece removed to prevent a reestablishment of contact. The observations were repeated with the same form of stimulation as before, with results shown by the broken line of this figure. For all tones from 290 to 2000 Hz there was a loss of sensitivity that averaged 20 db and reached a value of 34 db at 1000 Hz. For lower and higher tones there were slight variations, or small improvements, in the range between 2500 and 5000 Hz that are of doubtful significance. The considerable changes in the middle range show the conductive mechanism of this ear to be of valuable service in the reception of sounds.

In this same animal, measurements were made also by the use of vibratory stimuli before and after the columella sectioning. The vibrations were applied to the quadrate bone through a needle on its surface. In Fig. 6-73 the solid line shows the results with the ear intact and the broken line shows those obtained after clipping the columella.

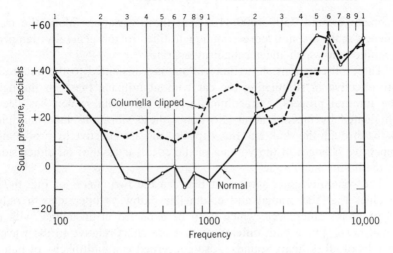

FIG. 6-72. Auditory sensitivity to aerial sounds in a specimen of *Holbrookia maculata* before and after interrupting the columella. For both curves the sounds were presented through a tube sealed over the skin in the ear region. From Wever, 1973b.

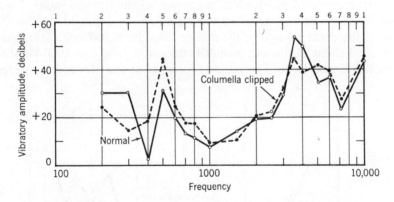

FIG. 6-73. Auditory sensitivity in *Holbrookia maculata* for vibratory stimulation of the quadrate before and after clipping the columella. From Wever, 1973b.

These two curves are similar, but there are some significant variations. As a result of the columella sectioning, there are losses at 400 and 500 Hz of 16 and 13 db, and gains at 3500 and 4000 Hz of 9 and 11 db, though at other points the variations are small and of doubtful validity. The larger changes reflect alterations in the resonance characteristics of the system. Evidently the region of 400 Hz is one of normal mechanical resonance, and loss of the columella has disturbed this resonance, producing a significant loss. A contrary effect appears in the high frequencies around 3500-4000 Hz. That these variations along the frequency scale are not more common testifies to the generally damped condition in which the mechanism operates.

These observations show that vibrations transmitted to the quadrate are effective in stimulating the ear without utilizing the path through the internal process, extracolumella, and columella, which has been eliminated by the operation. No doubt these vibrations take multiple paths through the skull and otic capsule, and are effective by producing repetitive changes in the volume of the cochlear cavities on either side of the basilar membrane.

The generally good agreement between the two curves of Fig. 6-73 signifies that the cranial and columellar pathways operate with only slight phase differences, which is to be expected in an animal with a small head, giving path differences that are short relative to the wavelengths of all ordinary sounds. Also concerned is a multiplicity of pathways by both cranial and columellar routes, and a summation of their effects is involved in producing the final stimulation.

Other experiments were carried out in a specimen of *Phrynocephalus maculatus*. The solid curve of Fig. 6-74 gives results obtained by vibratory stimulation of the quadrate when the middle ear was intact. Then the columella was clipped, and the measurements repeated. In this instance there was a loss over the middle range from 400 to 2500 Hz that averaged about 11 db, and included changes at 1000 and 1500 Hz that reached 19 and 17 db. This region appears to be one of broad resonance

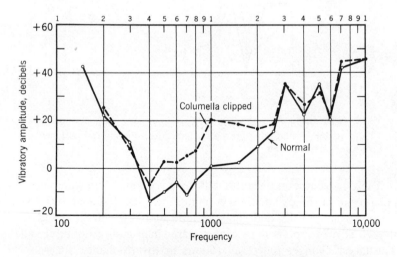

FIG. 6-74. Auditory sensitivity in *Phrynocephalus maculatus* for vibratory stimulation of the quadrate before and after clipping the columella. From Wever, 1973b.

of the columellar system, which augments the sensitivity of the ear to a significant degree.

Elsewhere the changes produced by cutting the columella are small and insignificant. Thus these results are in good agreement with the ones given for *Holbrookia*, indicating that in the absence of an ossicular connection between quadrate and columella the ear responds well to vibratory stimulation. These results are also in agreement in showing that the middle ear mechanism in these ears, though of abnormal form, contributes significantly to the reception of aerial sounds in some portions of the auditory range.

7. FAMILY IGUANIDAE:
THE IGUANID LIZARDS

The iguanids form one of the leading lizard families, containing about 700 species in some 65 genera. These lizards are mainly confined to the Western Hemisphere, and largely are found in North and South America, with several species in the West Indies.

A few iguanids live elsewhere, and all these are island dwellers, evidently being relict populations freed of competition with other lizards, particularly the agamids of similar life habits. It is remarkable that in only one place in the world, on Fiji Island, are iguanid and agamid species both present. A few island species occur also in the Western Hemisphere, and these include the giant marine form, *Amblyrhynchus cristatus*, on the Galapagos Islands. Others are halfway around the world: seven species belonging to two genera are on the island of Madagascar, and a single species lives on Fiji and Tonga in the South Pacific. The existence of these remote species has long been puzzling, but an explanation has been found in line with the theory of continental drift, through which the islands that they now inhabit can plausibly be regarded as once connected with South America (Avery and Tanner, 1971).

The species *Iguana iguana* of South America is represented in the frontispiece, and another species, *Sceloporus magister*, an inhabitant of semidesert areas of the southwestern United States, is shown in Fig. 7-1.

Because the iguanid family is large and varied, numerous attempts have been made to subdivide it. Certain more or less distinct phylogenetic lines are recognized, and it has sometimes been suggested that the anoline species, which make up a fourth or more of the total, might well constitute a subfamily. Systematic relationships are treated in some detail by Mittleman (1942), Savage (1958), Etheridge (1964), Avery and Tanner (1971), and Zug (1971). Following Zug's treatment, the 47 species of the present study will be subsumed under six categories: the anolines, basiliscines, iguanines, sceloporines, tropidurines, and Malagasians. These species, and the number of specimens of each, are as follows:

FIG. 7-1. The lizard *Sceloporus magister*.
Drawing by Anne Cox.

ANOLINES

Anolis carolinensis (Voight, 1832)—10 specimens; *Anolis chlorocyanus* (Duméril and Bibron, 1837)—2 specimens; *Anolis cristatellus* (Duméril and Bibron, 1837)—4 specimens; *Anolis cybotes* (Cope, 1862)—2 specimens; *Anolis distichus* (Boulenger, 1885)—1 specimen; *Anolis lionotus* (Cope, 1861)—3 specimens; *Anolis sagrei* (Duméril and Bibron, 1837)—1 specimen; *Anolis* sp.—3 specimens; and *Polychrus acutirostris* (Spix, 1825)—1 specimen.

BASILISCINES

Basiliscus basiliscus (Linnaeus, 1758)—2 specimens; *Basiliscus vittatus* (Wiegmann, 1828)—3 specimens; *Laemanctus longipes* (Wiegmann, 1834)—1 specimen.

IGUANINES

Ctenosaura acanthura (Shaw, 1802)—1 specimen; *Ctenosaura similis similis* (Gray, 1831)—1 specimen; *Ctenosaura* sp.—1 specimen; *Cyclura stejnegeri* (Barbour and Noble, 1916)—1 specimen; *Dipsosaurus dorsalis dorsalis* (Baird and Girard, 1852)—4 specimens; *Enyaliosaurus clarki* (Bailey, 1828)—2 specimens; *Iguana iguana* (Linnaeus, 1758)—5 specimens; *Sauromalus obesus* (Baird, 1858)—4 specimens.

SCELOPORINES

Callisaurus draconoides ventralis (Hall, 1852)—8 specimens; *Cophosaurus texanus* (Troschel, 1852)—2 specimens; *Crotaphytus collaris*

(Say, 1823)—16 specimens; *Crotaphytus insularis* (Axtell, 1972)—3 specimens; *Crotaphytus wislizenii wislizenii* (Baird and Girard, 1852)—5 specimens; *Holbrookia maculata* (Girard, 1851)—2 specimens; *Phrynosoma asio* (Cope, 1864)—1 specimen; *Phrynosoma cornutum* (Harlan, 1825)—1 specimen; *Phrynosoma coronatum* (Blainville, 1835)—4 specimens; *Phrynosoma m'callii* (Hallowell, 1852)—1 specimen; *Phrynosoma platyrhinos* (Girard, 1852)—4 specimens; *Phrynosoma solare* (Gray, 1845)—1 specimen; *Sceloporus clarkii* (Baird and Girard, 1852)—3 specimens; *Sceloporus cyanogenys* (Cope, 1885)—2 specimens; *Sceloporus graciosus* (Camp, 1916)—2 specimens; *Sceloporus magister* (Hallowell, 1854)—7 specimens; *Sceloporus occidentalis* (Baird and Girard, 1852)—5 specimens; *Sceloporus orcutti* (Stejneger, 1893)—3 specimens; *Sceloporus undulatus consobrinus* (Baird and Girard, 1854)—11 specimens; *Streptosaurus mearnsi* (Stejneger, 1894)—2 specimens; *Uma notata* (Baird, 1858)—4 specimens; *Urosaurus ornatus linearis* (Baird, 1857)—11 specimens; and *Uta stansburiana* (Baird and Girard, 1852)—4 specimens.

TROPODURINES

Leiocephalus carinatus armouri (Barbour and Shreve, 1935)—4 specimens; *Leiocephalus barahonensis aureus* (Cochran, 1934) × *oxygaster* (Schwartz, 1967)—1 specimen; *Plica plica* (Linnaeus, 1758)—1 specimen.

MALAGASIAN

Oplurus sebae (Duméril and Bibron, 1837)—1 specimen.

Measurements of sensitivity in terms of cochlear potentials were obtained on nearly all the above specimens, and the ears then prepared histologically for anatomical examination. For *Polychrus acutirostris, Phrynosoma asio*, and *Sceloporus graciosus*, experimental difficulties and the limited availability of specimens prevented the obtaining of reliable sensitivity curves, and these species are dealt with only anatomically.

ANATOMICAL OBSERVATIONS

EXTERNAL AND MIDDLE EAR STRUCTURES

Some iguanids have a superficial tympanic membrane, but more often this membrane lies a little below the surface, so that a short auditory meatus can be identified. Sometimes, as shown earlier for *Sceloporus magister* (Fig. 6-13), a row of scales extends a short distance from the anterior edge of the ear opening, partially overlying the tympanic membrane below. In Fig. 7-1 these scales hide the opening entirely. In some other species a single enlarged scale or a little flap of skin lies in this position. It is apparent that such structures serve in some measure to protect the meatus from the entrance of sharp objects and particles of debris as the animal moves forward in tall grass or burrows in loose sand and soil. However, no specific muscle for closing the ear aperture, as found in most of the geckos, has been observed in iguanids.

The tympanic membrane is usually oval, facing laterally and a little posteriorly. Its middle portion is commonly bulged outward slightly by the attachment to its inner surface of the pars inferior of the extracolumella.

The sound-conductive mechanism leading from the tympanic membrane to the oval window of the cochlea, as in all the reptiles, consists typically of two elements, an osseous columella and a cartilaginous extracolumella, as described earlier. A constant feature is the presence of an element known as the internal process, which extends from the main body of the extracolumella to an anchorage on the quadrate bone. This feature identifies the iguanid type of middle ear as found in this group and also in several other lizard families. In a few species with a degenerate conductive system the internal process is greatly modified but still is recognizable as a distinctive feature.

The effectiveness of the middle ear mechanism as a receiver and transmitter of sounds has already been fully treated for the iguanids, as represented by *Crotaphytus collaris* and *Sceloporus magister* (Chapter 6).

INNER EAR

As generally true among lizards, the cochlear duct of iguanids is a cavity with the form of an inverted pyramid, enclosed in the otic capsule by large masses of limbic tissue. In this family the portion of the duct occupied by the auditory papilla is relatively small, and the lagenar macula spreads broadly over the anterior and ventral regions.

As pointed out by Miller (1966a) from his observations on the form of the cochlear duct, the ear of iguanids, along with that of agamids, stands at the lower end of the series of lizard families in level of structural development. This low ranking is borne out by many other features, such as the size and degree of elaboration of the auditory papilla and the size of the hair-cell population. However, within the iguanid family there is a considerable range of variation.

The general features of the cochlea of iguanids and the relations to surrounding structures will first be described for a representative species, and then the variations occurring among the different groups will be considered. Finally a few species in the different groups will receive particular attention.

Ear of *Iguana iguana*. The species chosen for early examination is *Iguana iguana*, with one of the best developed ears in this family. This species has already been used to present the general anatomy of the reptilian inner ear in Chapter 3, and that section, along with the illustrations presented there, may well be consulted for contributory material.

A lateral view of the right inner ear region of *Iguana iguana* is presented in Fig. 7-2. This is a reconstruction drawing based on a series of sections cut in the sagittal plane and passing through the long axis of the auditory papilla. Above in this figure is the saccule (with its membranous walls considerably collapsed), and to the right, i.e. anteriorly, is a portion of the utricle. The dorsal wall of the cochlear duct is formed by Reissner's membrane, which separates this duct from the scala vestibuli. A number of capillaries line the vestibular surface of Reissner's membrane. In this view the scala vestibuli is seen only in part; it extends far laterally, toward the observer, where it is continuous with a large chamber, the perilymphatic cistern. This lateral extension of the scala vestibuli is enclosed by bony walls except for the oval window, which is covered by the footplate of the columella.

The larger portion of the cochlear duct is occupied by the lagenar macula. This organ consists of a dense layer of ciliated cells, overlaid first by a tectorial network and then by a mass of otolithic crystals. As seen here, this macula occupies the anterior wall of the cochlear duct and the ventral portion of its posterior wall.

The auditory papilla is located high on the posterior wall of the cochlear duct, supported by the basilar membrane, which covers an elongated oval opening in the thick limbic plate in this region.

The scala tympani is shown below and posterior to the auditory papilla. It leads through the round window, a membrane-covered opening in the posterolateral wall of the otic capsule, into the round window niche. This niche is an extension of the middle ear cavity and is filled with air. Accordingly, the round window membrane can bulge into this air cavity to permit vibratory displacements of the cochlear fluids.

As the drawing shows, a large mass of tissue, which bulges to form the limbic mound, lies anterior and dorsal to the papilla, and from the posteroventral side of this mound extends a thin membrane of tapered form, which is the tectorial membrane. This membrane ends in a tectorial plate lying close over the middle portion of the auditory papilla.

Further details of the auditory papilla have already been shown in Fig. 4-24, which gives a lateral view much as in Fig. 7-2 but rotated to the left. This picture represents the tectorial plate, about 80 μ long, lying over the ciliary tufts of about six rows of hair cells near the middle of the papilla, and shows the connections of the tips of these tufts to

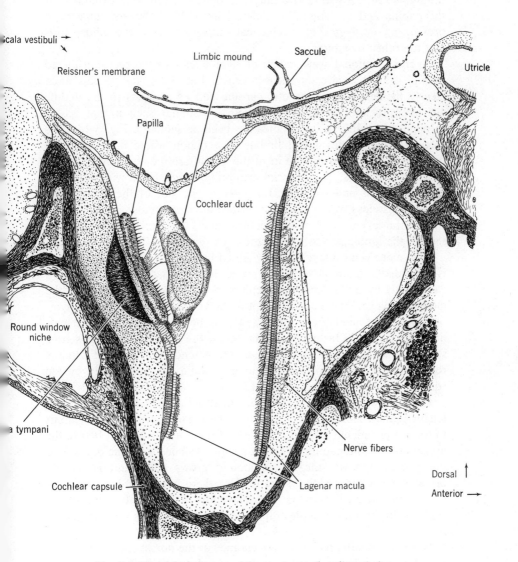

Scala vestibuli

Reissner's membrane

Limbic mound

Saccule

Utricle

Papilla

Cochlear duct

Round window
niche

a tympani

Cochlear capsule

Nerve fibers

Lagenar macula

Dorsal ↑

Anterior →

FIG. 7-2. The right inner ear of *Iguana iguana* in a lateral view.
A reconstruction based on serial sections. Scale 50×.

the underside of the plate. Also shown is an extensive mass of embedding material, consisting of a tangle of delicate fibrils that extends over the papilla and the adjoining ends of the limbic plate and appears to provide a low degree of mechanical linkage between the ciliary tufts and the other structures.

A cross-sectional view, taken at right angles to the one considered in Fig. 4-24, has already been shown in Fig. 3-11, and reveals more clearly the relations of the otic capsule and its contents to the middle ear laterally and to the brain within. A similar section that passes through the middle of the cochlea has been presented in Fig. 3-14. This section includes the medial limbus in a region where it is especially broad and thick. Shown here in further detail are the attachments to the tectorial structures. The tectorial membrane is connected with the tectorial plate at its anterior edge, runs free to the side of the limbus, and then passes close over the limbic surface along a sinuous course all the way to the lagenar end of the cochlear duct. No root process could be seen in the limbic notch, which here is extremely shallow.

At a more ventral level as represented in Fig. 3-12 the scala tympani is expanded into a relatively large recess, and is separated from the brain cavity only by the arachnoid reticulum. Somewhat farther ventrally the glossopharyngeal nerve, whose root is shown, passes out of the perilymphatic sac and finally penetrates the lateral bony walls to enter the middle ear space. Still farther ventrally, below the ampulla of the posterior semicircular canal, the perilymphatic sac extends to the round window region, where it is separated from the middle ear cavity only by the round window membrane (Fig. 3-13).

Let us now consider the variations in structure that appear along the length of the cochlea. The elevation of the medial portion of the limbus to form a prominent mound is evident at the dorsal end even before the appearance of the papilla. When at a short distance from the dorsal end the papilla appears, the limbic mound grows taller and noticeably broader. Near the middle of the cochlea there is a secondary elevation of this mound medial to the main elevation. At the ventral end of the cochlea this mound dwindles rapidly and flattens out as the papilla terminates.

When the papilla makes its appearance at the dorsal end of the cochlea, the tectorial membrane may be seen closely covering the peak of the limbic mound and a portion of its lateral face. A little farther ventrally this membrane leaves the mound and extends a short distance in the direction of the papilla. In this region also the membrane may be followed around the anteromedial surface of the mound through the limbic notch and continuing anteriorly to the region of the lagenar macula. This membrane is often difficult to trace further, but at times it may be seen to pull away from the limbic wall and become incorporated into the tectorial reticulum of the lagenar macula.

About the middle of the cochlea the tectorial membrane reaches the papilla and attaches to the tectorial plate, as already seen. The tips of the ciliary tufts of the hair cells are embedded in the undersurface of the plate, and for the most part appear to make this connection to nodular thickenings that correspond at least roughly to the transverse rows of hair cells. These connections are shown in Fig. 7-3.

Over the remaining portion of the cochlea, both dorsal and ventral to the tectorial plate, the ciliary tufts end freely, apart from tenuous connections with embedding material. A section just dorsal to the plate region is represented in Fig. 7-4.

In these last two figures the basilar membrane is shown anchored to the two edges of the limbus, with its greatly thickened portion, the fundus, located asymmetrically so that a narrow ribbon of thin membrane extends beyond the lateral side of the papilla. Over the surfaces of the limbus, immediately adjacent to the papilla, are border cells, usually in one or two rows on either side, and farther away is the usual inner covering of epithelial cells, here of rather simple character.

The papilla has the standard construction. Over the fundus is a thick array of supporting cells, which send long columnar processes upward between the hair cells to enclose their upper ends. The nuclei of the supporting cells are clearly evident, but their cell walls are not usually discernible with the staining procedure commonly used. Nerve fibers run in a thick strand through the medial limbus and penetrate the thin inner edge of the basilar membrane, entering the base of the papilla to course obliquely to the lower ends of the hair cells. These fibers lose their medullary sheaths just before entering the papilla, and their paths through the papilla to the hair cells are usually difficult to see.

The lengths of the ciliary tufts of the hair cells vary systematically along the cochlea, as best shown in Fig. 4-24. At the dorsal end these tufts are relatively short, with their longest cilia measuring around 6 μ. More ventrally they increase progressively in length until, just before the tectorial plate begins, they measure 17-21 μ. Coincident with the appearance of the tectorial plate, these tufts are markedly reduced in length, and here again measure about 6 μ. Ventral to the plate the tufts become elongated once more, and in this specimen were around 23 μ in length. At the ventral end the cilia again become reduced, some measuring 4 μ or less. Thus the range of variation over the cochlea in the length of the ciliary tufts exceeds fivefold.

The orientation of the ciliary tufts varies along the cochlea. Over the dorsal and ventral segments this orientation is regularly bidirectional, with most of the hair cells on the lateral side of the papilla having their

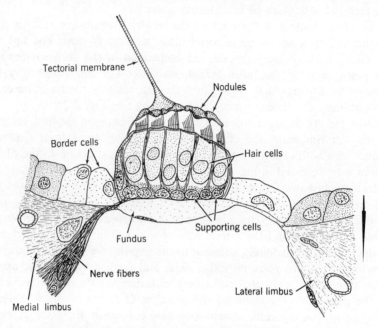

FIG. 7-3. The auditory papilla of *Iguana iguana* shown in cross section near the middle of the cochlea. From Wever, 1967a.

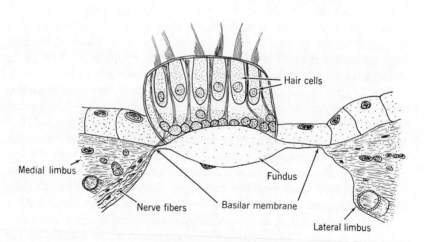

FIG. 7-4. The auditory papilla of *Iguana iguana* shown in cross section just dorsal to the tectorial plate. Scale 500×.

tufts facing medially (i.e., with the longest cilia, together with the kino-cilium, on the medial side), and contrariwise those cells on the medial side having their tufts facing laterally, as shown in Fig. 7-4. Usually the division between these two orientations is near the middle of the trans-verse row of hair cells, but occasionally a cell will depart from this or-derly arrangement and faces contrary to its neighbors.

Beneath the tectorial plate, as Fig. 7-3 has shown, the orientation of ciliary tufts is unidirectional, with all the tufts having their longest cilia on the lateral side. Usually, as indicated in this figure, a small number of stereocilia, together with the kinocilium, extend well beyond the main group and make the tectorial connections.

The various iguanid species included in this study will now be con-sidered, with particular attention to the forms of the tectorial mem-brane, the size and location of the tectorial plate, and the numbers and distribution of the hair cells. These features will be related to the six subdivisions of the family already referred to, with an indication of the degree to which the auditory structures follow these lines of distinction based on numerous other features.

Types of Tectorial Membrane. Among the Iguanidae the tectorial membrane takes two distinct forms, and each of these exhibits two subtypes.

Type I. — In most of the iguanids studied the tectorial membrane takes a peculiar elevated course not found elsewhere in any other liz-ards. The membrane arises from an attachment at the limbic notch, and then curls around in a wide arc to its insertion on the tectorial plate over the auditory papilla, with no contact with the limbic mound. Two sub-types may be distinguished according to the degree of elevation above the mound. In most species with this form of tectorial membrane there is a wide separation from the limbic mound, as in I*a* of Fig. 7-5, but in a few this separation is only slight, especially over the peak of the mound (I*b* of this figure).

Type II. — The second form of tectorial membrane found among iguanids is a direct type, much like the one most commonly encoun-tered in other lizard families. This membrane is in contact with the lim-bic mound, or even attached to it, and leaves this surface to run directly to the tectorial plate lying over the papilla. Again there are two sub-types: in one the membrane arises from the limbic notch, or passes through it, and continues around the surface of the mound in more or less close contact, and then leaves the mound just beyond the peak to run to the tectorial plate (II*a* in Fig. 7-5). In the other form the mem-brane either arises from the lateral slope of the limbic mound or arises farther back and then attaches to this lateral surface, and finally passes directly to the tectorial plate (II*b* in Fig. 7-5).

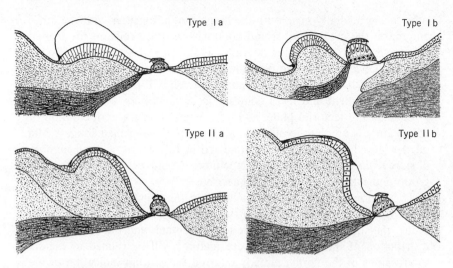

FIG. 7-5. Types of tectorial membranes in Iguanidae. The sketches are based on selected species: *Sceloporus clarkii* for Type Ia, *Anolis lionotus* for Type Ib, *Dipsosaurus dorsalis* for Type IIa, and *Ctenosaura acanthura* for Type IIb. Scale 200×.

A further representation of this IIb form of tectorial membrane is afforded in Figs. 7-6 and 7-7 for the species *Ctenosaura acanthura*. Figure 7-6 shows a cross section through a specimen at a point just dorsal to the appearance of the tectorial plate where the ciliary tufts have their usual free-standing form, with bidirectional ciliary orientation, and the end of the tectorial membrane is extending from its attachment to the side of the limbic mound but at this point does not reach the region of the papilla. The next figure shows this same specimen a little farther ventrally where the tectorial membrane extends farther and connects to the plate over the papilla. The ciliary tufts here are short and all are oriented laterally.

At times Types Ib and IIa may be difficult to distinguish, for in certain instances the elevation of Ib becomes minimal. The criterion is whether there is anywhere a place of contact with the mound, which makes the specimen belong to Type IIa.

Occasionally, as shown in Fig. 7-8 for *Enyaliosaurus clarki*, a strand of tectorial tissue runs from the peak of the limbic mound to Reissner's membrane and then extends anteriorly along this membrane to the lagenar region, where it becomes incorporated into the macular reticulum. It is doubtful that this variation of the Type IIa tectorial membrane has any functional significance.

A further variation in the form of the tectorial membrane is shown in Fig. 7-9 for a specimen of *Dipsosaurus dorsalis*. In this specimen an

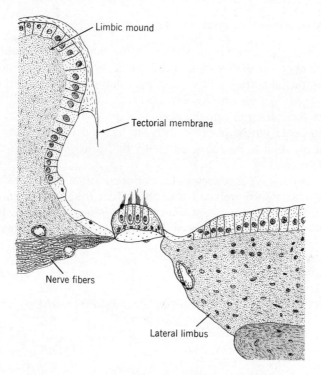

FIG. 7-6. The auditory papilla and tectorial membrane of *Ctenosaura acanthura* at a level dorsal to the tectorial plate. Scale 250×.

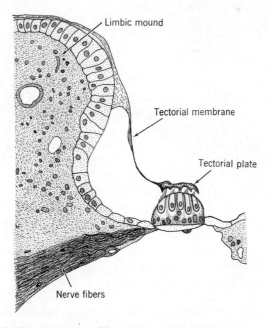

FIG. 7-7. The auditory papilla and tectorial membrane of *Ctenosaura acanthura* at the level of the tectorial plate. Scale 250×.

extension of the tectorial membrane to the limbic notch could not be found. Such an extension is the rule, usually with a definite connection to the epithelial surface within the notch, and often with a further continuation of the membrane anteriorly along the epithelial surface. Whether the extension actually occurs in specimens like the one shown here and could not be resolved with the light microscope must remain an open question to be answered finally by further study with the electron microscope.

In general the types of tectorial membranes found among the iguanids of this study conform well with the subfamilial groupings. All the sceloporines are of Type I*a* except the two *Crotaphytus* species, which are Type II*a*. All the iguanines are of Type II, though divided between sub-

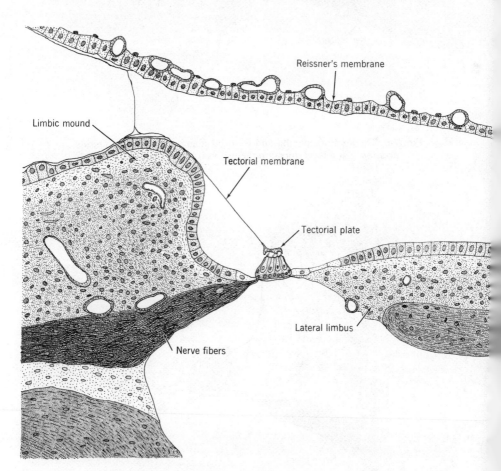

FIG. 7-8. The auditory papilla and tectorial membrane of
Enyaliosaurus clarki. Scale 200×.

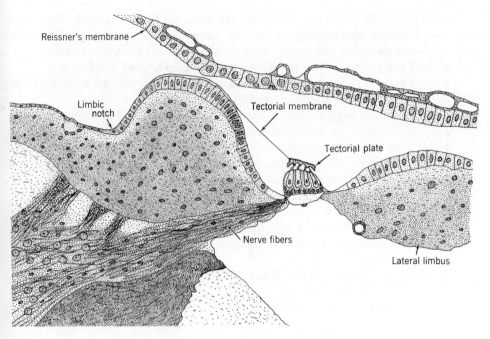

Reissner's membrane

Limbic notch

Tectorial membrane

Tectorial plate

Nerve fibers

Lateral limbus

FIG. 7-9. The auditory papilla and tectorial membrane of
Dipsosaurus dorsalis. Scale 200×.

types *a* and *b*. Among the basiliscines the two *Basiliscus* species are
Type I*b* and *Laemanctus* is Type I*a*. The anolines divide, with six
Anolis species showing the I*b* type, *Anolis cybotes* and *Polychrus
acutirostris* showing the II*a* type, and the unidentified *Anolis* showing
the II*b* type. The tropidurines also divide, with *Leiocephalus carinatus*
and *Leiocephalus barahonensis* showing the I*a* type and *Plica plica* the
II*b* type. These characters are indicated for all the species in Table 7-I.

Forms of Tectorial Plates.   The tectorial plates vary in size and in their
location along the auditory papilla. Figs. 7-10 to 7-15 represent the
lengths of the papillae and also the locations and extent of the tectorial
plates among this series of iguanids. The diagrams are arranged in
groups, and within each group the order is according to the length of
the papilla.

The upper part of Fig. 7-10 represents the nine anolines studied. The
eight *Anolis* species present a distinctive pattern, with the tectorial plate
close to the ventral end of a papilla of considerable length. *Polychrus
acutirostris*, however, shows a plate of good length somewhat dorsal to
the middle of a short papilla.

Below in Fig. 7-10 are represented the three basiliscines studied. The two *Basiliscus* species show a peculiar pattern, with a ventral papilla well separated from the larger dorsal one. The tectorial plate is located toward the ventral end of the dorsal papilla. *Laemanctus longipes* displays a more common type of structure, with an undivided papilla bearing its tectorial plate a little ventral of its midpoint.

Seven species of iguanines are represented in Fig. 7-11. Two of these, *Dipsosaurus dorsalis* and *Sauromalus obesus*, have moderate size papillae with their tectorial plates a little dorsal of the midpoints. The three *Ctenosaura* species, along with *Iguana iguana*, display greatly elongated papillae with tectorial plates of more than average size, disposed mostly near the middle, except in *Ctenosaura acanthura*, where it is shifted somewhat ventrally. The species *Enyaliosaurus clarki*, considered to be a derivative of the *Ctenosaura* type (Avery and Tanner, 1971), occupies an intermediate position in this iguanine series.

TABLE 7-I

MEAN HAIR-CELL POPULATIONS IN IGUANIDAE

Species	Number of Ears	Dorsal	Under Plate	Ventral	Total	Tectorial Membrane Type
Anolines						
Anolis carolinensis	6	131.0	22.8	3.3	157.2	I*b*
Anolis chlorocyanus	2	69.5	25.0	2.5	97.0	I*b*
Anolis cristatellus	4	147.7	23.5	1.2	172.5	I*b*
Anolis cybotes	2	176.0	34.0	4.0	214.0	II*a*
Anolis distichus	2	144.5	17.0	2.0	163.5	I*b*
Anolis lionotus	6	169.5	54.8	4.5	228.8	I*b*
Anolis sagrei	2	106.5	22.5	2.0	131.0	I*b*
Anolis sp.	6	111.5	24.3	5.8	141.7	II*b*
Polychrus acutirostris	1	12.0	43.0	30.0	85.0	II*a*
Basiliscines						
Basiliscus basiliscus	4	76.7	45.5	3.0+26.7	152.0	I*b*
Basiliscus vittatus	4	86.2	44.5	6.2+33.2	170.2	I*b*
Laemanctus longipes	2	96.0	37.5	44.0	177.5	I*a*
Iguanines						
Ctenosaura acanthura	2	136.0	36.5	99.5	272.0	II*b*
Ctenosaura similis	2	104.0	49.5	95.0	248.5	II*b*
Ctenosaura sp.	2	65.0	28.5	50.5	144.0	II*b*
Cyclura stejnegeri	1	33.0	32.0	74.0	139.0	II?
Dipsosaurus dorsalis	4	18.5	34.0	26.7	79.2	II*a*
Enyaliosaurus clarki	4	81.7	25.7	74.2	181.7	II*b*
Iguana iguana	6	125.1	55.2	111.7	292.0	II*a*
Sauromalus obesus	4	44.2	21.2	58.5	124.0	II*a*

TABLE 7-I (cont.)

MEAN HAIR-CELL POPULATIONS IN IGUANIDAE

Species	Number of Ears	Dorsal	Under Plate	Ventral	Total	Tectorial Membrane Type
Sceloporines						
Callisaurus draconoides ventralis	8	20.4	26.9	15.9	63.4	Ia
Cophosaurus texanus	4	22.7	33.5	22.7	79.0	Ia
Crotaphytus collaris	16	28.8	50.9	28.0	107.8	IIa
Crotaphytus wislizenii	10	13.5	23.8	19.6	56.8	IIa
Holbrookia maculata	4	16.0	23.7	20.2	60.0	Ia
Phrynosoma asio	2	28.5	8.5	25.0	62.0	Ia
Phrynosoma cornutum	2	15.5	9.5	17.0	42.0	Ia
Phrynosoma coronatum	6	19.0	8.2	18.7	45.8	Ia
Phrynosoma m'calli	2	17.0	19.0	13.0	49.0	Ia
Phrynosoma platyrhinos	6	16.2	20.8	15.0	52.0	Ia
Phrynosoma solare	2	23.5	15.5	22.0	61.0	Ia
Sceloporus clarkii	2	30.5	17.5	29.0	77.0	Ia
Sceloporus cyanogenys	1	35.0	11.0	42.0	88.0	Ia
Sceloporus graciosus	4	18.7	6.0	24.0	48.7	Ia
Sceloporus magister	6	32.5	15.3	31.0	78.8	Ia
Sceloporus occidentalis	8	28.1	9.5	27.9	65.5	Ia
Sceloporus orcutti	6	28.3	10.3	27.8	66.5	Ia
Sceloporus undulatus consobrinus	4	22.0	11.5	23.5	57.0	Ia
Streptosaurus mearnsi	4	21.0	18.5	31.0	70.5	Ia
Uma notata	4	25.0	17.8	25.2	68.0	Ia
Urosaurus ornatus linearis	6	21.3	10.3	22.8	54.5	Ia
Uta stansburiana	8	20.5	9.0	22.0	51.5	Ia
Tropidurines						
Leiocephalus barahonensis	2	41.0	14.0	38.0	93.0	Ia
Leiocephalus carinatus armouri	5	60.8	12.6	51.2	124.6	Ia
Plica plica	2	238.5	44.0	8.0	290.5	IIb
Malagasian						
Oplurus sebae	2	59.5	29.5	58.0	147.0	Ia
For 22 sceloporines	115	22.9	17.1	23.8	63.9	
For 24 other iguanids	77	97.3	32.4	38.1	167.8	
For all 46 species	192	61.7	25.1	31.3	118.1	

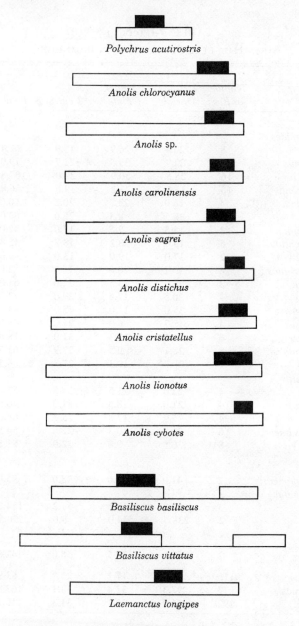

FIG. 7-10. Bar diagrams for nine species of anolines and three species of basiliscines. The open rectangles represent the auditory papilla and the black rectangles show the location and extent of the tectorial plate. In this and the following five figures the lengths are on the same scale, which is 125×. Dorsal is to the left, ventral to the right.

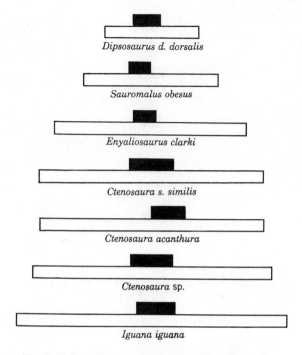

Dipsosaurus d. dorsalis

Sauromalus obesus

Enyaliosaurus clarki

Ctenosaura s. similis

Ctenosaura acanthura

Ctenosaura sp.

Iguana iguana

FIG. 7-11. Bar diagrams for seven species of iguanines.

The papillar patterns in the sceloporines are shown in the next three figures. Figure 7-12 represents six species of *Phrynosoma* with papillae of short to moderate length and rather small tectorial plates located near the middle of the papilla—though displaced a little dorsally in *Phrynosoma platyrhinos*.

The seven *Sceloporus* species studied, as Fig. 7-13 shows, have papillae of lengths varying from moderate to long. The tectorial plates vary around the midposition, and vary considerably also in length from very short in *Sceloporus graciosus* and *Sceloporus magister* to long in *Sceloporus undulatus consobrinus* and *Sceloporus clarkii*.

The remaining sceloporines, shown in Fig. 7-14, continue the pattern already seen, except for the species *Streptosaurus mearnsi* and *Uma notata*, in which the tectorial plate is particularly broad. Also in *Holbrookia maculata* and in *Uta stansburiana* this plate is displaced somewhat toward the dorsal end of the papilla and in *Streptosaurus mearnsi* very noticeably so.

The three tropidurines included in the study, represented in Fig. 7-15, show some variation also: the two *Leiocephalus* species have moderately long papillae with a narrow plate slightly ventral to the middle,

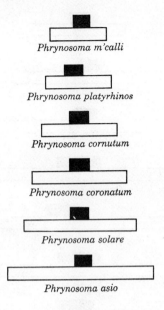

Phrynosoma m'calli

Phrynosoma platyrhinos

Phrynosoma cornutum

Phrynosoma coronatum

Phrynosoma solare

Phrynosoma asio

FIG. 7-12. Bar diagrams for six species of *Phrynosoma*.

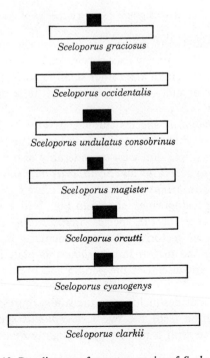

Sceloporus graciosus

Sceloporus occidentalis

Sceloporus undulatus consobrinus

Sceloporus magister

Sceloporus orcutti

Sceloporus cyanogenys

Sceloporus clarkii

FIG. 7-13. Bar diagrams for seven species of *Sceloporus*.

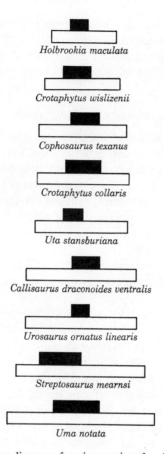

FIG. 7-14. Bar diagrams for nine species of sceloporines.

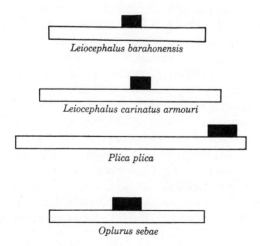

FIG. 7-15. Bar diagrams for three tropidurines and one Malagasian species.

and *Plica plica* shows a very long papilla with the tectorial plate close to the ventral end as found elsewhere only in the anoles.

The single representative of the Malagasian group, *Oplurus sebae*, shown in Fig. 7-15, has an average form of papilla with its tectorial plate in the middle.

HAIR-CELL POPULATIONS

In the counting of hair cells for the various iguanid specimens, account was taken of the numbers contained in the different cochlear segments: in the region dorsal to the tectorial plate, beneath this plate, and in the region ventral to the plate. These results showed individual variations, but a species pattern was evident, especially in the size of the population and more particularly in the number of hair cells beneath the tectorial plate. As expected, these numbers follow closely the spatial patterns seen in Figs. 7-10 to 7-15.

Table 7-I shows the average number of hair cells for each species, first separately for the three segments of the papilla and then for the total population.

Appearing first in this table are results for the nine anolines studied, with the *Anolis* species (except *chlorocyanus*) having relatively large hair-cell totals. The large majority of these cells are in the dorsal segment, as expected from its considerable length (see Fig. 7-10), and only a negligible number of cells is found in the ventral segment. What was noted for *Polychrus acutirostris* concerning the spatial pattern of the papilla is further borne out here: there is hardly any resemblance to the anoles. The dorsal segment contains few hair cells, the ventral segment a good many, and the total number falls below that of the other species in this group (though not far below *Anolis chlorocyanus*).

The basiliscines present a good sized total population of hair cells, with 20-26% under the tectorial plate. The distribution for *Laemanctus longipes* agrees well with that for the *Basiliscus* species, apart from the splitting off of the ventral portion of the structure in these ears.

The patterns for the iguanines already seen in Fig. 7-11 are followed closely by the hair-cell distribution data in Table 7-I. *Dipsosaurus dorsalis* has the smallest total population of 79.2, though the number of cells beneath the tectorial plate amounts to a surprising 43% of this total. *Sauromalus obesus*, with a total of 124, has only 21.2 cells or 17% of the total beneath this plate. *Iguana iguana* leads the group both in the total number of hair cells and in the number beneath the plate. Among other species examined, only *Plica plica*, one of the tropidurines, approaches this population size.

The sceloporines as a group show the lowest level of ear development in terms of hair-cell numbers. The 22 species examined present totals ranging from 42 in *Phrynosoma cornutum* to 107.8 in *Crotaphytus collaris*, and the average population is 63.9. This number is to be compared with the average of 118.1 for the whole iguanid group. The population size of 107.8 in *Crotaphytus collaris* approaches this general average, but no other sceloporine does so: the next in order is *Sceloporus cyanogenys* with an average of 88.0 hair cells, which is about three-fourths of the general average for Iguanidae.

The numbers of hair cells under the tectorial plate are of comparative interest also. This number averages 25.1 for the whole iguanid series, and is 17.1 for the sceloporines. Again *Crotaphytus collaris* with 50.9 leads the sceloporine group and is far ahead of the next in line, which is *Cophosaurus texanus* with 33.5. The lowest number under the plate is found in *Sceloporus graciosus* with 6.0, and only a little larger are the numbers 8.2 in *Phrynosoma coronatum* and 8.5 in *Phrynosoma asio*. The small numbers in the *Phrynosoma* species seem reasonable in view of the fact that this group shows several degenerative modifications of the sound receptive system, no doubt as a result of their adaptation to burrowing and sand swimming in a desert environment. But the comparatively large number of hair cells both in total and under the tectorial plate in the "greater earless lizard," *Cophosaurus texanus*, is surprising. As noted at another place, this condition probably represents a secondary compensation for the loss of the tympanic membrane.

COCHLEAR DIMENSIONS

The following descriptions of representative species among the several groups of iguanids will bring out the structural variations found and the level of differentiation in these ears.

ANOLINES

Anolis cristatellus. — The species *Anolis cristatellus* will be taken as representative of the anoles. In the specimen selected for this description the cochlear duct was about 820 μ long from its narrow connection to the saccule to its expanded lagenar end. At the dorsal end of the cochlear duct the limbic plate forms two mound-like elevations. As shown in Fig. 7-16, the main elevation is covered by tall columnar cells, and anterior to it is a secondary mound covered by a simple epithelial layer. Toward the middle of the cochlea the principal mound grows somewhat in height and breadth, while the other dwindles; toward the ventral end of the cochlea this secondary mound disappears altogether.

Reissner's membrane is thin over the posterior third of its extent, but anteriorly it becomes considerably thickened by expansion of its cochlear layer, whose cells take a tall columnar form.

The scala tympani is a blind tunnel at both dorsal and ventral ends, and only in the middle half of the cochlea is it open to the tympanic recess.

The tectorial membrane makes its appearance in the dorsal region, arising from the limbic notch. As shown in Fig. 7-17, it extends above the epithelial surface, turns in a wide arc anteriorly and laterally, and then runs close over the top of the limbic mound and along its postero-lateral surface; finally, in the ventral region, it leaves this surface and connects to the tectorial plate. This is an exceptional form of tectorial membrane, not conforming well to the types described above. It begins as a Type I membrane with a free course in the fluid as it curls around the anterior portion of the limbic mound, and then it approaches the mound and runs closely over its middle surface in the fashion of a Type II membrane, and finally extends to the tectorial plate. Because this membrane makes a close contact with the limbic mound, it meets the definition of a Type II structure and may be designated as Type IIa.

The basilar membrane in *Anolis cristatellus* is of medium length, measuring a little under 500 µ as shown in Fig. 7-18. It rises quickly to a maximum width of 130 µ, from which it narrows, at first rapidly and then more gradually, until it terminates at the ventral end. Its thickened portion, the fundus, takes a similar course, with a maximum of 72 µ reached abruptly, and with a slower decline along its course.

The thickness of the fundus, shown in part *b* of this figure, rises to a maximum near the dorsal end of the cochlea, and then takes an irregular course over the remaining region.

The form of the auditory papilla is much the same as that of the basilar membrane. Its area increases rapidly to a maximum near the dorsal end and then declines, with a little wavering, until it quickly terminates. The variation over its main course is only about 30%. The organ of Corti portion, shown by the broken curve in part *c* of the figure, follows the same course.

The longitudinal rows of hair cells exhibit remarkable uniformity, with 8 in a row everywhere except at the ends of the papilla. The hair-cell density (expressed as the number of cells per 20 µ section along the cochlea) is correspondingly uniform, quickly attaining a maximum near the dorsal end and falling off only moderately thereafter, with slight dips in the midregion.

Altogether this cochlea is relatively simple and undifferentiated. The variation in the width of the membrane, amounting to about 2.5-fold, is the only significant dimension that might make for a selective action of tonal frequencies. We must regard this ear, therefore, as only slightly graduated as far as local action of tonal stimuli is concerned. Any significant amount of tonal discrimination would have to be achieved through frequency following by the cochlear nerve fibers.

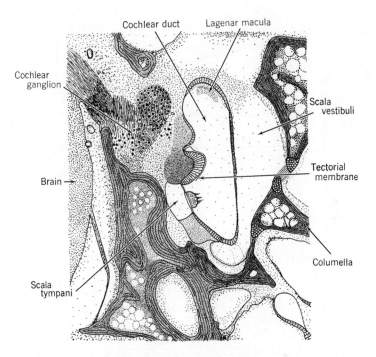

FIG. 7-16. The cochlear region in *Anolis cristatellus*. Anterior is above and lateral to the right.

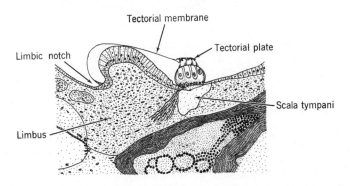

FIG. 7-17. The auditory papilla and tectorial membrane of *Anolis cristatellus*. Scale 150×.

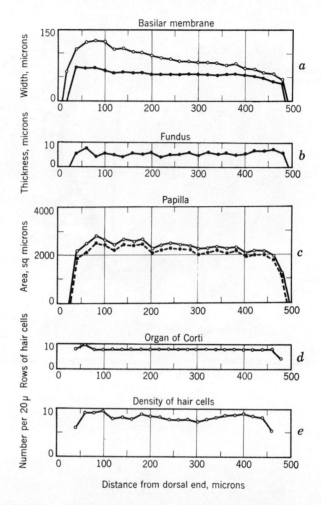

FIG. 7-18. Cochlear dimensions in *Anolis cristatellus*. In this graph and elsewhere (unless otherwise noted) the upper curve of part *a* represents the width of the basilar membrane and the lower curve the width of the thickened portion, the fundus. Part *b* represents the thickness of the fundus. The solid curve of part *c* shows the area of the papilla (including the fundus), and the broken curve shows the area of the cellular portion only (organ of Corti). Part *d* indicates the number of longitudinal rows of hair cells along the cochlea, and part *e* represents the density of these cells. The density is expressed as the number of hair cells for a section 20μ thick, and the curve shows a rolling average of these values. Each consecutive three values are averaged.

Polychrus acutirostris. — In *Polychrus acutirostris,* another species included among the anolines, there is a single limbic mound of moderate height, covered with low columnar epithelium. Reissner's membrane is of ordinary form, lacking the pronounced thickening in its anterior portion noted for *Anolis cristatellus.* The basilar membrane is very short, measuring only about 200 μ, as Fig. 7-19 indicates. Its width increases from the dorsal end to a ventral maximum of 124 μ. The width of the fundus follows the same form, and is a little over half that of the whole membrane. Thus a relatively wide strip of thin membrane runs along the lateral border of the papilla. The thickness of the fundus

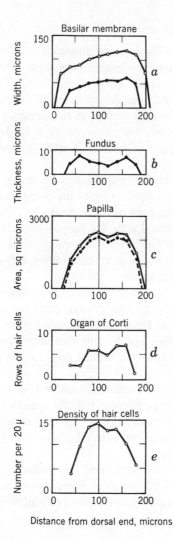

Distance from dorsal end, microns

Fig. 7-19. Cochlear dimensions in *Polychrus acutirostris.*

shows two maximums, with a thin region in the middle of the cochlea.

The auditory papilla has a simple, nearly symmetrical form, with a slight depression near its midpoint. The cellular portion (organ of Corti) follows the same form.

The number of rows of hair cells along the organ of Corti increases in two stages from 3 to 7, and then falls abruptly as the papilla ends. The density of the hair cells rises rapidly to a maximum in the middle of the papilla, and then falls away.

A comparison of this cochlea with that of *Anolis cristatellus* shows many differences among which the reversed form of the basilar membrane is perhaps the most striking.

BASILISCINES

Basiliscus basiliscus. — The basiliscines are considered to be a specialized side branch of the Iguanidae, containing only three genera, *Basiliscus*, *Corythophanes*, and *Laemanctus* (Etheridge, 1964). The inner ear in most respects exhibits a typical iguanid form, but in the *Basiliscus* species it differs from all others of this family in the presence of two separate openings in the limbic plate, each covered by a basilar membrane that bears an auditory papilla. The arrangement is shown diagrammatically, to scale, in Fig. 7-20.

As the figure shows, the two basilar membranes lie along an axis that runs nearly dorsoventrally. The dorsal membrane is the larger, and its papilla has about three times the area of the ventral one. The dorsal papilla bears a tectorial plate, located near its ventral end, as has already been indicated in Fig. 7-10. This plate is attached to the tectorial membrane, which arises over a broad area of the limbus and narrows greatly as it swings over the papilla to this connection.

The structural relations are seen in further detail in the cross-sectional view of Fig. 7-21, which passes through the tectorial plate region at the place indicated by the arrow at *a* in Fig. 7-20. The tectorial membrane arises by a root process in the limbic notch, runs well above the rather low mound as shown, and connects to the medial edge of the tectorial plate.

A cross section through the ventral papilla, shown in Fig. 7-22, cuts through a portion of the tectorial membrane, but this membrane does not approach the papilla here. The hair cells in this region, as also in the dorsal papilla everywhere outside the tectorial plate, send their ciliary tufts freely into the cochlear fluid.

A cross section taken through the portion of the limbus separating the two papillae, shown in Fig. 7-23, reveals the complete continuity of the limbic tissue, though this plate becomes greatly reduced in thickness here.

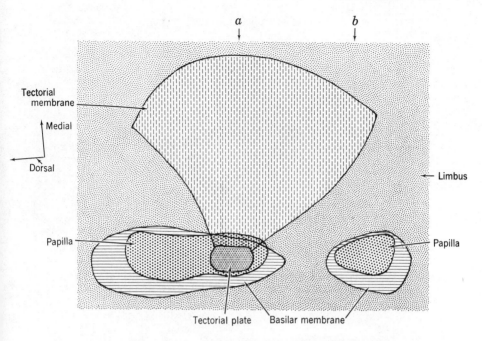

FIG. 7-20. A diagrammatic representation of the right limbic plate of *Basiliscus basiliscus* with locations of the auditory structures. The arrows at *a* and *b* indicate the levels of cross sections shown in the next two figures. From Wever, 1971a.

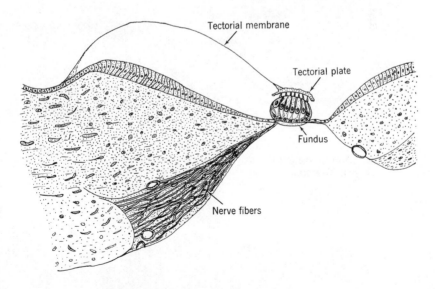

FIG. 7-21. A cross section through the dorsal cochlear region of *Basiliscus basiliscus*, passing through the plane shown at *a* in the preceding figure. From Wever, 1971a.

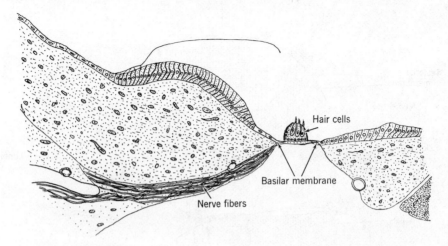

FIG. 7-22. A cross section through the ventral cochlear region of *Basiliscus basiliscus*, along the plane indicated by *b* in Fig. 7-20. From Wever, 1971a.

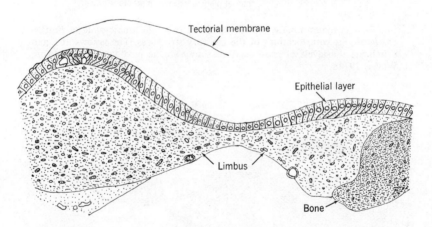

FIG. 7-23. A cross section through the isthmus separating dorsal and ventral basilar membranes in *Basiliscus basiliscus*. From Wever, 1971a.

The nerve fibers supplying the two papillae arise in a single ganglionic cluster anteromedial to the cochlea, run outward as the cochlear bundle, and in an area medial and posterior to the limbic mound, about opposite its peak, this bundle divides into two parts, a larger one passing to the dorsal papilla and another, much smaller, going to the ventral papilla.

Dimensions of the cochlear structures are shown for a specimen of *Basiliscus basiliscus* in Fig. 7-24. The width of the dorsal basilar membrane, on the left of part *a* of this figure, rises rapidly to a maximum of 120 μ, then falls somewhat sinuously to about half this value at the 300 μ point, after which the drop is very rapid toward zero. The ventral basilar membrane begins near the 400 μ point, rises rapidly to a peak of 100 μ at the 500 μ point, and falls even more rapidly to zero at the ventral end.

The width of the fundus varies in a similar fashion; for the dorsal structure it rises abruptly to a peak of about 60 μ, then declines rather gradually until near the ventral end it rounds off quickly. The ventral portion shows a rapid and then a more gradual increase to a maximum of about 50 μ, and then a rapid drop. The curve representing the thickness of the dorsal fundus shows double humps, first a rather flat one and then a broad triangular one. The ventral fundus exhibits only a moderate thickness throughout its short course.

The areas of the two papillae follow closely the form of the basilar membrane. The dorsal one rises rapidly to a maximum, decreases and then rises slightly to a secondary peak, and then rapidly falls away. The ventral papilla rises to a peak at the 500 μ point, and then quickly terminates.

In the dorsal papilla the number of longitudinal rows of hair cells varies around 5, reaches a maximum of 7 toward the end of this structure, and then rapidly diminishes. In the ventral papilla this number only varies between 2 and 3.

The duplex character of the cochlea of *Basiliscus basiliscus* is clearly reflected in all its structural features. The dorsal papilla is dominant, and the ventral one appears only as a minor adjunct.

How this duplexity arose is a matter of speculation. It seems likely that these lizards were derived from early iguanids with a single, continuous basilar membrane, and this membrane was divided through an invasion of limbic tissue. A minor degree of duplexity in the form of double maximums in the dimensions of the basilar membrane and papilla are sometimes seen in other iguanids, as in Figs. 7-26 and 7-28 for *Laemanctus longipes* and *Iguana iguana*, and occurs also in other families. A complete separation into dorsal and ventral portions is found elsewhere only in the Lacertidae and occasionally in the Varanidae, as described below (Chapters 11 and 13).

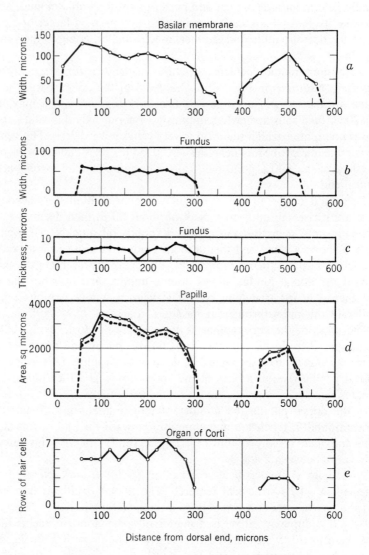

FIG. 7-24. Cochlear dimensions in a specimen of
Basiliscus basiliscus. From Wever, 1971a.

Functionally the division of the cochlea of *Basiliscus* into two parts seems to have little or no effect. There are no outstanding differences from other iguanids in degree of sensitivity or the form of the sensitivity curve.

Laemanctus longipes. — The duplexity of basilar membrane and papilla that is found in *Basiliscus basiliscus* and also in *Basiliscus vittatus* does not appear in *Laemanctus longipes*. In this species, as represented in Fig. 7-25, the limbic mound is low and broad, and the tectorial membrane arises from a root process in a wide depression anteromedial to this mound. This membrane swings around in a wide arc, and a little beyond the middle of the cochlea it connects to the inner edge of the tectorial plate.

Reissner's membrane shows an unusual structure. For the most part it is relatively thick, due to the presence in its cochlear layer of large, irregularly arranged cells, and through its middle region it contains a series of large fluid spaces of mostly elongate form. Over its dorsal half this membrane shows a region near the outer edge, and usually another at its inner edge also, where it is relatively thin and where its cochlear layer is made up of long, flat cells in an orderly series.

The orientation of the ciliary tufts of the hair cells is bidirectional throughout the cochlea, even beneath the tectorial plate where in most species this orientation is unidirectional.

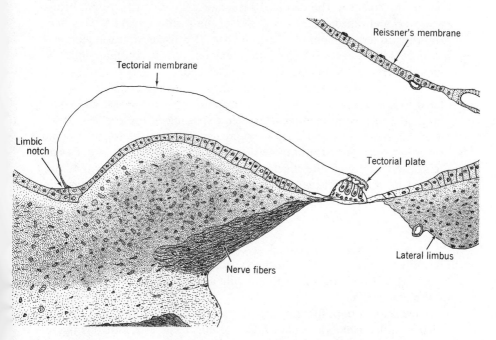

Fig. 7-25. The cochlear region of *Laemanctus longipes*. Scale 200×.

Dimensions of the cochlear structures are represented in Fig. 7-26. The width of the basilar membrane, shown by the upper curve in *a* of this figure, rises rapidly at the dorsal end to a maximum of 96 μ, then declines a little in the middle of the cochlea where it measures around 80 μ, rises sharply to a second maximum at 96 μ, and then falls rapidly to its ventral end. The length of the basilar membrane is 460 μ.

The fundus, represented by the lower curve in *a* of this figure, has a width about half that of the whole membrane, and follows the same form. The thickness of the fundus, shown in part *b*, shows two maximum regions, one in the dorsal part of the cochlea that is followed by a considerable decline through the middle of the cochlea, and a second in the ventral part that almost equals the dorsal one.

The auditory papilla is definitely bimodal in form; its area increases sharply to a maximum near the dorsal end, falls to a minimum in the middle of the cochlea, and then rises to a second, broader maximum in the ventral region. The two curves shown here, for the papilla as a whole and for its cellular portion, are close together and follow the same path because of the small size and relative uniformity of the fundus in this species.

The arrangement of hair cells (part *d*) shows considerable uniformity also. Five longitudinal rows appear at the dorsal end of the cochlea, vary between 6 and 3, then continue between 4 and 3 over the remainder of the cochlea. The density of the hair cells varies more markedly, however, and reflects the varying size of the auditory papilla. This density rises to a maximum of 13.5 cells per 20 μ in the dorsal region and at once falls away, and thereafter remains nearly constant until the papilla begins to terminate at the ventral end.

The bimodality seen especially in the size of the papilla may be related to the complete division seen in the *Basiliscus* species, and could be thought of as an approach to such a splitting of the cochlea. However, the bimodality seen here is no greater than may be found in some other iguanids, and also in more distant species among the anguids and teiids, so that no definite conclusions can be drawn about steps that may have led to the division of the cochlea in *Basiliscus* species and in Lacertidae.

IGUANINES

Iguana iguana. — A general description of the ear of *Iguana iguana* has already been given. A further view of this ear is presented in Fig. 7-27. This section runs near the middle of the cochlea where the tectorial plate appears.

Dimensional data for this ear are presented in Fig. 7-28. The width of the basilar membrane, shown by the upper curve in *a* of this figure, increases rapidly at first and then more gradually, with a few irregularities, to a maximum around 115 μ near the ventral end, after which this structure quickly terminates. The width of the fundus follows the same course, reaching a maximum of about 60 μ.

The thickness of the fundus shows a duplex form, reaching a flat maximum of considerable magnitude, around 17 μ, over the dorsal region, and then falling to 10 μ, and thereafter rising somewhat irregularly to a greater value around 19-24 μ over the ventral part of the cochlea.

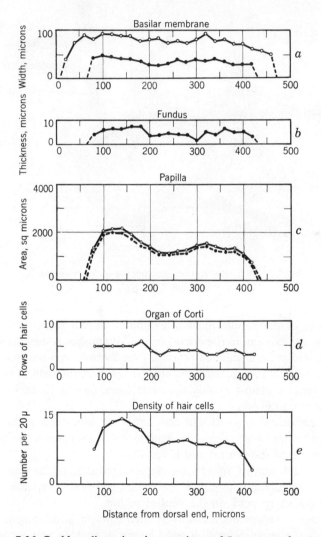

FIG. 7-26. Cochlear dimensions in a specimen of *Laemanctus longipes*.

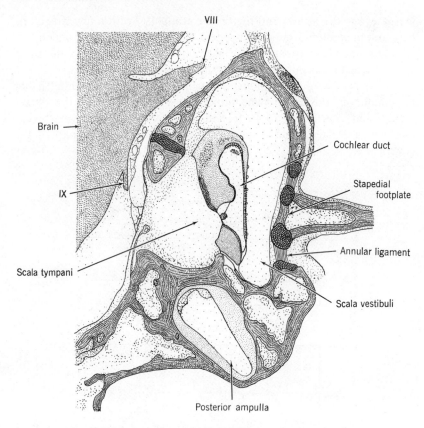

FIG. 7-27. A frontal section showing the auditory region in
Iguana iguana. Scale 15×.

The area of the papilla likewise exhibits a duplex form. It increases
rapidly to a maximum in the dorsal region, diminishes somewhat irregularly to a minimum at the 260 μ point, then ascends to a new maximum
in the far ventral region, after which it drops precipitously. The organ
of Corti follows almost exactly the same form, though its curve is well
separated from the other.

The longitudinal rows of hair cells, after an initial rise at the dorsal
end, show remarkable uniformity, numbering 7, except for two upward
deflections, until the ventral end is reached.

The hair-cell density reflects the duplex character of the papilla and
shows two maximums, one near the dorsal end and a broader region,
containing a sharp peak, in the ventral region.

This species represents a relatively high level of development of the
ear among the iguanids, with a good length of cochlea and a consider-

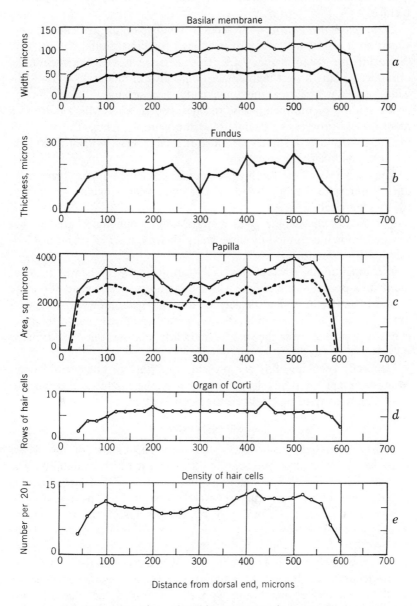

FIG. 7-28. Dimensions of cochlear structures in *Iguana iguana*.

able density of hair cells. There is relatively little differentiation of cochlear structures, though a moderate degree of duplexity complicates the picture.

Ctenosaura acanthura. — A second member of the iguanine group, *Ctenosaura acanthura*, was also examined for dimensional details, with the results presented in Fig. 7-29. This species has a basilar membrane of moderate length, just below 600 μ, and also of moderate width. This width increases rapidly at the dorsal end, varies around 90 μ and finally, at the middle of the membrane, reaches 100 μ, after which there is a more or less progressive decline over the ventral region. The width of the fundus follows a somewhat more regular course, with a maximum about the middle of the structure.

The thickness of the fundus presents an obviously duplex form, with dorsal and ventral mounds—the latter somewhat irregular—separated by a wide trough.

The area of the papilla gives a suggestion of bimodal character, but this pattern is interrupted by three sharp deflections. The organ of Corti follows this same form.

The longitudinal rows of hair cells run around 5 or 6 over most of the papilla except in the middle where they fall to 4. The general picture is one of uniformity. The density of the hair cells, however, shows a rapid rise to a high maximum in the dorsal region; the curve then declines to a plateau over the middle and early ventral region, and finally falls.

Generally speaking, this ear presents a picture of considerable uniformity, except for duplexity in the form of the fundus and a particularly high density of hair cells in the dorsal region.

SCELOPORINES

Sceloporus magister. — Much study has been made of the desert spiny lizard, *Sceloporus magister*, in part because of its availability. The general anatomical picture was presented earlier in Fig. 3-1, and details of outer and middle ear structures were shown in Figs. 6-13 and 6-17.

The inner ear of this lizard is of relatively simple form, as shown in the cross-sectional view of Fig. 7-30, taken from the middle of the cochlea. The limbic mound is substantial, and here the tectorial membrane arises from the limbic notch and takes a particularly high course to the tectorial plate over the papilla. This papilla is situated near the middle of the basilar membrane, with a wide strip of free membrane between it and the medial limbus. Over this part of the cochlea the fundus is very thin.

A sectional view just dorsal to the tectorial plate, where the ciliary tufts stand free, is shown in Fig. 7-31, and here the papilla is nearly round in cross section and the fundus is noticeably thickened. There are only two rows of hair cells in this region.

Dimensional details for this cochlea are presented in Fig. 7-32. Part *a* shows the length of the basilar membrane to be about 360 μ, and its form is a broad oval: blunt at both ends and with a nearly constant width of 98 μ over its middle portion. The variation in width is only about 25% over the main course of the structure.

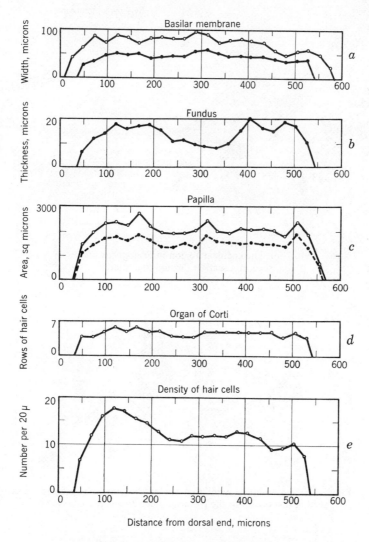

Fig. 7-29. Dimensions of cochlear structures in a specimen of *Ctenosaura acanthura*.

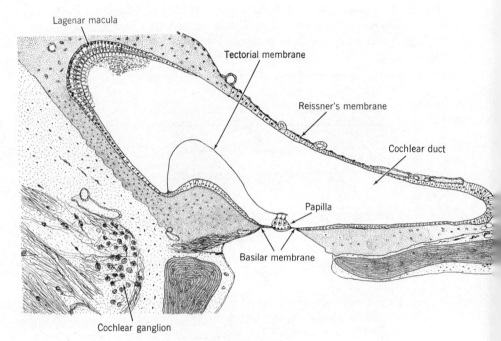

FIG. 7-30. The cochlear region in *Sceloporus magister*, seen in a section through the middle.

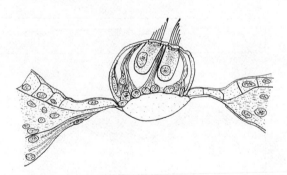

FIG. 7-31. The auditory papilla of *Sceloporus magister*, seen in cross section just dorsal to the tectorial plate. From Wever, 1965a.

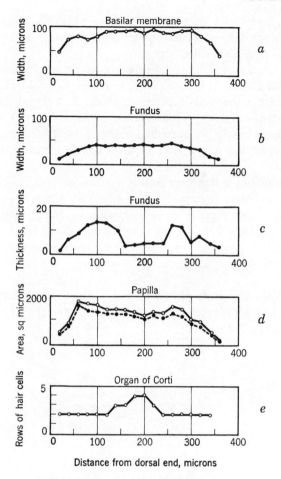

FIG. 7-32. Cochlear dimensions in a specimen of *Sceloporus magister*. From Wever, 1965a.

The fundus, shown at *b* in this figure, is 340 μ long and shows a rather flat middle portion and tapering ends. The fundus varies in thickness as shown in *c* of this figure, with dorsal and ventral regions attaining considerable thickness and the middle being uniformly thin.

The area of the papilla varies only moderately except for a depression over the midregion and rapidly tapering ends. The area of the organ of Corti follows this same form.

The papilla has only 2 longitudinal rows of hair cells at its two ends, as part *e* shows, whereas in the middle this number increases to 3 and then to 4 for a short distance.

This is a simple cochlea with little evidence of differentiation. The width of basilar membrane and fundus is nearly constant over the middle region, and varies only at the tapering ends. The thickness of the fundus exhibits the most striking variations, but by itself can hardly achieve any significant selectivity.

Uma notata. — Another relatively simple ear is encountered in the species *Uma notata*. This ear is represented in Fig. 7-33 in a cross section taken from the middle of the cochlea. A tectorial membrane of Type I*a* extends to a very small tectorial plate.

Dimensional details are presented in Fig. 7-34. The basilar membrane has a length of 360 μ, and over its main course it varies in width only about 27%.

The fundus has a length of 340 μ, and varies in width only slightly except for moderate tapering at the ends. The thickness of the fundus varies in the same manner as in *Sceloporus magister*, with the middle portion particularly thin, but the magnitude is much less.

The size of the papilla varies only moderately except for the rapidly tapering ends. The organ of Corti follows closely the form of the whole structure.

Two longitudinal rows of hair cells are present at dorsal and ventral ends, and the middle region has a constant number of 3. This condition is shown in Fig. 7-35.

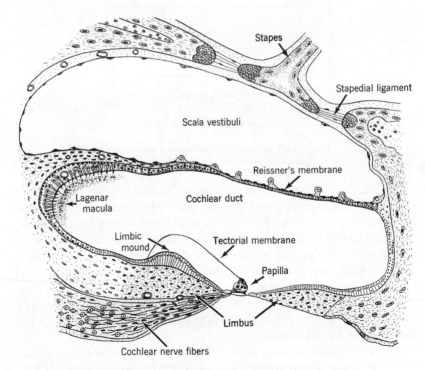

FIG. 7-33. The ear region in *Uma notata*. From Wever, 1965a.

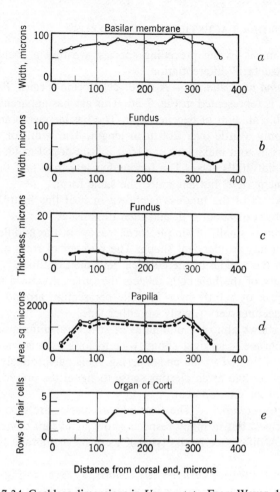

FIG. 7-34. Cochlear dimensions in *Uma notata*. From Wever, 1965a.

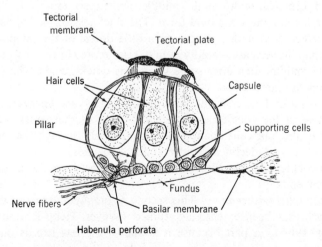

FIG. 7-35. The auditory papilla of *Uma notata*, seen in the region of the tectorial plate. Scale 800×. From Wever, 1965a.

Here again, as in the preceding species, we find a cochlea almost lacking in structural differentiation.

Phrynosoma platyrhinos. — A third sceloporine species, *Phrynosoma platyrhinos*, is represented in Fig. 7-36. This ear has apparently suffered a considerable amount of degeneration. The basilar membrane is short, measuring only a little over 200 μ in length. The width of the basilar membrane rises to a maximum of 116 μ a little dorsal to its midregion, and then rapidly falls away. The fundus has less than half the width of the whole membrane, but has much the same form.

The thickness of the fundus, as shown in *b* of this figure, rises to a small maximum of about 6 μ, and then diminishes.

The papilla is small, of simple form, widest in the middle. The cellular portion has an identical shape. The number of rows of hair cells rises to 5 in the middle of the cochlea, and then declines.

The density of the hair cells reflects the other structural features: it reaches a peak of a little over 10 cells per section around the 100 μ point, and declines very rapidly on either side.

This ear shows almost a minimum of structural differentiation. The basilar membrane is an oval about twice as long as it is wide, and differentiation is limited to the end tapering. This ear probably serves no further purpose than as an alerting sense to signal the presence of some occurrence in the environment.

Crotaphytus collaris. — The *Crotaphytus* species are grouped among the sceloporines, but in many respects show significant deviations from the others. A difference in the type of tectorial membrane has already been noted.

Dimensional data for *Crotaphytus collaris* are given in Fig. 7-37. The basilar membrane in this species is just over 300 μ long, reaches a maximum of 116 μ in width at the middle, and tapers rapidly at the ends. The fundus has the same oval form. The thickness of the fundus varies rather irregularly, with a maximum a little dorsal of the midpoint.

The papilla increases rapidly at first, rises more slowly to a maximum near the middle, then after some variations quickly ends. The organ of Corti has the same shape.

The rows of hair cells increase rapidly to 6, vary between 6 and 5 over most of the cochlea, and then fall away at the ventral end. The density of the hair cells changes only moderately through the middle of the cochlea, and declines rapidly at the two ends.

This cochlea exhibits little differentiation apart from the usual tapering at the ends.

Crotaphytus wislizenii. — This second *Crotaphytus* species, the leopard lizard, has been extensively studied (Wever, Hepp-Reymond, and Vernon, 1966), in part because it is one of the few lizards outside of the geckos in which vocalizations have been reported. Such observations, with tape recordings of the emitted sounds, were made by Clive D. Jorgensen and his associates at Brigham Young University, who provided our laboratory both with samples of the tapes and a number of specimens collected in desert areas of southern Nevada.

A general view of the auditory region in this species is given in Fig. 7-38, taken from a section through the middle of the cochlea. A tectorial plate is present here, served by a Type II*a* tectorial membrane. This plate and its relations to the hair cells are represented in further detail in Fig. 7-39, and dimensional data are shown in Fig. 7-40.

The basilar membrane is extremely short, only 180 μ. Its width is a little less than half the length, and the form is a simple oval. The fundus

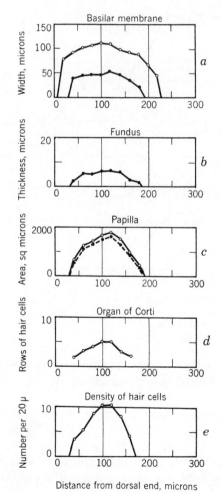

FIG. 7-36. Cochlear dimensions in *Phrynosoma platyrhinos.*

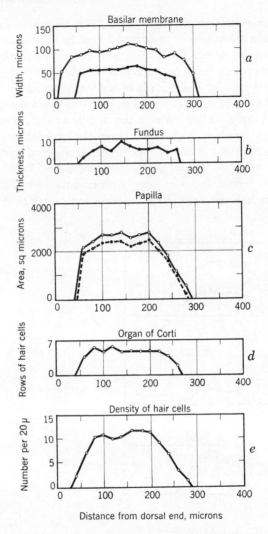

Fig. 7-37. Cochlear dimensions in
Crotaphytus collaris.

varies only slightly in width along its extent. The thickness of the fundus increases to a maximum a little dorsal of the middle, and then falls rapidly.

The area of the papilla presents a sharp maximum somewhat dorsal of the midregion, from which it falls rapidly on either side. The organ of Corti follows the same form.

The number of longitudinal rows of hair cells begins at 3, increases to 5 in the midregion, and then falls to 3 again at the ventral end. Hair cells are present over only 120 μ of the basilar membrane.

This inner ear is one of the simplest found, with little structural differentiation. Though this lizard produces vocal sounds, there is nothing in the construction of the ear to indicate any special developments of discriminatory capability. Yet, as will be shown, this ear displays relatively keen sensitivity as measured by the cochlear potentials, with the best region in the low tones from 300 to 700 Hz.

TROPIDURINES

Leiocephalus carinatus armouri. — In the species *Leiocephalus carinatus armouri*, as Fig. 7-41 shows, the basilar membrane is 435 μ long, and its width increases rapidly to a maximum of 100 μ in the dorsal region. Though the measurements show large variations, the trend is a decline to about half this width near the ventral end, after which the

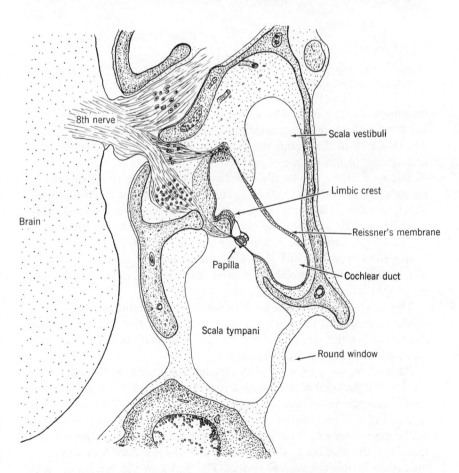

FIG. 7-38. The auditory region of *Crotaphytus wislizenii*.
From Wever, Hepp-Reymond, and Vernon, 1966.

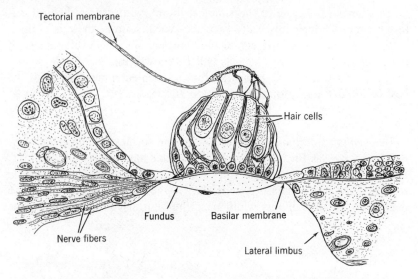

Tectorial membrane

Hair cells

Fundus

Basilar membrane

Nerve fibers

Lateral limbus

FIG. 7-39. The auditory papilla of *Crotaphytus wislizenii*, in a section through the tectorial plate. From Wever, Hepp-Reymond, and Vernon, 1966.

membrane quickly ends. The width of the fundus shows variations also, but with indications of a maximum near the middle of the membrane. The thickness of the fundus shows two maximums separated by an irregular trough in the middle region.

The area of the papilla also displays a duplex pattern. A regular increase at the dorsal end leads to a rounded maximum at the 100 μ point, which is followed by a narrow trough and then a rise to a second maximum at the 340 μ point. The organ of Corti has nearly the same form.

The longitudinal rows of hair cells are unusually few. There are only 3 such rows over most of the cochlea, increasing to 4 at scattered points. The density of the hair cells is greatest near the ends of the cochlea and noticeably less through the middle.

This is a relatively simple cochlea with only moderate variation in the basilar membrane, but with a duplex pattern in the thickness of the fundus and in the size of the papilla. This duplexity is not significantly reflected in the hair-cell distribution.

Plica plica. — A cross section through the region of the ear at the level of the tectorial plate is shown for a specimen of *Plica plica* in Fig. 7-42. Dimensional details for this ear are presented in Fig. 7-43.

The width of the basilar membrane increases rapidly to a maximum in the dorsal region, then somewhat irregularly decreases through the middle of the cochlea to a minimum from which it rises to a secondary maximum in the far ventral region. The width of the fundus is much more uniform, though the same general regions are recognizable. The

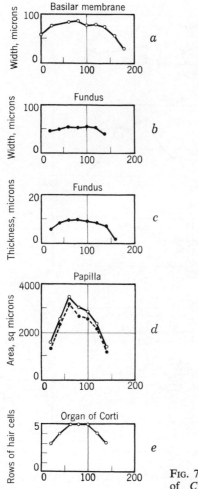

FIG. 7-40. Dimensional data for a specimen of *Crotaphytus wislizenii*. From Wever, Hepp-Reymond, and Vernon, 1966.

thickness of the fundus increases gradually, with a few variations, toward the ventral end.

The area of the auditory papilla increases rapidly to a flat maximum, falls slightly and then rises to a higher maximum in the middle of the cochlea, falls to a minimum, and then rises once more to a sharp maximum near the ventral end. The organ of Corti follows this same course.

The rows of hair cells show great uniformity; their number increases to 5 in the dorsal region, and then fluctuates between 5 and 6 over the remainder of the cochlea, almost to the end. The density of the hair cells follows a pattern closely resembling the area of the papilla, but with some enhancement toward the ventral end.

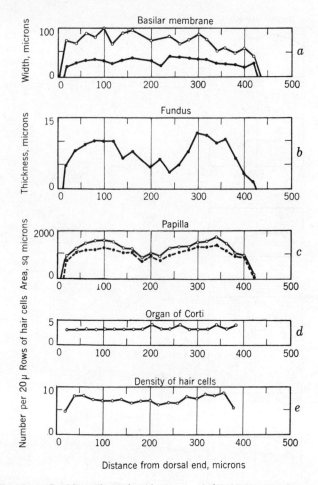

FIG. 7-41. Cochlear dimensions in *Leiocephalus carinatus armouri*.

MALAGASIAN

Oplurus sebae. — The only iguanid species from the island of Madagascar was *Oplurus sebae*, whose cochlear dimensions are represented in Fig. 7-44. The basilar membrane in this specimen is 385 μ long, rises rapidly to a maximum width of 98 μ in the dorsal region, and then falls away progressively to the ventral end. The width of the fundus varies a little less over its main course and ends a bit more abruptly. The thickness of the fundus shows a marked duplexity, with irregular maximum regions and a severe decline in the middle.

The size of the papilla shows this same duplex character. The area increases rapidly at the dorsal end to a rounded maximum, declines to a flat minimum in the middle, and then increases to a second maximum in the ventral region. The organ of Corti has a similar shape.

The longitudinal rows of hair cells increase from the dorsal end to a maximum of 7, then decline somewhat and run around 4 and 5 over the remainder of the cochlea until over the ventral region there is a progressive decline. The density of the hair cells likewise increases to a maximum in the early dorsal region, remains fairly steady over the middle of the cochlea, and then declines rapidly at the ventral end.

This cochlea follows a pattern that is now familiar among the iguanids: only a moderate variation in width of basilar membrane and fundus, a duplex form of fundus width and papilla size, and a relatively simple arrangement and distribution of the hair cells.

In summary, the iguanid lizards show only a moderate development of the cochlea, with slight to minimal differentiation of dimensional

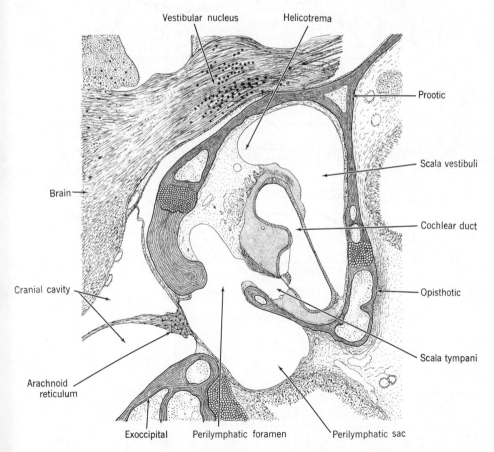

FIG. 7-42. A frontal section through the ear region of *Plica plica*.

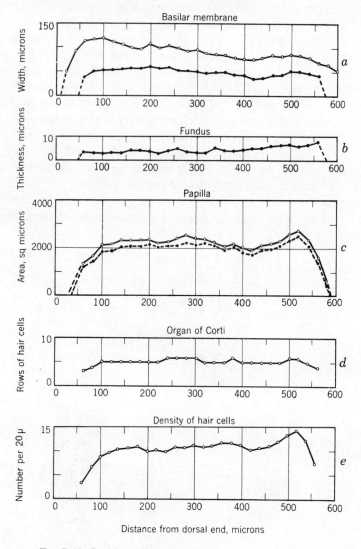

FIG. 7-43. Cochlear dimensions in a specimen of *Plica plica*.

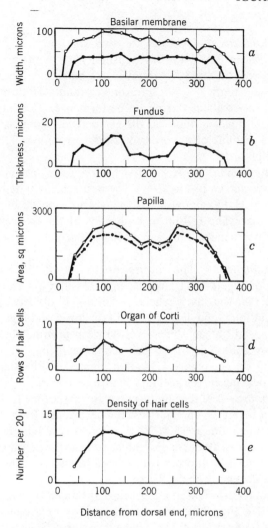

FIG. 7-44. Cochlear dimensions in a specimen of *Oplurus sebae*.

features. These ears no doubt are serviceable in the reception of sounds over a moderate frequency range, but give little evidence of more than a small amount of differentiation in terms of place of response over the basilar membrane.

AUDITORY SENSITIVITY

The results of the cochlear potential tests of sensitivity in the iguanids will be presented under the six groups of species identified earlier.

ANOLINES

Anolis carolinensis. — Curves for two specimens of the Carolina anole are shown in Fig. 7-45. Best sensitivity for these ears is indicated over a range from 500 to 3000 Hz, with the maximum at 1500 Hz. The best level of sensitivity is −59 db for one ear and −56 db for the other, representing an excellent degree of performance. The high-frequency roll-off is very rapid, around 60 db per octave, and that for the low tones is more gradual, of the order of 20 db per octave. This sensitivity is of a very high order.

Anolis chlorocyanus. — A sensitivity curve for a specimen of *Anolis chlorocyanus* is shown in Fig. 7-46. The best sensitivity is in the region of 400 to 1500 Hz, attaining levels of −44 to −46 db at three points within this range. The roll-off at both ends of the frequency range is about 20 db per octave. This ear displays very good sensitivity in the middle range.

Anolis cristatellus. — The species *Anolis cristatellus* is represented by the curve of Fig. 7-47. Unusually high sensitivity is shown within a limited range. The sensitivity reaches −76 to −79 db between 2000 and 3000 Hz, and though the decline is rapid for lower and higher tones, the performance is still at an acute level between 600 and 6000 Hz. The high-frequency roll-off is 40 db per octave, and that for the low frequencies is 20 db per octave. This is a highly sensitive ear, with relatively sharp peaking in the medium high frequencies.

Anolis cybotes. — A sensitivity curve for a specimen of *Anolis cybotes* is given in Fig. 7-48. The best region extends from 290 to 2500 Hz, with the maximum at 1500-2000, where a very high level of sensitivity of −60 db is attained. For higher tones the sensitivity declines rapidly, at a rate of 40 db per octave, and for tones at the low end of the range the decline is about the same.

Anolis distichus. — Observations on a specimen of *Anolis distichus* are presented in Fig. 7-49. The best sensitivity of around −37 db is attained at 600 and at 2000 Hz. This ear performs well over the whole range from 300 to 3000 Hz. Below 400 Hz the roll-off has a moderate rate of 20 db per octave, and above 2000 Hz it is around 30 db per octave.

Anolis lionotus. — The species *Anolis lionotus* is represented by two specimens as shown in Fig. 7-50. The sensitivity is best around 400 Hz, where it reaches −44 db. At higher frequencies there is a moderate decline, and then good maintenance of sensitivity for a little over two octaves until 2500 Hz is reached. Thereafter the decline is progressive at a rate of about 30 db per octave. The roll-off at the low end of the scale, below 400 Hz, has about the same rate. This type of ear appears to respond well in the low-frequency region from 400 to 1000 Hz, and still

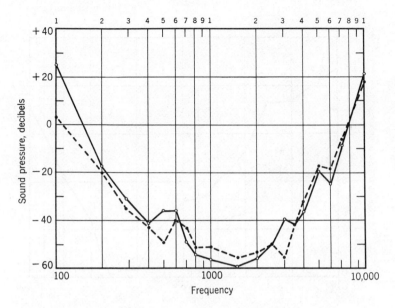

FIG. 7-45. Sensitivity curves for two specimens of *Anolis carolinensis*. Each curve indicates the sound pressure, in decibels relative to a reference level of 1 dyne per sq cm, required to produce a response at the round window of 0.1 μv.

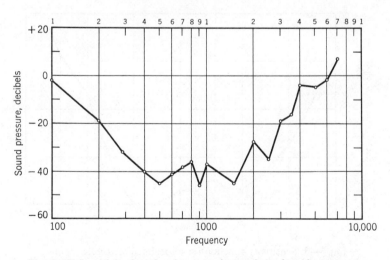

FIG. 7-46. Sensitivity function for a specimen of *Anolis chlorocyanus*.

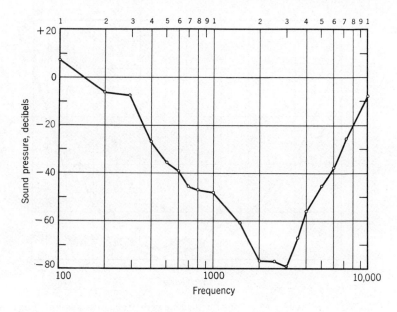

Fig. 7-47. Sensitivity function for a specimen of *Anolis cristatellus*.

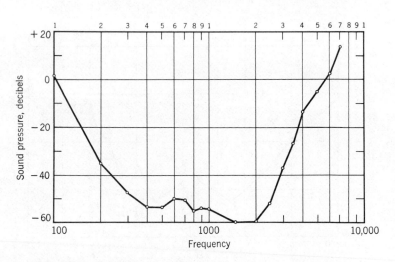

Fig. 7-48. Sensitivity function for a specimen of *Anolis cybotes*.

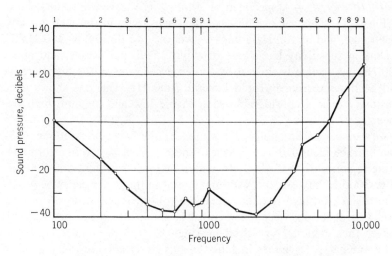

FIG. 7-49. Sensitivity function for a specimen of *Anolis distichus*.

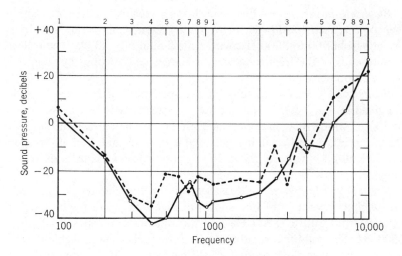

FIG. 7-50. Sensitivity curves for two specimens of *Anolis lionotus*.

performs effectively for somewhat higher tones, until the upper roll-off appears.

Anolis sagrei. — A specimen of *Anolis sagrei* gave the results shown in Fig. 7-51. This ear exhibits excellent sensitivity in the medium low-tone range, and at 400 Hz attains a level of −44 db. Below 300 Hz the decline in sensitivity has a rate of about 15 db per octave, and above 500 Hz the decline is slow at first, then accelerates in the higher range and becomes extremely rapid beyond 7000 Hz. This ear shows good sensitivity over a considerable span in the low and medium high frequencies.

Anolis sp. — The anolis represented by the two curves of Fig. 7-52 could not be identified as to species. These ears show good performance in the region of 600 to 4000 Hz, where two peaks around 500-700 and 2500-3000 Hz are evident, with maximums in the better ear of −49 and −46 db. At the high-frequency end the roll-off is about 40 db per octave, and at the low-frequency end it is about 20 db per octave. This species thus exhibits very good hearing.

In summary, the anoles included in this study all show very good to excellent sensitivity, with best performances in *Anolis carolinensis* and *Anolis cristatellus* in the medium high frequencies.

BASILISCINES

Basiliscus basiliscus. — Fig. 7-53 shows a sensitivity curve for a specimen of *Basiliscus basiliscus*, in which a region of moderate sensitivity extends from 400 to 3000 Hz, with a maximum of −23 db. The roll-off at both ends of the frequency scale is somewhat irregular, but approximates 20 db per octave.

Basiliscus vittatus. — Results for two specimens of *Basiliscus vittatus* are given in Fig. 7-54. In this species the sensitivity is best in the region of 500 to 4000 Hz, with a peak at 1500 Hz in one animal and at 2500 Hz in the other, and attaining maximums of −35 and −27 db, respectively. Roll-off rates are about 30 db per octave at both ends of the range.

Laemanctus longipes. — Measurements on a specimen of *Laemanctus longipes* gave the function shown in Fig. 7-55. Here the best region extends from 500 to 5000 Hz and the maximum reaches −44 db at 2500 Hz. The roll-off is rapid at the high-frequency end, reaching 50 db per octave, but is slow, around 15 db per octave, at the low-frequency end. This ear performs well in the middle-frequency range.

IGUANINES

Ctenosaura acanthura. — A specimen of *Ctenosaura acanthura* gave the sensitivity function presented in Fig. 7-56. Best responses are in a

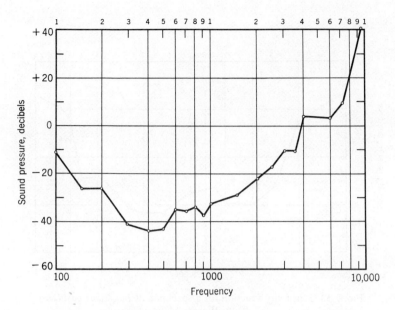

Fig. 7-51. Sensitivity function for a specimen of *Anolis sagrei*.

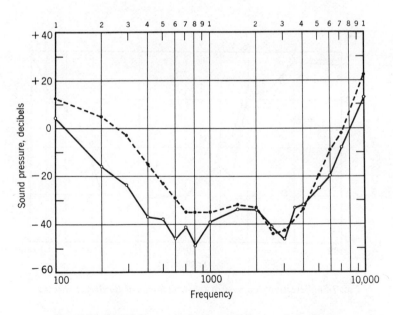

Fig. 7-52. Sensitivity curves for two specimens of *Anolis* sp.

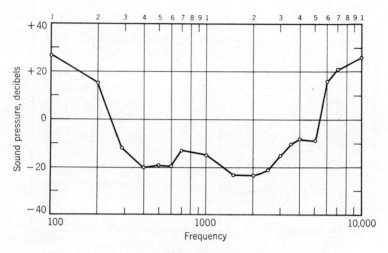

Fig. 7-53. Sensitivity function for a specimen of *Basiliscus basiliscus*.
From Wever, 1971a.

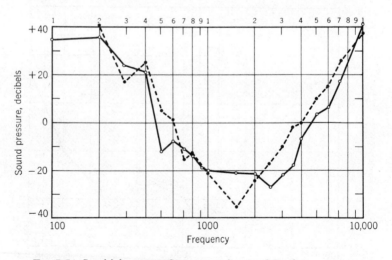

Fig. 7-54. Sensitivity curves for two specimens of *Basiliscus vittatus*.

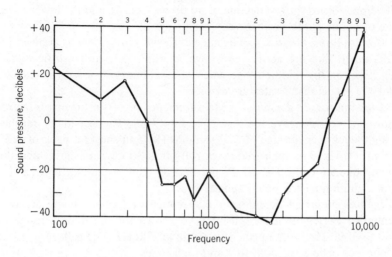

FIG. 7-55. Sensitivity function for a specimen of *Laemanctus longipes*.

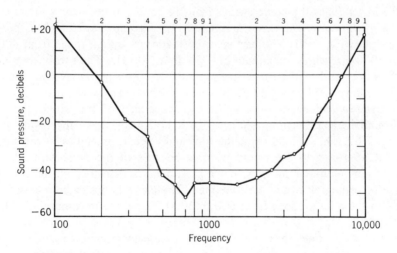

FIG. 7-56. Sensitivity function for a specimen of *Ctenosaura acanthura*.

broad region from 500 to 3000 Hz, with a peak of −53 db at 700 Hz and good sensitivity of the order of −45 db between 800 and 1500 Hz. The upper tones show a decline of 30 db per octave, and the lower tones one of about 25 db per octave.

Ctenosaura s. similis. — The species *Ctenosaura s. similis* showed results much like the above, as shown in Fig. 7-57, with a little higher sensitivity in the middle range where a maximum of −53 db is reached at 1000 Hz. Roll-off rates are similar.

Dipsosaurus d. dorsalis. — Measurements on a specimen of *Dipsosaurus d. dorsalis* gave the results of Fig. 7-58. The sensitivity reaches a high level over a range of 290 to 3000 Hz, and shows a peak of −57 db at 600 Hz. The high-frequency roll-off is 30 db per octave, and the low-frequency roll-off is 20 db per octave. This is an excellent ear for the medium high-frequency range.

Iguana iguana. — Results for a specimen of *Iguana iguana* are presented in Fig. 7-59. Best sensitivity is shown for the middle range of 500 to 3000 Hz, with a peak of −47 db at 700 Hz. The roll-off is about 25 db per octave for both low and high tones.

Sauromalus obesus. — Two specimens of *Sauromalus obesus* gave the curves of Fig. 7-60. Best response is in the middle range, increasing somewhat from 400 to 2500 Hz, where peaks of −34 and −39 occur. For higher tones the decline is rapid, around 40 db per octave, and for the lower tones the curve swings from 15 to 20 db per octave in slope.

Enyaliosaurus clarki. — A function for *Enyaliosaurus clarki* is presented in Fig. 7-61. The best sensitivity appears over a range from 400 to 3000 Hz, with a maximum of −50 db at 700 Hz. The roll-off rate at the upper end of the range averages about 20 db per octave, and that at the lower end about 30 db per octave. This is an excellent ear.

Cyclura stejnegeri. — A half-grown specimen of the Mono Island ground iguana *Cyclura stejnegeri* (weight 5.5 kg) gave the results of Fig. 7-62. This animal presented unusually great difficulties in the surgical approach to the round window because of the large protruding pterygomandibular muscles, but a reasonably satisfactory placement of the recording electrode was obtained. The curve indicates best sensitivity in the middle range from 600 to 3000 Hz, with a maximum of −30 db at 800 Hz.

These iguanines show a high degree of agreement, with good to excellent sensitivity over a broad area through the middle frequencies.

SCELOPORINES

A good many sceloporines were included in this study, reflecting their ready procurement during the early stage when the work was getting under way.

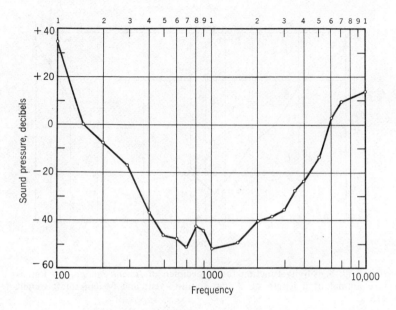

FIG. 7-57. Sensitivity function for a specimen of *Ctenosaura s. similis.*

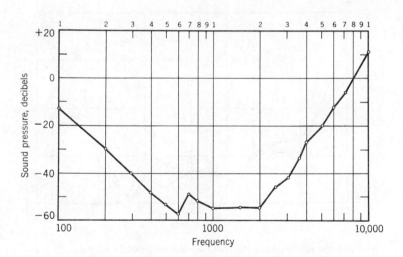

FIG. 7-58. Sensitivity function for a specimen of *Dipsosaurus d. dorsalis.*

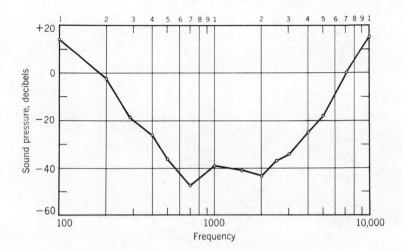

FIG. 7-59. Sensitivity function for a specimen of *Iguana iguana*. An imma-ture animal of a length of 17 cm snout-to-vent and 66 cm total; weight 115 g.

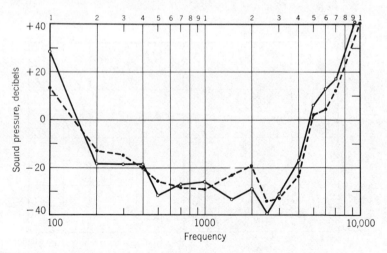

FIG. 7-60. Sensitivity functions for two specimens of *Sauromalus obesus*. After Wever, Crowley, and Peterson, 1963.

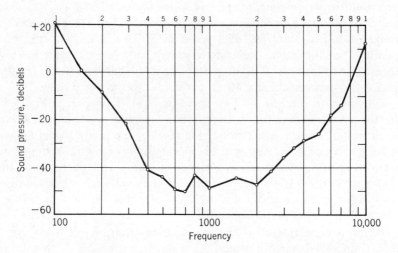

FIG. 7-61. Sensitivity function for a specimen of *Enyaliosaurus clarki*.

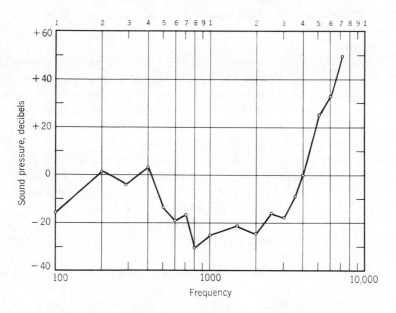

FIG. 7-62. Sensitivity function for a specimen of *Cyclura stejnegeri*.

Callisaurus draconoides ventralis. — Results for the zebra-tailed liz-ard, *Callisaurus draconoides ventralis,* are presented in Fig. 7-63. The better of the two functions represents a good level of sensitivity over a range of 400 to 3000 Hz, with levels of 32-34 db reached at several points. The other curve indicates relatively poor sensitivity for the low tones. The roll-off rate is 30 db per octave at the upper end of the range, and varies between 25 and 40 db per octave in the lower range.

Cophosaurus texanus. — Two curves were shown earlier in Fig. 6-56 for the greater earless lizard, *Cophosaurus texanus,* which are in agree-ment in indicating a remarkably good level of sensitivity for an animal lacking a tympanic membrane. The peculiarities of sound transmission in this ear were considered in Chapter 6. The sensitivity peaks in the region of 600-800 Hz, where the better of the two ears reaches −48 db. Over a wider range, from 400 to 2000 Hz, this ear presents a very serviceable level of performance.

Holbrookia maculata. — The lesser earless lizard, *Holbrookia macu-lata,* is another iguanid in which the tympanic membrane is absent and further alterations of the conductive mechanism are present. This spe-cies also has received special consideration in Chapter 6. As there shown in Fig. 6-62, the sensitivity is poor, and a maximum of only −8 db is reached in the middle range around 800 Hz. The region from 290 to 2000 Hz probably serves for limited yet still useful sound perception in this animal.

Phrynosoma species. — Sensitivity data were obtained for five species of *Phrynosoma,* the horned lizards. For two of these species, *Phryno-soma coronatum* and *Phrynosoma platyrhinos,* the results have already been presented in Chapter 6, in Figs. 6-58 and 6-59. Functions for the other three, *Phrynosoma cornutum, Phrynosoma douglassi,* and *Phryno-soma solare,* are shown in Figs. 7-64 to 7-66.

In all these species the sensitivity is poor, with best levels around 0 to −5 db in *Phrynosoma cornutum* and about −10 db in *Phrynosoma solare,* but reaching −20 db or better in *Phrynosoma coronatum* and *Phrynosoma platyrhinos.* The species *Phrynosoma douglassi* shows the worst performance, attaining only +20 db at one frequency when meas-ured at the standard temperature of 25° C, though decidedly better when measured at 35° C.

As suggested earlier, the poor performance of these *Phrynosoma* spe-cies evidently relates to a variety of conditions attributed to special adaptations. The tympanic membrane is usually thick and often scaly, and the parts of the columella are relatively massive. A curious anom-aly, seen in some of these species, is the presence of fluid about the round window. Also these animals are adapted to an especially high temperature, and measurements made at a temperature of 25° C no

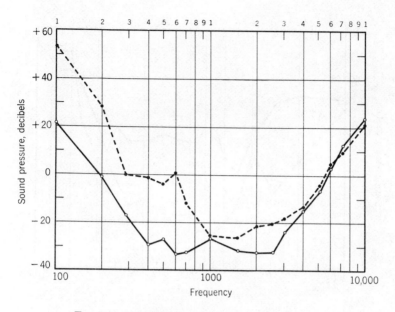

FIG. 7-63. Sensitivity functions for two specimens of
Callisaurus draconoides ventralis.

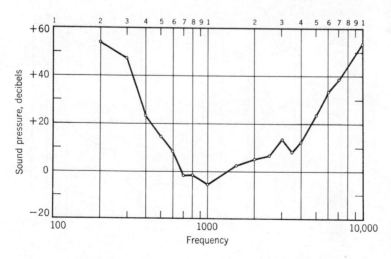

FIG. 7-64. Sensitivity function for a specimen of *Phrynosoma cornutum.*

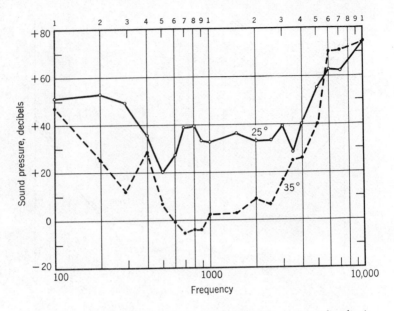

FIG. 7-65. Sensitivity functions for a specimen of *Phrynosoma douglassi*, obtained at two temperatures, 25° and 35° C.

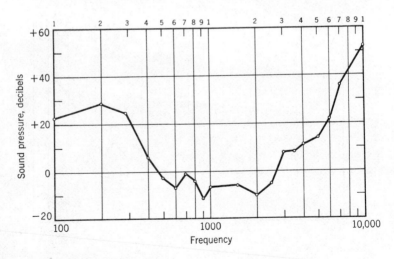

FIG. 7-66. Sensitivity function for a specimen of *Phrynosoma solare*.

doubt represent the performance of their ears less favorably than is true for most other lizards. The considerable improvement shown for *Phrynosoma douglassi* when the temperature is elevated to 35° C is evidence that such is the case. (For a general discussion of temperature relations, see Chapter 2.)

Sceloporus clarkii. — Several species of the spiny lizards of the genus *Sceloporus* were investigated. Results for the first of these, *Sceloporus clarkii*, are shown in Fig. 7-67. Three specimens are represented, all giving much the same picture of moderately good sensitivity in the middle range from 290 to 3000 Hz and maximums around 1500 Hz, though one curve has a secondary peak at 400 Hz. The peak values vary from −22 to −34. Roll-off rates for these ears are around 30 db per octave in the high frequencies and 20 db per octave in the low frequencies.

Sceloporus cyanogenys. — A sensitivity curve for a specimen of *Sceloporus cyanogenys* is shown in Fig. 7-68. This curve exhibits the best sensitivity over the range 500 to 4000 Hz, with a maximum of −25 db at 2500 Hz. The high-frequency roll-off is very rapid, around 60 db per octave, and is slower, averaging around 30 db per octave, in the low frequencies.

Sceloporus magister. — A number of measurements were made on the desert spiny lizard, *Sceloporus magister*. Results obtained on one batch of these animals are shown in Fig. 7-69. Four curves are presented showing levels of sensitivity that vary considerably, though in good agreement as to general shape. These indicate best responses over a range from 290 to 4000 Hz, and agree in indicating two regions of good sensitivity, one varying between 400 and 1000 Hz and the other between 2000 and 3000 Hz. The best curve reaches a peak of −32 db, but the others show only moderate to poor maximums. Roll-off rates at the high-frequency end of the range vary between 40 and 60 db per octave, and at the low-frequency end run around 20 db per octave. Further results on this species, obtained at another time on specimens from a different source, are shown in Fig. 7-70. These functions represent considerably better sensitivity, though they agree with the others in general form, and have at least a tendency toward the bimodality that the others show. The better of these functions reaches an excellent level of sensitivity of −56 db at 700 Hz. These curves have high-frequency roll-offs around 25-30 db per octave.

It is possible that these two sets of curves represent different subspecies of *Sceloporus magister*, but the places where the specimens were collected were unknown and identification beyond the species level was not made. Alternatively, the variations shown may only represent the range of individual differences among these animals.

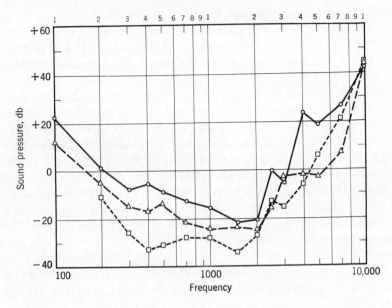

Fig. 7-67. Sensitivity curves for three specimens of *Sceloporus clarkii*.
From Wever and Peterson, 1963.

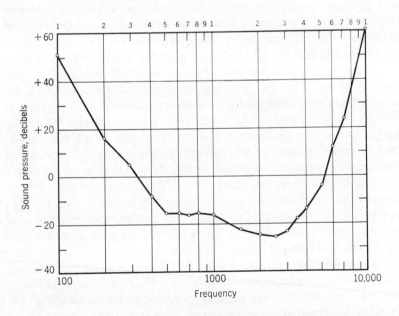

Fig. 7-68. Sensitivity function for a specimen of *Sceloporus cyanogenys*.

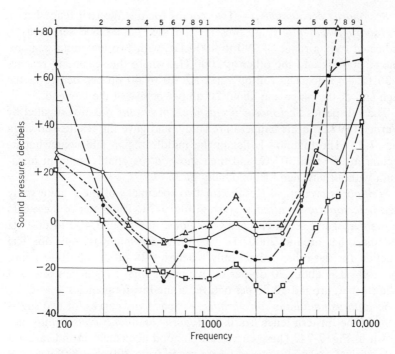

FIG. 7-69. Sensitivity curves for four specimens of *Sceloporus magister*. From Wever, Crowley, and Peterson, 1963.

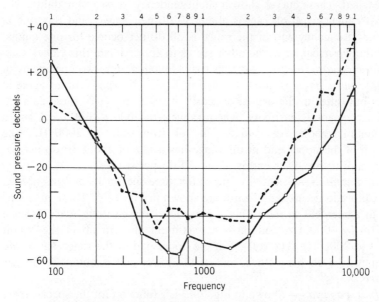

FIG. 7-70. Sensitivity functions for two additional specimens of *Sceloporus magister*.

Sceloporus occidentalis. — Two curves for the Western fence lizard, *Sceloporus occidentalis*, are presented in Fig. 7-71. Good sensitivity is indicated over a range of 290 to 4000 Hz, with two maximum regions, one at 600 Hz and the other at 2000 Hz, where the sensitivity reaches −36 to −43 db. The roll-off rates are 30 to 40 db per octave at the high end of the range and about 20 db per octave at the low end.

The subspecies *Sceloporus occidentalis biseriatus* was investigated by Werner (1972) in his temperature study, and gave the results shown in Fig. 7-72. The function is flat in the middle region and somewhat irregular at 25° and 30° C, and then shows a reasonably uniform maximum without any sign of bimodality at 35° C.

Sceloporus orcutti. — Results for two specimens of the granite spiny lizard, *Sceloporus orcutti*, are given in Fig. 7-73. The better of these ears shows very good sensitivity in two regions, one around 400 to 600 Hz and the other around 2000 Hz, where the maximum is −40 db. The function for the other animal only reaches −28 db at the 2000 point. The roll-off is about 20 db per octave in the high frequencies and is a little steeper, around 30 db per octave, in the low frequencies.

Sceloporus undulatus consobrinus. — Sensitivity curves for two specimens of the Eastern fence lizard, *Sceloporus undulatus consobrinus*, are shown in Fig. 7-74. These curves are in good agreement in indicating a high level of sensitivity over a broad range from 290 to 4000 Hz, with a clear maximum in one of these at 2000 Hz, where a level of −36 db is reached. These curves show a slight tendency toward bimodality. The high-frequency roll-off rate is about 30 db per octave, and this rate at the low-frequency end of the range is 30 db per octave for one animal and though irregular in the other seems to approximate this value.

Streptosaurus mearnsi. — Sensitivity functions for two specimens of *Streptosaurus mearnsi* are presented in Fig. 7-75. These curves agree in form but differ in the degree of sensitivity. The better of these ears exhibits very good sensitivity over a range from 600 to 4000 Hz, and a maximum response of −46 to −47 db from 1000 to 2000 Hz. The roll-off at the upper end of the range has a rate of about 30 db per octave, and at the lower end is about 20 db per octave.

Uma notata. — Sensitivity curves for three specimens of *Uma notata*, the Colorado desert sand lizard, are shown in Fig. 7-76. These functions are in agreement in indicating a good level of sensitivity over a range of 600 to 4000 Hz, and show a maximum of −36 db in the region 1000 to 2000 Hz. The roll-off at the upper end of the range has a rate of about 30 db per octave, and at the lower end is about 20 db per octave.

Uma scoparia. — Shown in Fig. 7-77 is a function for the species *Uma scoparia*, the Mojave sand lizard, and represents the averaged results

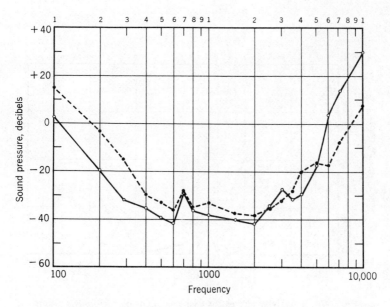

FIG. 7-71. Sensitivity functions for two specimens of *Sceloporus occidentalis*.

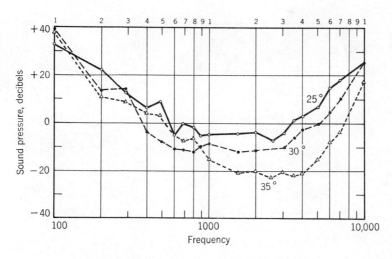

FIG. 7-72. Averaged sensitivity curves for *Sceloporus occidentalis biseriatus*, obtained at three temperatures. After Werner, 1972.

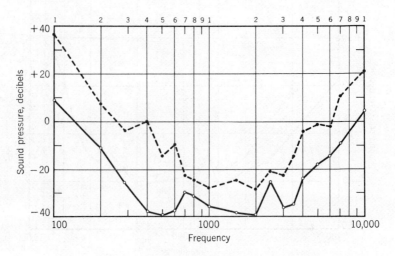

FIG. 7-73. Sensitivity curves for two specimens of *Sceloporus orcutti*.

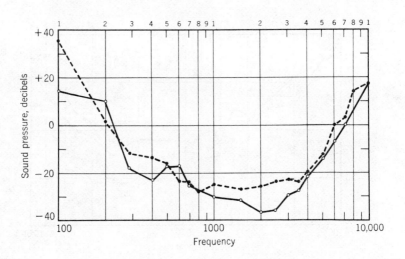

FIG. 7-74. Sensitivity curves for two specimens of
Sceloporus undulatus consobrinus.

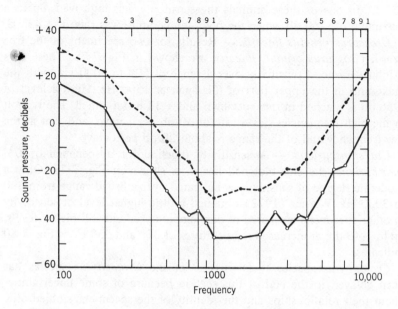

FIG. 7-75. Sensitivity curves for two specimens of *Streptosaurus mearnsi.*

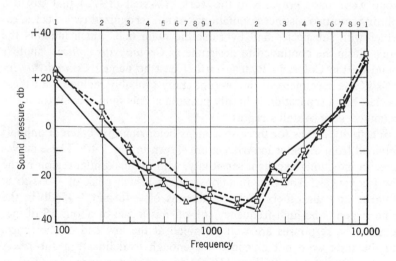

FIG. 7-76. Sensitivity curves for three specimens of *Uma notata.*
From Wever and Peterson, 1963.

obtained by Werner (1972) on four specimens at a temperature of 25° C. In one of these animals the sensitivity was improved, up to a maximum of −60 db, when the body temperature was raised to 43° C.

Urosaurus ornatus linearis. — Results for two specimens of the tree lizard, *Urosaurus ornatus linearis*, are shown in Fig. 7-78. These ears exhibit very good sensitivity over a range of 500 to 3000 Hz, with the maximum in the upper part of this span at 1500 Hz. Here a level of −36 db is reached in one specimen and −38 in the other. The roll-off in the high frequencies has a rate of 30 db per octave, and that at the low-frequency end of the range is about 20 db per octave.

Uta stansburiana. — A sensitivity function for a specimen of *Uta stansburiana*, the side-blotched lizard, is shown in Fig. 7-79. Only a moderate degree of sensitivity is indicated over a broad range from 290 to 3000 Hz. Werner (1972) obtained a little higher level of sensitivity in other specimens measured at this temperature, and still higher levels, up to −40 db, at increased temperatures of 30° and 35° C, as Fig. 7-80 indicates.

Crotaphytus species. — Discussion of the *Crotaphytus* species has been delayed to the end of this section because of some uncertainties about their relationships and the identity of the specimens studied. Zug (1971) placed these lizards in the sceloporine group, but others have questioned this position. Most of the present measurements on this group were made previous to the work of Axtell (1972) that led to a splitting into two distinct populations. Axtell recognized two species, together with subspecies and hybrids: the main one, which inhabits the Great Basin, he continued to designate as *Crotaphytus collaris*; another found on the Colorado Plateau farther east he named *Crotaphytus insularis*. The specimens of the present study probably include both these species, but a separation is hardly possible at this time, except for a few examined after Axtell's report.

Sensitivity curves for three specimens belonging to a batch of animals obtained from a dealer in Arizona are shown in Fig. 7-81. These curves agree in indicating excellent sensitivity, with the maximum region in the low frequencies. The best function reaches a high peak of −59 db at 400 Hz, and the others show maximums of −48 and −45 db in the same region. The high-frequency roll-off has a rate of about 35 db per octave. The responses are well sustained at the low end of the range, but the tests were not extended far enough to delineate a cut-off. A function obtained by Werner (1972) in a specimen then identified as *Crotaphytus collaris baileyi* is reproduced in Fig. 7-82, and shows excellent sensitivity in the range from 200 to 2000, with a maximum of −64 at 500 Hz. In this ear the roll-off rate for the high frequencies is 40 db per octave and that for the low frequencies is 15 db per octave.

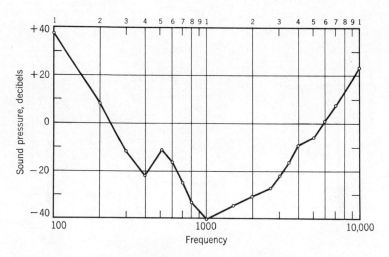

FIG. 7-77. An average sensitivity curve for four specimens of *Uma scoparia*.
Data from Werner, 1972.

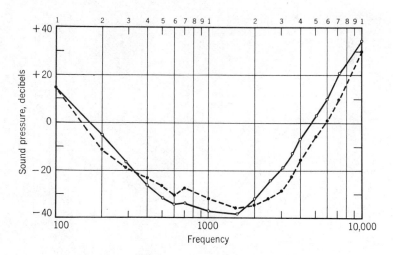

FIG. 7-78. Sensitivity curves for two specimens of
Urosaurus ornatus linearis.

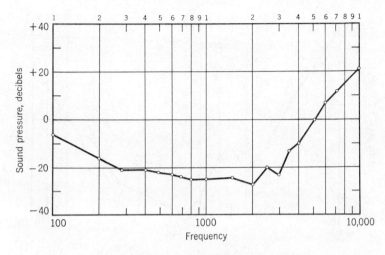

FIG. 7-79. Sensitivity function for a specimen of *Uta stansburiana* at a body temperature of 24° C. Observations by Y. L. Werner.

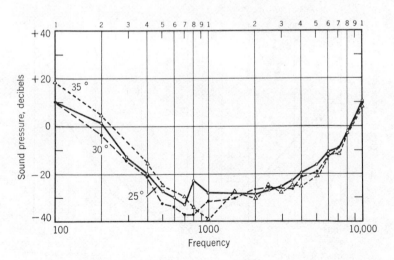

FIG. 7-80. Sensitivity curves for a specimen of *Uta stansburiana*, obtained at three body temperatures. Data from Werner, 1972.

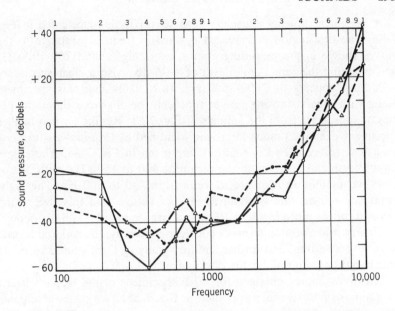

FIG. 7-81. Sensitivity curves for three specimens of *Crotaphytus collaris*.

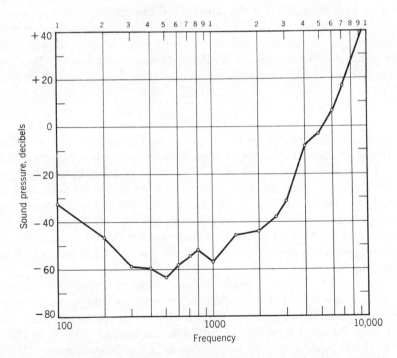

FIG. 7-82. Sensitivity function for a specimen of *Crotaphytus collaris baileyi*.
Data from Werner, 1972.

A specimen showing unusually keen sensitivity is represented in Fig. 7-83. The best region is displaced upward in frequency relative to the curves shown in the preceding graphs, centering around 700-1000 Hz and attaining the remarkable level of −96 db. Also a clear secondary maximum appears at 2500-3000 Hz. It is likely that this specimen, along with others showing similar functions, belongs to a different species or subspecies from the animals of Fig. 7-81. Results on two further specimens examined much later, and identified as *Crotaphytus insularis* on a basis of Axtell's key features, are given in Fig. 7-84. These ears show very high sensitivity in the region of 400 to 1500 Hz, with one of them outstanding in response to tones from 290 to 600 Hz. The high-frequency roll-off rate is about 35 db per octave, and this rate at the low end for the more regular curve has a large value of 50 db per octave.

Clearly, when seen in terms of cochlear sensitivity, the collared lizards are a diverse group, and further auditory study of them should be made with careful account of locality of origin.

Sensitivity curves obtained for three specimens of the leopard lizard *Crotaphytus wislizenii* are presented in Fig. 7-85. Two of these animals show very good sensitivity in the low-tone range, around 290-1000 Hz, with maximums of −43 and −47 db. The third curve is irregular, and has its maximum at 1000 Hz. The roll-off rate in the high frequencies is about 20 db per octave. In the low frequencies the responses are too scattered to determine a roll-off rate with any certainty, though one curve has a uniform slope of 20 db per octave. These animals produce high-pitched squeals when disturbed (Jorgensen, Orton, and Tanner, 1963).

TROPIDURINES

Leiocephalus species. — Three species of tropidurines were included in the study. One of these, *Leiocephalus carinatus armouri*, is represented by the two functions of Fig. 7-86. The better of these curves shows a broad region of good sensitivity extending from 290 to 4000 Hz, with a nearly flat maximum of −36 db. The other curve has nearly the same form, but the sensitivity is much less. The roll-off rates are about 40 db per octave in the high frequencies and vary for the two curves in the low frequencies, being 20 db per octave in one and 30 db per octave in the other.

Observations on a specimen of *Leiocephalus barahonensis*, further identified as a hybrid between the subspecies *aureus* and *oxygaster*, produced the results of Fig. 7-87. The sensitivity is very good, around −40 db, over the range from 400 to 2000 Hz, and reaches −43 db at 1500 Hz. The roll-off in the low frequencies is at a rate approaching 20 db per octave, and is around 30 db per octave above 3500 Hz.

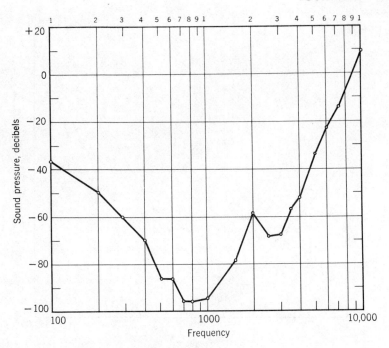

FIG. 7-83. Sensitivity function for a specimen of *Crotaphytus collaris*.

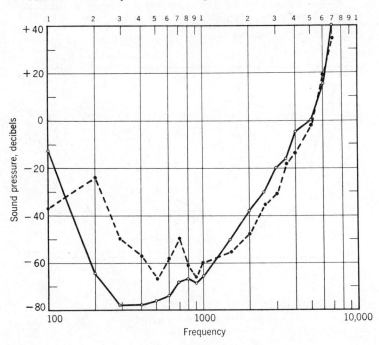

FIG. 7-84. Sensitivity functions for two specimens of *Crotaphytus insularis*.

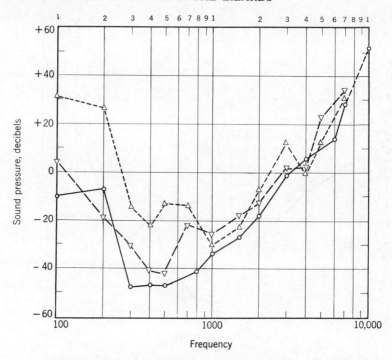

FIG. 7-85. Sensitivity curves for three specimens of *Crotaphytus wislizenii*. From Wever, Hepp-Reymond, and Vernon, 1966.

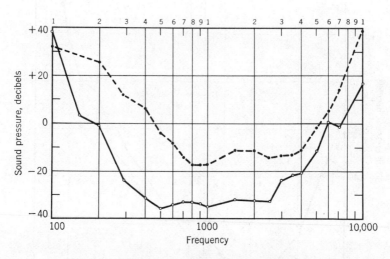

FIG. 7-86. Sensitivity functions for two specimens of *Leiocephalus carinatus armouri*.

Plica plica. — Measurements made on a specimen of *Plica plica* gave the results shown in Fig. 7-88. High sensitivity is evident within a narrow region around 700-1500 Hz, with a peak of −47 db at 1000 Hz. The roll-off rate is about 25 db per octave at the upper end of the range, and about 20 db at the lower end.

MALAGASIAN GROUP

Oplurus sebae. — A single specimen from the island of Madagascar was available for study. This was of the species *Oplurus sebae*, and gave the results shown in Fig. 7-89. The sensitivity is very good in the middle range from 700 to 2500 Hz, with a broad maximum centering at −40 db for the tone 1500 Hz. The roll-off rate is about 30 db per octave at both ends of the frequency range.

This sampling of the Iguanidae has revealed a considerable range of specific auditory capabilities. At the lower end of the distribution are certain of the sceloporines with tympanic membrane and other middle ear anomalies. The other sceloporine species exhibit great scatter, but the remaining species present performances that are mainly good to excellent. The *Anolis* species are on a high level, with *Anolis carolinensis* and *Anolis cristatellus* in the superior category. Among the basiliscines the two *Basiliscus* species are on a mediocre level and *Laemanctus longipes* is much better. The iguanines in general show good to excellent performance. The most common acoustic picture in this family of lizards is a broad range of sensitivity in the medium low to middle frequencies, with the maximum reaching −40 db or better.

HAIR-CELL NUMBERS AND SENSITIVITY

A study was made of the relations between iguanid hair-cell populations and auditory sensitivity as measured by the cochlear potentials. With a group of 36 specimens representing as many species, for which sensitivity measurements had been made and counts of hair cells were also available, a calculation of correlation coefficients was carried out in four different ways. Two measures of sensitivity were used, one representing maximum sensitivity—the lowest point (in decibels) reached by the response curve—and the other the "mean best sensitivity," obtained as the mean of all points on the curve within 10 db of this maximum. For example, the solid-line curve of Fig. 7-45 for *Anolis carolinensis* shows a maximum sensitivity of −59 db and a "mean best sensitivity," averaged over five points, of −55.2 db. Two populations of hair cells were used, the total number and the number beneath the tectorial plate.

The results of these calculations are shown in Table 7-II, where values are given for both the correlation coefficients (r) and a level of sig-

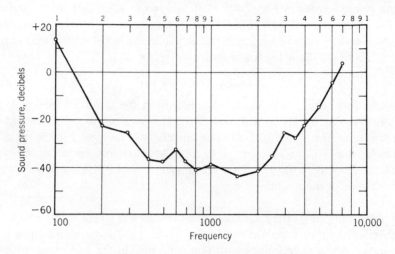

FIG. 7-87. Sensitivity function for a specimen of
Leiocephalus barahonensis aureus × *oxygaster*.

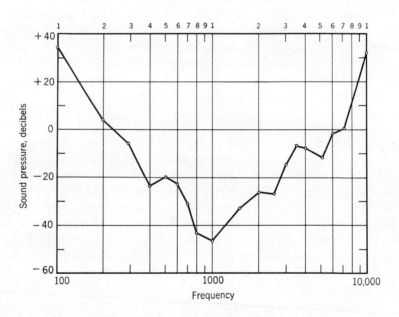

FIG. 7-88. Sensitivity function for a specimen of *Plica plica*.

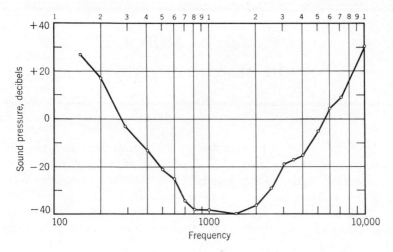

FIG. 7-89. Sensitivity function for a specimen of *Oplurus sebae*.

TABLE 7-II

SENSITIVITY AND HAIR-CELL POPULATIONS
IN IGUANIDAE
(Correlation Coefficients)

	Maximum Sensitivity	Mean Best Sensitivity
Total hair cells	$r = 0.49$ $p = 0.002$	$r = 0.46$ $p = 0.004$
Hair cells under tectorial plate	$r = 0.41$ $p = 0.01$	$r = 0.38$ $p = 0.02$

nificance (p). (The value of p represents the probability that the correlation obtained is due to chance.)

These results show a moderate positive relationship between hair-cell numbers and sensitivity, and the values are statistically significant. The two ways of designating sensitivity give much the same picture, and though the total population of hair cells yields slightly larger coefficients than the number of cells beneath the tectorial plate, the differences are hardly significant.

8. FAMILY AGAMIDAE:

THE AGAMID LIZARDS

The agamids may well be regarded as the typical lizards of the Old World, over which they are widely distributed, with strong preferences for the warmer regions. There are about 300 species in 34 genera (Wermuth, 1967). More than half of these species belong to just four genera, which are *Agama, Calotes, Draco,* and *Phrynocephalus.* The head of a specimen of a less common form, *Physignathus cocincinus,* is shown in Fig. 8-1.

FIG. 8-1. The head of a specimen of
Physignathus cocincinus.
Drawing by Anne Cox.

The agamids occupy a variety of habitats. Most are ground or rock dwellers, a good many are arboreal, and a few are semiaquatic. The genus *Draco,* with about 40 species, has the distinction along with the gecko *Ptychozoon* of including the only modern "flying" reptiles, with the capability of gliding through the air by extending the skin membranes along the sides of the body.

The present study included 13 species in 8 genera. These were *Agama agama* (Linnaeus, 1758)—4 specimens; *Agama agilis* (Olivier, 1807) —4 specimens; *Agama* cf. *cyanogaster* (Rüppell, 1835)—1 specimen; *Calotes mystaceus* (Duméril and Bibron, 1837)—3 specimens; *Calotes versicolor* (Daudin, 1802)—4 specimens; *Ceratophora stoddarti* (Gray, 1834)—5 specimens; *Ceratophora tennenti* (Gunther, 1861)—2 specimens; *Draco volans* (Linnaeus, 1758)—4 specimens; *Leiolepis belliana*

(Gray, 1827)—4 specimens; *Phrynocephalus maculatus* (John Anderson, 1872)—4 specimens; *Phrynocephalus* sp.—1 specimen; *Physignathus cocincinus* (Cuvier, 1829)—3 specimens; and *Uromastix hardwicki* (Gray, 1827)—5 specimens. All were first studied in terms of cochlear potentials and then prepared for histological examination.

ANATOMICAL OBSERVATIONS

EXTERNAL AND MIDDLE EAR STRUCTURES

In agamids the external ear is lacking. The tympanic membrane (except in deviant species in which it is absent) lies at the surface at the side of the head, depressed only by a little more than the thickness of the skin. This condition in *Agama agama* is shown in Fig. 8-2, where part *a* presents a general view of the head and *b* gives an enlarged view of the tympanic area. Elongated spines in patches around the membrane perhaps afford a little protection for the receptive surface.

The middle ear was studied in five species of agamids by Versluys (1898), who gave particular attention to the manner of enclosure of the tympanic cavity and described the forms of the columellar mechanism, which vary somewhat with species.

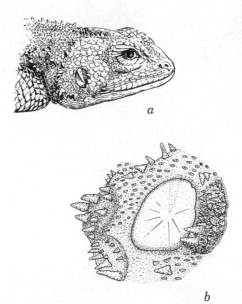

a

b

FIG. 8-2. The head of *Agama agama* in a side view at *a*, and details of the temporal region at *b*. Scales 0.5× for *a* and 3.5× for *b*.

The middle ear of *Agama agama* is represented in Fig. 8-3. As indicated, the tympanic membrane is firmly supported by a number of structures; anteromedially by the curved edge of the quadrate, anteriorly by the attached ends of the pterygoid and quadrate, anterolaterally by the mandibular articulation, ventrally by the retroarticular process, and posteriorly by a thick band of connective tissue running along the depressor mandibulae muscle. This muscle has three divisions as shown, and the main support for the tympanic membrane is provided by the fascial sheath along its anterior bundle.

The columellar mechanism and its attachments are shown in the figure. A long, slender osseous columella has its narrow footplate in the oval window, and is extended laterally by an extracolumella of simple form. As Versluys noted in a related species, *Agama colonorum*, the accessory processes of the extracolumella are absent. A slender ligament extending posterolaterally along the posteromedial edge of the tympanic

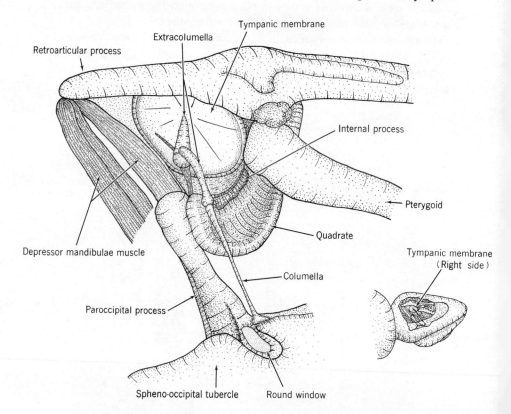

FIG. 8-3. The middle ear region of *Agama agama*, right side. The inset at the lower right indicates the angle of view, which is ventrolateral and slightly posterior.

membrane probably corresponds to the usual posterior process. An extracolumellar ligament runs from the end of the pars superior of the extracolumella to a cartilage that most likely is the intercalary, though Versluys described it differently in *Agama colonorum* and located it on the paroccipital process.

A lateral bracing of the columellar mechanism is provided by the internal process, a vane of cartilage inserted along the lateral rim of the quadrate.

In this species there is no true joint between columella and extracolumella; these two elements are firmly united by fibrocartilage.

Divergent Middle Ear Structures. As already noted, the tympanic membrane is absent in some of the agamids. This condition is characteristic of three groups, represented in the present study by five species: *Ceratophora stoddarti, Ceratophora tennenti, Draco volans, Phrynocephalus maculatus*, and *Phrynocephalus* sp. In the *Ceratophora* and *Phrynocephalus* species, this lack of a standard sound-receptive surface is further complicated by the forward intrusion of the large mass of the depressor mandibulae muscle, which serves to reduce the penetration inward of sounds acting on the skin of this region. These changes in the sound receptive mechanism can best be regarded as degenerative.

Other structural modifications in these animals appear to represent new adaptations tending to restore the lost auditory capabilities in some degree. In *Phrynocephalus* the pars inferior is greatly extended, and its lateral portion lies along the inner surface of the depressor mandibulae muscle. Also a narrow air cavity, which is a derivative of the middle ear space, is located just medial to the pars inferior and the muscle mass, contributing to the mobility of the skin and muscle tissue when driven by sound pressures applied to the external surface. In *Ceratophora* the end of the pars inferior inserts into a shallow depression in the inner surface of the depressor mandibulae muscle, in a manner that seems effective for the transmission of sound vibrations inward to the cochlea. These arrangements were considered in further detail in the section on conductive anomalies in Chapter 6.

Inner Ear

The cochlear duct of the Agamidae was described and pictured by Miller (1966a) as triangular and nearly as broad as it is long. The limbic plate is roughly circular in outline, with an oval opening covered by the basilar membrane.

The size of the basilar membrane and its papilla varies considerably among species. For the specimens of the present study the length of the basilar membrane varied from 500 μ in *Leiolepis belliana* to 210 μ in

Ceratophora stoddarti. In the same specimens the papilla varied in length from 460 to 130 μ.

A cross section of the cochlear duct in *Agama agama* is shown in Fig. 8-4. This section was taken at a point near the middle of the papilla, 160 μ from its dorsal end, and shows the hair cells with their ciliary tufts standing free. The kinocilia, each with a small group of stereocilia adjoining it in the tuft and about equally long, measured about 25 μ.

In the agamids a tectorial plate is present over a portion of the papilla, with its extent and location varying with species. Figure 8-5 shows for all the species studied the varying lengths of the papilla, with the position and length of the tectorial plates indicated. The first four species, including *Phrynocephalus maculatus, Phrynocephalus* sp., *Uromastix hardwicki,* and *Physignathus cocincinus,* have their tectorial plates nearly in the middle of the papilla. The others, save one, have them variously disposed toward the ventral end. The species *Leiolepis belliana* is unusual in having a very long plate covering the dorsal two-thirds of the papilla. Also a line of sallets covers the ventral portion of this papilla. These and other peculiarities of this species will be given special consideration presently. *Agama agama,* it will be noted, is one of the species with the tectorial plate at a far ventral position. A detailed study of cochlear structures was carried out in a few specimens selected to represent the range of variations among these species.

Agama agama. — A cross section of the same specimen of *Agama agama* shown in Fig. 8-4 is now presented in Fig. 8-6 at a point that

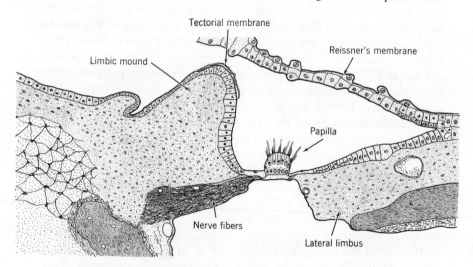

FIG. 8-4. A frontal section showing the auditory structures of *Agama agama* at a point a little beyond the middle of the cochlea.

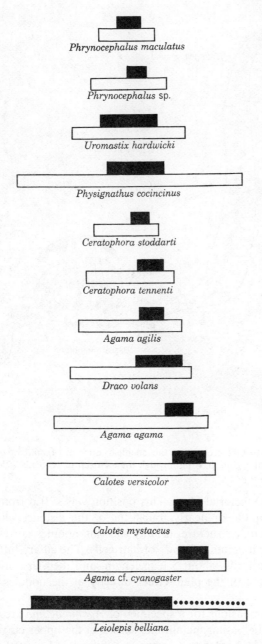

Fɪɢ. 8-5. Bar diagrams showing length of the auditory papilla (open rectangles) and of the tectorial plate (solid rectangles) for 13 species of Agamidae. The line of dots for *Leiolepis belliana* represents a series of sallets. Scale 125×.

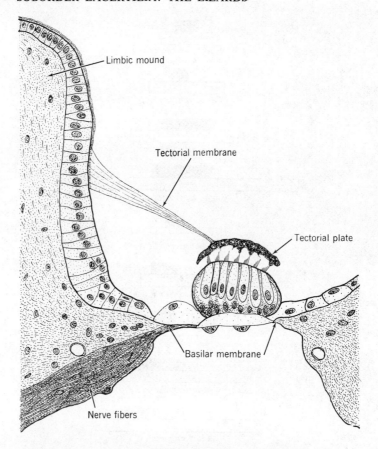

Limbic mound

Tectorial membrane

Tectorial plate

Basilar membrane

Nerve fibers

FIG. 8-6. A frontal section of the cochlear duct of *Agama agama* in the ventral region, passing through the tectorial plate. Scale 375×.

cuts across the tectorial plate. This position was 220 μ from the dorsal end of the papilla, or nearly four-fifths of the distance along it. This plate is a reticular structure, usually showing nodules corresponding in number to the longitudinal rows of hair cells. The ciliary tufts are short, with the longest cilia about 5 μ in length, and their tips are connected to the underside of the plate either directly at the nodules or through fine fibers.

Cochlear dimensions for a specimen of *Agama agama* are represented in Fig. 8-7. The basilar membrane, shown in the upper curve of part *a*, is 460 μ long and reaches its greatest width of 145 μ in the dorsal region, after which it declines progressively.

The fundus, which is the thickened portion of the basilar membrane, has much the same form, but has only about half the width of the whole

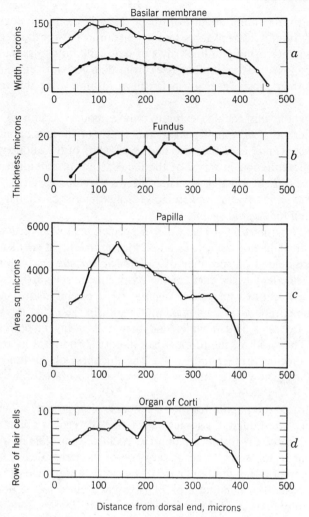

FIG. 8-7. Cochlear dimensions in a specimen of *Agama agama*. In part *a* the upper curve represents the width of the basilar membrane and the lower curve the width of the fundus.

membrane. Unlike most lizards, in which the fundus is located at the medial side of the membrane, this thickening in agamids is either near the middle (as shown in Fig. 8-4) or at the lateral side (as in Fig. 8-6).

The fundus attains a moderate thickness of 12-15 μ near the dorsal end of the papilla and varies only slightly over its remaining course, with a low maximum near its middle.

The size of the auditory papilla varies considerably along the cochlea. The area of this structure increases rapidly at the dorsal end, reaching a sharp maximum about a third of the way along the cochlea, and then falls rapidly, except for a minor plateau in the ventral region.

The number of longitudinal rows of hair cells fails to reflect papillar size very closely. This number rises from 5 to 8, with some irregularities, and then subsides, again in irregular fashion, over the ventral region.

As a whole this cochlea exhibits only a moderate amount of structural differentiation, manifested in the width of the basilar membrane and the size (and hence the mass and resistance) of the auditory papilla. Tonal differentiation in terms of the local action of stimuli would thus seem to be only of low degree. The other two *Agama* species, *Agama* cf. *cyanogaster* and *Agama agilis*, showed characteristics similar to those represented for *Agama agama*.

Calotes mystaceus. — The *Calotes* species exhibited structural features much as just described. Detailed measurements of cochlear dimensions were made in a specimen of *Calotes mystaceus*, and are shown in Fig. 8-8. The length and form of the basilar membrane are closely similar to these features in *Agama agama*. As *a* of this figure shows, the basilar membrane attains its maximum width in the early dorsal region, where it becomes 180 μ wide, and then falls away in corresponding fashion. The width of the fundus rises almost abruptly to its maximum, and then declines rather uniformly. The thickness of the fundus varies only slightly, showing a mild and somewhat irregular decline along its course.

The size of the papilla follows a function almost exactly like that of *Agama agama*, though it is somewhat smaller. It reaches a rounded peak about a third of the distance along the cochlea, then falls away, and in the ventral region shows a secondary rise of moderate degree.

The number of longitudinal rows of hair cells follows closely the form of the papilla. The maximum of 11 is reached close to the dorsal end, and ventrally there is an irregular decline marked by a secondary maximum.

Uromastix hardwicki. — The general form of the inner ear is much like that of *Agama agama* as described above, except for smaller size. Dimensional details for one of the specimens are shown in Fig. 8-9. This cochlea is small in all respects. The basilar membrane has a length of 325 μ, and varies little in width. The width rises to about 85 μ near the dorsal end and remains there, with only slight variations, up to the middle of the cochlea, after which there is a decline in two stages. The fundus follows a closely similar course. The thickness of the fundus shows only slight variations, reaching a maximum in the dorsal region and hardly changing through the middle region, until an irregular decline occurs over the ventral third.

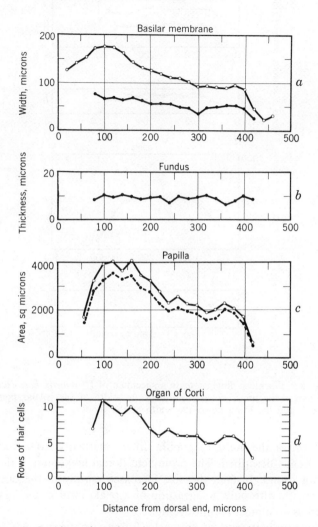

FIG. 8-8. Cochlear dimensions in a specimen of *Calotes mystaceus*. In part *a* the upper curve represents the width of the basilar membrane and the lower curve the width of the fundus. In part *c* the upper curve represents the area of the whole papilla, and the lower curve the area of the cellular portion (organ of Corti) only.

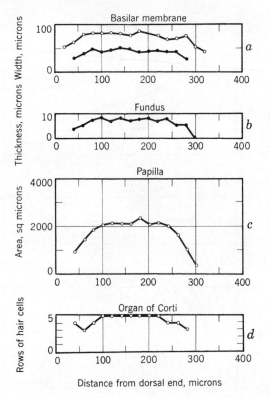

FIG. 8-9. Cochlear dimensions in a specimen of *Uromastix hardwicki*. In part *a* the upper curve represents the width of the basilar membrane, and the lower curve the width of the fundus.

The form of the auditory papilla differs markedly from that of the species already described. The prominent dorsal maximum of those species is not found. Instead the size increases rapidly at the dorsal end and levels off, with only a suggestion of a peak, until in the ventral region there is an almost abrupt decline.

The longitudinal rows of hair cells follow a pattern much like that of the papilla. A maximum of 5 rows prevails over the whole middle part of the cochlea, with a decline on either side.

Here is a cochlea of much simpler form than in the preceding species. There is hardly any differentiation except at the two ends of the cochlea, and it must be inferred that tonal discrimination in terms of place of action along the cochlea is minimal. This simplicity of *Uromastix* is exhibited in other ways, as in the caudal vertebrae and the trunk musculature (Werner, 1961).

Ceratophora tennenti. — Similar dimensional measurements were carried out in a specimen of *Ceratophora tennenti*, one of the agamid species with anomalies of the sound-conductive mechanism. As part *a* of Fig. 8-10 shows, the basilar membrane is very short, measuring only 270 μ in length. Its width rises progressively to a maximum near the middle of the cochlea, after which it declines rapidly. The width of the fundus shows a similar form. The thickness of the fundus was so slight that it could not easily be measured, and this feature is not shown. The size of the papilla increases to a sharp peak near the middle of the cochlea and then falls precipitously.

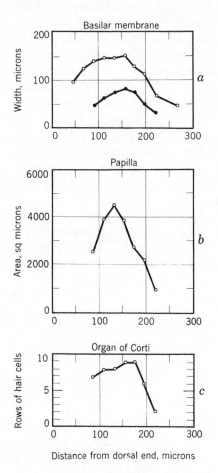

Fig. 8-10. Cochlear dimensions in a specimen of *Ceratophora tennenti*. The upper curve of part *a* represents the width of the basilar membrane, and the lower curve represents the width of the fundus.

The number of longitudinal rows of hair cells increases from 7 to 9, then falls abruptly.

This is a very short cochlea, with no significant differentiation.

Leiolepis belliana. — As already indicated, the ear of *Leiolepis belliana*, a terrestrial species of Southeast Asia, is highly developed and takes an unusual form. A representative specimen will be described in detail.

The papilla begins almost at the dorsal end of the basilar membrane, and from the beginning is covered by a tectorial plate. As shown in Fig. 8-11, taken at a point 40 μ from the end of the basilar membrane, the medial limbus is only moderately thickened and shows a small conical elevation of the epithelial covering.

The basilar membrane here is 96 μ wide, and the fundus, which is 38 μ wide and only moderately thickened, is located 15 μ away from the inner edge of the membrane. The supporting cells stand relatively high above the fundus, and send slender root processes down to the surface of the fundus, as well as the usual fine columns that pass upward between the hair cells to the papillar surface.

The papilla is sharply tipped in a medial direction and shows four rows of hair cells. For the most part these cells send their short ciliary tufts into shallow recesses in the lower surface of the tectorial plate.

Tectorial tissue is present over the peak of the limbic mound, but does not form a free membrane in this region.

The number of transverse rows of hair cells increases rapidly, until at a point 120 μ along the basilar membrane, as shown in Fig. 8-12, there are 11 rows of cells, all of uniform size, covered closely by the cap-like tectorial plate. The papilla in this region shows little of its former obliquity on the basilar membrane.

A further change occurs a little farther ventrally. The papilla becomes somewhat lozenge-shaped, with its medial portion raised. The hair cells become larger and fewer, stabilizing at 7 rows. Also at this point, which is 160 μ from the dorsal end, the tectorial membrane extends freely from the peak of the limbic mound (which has grown somewhat larger) and connects with the tectorial plate, as shown in Fig. 8-13. The supporting cells here for the most part rest closely over the fundus.

The form just described continues with little change for a considerable distance. At 260 μ from the dorsal end the papilla has become nearly symmetrical in form, and the number of rows of hair cells remains at 7.

Then in a region about 340 μ from the dorsal end another abrupt change occurs. The tectorial plate becomes narrower, and now has the form of a peaked mound rather than a uniform cap. At this point it

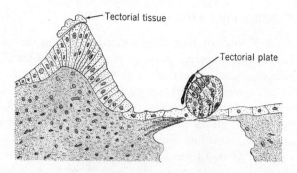

FIG. 8-11. The auditory papilla and its limbic support in *Leiolepis belliana* at a point close to the dorsal end. Scale 250×.

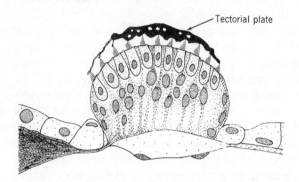

FIG. 8-12. The auditory papilla of the same specimen as in the preceding figure, a little farther ventrally. Scale 500×.

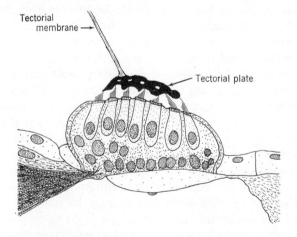

FIG. 8-13. The auditory papilla of the same specimen, taken near the middle of the cochlea where the tectorial membrane becomes attached to its plate. Scale 500×.

connects with 5 rows of hair cells located at the medial side of the papilla. Three other rows are present, one with a free-standing ciliary tuft, and two, the most lateral ones, with their tufts connected to a small sallet. This arrangement is represented in Fig. 8-14.

At a point 20 μ farther ventrally, represented in Fig. 8-15, the tectorial membrane is no longer attached to the plate. A reduced portion of the plate remains, however, making connection to the ciliary tufts of four rows of hair cells. In this region the sallet continues over the lateral portion of the papilla, connecting with three rows of hair cells.

Then, at a position 400 μ from the dorsal end, all traces of the tectorial plate are lost, and an enlarged sallet located over the middle of the papilla serves all 7 rows of hair cells. These cells are of shortened form, and the supporting cells once more are raised well above the fundus, sending their root processes down to its surface as described for the dorsal region. The fundus is somewhat reduced in width.

A little farther ventrally a medial inclination of the papilla becomes noticeable, giving much the same picture as described for the dorsal end. This inclination increases rapidly until at a point 460 μ from the dorsal end, as shown in Fig. 8-16, the angle with the basilar membrane is about 45°. At the same time the papilla becomes smaller, the number of longitudinal rows of hair cells falls to 3, and the sallet is much reduced. Thereafter the papilla quickly comes to an end.

This papilla is distinctive among all the agamids studied. The most outstanding feature is the presence of a row of sallets along with an extensive tectorial plate. Sallets have not been encountered in any other member of this family.

Another distinctive feature is the dorsal extension of the tectorial plate far beyond its connection to the tectorial membrane. In other species the presence of this plate is nearly coincident with its connection to the membrane and hence its secure anchorage to the limbus. However this long dorsal extension of the plate without such a connection probably has no significant effect upon function. The plate appears to have sufficient rigidity that the restraint imparted to its middle portion is readily communicated to the dorsal portion also, so that everywhere it imposes a high degree of immobility upon the tips of the ciliary tufts attached to it.

The row of sallets begins at the lateral edge of the papilla, as Fig. 8-14 shows, when the tectorial plate is still present, and then expands ventrally over the middle of the papilla after this plate has terminated, as Fig. 8-15 has indicated. These sallets are probably more or less independent bodies. There is no indication of an interconnection along their top surfaces such as occurs regularly for the sallets found in the gekkonid cochlea. It is possible that a connection runs along the bases of these sallets, but this point could not be determined with any certainty in the frontal sections available and needs to be checked in sagittal series, or by a surface study.

Ciliary Orientation Patterns.   In all agamids, including *Leiolepis belliana*, the orientation of the ciliary tufts beneath the tectorial plate is uniformly lateral: the longest cilia, including the kinocilium, are on the lateral side of the series. Elsewhere the pattern is bidirectional, but with a strong predominance of laterality. This condition is represented in Fig. 8-4 for *Agama agama*, where four of the six cells in the transverse row have a lateral orientation. Usually the most lateral cell, and perhaps the one next to it, will have a medial orientation, but in the next adjacent row the corresponding cells may face laterally. Frequently a single cell in a region where all others have a lateral orientation will show the contrary position. Thus the arrangement of the ciliary tufts in agamids is not as orderly as in most other lizard families. In *Leiolepis* the ciliary orientation is bidirectional in the region where sallets are present, as Figs. 8-14 and 8-15 portray.

Numbers of Hair Cells.   The degree of cochlear development in these and other agamid species studied is further indicated in the size of the hair-cell population. Counts of the hair cells were made, with separate records of the cells beneath the tectorial plate and elsewhere (where in

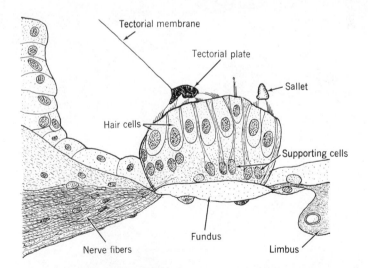

FIG. 8-14. The auditory papilla of the same specimen at a point about two-thirds of the distance along the cochlea where the tectorial plate is reduced and a sallet first makes its appearance. Scale 500×.

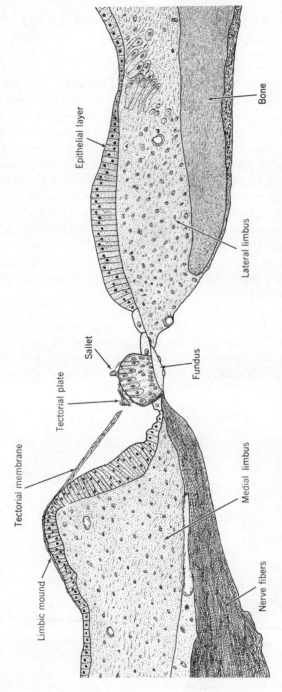

FIG. 8-15. The cochlear region of *Leiolepis belliana* in frontal section at a point far ventrally. Scale 200×.

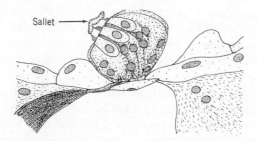

Sallet

FIG. 8-16. The auditory papilla of *Leiolepis belliana* in frontal section
at a point close to the ventral end. Scale 500×.

all but *Leiolepis belliana* the ciliary tufts stand free). Averages are given
in Table 8-I for the different species. The listing is in the order of magnitude of the totals.

The largest hair-cell population appears in *Leiolepis belliana* with an
average of 262.5, and *Physignathus cocincinus* is a close second with
an average of 257.6. Next appear the two *Calotes* species and two of the
Agama species, *Agama agama* and *Agama cyanogaster*, with averages between 248.2 and 222.5. Noticeably smaller populations occur in *Draco
volans* and *Agama agilis*, and a further decline is seen in *Uromastix hardwicki*, with a count of 101.9. The *Ceratophora* species show further
drops, and at the end of the series are the *Phrynocephalus* species, with
Phrynocephalus maculatus showing an average total of 55.8 hair cells.
The range of variation throughout this group is nearly fivefold.

TABLE 8-I

MEAN HAIR-CELL POPULATIONS
IN AGAMIDAE

Species	Number of Ears	Dorsal to Plate	Under Plate	Ventral to Plate	Total
Leiolepis belliana	6	0	199.2	63.3	262.5
Physignathus cocincinus	3	127.3	51.0	79.3	257.6
Calotes mystaceus	6	201.0	42.5	4.7	248.2
Agama agama	6	201.8	30.8	6.7	239.3
Calotes versicolor	6	187.3	47.0	4.0	238.3
Agama cyanogaster	2	184.0	24.5	14.0	222.5
Draco volans	6	82.2	61.7	6.5	150.3
Agama agilis	6	117.0	19.5	9.2	145.7
Uromastix hardwicki	7	17.0	60.4	24.4	101.9
Ceratophora stoddarti	10	58.6	24.6	3.6	86.8
Ceratophora tennenti	2	46.5	22.5	2.0	71.0
Phrynocephalus sp.	2	29.0	27.0	8.5	64.5
Phrynocephalus maculatus	6	23.5	23.8	8.5	55.8
Average of 13 species	—	98.1	48.8	18.1	165.0

The low position of the *Ceratophora* and *Phrynocephalus* species might be expected in view of the decline in effectiveness of their sound-conductive mechanisms, which must impair the serviceability of their ears, but the position of *Draco volans* is higher than might be expected on such grounds.

There is reason to believe, from the results of overstimulation experiments (described in Chapter 4), that the hair cells beneath the tectorial plate are by far the most sensitive, and are primarily responsible for an animal's detection of the faintest sounds. If this is the case, an evaluation of the general serviceability of the ear ought to focus on this part of the hair-cell population rather than the total number. With this criterion, most of the species in Table 8-I hold their rankings as shown, but there are some notable exceptions: *Draco volans* and *Uromastix hardwicki* gain markedly and become second and third after *Leiolepis belliana, Agama agilis* falls to the last position, and the other two *Agama* species lose considerable rank. A large gap opens between *Leiolepis belliana* and all the others because of the great length of its tectorial plate.

These structural relations will be considered further after an examination of sensitivity data on these lizard species.

SENSITIVITY FUNCTIONS

Representative results of the sensitivity tests are presented in a number of figures to follow. Figure 8-17 gives cochlear potential curves for two specimens of *Agama agama*. The sensitivity is very good over a middle range of frequencies from 400 to 3000 Hz, with the maximum at −44 db. A rate of decline in sensitivity of about 20 db per octave is present for the low tones, and a rate around 30 db per octave appears for the high tones.

Sensitivity curves for two specimens of *Agama agilis* are shown in Fig. 8-18. These results are generally similar to those just presented for *Agama agama*, though one of the curves indicates somewhat greater sensitivity in the low frequencies. Again the response is best in the middle of the range, with the maximum reaching −46 db. The roll-off rate for the low tones is around 30 db per octave, and is about the same for the high tones.

Results for two specimens of *Calotes mystaceus* are presented in Fig. 8-19. One of the curves is like the foregoing, with the best response in the middle region, where two maximums appear with a moderate reduction between. The other curve shows considerably less sensitivity. The roll-off rate is rapid, around 40 db per octave, at both ends of the solid-line curve.

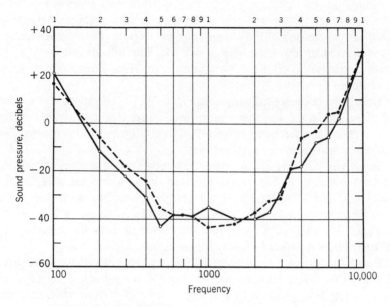

FIG. 8-17. Sensitivity curves for two specimens of *Agama agama*.

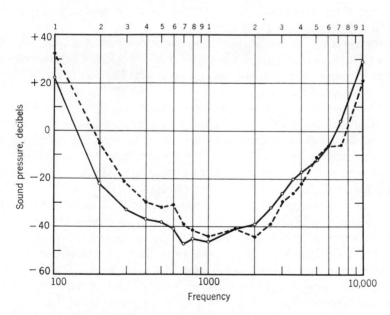

FIG. 8-18. Sensitivity curves for two specimens of *Agama agilis*.

Results for two specimens of *Calotes versicolor* are given in Fig. 8-20. The two functions are in general agreement, with maximum points at 3000 Hz, where one of the curves reaches −40 db. Both curves show a curious point of low sensitivity at 600 Hz. The roll-off rate for the low tones is around 30 db per octave, and for the high tones is about 40 db per octave.

Results were presented earlier, in Fig. 6-70, for two specimens of *Ceratophora stoddarti*. For these ears the sensitivity is poor, with the best region in the low frequencies from 200 to 600 Hz, and with progressively poorer sensitivity for higher tones.

Sensitivity data were also presented earlier for a specimen of *Ceratophora tennenti*, in Fig. 6-71. This ear shows a maximum region between 400 and 1500 Hz, where the best sensitivity is −4 db.

A sensitivity function for a specimen of *Draco volans* was presented in Fig. 6-66. This curve runs around 0 to −6 db over the range of 500 to 3000 Hz, but falls off rapidly for lower and higher tones.

Results for three specimens of *Phrynocephalus maculatus* were given in Fig. 6-68. These curves agree in showing a mediocre level of sensitivity, with best response in the region around 300 to 1000 Hz, and rapidly declining sensitivity for lower and higher tones, although one ear shows a secondary maximum at 2000 Hz. The best level attained by one of these ears was −20 db at 500 Hz; the other two ears reached −10 and −14 db for tones a little higher.

As Fig. 8-21 shows, an even poorer performance was found in another *Phrynocephalus* specimen whose species could not be determined. This ear reached a maximum of +4 db at 400 Hz, and fell off rapidly for lower and higher tones.

The poor level of hearing in the *Ceratophora*, *Draco*, and *Phrynocephalus* species is due in large part to the absence of a tympanic membrane along with other conductive anomalies, as discussed in Chapter 6. A deficiency of inner ear performance is also indicated, however—the sensitivity is still below average as assessed by vibratory stimulation. See, for example, the results on *Phrynocephalus maculatus* in Fig. 6-74.

In sharp contrast to the results just described are those obtained on the species *Leiolepis belliana*. The two ears represented in Fig. 8-22 display a superior level of sensitivity, with the maximum in the region of 500 to 1000 Hz. Although the sensitivity declines rapidly outside this octave range it is still very good from 200 to 3000 Hz. The maximum attained for one of these ears was −78 db, which is a remarkably high degree of sensitivity.

Good hearing was found also for the species *Physignathus cocincinus*, as represented in Fig. 8-23. The two curves have their best region of sensitivity in the range of 400 to 1000 Hz, where they reach levels of

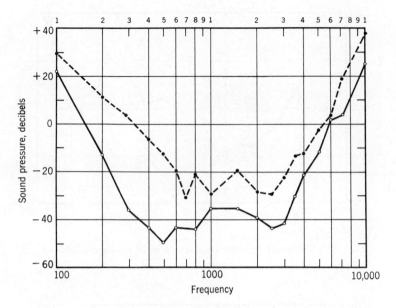

Fɪɢ. 8-19. Sensitivity curves for two specimens of *Calotes mystaceus*.

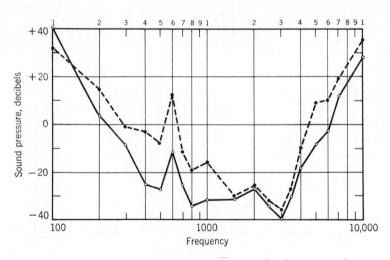

Fɪɢ. 8-20. Sensitivity curves for two specimens of *Calotes versicolor*.

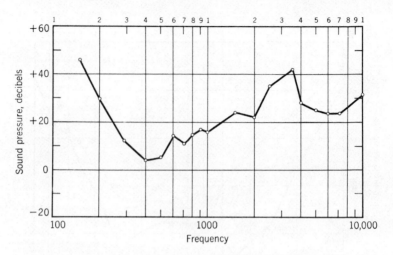

FIG. 8-21. Sensitivity function for a specimen of *Phrynocephalus* sp.

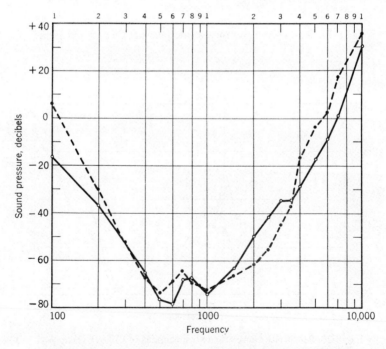

FIG. 8-22. Sensitivity curves for two specimens of *Leiolepis belliana*.

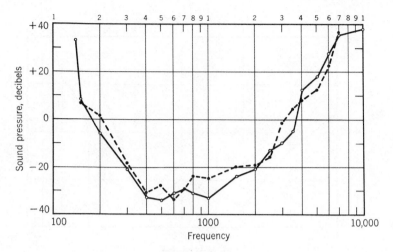

Fɪɢ. 8-23. Sensitivity functions for two specimens of *Physignathus cocincinus*.

−34 db. The decline of sensitivity is very rapid at the low end of the tonal range, and is around 30 db per octave at the high end. One of these specimens was tested also with vibratory stimulation. The tip of the mechanical vibrator was applied in one series to the outward bulge of the tympanic membrane over the end of the pars inferior, thereby driving the columellar system directly. In another series the cephalic condyle was exposed and the stimulating tip applied to its surface. The resulting curves (Fig. 8-24) are of similar form, with the one for pars inferior driving more sensitive for the low tones, especially around 400 Hz. There is close resemblance between these curves and those obtained by aerial stimulation as shown in Fig. 8-23, all these functions showing the best sensitivity in the middle range from 300 to 2000 Hz.

Curves for two specimens of *Uromastix hardwicki* are shown in Fig. 8-25, and represent very good sensitivity in the middle range. The better of these ears reaches −50 db at 1500 Hz, and the sensitivity is good over a range that includes two octaves below this and about half an octave above. The decline beyond 2000 Hz is rapid, of the order of 40 db per octave.

The above sample of agamid sensitivity covers a wide range, from the superior performance of *Leiolepis belliana* to the severely limited response of species with conductive anomalies. It should be pointed out that this sample was biased by a deliberate effort to obtain species with conductive impairments, in the interests of the special study of such anomalies in Chapter 6.

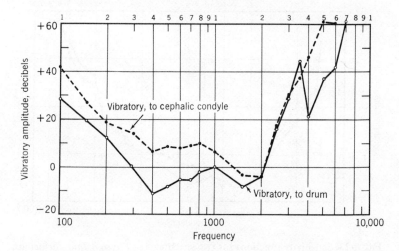

FIG. 8-24. Sensitivity functions for a specimen of *Physignathus cocincinus* with vibratory stimulation at the tympanic membrane over the pars inferior (solid line) and on the cephalic condyle (broken curve). Vibratory amplitude is in decibels relative to 1 mμ.

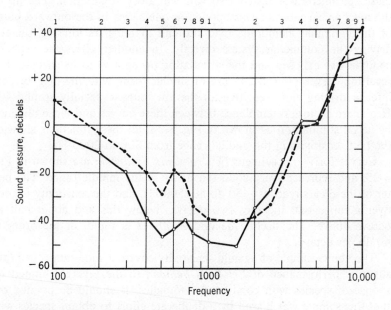

FIG. 8-25. Sensitivity functions for two specimens of *Urosaurus hardwicki*, with aerial stimulation.

Sensitivity and Hair-cell Populations. A positive correlation was found between auditory sensitivity in these agamid species and the size of the hair-cell population. This relationship was investigated by using both the maximum point in the sensitivity function and the mean of all values within 10 db of the maximum. Also these measures of sensitivity were related to hair cell numbers obtained in two ways, as the number of hair cells beneath the tectorial plate and the total number of hair cells in the papilla. Accordingly, four measures of correlation were obtained as indicated in Table 8-II.

TABLE 8-II

SENSITIVITY AND HAIR-CELL POPULATIONS
IN AGAMIDAE
(Correlation Coefficients)

	Maximum Sensitivity	Mean Best Sensitivity
Total hair cells	$r = 0.69$	$r = 0.70$
	$p = 0.013$	$p = 0.012$
Hair cells under	$r = 0.63$	$r = 0.66$
tectorial plate	$p = 0.027$	$p = 0.020$

All four correlation coefficients are high enough to be meaningful, and their reliability is indicated as good. It is a little surprising that the relationship of sensitivity to the total number of hair cells is even a little higher than to the number beneath the tectorial plate, though the difference is slight. This same condition existed in the Iguanidae, and no very satisfactory explanation was found for it in that family. In the Agamidae the correlation is considerably higher than was found among the iguanids, and this question becomes more pointed. It can still be urged that maximum sensitivity is determined by the action of a very small number of hair cells, and that a larger population of cells increases the probability that some of them will operate with great efficiency.

This survey of the agamid family shows these lizards as a somewhat varied group, with large differences in the general size of the inner ear and its complement of hair cells. Variations occur also in the middle ear, which in a number of species has undergone degenerative changes. In general these ears show little cochlear differentiation, and more than a very limited amount of frequency discrimination in terms of place of response along the basilar membrane seems unlikely. The sensitivity of this type of ear can be very good, however, as shown in a few species and especially in *Leiolepis belliana*. This last species is exceptional both in its degree of sensitivity and in cochlear structure.

9. FAMILY CHAMAELEONIDAE:

THE CHAMELEONS

The chameleons, represented in Fig. 9-1 by a specimen of *Chamaeleo jacksoni*, are lizards of distinctive form, readily recognized by the laterally compressed body and crested head, large mobile eyes with a small central aperture, long extensible tongue, clasping feet with the digits in a two and three division, and the absence of an external ear opening and tympanic membrane. Many species are distinguished also by prominent scale-covered flaps of skin in the occipital region.

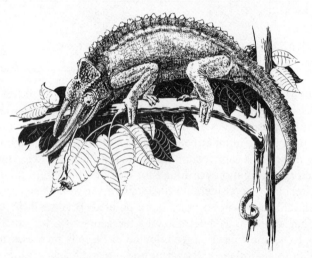

FIG. 9-1. A chameleon of the species *Chamaeleo jacksoni.*
Drawing by Anne Cox.

These animals are noted for their rapid color changes, though in this regard they are easily matched by other lizards, such as the anoles. The horns exhibited by the animal in the illustration represent an additional feature found in certain species, usually restricted to the males.

The absence of an ear opening is apparent in Fig. 9-2. A flat and slightly sunken area lies over a portion af the pterygoid bone and outlines an "auditory area," further defined below.

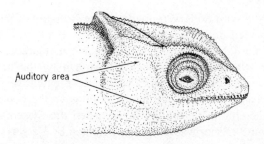

Fig. 9-2. The head of *Chamaeleo senegalensis*. From Wever, 1968b.

The specializations just mentioned and others less immediately obvious make this group aberrant among lizards and difficult to locate among the other families. However several investigators, beginning with Cope (1900), have suggested that the chameleons are an early offshoot of the agamid family (Camp, 1923; Broom, 1935; Brock, 1940; Malan, 1945).

The chameleons are mainly arboreal, though a few, notably the *Brookesia* species, live on the forest floor. Almost half the living species are restricted to the island of Madagascar, and most of the others are in Africa. Three species occur in southern Asia, and one, *Chamaeleo chamaeleon*, is scattered along the southern and eastern borders of the Mediterranean and on some of its islands; and even extends as far as the European mainland in Spain. Because this species is the one best known to Europeans, it is called the common chameleon—as witnessed by its former name of *Chamaeleo vulgaris*.

The chameleon family contains over 80 species assigned to the six genera *Chamaeleo, Brookesia, Rhampholeon, Microsaura, Leandria,* and *Evoluticauda*. The most extensive study of the group was made by Franz Werner in 1902.

Werner in his 1902 monograph identified 69 species in the genus *Chamaeleo*, and later in a checklist of 1911 he increased this number to 73, along with several subspecies. The few species outside this genus are mainly in the genera *Brookesia* and *Rhampholeon*—seven each according to Werner's early listing (see Mertens, 1966).

Werner attempted to organize the *Chamaeleo* species into distinct assemblages. He suggested seven groups containing 5 to 14 members each, plus a single species left by itself. This grouping was based mainly on external features such as scalation, the presence and size of the occipital lobes, scaly ridges along dorsal and ventral surfaces of the body, and the presence of "ornamental" characters, often restricted to the males, such as nose processes, horns, and heel spurs.

A few early observations were made on the structure of the chameleon ear, mainly carried out in connection with studies of the skull. The absence of an external ear opening and tympanic membrane, and the peculiar form of the columella, were well known by the middle of the nineteenth century. Parker in 1885 reported an extensive study of the skull of *Chamaeleo chamaeleon*, in which he gave the first detailed description of the columella, though he regarded this system as nonfunctional. This same species was extensively treated by Versluys a few years later (1898).

Specimens of other genera were included in some of the early descriptions. Siebenrock (1893) studied *Brookesia superciliaris* and Toerien much later (1963) described another member of this genus, *Brookesia marshallii*. *Microsaura pumila* was briefly dealt with by Parker, and more fully treated by Toerien. The related species *Microsaura ventralis* was described by Brock (1940) and by Engelbrecht (1951). Frank (1951) worked on *Rhampholeon platyceps*. These accounts deal with the auditory structures so briefly that little pertinent information can be gleaned from them. The main indications are of degenerative changes in these genera that are more extensive than in *Chamaeleo* species, such as the absence of a tympanic cavity in *Rhampholeon platyceps* and its reduction to a "mere vestige" in *Microsaura ventralis*. Extreme reduction of the columella is reported also in *Brookesia marshallii* and *Rhampholeon platyceps*.

The present study was carried out on 36 specimens belonging to 8 species and one subspecies, all in the genus *Chamaeleo*. These were *Chamaeleo chamaeleon* (Linnaeus, 1785)—2 specimens; *Chamaeleo chamaeleon calcarifer* (Peters, 1870)—5 specimens; *Chamaeleo dilepis dilepis* (Leach, 1819)—3 specimens; *Chamaeleo quilensis* (Bocage, 1866)—3 specimens; *Chamaeleo senegalensis* (Daudin, 1802)—8 specimens; *Chamaeleo ellioti* (Gunther, 1895)—3 specimens; *Chamaeleo höhnelii* (Steindachner, 1891)—5 specimens; *Chamaeleo jacksoni* (Boulenger, 1896)—5 specimens; and *Chamaeleo fischeri tavetensis* (Steindachner, 1891)—2 specimens.

The first five forms listed belong to Werner's Group 1, the next three are in his Group 3, and the last is in his Group 5.

Nearly all these animals were first tested by means of cochlear potentials, and then the ears were treated histologically and examined in anatomical detail. Some of the results of this study were reported earlier (Wever, 1968b, 1969a, b).

As already indicated, the chameleon ear presents certain peculiar features that must be regarded as deficiencies and a result of degenerative changes. Most obvious is the absence of an external ear opening and tympanic membrane. Also lacking in this ear is the round window, usually found in the wall of the perilymphatic sac and leading to the air cavity of the middle ear. This lack of a round window is a serious matter, and will be treated in detail presently.

ANATOMICAL OBSERVATIONS

THE SOUND-CONDUCTIVE MECHANISM

In the chameleons examined, the middle ear mechanism is contained in a definite tympanic cavity, separated from the pharyngeal region by a membrane except for a small oval opening. This opening in a specimen of *Chamaeleo quilensis* measured 0.5 × 2.0 mm. Removal of the floor of the mouth and then of this medial membranous wall provides a full view of the middle ear structures.

A still broader view is presented in Fig. 9-3, which shows the skull of a specimen of *Chamaeleo quilensis* in which the columellar structures were preserved.

The oval window is seen in the wall of the otic capsule, and is nearly filled by the columellar footplate. This footplate is relatively large and is held in place by an annular ligament that is especially heavy along its lateral edge and toward its anterior end.

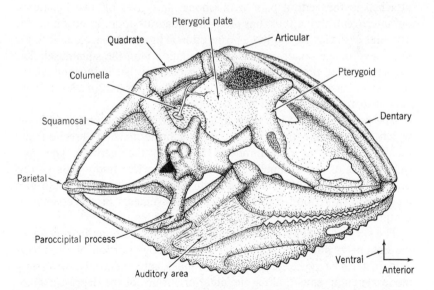

FIG. 9-3. The skull of *Chamaeleo quilensis*, oriented to show the otic region of the right side. The auditory area of the left side is indicated also. From Wever, 1968b.

Rising from the footplate is the shaft of the columella, which is osseous and noticeably stiff in this species. Beyond the osseous portion lies the extracolumella, whose processes are cartilaginous and somewhat flexible. In this species the extracolumella branches immediately to produce anterior and posterior processes.

It should be noted that the designations "anterior process" and "posterior process" as used for these species are based simply upon the spatial relations, and do not imply a correspondence with the similarly named components of the normal extracolumella of other lizards. Versluys attempted to homologize the extracolumellar processes in *Chamaeleo chamaeleon* with those of normal ears, but his identifications are almost certainly in error. The severe modifications of middle ear structure that have taken place in the chameleons, and the variations from one species to another (and to some degree within species) make this task an extremely hazardous one.

As shown in Fig. 9-4 for *Chamaeleo quilensis* and in Fig. 9-5 for *Chamaeleo senegalensis*, the anterior process bends forward and runs over the surface of a pterygoid plate, to which it is attached by ligamentary fibers. The posterior process continues the direction of the columellar shaft and extends to the close vicinity of the quadrate. Here this process expands as a flat plate and runs anteroventrally along the surface of the quadrate to which it becomes attached. The plate is loosely attached to the ventral part of this bone, and gives off two ligaments, an anteroventral one that runs along the quadrate to an insertion on the articular of the lower jaw and a posterodorsal ligament that extends to the upper end of the quadrate near its articulation with the squamosal. The extracolumellar plate is bounded posteriorly by the midportion of the depressor mandibulae muscle, but in these species is not closely connected to the muscle.

The pterygoid plate is a derivative of the wing of the pterygoid bone, which becomes greatly thinned out in a posteromedial direction and then is continued as a membrane to the edge of the quadrate. The anterior process of the columella connects with the free midportion of this plate. It was easily observed that slight displacements of the plate in a mediolateral direction produced rocking movements of the columellar footplate about an axis along its anterolateral edge.

The species *Chamaeleo d. dilepis* likewise has an anterior process of the columella that attaches firmly to the osseous portion of the pterygoid plate. The posterior process continues anteroventrally, with only moderate enlargement, along the anterior surface of the depressor mandibulae muscle. A narrow fin on the posterior part of this process lies in a shallow groove in the muscle and ends as a short ligament close to (but not connected with) the retroarticular process of the lower jaw.

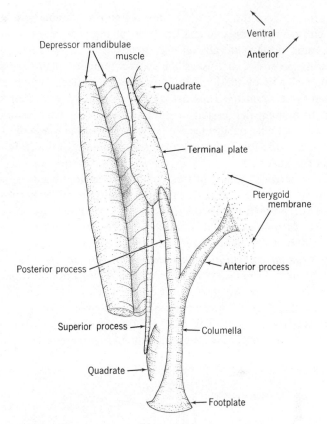

Fig. 9-4. The conductive mechanism of *Chamaeleo quilensis*. Normally the terminal plate lies deep in the depression along the depressor mandibulae muscle.

In *Chamaeleo chamaeleon calcarifer*, as in the foregoing species, the columella sends off anterior and posterior processes, but the anterior process makes contact only with the membranous portion of the pterygoid plate. The posterior process extends between the quadrate and the depressor mandibulae muscle, with contact along the surface of the medial part of this muscle. It never penetrates the muscle, though in the ventral region the medial end of the muscle curls around the cartilage somewhat to form a close contact.

In *Chamaeleo chamaeleon*, as shown in Fig. 9-6, there is an anterior process also, but this process arises at a deep ventral position, well beyond the osseous portion of the columella. In some specimens it runs for a considerable distance ventrally along the pterygoid surface, which remains membranous all this way. Only far beyond the end of the anterior process does the pterygoid plate contain bone, and here the bone is thin and occupies only a small portion of the surface. A posterior ligament runs posterolaterally at the quadrate. The terminal part of the posterior process expands as the extracolumellar plate and lies along the medial half of the depressor mandibulae muscle, much as just described for *Chamaeleo chamaeleon calcarifer*, and as in that subspecies it lies in a slight depression in the muscle surface near its ventral end, but never actually penetrates the muscle. This plate sends off a thin ligament (the superior ligament) that runs medially, usually reaching the end of the quadrate.

The differences found in this conductive structure among specimens of *Chamaeleo chamaeleon* are perhaps related to the general variability of this species as it occurs over its widely distributed and spotty habitat. Hoofien (1964) reported variability in several features in specimens from different regions of Israel and its environs.

In *Chamaeleo ellioti, Chamaeleo fischeri tavetensis, Chamaeleo höhnelii,* and *Chamaeleo jacksoni* there is no anterior process of the columella. The osseous portion of the columella continues far ventrally, reaching the close vicinity of the quadrate before it ends and the cartilaginous portion begins. This cartilaginous portion evidently corresponds to the posterior process and extracolumellar plate of other chameleons,

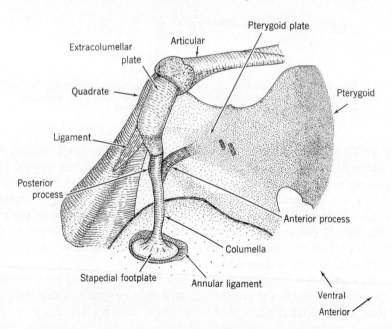

FIG. 9-5. The conductive mechanism of *Chamaeleo senegalensis*. The depressor mandibulae muscle has been removed. From Wever, 1968b.

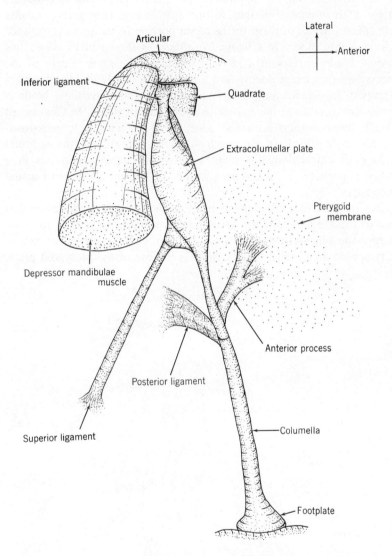

FIG. 9-6. The conductive mechanism of *Chamaeleo chamaeleon*. The lower portion only of the depressor mandibulae muscle is represented.

and extends between the quadrate bone and the depressor mandibulae muscle, with intimate relations to this muscle that vary greatly among these species. In all of them the terminal portion of the process expands to form a thin blade. In *Chamaeleo ellioti*, as shown in Fig. 9-7, this terminal blade turns nearly at right angles to the main course of the process and penetrates deeply into a notch that nearly divides the muscle in two. In *Chamaeleo fischeri tavetensis* a similar terminal blade is formed that penetrates the muscle to about half its depth. In *Chamaeleo höhnelii* the condition is similar except that the muscle is penetrated only about one-third. In *Chamaeleo jacksoni* (Fig. 9-8) the terminal cartilage or extracolumellar plate is relatively narrow, and lies in a shallow depression in the anterior surface of the muscle without actual penetration.

Thus we have identified in the chameleons two special strategies that seem to represent secondary developments tending to restore in some measure the sensitivity lost to these ears through the disappearance of the tympanic membrane. One of these is the use of the pterygoid plate,

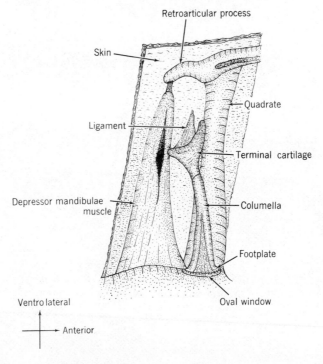

Fig. 9-7. The conductive mechanism of *Chamaeleo ellioti*. The terminal cartilage of the extracolumella has been pulled out of the deep notch in the depressor mandibulae muscle.

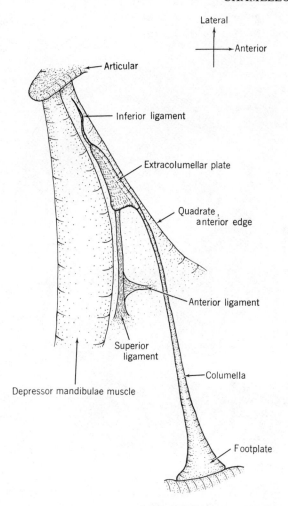

FIG. 9-8. The conductive mechanism of *Chamaeleo jacksoni.*

and the other the use of the mass of the depressor mandibulae muscle, to which extracolumellar attachments are made. Both these devices utilize the skin of the "auditory area" as a sound-receptive surface, and in these different ways convey the resulting vibrations to the columella and inner ear. The effectiveness of these means of sound reception will be brought out later in connection with cochlear potential measurements.

INNER EAR

The general orientation of the inner ear structures in chameleons is indicated in Fig. 9-9, which portrays the species *Chamaeleo chamaeleon.*

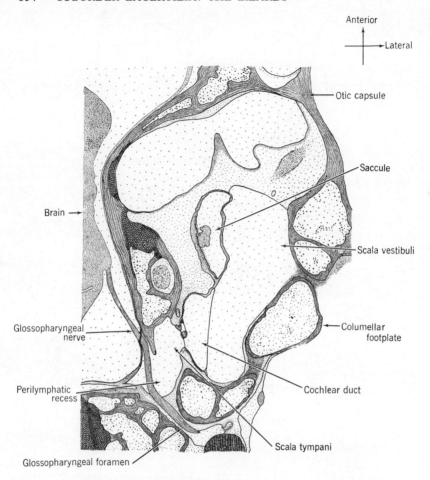

FIG. 9-9. A frontal section through the inner ear region of
Chamaeleo chamaeleon. Scale 30×.

The bony enclosure, containing the labyrinthine sense organs anteriorly
and the cochlear structures in its posterior part, is complete with relatively heavy walls except at two principal places and a third minor one.
The largest opening is the oval window in the posterolateral wall, and
contains the columellar footplate. A smaller opening appears at the posteromedial corner, and is occupied by the perilymphatic recess, a fluid
space that intrudes into the brain cavity and is separated from that cavity only by a part of the arachnoid membrane. The perilymphatic recess
leads posterolaterally to the glossopharyngeal foramen, and along this
pathway the glossopharyngeal nerve passes from its origin at the medulla to its exit into the middle ear cavity.

The chameleon inner ear is of extremely simple form. The limbic plate along the medial wall of the cochlear duct is only moderately thickened, as seen in Fig. 9-9 and further in Fig. 9-10, and its medial (or more strictly anteromedial) portion does not display a prominent elevation or crest, but only a slight thickening of the epithelial layer where the tectorial membrane arises.

A small mass of tectorial tissue lies on the medial limbus and sends forth usually two processes posteriorly, one that extends as a slender strand to the region of the auditory papilla, and another strand that runs close along the limbic surface for a short distance, or in certain instances continues to the medial end of the papilla and attaches on its surface.

The main tectorial strand connects to a tectorial structure over the hair cells of the papilla whose form varies considerably with species. In *Chamaeleo chamaeleon*, in which this tectorial structure is simplest among the species studied, the connecting strand continues above the papilla and gives off three or four branch fibers, which with a few fine cross filaments form a simple network or endbrush, as shown in Fig. 9-10. The tips of the ciliary tufts of the hair cells attach to the elements of this network.

Most of the other species show networks of somewhat more complex form, containing more long fibers and more numerous cross connections. *Chamaeleo chamaeleon calcarifer* (see Fig. 3-21 above), *Chamaeleo höhnelii, Chamaeleo jacksoni,* and *Chamaeleo ellioti* show such networks, with the complexity increasing roughly in the order named. In *Chamaeleo fischeri tavetensis* the structure has a fanlike form, with the terminal fibers spreading out from the connecting strand in the manner shown in Fig. 9-11.

The form displayed by *Chamaeleo senegalensis, Chamaeleo dilepis,* and *Chamaeleo quilensis* seems to be a modification of the simple dendritic type. In *Chamaeleo senegalensis,* as shown in Fig. 9-12, the terminal structure consists of a brushlike body with a top piece that is an extension and thickening of the connecting strand, and running from it are several thick strands that descend obliquely to the papillar surface. Many of these descending strands connect to single ciliary tufts. The more peripheral strands, however, often attach to more than one, sometimes to as many as four.

In a series of sections that passed through the ear of a specimen of *Chamaeleo senegalensis* in the sagittal plane, one section by a fortunate chance showed almost the whole papilla, and a drawing of this section, based on photomicrographs and study under the microscope, is presented in Fig. 9-13. The view is somewhat oblique, with the hair-cell surface of the papilla tipped a little toward the observer, and with its deeper portion containing numerous supporting cells. The focus is su-

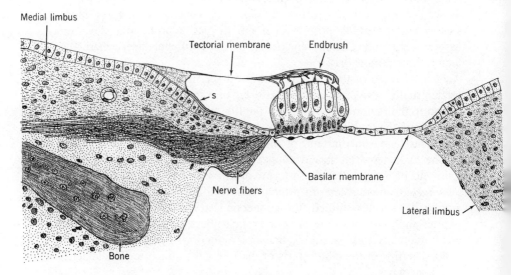

FIG. 9-10. The auditory papilla of *Chamaeleo chamaeleon*.
s represents a secondary strand of the tectorial membrane. Scale 300×.

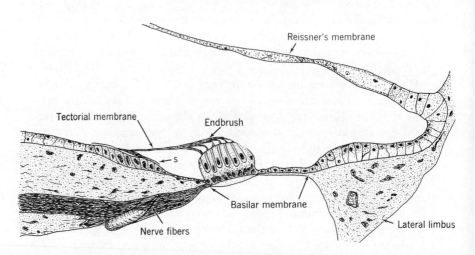

FIG. 9-11. The auditory papilla of *Chamaeleo fischeri tavetensis*.
s represents a secondary strand of the tectorial membrane. Scale 250×.

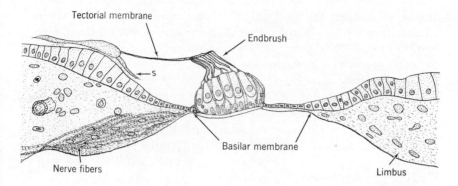

FIG. 9-12. The auditory papilla of *Chamaeleo senegalensis*, seen in cross section. *s*, secondary strand. Scale 250×.

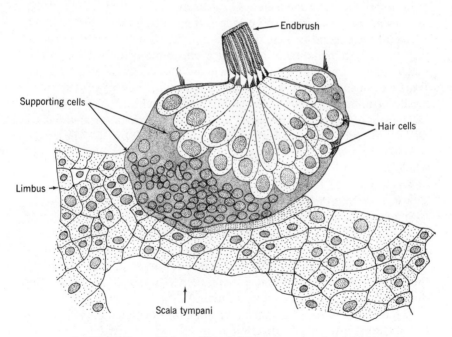

FIG. 9-13. The auditory papilla of *Chamaeleo senegalensis* in an obliquely lateral view. Scale 500×.

perficial, revealing the most lateral hair cells running obliquely upward to their attachments to the ends of the tectorial strands. Beneath these lateral hair cells are the basal ends of other, more deep-lying cells whose ciliated ends emerge in rows medial to the lateral row pictured here (and to be seen by adjusting the focus). Altogether six such rows could be made out, with the widest row containing seven cells. There were four other ciliary tufts outside this main group and without tectorial connections; two of these are shown in the figure. These tufts belong to cells far medial to the plane of the drawing.

In this specimen 17 vertical tectorial strands could be counted, and there were 41 hair-cell nuclei. Many of the tectorial strands were seen to connect successively to the tufts of two or three hair cells, and in at least one instance to four.

Cochlear Dimensions.   Detailed measurements of the dimensions of the cochlear structures were carried out in two species, *Chamaeleo senegalensis* and *Chamaeleo fischeri tavetensis*.

In a specimen of *Chamaeleo senegalensis*, represented in Fig. 9-14, the basilar membrane was 160 μ long and attained a maximum width of 132 μ a little beyond its midpoint. The fundus is considerably narrower than the basilar membrane, attaining a maximum of only 55 μ in the middle and decreasing rapidly at the ends. The thickness of the fundus (part *b* of this figure) varies irregularly up to 6 μ, and this structure was hardly identifiable except in the middle of the cochlea.

The size of the papilla increases rapidly to a maximum near the middle of the cochlea, where it reaches a value of 3400 sq μ. The cross-sectional area of the sensory portion is nearly as great as that of the whole structure.

The number of longitudinal rows of hair cells varies from 3 to 7 as shown, reflecting the simple form of the papilla.

Similar data are given for the species *Chamaeleo fischeri tavetensis* in Fig. 9-15. The length of the basilar membrane is slightly greater, reaching 180 μ. The width of this membrane reaches a maximum of 145 μ, and that of the fundus has a maximum of 71 μ, both exceeding these dimensions in *Chamaeleo senegalensis* by small amounts. The thickness of the fundus has a maximum of 10 μ.

The area of the papilla attains a maximum in the middle of the cochlea of 3000 sq μ, mainly determined by its sensory portion. This size is a little smaller than in *Chamaeleo senegalensis*, but hardly significantly so.

The number of longitudinal rows of hair cells varies from 2 to 7, representing a very simple sensory organ.

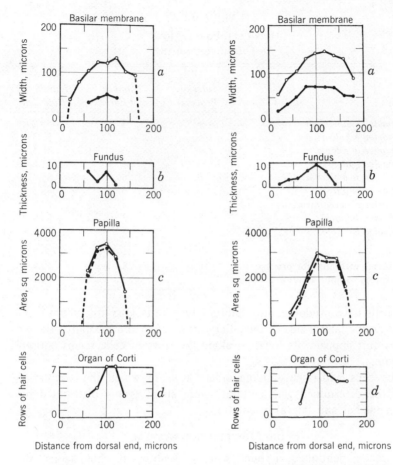

FIG. 9-14. Cochlear dimensions in a specimen of *Chamaeleo senegalensis*.

FIG. 9-15. Cochlear dimensions in a specimen of *Chamaeleo fischeri tavetensis*.

Numbers of Hair Cells. The size of the hair-cell population was determined in most of the specimens, and averages are reported in Table 9-I. These numbers varied as much between individual ears of a given species as between the species. The average for the 51 ears examined is 50.7.

Ciliary Orientation Pattern. In the chameleon papilla nearly all the hair cells have tectorial connections to their ciliary tufts, and only a few at the periphery of the structure have free-standing cilia. The tectorial

TABLE 9-I

HAIR-CELL POPULATIONS
IN CHAMAELEONIDAE

Species	Anterior Process	Number of Ears	Mean Number of Hair Cells
Chamaeleo chamaeleon	+	4	45.2
Chamaeleo c. calcarifer	+	7	57.3
Chamaeleo d. dilepis	+	6	48.5
Chamaeleo quilensis	+	5	51.6
Chamaeleo senegalensis	+	6	45.4
Chamaeleo ellioti	−	6	42.5
Chamaeleo fischeri tavetensis	−	4	57.7
Chamaeleo höhnelii	−	7	54.9
Chamaeleo jacksoni	−	6	53.2
All nine forms		51	50.7
Five forms with anterior process		28	49.9
Four forms without anterior process		23	51.7

hair cells have predominantly lateral ciliary tufts (the kinocilium is located on the lateral side of the ciliary array). When a medially oriented ciliary tuft appears, its location along the rows of cells seems entirely random.

The cells with free-standing ciliary tufts in which the orientation could be determined with certainty were about evenly distributed between medial and lateral orientations.

FLUID MOBILIZATION SYSTEM

As already mentioned, a round window is absent in chameleons. Yet their ears, though showing levels of sensitivity well below average for lizards in general, are not as profoundly deaf as they would be if no means at all were provided for mobilization of the cochlear fluids. A search therefore was made for a substitute pathway from scala tympani to middle ear region that would give the mobilization required, and such a path was found. It is, however, a somewhat narrow and hindered one.

As shown in Fig. 9-9, the scala tympani opens broadly in a posterior direction into a cavity in the exoccipital bone that extends medially to the brain space and is separated from this space only by a thin membrane that is a part of the arachnoid. This cavity is the perilymphatic recess, and it extends posterolaterally through the occipital bone with varying conformations in the different chameleon species. In *Chamaeleo chamaeleon calcarifer*, as shown in Fig. 9-16, it extends about halfway

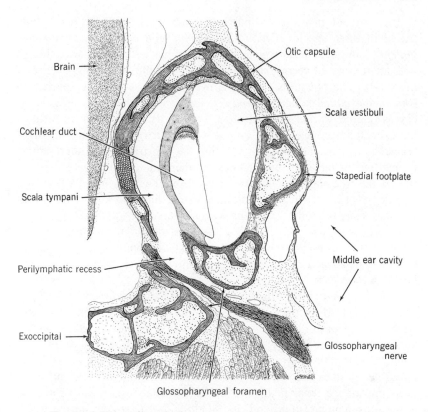

Brain

Otic capsule

Cochlear duct

Scala vestibuli

Scala tympani

Stapedial footplate

Perilymphatic recess

Middle ear cavity

Exoccipital

Glossopharyngeal
nerve

Glossopharyngeal foramen

FIG. 9-16. The perilymphatic recess and glossopharyngeal foramen of
Chamaeleo chamaeleon calcarifer.

through this bone as a relatively wide opening and then narrows greatly
over the remaining distance to the glossopharyngeal foramen.

This passage transmits the glossopharyngeal nerve, but is not entirely
filled by it. Fluid spaces occur alongside the nerve, and a reconstruction
of the region showed that these fluid spaces have enough continuity to
permit a surging of fluid back and forth under the influence of sound
pressures exerted at the columellar footplate.

This is a pathway for fluid mobilization that is in some sense intermediate between the usual round window type and a complete reentrant
fluid circuit as found in turtles, snakes, and others (see Chapter 4). A
fluid pathway leads from the scala tympani through the perilymphatic
duct and sac and then through the glossopharyngeal foramen to the middle ear space, and then is continued by a short aerial link in the middle
ear to the columellar footplate.

This path of pressure discharge through the ear has been carefully traced in three species, *Chamaeleo chamaeleon calcarifer, Chamaeleo quilensis*, and *Chamaeleo senegalensis*, and has been noted also, though without the use of reconstruction methods, in the remaining species of this study. The openness of the pathway clearly varies with species. In *Chamaeleo quilensis* and *Chamaeleo senegalensis* the perilymphatic recess is relatively wide through about two-thirds of the distance through the exoccipital bone, leaving only a short constricted portion. In other species this open passage is shorter and the constricted portion amounts to one-half or more of the thickness of the bone. Also variable are the size of the free space around the nerve in its passage and the presence and extent of nerve sheaths and other tissues that tend to obstruct vibratory movements along this path. Again *Chamaeleo quilensis* and *Chamaeleo senegalensis* show less of this obstruction than other species. Figure 9-17 represents the passage through the bone in a specimen of *Chamaeleo senegalensis*, and Fig. 9-18 shows the corresponding region in *Chamaeleo jacksoni*.

As already indicated, this substitute for the round window makes possible a mobilization of the cochlear fluids, but hardly with ideal efficiency. The vibratory movements of the fluids are not as free as in lizards with a round window covered with a thin, flexible membrane and bounded externally by an air space. The movements involve the interposed membranes and various tissue elements along the path, and incur mass impedance from this cause. Also frictional resistance is involved in the movements of fluid and tissues along the narrow enclosing walls. The frictional effect reduces the amplitude of vibratory motion and impairs the ear's sensitivity. The mass effect also reduces sensitivity, especially for the higher frequencies. The result is an ear that may perform reasonably well in the low frequencies, but can be expected to be lacking in the upper range.

The effectiveness of the chameleon ear in the reception of sounds can be judged from the results of a series of experiments in which the sensitivity was measured in terms of the cochlear potentials.

AUDITORY SENSITIVITY

In a few experiments measurements were made of the ear's sensitivity without special procedures, and the following figures show results for five species.

In Fig. 9-19 are given aerial sensitivity curves for two specimens of *Chamaeleo senegalensis*. These two curves agree in indicating best sensitivity in the low frequencies, in the region of 200 to 600 Hz, and a rapid decline in the higher frequencies.

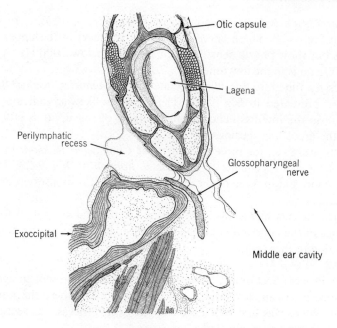

FIG. 9-17. The passage through the exoccipital bone in
Chamaeleo senegalensis.

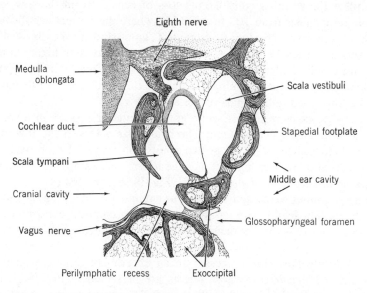

FIG. 9-18. The passage through the exoccipital bone in
Chamaeleo jacksoni.

Similar results for two specimens of *Chamaeleo quilensis* are presented in Fig. 9-20. These curves resemble the others in the upper frequencies but show poorer sensitivity for all tones below 2000 Hz. Again the best region is in the low tones.

Sensitivity functions for two specimens of *Chamaeleo fischeri tavetensis* are presented in Fig. 9-21. These curves likewise indicate best sensitivity in the low frequencies, but the level of sensitivity is still less than in the preceding species.

Results are given for three specimens of *Chamaeleo jacksoni* in Fig. 9-22 and for two specimens of *Chamaeleo höhnelii* in Fig. 9-23. These ears are poor, and in three of the specimens the best responses are in the high frequencies.

Further measurements of sensitivity were made along with surgical alterations of the conductive mechanisms designed to reveal the manner of operation of these ears.

Anterior Process Sectioning.   An anterior process is present in certain of the species, as already noted, and in a number of these the normal sensitivity curve was first obtained, and then this process was sectioned and the measurements repeated.

Shown in Fig. 9-24 are results obtained in a specimen of *Chamaeleo dilepis*. After the columella was freed from its connection with the pterygoid plate there was a considerable loss of sensitivity in the low and middle frequencies from 100 to 2500 Hz, where the normal ear showed the best reception. The loss was greatest, amounting to 41 db, at 400 Hz, and averaged 27 db over the range mentioned. For higher tones there were only moderate effects of this sectioning. Clearly the pterygoid mechanism in this species is highly effective in sound reception throughout the lower and middle part of the auditory range.

Fig. 9-25 shows results of sectioning the anterior process in a specimen of *Chamaeleo chamaeleon*. The result was a loss in the low-tone region that reached 17-18 db at two points and averaged 8 db over the range 100-800 Hz. For higher tones the losses were small and irregular. In a second specimen of this species (Fig. 9-26), the losses as a result of this sectioning were even smaller but extended over a broader range. It appears that in this species the pterygoid mechanism contributes to the sensitivity to a varying degree in different specimens.

Sectioning the Posterior Process.   In an experiment on a specimen of *Chamaeleo senegalensis*, whose results are presented in Fig. 9-27, the normal curve was first obtained, then the posterior process was sectioned, and finally the anterior process was sectioned also. The first of these operations gave the function indicated by the short-dashed curve,

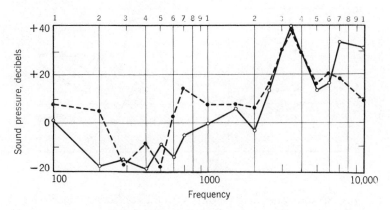

FIG. 9-19. Sensitivity functions for two specimens of *Chamaeleo senegalensis*, with aerial stimulation. From Wever, 1968b.

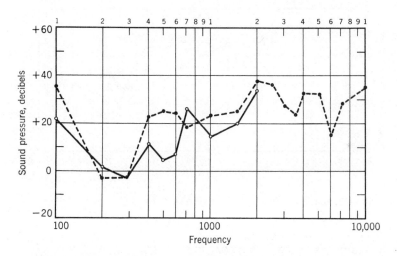

FIG. 9-20. Sensitivity functions for two specimens of *Chamaeleo quilensis*, with aerial stimulation.

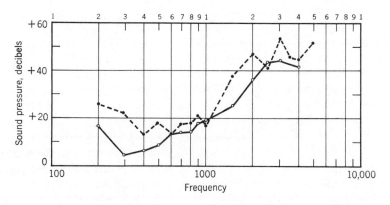

FIG. 9-21. Sensitivity functions for two specimens of *Chamaeleo fischeri tavetensis*.

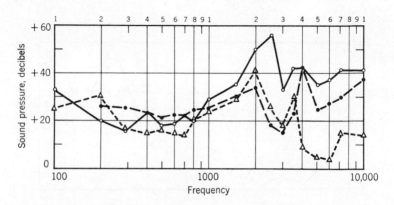

FIG. 9-22. Sensitivity functions for three specimens of *Chamaeleo jacksoni*. From Wever. 1969b.

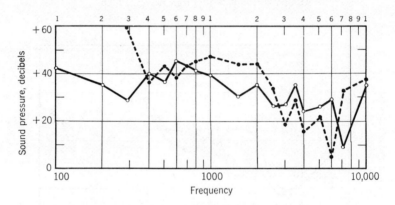

FIG. 9-23. Sensitivity functions for two specimens of *Chamaeleo höhnelii*.

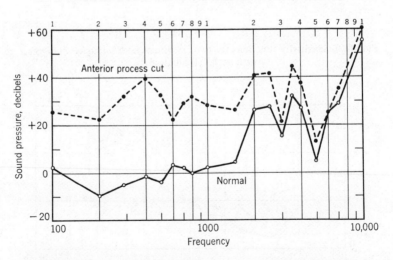

FIG. 9-24. Sensitivity functions for a specimen of *Chamaeleo dilepis* under normal conditions (solid line) and after sectioning the anterior process (broken line).

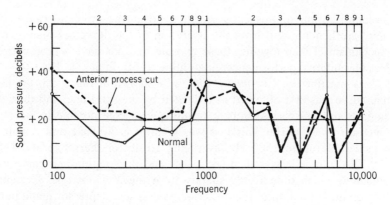

FIG. 9-25. Sensitivity functions for a specimen of *Chamaeleo chamaeleon* before and after sectioning the anterior process.

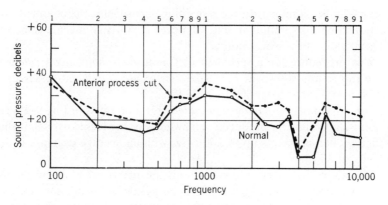

FIG. 9-26. A second specimen of *Chamaeleo chamaeleon* in which the anterior process was sectioned.

in which large losses occur for the tones 200 and 290 Hz, and moderate losses for higher tones up to 3000 Hz. The decline for 200 Hz amounts to 30 db, that for 290 Hz is 18 db, and for the range 400-2500 it averages 8 db. For still higher tones the results are irregular, or even show reversals. It appears that in this species the transmission through the depressor mandibulae muscle and the terminal cartilage of the extracolumella has a moderately beneficial effect over a good deal of the range up to 2500 Hz, and has a strong effect around 200-290 Hz. In the next step of the procedure, when the anterior process was sectioned, the loss for the low and middle tones was great as shown by the long-dashed curve, amounting to 45 db at 600 Hz and averaging 26 db for the range up to 2000 Hz. It is clear that the pterygoid mechanism makes the major contribution to this ear's sensitivity over the principal range, but there are benefits also from the operation of the depressor mandibulae mechanism. Taken together these two mechanisms have a profound effect on auditory performance, amounting to 50 db at 600 Hz.

In a specimen of *Chamaeleo dilepis*, as shown in Fig. 9-28, the normal curve was first obtained as indicated by the solid line, and then the anterior process was sectioned, resulting in a considerable loss for all tones up to 1500 Hz. This loss was greatest at 400 Hz and remained considerable up to 1500 Hz, averaging 16 db for the range 300-1500 Hz. The columella was then sectioned, and the long-dashed curve was obtained, which differs only slightly from the previous one. Evidently the conductive action of the posterior process and depressor mandibulae muscle, which is excluded by this second operation, was contributing little to sound reception in this ear.

Columella Sectioning.   In the remaining experiments the columella was sectioned near its midpoint, and whatever conductive mechanisms were present were put out of action.

This procedure for a specimen of *Chamaeleo chamaeleon calcarifer* gave the results shown in Fig. 9-29. There are significant losses up to 28 db over a range from 600 to 2000 Hz.

. For a specimen of *Chamaeleo quilensis*, represented in Fig. 9-30, sectioning the columella produced somewhat varying losses over the whole range of measurements up to 2000 Hz, with an average decrease in sensitivity of 17 db.

The above two species possess both the pterygoid plate mechanism and an extracolumellar plate running alongside the depressor mandibulae muscle, and the losses after sectioning the columella represent the elimination of both sound-conductive systems.

In the next three species in which this columellar sectioning was carried out the anterior process is lacking, and the reception of sounds must depend on transmission through the depressor mandibulae muscle and the posterior process. The results for a specimen of *Chamaeleo ellioti* are shown in Fig. 9-31. The loss of sensitivity is general, with an average of 12 db over the range up to 1500 Hz. Somewhat less regular results were obtained for a second specimen of this species, as shown in Fig. 9-32; but for the lowest tones and those between 800 and 3500 Hz, the losses were significant.

Results of this procedure for a specimen of *Chamaeleo höhnelii* are shown in Fig. 9-33. The curve after the sectioning takes much the same course as before, though there are differences at a few points. It appears that the action of the columellar system in this species contributes little to the reception of aerial sounds.

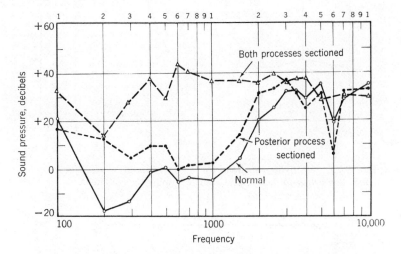

FIG. 9-27. Sensitivity functions for a specimen of *Chamaeleo sene-galensis* under three conditons, with the ear normal (solid line), after the posterior process was sectioned (short-dashed line), and after the anterior process also was sectioned (long-dashed line). From Wever, 1968b.

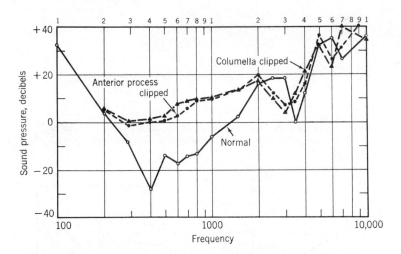

FIG. 9-28. Sensitivity functions for a specimen of *Chamaeleo dilepis* in which a normal curve was obtained (solid line), the anterior proc-ess was sectioned (short-dashed line), and finally the columella was sectioned also (long-dashed line).

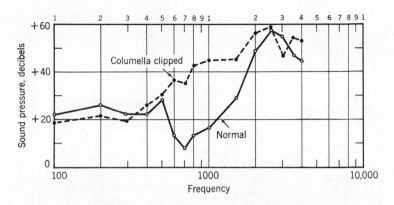

FIG. 9-29. Sensitivity functions for a specimen of *Chamaeleo cha-maeleon calcarifer* before and after sectioning the columella.

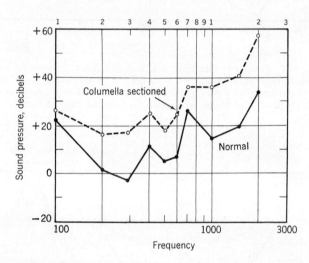

FIG. 9-30. Sensitivity functions for a specimen of *Chamaeleo quilensis* before and after sectioning the columella. From Wever, 1968b.

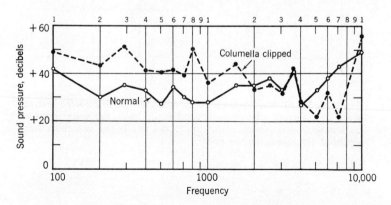

FIG. 9-31. Sensitivity functions for a specimen of *Chamaeleo ellioti* before and after sectioning the columella.

The third experiment in this series, carried out on a specimen of *Chamaeleo jacksoni*, gave results shown in Fig. 9-34. In this instance the losses after sectioning the columella are small and limited to a few frequencies, and occur mostly in the medium high-tone range. These changes are of doubtful significance. In this species the anterior process is lacking, and the terminal cartilage of the extracolumella makes only superficial contact with the depressor mandibulae muscle. Evidently the transmission of sound to this ear is merely by bone and tissue conduction.

Vibratory Stimulation

The action of the chameleon ear was further studied by the use of vibratory stimulation applied in a number of ways.

In one experiment the vibrator with a plastic tip of 2 mm diameter was used to explore the surface of the head and especially the "auditory area" as represented in Fig. 9-2. With a stimulating frequency of 600 Hz and a constant input level, the cochlear potential output was observed with numerous locations of the vibrating tip. The stimulation was only moderately effective at points along the edge of the lower jaw over the articular bone. The most advantageous position was at the dorsal end of the "auditory area," close to the end of the quadrate bone. The high degree of effectiveness of stimulation in this region is probably due to the ready conduction of vibrations to the otic capsule by way of the quadrate and the paroccipital process of the exoccipital bone. Of intermediate effectiveness was application of the vibrator to the middle of the auditory area where it contacted skin and the thin muscle layer lying over the pterygoid plate.

In a number of experiments both aerial and vibratory stimulation was used on the same ear, often with the vibrator in contact with the cut end of the columella so that the driving was "axial," along the line of the columellar shaft. Results of such observations on a specimen of *Chamaeleo quilensis* are shown in Fig. 9-35. This ear with aerial stimulation exhibits good sensitivity in a narrow region within the low frequencies, and then takes a somewhat wavering course showing poor sensitivity in the high frequencies. Vibratory stimulation gave a more uniform function through the low and middle frequencies, and then showed some sharp variations in the upper range. Such variations in the high frequencies are common with vibratory stimulation, and no doubt are mainly due to complex resonances.

A direct comparison of aerial and vibratory functions is difficult to make because the two forms of stimulation are expressed in different physical terms: in terms of sound pressure for aerial sounds and in terms of amplitude of motion for vibratory stimuli. These different di-

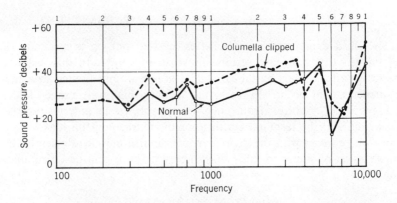

FIG. 9-32. Effects of columella sectioning in a second specimen of *Chamaeleo ellioti.*

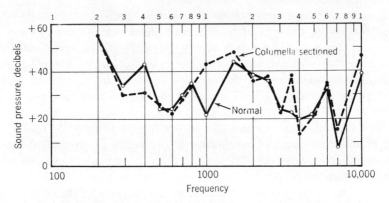

FIG. 9-33. Sensitivity functions for a specimen of *Chamaeleo höhnelii* before and after columella sectioning. From Wever, 1969b.

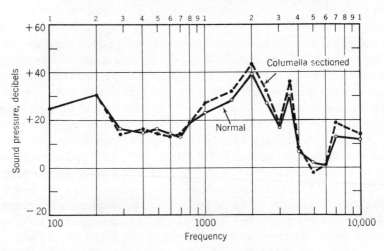

FIG. 9-34. Sensitivity functions for a specimen of *Chamaeleo jacksoni* before and after columella sectioning. From Wever, 1969b.

mensions of sounds can be equated only when the impedances of the vibrating structures are known. In an aerial medium where the impedance is determinable and is constant with frequency we can say that for sounds of constant pressure the amplitude of movement of the air particles varies inversely with frequency. The ear's impedance, however, is not known for either aerial or vibratory stimulation, and most certainly is not constant with frequency. For aerial stimulation this impedance can be assumed to increase with frequency in a general way, but not in a uniform manner because of the presence of resonances in the moving structures. For vibratory stimulation the ear's impedance varies also, but in a manner very different from that for aerial stimulation because of the presence of the vibrator closely coupled to the structures. However, with the vibrator used here, for which the mechanical impedance was extremely great, the impedance characteristics of the driven structures can be neglected; they are small in relation to the impedance of the vibrator itself.

Because of these limitations, a comparison of aerial and vibratory functions like those of Fig. 9-35 becomes largely empirical; we have to discover the forms of the two kinds of curves and then to assess any marked variations from the usual picture.

The limitation just indicated does not impair the use of vibratory stimulation in comparisons between different animals, or comparisons involving different ways of applying these stimuli.

Figure 9-36 compares the sensitivity curves obtained with vibratory stimulation in *Chamaeleo quilensis* and *Chamaeleo senegalensis*, with the vibrating needle applied axially to the cut end of the columella. With this form of stimulation *Chamaeleo senegalensis* maintains the considerable superiority of sensitivity already shown in the tests with aerial sounds.

The axial mode of stimulation of the cut end of the columella with a vibrating needle, as already described, is the most convenient one to carry out, because the vibrator tip has only to be inserted into the middle ear cavity and brought to the proper position with a manipulator. The columellar footplate is then caused to move in and out of its window like a piston in a cylinder.

There is reason to believe, however, that this piston movement is not the natural mode of action of the system in those chameleon species with the pterygoid mechanism. The anatomical relations in these ears, as pictured in Fig. 9-5, suggest that a rocking action is produced at the footplate when the pterygoid plate is vibrated in and out. If the footplate is firmly held at one edge of the oval window, and is relatively free to move at the opposite edge, then it will operate like a hinged cover. Then if this plate has a symmetrical form (which is at least ap-

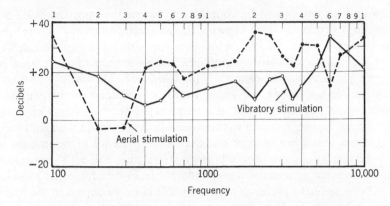

FIG. 9-35. Sensitivity functions for a specimen of *Chamaeleo quilensis* obtained by aerial and vibratory stimulation. From Wever, 1968b.

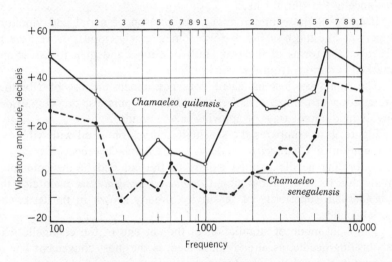

FIG. 9-36. Sensitivity for vibratory stimulation in *Chamaeleo senegalensis* and *Chamaeleo quilensis*. From Wever, 1968b.

proximately the case), the volume displacement of the cochlear fluids will be the same as if the whole plate moved uniformly at the amplitude imparted to its center.

In an experiment whose results are presented in Fig. 9-37, the columella of a specimen of *Chamaeleo senegalensis* was driven in two ways, first axially and then mediolaterally, i.e., with the vibrator applied in a direction perpendicular to the columellar axis and in such a way as to produce motion about the anterolateral border of the columellar footplate. The results show that the mediolateral method of driving is a very effective one, producing a curve showing sensitivity about 20 db greater than that afforded by axial driving in the frequency region from 100 to 700 Hz. For higher tones this advantage is lost, and the two functions are hardly distinguishable. Most likely at these high frequencies the motion of the footplate is different.

In Fig. 9-38 results are given for an experiment on a specimen of *Chamaeleo jacksoni* in which vibratory stimuli were delivered in two ways, to the side of the head over the quadrate bone and at a point on the remaining portion of the sectioned columella. The driving of the columella was mediolateral, at right angles to its shaft. It is somewhat surprising that head stimulation was the more effective, in an average amount of about 8 db over the range up to 1000 Hz. For higher tones up to 2500 Hz this relation is reversed, and the response is highly irregular at the upper end of the range.

The low effectiveness of this form of columellar driving in *Chamaeleo jacksoni* is not fully explained, but a suggestion arises from an examination of the footplate in the oval window of this species. This footplate appears to be more centrally located in its window than in other chameleon species, with a relatively wide annular ligament. There is a possibility that with this arrangement a lateral displacement of the shaft tends to produce a simple rocking of the footplate about its center, so that one edge is raised while the other is depressed, and little displacement of the cochlear fluid occurs. Further experiments will be necessary to test this possibility.

Another experiment, carried out on a specimen of *Chamaeleo höhnelii*, was concerned with the effects of sectioning the columella on responses to vibratory stimuli applied to the side of the head over the ventral end of the quadrate bone. As shown in Fig. 9-39, this sectioning produced a moderate reduction in sensitivity, amounting to about 6 db on the average, over the range from 100 to 2500 Hz. In this species the terminal cartilage of the extracolumella is inserted about halfway into the mass of the depressor mandibulae muscle, and a likely pathway of transmission of vibrations applied over the end of the quadrate is through the retroarticular process to the muscle, then by way of the terminal cartilage and columella to the inner ear. Obviously other bone-conduction pathways are nearly as effective, as eliminating this one through the columella causes only a slight loss of sensitivity.

Peripheral Determinants of Chameleon Sensitivity. The variations in auditory sensitivity among the different chameleon species are clearly related to the forms of special sound-receptive and conductive mechanisms developed in these animals. The most sensitive species are *Chamae-*

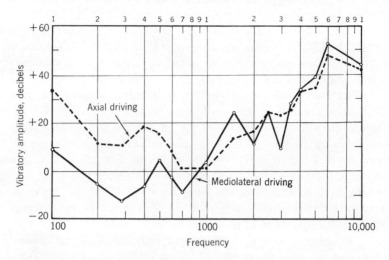

Fig. 9-37. Vibratory stimulation in a specimen of *Chamaeleo sene-galensis* produced by two methods of driving the columella: in line with the columellar axis and at right angles to this axis. From Wever, 1968b.

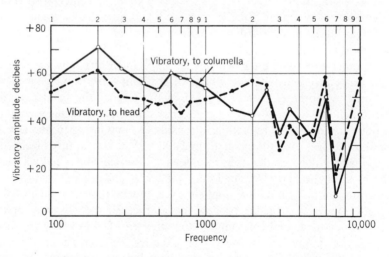

Fig. 9-38. Vibratory stimulation in a specimen of *Chamaeleo jacksoni* produced by two methods, with the vibrating tip applied to the head at the quadrate bone and to the side of the columellar shaft. From Wever, 1969b.

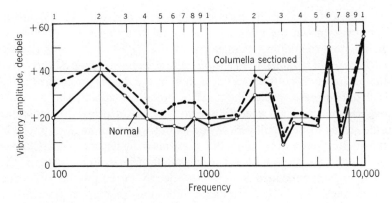

FIG. 9-39. Sensitivity curves in a specimen of *Chamaeleo höhnelii* with vibratory stimulation before and after sectioning the columella. From Wever, 1969b.

leo d. dilepis, Chamaeleo quilensis, and *Chamaeleo senegalensis* (which may be referred to as the *dilepis* group), showing for the specimens studied an average sensitivity of −10.6 db. These species have an anterior process arising from the stiff part of the columella immediately distal to its point of transition from bone to cartilage, and the anterior process inserts on the osseous portion of the pterygoid plate. In *Chamaeleo chamaeleon* and *Chamaeleo chamaeleon calcarifer* the sensitivity is considerably less, with a mean value of +3.5 db, and in these animals the anterior process arises farther distally, from the cartilaginous extension of the columella, and its insertion is on the membranous portion of the pterygoid plate. In *Chamaeleo chamaeleon* the osseous portion of this plate does not extend as far medially as in the other species, and the anterior process remains particularly remote from it.

It should be mentioned that in one specimen of *Chamaeleo chamaeleon calcarifer* the anterior process was lacking on one side, and this ear had much reduced sensitivity, measured as +28 db. This ear is considered as anomalous, and has not been included in the sensitivity averages for this species.

In all the above species the posterior process of the columella shows superficial relations with the depressor mandibulae muscle, running between its middle portion and the quadrate, but without any direct connections with the muscle. The posterior process expands to form a plate lying close over the surface of the quadrate, and from this plate ligaments extend for varying distances toward the two ends of the quadrate. These attachments seem to constitute a distal anchorage of the columellar system that serves as a pivot line, so that, when the pterygoid plate vibrates carrying the anterior process in and out, the columellar system twists about the pivot and the footplate rocks in its window.

In the four remaining chameleon species—*Chamaeleo ellioti, Chamaeleo höhnelii, Chamaeleo jacksoni*, and *Chamaeleo fischeri tavetensis* —the sensitivity is considerably less than in the *dilepis* group, averaging +18.1 db. In these species the anterior process is absent, and an alternative means of sound reception involving the depressor mandibulae muscle seems to be employed. When sound waves impinge on the skin of the auditory area over the middle portion of the depressor mandibulae muscle this mass of tissue is set in vibration. The terminal blade of the extracolumella, embedded more or less deeply in the muscle, is carried along in the action and transmits the vibrations to the footplate in the oval window.

Support for this inference comes from the observation that the losses of sensitivity that occur in *Chamaeleo ellioti, Chamaeleo höhnelii*, and *Chamaeleo jacksoni* as a result of sectioning the columella decrease progressively in the order named, and this order corresponds to the decreasing depth of penetration of the depressor mandibulae muscle by the extracolumellar process in these three species. (The species *Chamaeleo fischeri tavetensis* cannot be located in this series because the procedure of columella sectioning was not carried out on these specimens.)

There is reason to believe that the columellar mechanism in the *dilepis* group involves a lever action, with the anterior process forming the primary arm of the lever and the footplate forming the secondary arm. The arrangement is that of a reducing lever, and the decrease in amplitude is accompanied by a corresponding increase in sound pressure, which is favorable to a transfer of vibratory energy from an aerial medium.

The action of the second mechanism involving the depressor mandibulae muscle is more likely a direct plunger effect, with the movements of the superficial mass of tissue communicated to the columella so that its footplate moves in and out of the oval window, without any lever action.

The evidence shows that this second receptive system is much less effective than the first. Eliminating the pterygoid plate mechanism by cutting the anterior process produces an average loss of responsiveness of about 22 db, whereas eliminating the depressor mandibulae mechanism gives an average loss of about 7 db. The average transmission difference between the two mechanisms is thus about 15 db.

Also varying with species is the degree of patency of the pathway through the glossopharyngeal foramen. The impression gained from examining these pathways in serial sections is that they are somewhat more open in the *dilepis* group than in the others. No measurements of the acoustic impedance of this pathway have been made in the different species—although such determinations are possible.

The observation that the *dilepis* group differs from the other group in overall sensitivity by about 28 db, whereas the columella sectioning experiments indicate transmission differences of 15 db between the two, leaves a considerable variation still to be accounted for. At least a part of this difference can be ascribed to impedance in the glossopharyngeal pathway.

Sensitivity does not appear to be related to the size of the hair-cell population in these animals. This size varies among individuals of a species about as widely as across species, and the species averages are remarkably similar. As shown in Table 9-I, the mean of 53 specimens of all species studied was 50.7 hair cells. For species possessing an anterior process this average was 49.9 and for those without this process it was 51.7—an insignificant variation.

It will be noted that the better hearing species of chameleons are in F. Werner's Group 1, and the others are in his Groups 3 and 5. It would be of interest to investigate still other members of the genus *Chamaeleo*, and species belonging to other genera as well, to discover how widespread are the two special devices for sound reception here noted, and whether still other mechanisms have been developed among the chameleons.

10. ANGUID LIZARDS AND
THEIR RELATIVES

Grouped together as the Anguoidea are three lizard families—one of moderate size, the Anguidae, and two others that are very small, the Anniellidae and Xenosauridae.

FAMILY ANGUIDAE: THE ANGUID LIZARDS

The lizards of the family Anguidae form an assemblage of over 70 species of scattered distribution and diverse body types. The living forms belong to three subfamilies, the Anguinae, Gerrhonotinae, and Diploglossinae. The Anguinae are limited to a single genus, *Anguis*, occurring in Europe. The Gerrhonotinae include five genera: *Ophisaurus, Gerrhonotus, Abronia, Barisia*, and *Elgaria*, though some authorities subsume the last three under *Gerrhonotus*. Most of these species are found in America, especially in the western United States and Mexico, but some of the *Ophisaurus* species, along with *Anguis*, occur in the Old World. The Diploglossinae include four or five genera sparingly distributed in parts of the United States, Mexico, Central America, and the West Indies.

These lizards show wide variations in body types. *Anguis* is a legless form, known in Europe as the slowworm. The several species of *Ophisaurus* are also legless and snakelike, and are commonly known as glass snakes or glass lizards because of the fragility of the long tail. There are a number of species of *Gerrhonotus*, usually called alligator lizards because of their strong bodies and very short legs. One of these, the foothill alligator lizard *Gerrhonotus multicarinatus*, is represented in Fig. 10-1. The Diploglossinae, popularly known as galliwasps, have more typical body forms.

Available for study were 37 specimens representing seven species belonging to four of the above genera. These were the following: *Anguis fragilis* (Linnaeus, 1758)—3 specimens; *Barisia gadovii* (Boulenger, 1913)—6 specimens; *Gerrhonotus multicarinatus* (Blainville, 1835)—19 specimens; *Ophisaurus apodus* (Pallas, 1775)—2 specimens; *Ophisaurus attenuatus longicaudus* (McConkey, 1952)—1 specimen; *Ophisaurus compressus* (Cope, 1900)—1 specimen; and *Ophisaurus ventralis* (Linnaeus, 1766)—5 specimens.

FIG. 10-1. The alligator lizard *Gerrhonotus multicarinatus.*
Drawing by Anne Cox.

Earlier studies of the anguid ear include an extensive series of sensitivity measurements in terms of cochlear potentials carried out by Crowley (1964) on the alligator lizard *Gerrhonotus multicarinatus*, and the treatment of a number of species by both anatomical and electrophysiological methods by Wever (1971d). The alligator lizard also received special attention in a description of ultrastructural details of the reptile ear by Mulroy (1968, 1974) and in experiments on intracellular responses by Weiss, Mulroy, and Altmann (1974) and by Mulroy, Altmann, Weiss, and Peake (1974).

ANATOMICAL OBSERVATIONS

OUTER AND MIDDLE EAR STRUCTURES

In all species examined except *Anguis fragilis*, an external ear opening is present, though in the *Ophisaurus* species this opening is small. Mention has already been made in Chapter 6 of the variable form of the external ear opening in *Anguis fragilis*, with the western European specimens lacking this opening and those from Russia and Hungary showing at least a minute opening. In all specimens seen in the present study, this opening is absent, along with other modifications of the conductive mechanism.

Also treated in Chapter 6 is the question of meatal muscles and their actions in anguids, with the indication that these muscles are of doubtful service in closure of the external ear opening in these animals.

In all anguids studied except *Anguis fragilis* the conductive mechanism is of normal form, as shown in Fig. 10-2. This mechanism is of the scincid type, lacking the internal process that in many other lizards, such as the iguanids, runs to the quadrate and serves as a stiffening vane. The usual fold of mucous membrane extends from quadrate to columella and provides a measure of support.

INNER EAR

In anguids the basilar membrane lies over an elongated opening in the limbus and sustains an auditory papilla of moderate size. The limbic plate is somewhat thickened on the lateral side, and much more so on the medial side to form a prominent elevation as may be seen in Fig. 10-3. This figure presents a cross section of the cochlea of *Gerrhonotus multicarinatus*, and shows the tectorial membrane extending from the lateral side of the limbic crest close to its peak.

The general orientation of the anguid ear and its relation to adjacent structures are portrayed in the reconstruction drawing of Fig. 10-4. This drawing was based upon a series of sagittal sections. The peripheral portions of the drawing represent a particular section in which the otic capsule had been entered, and various structures—including saccule, utricle, and round window niche, as well as the cochlea and lagena—had been revealed. In the middle of the drawing the cochlear structures have been depicted in further depth by reference to additional sections, so that the auditory papilla appears in its setting in a hollow region of the limbus. This papilla lies on the outer side of the basilar membrane, which covers and thus obscures an oval opening that extends all the way through the limbic plate. Over the main surface of the papilla the hair cells have long ciliary tufts extending dorsolaterally and standing free of any coverings. At the lateroventral end of the papilla (the end turned toward the observer), the hair cells are surmounted by a tectorial plate, a sheet of tissue containing several perforations large and small, and to whose underside the short ciliary tufts of this region are attached. A tapered tectorial membrane extends from an elevated part of the limbus anterior and lateral to the papilla, and attaches to one edge of the tectorial plate.

The footplate of the columella is located lateral to the section shown, and when vibrated by sounds exerts its pressures on a large body of fluid in the perilymphatic cistern and scala vestibuli, which is continuous with the cavity shown in the upper part of this drawing. The effects are transmitted through Reissner's membrane to the basilar membrane, which is free to move because it is bounded by the scala tympani, not shown in the drawing but lying medial and posterior to this membrane and connecting through the flexible round window membrane to the air of the round window niche.

A further view of the auditory papilla of this species is given in Fig. 10-5, taken from a section that passed through the ventral end of the papilla where the tectorial plate is located, and then continued obliquely in a dorsal direction across the ends of the hair cells and the ciliary tufts. The ventral hair cells may be seen with their tips pointed toward the

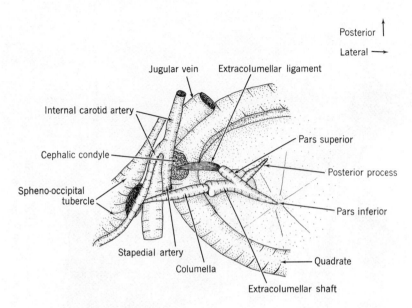

FIG. 10-2. The middle ear mechanism of *Gerrhonotus multicarinatus*.

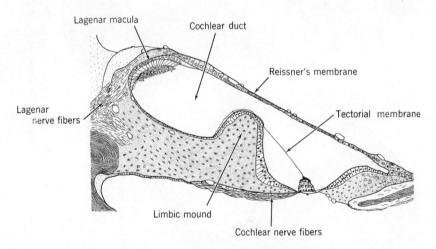

FIG. 10-3. A frontal section through the cochlear duct of
Gerrhonotus multicarinatus. Scale 75×.

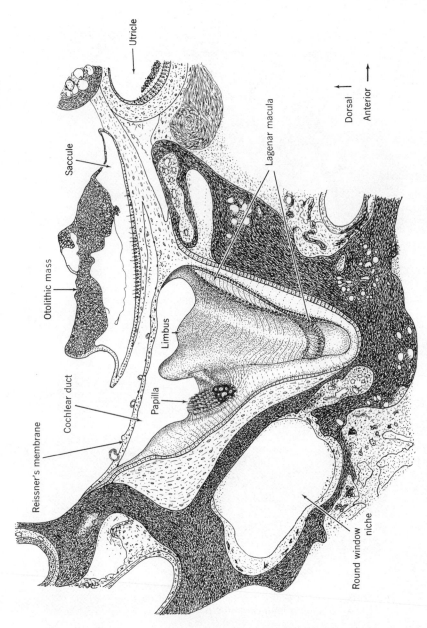

Utricle

Dorsal

Anterior →

Lagenar macula

Saccule

Otolithic mass

Limbus

Cochlear duct

Papilla

Reissner's membrane

Round window niche

Fig. 10-4. A reconstruction of the ear of *Gerrhonotus multicarinatus* in a lateral view, based on serial sections. Scale 50×. From Wever, 1971d.

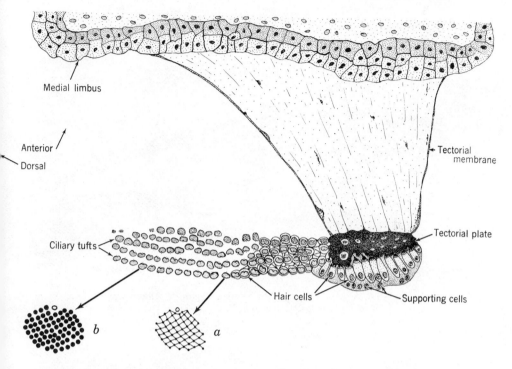

Fɪɢ. 10-5. A section through the auditory papilla of *Gerrhonotus multicarinatus* producing an oblique surface view.

tectorial plate, and in many instances extending into round perforations in this plate. For the more dorsal hair cells (those to the left of the drawing) the ciliary array is seen endwise, with its field of stereocilia and the single kinocilium lying in a notch on one side of the array. When the section passes close to the hair-cell surface (as at *a*) the cilia appear to be connected by delicate filaments, but when this section is farther from the surface and the cilia are cut through closer to their outer ends (as at *b*) they are much thicker in diameter and seem to be independent.

In the ciliary tufts illustrated, the number of stereocilia was counted as 58 at *a* and 71 at *b*. Mulroy (1968a) reported the number of stereocilia in this species as 55-70.

Other anguid species have much the same general form of inner ear structure as just indicated. A cross section through the cochlea of *Anguis fragilis* is presented in Fig. 10-6, and shows the auditory papilla near its middle where it bears a tectorial plate with a connection to the limbic mound through the tectorial membrane. The form of the papilla at a more dorsal position in this same species is represented in Fig. 10-7.

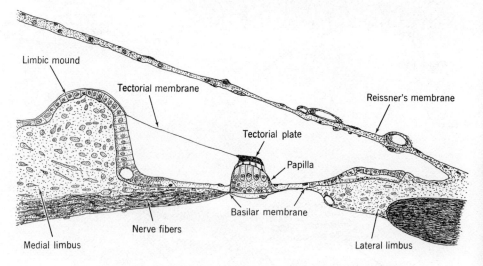

FIG. 10-6. A section through the cochlear region of *Anguis fragilis*.
Scale 200×. From Wever, 1971d.

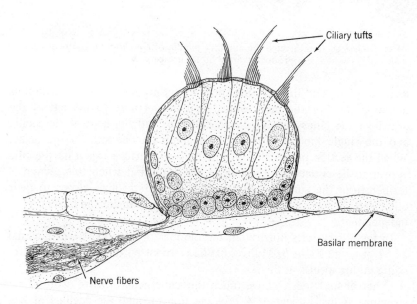

FIG. 10-7. The auditory papilla of *Anguis fragilis* in cross section through
its dorsal region. Scale 835×. From Wever, 1971d.

Similar sections are shown for the species *Ophisaurus ventralis*, with the midcochlear region represented in Fig. 10-8 and a more dorsal region shown in Fig. 10-9.

As was indicated for *Gerrhonotus multicarinatus* in Fig. 10-4, there is only one region of the anguid cochlea in which a tectorial plate is present, and the tectorial membrane extending from the limbus narrows down to make an attachment to it. The location shown is at the ventral end of the cochlea, but this location varies wtih species. In *Ophisaurus* species this plate is more toward the middle of the cochlea, and in *Ophisaurus attenuatus longicaudus* it is almost exactly in the middle. In *Anguis fragilis* the location is somewhat variable around the middle, and sometimes it is displaced a little toward the dorsal end. Figure 10-10 represents the lengths of the auditory papilla in all the species studied, with the location and extent of the tectorial plate for each form. Two specimens of *Anguis fragilis* are portrayed to show the range of variations found in this species.

The ciliary tufts in anguids show a systematic variation in form according to location along the papilla. When these tufts lie beneath the tectorial plate, they are relatively short and taper rapidly, as shown in *a* of Fig. 10-11, and when they are free-standing they are twice to four times as long, and either taper gradually as shown in *b* of this figure or have a tapered portion extended by a group of three to five long fibers as in *c*. Type *b* is usually encountered close to the tectorial plate, and evidently represents a transitional form. The short form has already been seen in Fig. 10-8, and the long, continuously tapered form in Fig. 10-7.

Cochlear Dimensions. Detailed measurements of the magnitude of cochlear structures were carried out in specimens of *Gerrhonotus multicarinatus* and *Ophisaurus ventralis*.

Gerrhonotus multicarinatus. — Results for one of the alligator lizards are presented in Fig. 10-12. This ear has a basilar membrane 420 μ long, with moderate variations of width along its course. As *a* of this figure shows, the width increases sharply at the dorsal end and rises to a maximum of 103 μ in the mid-dorsal region. It then declines to a minimum, and thereafter increases somewhat irregularly to a second maximum in the ventral region where the tectorial plate appears. The variation in width over the main course of the membrane is about 68%.

The width of the fundus (part *b*) varies only moderately over its main course, in the amount of 60%. From the maximum reached in the dorsal region the width falls to a minimum at the 260 μ point, and then a second flat maximum appears in the ventral region.

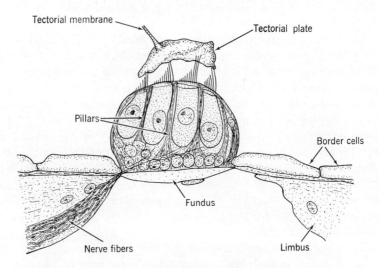

Fig. 10-8. A section through the auditory papilla of *Ophisaurus ventralis*, passing through its midregion. Scale 640×. From Wever, 1971d.

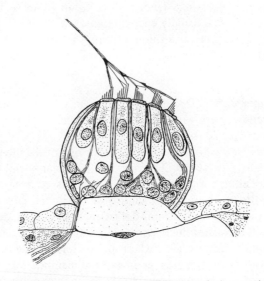

Fig. 10-9. The auditory papilla of *Ophisaurus ventralis* in a section dorsal to the tectorial plate. From Wever, 1965a.

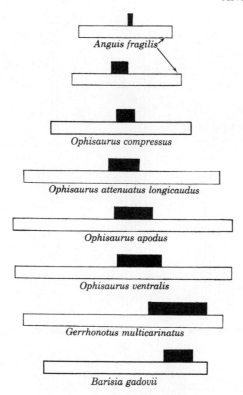

Fig. 10-10. Representation of the length of the auditory papilla (open rectangles) and the location and extent of the tectorial plate (solid rectangles) for seven species of anguids. Scale (length only) 125×.

The thickness of the fundus (part c) varies about 53%, in a rather irregular manner.

The size of the papilla (solid line of part d) varies a little over twofold from a high value in the dorsal region through a barely noticeable minimum to a second maximum in the ventral region. The organ of Corti (broken line of part d) follows almost exactly the same course.

The number of longitudinal rows of hair cells (part e) varies from 4 at the dorsal end to a maximum of 6 soon reached in this region, then declines to 5 and finally to 4 in the remainder of the cochlea.

This ear is peculiar in that it is reverse-tapered in comparison with most lizard species. The basilar membrane shows a considerable decline in width from dorsal to ventral regions, but with a strong suggestion of bimodality. The fundus reflects this same form. The auditory papilla exhibits this variation, and the number of rows of hair cells also bears it out.

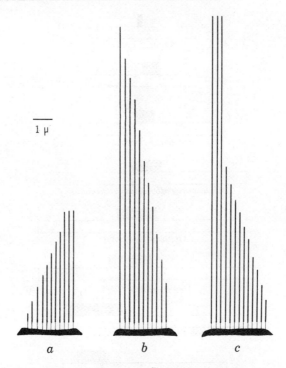

1 μ

a   *b*   *c*

Fig. 10-11. Forms of ciliary tufts. (From a
specimen of *Ophisaurus compressus*).

This ear exhibits a moderate amount of variation in dimensional characteristics. The fundus varies rather irregularly in thickness, but its width
tends to follow the variations in the basilar membrane as a whole and
in the size of the papilla. Altogether these changes would seem to favor
the action of low tones in the dorsal part of the cochlea and the action
of high tones in the ventral part. However, the ventral part in which a
tectorial plate covers many of the hair cells is but slightly differentiated,
and it may be presumed that these cells tend to work as a unitary group.
Weiss, Mulroy, and Altmann (1974), in recording single fiber responses
to clicks in this species, obtained evidence indicating a localization of
high-frequency responses in the dorsal region of the cochlea and low-
frequency responses in the ventral region.

Ophisaurus ventralis. — Dimensional data for *Ophisaurus ventralis*
are presented in Fig. 10-13. The basilar membrane is 488 μ in length,
and rises a bit irregularly to a maximum width in the middle region and
then falls toward the ventral end. The variation in width over the main
course is 53%.

The fundus (part *b*) has much the same form, and varies in width by 50%. However, as shown in part *c*, the thickness of the fundus varies considerably and presents two distinct modes, a larger one in the dorsal region and a very symmetrical one located ventrally. The variation between the dorsal hump and the low point near the middle is 3.3-fold.

The papilla is strongly duplex, with the dorsal part attaining a maximum area noticeably greater than the ventral one, and exceeding the size of the intermediate region by 2.3-fold.

The rows of hair cells reflect the dual character of the papilla. Their number rises rapidly from 2 to 5 in the dorsal region, where it remains for some distance, then this number falls to 3 in the intermediate region. Thereafter the number rises in two stages, attaining 6 at one point and then stabilizing at 5 over the remainder of the cochlea.

This ear shows a moderate amount of variation in the widths of the basilar membrane and fundus, and these are such as to favor the action of the higher tones at the two ends of the cochlea. Yet these variations are cut across by the large and rapid changes in the thickness of the fundus and the size of the sensory structures as represented by the dorsal and ventral enlargements. The changes in the fundus have two contrary effects: its increase in thickness adds to the stiffness of the moving structures thus favoring the response of high tones, whereas this increase augments the mass, favoring the response of low tones. The two effects therefore are counteractive, though it may be suggested that the stiffness effect is predominant. Hence the changes seen in the fundus are such as to improve high-tone response in both dorsal and ventral regions. In these same regions, however, the size of the auditory papilla, with its large mass changes, is in a contrary direction. The net effect of these mass and stiffness factors is probably a spreading out of the distribution of response caused primarily by width variations in basilar membrane and fundus. Hence this ear may be expected to exhibit two areas of differentiation, one in the dorsal part of the cochlea and another in the ventral part, with each part favoring the high tones in their end regions. The tectorial plate lies in a region a little beyond the middle of the cochlea, where the basilar membrane and papilla have their greatest width and where, contrariwise, the thickness of the fundus and the size of the auditory papilla are least. Probably this region serves for the response to tones of intermediate frequency, and may account for the peculiar bimodal form of the sensitivity function seen in Fig. 10-20.

Observations on the varying width of the basilar membrane were made also in representatives of the other species, and the results are shown in the outline sketches of Fig. 10-14. In these outlines the minor variations are smoothed out, and symmetry about a longitudinal axis is assumed. The dorsal end of the membrane is to the left.

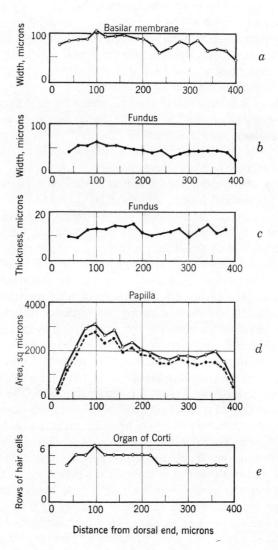

FIG. 10-12. Cochlear dimensions in a specimen of *Gerrhonotus multicarinatus*. From Wever, 1965a.

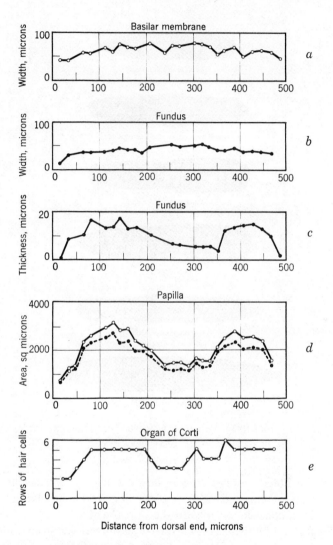

FIG. 10-13. Cochlear dimensions in a specimen of *Ophisaurus ventralis*. From Wever, 1965a.

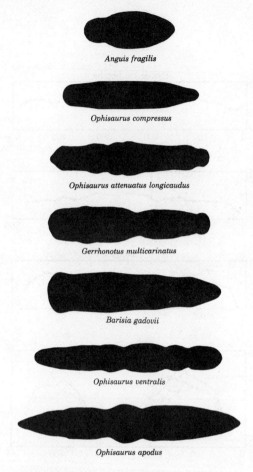

Anguis fragilis

Ophisaurus compressus

Ophisaurus attenuatus longicaudus

Gerrhonotus multicarinatus

Barisia gadovii

Ophisaurus ventralis

Ophisaurus apodus

Fig. 10-14. Forms of the basilar membrane
in seven species of Anguidae. Scale 100×.

As will be noted, *Anguis fragilis* has a much shorter basilar membrane than the others, and its form is peculiar. It has an almond shape disturbed by a constriction toward the dorsal end.

In all these forms there is relatively little differentiation in width except for tapering at the ends. For *Anguis fragilis, Ophisaurus compressus, Ophisaurus attenuatus longicaudus,* and *Ophisaurus apodus* the widest point is near the middle. In *Ophisaurus ventralis* there are two wide regions, one on either side of the midpoint. For *Gerrhonotus multicarinatus* and *Barisia gadovii* the form is asymmetrical, with the widest region toward the dorsal end and a progressive narrowing ventrally, though disturbed by a marked constriction in *Gerrhonotus multicari-*

natus. On the whole the anguids exhibit only a moderate degree of differentiation in terms of basilar membrane width.

Numbers of Hair Cells.   The hair-cell population varies considerably among anguid species as shown in Table 10-I. This table shows species averages that include all specimens that could be counted reliably, and represents separately the hair cells with free-standing cilia and those cells beneath the tectorial plate.

TABLE 10-I

MEAN HAIR-CELL POPULATIONS

IN ANGUIDAE

Species	Number of Ears	Number of Hair Cells		
		Free-standing	Under Plate	Total
Anguis fragilis	6	35.3	10.0	45.3
Barisia gadovii	8	91.0	20.1	111.1
Gerrhonotus multicarinatus	9	126.1	31.4	157.6
Ophisaurus apodus	4	166.0	24.7	190.7
Ophisaurus attenuatus longicaudus	2	186.0	28.0	214.0
Ophisaurus compressus	2	124.5	15.0	139.5
Ophisaurus ventralis	5	149.8	27.4	177.2
Average of 7 species	—	125.5	22.4	147.9

The hair-cell population in *Anguis fragilis* is relatively small, with an average total of 45.3. This number is consistent with the low level of development (or more likely the degree of degeneration) of this ear. *Barisia gadovii* is next in order with a mean of 111.1, nearly two and a half times as many. The remaining species show significantly larger populations, from *Ophisaurus compressus* with a mean of 139.5 to *Ophisaurus attenuatus longicaudus* with a mean of 214.0 hair cells. *Gerrhonotus multicarinatus* stands between these two with 157.6 hair cells on the average. These numbers are moderate in relation to hair-cell populations of many other lizards such as geckos and varanids, but compare favorably with most iguanids and agamids in which the populations vary around 60 to 100. The anguids therefore must be placed below the middle of the general series of lizards as regards cochlear development, but (except for *Anguis*) not at the lower end.

The relative numbers of hair cells beneath the tectorial plate vary with species in a suggestive manner. In *Anguis*, *Barisia*, and *Gerrhonotus* species, the number of these cells with tectorial restraints is about one-fifth of the total, whereas in the *Ophisaurus* species it is significantly less, from 10.8% in *Ophisaurus compressus* to 15.5% in *Ophisaurus ventralis*.

Ciliary Orientation Patterns.   In all regions of the cochlea away from the tectorial plate, the ciliary orientation is bidirectional, and with a nearly equal division between medially directed ciliary tufts on the lateral side of the auditory papilla and laterally directed ones on the medial side. Beneath the tectorial plate these tufts are strictly unidirectional, with all the tufts laterally oriented.

In line with evidence obtained from overstimulation studies and an analysis of cochlear patterns in the geckos, it is a reasonable assumption that the maximum sensitivity of the anguid ear is largely determined by that part of the cochlea covered by the tectorial plate. The presence of unidirectional ciliary orientation here should contribute to the sensitivity.

AUDITORY SENSITIVITY

Sensitivity curves in terms of cochlear potentials were obtained for most of these specimens. Results for *Anguis fragilis* have already been shown in Fig. 6-64. This ear exhibits best sensitivity at three points within the range of 400 to 2000 Hz, beyond which the sensitivity falls off rapidly. Even at the best point at 400 Hz the sensitivity is relatively low, and the performance of this ear must be characterized as poor.

Curves for two specimens of *Barisia gadovii* are given in Fig. 10-15. The sensitivity is excellent in the middle range, with a tendency toward bimodality. The better of these ears reaches −56 db in the region 400-700 Hz and −48 db in the region 1500-2000 Hz. The roll-off beyond the sensitive region is about 40 db per octave for the low tones and 25 db per octave for the high tones.

Results for three specimens of *Gerrhonotus multicarinatus* are presented in Fig. 10-16. These curves are in good agreement in showing the best sensitivity in the region of 400-1000 Hz, with the maximum points varying between −62 and −78 db over this region. In the low tones the roll-off is around 30 db per octave, and more rapid, approaching 40 db per octave, in the high tones. Within the medium low-tone range the sensitivity is outstanding.

Observations made on this species by Crowley gave the curves of Fig. 10-17, which show the same form, and though the sensitivity is somewhat less, it is still of high degree.

Curves for two specimens of *Ophisaurus apodus* are given in Fig. 10-18. These curves are in fair agreement in indicating a moderate degree of sensitivity in the region of 500 to 3500 Hz, where levels between −10

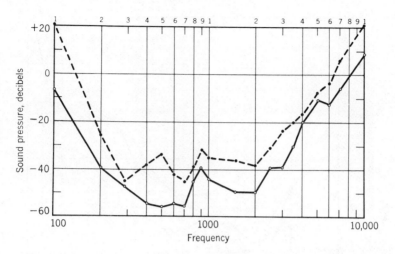

FIG. 10-15. Sensitivity curves for two specimens of *Barisia gadovii*.

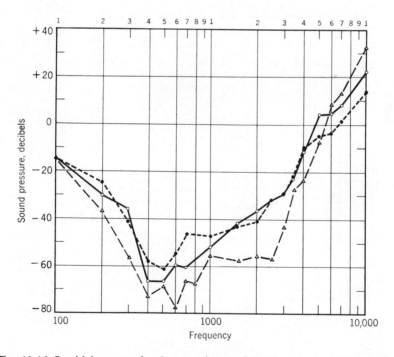

FIG. 10-16. Sensitivity curves for three specimens of *Gerrhonotus multicarinatus*.

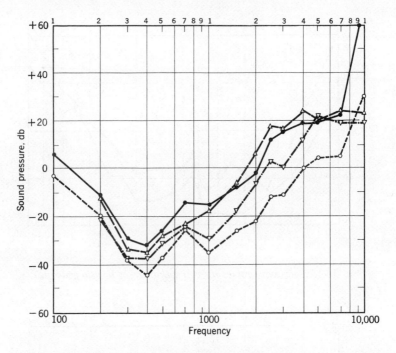

Fig. 10-17. Sensitivity curves for four specimens of *Gerrhonotus multicarinatus* obtained by Crowley (1964).

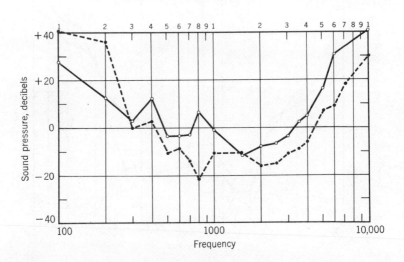

Fig. 10-18. Sensitivity curves for two specimens of *Ophisaurus apodus*. From Wever, 1971d.

and −22 db are reached. The roll-off at the low end of the range is irregular around 20 db per octave, and at the high end is about 30 db per octave. The general performance must be regarded as limited.

Results for a specimen of *Ophisaurus attenuatus longicaudus* are shown in Fig. 10-19. The best region of sensitivity is between 600 and 3000 Hz, with a peak of −34 db at 800 Hz. The low-frequency roll-off is 30 db per octave, and the high-frequency roll-off is about 50 db per octave.

Sensitivity curves for two specimens of *Ophisaurus ventralis* are shown in Fig. 10-20. These curves agree in indicating very good sensitivity in the region of 290-800 Hz, with peaks of −37 and −41 db. Then, after a sharp decline in sensitivity at 1000 Hz, a further region appears between 1500 and 3000 Hz where the response is also good. The roll-off at the low end of the range is rapid, around 40 db per octave, and is only a little less at the high end.

Some further tests were made on a specimen of *Gerrhonotus multicarinatus* by the use of vibratory stimulation. This ear was stimulated in two ways, with the vibrator applied to the quadrate, which was exposed at its posterodorsal corner, and then to the retroarticular process of the mandible. The results are shown in Fig. 10-21. These two curves are generally similar, showing best sensitivity in the middle range of frequencies, but more sharply located around 600 Hz for quadrate stimulation and more uniform between 400 and 2000 Hz for mandibular stimulation. These results may be compared with the aerial curve for the same ear as shown in Fig. 10-22, which presents the maximum sensitivity in the same region of frequency. These results are in agreement with more extensive studies on other lizard species indicating a major role taken by the inner ear in determining the form of the sensitivity function.

The sensitivity tests reveal a wide range of auditory capability among the anguids, ranging from poor in *Anguis fragilis* and *Ophisaurus apodus* to excellent in *Barisia gadovii* and *Gerrhonotus multicarinatus*. Further study is needed on the *Ophisaurus* species, especially on *Ophisaurus attenuatus longicaudus* and *Ophisaurus compressus*, for which only single specimens were available. The deep location of the round window in these species makes the placing of the electrode particularly difficult, and for the *Ophisaurus compressus* specimen the sensitivity measurements were unsatisfactory and no results are presented here.

The anguid ear appears to be mainly serviceable for sound detection, with a wide range of effectiveness among the different species. Dimensional variations along the cochlea are not outstanding, and cochlear tuning in these terms does not appear to be a significant feature. The presence of the heavy tectorial plate in one region may favor the re-

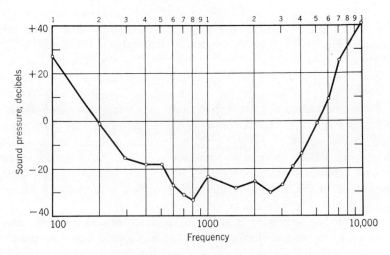

FIG. 10-19. A sensitivity function for a specimen of
Ophisaurus attenuatus longicaudus. From Wever, 1971d.

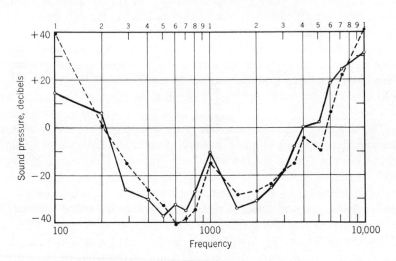

FIG. 10-20. Sensitivity curves for two specimens of *Ophisaurus ventralis*.
From Wever, 1971d.

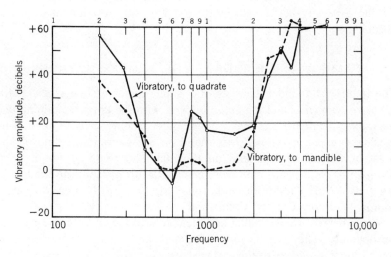

Fig. 10-21. Sensitivity for vibratory stimulation in a specimen of
Gerrhonotus multicarinatus.

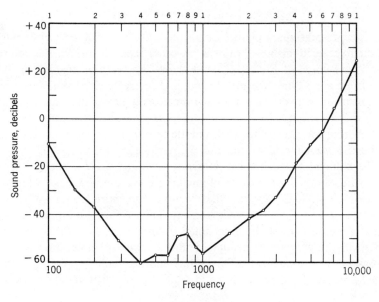

FIG. 10-22. An aerial sensitivity curve for the same ear as in
the preceding figure.

sponse of the low frequencies there, and may account in some measure for the generally better sensitivity found for the low and middle frequencies.

FAMILY ANNIELLIDAE: THE LEGLESS LIZARDS

The family Anniellidae consists of a single genus of two species, *Anniella geronimensis* and *Anniella pulchra*, with the latter containing two subspecies—*pulchra* and *nigra*—that differ mainly in coloration. The species *Anniella geronimensis* is found on San Geronimo Island off the coast of Baja California and on the mainland south of Punta Baja. *Anniella p. pulchra* is found from central California south to the northern end of Baja California, and *Anniella p. nigra* is limited to a small area in Monterey County, California.

These lizards are wormlike and slender, with small eyes and without ear openings. They inhabit pine forest and sandy areas, where they burrow in loose soil close to the surface.

Six specimens of *Anniella p. pulchra* (Gray, 1852) were studied. All were of adult size, 10-13 cm in snout-to-vent length and 14-22 cm in total length, with a body diameter of 4-5 mm.

ANATOMICAL OBSERVATIONS

This species presents a picture of serious reduction in the receptive and conductive mechanisms of the ear. A cross section through the ear region is presented in Fig. 10-23 and shows the essential features of this ear. There is no tympanic membrane, and the scaly skin continues without interruption over the ear area. Immediately below the skin layer are large masses of the depressor mandibulae muscle and a long forward extension of the constrictor colli muscle that together form a substantial barrier between the inner ear and the periphery.

The columellar mechanism is greatly modified, though the two main parts, the columella and extracolumella, are represented. The columella is reduced to a large footplate that occupies the entire lateral wall of the otic capsule and a strong headpiece extending from its midportion. The extracolumella consists of a cartilage of conical form with its base attached to the columellar headpiece and its rounded apex inserted in a depression in the quadrate. This apical end no doubt represents the internal process of the extracolumella, but other extracolumellar processes are lacking.

Another picture of the middle ear structures and their relations to the bones of the skull is presented in Fig. 10-24. The columella and extracolumella are seen wedged between the quadrate and prootic, with the large rounded expansion of the columella, constituting the footplate, inserted in the oval window. The muscle and skin layers normally covering these structures have been dissected away.

Figure 10-23 shows the outline of a flattened sac lateral to the extra-columella and quadrate, which resembles the accessory air cavity often encountered in divergent types of middle ears (see Chapter 6). In this species, however, there is no connection between the cavity of this sac and the middle ear cavity, and the absence of such a connection makes it practically certain that this sac is fluid filled. If this is the case, the sac cannot assist the mobilization of the skin and tissues lateral to it, as an accessory air space does in other species. Therefore the transmission of sounds from the outside to the middle ear can occur in this species only by bulk conduction through the tissues, involving the skin and the muscle mass below. Particularly concerned in this bulk movement is the anterior lobe of the depressor mandibulae muscle, which is in contact with the outer surfaces of the extracolumella and the head of the columella.

When the columellar footplate is exposed to vibratory pressures exerted by the superficial tissues, this body is able to move only if there is some yielding area medial to it. In most ears a round window serves this purpose, but in *Anniella* this window is absent. An alternative method of mobilization is employed, consisting of a reentrant fluid circuit as indicated by the arrows in Fig. 10-23.

As the columellar footplate moves inward, a fluid displacement is transmitted by a bulging of Reissner's membrane, continued in the cochlear duct by a corresponding bulging of the basilar membrane, and propagated further through the scala tympani and the perilymphatic sac and across the arachnoid membrane into the cranial cavity. Within the cranial cavity the discharge passes anteriorly and then turns laterally through fluid spaces adjacent to the facial nerve, and finally reaches the anterior and anterolateral surface of the footplate, thus completing the circuit.

A reentrant fluid circuit of this kind serves as the means of fluid mobilization in the cochleas of snakes, turtles, and amphisbaenians, but is rare in lizards. It serves its purpose well for the low tones, but imposes considerable impedance to the transmission of the high tones, especially when the fluid pathway is as narrow and circuitous as it is in *Anniella*.

Because the basilar membrane is involved in this surging motion of fluids in response to sound pressures, the sensory cells borne by this membrane are stimulated, and the sounds are heard. This hearing has limits imposed by the fluid circuit and also by further characteristics of the inner ear structures.

Inner Ear.   In *Anniella* the cochlea is short, and the basilar membrane is small and little differentiated. It has a pear shape as shown in part *a* of Fig. 10-25. This membrane is less than 200 μ in length and is half as wide at its maximum near the ventral end.

Detailed measurements of the width of the basilar membrane are represented in *a* of Fig. 10-26, together with dimensions of its middle portion, the fundus. The thickness of the fundus varies along the cochlea in a curiously bimodal manner, as *b* in this figure indicates. The auditory papilla, lying upon the basilar membrane, has a broadly peaked form as shown in *c* of this figure. The hair cells form two rows at the dorsal end of the cochlea, then increase rapidly to 7 rows in the ventral region.

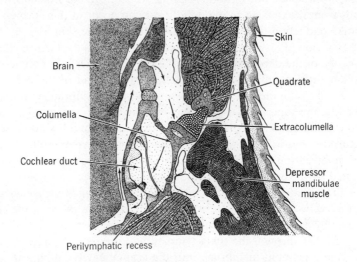

FIG. 10-23. A frontal section through the ear region of *Anniella pulchra*. Arrows indicate the path of vibratory fluid flow through the cochlea. From Wever, 1973b.

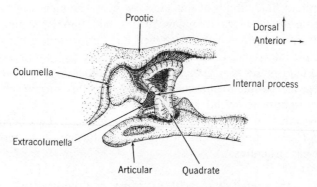

FIG. 10-24. A reconstruction drawing showing the conductive structures of the ear of *Anniella pulchra*.

FIG. 10-25. Part *a*, the form of the basilar membrane of *Anniella pulchra*. Part *b*, the length of the basilar membrane and location of the tectorial plate. Scale 100× (length only for part *b*).

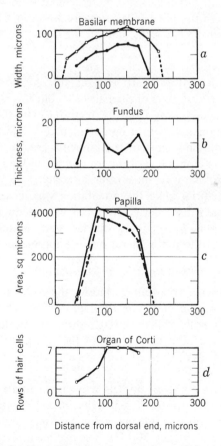

FIG. 10-26. Cochlear dimensions in a specimen of *Anniella pulchra*.

The number of hair cells was determined in 8 ears of 4 animals, and ranged from 43 to 81 with an average of 63.2.

A tectorial membrane arises from a sharply elevated mound on an anterior portion of the limbus, and extends to a plate over the papilla, as Fig. 10-23 has shown and Fig. 10-27 represents in further detail. This plate lies over the hair-cell surface of the papilla only in a limited region located a little beyond the middle of the cochlea, as part *b* of Fig. 10-25 indicates. Elsewhere the ciliary tufts of the hair cells extend freely into the endolymph of the cochlear duct. A cross-sectional view of the auditory papilla, at a level at which the tectorial plate has only just appeared, is given in Fig. 10-28. Of interest is the short form of the ciliary tufts beneath this plate, and the extended form of most of these tufts laterally where the plate fails to reach. Just ventral to this section the plate covers all the hair cells, and all the ciliary tufts have the short form.

Ciliary Orientation. Most lizard ears exhibit great regularity in the orientation of the ciliary tufts of the hair cells, but this is hardly the case in *Anniella*. In regions away from the tectorial plate there is usually a degree of bidirectionality, with a predominance of lateral (strictly posterolateral) orientation—most of the kinocilia are located on this side of the ciliary tuft. Yet in a few of the transverse rows all the ciliary tufts were found to face one way, most often laterally.

Beneath the tectorial plate the orientation was most commonly lateral, but occasional tufts departed from this pattern. In one instance the tufts beneath the plate were equally divided, with the more medial ones facing laterally and the more lateral ones facing medially.

COCHLEAR POTENTIAL STUDIES

Because of the absence of a round window, the cochlear potentials were recorded with a needle electrode insulated except at the tip and inserted in a small hole drilled through the roof of the saccule. This electrode was in the perilymph space of the left saccule. Sounds were presented through a tube sealed over the ear region at the left side of the head.

Sensitivity curves for two animals are shown in Fig. 10-29. These curves indicate a low degree of sensitivity, mostly around 0 db, over a broad range from 100 to 1500 Hz, going to a maximum of −14 db in one animal for the frequency 300 Hz. The decline at the two ends of the range is only moderately rapid, around 10 db per octave for the low tones in one animal in which the tests extended to 40 Hz, and around 20 db per octave for the high tones.

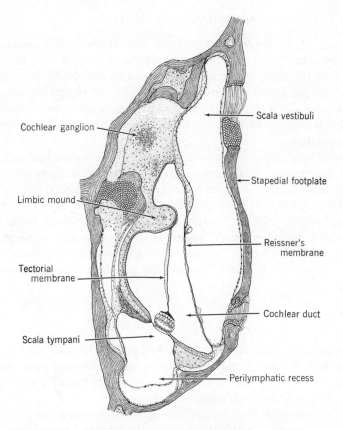

FIG. 10-27. The cochlear duct of *Anniella pulchra* at the level of the
tectorial plate, in a frontal section.

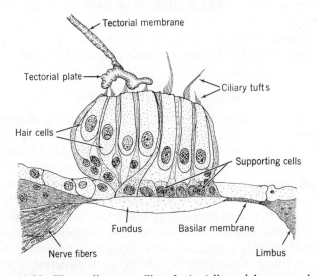

FIG. 10-28. The auditory papilla of *Anniella pulchra*, seen in a
frontal section at a point near one edge of the tectorial plate. Scale
500×.

In one of the specimens sensitivity measurements were made with vibratory stimulation, in which the stimulus was applied to the dorsal surface of the skull. The results are shown in Fig. 10-30, and indicate a primary maximum of sensitivtiy around 2000 Hz and a secondary maximum at 7000 Hz. The level of sensitivity is poor, and is consistent with the limitations of inner ear structure and the small size of the hair-cell population.

Anniella pulchra is a burrowing form in which the ear has undergone severe degenerative modifications that have produced a loss of sensitivity and serviceability. And yet this organ has been retained, even though it performs poorly. There is even evidence that secondary changes have occurred to assist the ear's response, and to compensate in some measure for the closure of the ear opening and other changes consequent to it. One of these secondary events is the appearance, or at least the retention, of a reentrant pathway providing a mobilization of the cochlear fluid.

FAMILY XENOSAURIDAE

This small family is divided by McDowell and Bogert (1954) into three subfamilies, one represented by an extinct genus *Melanosaurus*, another including a little-known form *Shinisaurus crocodilus* of China, and a third containing the single genus *Xenosaurus*.

Xenosaurus contains three species—*grandis, newmanorum,* and *platyceps*—all occurring within a limited range in central Mexico and Guatemala (King and Thompson, 1968). The species *Xenosaurus grandis* is separated into five subspecies, largely on the basis of color patterns. These animals are most often found in dry and rocky areas, but occasionally in other localities, including quasi-rainforests.

Four specimens of *Xenosaurus grandis* (Gray, 1856) were available for study.

ANATOMICAL OBSERVATIONS

Middle Ear.   As shown in Fig. 10-31, the conductive mechanism of *Xenosaurus grandis* departs from normality only in one significant respect: in the absence of a tympanic membrane and the use of the skin as the sound-receptive surface. This skin has a dense dermal layer and a thick subdermal layer, and the insertion piece of the extracolumella is applied to a particularly thick portion of this skin. A notable feature is the presence of a large tympanic recess, an air space derived from the middle ear cavity, which bounds the inner surface of this lateral area of skin, thereby defining a tympanic area and giving it a degree of mobility. The depressor mandibulae muscle lies well posterior to this tym-

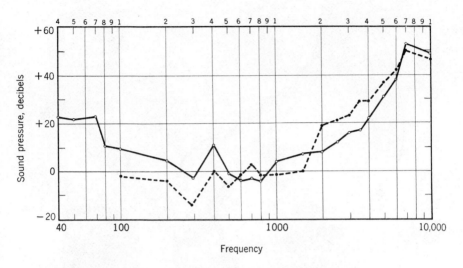

FIG. 10-29. Aerial sensitivity curves for two specimens of *Anniella pulchra*.
From Wever, 1973b.

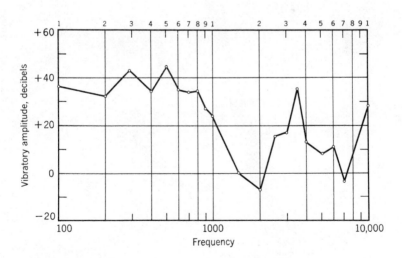

FIG. 10-30. Vibratory sensitivity in a specimen of *Anniella pulchra*.

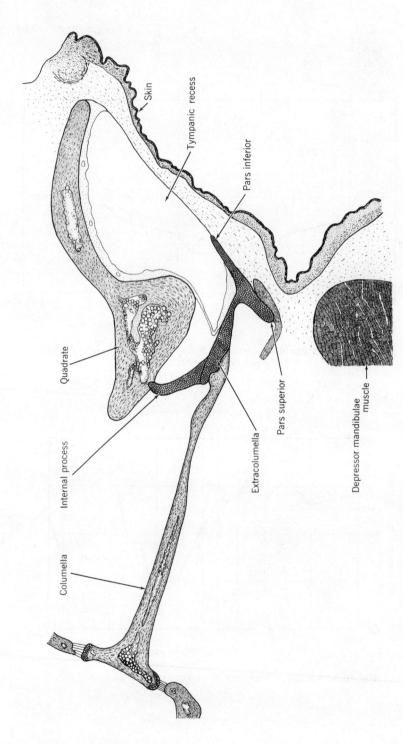

Fig. 10-31. The middle ear structures of *Xenosaurus grandis*. From Wever, 1973b.

panic area, and does not form an obstruction to the passage of sounds to the ear structures as it does in many other lizards that have lost the tympanic membrane.

The end of the pars superior, to which the extracolumellar ligament is attached, lies at the edge of the tympanic area and evidently forms an axis about which the free portion of the skin can move in and out under the influence of sound waves. The internal process running off from the medial end of the extracolumella to the quadrate is long and flexible, and its connection to the quadrate appears to be sufficiently loose to permit vibratory movements with little or no restraint.

Inner Ear.   The cochlea of *Xenosaurus* is exceedingly simple. Its basilar membrane is an oval about twice as long as it is wide, and resting on its surface is an auditory papilla of similarly oval form but considerably smaller size. These structures are represented in the reconstruction drawing of Fig. 10-32. The papilla lies close to the inner edge of the basilar membrane and leaves a wide area of free membrane on the outer side. The hair cells appear in longitudinal rows varying in number from 3 to 10 over the extent of the papilla.

A tectorial membrane is present on the surface of the medial limbus over a region almost as extensive as the basilar membrane, but in only one place does it leave this surface and extend laterally over the papilla. Accordingly this membrane makes connection, through a tectorial plate, with the ciliary tufts of only a few hair cells, those belonging to two or three transverse rows in the middle of the papilla. Also this plate does not quite cover all the cells of these rows, but stops short before the most lateral ones are reached. Figure 10-33 shows these relations together with other details.

The ciliary tufts beneath the tectorial plate are relatively short, but the others, which end freely, contain cilia two or three times as long. As suggested earlier, this greater length enlarges the surface that the tufts present to the cochlear fluid and increases the restraining force acting upon them, thereby improving the delicacy of response of their hair cells. The maximum sensitivity attained by such an ear is probably determined by those few hair cells whose tufts are contacted by the tectorial plate and operate effectively enough to excite their nerve fibers in response to the fainter sounds, and the remaining elements contribute to the neural discharge only at higher levels of stimulation. This arrangement extends the ear's dynamic range.

Numbers of Hair Cells.   The hair cells were counted in six ears, and ranged from 56 to 65, with a mean of 61.6. In one of these ears there were 11 cells (portions of two rows) with ciliary attachments to the tectorial plate.

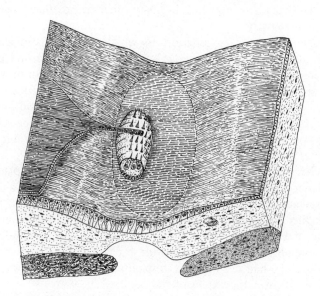

FIG. 10-32. The inner ear of *Xenosaurus grandis*, drawn from a reconstruction model. The view is dorsoposterior and lateral. Scale 125×.

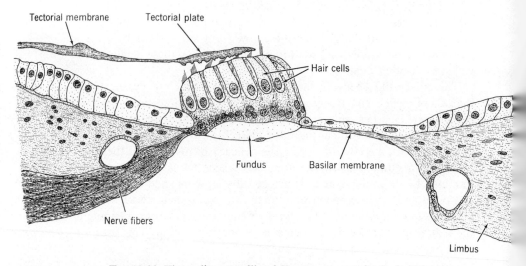

FIG. 10-33. The auditory papilla of *Xenosaurus grandis*. Scale 375×.

The Ciliary Orientation Pattern. Over most of the papilla, away from the tectorial plate, the ciliary orientation is bidirectional. Often the arrangement is the common one in which the ciliary tufts of the most medial hair cells face laterally (the longest stereocilia and the kinocilium are on the lateral side) and the lateral hair cells have the reverse orientation. Frequently, however, there are departures from this pattern, including the appearance of a tuft that is reversed from its expected position, or there are regular alternations along a transverse row.

For the ciliary tufts beneath the tectorial plate a determination of the orientation pattern was often difficult, but the tendency was certainly toward a lateral orientation. In one specimen all the tufts in contact with the plate had this lateral position. In another specimen the two tufts that could be seen plainly had a lateral orientation and four others probably had this same orientation, though the differentiation in length of the cilia was so slight as to prevent any firm decision. Two other tufts in this same transverse row that were not covered by the tectorial plate had the reverse orientation.

Cochlear Dimensions. Dimensional details for this cochlea are given in Fig. 10-34. In part *a* are shown the width of the basilar membrane (open circles) and of the fundus (filled circles). The basilar membrane attains a width of 175 μ in its midregion, and has a length of a little under 300 μ. The fundus is less than half as wide and has half the length. The thickness of the fundus (part *b*) shows only slight variations apart from its tapered ends. The papilla has a simple mound-like form as already indicated in Fig. 10-32. In this specimen the longitudinal rows of hair cells varied from 4 at the dorsal end to 6 at the ventral end.

COCHLEAR POTENTIAL STUDIES

Potentials in response to sounds were recorded from the round window in the usual way except that, in the absence of a tympanic membrane, the sound tube was sealed over a large area of skin at the side of the head. The results of these tests varied considerably in the different specimens, due in part to poor physiological condition in one or two of them, but perhaps reflecting individual differences of structure also. The first animal obtained, whose sensitivity curve was published earlier (Wever, 1973b), was obviously in a moribund state when received and tested, though the specimen proved valuable for the investigation of middle ear structures. A sensitivity curve for another animal that seemed to be in excellent condition is presented in Fig. 10-35. For this ear the sensitivity is best in a region from 290 to 1000 Hz, where it is nearly uniform around −26 db, and then falls off at the rapid rate of about 40 db per octave in the low frequencies and 30 db per octave in the high fre-

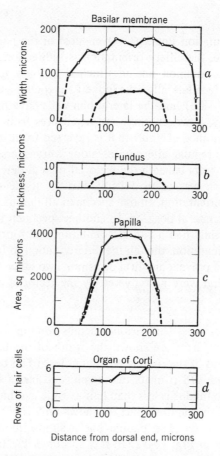

Fig. 10-34. Cochlear dimensions in *Xenosaurus grandis*.

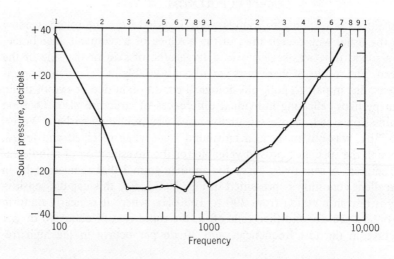

Fig. 10-35. A sensitivity curve for a specimen of *Xenosaurus grandis*.

quencies. The performance is remarkably good for an ear lacking a tympanic membrane.

The service of this ear as an acoustic receptor seems to be limited, however. The cochlea is not only small but differentiated only in slight degree. The discrimination of tones by the use of spatial cues will be negligible, and whatever discrimination exists must be in terms of the frequency of neural discharge. With the few receptor elements and nerve fibers present, this capability also must be severely limited. Thus it seems doubtful that this animal is able to utilize acoustic signals for anything but the simplest purposes, perhaps as a gross means of alerting it to possible dangers.

11. THE VARANIDAE AND
HELODERMATIDAE: MONITORS AND
BEADED LIZARDS

FAMILY VARANIDAE: THE MONITOR LIZARDS

The varanids occur in a few regions of the Eastern Hemisphere: the East Indies, Africa, and Australia. There is only one genus, *Varanus*, and the included species, about 28 in number, show great uniformity. They are large-bodied animals, with a long neck, strong legs, and a powerful laterally compressed tail much like that of a crocodile. The size varies with species from a small 20-cm type in Australia to the famous Komodo dragon of the East Indies that may attain a length of 3 meters and a weight of 100 kg or more. An immature specimen of *Varanus bengalensis nebulosus* is pictured in Fig. 11-1.

Fig. 11-1. The monitor lizard *Varanus bengalensis nebulosus*.
Drawing by Anne Cox.

The habitat of these lizards varies; some live in trees, some frequent moist areas near streams, others roam the land where they dig or appropriate burrows, and many are active in all these places. They live on a variety of prey, including fish, crustacea, frogs, birds, small mammals, and eggs of various kinds.

Three species were available for the present study, all represented by specimens of medium to small size. These were *Varanus bengalensis nebulosus* (Gray, 1831)—14 specimens; *Varanus griseus* (Daudin, 1803)—3 specimens; and *Varanus niloticus* (Linnaeus, 1762)—4 specimens.

ANATOMICAL OBSERVATIONS

MIDDLE EAR

In general the sound conductive mechanism of *Varanus* is of the standard iguanid type. Two features are of special interest, however. The end of the internal process is not anchored in a depression or groove in the surface of the quadrate as is usually the case, but simply lies on this surface, loosely held by mucous membrane. Also, as Fig. 11-2 indicates, a small ligament runs from one side of the internal process to an anchorage on the cephalic condyle. This arrangement allows great mobility of the middle portion of the conductive apparatus in its vibratory response to sounds, but presumably affords a protective restraint of displacements of large amplitude, such as might be produced by blows on the head or the like that might be damaging to the ear.

INNER EAR

In *Varanus* the cochlear duct is large and elongated. The limbus is thick and heavy, and its medial portion is only slightly elevated. A tectorial membrane arises from the limbic epithelium far medially at its junction

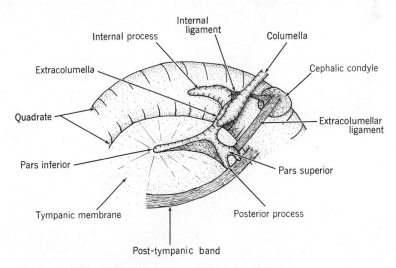

FIG. 11-2. The middle ear of *Varanus bengalensis nebulosus* in a ventromedial view, showing the inner surface of the tympanic membrane, the extracolumella, and the medial portion of the columella. Scale 4.5×.

with Reissner's membrane, runs over the elevated portion of the limbus in close contact with it, and then leaves this surface at almost a grazing angle along its lateral declivity, as shown in Fig. 11-3 for a specimen of *Varanus griseus*.

The tectorial membrane ends in a tectorial plate that lies over the papillar hair cells. The form of this plate varies with species, and also varies somewhat along the cochlea as will be described presently.

The varanid cochlea has a duplex character varying in degree in different specimens. There is a marked narrowing of the basilar membrane in a region that in all three species studied occurs about two-thirds of the distance from the dorsal end. In this narrow region there is a distinct reduction of the auditory papilla and of its hair cells or (in one instance) their complete disappearance.

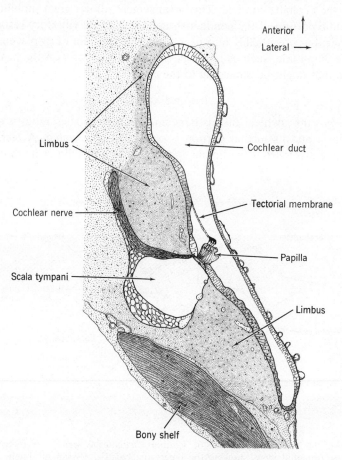

Fɪɢ. 11-3. The auditory papilla of *Varanus griseus* in a frontal section.

These general relations are shown in Fig. 11-4, which is based on a reconstruction of a specimen of *Varanus bengalensis nebulosus*. Here the medial limbus is on the left and the dorsal end of the auditory papilla is below. Reissner's membrane is represented by the shaded band to the left of the papilla. The basilar membrane is largely hidden beneath the papilla, though its edge appears to the right (indicated by simple hatching).

At the narrowest part of the papilla the hair cells are usually absent for one or more sections. In this region the tectorial plate is absent, and

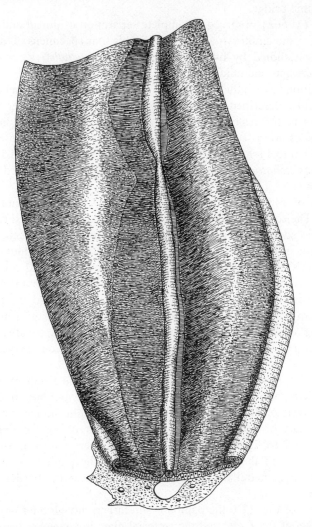

FIG. 11-4. A reconstruction drawing of the auditory papilla of
Varanus bengalensis nebulosus. Scale 50×.

the tectorial membrane is greatly shortened though usually still visible on the limbic surface.

In the right ear of one of the specimens of *Varanus griseus* there was a complete separation of dorsal and ventral portions of the papilla. In the region of this hiatus the two parts of the limbus were continuous and the basilar membrane was absent. This condition is represented in Fig. 11-5. The epithelial layer continues over the limbus in this region without any suggestion of a papilla, though with a conspicuous thickening. This figure also shows a capillary running over the tympanic surface of the limbic bridge.

Miller (1966a) observed a complete separation of dorsal and ventral segments of the auditory papilla in two of his four specimens of varanids; the species studied by him included *Varanus nuchalis*, *V. punctatus*, *V. salvator*, and one unidentified.

In addition to its reduction and near or complete disappearance at one point as just described, the auditory papilla undergoes other striking alterations in form along its course. These changes are best indicated in a series of cross-sectional views taken at intervals along the cochlea. Figures 11-6 to 11-8 present nine such views for a specimen of *Varanus bengalensis nebulosus* at approximately equal intervals from dorsal to ventral ends.

Part *a* of Fig. 11-6 shows the structures at the dorsal end of the cochlea. The scala tympani does not extend this far dorsally; the papilla lies on the edge of the limbus. The papilla is wedged between the tall border cells of the epithelial layer, and here contains 4 longitudinal rows of hair cells. The supporting cells are numerous and send up a good many fine columnar processes between the hair cells to the papillar surface. The tectorial membrane is a coarse network, and the tectorial structure over the hair cells at this point is little more than a moderate thickening of the membrane. Reissner's membrane is a rather thick, double-layered structure, with numerous capillaries on its vestibular surface.

In part *b* of this figure, representing a section a little distance ventrally, the scala tympani has appeared. The papilla here is tall and moderately wide and rests on the basilar membrane with its thin fundus. There is a tectorial plate of definite form, and the number of rows of hair cells has increased to 7. A thick band of nerve fibers passes through the medial limbus to the anteromedial corner of the papilla, where these fibers lose their medullary sheaths and pass between the border cells to the papilla.

In part *c* of Fig. 11-6 the papilla is still larger, though the rows of hair cells are about the same, and the tectorial plate is somewhat heavier.

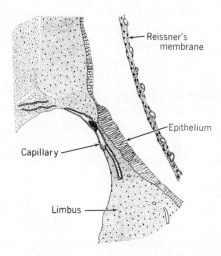

FIG. 11-5. The hiatus between dorsal and ventral portions of the auditory papilla in a specimen of *Varanus griseus*.

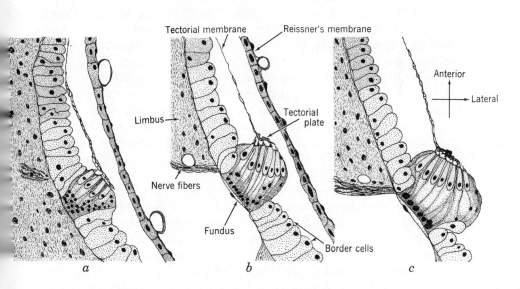

FIG. 11-6. Three frontal sections of the
auditory papilla of *Varanus bengalensis nebulosus* in the
dorsal region of the cochlea. Scale 250×.

This plate over most of the dorsal region is a heavy reticular structure with considerable thickness. A careful examination reveals several perforations or thin areas, and on the underside a number of pits with sharp projecting edges. The tips of the ciliary tufts either enter these pits and make contact with their inner surfaces or attach to the projecting portions.

Section *d* as shown in Fig. 11-7 represents the structures still farther ventrally. Here the papilla shows a curious enlargement on its lateral side and a slight reduction in the rows of hair cells. The number of columnar processes is greatly increased, and many of these are present in the elevated portion of the papilla. This picture also shows the courses of many of the nerve fibers passing through the lower part of the papilla to the hair cells. These fibers are not usually seen with the stains employed, but fortuitously showed up well in this instance. The fibers seen were more than ten times as numerous as shown in the drawing. The tectorial membrane is considerably thinner than in more dorsal regions, and the tectorial plate remains thick and heavy.

Section *e* is from a still more ventral region, approaching the end of the dorsal segment of the cochlea. The papilla is further enlarged, mainly by the upward and lateral extension of its lateral side. The longitudinal rows of hair cells are reduced to four. The supporting cells and their columnar processes are still numerous. The tectorial membrane is much as before, though its plate is narrowed in harmony with the reduced rows of hair cells.

In *f* of this figure is a section through the middle of the hiatus between dorsal and ventral segments. The papilla is reduced to a small framework containing supporting cells and their processes, but devoid of hair cells. The basilar membrane continues beneath the narrow base of the papilla, but a fundus cannot be discerned. The tectorial membrane extends toward the papilla, but a tectorial plate is lacking and no peripheral contact is made.

Beyond the hiatus as shown in Fig. 11-8 the cochlear structures are again prominent. In *g*—a location about a third of the distance between hiatus and ventral end—the papilla is wide, with 9 rows of hair cells, and rests on a broad basilar membrane with a thin fundus. The tectorial membrane is a little thicker and shows its reticular structure clearly. The tectorial plate has a dense reticular form.

Still farther ventrally, near the middle of this ventral segment as shown in *h*, the papilla is taller, with its lower portion greatly compressed between the many border cells, and with a moderately increased number of hair cells. The thin columnar processes are especially numerous. The tectorial plate here, as throughout the ventral segment, consists of a loose network in contrast to the relatively dense body over most of the dorsal segment.

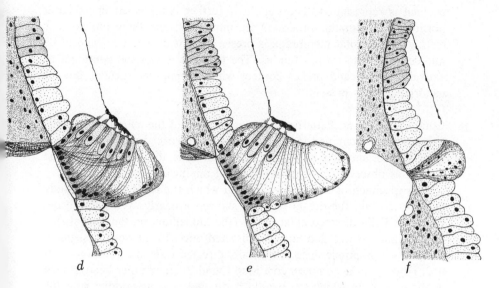

FIG. 11-7. Three views of the same specimen as in the preceding figure, taken from the middle of the cochlea. Scale 250×.

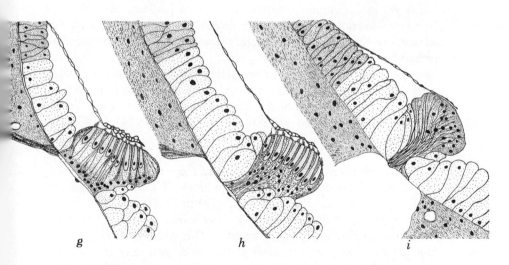

FIG. 11-8. Three views of the same specimen as in the two preceding figures, taken from the ventral region of the cochlea, beyond the hiatus. Scale 250×.

In *i*, the final picture of this series, the extreme ventral end of the cochlea is represented. The papilla is further compressed in its basal portion, the columnar processes are numerous, and there are no hair cells. A thin basilar membrane is present, but it does not thicken in its middle portion to form a fundus. The tectorial membrane continues to extend outward and makes contact with the papillar surface, but no tectorial plate is present.

Ciliary Orientation Patterns.   The orientation of the ciliary tufts of the hair cells was examined in all specimens, and worked out in detail in a few. This orientation is bidirectional, but fails to show a systematic pattern as observed in many other lizard families.

In a specimen of *Varanus griseus*, in which the most extensive study was carried out, the dorsal segment showed a nearly even balance between medially directed ciliary tufts (the kinocilium on the medial side of the ciliary array) and laterally directed ones. In the ventral segment there was a preponderance of medially directed tufts in a ratio of about 3:2. However, the common condition found in many other lizards, such as the geckos, in which the papilla is divided near its middle, with the medially directed tufts on the lateral side and the laterally directed ones on the medial side, does not hold regularly for this ear, though it occasionally occurs as was shown in Fig. 11-6*a* above.

And yet the orientation pattern is far from random: the tufts of a given row sometimes show the same orientation, or more often all face the same way except one, and in other rows other regular arrangements are found.

Further study of the orientation pattern in the varanids is needed, preferably with the scanning electron microscope, which provides a more comprehensive view of the whole papillar surface. Some limited observations of this kind were made by Miller (1974) in *Varanus exanthemicus*.

Cochlear Dimensions.   A detailed study of the dimensions of the cochlear structures was carried out on a specimen of *Varanus bengalensis nebulosus*, with results shown in Figs. 11-9 to 11-11.

The upper curve of part *a* in Fig. 11-9 shows the varying width of the basilar membrane along the cochlea. This width quickly rises to a value around 100 μ and remains at that level with only minor variations over the whole dorsal segment. It drops precipitously to about 10 μ at the hiatus, which occurs at a point 1500 μ from the dorsal end, almost exactly two-thirds of the distance along the cochlea. Beyond this point the width increases rapidly to a new maximum around 125 μ, which is maintained with only a slight decline over most of the ventral segment.

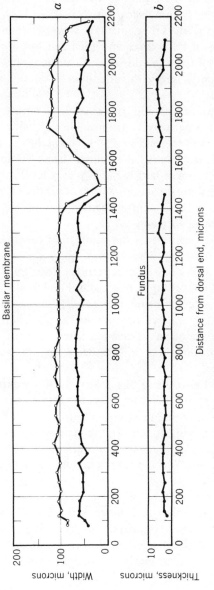

Fig. 11-9. Cochlear dimensions in a specimen of *Varanus bengalensis nebulosus*. Part *a* represents the width of the basilar membrane (upper curve) and of the fundus (lower curve). Part *b* represents the thickness of the fundus.

The width of the fundus, shown by the lower curve of this figure, varies in practically the same way as the whole membrane, except that the fundus disappears in the region of the hiatus.

The thickness of the fundus, shown in *b* of Fig. 11-9, is slight and varies but little. In the dorsal segment it wavers around a value of about 3 μ over most of its course, and then reaches about twice this value at a point 1300 μ from the dorsal end. At the hiatus it disappears. In the ventral segment it grows a little thicker, attaining 5-6 μ over most of that segment.

The size of the auditory papilla, represented as an area in part *c* of Fig. 11-10, increases progressively from an initial value around 2000 sq μ to a maximum near the end of this segment of about 9000 sq μ, an increase of 4.5-fold, then in the hiatus drops abruptly to a low value. In the ventral segment it rises rapidly to a level around 5000 sq μ, which is held with some variation, and finally falls off rapidly at the end. The maximum size in the ventral segment is a little more than half that attained in the dorsal segment.

Part *d* of Fig. 11-10 represents the number of longitudinal rows of hair cells along this cochlea. The number quickly rises to 6 and remains at this value, with occasional variations, over half the length of the cochlea. Beyond the 1000 μ point this number drops in varying steps to 4, then in the region around 1500 μ there are several sections with a single hair cell, or at two points none at all. Thereafter a rapid and progressive increase occurs, reaching a new level of 11 rows, followed by an irregular fall to 8 or 9, and then a final drop as the papilla ends.

The density of the hair cells in this specimen of *Varanus bengalensis* is shown in Fig. 11-11 as increasing rapidly to a maximum near the dorsal end, and then declining progressively to the end of this segment, where it quickly falls to almost zero. In the ventral segment the increase is rapid to a somewhat irregular maximum area, and falls precipitously again as the papilla ends. The ventral maximum is nearly twice as large as the dorsal one.

These density changes closely follow the row structure of the papilla, but depart markedly from the areal dimension, especially along the dorsal segment. As noted earlier, the papilla is distorted by the appearance of a large hump on its lateral side, which seems to crowd the hair cells medially, and increasingly so along the dorsal segment until a maximum is reached near its end.

Numbers of Hair Cells.   Counts of hair cells were made in both ears of five specimens of *Varanus bengalensis nebulosus* and in two specimens each of the other two species, and averages for these three species are

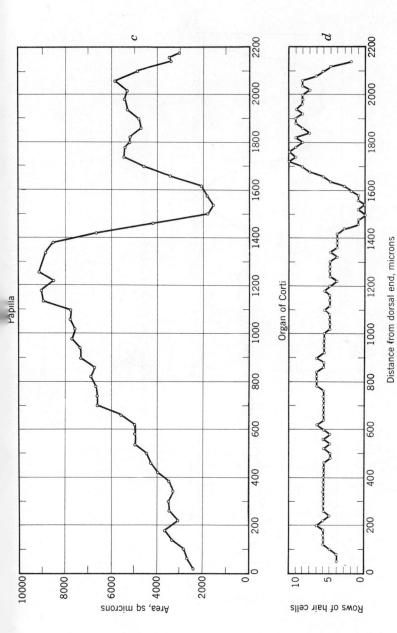

FIG. 11-10. Cochlear dimensions in *Varanus bengalensis nebulosus*. Part *c* shows the area of the auditory papilla. Part *d* shows the varying number of longitudinal rows of hair cells along the cochlea.

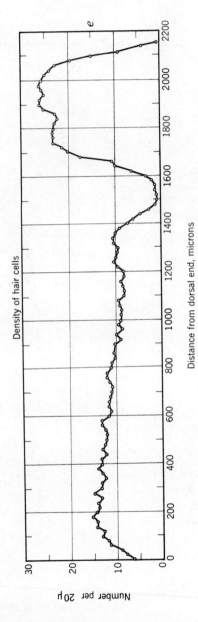

FIG. 11-11. Hair-cell density in *Varanus bengalensis nebulosus*.

given in Table 11-I. The numbers are given separately for the dorsal and ventral segments, with the totals in the last column.

A large population of hair cells is present in *Varanus bengalensis nebulosus* (averaging 1285), of which nearly two-thirds are in the dorsal segment. In *Varanus niloticus* the average number is 902, with nearly the same proportion in the dorsal segment; and in *Varanus griseus* this average is 424, with a little greater share in the dorsal segment. The variation in total population thus is about threefold among these species.

TABLE 11-I

MEAN HAIR-CELL POPULATIONS
IN VARANIDAE

Species	Number of Ears	Dorsal Segment	Ventral Segment	Total
Varanus bengalensis nebulosus	10	802	483	1285
Varanus niloticus	4	554	348	902
Varanus griseus	4	291	133	424

COCHLEAR POTENTIAL STUDIES

Sensitivity curves for two specimens of *Varanus bengalensis nebulosus* are presented in Fig. 11-12. A high degree of sensitivity is indicated in the medium low to medium high frequencies, with moderate differences between the two animals shown. The maximum for one animal is at 600 Hz, where the curve reaches −45 db. For the other animal there are two maximums, one at 800 and the other at 2500 Hz, with the better of these reaching −49 db. Beyond the middle-tone region the sensitivity declines at rates of about 20 db per octave at the low end and about 30 db per octave at the high end.

Results for two specimens of *Varanus griseus* are given in Fig. 11-13. This species shows the best sensitivity around 500-1000 Hz, where the very acute level of −60 db is reached. At the two extremes of the frequency scale the sensitivity declines in much the same fashion as in the preceding species.

Two specimens of *Varanus niloticus*, whose cochlear potential curves are represented in Fig. 11-14, gave somewhat lower levels of sensitivity than the ones just shown. One specimen had maximums around −40 db and the other reached only −20 db. Again the best region is in the middle tones, around 400-800 Hz in one animal and 700-1500 Hz in the other.

One of the specimens of *Varanus bengalensis nebulosus* was used for a study of sensitivity to vibratory stimulation. First, for purposes of com-

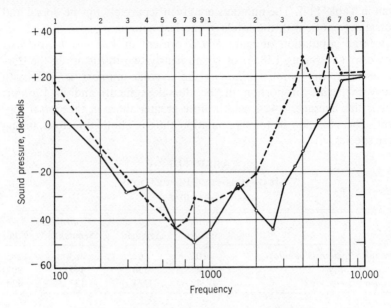

FIG. 11-12. Sensitivity functions for two specimens of
Varanus bengalensis nebulosus.

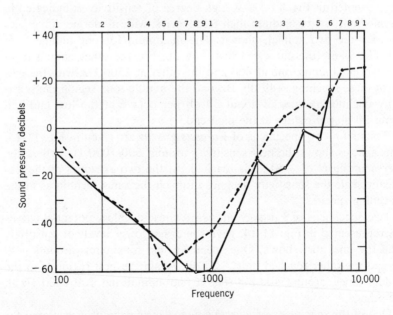

FIG. 11-13. Sensitivity functions for two specimens of *Varanus griseus.*

parison, an aerial response curve was obtained in the usual way, with the sounds applied through a tube sealed over the external auditory meatus, and the curve of Fig. 11-15 was obtained. Then, with the recording electrode still in position on the round window membrane, the ear was stimulated with a mechanical vibrator in two different ways. In one series the tip of the vibrating needle was located on the tympanic membrane over the end of the pars inferior, thus driving the columellar system directly, and gave the results shown by the solid curve of Fig. 11-16. In a second series the vibrating needle was applied to the exposed surface of the quadrate bone, with results indicated by the broken curve of this figure.

It is evident that quadrate stimulation shows poorer sensitivity to the extent of about 20 db in the region of frequencies where this ear performs best, and also this maximum region is considerably narrower than for stimulation of the pars inferior. Sound vibrations reaching the inner ear by way of the columellar system are significantly more effective than those conducted through the quadrate and other bones of the skull.

A comparison with the sensitivity curve of Fig. 11-15 may now be made, though, as mentioned earlier, this comparison of aerial and vibratory curves must be made with some reservation because not only are the conductive pathways different but there are differences in the physical quantities by which the stimuli are measured: one is in units of sound pressure and the other in units of vibratory amplitude.

Taken grossly, the aerial and vibratory functions are in agreement in showing the best sensitivity as located broadly in the middle frequencies from 400 to 2000 Hz, which clearly can be attributed to basic characteristics of the inner ear. The aerial function indicates maximum sensitivity at the lower end of this range, which can be considered as due in part to the greater effectiveness of the middle ear mechanism in this low-frequency range.

These observations show excellent auditory sensitivity in these varanid species. The medium low and middle frequencies are favored, and this is true for both aerial and vibratory stimulation, though the vibratory curves are flatter in this region. The general picture is compatible with the observation of a very large population of hair cells and only a slight amount of differentiation along the basilar membrane.

FAMILY HELODERMATIDAE: THE BEADED LIZARDS

The beaded lizards are so called because of the nodular form of their scales and their orderly arrangement over much of the body surface. These scales are seen in the neck and throat regions in Fig. 11-17, along

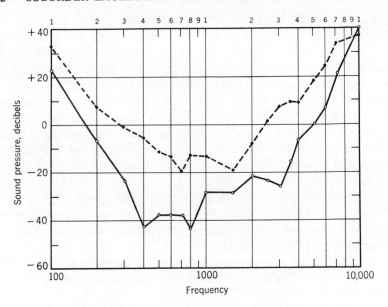

FIG. 11-14. Sensitivity functions for two specimens of *Varanus niloticus*.

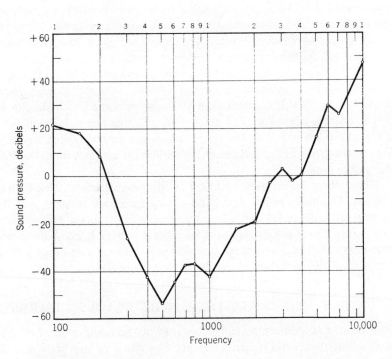

FIG. 11-15. An aerial sensitivity curve for a specimen of *Varanus bengalensis nebulosus* whose vibratory functions are shown in the next figure.

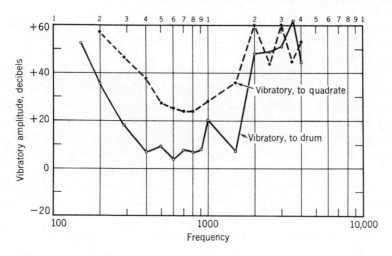

FIG. 11-16. Sensitivity in a specimen of *Varanus bengalensis nebulosus* for vibratory stimulation, with the vibrator point applied to the tympanic membrane over the pars inferior (solid line) and applied to the quadrate bone (broken line). Compare with the aerial curve for this same ear in the preceding figure.

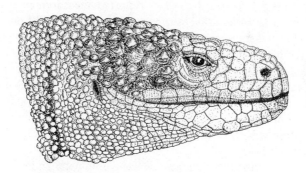

FIG. 11-17. The head of *Heloderma horridum*,
redrawn from Bogert and Martín del Campo (1956),
after Duméril, Bocourt, and Mocquard (1874).

with greatly enlarged scales on the dorsal surface of the head. These are large-bodied, sturdy animals, attaining body lengths around 50-90 cm and weights of 3-4 kg. They are generally slow-moving, but are capable of swift lunges in offense and defense, as the unwary may discover.

These lizards belong to a single genus, *Heloderma*, containing two species, *horridum* and *suspectum*, the first of these with three subspecies and the second with two (Bogert and Martín del Campo, 1956).

The helodermatids are limited to scattered regions of the southwestern United States and Mexico, with the *suspectum* species occurring in

Arizona and its border areas, including the northwestern corner of Mexico, and the *horridum* species mainly distributed along the Pacific coastline of Mexico.

The habitat is varied and includes tropical forest and coastal plains as well as shrubby areas and semiarid rocky regions. These animals feed on a variety of small mammals and on the eggs of ground-nesting birds.

They are unique in being the only venomous lizards. Venom glands are present as modified salivary glands in the lower jaw adjacent to deeply grooved teeth. The most prominent of these teeth have cutting edges adjacent to the grooves, and the venom is worked into the tissues of the enemy or prey as it is seized and repeated biting pressures are applied. The bite can be serious in man, but accounts are confused on the matter of fatalities from this cause.

Two specimens of *Heloderma horridum* (Wiegmann, 1829) were studied.

ANATOMICAL STRUCTURES

External and Middle Ear.   As shown in Fig. 11-17, the external ear opening is an elongated oval only slightly inclined from the vertical. The tympanic membrane, of a broad oval form, is not usually visible from the outside. Versluys (1898) reported measurements of the size in a specimen of *Heloderma suspectum* as 6 mm in width and 9 mm in height. The present specimens of *Heloderma horridum* had somewhat smaller tympanic membranes, one measuring 4.5 by 6.2 mm.

The middle ear is of the iguanid type, with a short internal process running to a notch in the edge of the quadrate, as shown in Fig. 11-18. The columella is a slender bony rod, which in one specimen measured 7.8 mm from the footplate to its junction with the extracolumella. The latter process was 3.8 mm long from this junction to the plane of the tympanic membrane.

Versluys described the junction between columella and extracolumella in *Heloderma suspectum* as a joint with a definite cleft, the surface on the columella side being concave and fitting a convex surface at the extracolumellar shaft. In the two *Heloderma horridum* specimens studied no joint was found. As Fig. 11-18 shows, the inner end of the bony columellar shaft is continued for a short distance by cartilage, and then this goes over into a large rounded mass of fibrocartilage, which in turn leads to the hyalin cartilage of the extracolumellar shaft. These parts are closely joined with no physical discontinuity, though on the outer or extracolumellar side there is a clear demarcation in the two forms of cartilage. The cartilage on the inner side of this junction is of a mixed type with a gradual transition to the mass of fibrocartilage.

The footplate end of the columella shows only a minor expansion to

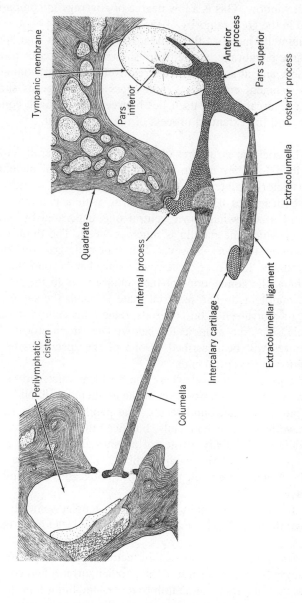

FIG. 11-18. The middle ear mechanism of *Heloderma horridum*, reconstructed from serial sections. Scale 10×.

a width of about 0.6 mm in its osseous part, which is about three times the diameter of the shaft along its midportion. Rounded edges of cartilage extend this osseous end to 0.84 mm, so that it fills about half the area of the oval window. This is a low degree of coverage in comparison with that found in the great majority of ears.

The extracolumella includes the standard four processes that spread over the tympanic membrane, as the figure shows. The posterior process is well developed, and attached to it is a relatively long extracolumellar ligament. This ligament embodies a large mass of cartilage as shown, and ends in the intercalary that itself lies between the cephalic condyle of the quadrate and the extended process of the paroccipital.

Inner Ear.   The cochlear structures of *Heloderma* are generally similar to those described for *Varanus*, in that there is a tectorial membrane extending to a tectorial plate over the greater part of the auditory papilla, and these structures have a duplex form, with a marked dorsal development and a somewhat lesser ventral one, separated by a wide middle region in which the papilla is much reduced. The duplexity is not as striking as in *Varanus*, however. The hair cells are continuous, and although reduced to two or three rows somewhat beyond the middle of the cochlea, the connections of their ciliary tufts to the tectorial membrane are maintained. The tectorial plate is much the same as in *Varanus* near its dorsal end, but over its remaining course it differs greatly, becoming clawlike as indicated below. The varying form of the cochlear structures will be described in one of the specimens as it is followed from dorsal to ventral ends.

The papilla appears at a level where the cochlear duct and saccule first become separated by a constriction of the epithelial wall, and where the scala tympani is a small blind recess. The papilla increases rapidly in size, with several rows of hair cells, but here these cells have only free-standing ciliary tufts. Only a trace of a tectorial membrane can be discerned on the surface of the medial limbus, where there is a slight elevation of the limbic plate itself and a more noticeable thickening of its epithelial layer.

The papilla has a relatively thin fundus, and occupies somewhat less than half the width of the basilar membrane, well displaced to its medial side.

A little farther along the cochlea a thin tectorial plate lies close over the hair cells along the main portion of the papillar surface, and connects to a tectorial membrane that can be followed over the limbus all the way to the lagenar end of the cochlear duct, where it seems to arise near the root of Reissner's membrane. The tectorial membrane is extremely thin over the low limbic elevation, and leaves this surface at a grazing angle as it runs to its plate structure over the papilla. This condition is represented in Fig. 11-19.

A little farther ventrally this plate thickens and exhibits the deep, cavernous recesses on its undersurface found in many other reptilian species. The caverns are formed by extended thornlike processes that come close to the papillar surface, mainly at points between the ends of the hair cells.

Still farther ventrally the papilla grows smaller, with a reduction in the rows of hair cells and also in the size of the tectorial plate. This plate also becomes more clawlike, as shown in Fig. 11-20.

More ventrally still, beyond the middle of the cochlea, the number of rows of hair cells is reduced to 3 and finally to 2 in a narrow region, and the tectorial plate is reduced to two or three narrow prongs that seem only to be a continuation of the rather thickened tectorial membrane.

Farther toward the ventral end of the cochlea the papilla shows a secondary enlargement, with the number of rows of hair cells increasing rapidly to 5 and 6. The tectorial plate structure also changes, with a definite clawlike form appearing once more, and then comes a dwindling to a small plate-like process at the end of an extremely thin membrane. Finally, at the ventral end, the tectorial membrane disappears, and the papilla terminates.

A curious feature, seen in three of the four ears examined, is the appearance at the extreme dorsal end of the papilla, just as the tectorial plate first extends across the middle rows of hair cells, of a small sallet that lies over the medial edge of the papilla and makes connection with the ciliary tufts of the two most medial rows of hair cells. This structure could be followed only a short distance of 60-90 μ, and then it disappeared. Probably it is of little or no functional significance.

The total number of hair cells in the four ears examined varied from 281 to 325, and averaged 311.3. In one of these ears there were 184 cells in the dorsal region up to the minimum, which in this ear appeared about two-thirds of the distance along the cochlea, and 136 cells in the ventral region.

The ciliary orientation of the hair cells is bidirectional, but with a bias toward laterality. In general, the more medial cells beneath the tectorial plate have a lateral ciliary orientation and one or two rows at the lateral edge of the plate show a medial orientation. Often, and more noticeably so toward the ventral end of the cochlea, the tectorial plate fails to extend to the most lateral row of hair cells, and these free-standing cells show medial ciliary orientation.

Dimensional Measurements. — The changing forms of the cochlear structures are most easily followed in Fig. 11-21. Shown in the upper curve of part *a* is the varying width of the basilar membrane, in which

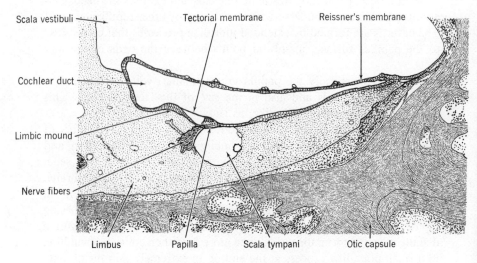

FIG. 11-19. A frontal section through the cochlear region of *Heloderma horridum*. Scale 50×.

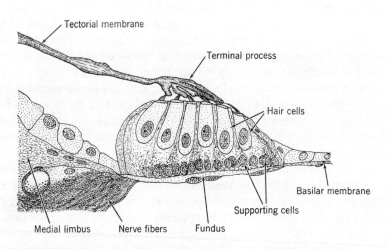

FIG. 11-20. The auditory papilla in *Heloderma horridum*. Scale 500×.

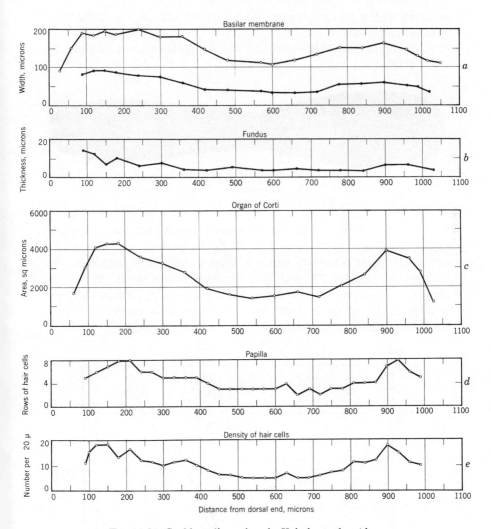

FIG. 11-21. Cochlear dimensions in *Heloderma horridum*.

a bimodal form is evident. The width of the fundus, represented by the lower curve, is markedly less, especially near the middle of the cochlea, but presents the same form. As shown in part *b* of this figure, the fundus is particularly thin, especially over the midregion.

The size of the papilla, shown in part *c*, exhibits the duplexity already seen in the basilar membrane, with a large maximum near the dorsal end and a smaller one far ventrally.

The arrangement of the hair cells, represented in part *d* as the row structure and in part *e* in terms of hair-cell density, closely follows the general dimensions.

A representation of the form of the basilar membrane is given in Fig. 11-22.

FIG. 11-22. The form of the basilar membrane in *Heloderma horridum*. Scale 100×.

Auditory Sensitivity. Cochlear potential functions for the two specimens studied are presented in Fig. 11-23. *Heloderma horridum* shows a high degree of sensitivity within a narrow range in the medium low frequencies, between 300 and 1000 Hz. Within this range one animal attained a maximum of −46 db, and the other reached the remarkably acute level of −67 db. Probably this species possesses serviceable hearing over about three octaves of the low and middle range.

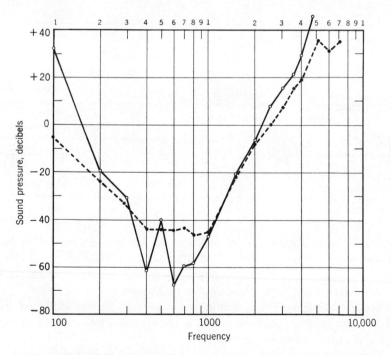

FIG. 11-23. Auditory sensitivity for two specimens of *Heloderma horridum*.

12. FAMILY TEIIDAE:
THE TEIID LIZARDS

The teiids form a group of about 40 genera including some 200 species limited to the Western Hemisphere. These lizards are most abundant in the tropical regions of South America, and it is believed that the family originated there. Some members later spread north and south, with only a few making their way along the land bridge into Central America and Mexico, and a single genus, *Cnemidophorus*, entering the United States. These lizards are almost exclusively land dwellers, but are found in a variety of localities, including forests, deserts, and mountains as well as open fields. Some are burrowers, and a few are semiaquatic, notably the caiman lizard *Dracaena* that lives at water's edge feeding mainly on snails. A common form is the six-lined race runner, *Cnemidophorus sexlineatus*, from the southeastern and central part of the United States, pictured in Fig. 12-1.

FIG. 12-1. The six-lined race-runner, *Cnemidophorus sexlineatus*.
Drawing by Anne Cox.

Seven species were available for the present study, including *Ameiva ameiva* (Linnaeus, 1758)—10 specimens; *Ameiva lineolata privigna* (Schwartz, 1966)—1 specimen; *Cnemidophorus lemniscatus* (Gray, 1845)—5 specimens; *Cnemidophorus tessellatus aethiops* (Cope, 1898)—3 specimens; *Cnemidophorus tigris gracilis* (Baird and Girard, 1852)—7 *specimens; Cnemidophorus velox* (Springer, 1927)—4 specimens; and *Tupinambis nigropunctatus* (Gray, 1845)—4 specimens.

ANATOMICAL OBSERVATIONS

MIDDLE EAR

The conductive mechanism of teiids is of the iguanid type, with the general form as represented in Fig. 12-2 for *Ameiva lineolata*. In this species the posterior process is present, but an anterior process was not seen. A particularly large extracolumellar ligament arises from the cephalic condyle and inserts on the pars superior.

INNER EAR

The cochlea is well developed. The limbus is greatly thickened in its anterolateral portion, and curls around in a lip resembling the one found in gekkonids and pygopodids. As shown in Fig. 12-3, which is a cross section from a specimen of *Cnemidophorus velox* near the dorsal end of the auditory papilla, this limbic lip is particularly thick—much more so than in the other two families mentioned. At this point along the cochlea the lip is only moderately curled, but farther ventrally the curvature in-

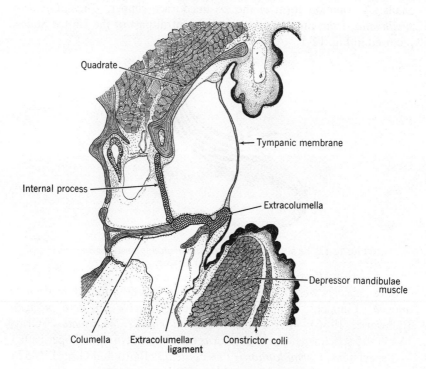

FIG. 12-2. The middle ear of *Ameiva lineolata*,
drawn from a series of sections. Scale 40×.

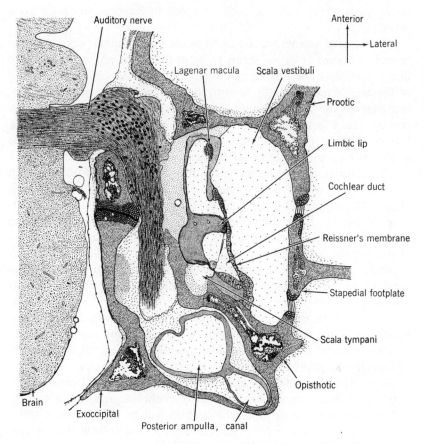

Fig. 12-3. The auditory region of *Cnemidophorus velox*, in a frontal section
near the dorsal end of the cochlea. Scale 50×.

creases and the outer end of the lip comes to lie almost directly over the
papilla as it commonly does in the geckos and pygopodids.

At its anterior end, as may be seen in the figure, the limbus is extended as a narrow strip that runs forward, then bends laterally and
turns back as a loop to form the enclosure of the lagenar portion of the
cochlear duct. The terminal end of the loop serves as the forward anchorage of Reissner's membrane, which stretches across in a posterolateral direction to its second anchorage on the opisthotic bone. The figure
shows this membrane in close contact with the lateral crest of the limbic
lip, and this condition was often observed. Whether a contact is made in
life is uncertain, but the point probably lacks any functional significance.
Reissner's membrane even when foreshortened by this contact, and even
when it is thick as in this species, will still have sufficient flexibility to

move freely along with the cochlear fluids in response to sound pressures.

The tectorial membrane arises from the anterolateral surface of the limbus. Its point of origin seems to be at the notch just anterior to the heavy portion of the limbus, as observed in other species. A thickening of the membrane is usually seen here, with indications of a secure attachment to the limbic surface. In this species a very thin membrane can be traced farther forward along the limbic extension and into the lagenar macula. Thus continuity exists between the tectorial membrane of the cochlea and the tectorial reticulum of the lagena.

The tectorial membrane can nearly always be followed from the notch anterolaterally to the posterolateral edge of the limbic lip, where it leaves the surface and extends to the auditory papilla. At times a definite thickening of the membrane may be seen at the point of leaving the surface. In this species a further thickening of the membrane appears in the middle of its free portion, as indicated in Fig. 12-3 for *Cnemidophorus velox*. This thickening is slight in comparison with that seen regularly in the diplodactyline species and in *Thecadactylus rapicaudus* among the geckos.

The basilar membrane and auditory papilla present a degree of duplexity along the cochlea. The width of the membrane shows two maximums, one near the dorsal end and the other far ventrally, and correspondingly the papilla undergoes striking alterations in size.

Along with these changes are variations in the hair-cell connections, differing in detail among species and in smaller degree within species.

The varying structure along the cochlea will be followed in a specimen of *Cnemidophorus velox*. In this ear the hair cells at the extreme dorsal end of the cochlea have their tufts ending freely, without any tectorial connections. Just ventral to this small region is a single sallet making contact with the hair cells in two transverse rows. The form of this sallet is a flattened sphere much as was shown in Fig. 4-14*b* for a related species. Then still more ventrally a plate appears, attached to the tectorial membrane. This plate continues along most of the cochlea, but varies considerably in form over its course. Some of these variations are seen in Figs. 12-4 and 12-5, which show sections taken from two regions, one mid-dorsal and the other toward the ventral end of the cochlea.

In the narrowest part of the basilar membrane, beyond the dorsal maximum, the tectorial plate in *Cnemidophorus velox* is reduced to a minute fragment, whereas in the maximum region (as in Fig. 12-5) it covers a large part of the papillar surface. In the far ventral region this plate loses its connection to the tectorial membrane. A little farther ventrally it is replaced by rounded sallets much like the one found at the dorsal end. At the extreme ventral end in this specimen the hair tufts end freely, without any overlying structures. These relations vary in detail among individual specimens.

Somewhat similar variations along the cochlea are shown for a specimen of *Cnemidophorus tessellatus aethiops* in the cross-sectional views of Figs. 12-6 to 12-8. In the first of these the cochlear structures are shown in a section through the peak of the dorsal region, where the limbic lip is well developed and the tectorial membrane runs from its tip to a substantial tectorial plate. More ventrally the papilla grows

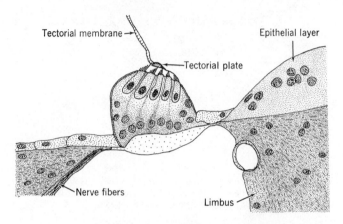

FIG. 12-4. The auditory papilla of *Cnemidophorus velox* in the dorsal region of the cochlea. Scale 500×.

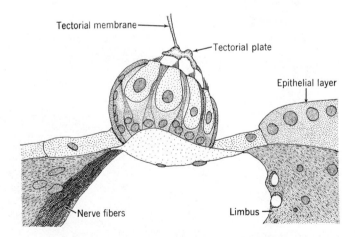

FIG. 12-5. The auditory papilla of *Cnemidophorus velox* in the ventral region of the cochlea. Scale 500×.

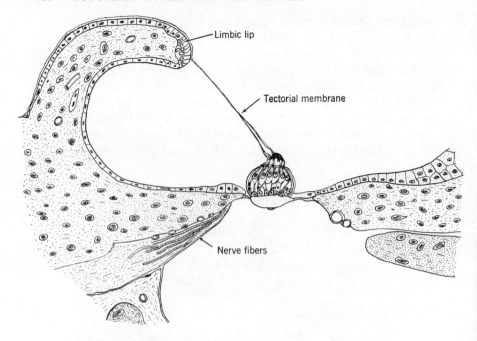

FIG. 12-6. The auditory region of *Cnemidophorus tessellatus aethiops*, seen at its peak in the dorsal portion of the cochlea. Scale 250×. From Wever, 1965a.

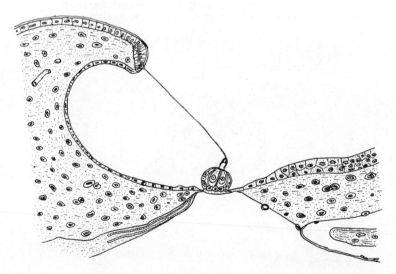

FIG. 12-7. The auditory region of the same specimen as in the preceding figure, seen at the minimum between dorsal and ventral expansions of the papilla. Scale 250×. From Wever, 1965a.

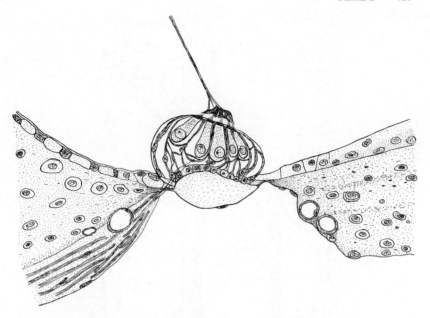

Fɪɢ. 12-8. The auditory region of *Cnemidophorus tessellatus aethiops* seen at its peak in the ventral portion of the cochlea. This is the same ear shown in the two preceding figures. Scale 250×. From Wever, 1965a.

smaller and reaches a minimum, as shown in Fig. 12-7, about a third of the distance along the cochlea. Here only two rows of hair cells remain, and their ciliary tufts connect to a much reduced tectorial plate. Farther ventrally the papilla grows rapidly in size and reaches a new maximum, much larger than the first, at a point toward the ventral end. The structure here has the form shown in Fig. 12-8. The tectorial plate is again substantial, though the rows of hair cells number no more than 5.

A more comprehensive view of the varying structures along the teiid cochlea was obtained in a specimen of *Ameiva ameiva* in which a longitudinal orientation gave sections passing through the entire auditory papilla as shown in Fig. 12-9. Here a single sallet appears at the dorsal end, then two rows of free-standing hair cells, followed by a tectorial plate that continues over most of the cochlea. As the figure shows, the tectorial membrane does not extend to the most dorsal portion of this plate, but its restraining influence is transmitted along a connecting strand of tissue. Ventrally a similar condition prevails with a large sallet contributing its effects.

Cochlear Dimensions.   Measurements of cochlear structures were made in three species, *Cnemidophorus tessellatus aethiops, Cnemidophorus tigris gracilis*, and *Tupinambis nigropunctatus*. The first set of measurements is shown in Fig. 12-10.

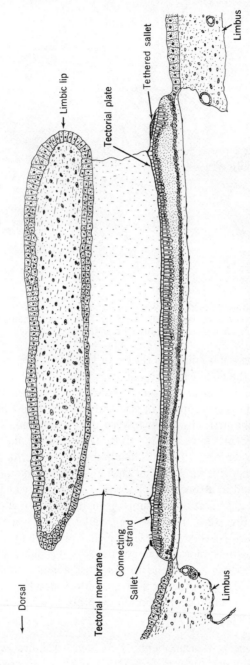

FIG. 12-9. A longitudinal view of the auditory papilla of *Ameiva ameiva*, based upon a series of sagittal sections. Scale 125×. From Wever, 1971c.

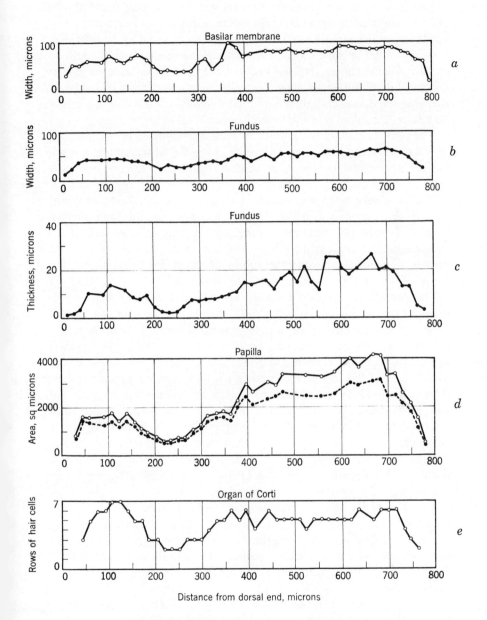

FIG. 12-10. Cochlear dimensions in a specimen of
Cnemidophorus tessellatus aethiops. From Wever, 1965a.

Part *a* of this figure shows the varying width of the basilar membrane over its length of nearly 800 μ. The width is duplex, with one maximum appearing in the early dorsal region and another extending broadly over the ventral half of the cochlea. The width of the fundus (part *b*) follows much the same course. The thickness of the fundus, represented in part *c* of this figure, presents a clear duplexity, with a maximum of moderate size near the dorsal end, then a marked thinning, followed by an irregular rise to a maximum in the ventral region.

The sensory structures exhibit the same variations over the cochlea, with a moderate elevation of the papilla in the dorsal region and a much higher and broader one over the ventral region. The rows of hair cells follow a corresponding pattern.

Similar observations were made on a second species, *Cnemidophorus tigris gracilis*, as shown in Fig. 12-11. In this species the same pattern is seen: a moderate dorsal development and a much greater ventral development. A curious feature is a reduction in the number of hair-cell rows in a region just beyond the middle of the ventral segment, so that the whole curve (part *e*) presents three maximums.

Cochlear dimensions were studied also in a specimen of *Tupinambis nigrolineatus* as shown in Figs. 12-12 and 12-13. In *a* of the first of these figures, the basilar membrane is seen as about 1.5 mm long. Its width varies only moderately along its extent. After quickly attaining a width around 150 μ close to the dorsal end, it narrows to 130 μ a little over a third of the way along the cochlea, then remains nearly constant through the middle region, rises somewhat toward the ventral end, and then terminates. The width of the fundus takes a similar course. The thickness of the fundus (part *b* of this figure) exhibits somewhat greater variations. There is a sharp rise to a maximum in the dorsal region, then a fairly progressive decline to a minimum around the 600 μ position, a rise to a second maximum near the 1100 μ point, and finally a moderate decline until the ventral end is reached.

The area of the auditory papilla, represented in part *c* of Fig. 12-13, shows much the same variations. There is a rather sharp maximum in the dorsal region, a minimum a third of the way along the cochlea, and a second maximum in the ventral region. The ventral maximum is smaller than the dorsal one, contrary to the condition seen in *Cnemidophorus tigris gracilis*.

The rows of hair cells, seen in *d* of this figure, reach a maximum of 10 in the dorsal region, then exhibit remarkable uniformity around 7 through the middle of the cochlea, and finally vary between 6 and 7 over most of the more ventral region.

The density of the hair cells follows closely the row structure.

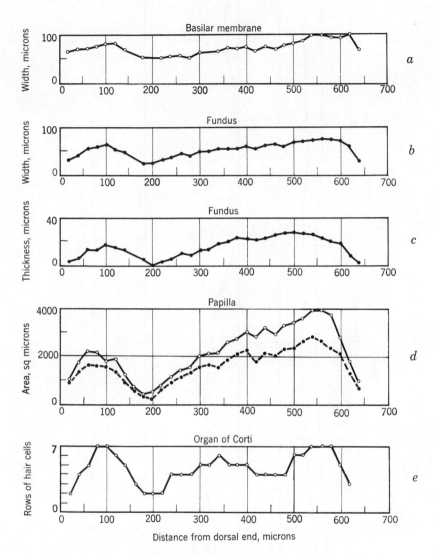

FIG. 12-11. Cochlear dimensions in a specimen of
Cnemidophorus tigris gracilis.

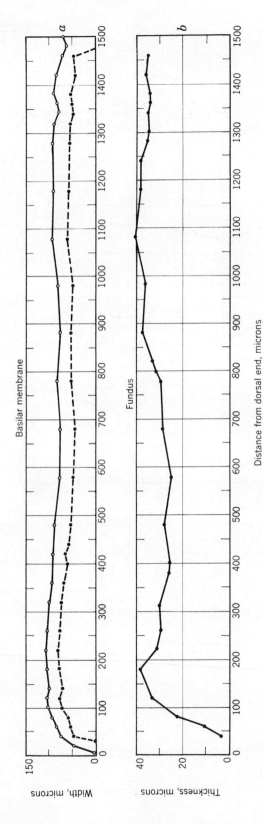

FIG. 12-12. Cochlear dimensions in a specimen of *Tupinambis nigrolineatus*, showing characteristics of the basilar membrane and fundus.

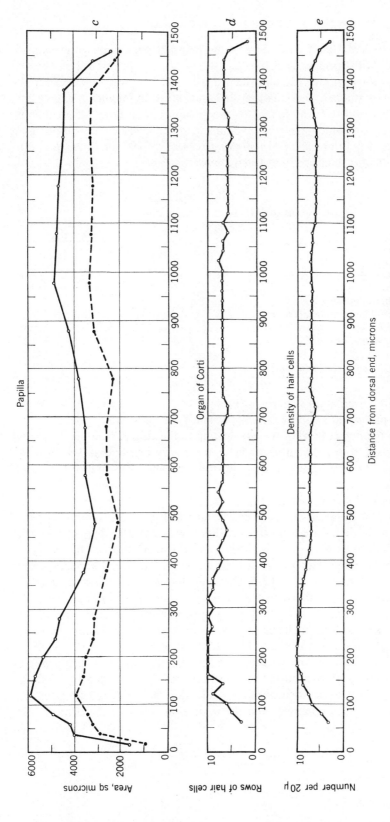

Fig. 12-13. Cochlear dimensions in the same specimen as in the preceding figure, with characteristics of the auditory papilla and the hair-cell distribution.

Numbers of Hair Cells.　Size of the hair-cell population varies considerably with species, as Table 12-I shows. This population is smallest in *Cnemidophorus velox*, for which a mean of 255 was obtained for the five ears studied, and is largest in *Tupinambis nigropunctatus* for which a mean of 1167 was found for the group of four ears. The remaining species for which data were obtained gave intermediate numbers: a mean of 369 for *Cnemidophorus tigris gracilis*, 460 for *Cnemidophorus lemniscatus*, and 551 for *Ameiva ameiva*.

TABLE 12-I

MEAN HAIR-CELL POPULATIONS
IN TEIIDAE

Species	Number of Ears	Mean
Ameiva ameiva	5	550.6
Ameiva lineolata	2	273.0
Cnemidophorus lemniscatus	4	459.7
Cnemidophorus tessellatus aethiops	2	359.5
Cnemidophorus tigris	6	369.2
Cnemidophorus velox	5	254.6
Tupinambis nigropunctatus	4	1167.0

Ciliary Orientation Patterns.　This pattern will be described for a specimen of *Cnemidophorus velox* in which it was examined in detail.

In the dorsal third of the cochlea, up to the constriction that separates this part from the ventral segment, the ciliary orientation is almost exclusively lateral. The picture uniformly presented was that of Fig. 12-4, except in a single transverse row where the extreme lateral cell had its ciliary tuft facing medially.

The orientation was very nearly lateral also in the ventral segment. The main departure from this condition was found in a region near the middle of this segment, where for a short distance most of the hair cells on the extreme lateral ends of their transverse rows had a medial orientation, as shown in Fig. 12-5. Occasionally also one or two cells in the neighboring rows showed such medial orientation. Still farther ventrally, toward the end of the papilla, this appearance of medially facing hair tufts became less frequent. In this region the extreme lateral cells usually alternated between medial and lateral orientations.

Miller (1973b) on the basis of scanning electron microscope studies of *Ameiva ameiva* reported 691 hair cells in all, of which about 600 had their ciliary tufts oriented in the direction designated here as lateral (actually posterolateral), and the remaining cells had the contrary orientation. He counted about 100 stereocilia in each tuft, with the five longest of these standing adjacent to the kinocilium.

In similar observations on *Tupinambis teguixin*, Miller found a total of about 1400 hair cells, with a complex and largely obscure pattern of ciliary orientation, though on the whole he observed an approximate balance between the two directions of orientation. A striking feature is the frequency with which the direction may change in a single transverse row, which can be as many as four times.

The kinocilia of all the teiids studied were found to have enlargements of the tips forming bulbs about twice the diameter of their stalks. This feature is similar to that seen regularly in geckos, though the enlargements are not as prominent. Miller's photomicrographs in his scanning electron microscope study of *Ameiva ameiva* show these enlarged tips.

COCHLEAR POTENTIAL STUDIES

Sensitivity curves for two specimens of *Ameiva ameiva* are shown in Fig. 12-14. A high degree of sensitivity is indicated over a broad range from 300 to 4000 Hz, where the standard response requires only −50 db or less, and with the maximum around 1500-3000 Hz. At the two ends of this range the sensitivity declines at a rate of about 30 db per octave.

A sensitivity curve for a specimen of *Ameiva lineolata* is shown in Fig. 12-15. This function is bimodal, with a sharp maximum of good sensitivity at 400 Hz, and then a second maximum of much greater sensitivity in the region of 1500-2500 Hz, where an acuity of −68 db is attained. In the upper frequencies the decline of sensitivity is at a rate of 34 db per octave, and in the low frequencies it is at a rate of about 22 db per octave.

Results for two specimens of *Tupinambis nigropunctatus* are given in Fig. 12-16. Here also the sensitivity is broad, but less so than in the preceding species. The sensitivity attains −46 to −50 db at its maximum in the region of 600-1000 Hz, which represents a notable acuity. The decline from this region is at a low rate, around 10 db per octave, at the lower end of the frequency scale and about 30 db per octave at the upper end.

The sensitivity was found to vary considerably among the four *Cnemidophorus* species. In *Cnemidophorus lemniscatus*, results for which are presented in Fig. 12-17, the sensitivity is high, around −40 db, for the middle frequencies from 500 to 3000 Hz, and declines at rates of 10-20 db per octave at the upper end. This is evidently a very serviceable ear in the middle range.

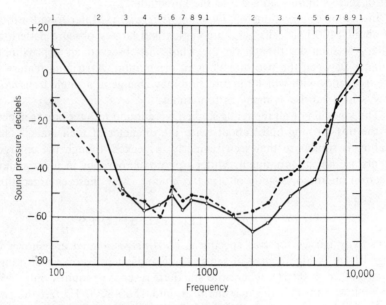

FIG. 12-14. Sensitivity curves for two specimens of *Ameiva ameiva*.

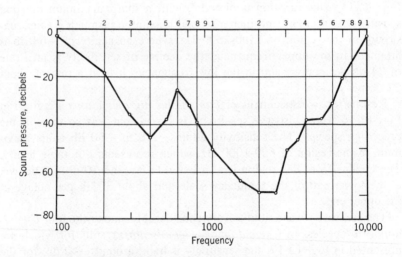

FIG. 12-15. Sensitivity curves for a specimen of *Ameiva lineolata*.

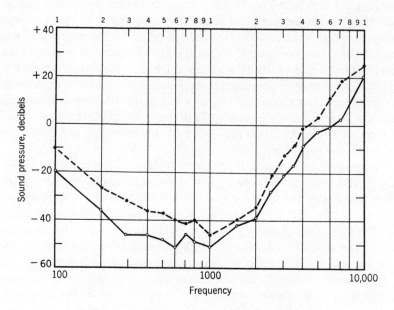

FIG. 12-16. Sensitivity curves for two specimens of *Tupinambis nigropunctatus.*

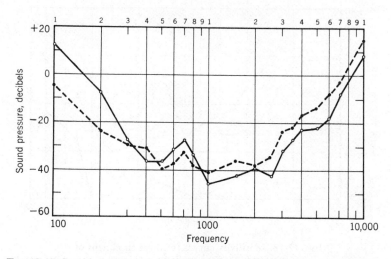

FIG. 12-17. Sensitivity curves for two specimens of *Cnemidophorus lemniscatus.*

Sensitivity data for three specimens of *Cnemidophorus tessellatus aethiops* are shown in Fig. 12-18. Two of these ears performed poorly, but one of them showed good sensitivity, attaining −40 db at several points between 700 and 4000 Hz. The fall-off in the low frequencies was only about 10 db per octave, and in the high frequencies it showed the more common rate of 30 db per octave (though sometimes irregular).

In *Cnemidophorus tigris gracilis* the sensitivity is excellent in the middle frequencies, as shown for two ears in Fig. 12-19. One of these ears attained the remarkable level of −75 db for 1500-2000 Hz. The roll-off in these ears is about 20 db per octave for the low tones and 30 db per octave for the high tones.

Curves for three specimens of *Cnemidophorus velox* are given in Fig. 12-20, and here we find the highest degrees of sensitivity among the teiids tested. For two of these ears the levels approach −80 db in two regions, at 500 and 1500 Hz, with somewhat less sensitivity between. The third specimen goes to −76 db at 2000 Hz in a somewhat irregular curve. The roll-offs are around 30 db per octave at both ends of the sensitive region. This degree of sensitivity is unusual in lizard ears.

In summary, the teiids examined show good to excellent auditory sensitivity in a moderate to broad range in the middle of the frequency scale.

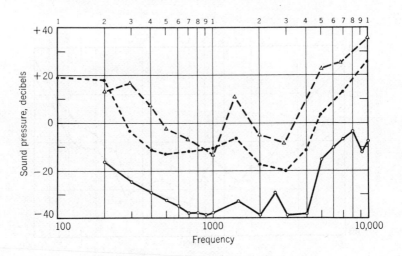

FIG. 12-18. Sensitivity curves for three specimens of
Cnemidophorus tessellatus aethiops.

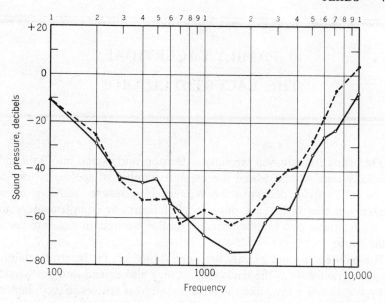

Fig. 12-19. Sensitivity curves for two specimens of
Cnemidophorus tigris gracilis.

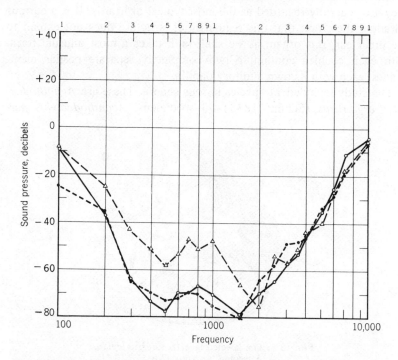

Fig. 12-20. Sensitivity functions for three specimens of *Cnemidophorus velox.*

13. FAMILY LACERTIDAE:
THE LACERTID LIZARDS

The lacertids are common throughout Europe, and extend into Asia and especially into Africa, where the greatest number of species is found. No lacertid species occurs in the Western Hemisphere, and it is often pointed out that in a sense this group of lizards is complementary to the teiids: these two families occupy similar habitats in separate parts of the world.

The habitats are not exactly the same, however; the lacertids mainly populate grasslands as the teiids do, but they also extend into bushy and rocky areas, and a few, like the *Acanthodactylus* species, have adapted to desert environments.

The body form of lacertids is highly regular, without loss of limbs or special adaptations of more than moderate degree. Partly for this reason they are generally regarded as the most typical of lizards. It is a curious paradox, therefore, that this regularity of structure does not extend to the inner ear; this organ, as we shall find, takes a most unusual form, with each cochlea containing two completely separate basilar membranes, each with its own auditory papilla.

The study included 11 species in four genera. These are *Acanthodactylus erythrurus* (Schinz, 1833)—3 specimens; *Acanthodactylus par-*

FIG. 13-1. The jeweled lacerta, *Lacerta lepida*.
Drawing by Anne Cox.

dalis (Lichtenstein, 1823)—6 specimens; *Acanthodactylus scutellatus* (Ardouin, 1829)—3 specimens; *Eremias argus* (Peters, 1869)—2 specimens; *Eremias velox persica* (Blanford, 1874)—3 specimens; *Lacerta agilis* (Linnaeus, 1858)—7 specimens; *Lacerta lepida* (Daudin, 1802) —4 specimens; *Lacerta muralis* (Laurenti, 1768)—7 specimens; *Lacerta sicula* (Rafinesque, 1810)—3 specimens; *Lacerta viridis* (Laurenti, 1768)—6 specimens; and *Psammodromus algirus* (Linnaeus, 1758)—5 specimens. A drawing of the jeweled lacerta is shown in Fig. 13-1.

ANATOMICAL OBSERVATIONS

MIDDLE EAR

The middle ear mechanism in the Lacertidae is of the regular iguanid type, with a long, thin internal process. As seen in Fig. 13-2, this mechanism exhibits no peculiarity except an interruption of the hyalin cartilage of the extracolumellar shaft by a mass of fibrocartilage.

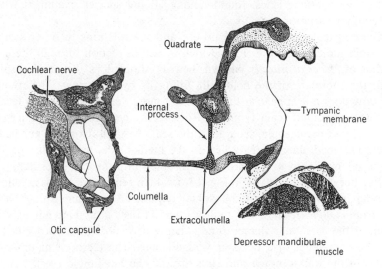

FIG. 13-2. The middle ear mechanism in *Lacerta muralis*. A reconstruction based on serial sections. Scale 37×.

INNER EAR

The duplex character of the lacertid inner ear has already been mentioned. There are two elongate openings in the limbic plate lying end to end, each covered by a separate basilar membrane, and each of these membranes is surmounted by an auditory papilla.

This duality of the inner ear in lacertids has long been known. It was observed by Deiters in 1862 in the species *Lacerta agilis*, and confirmed soon afterward by Clason (1871) in this species and perhaps in *Lacerta ocellata*. At this early date few lizard species had been examined, and the condition in *Lacerta* was regarded as typical. Retzius, who studied 11 different lizard species, seems to have been the first to recognize the duplexity in *Lacerta* as an unusual characteristic, though he found a similar condition in the varanid species *Psammosaurus caspius*. His remaining eight species of lizards had single basilar membranes.

Miller in 1966, in a general study of the cochlear duct of lizards, examined nine species of lacertids belonging to five genera, in all of which this dual structure was present. One of his pictures of the limbic plate is reproduced in Fig. 13-3. A detailed description of this ear was made for the species *Eremias velox persica* (Wever, 1967b), and a further treatment of the species *Eremias argus* appeared soon afterward (Wever, 1968a).

The two sensory structures of lacertids will be referred to as the dorsal and ventral basilar membranes and papillae. They exhibit a number of distinctive characteristics, found among all the species examined with only minor variations.

General features of the lacertid ear and the relations to other structures are indicated in Fig. 13-4, which is a cross section through the ear region of *Lacerta muralis*. Shown, heavily stippled, is the otic capsule with the footplate of the columella filling its oval window. The round window does not appear at this level, but more ventrally, where it leads into the expanded scala tympani. Reissner's membrane extends across the cavity of the capsule, separating the scala vestibuli from the cochlear duct. The medulla oblongata lies on the medial side, and sends a large trunk of the eighth nerve through a foramen into the otic capsule.

Further details are given in Fig. 13-5, which represents a cross section through the dorsal end of the dorsal papilla in a specimen of *Psammodromus algirus*. Here the auditory papilla rests at a sharp tilt on the peak of the triangular shaped fundus between well-defined border cells, and bears a sallet of a peculiar wedge shape. The tectorial membrane arises in the limbic notch and runs around the tall limbic crest, but at this level only begins its extension toward the papilla.

The general features of this duplex cochlea are shown in the reconstruction drawing of Fig. 13-6. Here the right ear of a specimen of *Eremias argus* is seen in a dorsolateral view. A portion of the limbic plate is shown much thickened on its medial side and bearing a prominent elevation, which begins dorsally with a large round hill and tapers away to a ridge in the ventral direction. Lateral to this elevation the plate is greatly thinned out to form a depression that contains the two

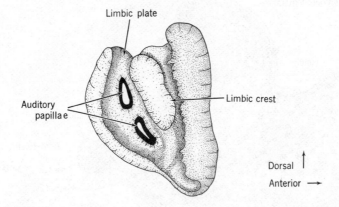

Fig. 13-3. The cochlear duct of *Acanthodactylus cantoris*, showing the two basilar membranes and their auditory papillae. Redrawn from Miller, 1966a, plate IX.

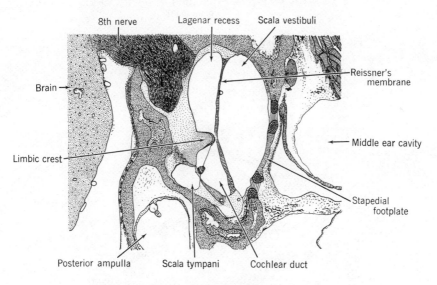

Fig. 13-4. A frontal section through the ear region of *Lacerta muralis*. Scale 75×.

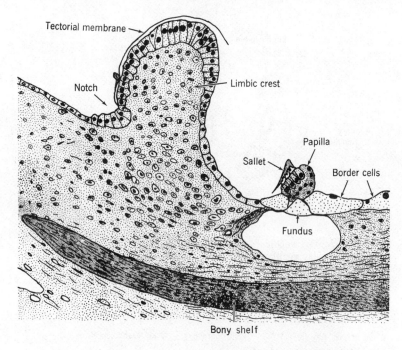

FIG. 13-5. The medial limbic region and auditory papilla of *Psammodromus algirus*. Scale 500×.

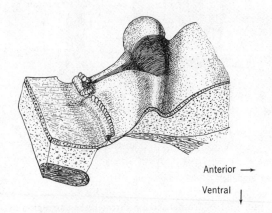

FIG. 13-6. A reconstruction drawing of the right cochlea of *Eremias argus*. From Wever, 1968a.

elongated openings in the limbic plate that were shown in Fig. 13-3. Here, however, these openings and the basilar membranes covering them are not seen, as they are obscured by the two papillae. The dorsal papilla is the shorter one (uppermost in the drawing), and the narrow end of the tectorial membrane extends to its more ventral portion. A few sallets are present on the dorsal half of this papilla, and others appear in a long row along the ventral papilla.

Features of this cochlea are further portrayed in a series of five drawings (Figs. 13-7 to 13-11), representing cross-sectional views of the cochlear region. Figure 13-7 shows a section through the dorsal portion of the dorsal papilla. The limbic crest is high and the fundus is thick, with the auditory papilla surmounted by a sallet with a wide base. No tectorial membrane is seen here.

In Fig. 13-8 is shown a section farther along, in the ventral half of this papilla. A tectorial membrane is present, arising from the limbic crest and extending to a plate that rests on the tips of the ciliary tufts of the hair cells.

The next drawing, shown in Fig. 13-9, shows a section passing through the separation region, or bridge, between the two papillae. The tectorial membrane continues to be seen, though abbreviated and ending freely in the cochlear fluid. The epithelial layer extends without interruption over the region of the limbus formerly occupied by the dorsal papilla.

Figure 13-10 shows a section still farther ventrally, through the most dorsal part of the ventral papilla. There are two rows of hair cells, whose ciliary tufts are surmounted by a tall, slender sallet. The tectorial membrane now appears only over the peak portion of the limbic crest.

A final view of this ventral papilla is given in Fig. 13-11, taken from a point near its ventral end. The papilla rests on the edge of a fundus of triangular cross section, and has taken a strongly oblique position, with its hair cells facing toward the medial limbus. The limbic crest is greatly reduced, and the tectorial membrane is absent.

A closer examination of the changing auditory structures in lacertids reveals a few further details. The following description is for a specimen of *Lacerta lepida*.

In this specimen the dorsal basilar membrane is 280 μ long and the ventral membrane 290 μ long, with a bridge between of 100 μ. The two sensory structures are entirely separate, with no continuity of basilar membrane or papillar tissue across the bridge. On the anterior surface (the one facing the cochlear duct) there is an extension of the border cells at the edges of the bridge, and in the thicker portion of the bridge the surface is covered with the regular epithelial layer of the area. On the posteromedial surface (the one facing the scala tympani) a capillary passes between medial and lateral parts of the limbus. In other lizards

FIGS. 13-7 TO 13-11. Views of the papillar region of *Eremias argus* at five locations along the cochlea from dorsal to ventral ends. From Wever, 1968a.

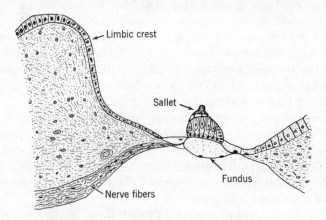

FIG. 13-7

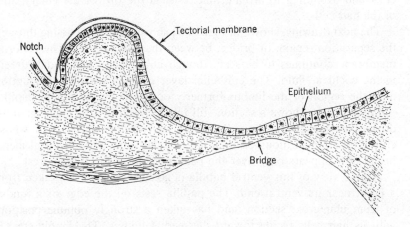

FIG. 13.8.

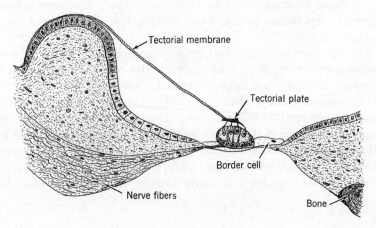

FIG. 13.9

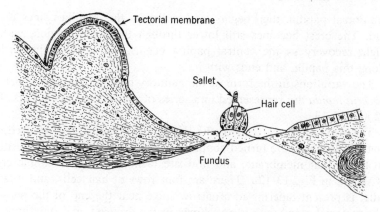

FIG. 13-10.

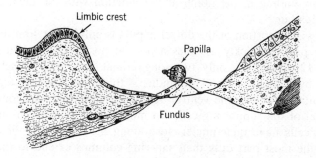

FIG. 13-11.

such a capillary is often seen crossing over in this middle region of the cochlea along the surface of the fundus itself.

Over the dorsal half of the cochlea, up to the bridge, the limbic plate is well supported over the whole posterior half of its extent by a thin vane of bone extending anteromedially as a part of the opisthotic. More ventrally, at the bridge region, this vane is much shorter, and thereafter it underlies only the lateral portion of the limbus. The anterior half of the plate is attached dorsally to light connective tissue leading to the anterior bony walls of the cochlear duct. Then from a point about a third of the distance along the cochlea it attaches directly to this bony wall.

The middle part of the limbic plate is elevated as the limbic crest. This crest varies a little in form among species, but in general is about twice as high as it is wide. In the *Lacerta lepida* specimen studied, it rises sharply to a height of 194 μ just before the basilar membrane begins, increases further to a maximum around 226 μ near the middle of the dorsal papilla, then begins to diminish as this papilla comes to an end. The crest becomes still lower through the bridge region, with a slight recovery as the ventral papilla begins, and then drops farther along this papilla and ends with it.

The variations in the form of the auditory papilla in this specimen of *Lacerta lepida* are represented in a series of seven drawings in Figs. 13-12 and 13-13.

The extreme dorsal end of the dorsal papilla is sharply twisted, so that the hair cells are turned about 40° away from their normal position on the basilar membrane, and now are facing toward the limbic crest as shown in Fig. 13-12a. There are four rows of hair cells, and a large sallet is present tapering to a narrow spike near the end of the papilla. The tectorial membrane is not shown in the drawing, and only appears as a thin line running along the anteromedial side of the limbic crest, apparently ending in the notch at the junction with the main body of the limbic plate.

This twisted portion of the dorsal papilla is only about 40 µ long, and then the papilla quickly assumes a normal position as shown in *b* of Fig. 13-12. There are only two longitudinal rows of hair cells here, covered by a small sallet. The fundus is thick and wide. A feature of this papilla, as well as the ventral one, is the relatively high location of the nuclei of the supporting cells. In most lizard cochleas the majority of these cells have their nuclei close along the surface of the fundus, and for the most part only their tapering columns extend to the upper regions of the capsule.

A short distance ventrally, near the middle of this papilla, the sallets come to an end and the hair cells have free-standing ciliary tufts, as shown in *c* of Fig. 13-12. At this point the tectorial membrane continues to lie along the anteromedial side of the limbic crest, ending near its peak.

A little farther ventrally, as shown in *d* of Fig. 13-12, a tectorial plate appears over the hair cells, and the tectorial membrane connects with it. Four transverse rows of hair cells are present, and the tectorial plate has projecting prongs on its underside to which the ciliary tufts often make their attachments. The fundus here has taken an asymmetrical form, and is thicker on the anteromedial side. This form continues to the end of the dorsal papilla.

Beyond the bridge region the ventral papilla begins as a tall, slender structure with 2 rows of hair cells, surmounted by a conical sallet as shown in *e* of Fig. 13-13. The fundus is relatively thin. A tectorial membrane is present only over the limbic crest, and throughout this ventral papilla has no relations to its hair cells.

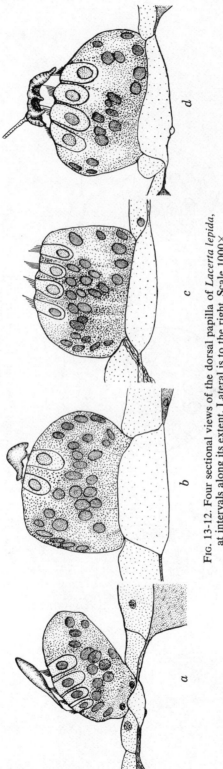

Fig. 13-12. Four sectional views of the dorsal papilla of *Lacerta lepida*, at intervals along its extent. Lateral is to the right. Scale 1000×.

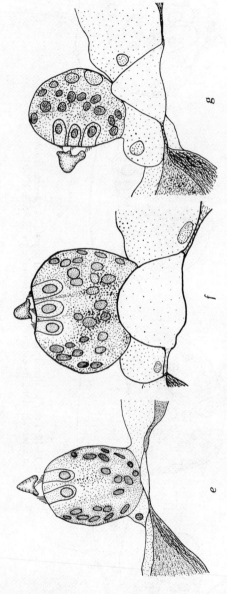

Fig. 13-13. Three sectional views of the ventral papilla of *Lacerta lepida* along its extent. Scale 1000×.

A little farther ventrally, as shown in *f* of Fig. 13-13, the number of transverse rows of hair cells increases to 3, the sallet is more regular in form, and the fundus becomes thick and rounded. At this point the papilla has a nearly upright position on the basilar membrane.

From here on the form of the structure undergoes progressive changes. The fundus becomes more triangular in cross section, and the papillar capsule, embraced by two large border cells, assumes a more and more oblique position, facing in the direction of the limbic crest. Part *g* of Figure 13-13 gives the picture at a point close to the ventral end of this papilla, when the angle of the hair cells relative to the basilar membrane is about 70°. At the very end this angle becomes practically 90°, and the papilla appears precariously perched on its side over the rounded corner of the fundus.

Relative to the basilar membrane, which in these species seems to maintain a nearly uniform position within the cochlear duct, each auditory papilla takes an upright position in its more middle course, adjacent to the separating bridge, but at the ends is sharply twisted dorsomedially. How these positional peculiarities affect the functioning of the receptor is an intriguing question. The twisted portion in each papilla is provided with sallets, and the oblique position would seem to convert the up-and-down movements of the basilar membrane into shearing forces at the hair cells, between the sallet and the ciliary tufts. Such shearing forces are considered by Békésy and others as more effective in stimulation than simple pressures, and the twisted position of the ends of these papillae may be a special means of improving the sensitivity. The twisted dorsal end of the dorsal papilla in this specimen includes only 24 out of the 89 hair cells in this dorsal papilla, or 27%, but the twisted ventral end of the ventral papilla includes 39 out of 66 cells, or 59%, and thus a substantial part of the whole. An improvement of sensitivity for these cells may well be significant in the performance of this ear.

Cochlear Dimensions.   Detailed measurements of the cochlear structures were made in two species, *Eremias argus* and *Lacerta lepida*. The results for the first of these are presented in Fig. 13-14. In *a* of this figure are shown the varying widths of the two basilar membranes, plotted along a scale beginning with the dorsal end of the dorsal membrane and extending to the ventral end of the ventral membrane, a distance of 500 μ. The dorsal membrane is 180 μ long and the ventral one 265 μ long. The dorsal membrane is the wider one, reaching 100 μ in its midregion. The ventral membrane increases rapidly to a maximum width of 80 μ, after which it slowly narrows to 60 μ and ends bluntly.

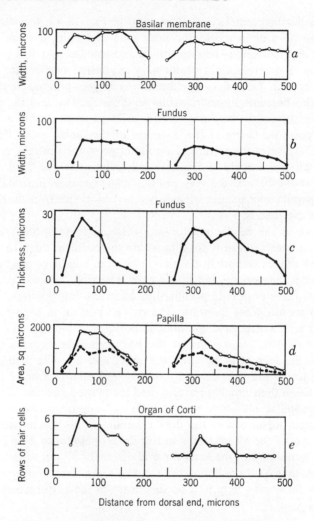

FIG. 13-14. Cochlear dimensions in a specimen of *Eremias argus*.
From Wever, 1968a.

Among the four ears of this species in which measurements were made, the bridge between the two basilar membranes varied in length from 40 to 80 μ. In the specimen shown here this length was 40 μ.

The width of the fundus, shown in *b* of Fig. 13-14, follows much the same course as the basilar membranes in both dorsal and ventral regions, but is about half as wide.

In the dorsal papilla, as part *c* of this figure shows, the fundus increases sharply in thickness to a maximum of 28 μ, then falls rapidly.

In the ventral papilla the fundus likewise thickens rapidly, reaches a maximum of 23 μ, and then falls off progressively to the ventral end.

The cross-sectional area of the two papillae is shown in Fig. 13-14d. The solid curves represent the whole moving structure, including the papillar capsule and the fundus on which it lies, whereas the broken curves represent the capsular part (organ of Corti) only. In the dorsal region the curves are nearly symmetrical; the sensory structure reaches a maximum near the middle of the basilar membrane and then falls away. In the ventral region these structures rise rapidly to a maximum and then decline more gradually.

The number of rows of hair cells is plotted in e of this figure. In the dorsal papilla are 3 rows at the dorsal end, then 6, and thereafter the number falls progressively to 3 just before the papilla ends. In the ventral papilla this number begins with 2, rises to 4 in a single section, then falls to 3 and 2 in succeeding regions.

The number of hair cells in each papilla of this ear was 51, for a total of 102. In the other ear of this same specimen these numbers were 47 for the dorsal papilla and 48 for the ventral one, for a total of 95.

Dimensional data for a specimen of *Lacerta lepida* are represented in Fig. 13-15. In a of this figure the varying widths of the two basilar membranes are shown by the curves with open circles, and the widths of the fundus portion of these membranes are shown by the curves below. For the dorsal papilla the basilar membrane width increases rapidly at first, attains a maximum of 110 μ, and then progressively declines. For the ventral papilla this membrane rises rapidly to a maximum, drops to a lower level that is maintained with only a moderate decline over most of this papilla, and then quickly falls away. In both papillae the fundus takes a somewhat similar course.

The thickness of the fundus, seen in part b of this figure, increases sharply in the dorsal papilla, then rises more slowly to a maximum, after which it declines rapidly. In the ventral papilla the course is similar.

Part c of Fig. 13-15 represents the varying cross-sectional areas of the two papillae, again shown for the entire structure (solid line) and for the organ of Corti only (broken line). For the dorsal papilla both curves rise to a large value that is maintained through the midregion, with a rapid decline near the end. For the ventral papilla the rise to a maximum is particularly rapid, and thereafter the decline is progressive to the end of this structure.

The arrangement of hair cells is shown in part d of this figure. For the dorsal papilla there are at first 4 rows, then a sharp rise to 7, and rapid changes thereafter between 3 and 6. For the ventral papilla the variations are only between 2 and 3 over the whole course of the structure.

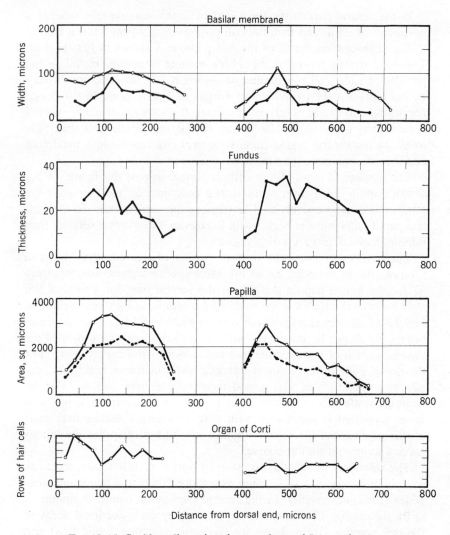

FIG. 13-15. Cochlear dimensions in a specimen of *Lacerta lepida*.

In the right ear of the specimen of *Lacerta lepida* there were 89 hair cells in the dorsal papilla, 42 of them served by sallets and 34 lying beneath the tectorial plate, with the remaining 13 cells in the transitional region between. The ventral papilla contained 66 hair cells, all overlaid by sallets.

Hair-cell populations for all species studied are represented in Table 13-I. This table shows for each species the distribution between the two papillae, and the total number of hair cells. These totals vary

TABLE 13-I

MEAN HAIR-CELL POPULATIONS
IN LACERTIDAE

Species	Number of Ears	Dorsal Papilla	Ventral Papilla	Total
Acanthodactylus erythrurus	3	55.3	37.7	93.0
Acanthodactylus pardalis	6	61.3	32.7	94.0
Acanthodactylus scutellatus	3	39.3	33.7	73.0
Eremias argus	4	41.5	50.2	91.8
Eremias velox persica	6	61.9	38.1	100.0
Lacerta agilis	8	54.9	50.4	105.2
Lacerta lepida	6	82.0	64.1	146.1
Lacerta muralis	8	63.3	46.3	109.6
Lacerta sicula	4	50.5	41.0	91.5
Lacerta viridis	8	75.4	60.1	135.5
Psammodromus algirus	8	54.8	51.6	106.4
Average of 11 species	—	58.2	46.0	104.2

moderately with species, from 73.0 in *Acanthodactylus scutellatus* to 146.1 in *Lacerta lepida*; the average for all species is 104.2. This is a relatively small size of population, below the general average found for iguanids (see Table 7-I), though well above that for the sceloporines. The relation between the numbers of hair cells in the two papillae is also of interest. In general the dorsal papilla contains the larger number by about 20%, but this preponderance is sometimes greater, especially in *Acanthodactylus pardalis*, and in one species, *Eremias argus*, the relationship is reversed.

Ciliary Orientation Patterns. The orientation of the ciliary tufts of the hair cells was carefully examined in one specimen each of the species *Acanthodactylus pardalis, Eremias velox persica, Lacerta lepida, Lacerta muralis*, and *Psammodromus algirus*, and briefly scanned in all other specimens. This pattern appears to be uniform in all species. The orientation in the dorsal papilla is mainly lateral as Fig. 13-12 shows. In some transverse rows (as in *c* and *d* of Fig. 13-12) the most lateral cell has its ciliary tuft facing medially, and this orientation was observed on one or two occasions for interior cells in the row. The condition in the dorsal and ventral parts of this dorsal papilla appears to be the same, though observation was particularly difficult near the dorsal end because here the sallet is large and lies so closely over the hair cells as to obscure the tufts in many instances. When a determination of the ciliary orientation could be made with certainty it was almost invariably lateral.

In the ventral papilla the orientation is bidirectional, with the outside cells always facing one another (as in Fig. 13-13). When there were three rows, the orientation of the middle cell was often obscure; but when it could be made out clearly it was nearly always lateral. Thus this pattern, though bidirectional, seems to be biased somewhat toward laterality. Miller (1974) in a study of *Lacerta viridis* and *Lacerta galloti* reported generally bidirectional ciliary orientation, with zones of unidirectionality.

The kinocilium was often found to bear a small spherical bulb at its tip, just as has been observed in teiids and gekkonids. This observation could be made only in certain sections, because this bulb is small and usually is closely applied to the lower surface of the sallet or tectorial plate. Nevertheless a few certain observations were made in the presence of these tectorial structures, though the clearest views of the form of the kinocilial tip were obtained in the middle of the papilla where a tectorial covering is usually absent for a short distance. In one such instance the kinocilium could be seen to arise adjacent to the longest stereocilia and then to wind partway around the tuft to end with its round bulb on the opposite side.

COCHLEAR POTENTIAL STUDIES

The auditory sensitivity of all these specimens was measured in terms of cochlear potentials. Figure 13-16 presents a curve for a specimen of *Acanthodactylus erythrurus* that shows good sensitivity in the midfrequency region, with the peak reaching −46 db at 1000 Hz. The sensitivity falls off at a rate of about 30 db per octave in the lower frequencies and 40 db per octave in the higher frequencies.

Results for two specimens of *Acanthodactylus pardalis* are given in Fig. 13-17. These curves show about the same peak sensitivity as the preceding species, but good response appears over a broader range, from 300 to 4000 Hz, and beyond this range the roll-off is relatively rapid, about 40 db per octave for the low tones and 60 db per octave for the high tones.

The species *Acanthodactylus scutellatus*, as represented in Fig. 13-18, also shows a broad range of good sensitivity, though with some irregularities. There is a region of reduced sensitivity around 800 to 2500 Hz, and a secondary maximum that is especially sharp in one specimen at 3000 Hz.

In general, all three *Acanthodactylus* species exhibit good sensitivity for the middle tones.

Sensitivity curves for two specimens of *Eremias argus* are presented in Fig. 13-19. These curves show two regions of best sensitivity, one

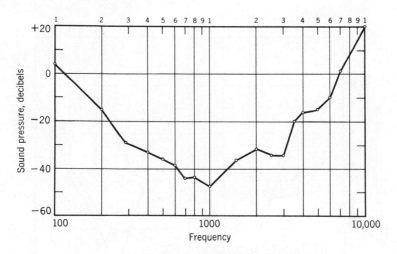

FIG. 13-16. A sensitivity function for a specimen of
Acanthodactylus erythrurus.

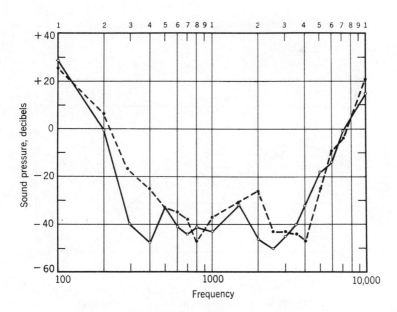

FIG. 13-17. Sensitivity functions for two specimens of
Acanthodactylus pardalis.

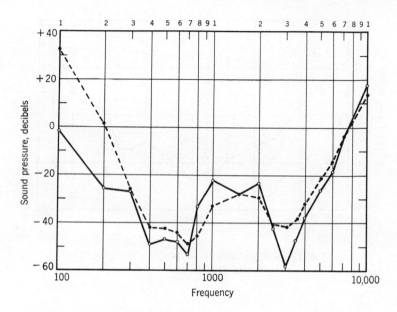

FIG. 13-18. Sensitivity functions for two specimens of
Acanthodactylus scutellatus.

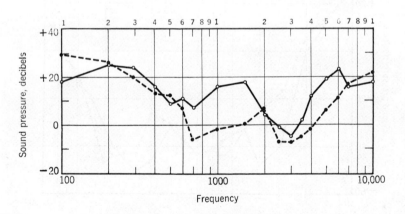

FIG. 13-19. Sensitivity functions for two specimens of
Eremias argus. From Wever, 1968a.

around 700 Hz and the other around 3000 Hz, with one animal some-what better than the other, but neither exhibiting good auditory capability.

Results for three specimens of *Eremias velox persica* are represented in Fig. 13-20. The best of these animals only reaches +22 db at 1000 Hz, and over most of the middle range these ears require sound pressures of 30-40 db above the reference level of 1 dyne per sq cm—which means absolute levels of 30-100 dynes per sq cm—to elicit the standard response of 0.1 μv, a very poor performance.

Results for two specimens of *Lacerta agilis* are given in Fig. 13-21. Sensitivity is good, around −40 db, for tones between 600 and 3000 Hz, with peak points at −44 and −45 db. Roll-off rates are rather high at the extreme frequencies. These ears display good function over a broad middle range.

The two curves for *Lacerta lepida*, shown in Fig. 13-22, have much the same form as the preceding species, but the sensitive range is shifted downward along the frequency scale. Response is excellent from 300 to 2000 Hz, with a peak region around 500-600 Hz that in one speci-men reaches the very acute level of −56 db. Roll-off rates vary at the low-frequency end from 20-40 db per octave, and are higher, around 40-50 db per octave, at the upper end of the range.

Results of observations on *Lacerta muralis* are shown in Figs. 13-23 and 13-24. These two sets of data are for specimens from separate batches, obtained from different sources. The curves agree closely in the high frequencies, but differ considerably in the low frequencies.

One animal, represented by the solid line of Fig. 13-23, shows excel-lent sensitivity in the low-tone range around 500-800 Hz, and good sen-sitivity for a wide range from 300 to 3500 Hz. The other three ears perform well between 2000 and 4000 Hz, but suffer a progressive de-cline toward the low-frequency end of the scale. The reasons for these variations are not known.

Results for two specimens of *Lacerta sicula* are given in Fig. 13-25. In these ears we find good sensitivity in the middle range between 700 and 4000 Hz, with peaks at 800 Hz for both and at 2500 Hz also for one of them. Roll-off rates are high, with one exception, at both ex-tremes of the frequency range.

Sensitivity curves for two specimens of *Lacerta viridis* are given in Fig. 13-26, for which the sensitivity is good in the range from 400 to 3500 Hz, with one specimen showing sharp peaks at 500 and 2000 Hz where the sensitivity attains significant levels of −50 and −55 db. Roll-off rates are about 30 db per octave at the low-frequency end and about 40 db per octave at the high-frequency end.

Results are given for two specimens of *Psammodromus algirus* in Fig. 13-27. These two ears exhibit excellent sensitivity in the mid-fre-

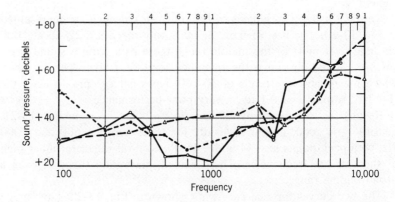

FIG. 13-20. Sensitivity functions for three specimens of *Eremias velox persica*.

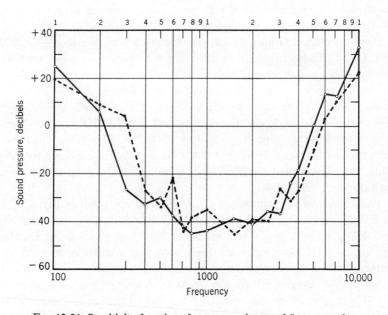

FIG. 13-21. Sensitivity functions for two specimens of *Lacerta agilis*.

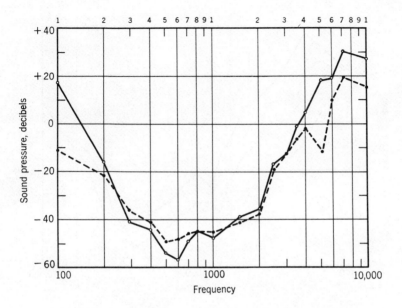

FIG. 13-22. Sensitivity functions for two specimens of *Lacerta lepida*.

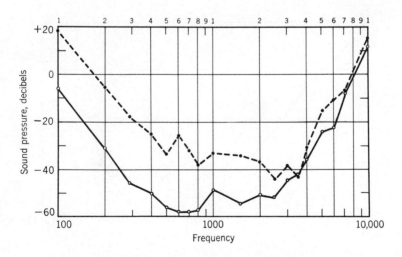

FIG. 13-23. Sensitivity functions for two specimens of *Lacerta muralis*.

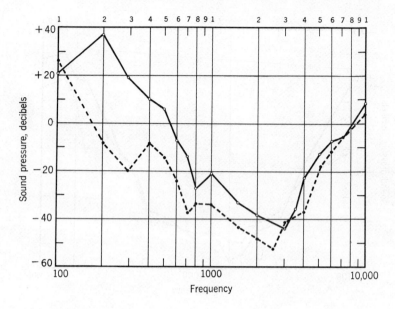

FIG. 13-24. Sensitivity functions for two additional specimens of *Lacerta muralis*.

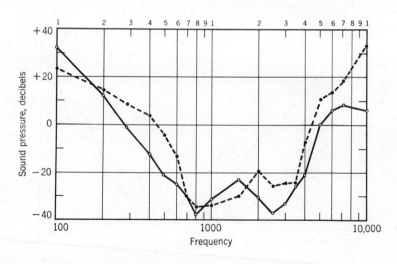

FIG. 13-25. Sensitivity functions for two specimens of *Lacerta sicula*.

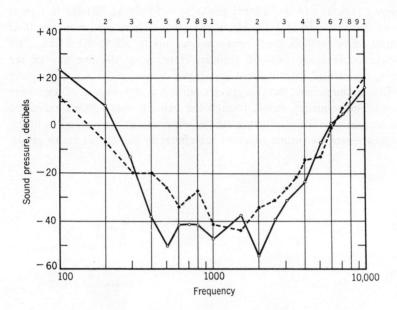

FIG. 13-26. Sensitivity functions for two specimens of *Lacerta viridis.*

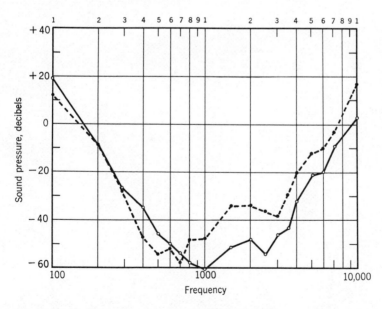

FIG. 13-27. Sensitivity functions for two specimens of *Psammodromus algirus.*

quency region, with prominent peaks of −58 db at 700 Hz in one of them and −61 db at 1000 Hz in the other. These ears maintain their acuity fairly well as the frequency rises up to about 3000 Hz, after which the response falls off rapidly. Here, as at the low end of the scale, the roll-off rate is about 40 db per octave.

In summary, most lacertid species possess a high degree of sensitivity in the midfrequency range, though the two *Eremias* species examined are seriously deficient. Further studies on *Eremias* are needed to determine whether the results reported are characteristic of this entire genus.

14. FAMILY GEKKONIDAE:
THE GECKOS

The geckos form a large group of 82 genera and more than 650 species, along with perhaps 175 subspecies, distributed throughout the warmer regions of all the continents (except Antarctica) and on numerous islands. A small species from the southwestern United States, *Coleonyx variegatus*, is shown in Fig. 14-1*a*, and the head of the large East Asian form, *Gekko gecko*, is shown in Fig. 14-1*b*.

FIG. 14-1*a*. The banded gecko, *Coleonyx variegatus*.
Drawing by Anne Cox.

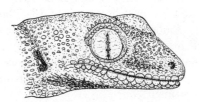

FIG. 14-1*b*. The head of *Gekko gecko*.

These lizards are adapted to a variety of habitats, but the large majority are terrestrial; a number live in bushes and trees, and several have adopted human habitations of both primitive and modern types. No gecko species is definitely aquatic, though certain ones have been observed to enter the water on occasion.

Most geckos are nocturnal or crepuscular, though all are considered to have developed from earlier lizards of diurnal habit. It is supposed that the adoption of nocturnality made more effective their practice of secretiveness, in permitting them to avoid the majority of predators that are active by day. Accompanying this change of habit were numerous alterations of eye structure, which enabled these animals to continue their use of vision in capturing prey even in dim light. The cones of the retina changed to rods, and a vertical pupil was acquired that in strong light could be reduced to a slit or a series of pinholes to protect the sensitive retina (Underwood, 1954). A few geckos, considered more primitive, have retained the retinal cones and a round pupil; these are active by day.

Most geckos lack eyelids, and the cornea is protected by a transparent covering, called the spectacle or brille; but in the eublepharines true eyelids are present.

Most geckos possess a voice, and the males at least produce chirps or calls that appear in many instances to have the function of proclaiming territorial rights and repelling other males. Some species, such as *Gekko gecko*, also emit loud cries as a part of their menacing behavior in the presence of an intruder. Chorusing behavior has sometimes been reported, but its function is obscure. A number of species utter a sharp squeak on being caught, but otherwise appear to be silent. Others, including all the sphaerodactylines, are reported to be mute (Kluge, 1967).

Despite these variations, there seems to be no doubt about what is a gecko. Yet classification within this group has posed something of a problem, and a number of systematic arrangements have been proposed (Underwood, 1954; Kluge, 1967). The arrangement developed by Underwood was based largely on eye characteristics, and especially the pupil. The one worked out by Kluge utilized 18 different features. The system adopted here follows Kluge, in which a single family is recognized and is divided into four subfamilies: the Eublepharinae, Gekkoninae, Diplodactylinae, and Sphaerodactylinae. As will be seen, the variations found in the structure and performance of the ear fit well into this systematic arrangement.

The gekkonid ear received early consideration from Shute and Bellairs (1953), who pointed to a number of features that they regarded as distinctive and as having systematic significance. They took particular note of the peculiar form of the limbus, whose medial portion is elevated into a curved lip that lies over the auditory papilla and suspends the tectorial membrane. However, this form of limbic elevation is not exclusive to the geckos, but occurs also in two other lizard families, the Pygopodidae and Teiidae.

A more distinctive feature of the gekkonid ear, though shared also with the Pygopodidae, is the use of two different stimulating mechanisms in a peculiar patterning of the hair cells along the auditory papilla (Wever, 1965a). This papilla is segmented, with a dorsal portion containing hair cells whose ciliary tufts are attached to a tectorial membrane and a ventral portion divided longitudinally so that the hair cells on one side continue the tectorial attachments whereas those on the other side are surmounted by a line of sallets that make the ciliary connections.

Shute and Bellairs noted also the great elongation of the gekkonid cochlear duct and the corresponding extension of the basilar membrane and its sensory structures. These structures are graduated along the cochlea to an extent exceeding that found in any other lizards, and in some species the degree of differentiation approaches that attained in the ears of the advanced mammals. Details of this development of the cochlea were worked out early for a few species (Wever, 1965a, 1967b) and presented in a further report (Wever, 1974b) for a group of 36 species.

The sensitivity of the gecko ear in terms of its electrical potentials has been determined in a number of experiments beginning with *Gekko gecko* (Wever, Vernon, Peterson, and Crowley, 1963; Hepp-Reymond and Palin, 1968; Werner and Wever, 1972), then extended to *Coleonyx variegatus* and *Hemidactylus turcicus* (Wever, Peterson, Crowley, and Vernon, 1964), to *Ptyodactylus hasselquistii* (Wever and Hepp-Reymond, 1967), and *Eublepharis macularius* (Werner and Wever, 1972), and finally was covered in 34 species (Wever, 1974b). Limited observations have been made also on the auditory nerve responses of a few forms in response to acoustic stimuli (Suga and Campbell, 1967; Hepp-Reymond and Palin, 1968; Campbell, 1969; Manley, 1972a).

ANATOMICAL OBSERVATIONS

All these observations agree in showing that the ear of geckos, though varying considerably with species, presents a well-developed structure and operates with a high degree of effectiveness in response to sounds. Evidence is now available on the structure and function of the ear in 45 species belonging to 26 genera as listed in Table 14-I. The number of specimens examined was about 230, nearly all of which were studied first by means of the cochlear potentials, and then the ears prepared for anatomical examination in serial sections.

OUTER AND MIDDLE EAR STRUCTURES

In all the geckos there is an external ear opening with the tympanic membrane lying relatively deep below the surface, so that a definite auditory meatus is present. In the majority of species, as noted in Chapter

TABLE 14-I

SPECIES OF GEKKONIDAE STUDIED

Eublepharinae
Coleonyx brevis (Stejneger, 1893)—1 specimen
Coleonyx variegatus (Baird, 1858)—28 specimens
Eublepharis macularius (Blyth, 1854)—11 specimens
Hemitheconyx caudicinctus (A. Duméril, 1851)—4 specimens

Gekkoninae
Aristelliger praesignis (Hallowell, 1857)—1 specimen
Cyrtodactylus kotschyi orientalis (Stepanek, 1937)—7 specimens
Geckonia chazaliae (Mocquard, 1895)—2 specimens
Gehyra variegata (Duméril and Bibron, 1836)—11 specimens
Gekko gecko gecko (Linnaeus, 1758)—26 specimens
Gekko monarchus (Duméril and Bibron, 1836)—2 specimens
Gekko smithi (Gray, 1842)—4 specimens
Hemidactylus brookii (Gray, 1844)—1 specimen
Hemidactylus brookii angulatus (Hallowell, 1854)—6 specimens
Hemidactylus echinus (O'Shaughnessy, 1875)—6 specimens
Hemidactylus mabouia (Moreau de Jonnes, 1818)—3 specimens
Hemidactylus sp.—4 specimens
Hemidactylus triedrus (Daudin, 1802)—2 specimens
Hemidactylus turcicus (Linnaeus, 1758)—3 specimens
Homopholis fasciata (Boulenger, 1890)—6 specimens
Lygodactylus picturatus (Peters, 1870)—2 specimens
Pachydactylus sp.—1 specimen
Phelsuma dubia (Boettger, 1881)—2 specimens
Phelsuma madagascariensis (Gray, 1831)—5 specimens
Phyllodactylus xanti (Cope, 1864)—3 specimens
Ptyodactylus hasselquistii guttatus (von Heyden, 1827)—20 specimens
Ptyodactylus hasselquistii puiseuxi (Boutan, 1893)—1 specimen
Stenodactylus orientalis (Blanford, 1876)—3 specimens
Stenodactylus s. sthenodactylus (Lichtenstein, 1823)—2 specimens
Tarentola mauritanica (Linnaeus, 1758)—1 specimen
Teratolepis fasciata (Blyth, 1853)—1 specimen
Teratoscincus scincus (Schlegel, 1858)—6 specimens
Thecadactylus rapicaudus (Houttuyn, 1782)—5 specimens

Diplodactylinae
Diplodactylus elderi (Stirling and Zietz, 1893)—2 specimens
Diplodactylus vittatus (Gray, 1832)—6 specimens
Diplodactylus williamsi (Kluge, 1963)—3 specimens
Hoplodactylus pacificus (Gray, 1842)—6 specimens
Lucasium damaeum (Lucas and Frost, 1895)—6 specimens
Oedura marmorata (Gray, 1842)—1 specimen
Oedura ocellata (Boulenger, 1885)—5 specimens
Oedura tryoni (de Vis, 1884)—2 specimens

Sphaerodactylinae
Gonatodes albogularis fuscus (Hallowell, 1855)—4 specimens
Gonatodes ceciliae (Donoso-Barros, 1966)—5 specimens
Gonatodes sp.—4 specimens
Gonatodes vittatus (Lichtenstein, 1856)—1 specimen
Sphaerodactylus f. fantasticus (Duméril and Bibron, 1836)—5 specimens
Sphaerodactylus lineolatus (Lichtenstein, 1856)—2 specimens
Sphaerodactylus monensis (Meerwarth, 1901)—4 specimens

6, a muscle is present in the meatal wall whose contractions produce a closure of the orifice.

As noted above, the meatal closure muscle occurs in two forms, one in the eublepharines and gekkonines and another in the diplodactylines. This muscle is absent in the sphaerodactylines and in three species (the day geckos) among the gekkonines examined.

As noted in Chapter 6, the middle ear of geckos is distinctive, exhibiting two features not found together elsewhere except in Pygopodidae. These are the existence of an extracolumellar muscle and the absence of an internal process to support the middle portion of the columella. The middle ear structure has been studied in detail in only two species, *Gekko gecko* and *Eublepharis macularius*, as described in Chapter 6, but in other gecko species it shows much the same form, and the presumption is that in these others the mechanism serves in sound reception with about equal effectiveness.

INNER EAR

The geckos display an advanced level of cochlear development, though there are considerable variations among species. This advance is represented in a number of structural features, and also is manifested in the functioning of this ear.

As observed earlier by Shute and Bellairs, the basilar membrane of gekkos is extended into a long, ribbonlike form, and is graduated in width over this extent. Most often the membrane is narrowest at the dorsal end, increases gradually for some distance, and then broadens rather rapidly to a maximum near the ventral end. In a few species this progressive taper is disturbed by a constriction in the mid-dorsal region, usually of moderate amount but quite marked in certain instances. In other species there are more striking departures from the usual form, as will be noted.

The hair cells are distributed over the papillar surface in rather orderly arrays. Usually there are reasonably well-defined arrangements into transverse and longitudinal rows, though in some regions the cells are crowded into partially interdigitating arrays. This patterning is difficult to make out in detail in serial sections, though by careful examination of successive sections the general form can be perceived. It is more readily discerned in the surface view afforded by the scanning electron microscope, provided of course that the preparation processes have not caused undue distortion of the tissues. Miller's excellent photomontage of the papillar surface in *Gekko gecko* (1973a) shows the hair cells in well-defined transverse and longitudinal rows, with only moderate waverings of their courses, especially over the ventral half of the papilla and on its outermost side. No doubt the row arrangement varies with species, and in lesser degree among individual specimens.

The papilla begins at its dorsal end with two or three longitudinal rows of hair cells, and other rows are added more ventrally as the structure widens, until in some species there are as many as 12-16 rows in the ventral expansion before a rapid termination occurs.

The distribution of the hair cells with respect to their two means of ciliary restraint—by attachment to the tectorial membrane or to sallets—has already been mentioned. In the dorsal portion of the cochlea, this attachment is exclusively to the tectorial membrane, whereas in the ventral portion in which the sallets appear there is a separation into two groups, one lying along the more medial side of the cochlea in which the tectorial connections continue and a lateral group in which the connections are made by sallets.

The ciliary tuft in geckos has the usual structure, consisting of a number of stereocilia of graduated lengths and a single kinocilium at one end of the array adjacent to the longest stereocilia. In *Gekko gecko* there are about 42 stereocilia and in *Coleonyx variegatus* and *Eublepharis macularius* about 32 (Miller, 1973a, b). The kinocilium in geckos bears a bulbous enlargement at its tip, noted by Miller in *Gekko gecko* as nearly spherical and measuring 1.4 μ in diameter.

The tips of the ciliary tufts are firmly attached to the overlying tectorial structures, either to the tectorial membrane or fibrous extensions from it or to the undersurface of the sallet or its extensions. This attachment is made by the kinocilial bulb, sometimes by itself but more often reinforced by a few of the longest stereocilia. According to Miller, this reinforcement in *Gekko gecko* is provided by five stereocilia that surround the kinocilial bulb and appear to have their upper ends attached to it.

Usually there is one sallet for each transverse row of hair cells, but sometimes a sallet serves the cells of two or three contiguous rows. Miller (1973b) noted a tendency for the sallets of *Eublepharis macularius* to divide at their lower ends and to connect to two adjacent transverse rows of hair cells.

A cross-sectional view of the inner ear region of a specimen of *Eublepharis macularius* is presented in Fig. 14-2. Shown is the great thickening of the limbus and its elevation into a prominent lip from which the tectorial membrane extends.

Cochlear Segmentation. The dorsal and ventral portions of the auditory papilla as already defined are separated by a transitional zone. In this zone the tectorial connections continue and sallets have not yet appeared, but one or two rows of hair cells are added on the lateral side of the array with ciliary tufts that end freely, without any form of restraint. These will be referred to as "pre-sallet cells," because a little more ventrally sallets appear above these rows and the hair tufts become attached to them.

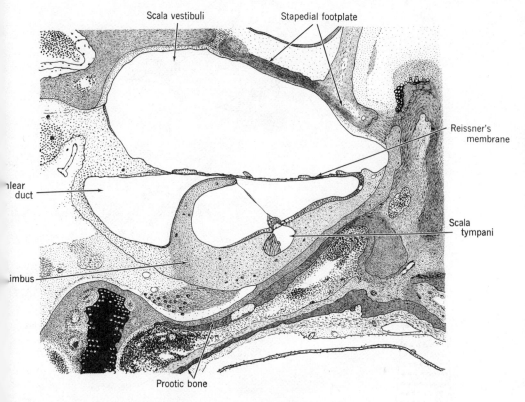

Scala vestibuli   Stapedial footplate

Reissner's
membrane

cochlear
duct

Scala
tympani

limbus

Prootic bone

Fig. 14-2. A frontal section through the inner ear region of
Eublepharis macularius, near the dorsal end of the cochlea. Scale 50×.

The pre-sallet cells have certain peculiarities besides bearing free-standing cilia. Their ciliary tufts often extend at a sharp lateral angle to the body of the cell, they have relatively few stereocilia and these show only limited graduation in length, and perhaps they often lack a kinocilium. Hair cells with free ciliary tufts have not been noted elsewhere along the auditory papilla except occasionally at its very ends.

The end of the transitional zone is marked by the appearance of sallets to make the connections to the ciliary tufts of the added lateral rows. A little more ventrally the sallets take over the connections also of hair cells in the rows toward the middle of the array, leaving only the most medial rows to be served by the tectorial membrane.

The sharing of the ciliary connections between tectorial and sallet structures varies somewhat along the cochlea, and also varies with species. Usually the sallets make more than half the connections.

The patterning of the gekkonid cochlea is represented in detail in Fig. 14-3, in which an outline is given of the basilar membrane of

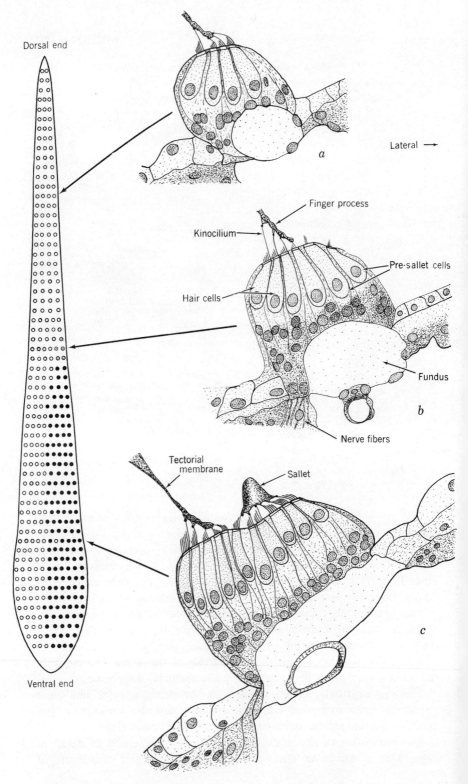

Dorsal end

Lateral →

Finger process

Kinocilium

Pre-sallet cells

Hair cells

Fundus

a

b

Nerve fibers

Tectorial membrane

Sallet

c

Ventral end

FIG. 14-3. The auditory papilla of *Eublepharis macularius*. On the left is the basilar membrane in outline showing the row structures of the hair cells; and on the right are three cross-sectional views of the auditory papilla at three cochlear regions. Scale for outline, 100×; for cross-sectional views, 400×. From Wever, 1974b.

Eublepharis macularius, and with the varying numbers of rows of hair cells indicated by the small circles. Open circles indicate hair cells with tectorial membrane connections, solid circles indicate those with sallet connections, and circles containing a dot represent pre-sallet cells in the transitional zone. About half the actual transverse rows are portrayed, but all the longitudinal rows in this particular specimen are represented. As will be noted, the number of longitudinal rows begins at the dorsal end with 2, increases to 5 at the beginning of the transitional zone, rises to 6 in that zone, and finally reaches 11 in the ventral region.

On the right of this figure are three drawings of the auditory papilla in cross section, one for each segment, to show the varying form of the structures and the ciliary connections. For these three sectional views, anterior is above and lateral is to the right.

The kinocilial bulb in *Eublepharis macularius* is relatively large, and is shown in the drawings in all instances in which it was clearly evident in the sections. A number of instances are shown in which the length of the kinocilium exceeded that of all the stereocilia, and this cilium made the attachment to the tectorial membrane or sallet by itself.

Ciliary Orientation Patterns.   Especially to be noted in Fig. 14-3 is the orientation of the ciliary tufts. In *a* these tufts all have the same orientation, with the kinocilium on the lateral side of the array. This orientation was found for all the hair cells of the dorsal segment in this lizard species. Within the transitional zone this lateral facing of the ciliary tufts was the rule also, as shown in *b* of this figure, but instances were found in which the contrary orientation occurred. This reverse orientation was common for the pre-sallet cells, although for many of these cells the orientation was uncertain because the tuft was poorly developed, having few stereocilia and often no discernible kinocilium.

The transitional zone in *Eublepharis* is comparatively short, and in the specimen shown here it includes only three rows of hair cells. It begins with a row containing four cells with tectorial membrane connections and one cell (the pre-sallet cell) that ends freely. The tufts of the first four cells are oriented laterally, and the one for the pre-sallet cell has a medial orientation. The next two rows of hair cells are similar except that there are two pre-sallet cells. Thereafter the sallets appear and the ventral segment begins.

It appears that the transitional zone is more closely related to the dorsal segment than to the other, for the majority of its cells continue the ciliary pattern of that segment, and the pre-sallet cells, without a tectorial attachment and perhaps lacking a kinocilium in many instances, must be considered as minimally functional. In some other species, however, the transitional zone is more extensive and the arrangement of its cells more complex.

Within the ventral segment, as represented by *c* of Fig. 14-3, there are two subpatterns, one for the hair cells served by the tectorial membrane and another for those cells served by sallets. For each of these subpatterns there are two orientations. In each, the more laterally disposed cells have the kinocilia on the medial side of their tufts, and the more medially disposed have them on the lateral side. This arrangement in the ventral region of the cochlea can be designated as a doubly bidirectional orientation, as contrasted with unidirectional orientation in the dorsal region.

Within each of the ventral subpatterns *Eublepharis macularius* shows a nearly even distribution of hair cells with laterally and medially oriented ciliary tufts. When two longitudinal rows are present in a subpattern, one tuft faces laterally and the other medially, and when three longitudinal rows are present the two outermost are thus oriented while the middle one alternates in direction regularly from one transverse row to the next, with only rare exceptions. When four or more rows are present the same rule holds, but exceptions are more common, especially for tufts near the middle of the array. Only rarely does the outermost tuft in a bidirectional array depart from the rule of facing toward the middle of the array.

A determination of the ciliary orientation patterns was made for most of the species listed in Table 14-I, with results that closely follow subfamily lines, but with certain exceptions.

In all species examined the ventral segment of the auditory papilla showed a doubly bidirectional pattern as just described. Species variations occurred only in the dorsal and transitional segments. These variations will now be shown, along with some general indications of the dimensions of the basilar membrane that will be preliminary to a more detailed treatment later.

For the portrayal of the regional variations of ciliary orientation in the different species, a schematic representation of the basilar membrane and its papilla will be used with four forms of line shading to represent the four patterns, together with dot shading to represent the ends of the basilar membrane that are not covered by the papilla. These schemes are defined in Fig. 14-4.

In all four of the Eublepharinae examined, the dorsal segment showed a unidirectional orientation as just described for *Eublepharis macularius*. Figure 14-5 presents schematic representations for these species. As will be seen, the basilar membranes of the two *Coleonyx* species are closely similar in form. The one for *Coleonyx brevis* was about 0.76 mm long and the one for *Coleonyx variegatus* about 1.0 mm long. For each, the dorsal segment, together with the transitional zone, occupies about a third of the total length. Along the dorsal segment there is only a

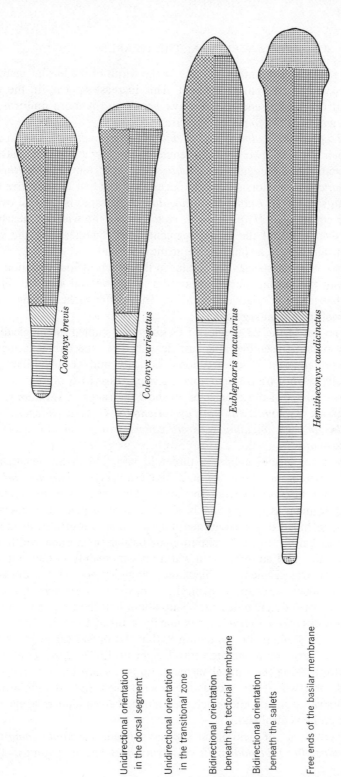

Coleonyx brevis

Coleonyx variegatus

Eublepharis macularius

Hemitheconyx caudicinctus

Fig. 14-5. Basilar membrane outlines for four species of Eublepharinae showing the segmental patterns. Scale 100×. From Wever, 1974b.

Unidirectional orientation in the dorsal segment

Unidirectional orientation in the transitional zone

Bidirectional orientation beneath the tectorial membrane

Bidirectional orientation beneath the sallets

Free ends of the basilar membrane

Fig. 14-4. Line shading scheme used to represent the ciliary patterns along the basilar membrane.

slight taper, and the main increase in the width of the basilar membrane occurs within the ventral segment. This increase appears in the region toward the ventral end in *Coleonyx brevis* and is more uniform along this segment in *Coleonyx variegatus*.

Eublepharis macularius and *Hemitheconyx caudicinctus* form a second pair. In these two the total length is considerably greater, attaining 1.31 mm in *Eublepharis macularius* and 1.42 mm in *Hemitheconyx caudicinctus*. The dorsal segment occupies about half the effective length in both species. In *Eublepharis macularius* the taper is gradual over the whole cochlea. In *Hemitheconyx caudicinctus* the width is nearly uniform over much of the dorsal segment, grows greater over the ventral region, and shows a sharp bulge at the very end.

Three of these species are similar in that the principal variation in the width of the basilar membrane occurs in the ventral segment, and the dorsal segment is relatively uniform. They differ most strikingly in the length of the dorsal segment.

In 16 out of the 23 species of Gekkoninae studied histologically, the ciliary pattern resembles that just described for the eublepharines: the ciliary orientation is unidirectional dorsally and doubly bidirectional ventrally, with a transitional zone between. Fig. 14-6 presents cochlear patterns for six of these, selected to show the range of variations. These species are *Cyrtodactylus kotschyi orientalis, Gehyra variegata, Hemidactylus brookii angulatus, Teratoscincus scincus, Gekko gecko,* and *Gekko smithi.*

These six species exhibit a threefold variation in the length of the basilar membrane, from 0.7 mm in *Cyrtodactylus kotschyi orientalis* to 2.1 mm in *Gekko smithi*. In the first three species of the group, the dorsal segment is comparatively short, and even when the transitional zone is included this portion makes up less than a fourth of the cochlea. There is hardly any differentiation in basilar membrane width within this portion, and an increase in width appears mainly in the ventral half of the ventral segment. In the three remaining species the dorsal segment is much more prominent. In *Teratoscincus scincus* it (together with the transitional zone) makes up about half the total, and in *Gekko gecko* and *Gekko smithi* approximately a third. In these three species there is only slight differentiation within the dorsal segment, and considerable variation along the ventral segment. In *Teratoscincus* the taper is uniform along the ventral segment, up to a sharp bulge at the ventral end. In the two *Gekko* species the taper is slight over the initial part of the ventral segment, then increases along the remainder, again with a bulge at the ventral end.

In seven of the gekkonine species studied, *Tarentola mauritanica, Thecadactylus rapicaudus, Ptyodactylus hasselquistii guttatus, Terato-*

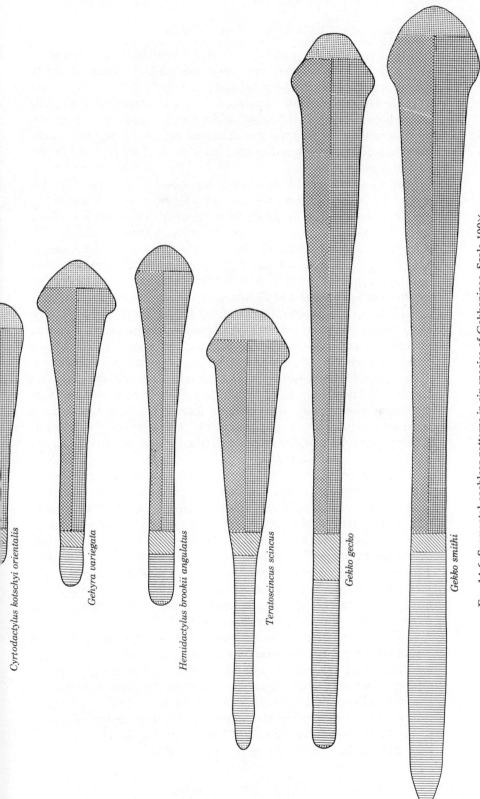

Cyrtodactylus kotschyi orientalis

Gehyra variegata

Hemidactylus brookii angulatus

Teratoscincus scincus

Gekko gecko

Gekko smithi

Fig. 14-6. Segmental cochlear patterns in six species of Gekkoninae. Scale 100×.

lepis fasciata, Phelsuma dubia, Aristelliger praesignis, and *Phelsuma madagascariensis*, the ciliary orientation departs from the usual one in certain respects, and will be described presently.

In the remaining two subfamilies of Gekkonidae, the Diplodactylinae and Sphaerodactylinae, the ciliary orientation pattern in the dorsal segment departs markedly from that described above: this segment exhibits a bidirectional pattern like that seen in the two portions of the ventral segment.

Outlines of the basilar membrane are shown in Fig. 14-7 for five species of diplodactylines: *Diplodactylus elderi, Lucasium damaeum, Diplodactylus vittatus, Hoplodactylus pacificus*, and *Oedura ocellata*. In all five the dorsal segment is comparatively short and is followed by a transitional zone of considerable length—in three of these species longer than the dorsal segment itself. These two parts together, however, make up only a fourth or less of the total length of the papilla. In *Diplodactylus elderi* and *Diplodactylus vittatus* there is a moderate amount of taper in the dorsal segment, but little in the remaining species. In *Lucasium damaeum* the transitional zone is of uniform width, but in the others there is an appreciable increase. Altogether the dorsal and transitional portions of the cochlea exhibit only a moderate amount of differentiation, and the principal variation is in the ventral segment. This variation occurs mainly in the ventral half of the ventral segment, with the dorsal half showing little change. Indeed, all these species have slight constrictions in the early part of the ventral segment, though this is barely perceptible in all but *Diplodactylus vittatus* and *Hoplodactylus pacificus*.

In these species, and especially in *Lucasium damaeum* and *Oedura ocellata*, the pre-sallet cells continue through a number of transverse rows. *Oedura ocellata* is notable also in that the dorsal ciliary orientation pattern, though bidirectional, is strongly biased toward laterality: most of the kinocilia are on the lateral side of their tufts, and indeed only a few of the most lateral hair cells have medially facing kinocilia. This same condition was observed in *Oedura tryoni*. A tendency of this sort was found also in *Diplodactylus elderi*, but the other species shown in Fig. 14-7 present more of a balance between laterally and medially directed ciliary tufts in the dorsal segment.

Ciliary orientation patterns for four species of Sphaerodactylinae are presented in Fig. 14-8. All these have relatively short basilar membranes, with the dorsal segment and the transitional zone especially curtailed. Also there are only moderate variations in the width of the basilar membrane. Barely noticeable constrictions appear in the initial part of the ventral segment in the first three species, and a more obvious distortion in the fourth, *Gonatodes ceciliae*. The cochlea of *Sphaerodactylus f. fantasticus* is the shortest among the geckos studied.

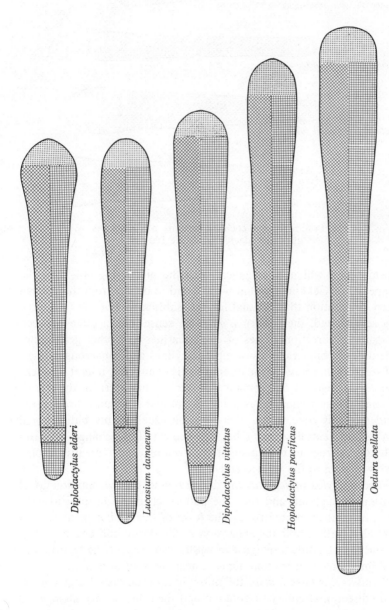

Diplodactylus elderi

Lucasium damaeum

Diplodactylus vittatus

Hoplodactylus pacificus

Oedura ocellata

Fig. 14-7. Segmental cochlear patterns in five species of **Diplodactylinae**. Scale 100×.

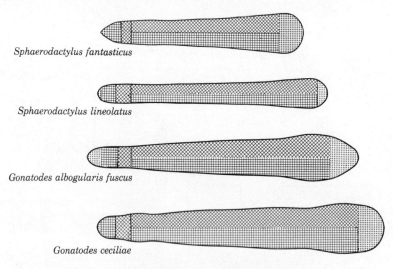

Sphaerodactylus fantasticus

Sphaerodactylus lineolatus

Gonatodes albogularis fuscus

Gonatodes ceciliae

FIG. 14-8. Segmental cochlear patterns in four species
of Sphaerodactylinae. Scale 100×.

Consideration will now be given, with the aid of Fig. 14-9, to the
seven species of Gekkoninae that were found to depart from the pattern
of ciliary orientation that is usual for this subfamily. For the first three
species represented, the pattern is in some sense intermediate between
the dorsally unidirectional ones of eublepharines and other gekkonines
and the dorsally bilateral ones of diplodactylines and sphaerodactylines.

In *Tarentola mauritanica* the orientation is completely lateral throughout the dorsal segment and remains so over the first third of the transitional zone. Then it changes to bidirectional and continues through the
remainder of this zone, going over into the usual doubly bidirectional
pattern as the ventral segment begins. There is little change in width
over the dorsal region, then a slight constriction, and thereafter a progressive widening to the ventral end.

In *Thecadactylus rapicaudus* the dorsal segment shows unidirectional
orientation throughout, and then the pattern changes abruptly to bidirectional in the transitional zone. The form of this basilar membrane is
unusual also, with a bulbous expansion at the dorsal end and a reverse
taper, and then a gradual increase in width over most of the ventral segment. A moderate constriction appears near the ventral end.

In *Ptyodactylus hasselquistii* the initial half of the dorsal segment exhibits unidirectional ciliary orientation and then this pattern changes to
bidirectional for the remainder of this segment. This bidirectional pattern continues throughout the transitional zone, and finally goes over into the typical ventral pattern. The basilar membrane is nearly uniform

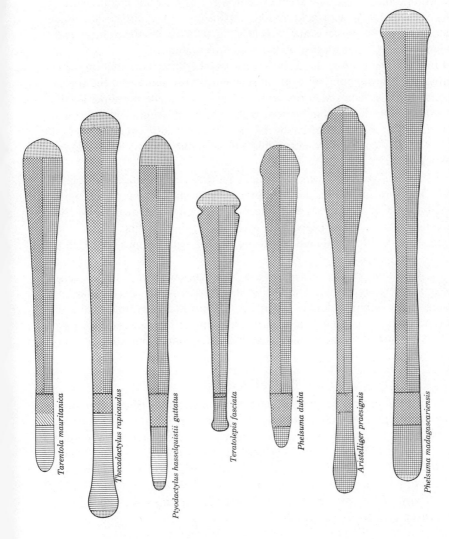

Tarentola mauritanica

Thecadactylus rapicaudus

Ptyodactylus hasselquistii guttatus

Teratolepis fasciata

Phelsuma dubia

Aristelliger praesignis

Phelsuma madagascariensis

FIG. 14-9. Segmental cochlear patterns in seven divergent species of Gekkoninae.

in width throughout the dorsal segment, increases along the transitional zone, then shows a marked expansion early in the ventral segment, and finally undergoes a somewhat wavering increase toward the ventral end.

The remaining species exhibit bidirectional ciliary orientation throughout the cochlea. In *Teratolepis fasciata* the dorsal segment is rather short, and the transitional zone extremely so. The basilar membrane has an unusual form, with a knob-like expansion at the dorsal end, then undergoes a progressive and accelerating increase to the ventral end, where a notch appears just as the papilla terminates.

In *Phelsuma dubia*, as in the foregoing, the ciliary orientation is bidirectional throughout. The dorsal segment is very short, and the transitional zone is about twice its length. There is a moderate taper in the dorsal region, then a constriction, with considerable uniformity through the middle region, followed by an increase in width over the ventral third of the cochlea. A noticeable enlargement appears close to the ventral end.

In *Aristelliger praesignis* the ciliary pattern is bidirectional throughout. The form of the basilar membrane is unusual, showing a club-like enlargement at the dorsal end, then a progressive narrowing until the middle of the cochlea is reached, after which there is first a moderate increase in width and then a rapid one, ending with a sharp constriction as the papilla terminates.

In *Phelsuma madagascariensis* the ciliary orientation pattern is also completely bidirectional. The basilar membrane in this species, like the foregoing, exhibits a double taper, but a more nearly symmetrical one.

These last four species, though classed among the Gekkoninae, have ciliary orientation patterns like those of the Diplodactylinae and Sphaerodactylinae.

Topical Treatment

The discussion thus far has brought out a number of general relations among the geckos, and especially the segmental form of the auditory papilla and its variations in the different groups. Further consideration will be given to these and other structural features, with particular regard to the question whether these ears are differentiated in some degree for selective response to tonal frequencies over their auditory range. Such selectivity of the ear has two important consequences for function: it improves the sensitivity within the range favored by the special tuning, and it presents the possibility of tonal discrimination, which can greatly enhance the information-giving characters of sounds. All ears possess this discrimination in some degree—indeed an ear would be entirely useless without it—but this capability varies greatly. Among the lizards the geckos appear to present the greatest possibilities of a large measure of aural selectivity and discrimination.

Bearing on this question are the structural features already noted: the considerable elongation of the basilar membrane, variations in its width along its extent, and the different forms of ciliary orientation and restraint. Other features generally regarded as significant in the selective response of an ear are the variations in stiffness of the vibrating structures and their varying mass.

For the gecko ear the following features will now be considered in some detail: the width of the basilar membrane, the width and thickness of its middle portion (the fundus), and the varying cross-sectional area of the organ of Corti—the encapsulated part of the papilla with its supporting cells and columns by which the hair cells are maintained. Also to be noted in this connection are the distributions of the hair cells: the varying density of these cells in the different cochlear regions and their allocation to tectorial and sallet systems of stimulation.

EUBLEPHARINAE

The eublepharines, distinguished among geckos by the presence of eyelids and the absence of a spectacle, form a relatively small assemblage with wide distribution in about seven separate areas around the world. One group, consisting of several species of the genus *Coleonyx*, is in the Western Hemisphere, in the southern part of North America and in Central America, and others are in Africa and the Orient.

Eublepharis macularius. — General features of the ear of *Eublepharis* have been shown in Figs. 14-2 and 14-3. Dimensional data are represented in Figs. 14-10 to 14-12. The upper curve in *a* of Fig. 14-10 indicates the varying width of the basilar membrane, and shown along with it is the width of the fundus. These structures exhibit a gradual increase over their dorsal region and then a more rapid rise to a maximum in the ventral region. From the 50 μ point to the maximum the width of the basilar membrane increases from 24 to 160 μ, a variation of 6.7-fold. The width of the fundus varies by about tenfold over this same course.

Shown in *b* of this figure is the thickness of the fundus, a structure that provides stiffness for the vibrating system. This thickness increases slowly from a low value at the dorsal end to a maximum around the middle of the cochlea, and then falls again toward the ventral end. The range of variation is a little over threefold.

Figure 14-11 presents two curves indicating the varying bulk of the sensory structures along the basilar membrane. The solid curve includes both the organ of Corti and the fundus on which it rests, whereas the broken curve is for the cellular mass only. The variation for the uppermost curve is about 7.4-fold, and that for the lower one about fivefold, again reckoning from the 50 μ point to the maximum.

The uppermost graph of Fig. 14-12 shows the varying numbers of longitudinal rows of hair cells along the basilar membrane. These are

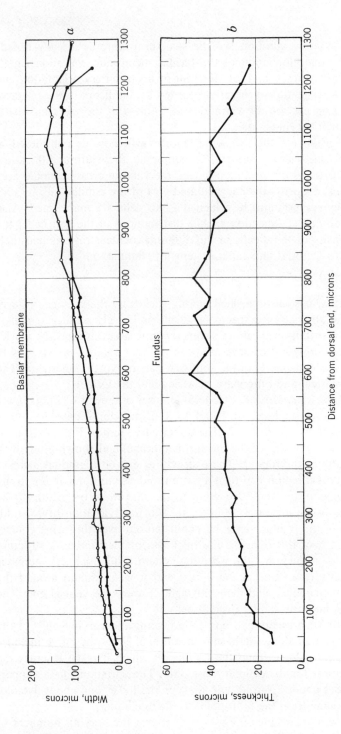

FIG. 14-10. Cochlear dimensions in *Eublepharis macularius*: basilar membrane and fundus. From Wever, 1974b.

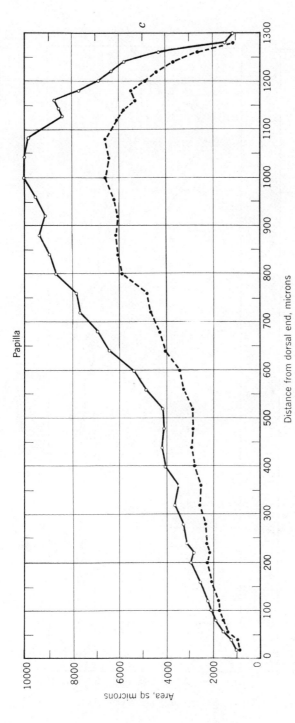

Fig. 14-11. Cochlear dimensions in *Eublepharis macularius*: size of auditory papilla. The solid line represents the whole structure, and the broken line the sensory
portion only. From Wever, 1974b.

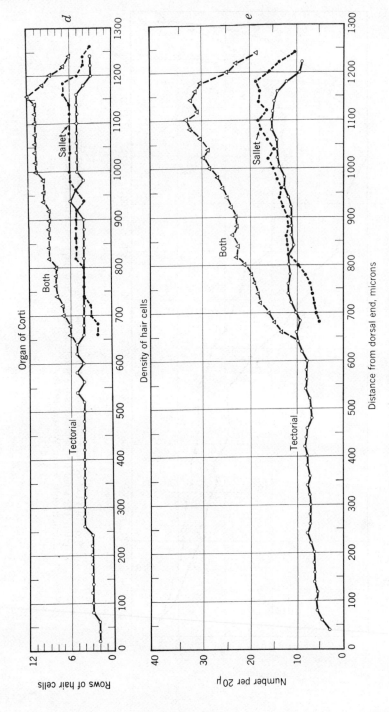

Fig. 14-12. Cochlear dimensions in *Eublepharis macularius*: hair-cell distribution. From Wever, 1974b.

the cells seen singly (i.e., one for each row) in transverse views like those of Fig. 14-3. Over the dorsal half of the cochlea these cells connect to the tectorial membrane exclusively. Two rows appear at the dorsal end, increase to 3 and then 4, and vary between 4 and 5 toward the end of this long dorsal segment. These rows then stabilize at 4 as the sallets appear, and after some variations go to 5 for a distance and finally fall away as the papilla terminates.

The sallets appear at the middle of the cochlea, first serving two rows of hair cells, and then increasing this number until it equals and finally exceeds the number served by the tectorial membrane, reaching seven a little before the papilla ends. In this figure the uppermost branch of the curve for the ventral region represents the rows served by both tectorial membrane and sallets. This combined curve shows a progressive rise until it includes 11, and even 12 rows at one point, after which it quickly subsides.

The varying density of the hair cells in the different cochlear regions is represented in the lower portion of this figure. What is portrayed here and in subsequent figures of this kind is lineal density: the number of hair cells per unit of length. The dorsal segment with tectorial connections to the hair cells shows much the same character as the row curve above, with a moderate rise along its course. The ventral segment, however, displays significant increases in density for both the tectorial and sallet cells. This occurs because toward the ventral end there is not only an increase in the number of rows of cells but also a closer packing of these cells in their rows. The increase is of about the same order of magnitude for both tectorial and sallet-connected cells and the effect is an increase in cell density at the ventral peak of more than ninefold from that near the dorsal end, or an increase of more than fourfold from the number prevailing over most of the dorsal segment.

Coleonyx variegatus. — The auditory papilla of *Coleonyx variegatus* is shown in two regions of the cochlea in Figs. 14-13 and 14-14. The first of these sectional views, taken close to the dorsal end, shows two rows of hair cells with their ciliary tufts connected directly to the overhanging tectorial membrane. The second view presents the cochlear structures in their full development in the ventral region with 10 rows of hair cells, 4 connected to the tectorial membrane and 6 surmounted by a sallet. In this ventral region the hair cells are long and closely packed, and nearly evenly divided between tectorial and sallet connections. The sallet has an unusual "top hat" form.

Dimensional data for this species were worked out earlier (Wever, 1965a). As represented in Fig. 14-15, the variation in the width of the basilar membrane is about fivefold, measured from the 50 μ point to the maximum, and that for the fundus is about eightfold. This width

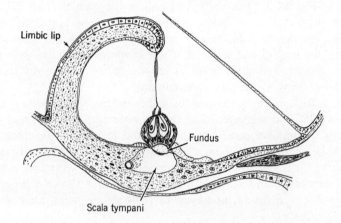

FIG. 14-13. A frontal section of the auditory region in *Coleonyx variegatus*, from the dorsal part of the cochlea. Scale 250×. From Wever, 1965a.

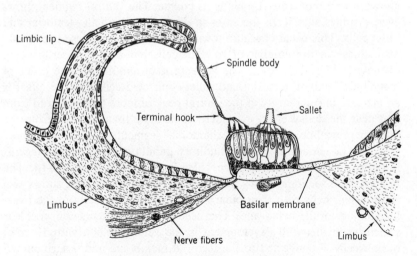

FIG. 14-14. A frontal section of the auditory papilla of *Coleonyx variegatus*, from the ventral region of the cochlea. Scale 200×. From Wever, 1967b.

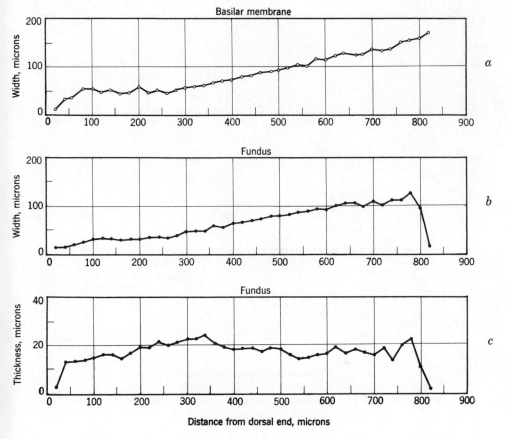

Fig. 14-15. Cochlear dimensions in *Coleonyx variegatus*: basilar membrane and fundus. From Wever, 1965a.

variation is a little less than that just shown for *Eublepharis macularius*. The thickness of the fundus varies about twofold, again less than in *Eublepharis*.

The sensory structures on the basilar membrane of *Coleonyx variegatus*, represented in Fig. 14-16, show about the same form of variation as in *Eublepharis macularius*.

The rows of hair cells were represented in this early study only in total, for tectorial and sallet-connected cells combined, but this curve agrees well with the total curve for these cells in *Eublepharis macularius*. Thus it appears that the cochleas of these species of eublepharines display close similarities of patterning, though the basilar membrane in *Coleonyx* is much the smaller.

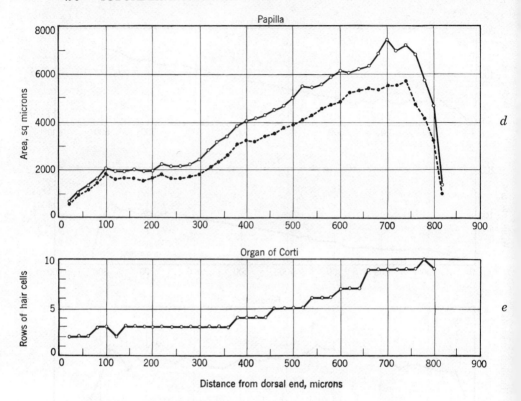

FIG. 14-16. Cochlear dimensions in *Coleonyx variegatus*: the auditory papilla
and rows of hair cells. From Wever, 1965a.

Hemitheconyx caudicinctus. — Dimensional measurements were
made also in a third eublepharine, *Hemitheconyx caudicinctus.* The
form of the basilar membrane in this species was shown in Fig. 14-5.
The distribution of the hair cells is represented in Fig. 14-17. As will
be noted, the hair cells make connections to the tectorial membrane
exclusively over the dorsal half of the cochlea. The number of longitudinal rows varies between 3 and 4 over this long course. On the
appearance of the sallets these tectorial rows continue for a way, then
rise to 5 and 6 in the ventral region. The number of rows served by
sallets, once these elements appear, increases rapidly from 1 to 4, then
varies between 5 and 6 through the ventral region.

The density of the hair cells remains remarkably constant over the
dorsal half of the cochlea, and then in the ventral half rises progressively to a maximum, after which the papilla quickly ends. This increase in
density is contributed to about equally by both tectorial and sallet-

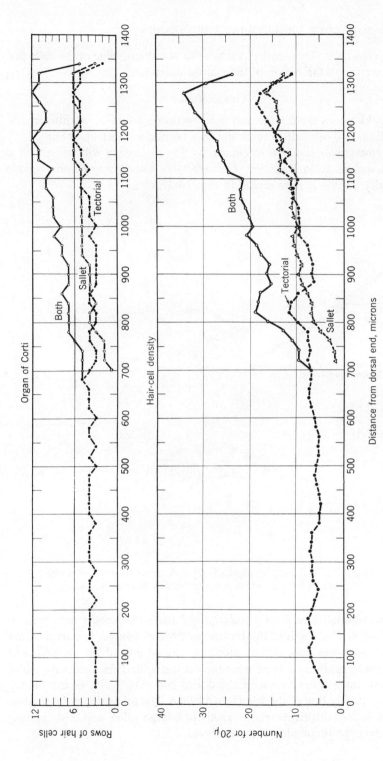

FIG. 14-17. Hair-cell distribution in *Hemitheconyx caudicinctus*.

connected cells. The density reaches the very high value of 34 cells per 20 µ section in the region of the ventral maximum.

GEKKONINAE

The gekkonines are a large and heterogeneous group of worldwide distribution, and difficult to represent in a limited number of examples.

Cyrtodactylus kotschyi orientalis. — The discussion will begin with a small species from Palestine, *Cyrtodactylus kotschyi orientalis*, whose auditory papilla is represented in Fig. 14-18.

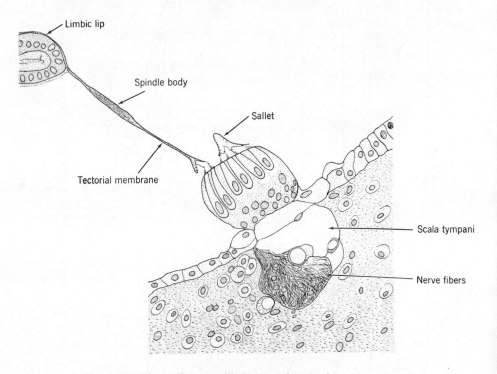

FIG. 14-18. The auditory papilla in *Cyrtodactylus kotschyi orientalis*, seen in frontal section near the dorsal end. Scale 400×.

Dimensional data for *Cyrtodactylus kotschyi orientalis* are given in Fig. 14-19. The width of the basilar membrane (shown in part *a* of the figure) increases only slightly along the dorsal part of the cochlea, and then rises rapidly to a peak near the ventral end. This maximum width is about three times that near the dorsal end. The width of the fundus follows closely that of the whole membrane. The thickness of the fundus, shown in *b* of this figure, rises quickly to a large value, and then changes little over the remainder of the cochlea.

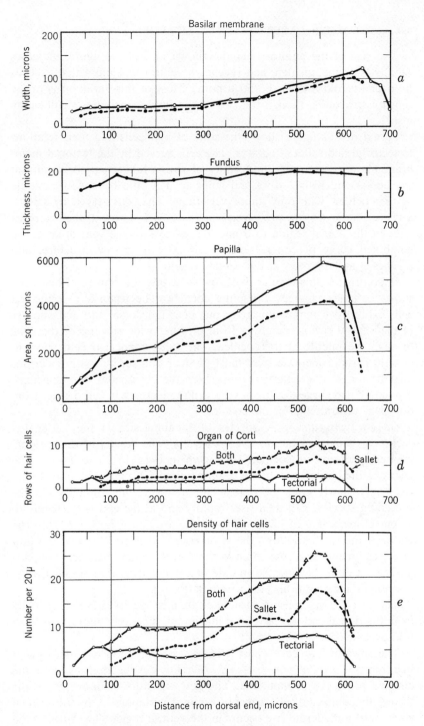

FIG. 14-19. Cochlear dimensions in *Cyrtodactylus kotschyi orientalis.*
From Wever, 1974b.

The size of the auditory papilla, shown in *c* of the figure, increases rapidly over the cochlea, and reaches a ventral maximum that exceeds by over fivefold that at the 50 μ point. Most of this variation is in the sensory structures, though the size of the fundus increases significantly also.

Part *d* of this figure shows the rows of hair cells and their relations to tectorial and sallet structures. The cells served by the tectorial membrane show considerable uniformity of row arrangement, beginning at the dorsal end with 2 rows and going to 3 a little beyond the middle of the cochlea, with only minor variations. The cells served by sallets, however, show a slow, progressive increase from 1 row at their appearance in the dorsal region to 6 and 7 rows toward the ventral end. The combined curve of course reflects this growth of row structure, and reaches 9 and 10 rows in the ventral region.

In contrast to this regularity of row structure, the density of hair cells shows complex changes along the cochlea. The tectorially connected cells exhibit two maximum areas, one near the dorsal end and another in the ventral region, resulting from a rather wide spacing of the cells through the middle of the cochlea. The sallet-connected cells show in general a rapid increase over most of their course, but with a curious interruption in the ventral region where the cells show wide spacings. The combined curve exhibits a very striking increase in cell density from dorsal to ventral ends of the cochlea.

Gehyra variegata. — A species of medium size is the variegated gecko, of wide distribution over Australia. The results of dimensional measurements on its cochlea are shown in Figs. 14-20 to 14-22. The width of the basilar membrane begins around 60 μ near the dorsal end, rises only slowly over the dorsal half of the cochlea much as in the preceding species, and then rises rapidly and at the end very steeply to a ventral maximum of 225 μ. The width variation is thus 3.7-fold. The width of the fundus follows the same course. The thickness of the fundus rises quickly to a maximum in the dorsal region, then falls continuously to a point beyond the middle of the cochlea, after which it rises to another maximum and then terminates.

The size of the sensory structures in this species increases rapidly from the dorsal end to the ventral maximum, increasing more than fourfold over its course (Fig. 14-21).

The rows of hair cells (part *d* of Fig. 14-22) show a pattern much like that of *Cyrtodactylus kotschyi orientalis*, but the changes along the cochlea are more pronounced. The row structure is uniform for the tectorially connected cells, except for minor variations in the transitional zone, until a moderate rise occurs in the ventral region. The rows of the sallet cells soon exceed the tectorial ones, and remain well above them

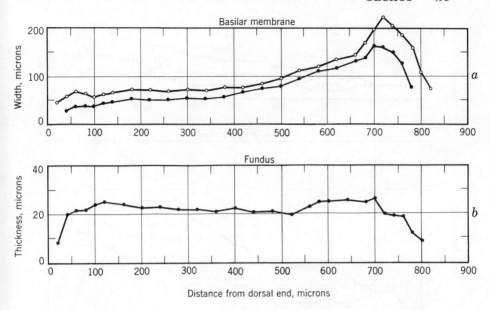

FIG. 14-20. Dimensions of the basilar membrane and fundus in *Gehyra variegata.*

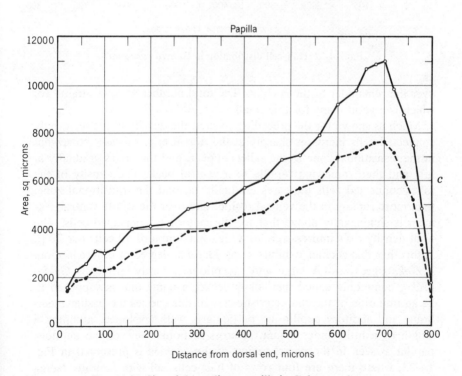

FIG. 14-21. Size of the auditory papilla in *Gehyra variegata.*

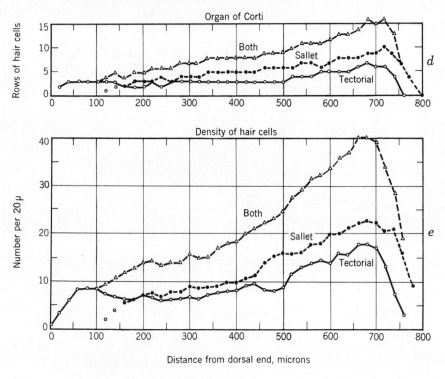

FIG. 14-22. Hair-cell distribution in *Gehyra variegata*.

over the remainder of the cochlea. The total number of rows attains 16, which is a good many for this small ear.

Much as shown for the preceding species, the density of the tectorially connected cells increases sharply at the dorsal end, subsides somewhat in the transitional zone as the sallets appear, and then rises gradually at first and then somewhat rapidly to a ventral peak. The density of the sallet-connected cells increases gradually beyond the transitional zone, then more rapidly in the ventral region, and over the whole ventral segment is significantly above that for the tectorially connected cells. The total density rises impressively to a high maximum of 40 cells per 20 μ. Altogether this cochlea exhibits a considerable degree of differentiation.

Gekko gecko. — A large and exceptionally hardy species is *Gekko gecko*, commonly called the Tokay gecko, a name that is supposed to be an imitation of the characteristic cry. In this species the basilar membrane and auditory papilla are particularly well developed, attaining a length of a little under 2 mm. A cross-sectional view of the auditory papilla as seen in the dorsal region of the cochlea is presented in Fig. 14-23, where there are four rows of hair cells, all with laterally facing

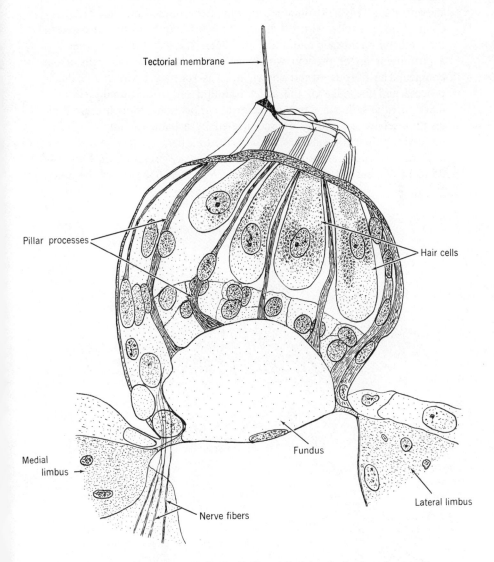

FIG. 14-23. The auditory papilla in *Gekko gecko*, seen in the dorsal segment of the cochlea. Scale 500×. From Wever, 1965a.

ciliary tufts, and all connected to a fibrous network from the tectorial membrane. Farther ventrally, near the middle of the cochlea, the picture shown in Fig. 14-24 is obtained in which the duplex manner of ciliary connections prevails, with the hair cells nearly equally divided between the tectorial membrane and a line of sallets. The tectorial membrane has a prominent finger process with which the innermost hair cells make contact. The fibrous strand shown here as running between the finger process and the edge of the sallet is found only occasionally; for the most part the sallets are free of tectorial connections, though these bodies themselves are closely interconnected by a band running longitudinally along their top surfaces, as described earlier.

Details of cochlear structure in this species are presented in Figs. 14-25 to 14-27, based on an earlier report (Wever, 1965a). It will be noted

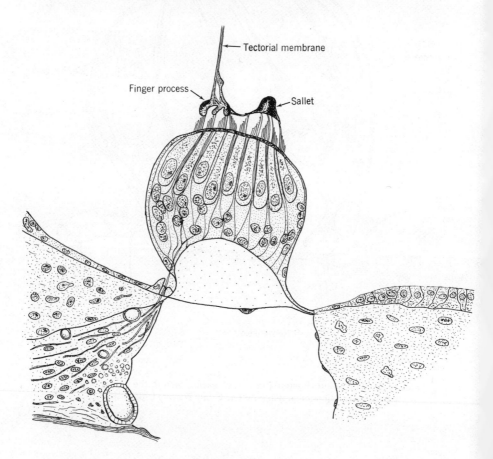

FIG. 14-24. The auditory papilla of *Gekko gecko*, seen in the ventral segment. Scale 500×. From Wever, 1965a.

that the abscissa scale is condensed compared with foregoing figures of this kind. The basilar membrane in the specimen studied was 1.84 mm long, and varied in width along its extent as shown in *a* of Fig. 14-25. This width rises progressively from the dorsal end, at first slowly and then at an accelerated rate, to a prominent maximum at the ventral end. The range from a point close to the dorsal end to the ventral maximum is 3.3-fold. The width of the fundus, shown in *b* of this figure, follows a similar course, though the ventral maximum is less sharp; this range of variation is about threefold. The thickness of the fundus, represented in *c* of this figure, increases also along the cochlea, but with considerable irregularities, and shows a range of variation of only 1.4-fold.

The auditory papilla, shown in Fig. 14-26, increases in size at a rapid rate and attains a sharp maximum near the ventral end, then almost abruptly terminates. Both the complete structure (solid line) and the cellular portion (broken line) show much the same form. The variation in area along the cochlea is about threefold.

The number of rows of hair cells, shown in Fig. 14-27, begins at 2, rises moderately in the dorsal region, increases in rapid steps to a level around 6-10 through the middle of the cochlea, and continues to rise in the ventral region to a maximum of 14-15, after which the papilla quickly ends. In this figure the tectorial and sallet cells are not treated separately, but the division between them is approximately equal, as shown in Fig. 14-24. The number of hair cells counted in serial sections is about 1600. Miller (1973a), using surface preparations, obtained counts around 2000 in his specimens (see the discussion in Chapter 2).

Gekko smithi. — A species resembling the more familiar Tokay gecko, but somewhat larger and of more slender body type, is *Gekko smithi*. Its cochlear development exceeds that of any other gecko, and specimens were studied in particular detail.

In one of the specimens the basilar membrane was 2.1 mm in length, and the number of hair cells was 2137 on the left side and 2078 on the right.

As shown in Fig. 14-6, the auditory papilla shows the typical gekkonid segmentation, with a dorsal portion making up about a third of the total length, and a short transitional zone leading into the extensive ventral segment.

In this species the terminal structures of the tectorial membrane exhibit some peculiar variations along the cochlea. At the extreme dorsal end this membrane runs around the edge of the limbic lip, comes free, immediately thickens to form a small spindle body, and then continues its free course to the papilla as an exceedingly thin membrane. Close over the papilla it shows a marked thickening, which includes a small densely staining area; this is the "rod" earlier described in *Gekko gecko*

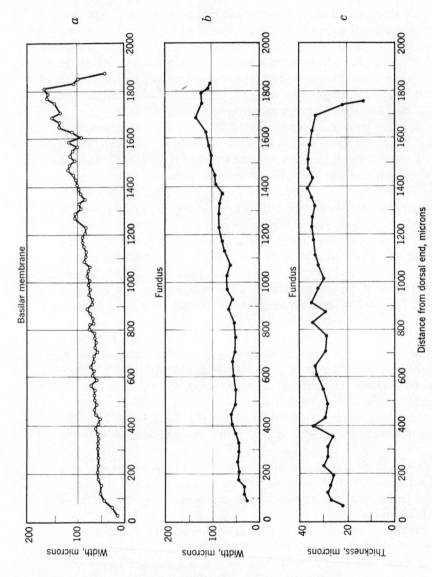

Fig. 14-25. Dimensions of the basilar membrane and fundus in *Gekko gecko*. Note the compression of the abscissa scale. From Wever, 1965a.

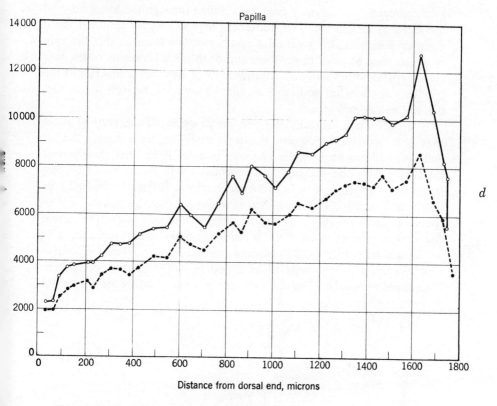

d

FIG. 14-26. Size of the auditory papilla in *Gekko gecko*. From Wever, 1965a.

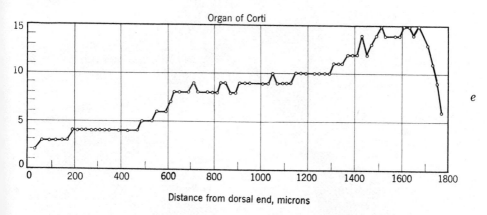

e

FIG. 14-27. Hair-cell distribution in *Gekko gecko*. From Wever, 1965a.

as a characteristic stiffening member running through the lower edge of the tectorial membrane. The thickening here is roughly spindle-shaped in cross section, and it gives off a number of fine fibers in the direction of the papilla. Most of these fibers approach the ciliary tufts of the hair cells, expand somewhat, and connect to their tips, but a few run to the papillar capsule itself and appear to make connections with it just medial to the hair-cell area.

A little farther ventrally the rod increases in size, becoming more triangular in cross section, and the fibers extending from it grow thicker and fuse to form an irregular chain of nodules lying immediately over the ciliary tufts and making the connections with them. This structure is the finger process, seen in more regular form still more ventrally, as shown in Fig. 14-28. The number of nodules along the "finger" corresponds at least roughly with the number of hair-cell rows. The actual connection appears to be made by the kinocilium, which terminates in a spherical bulb, often visible as shown in the drawing when the contact is with a thorn process sent off by a nodule, but sometimes difficult to differentiate visually when it fuses to the lower surface of the finger process.

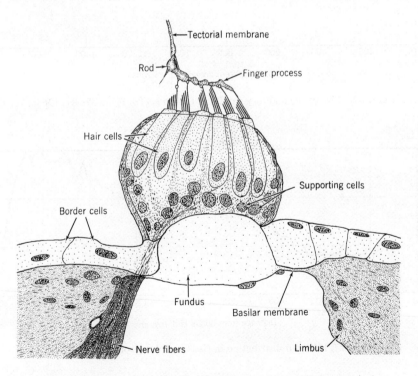

FIG. 14-28. The auditory papilla of *Gekko smithi*, seen in frontal section in the dorsal segment. Scale 500×.

Near the middle of the papilla, when sallets are present and serve for the connections to the more lateral hair cells, the finger processes become reduced. At the same time the rod structure expands, takes a pyramidal form, and its base makes connections to the most medial ciliary tufts. This stage is shown in Fig. 14-29. For a short distance two terminal bodies lie over this medial side of the papilla, one the expanded rod structure and the other the abbreviated finger process, along with the sallet on the lateral side. Then a little more ventrally the two medial bodies fuse to form a tectorial plate, seen in Fig. 14-30. This plate structure continues for a considerable distance ventrally, then becomes reduced somewhat and takes on a simple cuplike form, shown in Fig. 14-31.

The sallets in this species are comparatively small. They begin in the initial part of the ventral segment as little hillocks (Fig. 14-29), grow somewhat taller and broader in the middle of the cochlea, and then farther ventrally become still broader and flattened in conformity with the increased number of hair cells that they cover.

Details of papillar dimensions are represented for *Gekko smithi* in Figs. 14-32 to 14-36. In these figures the abscissa scale is the same as for preceding species (except *Gekko gecko*), and the length is shown in two segments.

The varying width of the basilar membrane is represented by the solid curve of Fig. 14-32. This membrane increases rapidly at the dorsal end to a value around 90 μ, increases slowly to nearly 100 μ, and then shows a gradual narrowing over a considerable distance, until somewhat beyond the middle of the cochlea a moderately rapid rise appears, which becomes almost abrupt to produce a maximum near the ventral end. Thereafter the width declines rapidly and the papilla ends. The variation in width is about threefold. The width of the fundus, shown in this same figure by the broken curve, follows much the same course, except that the maximum at the ventral end is much less prominent. The variation of this structure is only about twofold.

The thickness of the fundus is shown in Fig. 14-33, and throughout the cochlea is relatively great. This thickness begins at the dorsal end around 30 μ and maintains this level with remarkable uniformity throughout the dorsal half of the cochlea, then rises rapidly to 40 μ and varies around this value almost to the ventral end. The differentiation shown in this structure is only slight.

The varying area of the auditory papilla is represented in Fig. 14-34. The entire structure (solid line) shows a rapid increase at the dorsal end, then a slower rise to a maximum that reaches 14,000 sq μ in the far ventral region, after which the papilla quickly terminates. The variation in area, from the level at the 100 μ point to the maximum, amounts to about 3 times.

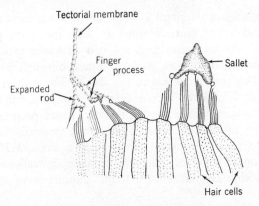

FIG. 14-29. The tectorial structure of *Gekko smithi*
near the middle of the cochlea.

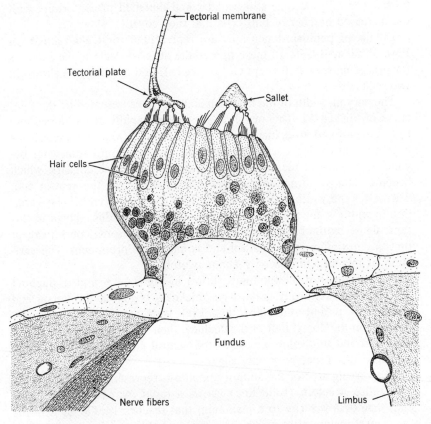

FIG. 14-30. The auditory papilla of *Gekko smithi*, seen at a point
a little beyond the middle of the cochlea. Scale 500×.

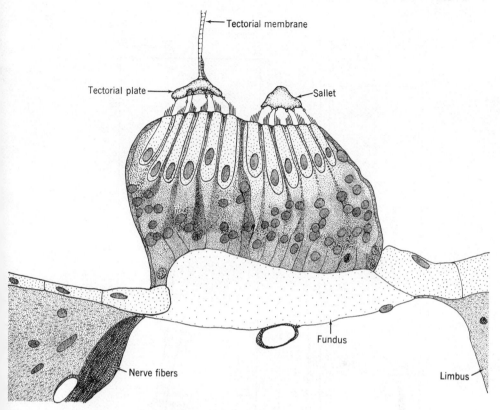

Fig. 14-31. The auditory papilla of *Gekko smithi*, seen at a point near its ventral end. Scale 500×.

The row structure along the basilar membrane is represented in Fig. 14-35 for hair cells with tectorial and sallet connections, considered separately and in total. For both types there is considerable uniformity. The number of rows begins at 4 at the dorsal end, increases to 5 and remains there, with some variations, through the dorsal segment, continues at that same level over the first half of the ventral segment, and then increases to 6 over most of the remainder of the cochlea. The rows served by sallets begin with 1 and 2, increase steadily to 4 and 5, reach 6 about the middle of the ventral segment, and vary around 7 and then 8 over the remainder of this segment. The total number of rows shows a wavering increase over the ventral segment, finally reaching 14 and 15 in the far ventral region.

The hair-cell density (Fig. 14-36) closely follows the row structure. This density for the tectorially connected cells varies only moderately

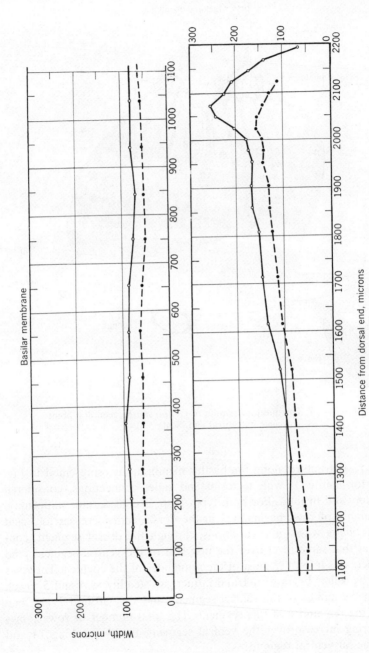

Basilar membrane

Distance from dorsal end, microns

Width, microns

Fig. 14-32. Width of the basilar membrane in *Gekko smithi*.

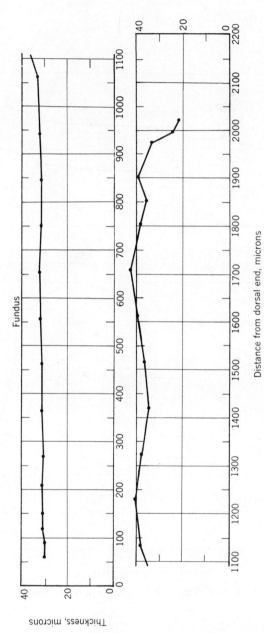

Fundus

Thickness, microns

Distance from dorsal end, microns

Fig. 14-33. Thickness of the fundus in *Gekko smithi*.

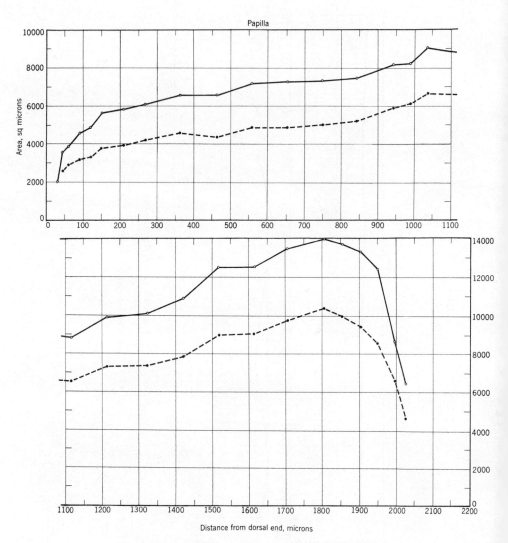

FIG. 14-34. Area of the auditory papilla in *Gekko smithi*.

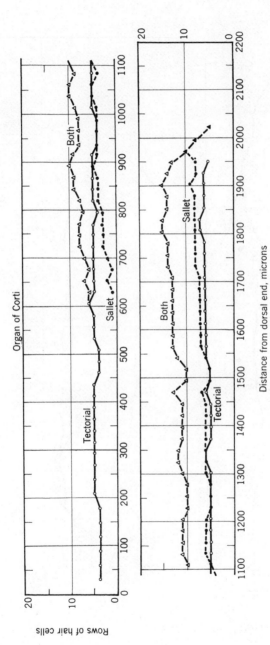

Fig. 14-35. Rows of hair cells in *Gekko smithi*.

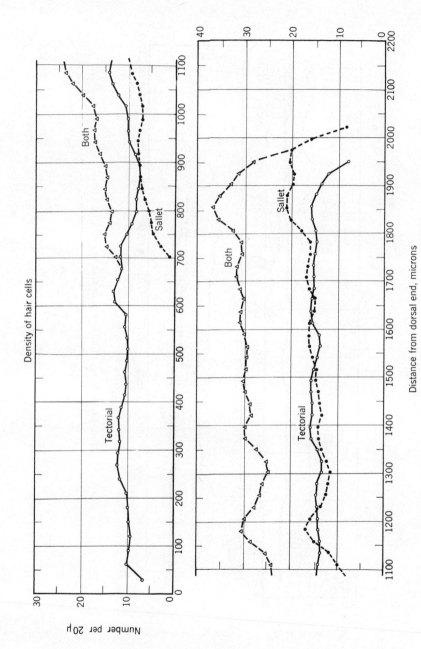

Density of hair cells

Number per 20μ

Distance from dorsal end, microns

Fig. 14-36. Density of hair cells in *Gekko smithi*.

over the dorsal segment, falls noticeably at the beginning of the ventral segment as the sallets take over their share, then a little thereafter, near the middle of the cochlea, these tectorial cells show an unusual preponderance for a short distance, and finally reach a wavering level around 15 cells per 20 μ section. The sallet-connected hair cells are less numerous than the others until the middle of the cochlea is passed, then vary around the same level until, near the ventral end, they considerably outnumber the others. As usual, only sallet-connected cells are found over the terminal part of the papilla.

DIPLODACTYLINAE

The diplodactylines are restricted to Australia and the immediately adjacent islands, and show an extensive radiation within the genus *Diplodactylus*.

Diplodactylus elderi. — This small species, with an adult length of about 70 mm, has a single habitat, the porcupine bush, and is popularly known as the porcupine bush gecko.

A cross section of the auditory papilla (Fig. 14-37) taken from the middle of the cochlea, shows the division of the hair cells into tectorial and sallet regions characteristic of the ventral segment. The tectorial membrane branches at its distal end, and the tips of the ciliary tufts connect directly to the end twigs or to thin filamentary extensions of them. On the lateral side the connections are made directly to the lower wings of the sallet.

As in all diplodactylines, and in no other geckos so far observed except the gekkonine *Thecadactylus rapicaudus*, the tectorial membrane is greatly thickened in its middle portion. The membrane here consists of the usual relatively dense reticular material along its medial edge, and then bulges laterally as a particularly tenuous network. This thickening, as shown in the figure, begins near the upper end of the spindle body, widens rapidly, and extends to a point close to the papilla, thus covering most of the free portion of the membrane. The main filaments of the network run somewhat obliquely in a lateral direction, and vary in length so that the lateral edge is irregularly serrated.

The function of this thickening of the tectorial membrane is not clear. It may serve to add mass to the membrane and increase the friction involved in its movements in the cochlear fluid, and thereby increase its restraint of the ciliary tufts.

Dimensional measurements on *Diplodactylus elderi* gave the results shown in Figs. 14-38 and 14-39. The width of the basilar membrane (upper curve in *a* of Fig. 14-38) increases progressively from the dorsal end, with moderate acceleration in the ventral region, to a maximum close to the ventral end. The range of variation is about 2.7. The width

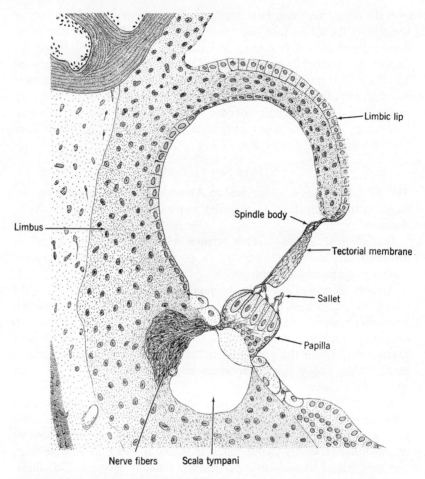

FIG. 14-37. Section through the cochlea of *Diplodactylus elderi*. Scale 250×.

of the fundus follows much the same course. The thickness of the fundus changes more markedly along the cochlea, rising a little irregularly to a relatively large value in the ventral region, with a range of variation exceeding twofold.

The area of the papilla, shown in Fig. 14-39, increases greatly toward the ventral end of the cochlea, with a prominent maximum approaching 7000 sq μ for the total structure and reaching 4400 sq μ for the cellular portion.

The distribution of the hair cells reflects these dimensional changes. The rows of hair cells (part c in Fig. 14-38) increase by fairly regular steps from 2 or 3 at the dorsal end to 8-10 in the far ventral region. The hair-cell density (e of Fig. 14-39) increases rapidly and at a rather

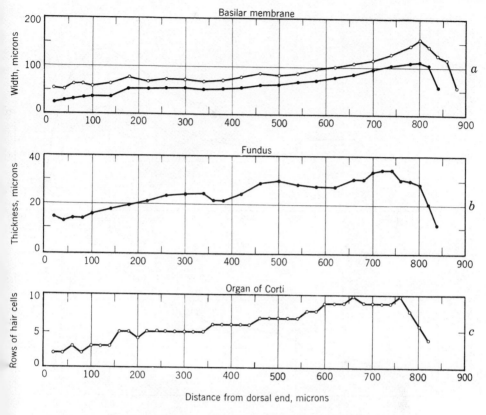

FIG. 14-38. Cochlear dimensions in *Diplodactylus elderi*.

uniform rate to a sharp maximum of 35 cells per 20 μ near the ventral end, and then quickly declines as the papilla ends.

This small cochlea exhibits a considerable degree of differentiation.

Lucasium damaeum. — A little larger diplodactyline species is the beaded gecko, *Lucasium damaeum*, a ground-dwelling form. Dimensional data on this ear are presented in Figs. 14-40 to 14-42. As shown by the upper curve of Fig. 14-40, the width of the basilar membrane increases from 45 μ near the dorsal end to 135 μ at the ventral maximum, a threefold variation. The width of the fundus follows the same course. The thickness of the fundus, at first about 14 μ, rises sharply in the early dorsal region and nearly uniformly thereafter to a maximum of 34 μ, a variation of 2.5-fold.

The size of the sensory structures, shown in Fig. 14-41, increases steadily by about sixfold to a maximum at the ventral end of about 7500 sq μ.

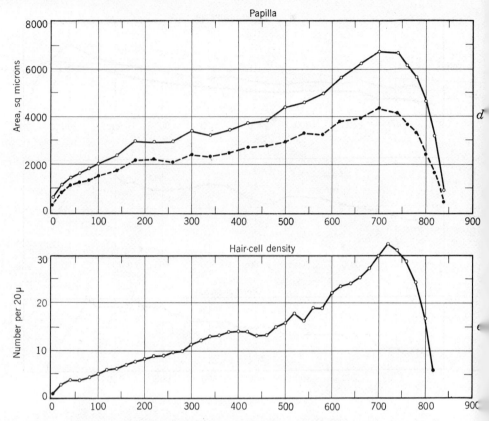

FIG. 14-39. Papillar area and hair-cell distribution in *Diplodactylus elderi.*

In *c* of Fig. 14-40 the rows of hair cells are represented separately for tectorial and sallet-connected cells, and also in combination. For the tectorial group the number of rows increases in the dorsal segment from 2 to 3 or 4; in the ventral segment the number at first varies between 3 and 4 and then between 4 and 5. For the sallet-connected cells there are only 2 rows through the initial part of the ventral segment, then between 2 and 3, and finally in the far ventral region mostly 4 rows. Thus the rows of sallet-connected cells are relatively fewer than in most gecko ears. This condition is in contrast to that found in the eublephar-ines and gekkonines, in which the rows of sallet-connected cells usually outnumber the others. The total rows show in general a regular increase up to the ventral maximum.

Some transitional cells are represented in this figure by the dotted circles around the 200 μ point.

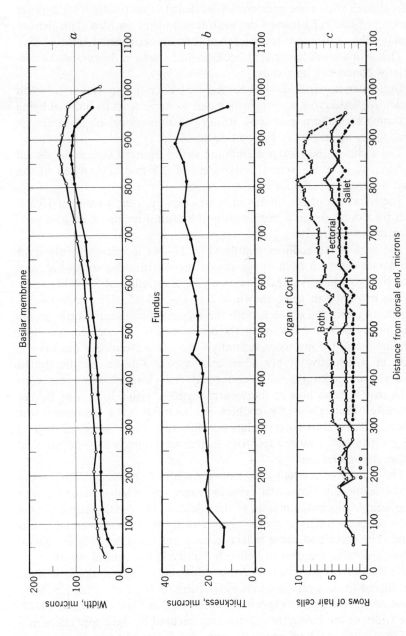

Fig. 14-40. Cochlear dimensions in *Lucasium damaeum*. From Wever, 1974b.

Hair-cell density, shown in Fig. 14-42, largely reflects the row structure, though with some acceleration of density that results from a closer packing of hair cells toward the ventral end of the cochlea. This density is noticeably greater for tectorial than for sallet-connected cells.

This is a well-differentiated cochlea, but shows a dominance of tectorially connected hair cells.

Diplodactylus vittatus. — A further diplodactyline species is the wood gecko, *Diplodactylus vittatus*, which gets its name from its habit of living in rotting logs and tree stumps. Results of dimensional observations on this species are shown in Figs. 14-43 to 14-45.

The width of the basilar membrane varies from 50 μ near the dorsal end to 140 μ at the ventral maximum, or 2.8-fold. The width of the fundus varies by 3.5-fold over this range.

The thickness of the fundus increases rapidly from a value of 13.5 μ near the dorsal end to a maximum in the ventral region of 35 μ, a variation of 2.6-fold.

The area of the auditory papilla (Fig. 14-44) increases rapidly from the dorsal end to a peak in the dorsal region, subsides somewhat, and then rises to a large ventral maximum, for a variation of 5.6-fold.

The rows of hair cells, shown in *d* of Fig. 14-45, show considerable regularity along the cochlea, with the tectorially connected cells occurring mostly in 3 or 4 rows over the dorsal half of the cochlea, and then forming 5 and 6 rows more ventrally. There is a prominent transitional zone in which several pre-sallet cells appear (shown by the dotted circles), and then the sallet-connected cells form 2 rows that continue over a large part of the ventral segment, rising to 3 rows farther ventrally. Throughout the cochlea the tectorial rows outnumber the others. The combined curve shows a slow rise along the dorsal half of the cochlea and a more rapid rise to the maximum, which occurs well before the ventral end.

The density of the hair cells shows much less regularity. The density of the tectorially connected cells increases somewhat erratically, with a noticeable dip in the middle of the cochlea. The density curve for the sallet-connected cells is more regular, rising to a broad ventral maximum. The combined curve reflects these local variations, but in general shows a notable increase in hair-cell density in the ventral region.

Hoplodactylus pacificus. — A New Zealand species of diplodactyline, considered to be a relatively primitive form, is *Hoplodactylus pacificus*. Dimensional data on this species are presented in Figs. 14-46 and 14-47. The width of the basilar membrane, represented by the upper curve in *a* of Fig. 14-46, increases regularly from the dorsal end to a broad maximum near the ventral end, a change of nearly fourfold. The width of the fundus follows the same course. The thickness of the fundus, shown

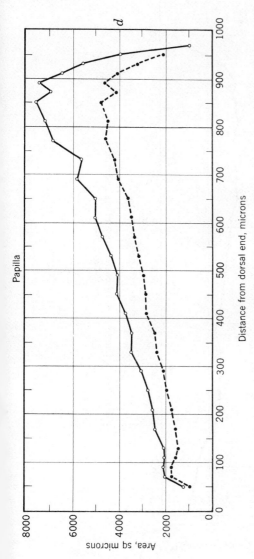

FIG. 14-41. Size of the auditory papilla in *Lucasium damaeum*. From Wever, 1974b.

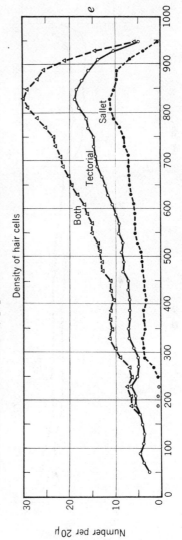

FIG. 14-42. Hair-cell density in *Lucasium damaeum*. From Wever, 1974b.

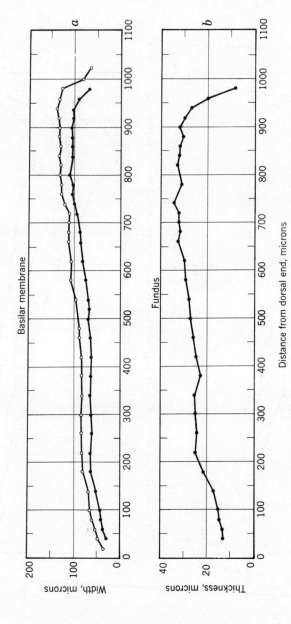

FIG. 14-43. Dimensions of basilar membrane and fundus in
Diplodactylus vittatus.

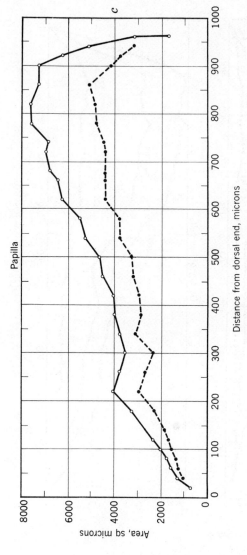

FIG. 14-44. Size of the auditory papilla in *Diplodactylus vittatus*.

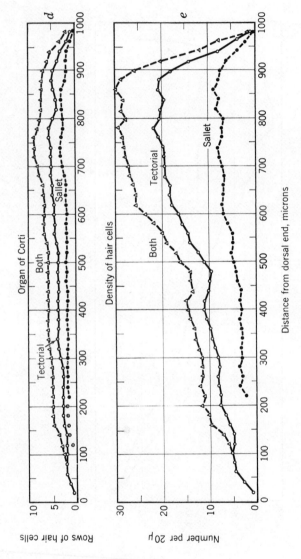

Fig. 14-45. Hair-cell distribution in *Diplodactylus vittatus*.

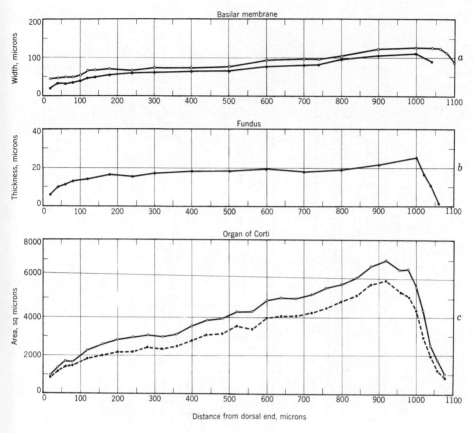

FIG. 14-46. Cochlear dimensions in *Hoplodactylus pacificus*.

in *b* of this figure, increases moderately to a ventral peak, with a range around twofold.

The area of the papilla, shown in *c* of this figure, increases steadily along the cochlea to a maximum of about 7000 sq μ near the ventral end, and its cellular portion follows a closely similar course. The variation is considerable, exceeding threefold.

The hair-cell distribution in this cochlea is represented in Fig. 14-47. The tectorially connected hair cells show considerable uniformity, with 3 rows over most of the dorsal portion of the cochlea, increasing to 4 and 5 rows in the ventral portion. The sallet-connected cells have generally fewer rows, from 2 in the dorsal third of the cochlea and increasing to 3 and 4 over most of the ventral region, but reaching 5 and 6 near the ventral end. The combined picture shows only a moderate increase along the cochlea.

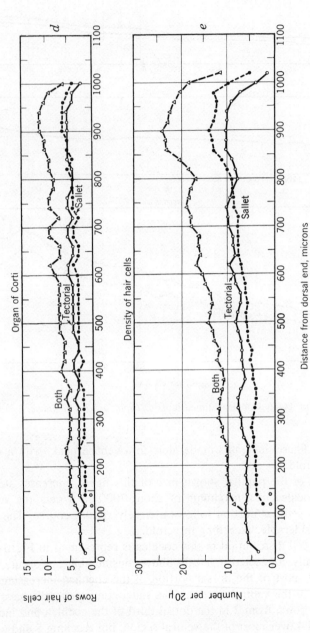

FIG. 14-47. Hair-cell distribution in *Hoplodactylus pacificus*.

The pattern of hair-cell density largely reflects the row structure. The density of tectorially connected hair cells rises slowly and somewhat irregularly and then takes a moderate spurt in the far ventral region. The density of sallet-connected cells takes a similar course, but is consistently lower over the dorsal three-fourths of the cochlea, then rises well above that of the tectorial cells in the far ventral region. This preponderance of sallet-connected cells extends over a greater area at the ventral end of the cochlea than in other diplodactylines examined.

Oedura ocellata. — A diplodactyline of particular interest is the eyed velvet gecko, *Oedura ocellata*, a relatively large species living in the Pilliga Scrub of New South Wales and Queensland, Australia. Fig. 14-48 is a cross section through the auditory region at a level close to the middle of the cochlea. The thickening of the tectorial membrane is a prominent feature, and at this level the membrane ends in an expanded foot.

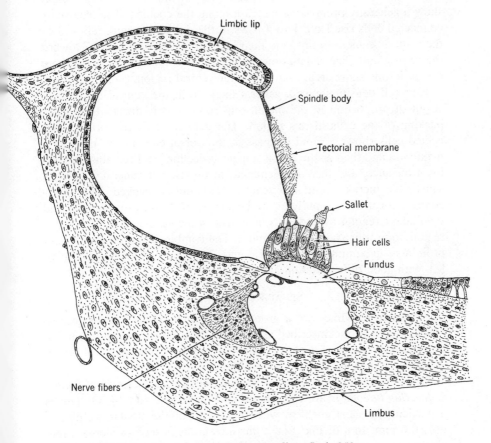

FIG. 14-48. The cochlear region in *Oedura ocellata*. Scale 250×.

Dimensional observations on this species are shown in Figs. 14-49 to 14-51. As *a* of Fig. 14-49 indicates, the width of the basilar membrane increases in a wavering fashion to a broad ventral maximum, with a range of variation of about twofold. The width of the fundus follows the same course. The thickness of the fundus increases near the dorsal end over a distance of about 400 μ, then remains relatively uniform over the whole middle part of the cochlea, rising slightly to a flat maximum near the ventral end.

The area of the papilla, shown in Fig. 14-50, changes only a little over a short distance at the dorsal end, then increases progressively to a prominent maximum, beyond which the papilla ends bluntly. The cellular portion of the papilla increases less rapidly and shows a much lower and flatter maximum at the ventral end. The areal differentiation is about fourfold for the entire structure.

The rows of hair cells in this species, represented in *d* of Fig. 14-51, show a generally progressive increase along the cochlea. The tectorially connected cells rise from 1 to 4 rows in the dorsal segment, fall to 3 as the ventral segment begins, and then increase by steps to 6 and 7 toward the ventral end. The sallet-connected cells show fewer rows everywhere except in one short area, even in the far ventral region.

Hair-cell density shows a general increase in the ventral region that is out of proportion to the increase in rows, and is due to the tighter packing of the cells in this region. The density of the tectorially connected cells shows a rapid increase at the dorsal end, passes through a maximum and then as the sallets appear it declines and remains uniform for a distance, and thereafter changes in bursts, but generally shows a substantial increase until the ventral maximum is reached. The sallet-connected cells rise rapidly at the beginning of the ventral segment and then take a regular ascending course, with some variations in the ventral region, to produce a broad maximum. The combined curve shows a very prominent maximum at the ventral end. This is a well-developed cochlea, with good differentiation along its extent.

SPHAERODACTYLINAE

The sphaerodactylines are a group of relatively small, diurnal geckos, distributed over circumscribed areas of the New World tropics, including the southern parts of Central America, northwestern parts of South America, and especially the West Indies.

Sphaerodactylus lineolatus. — Measurements on a specimen of *Sphaerodactylus lineolatus* are represented in Figs. 14-52 and 14-53. This is a particularly short cochlea, with the length of the basilar membrane only 0.6 mm. In *a* of Fig. 14-52 this membrane is seen to increase regularly to a maximum about double that near the dorsal end. The width

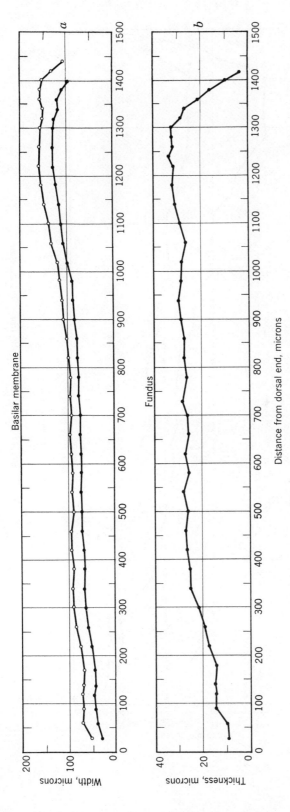

FIG. 14-49. Dimensions of basilar membrane and fundus in *Oedura ocellata*.

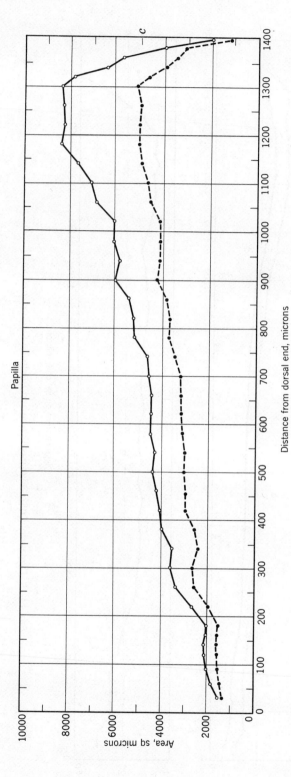

Fig. 14-50. Size of the auditory papilla in *Oedura ocellata*.

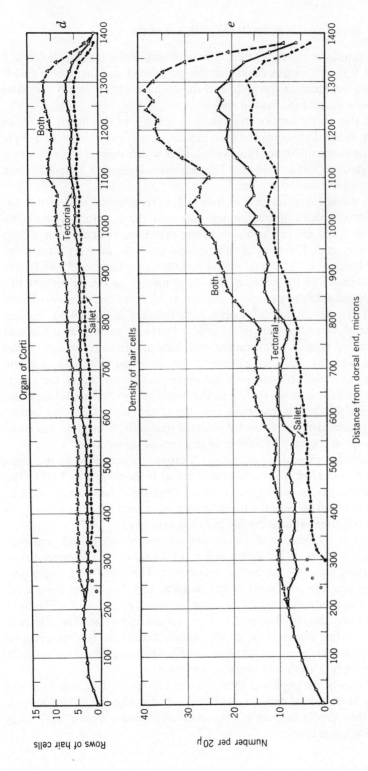

Fig. 14-51. Hair-cell distribution in *Oedura ocellata*.

of the fundus is closely similar. The thickness of the fundus, shown in *b* of this figure, increases rapidly at the dorsal end and then more slowly to a flat maximum in the middle of the cochlea, after which it decreases somewhat toward the ventral end.

The size of the sensory structures, shown in *c* of this figure, increases rapidly at the dorsal end and then remains nearly constant, with a slight decrease in the middle of the cochlea, after which there is a rise to a ventral maximum and a rapid fall. The maximum is about 3.3 times as large as the value at the 50 μ point.

The arrangement of rows of hair cells is represented in *d* of this figure. For the tectorially connected hair cells the pattern is remarkably uniform: there are just two rows along the entire cochlea, then a drop at the very end. The rows of sallet-connected cells, on the contrary, increase by steps in the dorsal region and then vary between 5 and 6 over the remainder of the cochlea. The total pattern of course reflects this form of the sallet rows, with a maximum of 8 reached at three points.

Hair-cell density for this species is represented in Fig. 14-53. For the tectorially connected cells this density increases rapidly at the dorsal end, passes through a maximum and falls as the sallets begin in the ventral segment, and then remains nearly constant, with only a gentle decline, over the remainder of the cochlea. The density of the sallet-connected cells changes rapidly in the initial part of the ventral segment, then rises slowly to a maximum near the ventral end. A striking feature is the greater density of sallet-connected hair cells throughout the cochlea except at the very beginning of the ventral segment.

Gonatodes albogularis fuscus. — A cross-sectional view of the auditory region of *Gonatodes albogularis fuscus* is shown in Fig. 14-54. The level is near the middle of the cochlea, where the ciliary tufts of the hair cells are served by both the tectorial membrane and a line of sallets, and the ciliary orientation is doubly bidirectional.

Cochlear dimensions in *Gonatodes albogularis fuscus* are presented in Figs. 14-55 to 14-57. Part *a* of Fig. 14-55 shows, in the upper curve, the varying width of the basilar membrane. The curve indicates a progressive increase over most of the cochlea, and then in the ventral region a rise to a maximum and a rapid decline. The range of variation is a little over twofold. The width of the fundus is similar. The thickness of the fundus, shown in *b* of this figure, takes a somewhat irregular course, with the principal maximum in the ventral region. This structure exhibits only a moderate range of variation.

The area of the papilla is shown in *c* of this figure. This area increases rapidly at the dorsal end, tends to level off ventrally and attains a maximum, then finally falls away. The cellular portion (broken line) takes much the same course.

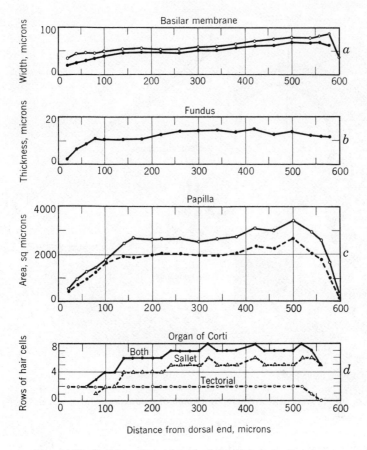

FIG. 14-52. Cochlear dimensions in *Sphaerodactylus lineolatus*.
From Wever, 1974b.

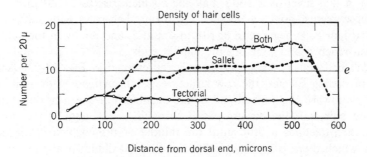

FIG. 14-53. Hair-cell density in *Sphaerodactylus lineolatus*.

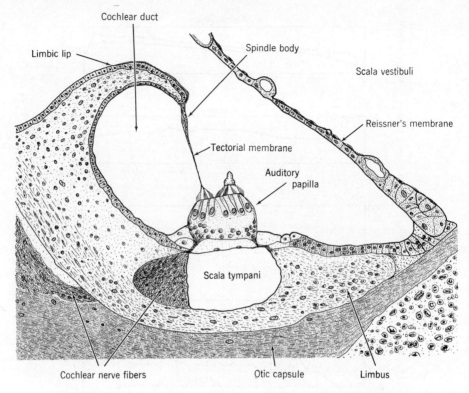

FIG. 14-54. The cochlear region in *Gonatodes albogularis fuscus*. Scale 250×.

The rows of hair cells (Fig. 14-56) vary considerably in number along this short cochlea. The rows with tectorially connected cells reach 3 for a short distance at the dorsal end, fall to 2 as the sallets begin and remain at that level over about half the cochlea, rise to 3 and then to 4 at three points, after which they fall away. The rows with sallet-connected hair cells begin with 3, rise to 4 and 5, and finally in the ventral region reach 6 and 7 before the papilla ends.

Hair-cell density in this cochlea is represented in Fig. 14-57. As in the preceding species, the row structure is reflected but with additions of cells toward the ventral end of the cochlea. For the tectorially connected hair cells the density passes through a maximum at the dorsal end, then declines a little until the middle of the cochlea is reached, after which there is a considerable rise to a maximum near the ventral end. The density of the sallet-connected hair cells is greater than that of the tectorial cells except at the beginning of the ventral segment.

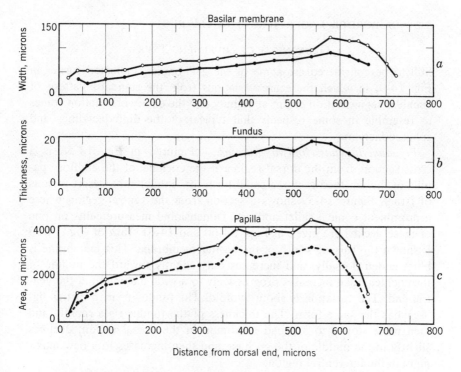

FIG. 14-55. Cochlear dimensions in *Gonatodes albogularis fuscus.*

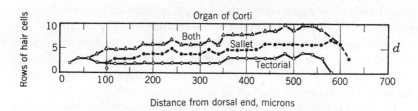

FIG. 14-56. Hair-cell distribution in *Gonatodes albogularis fuscus.*

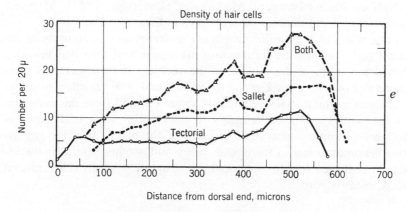

FIG. 14-57. Hair-cell density in *Gonatodes albogularis fuscus.*

Gekkonidae of Divergent Types

Still to be considered are some of the gekkonines, indicated earlier in Fig. 14-9, in which there are variations from the standard pattern of cochlear segmentation of this subfamily and the ciliary orientation comes to resemble in some respects that typical of the diplodactylines and sphaerodactylines.

Ptyodactylus hasselquistii guttatus. — Pictured in Fig. 14-58 is a cross section, from the dorsal region of the cochlea, of the auditory papilla of the fan-toed gecko, a small form that inhabits the desert regions of Israel. Figure 14-59 shows a section from the ventral region where a prominent conical sallet appears. Dimensional measurements on one of these geckos are given in Figs. 14-60 and 14-61. In *a* of Fig. 14-60 is shown the varying width of the basilar membrane. This membrane is blunt-ended dorsally, and increases only slightly toward the middle of the cochlea, then increases progressively to a wide maximum at the ventral end. The variation is about twofold. The fundus (part *b* of this figure) has the same form. The thickness of the fundus rises rapidly and then more slowly to a broad maximum in the dorsal region, declines slightly in the middle of the cochlea, and then increases to a new maximum in the far ventral region.

The variations in size of the auditory papilla are shown in *d* of Fig. 14-61. The whole structure (solid line) increases rapidly in area at the dorsal end, then more slowly through the middle of the cochlea, and rapidly once more to a maximum at the ventral end. The cellular portion (broken line) takes a similar course except for a smaller rise in the ventral region. The rows of hair cells (for tectorial and sallet-connected cells combined) increase by somewhat irregular steps all along the cochlea until a maximum is reached near the ventral end.

This cochlea does not show any special dimensional features except a slight tendency toward bimodality.

Phelsuma madagascariensis. — This gecko lives in coconut palms, chiefly on Pemba Island off Madagascar (Loveridge, 1947). Measurements of cochlear dimensions are shown in Figs. 14-62 to 14-64.

The width of the basilar membrane increases almost abruptly at the basal end, rises a little further to a maximum, and then declines to a broad minimum in the middle of the cochlea. It then increases once more and reaches a peak at the far ventral end. The variation from the minimum to the dorsal maximum is about 1.5, and from the minimum to the ventral maximum is nearly threefold. The width of the fundus follows a similar course. The thickness of the fundus likewise exhibits two maximums, one in the dorsal region and another ventrally; the reduction between is only slight, and this structure is hardly differentiated in an effective way.

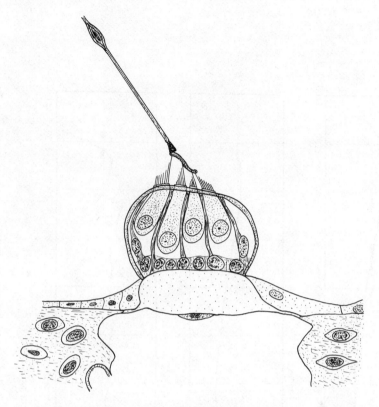

FIG. 14-58. The auditory papilla of *Ptyodactylus hasselquistii guttatus*, seen in the dorsal segment. Scale 600×.

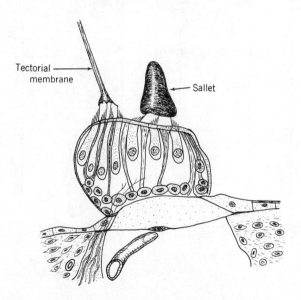

FIG. 14-59. The auditory papilla of *Ptyodactylus hasselquistii guttatus*, seen in the ventral segment. Scale 500×.

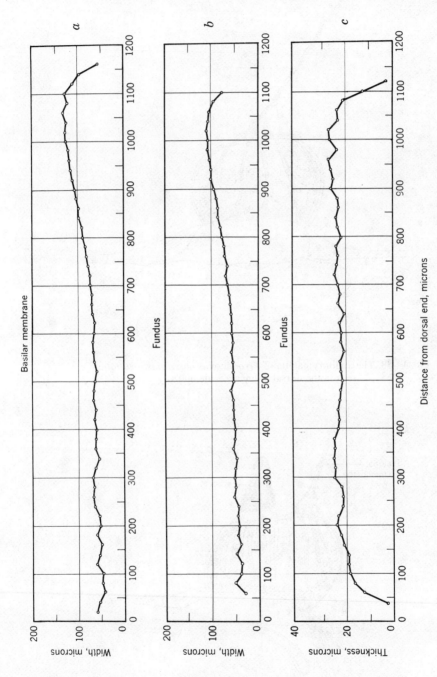

FIG. 14-60. Cochlear dimensions in *Ptyodactylus hasselquistii guttatus*: basilar membrane and fundus. From Wever, 1967c.

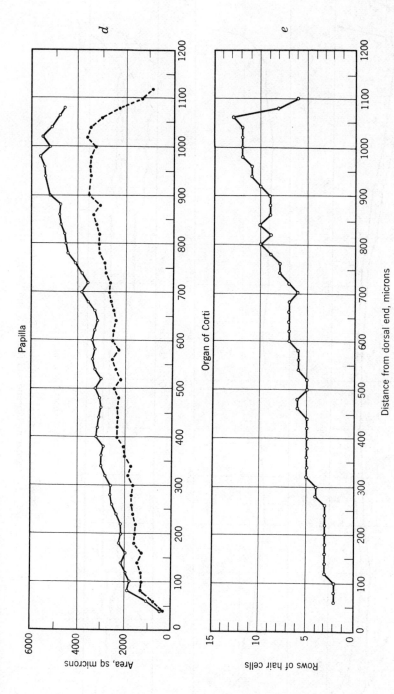

Fig. 14-61. Cochlear dimensions in *Ptyodactylus hasselquistii guttatus*: papillar size and hair-cell distribution. From Wever, 1967c.

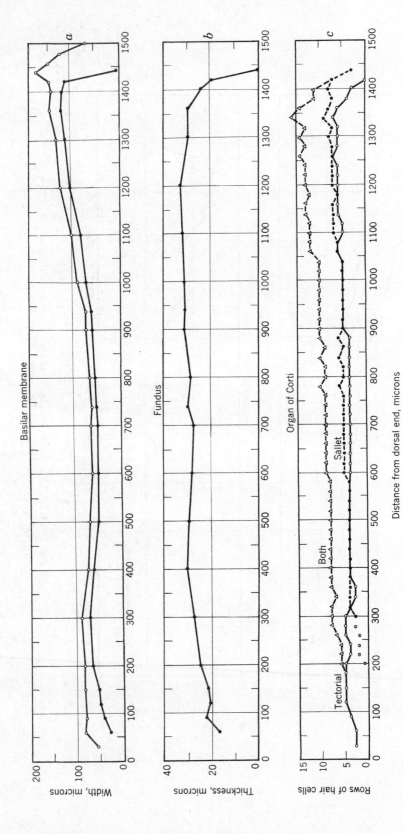

FIG. 14-62. Cochlear dimensions in *Phelsuma madagascariensis*: basilar membrane and fundus. From Wever, 1974b.

The size of the auditory papilla is represented in Fig. 14-63. This structure increases rapidly in area at the dorsal end and reaches a secondary maximum from which it declines a little, then increases once more at an accelerating rate to a high maximum at the ventral end. The cellular portion of the structure shows a similar form. The variation from near the dorsal end to the maximum is about 4.5-fold.

The rows of hair cells are represented in part c of Fig. 14-62. Those rows with tectorial connections increase to 5 at the dorsal end, vary somewhat through the transitional zone, and then remain constant at 4 through the middle of the cochlea. In the ventral region this number increases to 5 and 6 and even 7 before the papilla ends. The sallet rows quickly stabilize at 4, then increase to 5 through the middle of the cochlea, and reach 6 over most of the far ventral region. For the most part, the tectorial and sallet rows are well balanced, but in two regions there are one or two more sallet rows. The density pattern is represented in Fig. 14-64. There is a rapid increase for the tectorially connected cells at the dorsal end to a maximum of 13, then an irregular decline followed by considerable wavering with a general upward trend through the remainder of the cochlea until the far ventral region is reached. In this ventral region there is a rapid rise to a maximum before the terminal decline. The sallet-connected cells show a steady increase in density in the early part of the ventral segment and then a fluctuating upward course over which the sallet group shows for the most part a preponderance over the tectorially connected cells. The combined function shows a steady increase to a high maximum at the ventral end.

The remaining species of gekkonids have been examined in a general way, without detailed measurements but sufficiently to show that they follow the same basic plan as the ones represented here. The eublepharine and gekkonine cochleas indicate considerable amounts of differentiation, the diplodactylines somewhat less, and the sphaerodactylines the smallest degree.

AUDITORY SENSITIVITY

Now to be considered is the relative effectiveness of the gecko ear in sound reception as represented by the electrical potentials of the cochlea. Sensitivity curves were obtained by the standard method, with recording from the round window in nearly all the species listed in Table 14-I, and often for several specimens of each species.

An examination of the results of these measurements made it immediately clear that sensitivity in the geckos presents a limited number of distinctive patterns, and that a grouping of these patterns follows subfamily lines to a considerable degree. Accordingly, the sensitivity func-

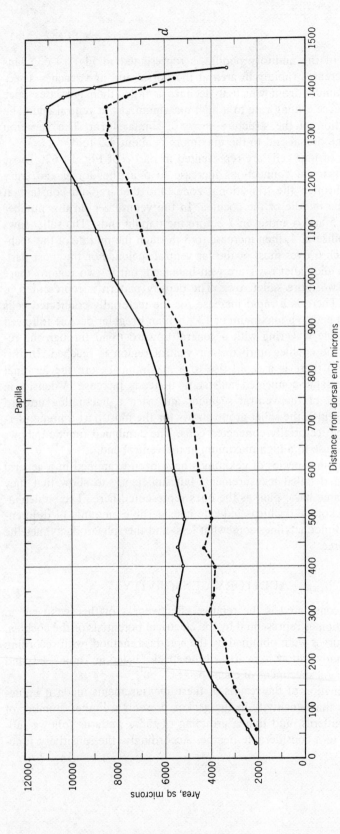

Papilla

Distance from dorsal end, microns

Area, sq microns

Fig. 14-63. Size of the auditory papilla in *Phelsuma madagascariensis*.
From Wever, 1974b.

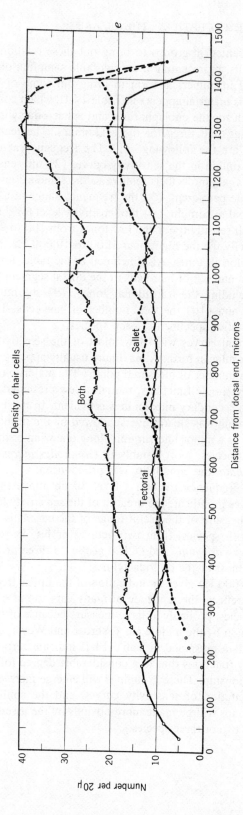

Fig. 14-64. Hair-cell density in *Phelsuma madagascariensis*. From Wever, 1974b.

tions will be presented in groups to bring out these relationships. Also some of the results, representing a considerable sampling of the whole, will be presented in numerical form to bring out some of the specific relationships. This material appears in Table 14-II, which contains seven sections, one each for the eublepharines and sphaerodactylines, two for the diplodactylines, and three for the gekkonines. The features shown for each species are the following: (1) The frequency at the peak of the principal maximum in the sensitivity curve, (2) the sensitivity level at this peak expressed in decibels relative to the standard level of sound pressure of 1 dyne per sq cm, (3) the frequency range at a level 30 db above the principal maximum, (4) the sensitivity level for the tone 4000 Hz, (5) the mean sensitivity for the low tones from 100 to 900 Hz, (6) the mean sensitivity for the high tones 1000 to 10,000 Hz, and (7) the mean sensitivity for all tones. Also included in the table, for later reference, are (8) the number of hair cells in the dorsal segment of the auditory papilla (including the transitional zone), (9) the number in the ventral segment, and (10) the total number of hair cells. The data are for one representative specimen of each species.

As the individual curves will show, most of the sensitivity functions are reasonably regular in form. A maximum usually appears as a definite trend involving a number of measured points. The majority of specimens showed relatively simple functions, though a few presented a degree of bimodality and a secondary maximum was evident. In most of these the secondary maximum was clearly overshadowed by the primary one, and it appeared only as a minor interruption along one wing of the sensitivity curve. In a few species, most notably in *Gonatodes albogularis fuscus* and *Oedura ocellata*, the bimodality was pronounced.

The range of frequencies within a level 30 db above the maximum (column 3) serves to indicate the breadth of the sensitivity function, and provides some measure of the useful range of the ear. This 30-db range varies greatly with species, from two octaves or less in some species, such as *Coleonyx variegatus* and *Gekko gecko*, to three or four octaves in others, as in most of the diplodactylines.

The level at 4000 Hz gives an indication of the upper frequency limit of the ear, and reflects the tendency of many ears to cut off sharply in the high frequencies, probably in large measure because of limitations in sound transmission by the middle ear (Werner and Wever, 1972).

The sample data presented in Table 14-II indicate certain groupings of the sensitivity functions that to a considerable degree follow subfamily lines of relationship. These groupings will emerge more clearly in the following presentation of sensitivity curves, and the similarities within particular groups will be made more obvious by the assembly in some of the figures of two or three species.

EUBLEPHARINAE

Sensitivity curves for two eublepharine species, *Coleonyx brevis* and *Hemitheconyx caudicinctus*, are presented in Fig. 14-65. The curve for *Coleonyx brevis* shows maximum sensitivity in the region of 600-1000 Hz, where the average sound pressure required for the standard response was −67 db. For lower and higher tones the sensitivity falls off sharply. For *Hemitheconyx caudicinctus* the region of best sensitivity lies between 500 and 1000 Hz, and the maximum attained is around −88 db.

For *Coleonyx variegatus* sensitivity curves were presented earlier (Wever, Peterson, Crowley, and Vernon, 1964) showing considerable variations among individual ears, with the maximum in the region of 500-1500 Hz. A function obtained more recently is shown in Fig. 14-66, and indicates best sensitivity in this same region of frequencies, with an average of −67 db in the peak area.

The species *Eublepharis macularius* was extensively studied by Werner and Wever (1972) in connection with middle ear measurements, and several functions for this species were presented in Chapter 6. One of these curves is included in Fig. 14-66 and shows best sensitivity around 400-600 Hz, peaking at −76 db.

The four eublepharines examined all display a high level of sensitivity within a narrow range in the medium low frequencies. The sensitivity declines at a rapid rate on either side of the peak region, yet remains within the −30 db level for a considerable range in both directions. Accordingly these ears must be regarded as functioning effectively over a range of four or five octaves, and as being extremely sensitive within a narrow portion of this range.

GEKKONINAE

The gekkonines present the greatest diversity of any of the gecko subfamilies. Their sensitivity curves fall into three groups. Two of these groups are alike in that the best sensitivity is in the low and middle frequencies, but they differ in the breadth of the sensitive band. The third group is distinctive in that the best sensitivity is in the high frequencies.

Group A: Low-frequency Ears. This group includes 9 species: *Gekko gecko, Gekko smithi, Hemidactylus brookii angulatus, Hemidactylus echinus, Hemidactylus* sp., *Hemidactylus triedrus, Pachydactylus* sp., *Phelsuma madagascariensis*, and *Teratoscincus scincus*. Sensitivity functions for three of these species are shown in Fig. 14-67.

Gekko smithi attains a very high peak of sensitivity, measuring −83 db at 400 Hz, and is still excellent over a range of 100-1500 Hz. For tones above 2000 Hz this ear is relatively poor, with a decline at the

TABLE 14-II
SENSITIVITY DATA

Species	Peak Frequency	Peak Sensitivity	30-db Range	4000 Hz Level	Mean Low Tones	Mean High Tones	Mean All Tones	Number of Hair Cells		
								Dorsal Segment	Ventral Segment	Total
Eublepharinae										
Coleonyx brevis	700	−70	320-2200	−33	−48.5	−19.9	−33.4	49	350	399
Coleonyx variegatus	700	−77	400-1300	−11	−70.0	−41.0	−54.7	68	446	514
Eublepharis macularius	1000	−100	240-3400	−64	−82.4	−40.7	−60.5	222	745	967
Hemitheconyx caudicinctus	1000	−88	220-1800	−27	−72.4	−30.5	−50.4	212	682	894
Gekkoninae, Group A										
Gekko gecko	600	−64	220-830	−23	−71.1	−30.5	−50.8	202	1423	1625
Gekko smithi	400	−83	120-1100	−13	−63.6	− 3.2	−33.4	414	1664	2078
Hemidactylus brookii angulatus	500	−56	140-1700	−16	−41.9	−10.5	−24.4	51	779	840
Hemidactylus echinus	500	−64	250-1900	−16	−44.2	−17.7	−30.3	50	592	642
Hemidactylus sp.	500	−34	300-2000	+ 6	−18.0	+ 4.6	− 4.5	65	587	652
Hemidactylus triedrus	500	−68	240-2300	−21	−48.4	−16.6	−31.8	97	692	789
Pachydactylus sp.	800	−57	270-2900	−18	−40.1	−12.2	−25.4	88	706	794
Phelsuma madagascariensis	600	−58	290-2200	−13	−37.8	−14.3	−25.4	127	1475	1602
Teratoscincus scincus	800	−90	200-2600	−40	−76.7	−36.0	−55.3	200	692	892
Gekkoninae, Group B										
Aristelliger praesignis	800	−39	100-10000	−24	−27.0	−30.7	−29.1	110	962	1072
Gehyra variegata	600	−58	230-5000	−27	−40.1	−21.4	−30.3	50	746	796
Gekko monarchus	500	−49	185-4000	−18	−32.4	−10.0	−21.2	96	687	783
Homopholis fasciata	1000	−58	270-4400	−34	−37.6	−27.4	−32.2	53	673	726
Ptyodactylus hasselquistii guttatus	1000	−80	280-6800	−53	−57.2	−48.4	−52.6	94	541	635
Stenodactylus s. sthenodactylus	1000	−36	220-7700	−25	−19.4	−17.1	−18.1	19	327	346
Tarentola mauritanica	600	−73	270-6000	−57	−50.8	−50.9	−50.8	85	727	812
Thecadactylus rapicaudus	400	−56	220-4300	−33	−44.2	−17.2	−30.3	171	1034	1205

TABLE 14-II (cont.)
SENSITIVITY DATA

Species	Peak Frequency	Peak Sensitivity	30-db Range	4000 Hz Level	Mean Low Tones	Mean High Tones	Mean All Tones	Number of Hair Cells Dorsal Segment	Number of Hair Cells Ventral Segment	Number of Hair Cells Total
Gekkoninae, Group C										
Cyrtodactylus kotschyi orientalis	3000	−48	670-6100	−40	−11.1	−26.5	−19.2	19	411	430
Geckonia chazaliae	1500	−50	230-5200	−47	−29.4	−29.0	−29.2	64	543	607
Hemidactylus mabouia	2500	−40	450-7500	−26	− 7.7	−26.3	−17.9	41	623	664
Phelsuma dubia	2000	−44	330-7500	−24	−13.9	−28.0	−21.3	35	789	824
Phyllodactylus xanti	2000	−48	180-6100	−38	−29.6	−32.7	−31.2	36	507	543
Teratolepis fasciata	2500	−29	800-7000	−20	+10.3	−14.1	− 2.5	40	390	430
Diplodactylinae, Group A										
Diplodactylus elderi	3500	−26	1000-7400	−23	+20.1	−10.1	+ 4.3	31	519	550
Diplodactylus vittatus	5000	−36	540-4600	−34	− 3.8	−24.3	−14.6	49	749	798
Diplodactylus williamsi	2500	−34	400-6300	−26	+ 1.0	−13.8	− 6.8	28	763	791
Hoplodactylus pacificus	3500	−35	360-8200	−27	−11.8	−20.3	−16.3	45	650	695
Lucasium damaeum	3000	−23	400-7000	−14	+ 4.7	− 8.7	− 2.4	51	532	583
Diplodactylinae, Group B										
Oedura marmorata	500	−56	200-3500	−26	—	—	—	—	—	1257
Oedura ocellata	400	−54	200-5000	−38	−39.6	−21.3	−29.4	100	1157	1204
Oedura tryoni	400	−55	200-4600	−38	−40.3	−22.1	−30.7	85	1119	
Sphaerodactylinae										
Gonatodes albogularis fuscus	3500	−54	500-6500	−52	−15.6	−32.0	−24.2	27	476	503
Gonatodes ceciliae	5000	−23	440-7000	−22	+10.9	−10.5	− 0.4	12	657	669
Gonatodes vittatus	5000	−30	700-1000	−30	+ 8.8	−20.9	− 6.8	16	489	505
Sphaerodactylus fantasticus	1500	−20	150-4000	−11	+ 1.9	−12.9	− 6.6	18	339	357
Sphaerodactylus lineolatus	4000	−42	600-8700	−25	− 5.1	−27.3	−16.8	17	374	391
Sphaerodactylus monensis	1500	−23	150-3500	+ 9	− 2.9	+ 3.3	+ 0.7	14	280	294

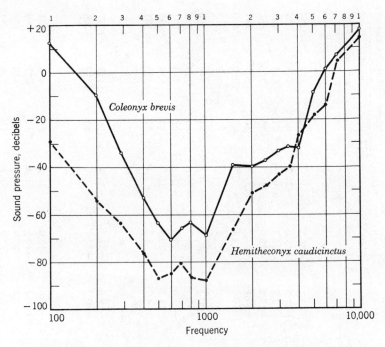

FIG. 14-65. Sensitivity functions for two species of Eublepharinae: *Coleonyx brevis* and *Hemitheconyx caudicinctus*. From Wever, 1974b.

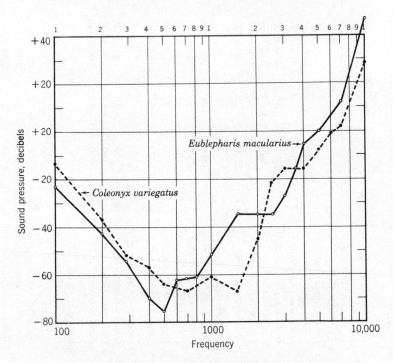

FIG. 14-66. Sensitivity functions for *Coleonyx variegatus* and *Eublepharis macularius*.

rate of 40 db per octave. Two additional curves for this species are presented in Fig. 14-68, and agree closely with the one just seen. Although one of these curves fails to reach the remarkably high level of better than −80 db presented by the others, it still exhibits a peak response of −65 db.

Referring to Fig. 14-67 once more, *Hemidactylus triedrus* presents the best sensitivity at 500 Hz, where the level reached is −68 db, and over a range from 300-2000 Hz the response is excellent. In the uppermost frequencies the roll-off for this ear is 40 db per octave.

Phelsuma madagascariensis, whose function is also included in Fig. 14-67, has a peak at 600 Hz, where it reaches a level of −59 db, and this ear still shows excellent performance between 400 and 2000 Hz. The high-frequency roll-off is about 35 db per octave.

These three species are in good agreement in the location of the peak of sensitivity in the region of 400-600 Hz, and in the attainment of very high acuity levels. They agree also in the form of the decline of sensitivity in the range above 1000 Hz. These species differ, however, in their sensitivity to the low frequencies, those below their peak regions.

Sensitivity curves for three additional gekkonines are presented in Fig. 14-69. The species *Hemidactylus brookii angulatus* shows peak sensitivity of −57 db at 500 Hz, and excellent sensitivity over the range 200-1000 Hz. The high-frequency roll-off has a rate of about 50 db per octave.

Hemidactylus echinus has a peak sensitivity of −64 db at 500 Hz, and excellent sensitivity in the range of 300-1500 Hz. The high-frequency roll-off is about 50 db per octave.

The *Pachydactylus* species also has its peak at 500 Hz, and the same level of sensitivity of −57 db as one of the *Hemidactylus* specimens. The range of excellent sensitivity is shifted upward a little relative to that of the other two species, and extends from 400 to 2500 Hz. The high-frequency roll-off is more rapid, around 55 db per octave.

These three functions show considerable similarity, taking much the same course despite local irregularities.

The sensitivity of *Gekko gecko* has been determined by a number of investigators, as mentioned earlier. Some of the best determinations were made by Hepp-Reymond and Palin (1968), and by Werner (1972) in the course of his temperature studies. Other curves have been reported by Werner and Wever in the middle ear study referred to in Chapter 6. Two curves from the latter report are reproduced in Fig. 14-70. These curves show maximum sensitivity of −62 to −64 db in the region of 400-600 Hz, and still excellent response over a range of two octaves from 200 to 800 Hz. A moderate level of sensitivity is sustained between 1000 and 4000 Hz, but for higher tones and those below 400 Hz the decline is rapid.

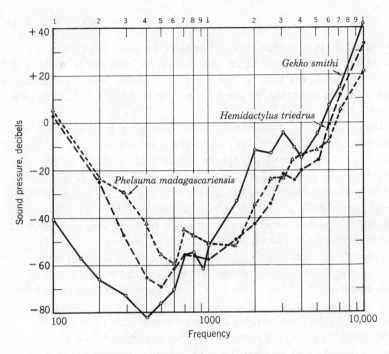

FIG. 14-67. Sensitivity functions for three species of Gekkonidae belonging to Group A. From Wever, 1974b.

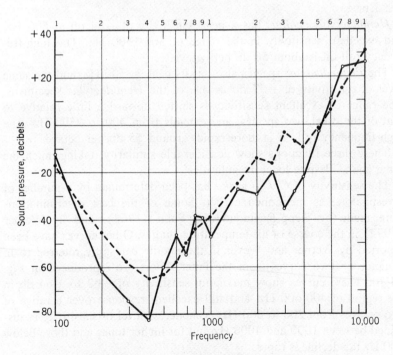

FIG. 14-68. Sensitivity curves for two additional specimens of *Gekko smithi*.

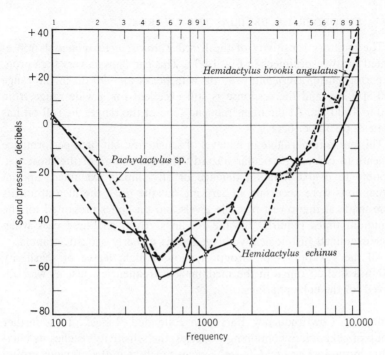

FIG. 14-69. Sensitivity functions for three additional species of Gekkoninae.

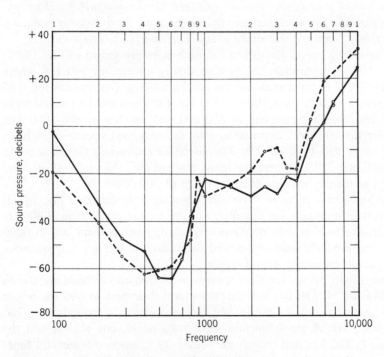

FIG. 14-70. Sensitivity functions for two specimens of *Gekko gecko*.
Data from Werner and Wever, 1972.

The auditory sensitivity of the lizard *Teratoscincus scincus* is represented for two specimens in Fig. 14-71. This ear shows a very high order of performance, averaging −86 db with peaks at −90 db in the range 400-800 Hz, and the response is still excellent for a wide range from 100 to 4000 Hz. At the high-frequency end of the range, the roll-off has a rate of 60 db per octave.

This ear is remarkable in its level of sensitivity, but such performance appears to carry a particular hazard. This ear is one of the most susceptible to overstimulation damage of all animals examined. Special precautions were necessary in carrying out the tests; the measurements were made at the lowest possible levels and the tones presented for the minimum times required for the readings. Even so, these ears when later examined histologically showed signs of early acoustic trauma.

All the gekkonines of Group A show a high degree of sensitivity within a limited range in the medium low frequencies, just as was observed for the eublepharines.

Group B: Low-frequency Ears with Extended Range. In a further group of gekkonines containing 8 species, the sensitivity reaches its maximum in the low and middle frequencies much as in the previous group, but falls off less rapidly in the upper frequencies. This group includes *Aristelliger praesignis, Gehyra variegata, Gekko monarchus, Homopholis fasciata, Ptyodactylus hasselquistii guttatus, Stenodactylus s. sthenodactylus, Tarentola mauritanica,* and *Thecadactylus rapicaudus.*

Sensitivity curves for three of these species are shown in Fig. 14-72. For *Gekko monarchus* the best sensitivity appears at 500 Hz, where a level of −49 db is attained, and good sensitivity (response at the −30 db level) extends from 300 to 1500 Hz. For *Homopholis fasciata* there are two peak regions, one at 400-500 and another at 800-1000 Hz, where the sensitivity approaches −60 db, and good sensitivity includes the range 300-4000 Hz. For *Thecadactylus rapicaudus* there are peaks at 400 and 700 Hz where the levels reached are −52 and −49 db, and good sensitivity appears over the range of 300-2500 Hz.

A sensitivity curve for *Aristelliger praesignis* is shown in Fig. 14-73. This is an unusually flat function, with the best response around 800 Hz where the level is −40 db. Good sensitivity extends from 200 to 6000 Hz, but the tests were not carried far enough in the upper frequencies to indicate the upper roll-off region.

Sensitivity curves for three specimens of *Gehyra variegata* are shown in Fig. 14-74. The two best functions have their peak at 600 Hz, where the level approaches −60 db, and excellent sensitivity extends from 300 to 4000 Hz. A third function shows two peaks, one at 500 and the other at 800 Hz, and though the sensitivity is lower, this ear still functions well over the range 400-1500 Hz.

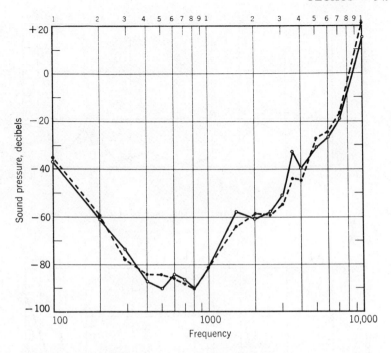

FIG. 14-71. Sensitivity functions for two specimens of *Teratoscincus scincus*.

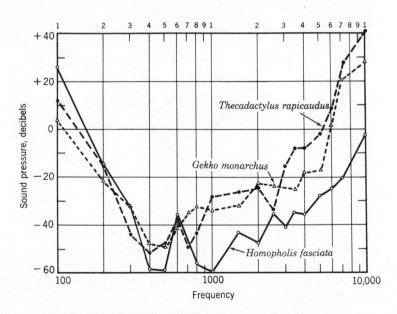

FIG. 14-72. Sensitivity functions for three species of gekkonines
belonging to Group B. From Wever, 1974b.

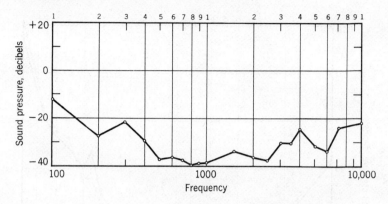

FIG. 14-73. Sensitivity function for a specimen of *Aristelliger praesignis*.

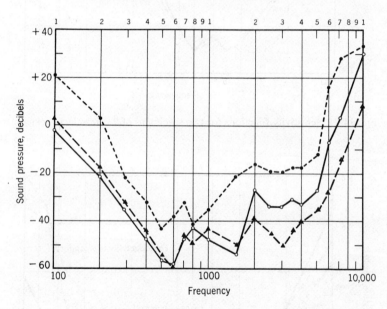

FIG. 14-74. Sensitivity functions for three specimens of *Gehyra variegata*.

The fan-toed gecko *Ptyodactylus hasselquistii guttatus* was studied earlier by Wever and Hepp-Reymond (1967) (wrongly identified then as the subspecies *puiseuxi*), and two of the functions reported at that time, along with a third, are presented in Fig. 14-75. These curves show best sensitivity in the region of 500-1000 Hz, where the level approaches −80 db. The performance of this ear is still excellent over a range from 200 to 3000-10,000 Hz. This is a remarkably sensitive species, rivaling

the performance of *Hemitheconyx caudicinctus*, and yet not as susceptible to acoustic trauma as that species; extreme sound levels do produce severe damage, however. It is of interest that the vocalizations
produced by males of this species, as reported by Frankenberg (1974),
which consist of a series of click sounds, have frequency components
mainly falling in the high-frequency portion of the sensitivity range, between 1000 and 5800 Hz.

Sensitivity curves for two specimens of *Stenodactylus s. sthenodactylus* are given in Fig. 14-76. The better of these curves reaches −36 db
at 1000 Hz and has a second peak at nearly that level at 2000 Hz. Good
sensitivity is found over a range from 500 to 2000 Hz, and the performance is fairly well maintained up to 5000 Hz. The other specimen
showed very poor sensitivity, and no doubt had suffered injury of some
kind.

A sensitivity function for *Tarentola mauritanica* is presented in Fig.
14-77. This species shows a sensitivity peak at 600 Hz that reaches a
level of −74 db, and the performance is excellent over a wide range,
from 300 to 7000 Hz.

Group C: High-frequency Ears. In sharp contrast to the gekkonines
just considered is a group in which the sensitivity is relatively poor for
the low tones and reaches its maximum in the high-frequency region.
These species are *Cyrtodactylus kotschyi orientalis, Geckonia chazaliae,
Hemidactylus mabouia, Phelsuma dubia, Phyllodactylus xanti*, and *Teratolepis fasciata*. Sensitivity functions for three of these are presented in
Fig. 14-78. For *Cyrtodactylus kotschyi orientalis* the best sensitivity is
at 3000-3500 Hz, and is still good within the range 2000-5000 Hz, but
falls off rapidly for higher and lower frequencies. *Phyllodactylus xanti*
shows a very regular function with a broad region of excellent sensitivity between 1500 and 4000 Hz, and good response over a wide range
from 400 to 5000 Hz. The species *Teratolepis fasciata* shows poorer
sensitivity, with a broad maximum in the region 1500-3500, and rapidly declining sensitivity elsewhere. These curves are similar at the high-
frequency end of the range studied, but differ greatly in the low frequencies.

In the species *Geckonia chazaliae* the observations were made at two
body temperatures, 25° and 27° C, as shown in Fig. 14-79. At the
standard temperature this ear shows excellent sensitivity in the medium
high tones, from 700 to 4000 Hz, with a broad maximum around 1500
Hz. For lower and higher tones the decline is rapid, with a roll-off rate
at the upper end of the scale of about 40 db per octave. When the body
temperature was increased to 27° C the sensitivity was increased
throughout the range, reaching −60 db in the maximum region.

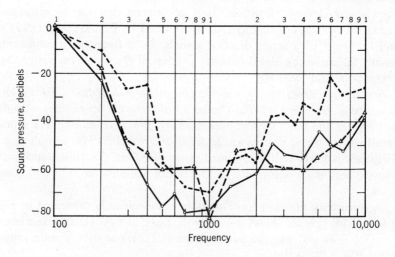

FIG. 14-75. Sensitivity functions for three specimens of
Ptyodactylus hasselquistii guttatus.

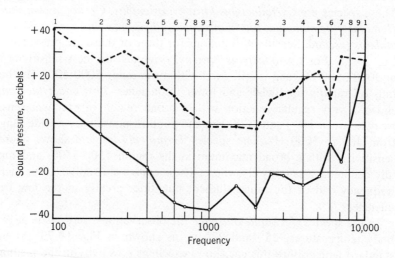

FIG. 14-76. Sensitivity functions for two specimens of
Stenodactylus s. sthenodactylus.

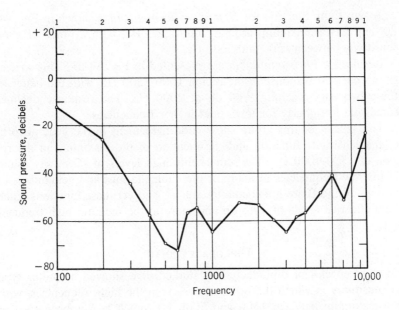

FIG. 14-77. A sensitivity function for *Tarentola mauritanica.*

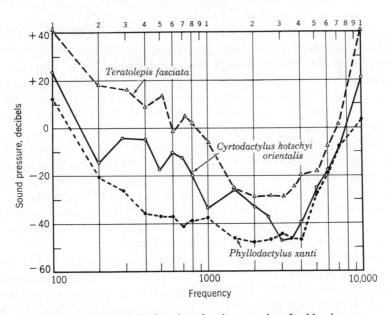

FIG. 14-78. Sensitivity functions for three species of gekkonines
of Group C, showing high-frequency sensitivity. From Wever, 1974b.

A sensitivity function for *Hemidactylus mabouia* is shown in Fig. 14-80, and has a maximum around 2000-2500 Hz, with a range of good sensitivity between 1500 and 5000 Hz.

Results for *Phelsuma dubia* are presented in Fig. 14-81. This species shows the best response in the region 1500-3500 Hz, with the better of these two ears reaching −44 db at 2000 Hz. The response declines somewhat irregularly for lower and higher frequencies.

The species of this group show good agreement in the general form of the sensitivity function and the location of the maximum in the region of 1500-4000 Hz. The remarkably high levels of sensitivity found in the low frequencies in many of the eublepharines and gekkonines of Groups A and B are not encountered here, and yet these functions reach fully effective levels of −40 db and more in the medium high frequencies.

DIPLODACTYLINAE

Of eight species of diplodactylines studied, five showed a similar type of sensitivity in which the best response is in the high frequencies, with the maximum between 1500 and 5000 Hz, much as has been seen in the third gekkonine group. Sensitivity functions for three of these species—*Diplodactylus vittatus, Hoplodactylus pacificus,* and *Lucasium damaeum*—are shown in Fig. 14-82.

In *Diplodactylus vittatus* the sensitivity peak is at 5000 Hz, where it reaches a level of −37 db, and good sensitivity extends over the range 1500-6000 Hz. The decline in the upper frequencies is particularly rapid, about 60 db per octave. *Hoplodactylus pacificus* shows a similar function, with about the same range of good sensitivity, though with some irregularity in the region of the maximum. The best response is at 3500 Hz and the level here is −36 db.

The species *Lucasium damaeum* is less sensitive, reaching only −24 db in the region 2500-3000 Hz. Additional curves for this third species are shown in Fig. 14-83, and their general agreement indicates that the relatively poor performance of this ear is a species characteristic.

In the species *Diplodactylus elderi*, as shown in Fig. 14-84, a maximum appears at 3500 Hz, and reaches a level of −26 db. This ear is only moderately sensitive within a narrow range.

The larger species *Diplodactylus williamsi* is represented in Fig. 14-85. The sensitivity is fairly good for two of the specimens studied, though poor for a third. The better ears show good response within a range of 1500-4000 Hz, with the best points around 2500-3000, where a level of −34 db is attained.

Three other diplodactyline species were examined, all in the genus *Oedura*, and revealed a strikingly different form of sensitivity function.

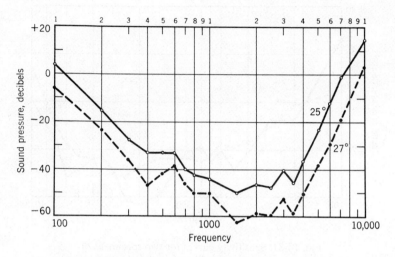

FIG. 14-79. Sensitivity functions for a specimen of *Geckonia chazaliae*, taken at two body temperatures.

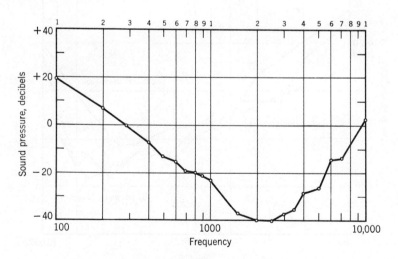

FIG. 14-80. A sensitivity curve for *Hemidactylus mabouia*.

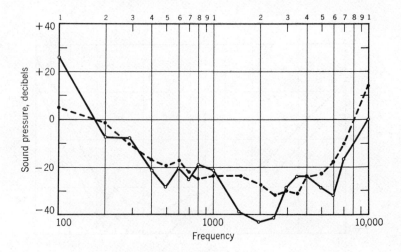

FIG. 14-81. Sensitivity curves for two specimens of
the day gecko *Phelsuma dubia*.

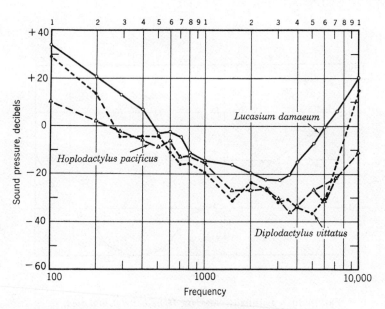

FIG. 14-82. Sensitivity functions for three species of Diplodactylinae.
From Wever, 1974b.

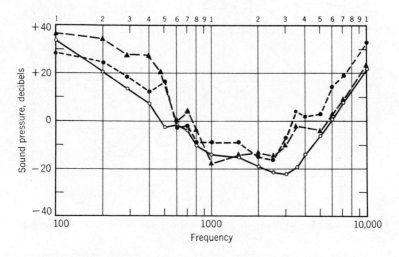

FIG. 14-83. Sensitivity curves for three additional
specimens of *Lucasium damaeum*.

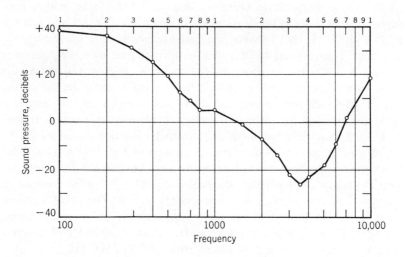

FIG. 14-84. A sensitivity function for *Diplodactylus elderi*.

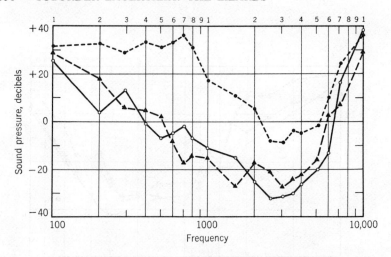

Fig. 14-85. Sensitivity functions for three specimens of *Diplodactylus williamsi*.

This function for the marbled velvet gecko, *Oedura marmorata*, is presented in Fig. 14-86, and has its most sensitive region at 400-500 Hz, where a level of −56 db is reached. The sensitivity is well maintained, varying around −40 db, in the higher frequencies up to 2500 Hz.

The eyed velvet gecko, *Oedura ocellata*, gave results shown in Fig. 14-87. These three curves present a degree of bimodality, with a high degree of sensitivity in two regions, one around 400-700 Hz and another at 2500-3000 Hz, but between the performance is only moderately diminished. Thus this ear has excellent sensitivity over a wide range, from around 200-400 Hz to about 4000 Hz. In the higher frequencies the three curves are in close agreement in showing a rapid decline at the rate of 40-45 db per octave, and in the low frequencies the decline is rapid also, though the three specimens vary considerably in this region.

A somewhat different form of sensitivity function is presented by Tryon's velvet gecko, *Oedura tryoni*, as seen in Fig. 14-88. This function shows a tendency toward duplexity also, but the low-frequency region is greatly emphasized, with peaks around −70 db in contrast to ones of −54 and −56 db in the secondary region. The two specimens differ in the locations of the primary peaks, which are an octave apart, but agree in the location of the secondary peak at 2000 Hz. The sensitivity is excellent over a wide range, from 300 to 2500 Hz.

These *Oedura* species represent a divergent form of auditory functioning among the diplodactylines, in which the degree of sensitivity is unusually high and extends over a wide range, with best performance in the low frequencies. The form of the function is similar to that seen in the gekkonines of Group B.

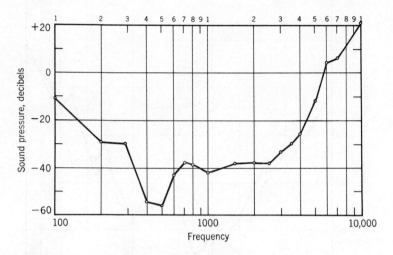

FIG. 14-86. A sensitivity function for a specimen of *Oedura marmorata*.

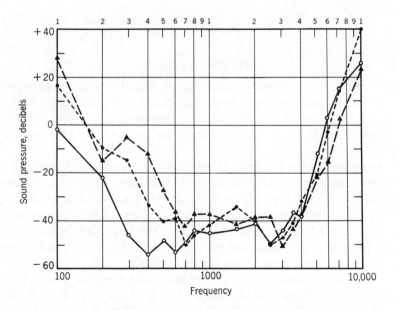

FIG. 14-87. Sensitivity functions for three specimens of *Oedura ocellata*.

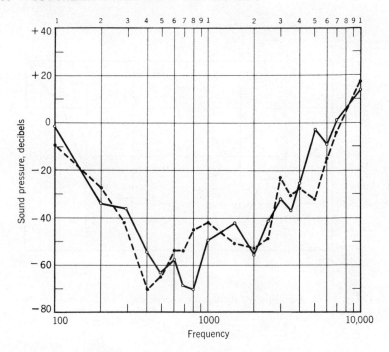

FIG. 14-88. Sensitivity functions for two specimens of *Oedura tryoni*.
From Wever, 1974b.

SPHAERODACTYLINAE

The species of sphaerodactylines studied were *Gonatodes albogularis fuscus, Gonatodes ceciliae, Gonatodes vittatus, Gonatodes* sp., *Sphaerodactylus f. fantasticus, Sphaerodactylus lineolatus,* and *Sphaerodactylus monensis.* All show a form of sensitivity that centers in the high frequencies.

Sensitivity curves for three of these species are presented in Fig. 14-89. *Gonatodes albogularis fuscus* shows a primary region of sensitivity around 3500 Hz, along with a secondary peak at 1000 Hz. In the primary region the sensitivity reaches −55 db, and excellent response is obtained over a range of 2500-5000 Hz. The secondary region shows very good sensitivity also, and this ear can be considered as performing well over the entire range from 700 to 6000 Hz.

The species *Gonatodes ceciliae* is much less sensitive, with a peak region around 3000-5000 Hz, and the best level reaching only −24 db.

The function for *Sphaerodactylus lineolatus* is intermediate between the other two species shown in this figure, and indicates excellent hearing within a limited range around 4000-5000 Hz. Within a somewhat wider range, from 3000-7000 Hz, this ear shows good performance.

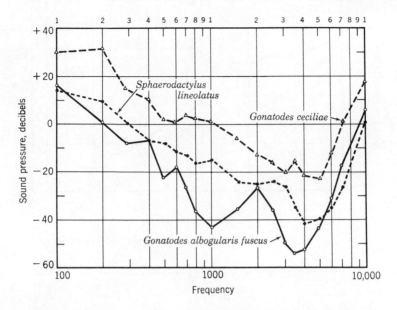

Fig. 14-89. Sensitivity functions for three species of Sphaerodactylinae.
From Wever, 1974b.

The species *Gonatodes vittatus* gave the curve shown in Fig. 14-90. The best sensitivity, at a level of −30 db, appears at 3500-5000 Hz, and a fair degree of response occurs over a range of 2000-6000 Hz.

Another *Gonatodes* collected in Trinidad but not identified as to species gave the results shown in Fig. 14-91. The best response is in the region of 2500-6000 Hz, with a peak at 4000 Hz, where the level reaches −34 db.

The species *Sphaerodactylus f. fantasticus*, represented in Fig. 14-92, shows only a fair degree of sensitivity, with a peak at 1500 Hz that reaches a level of only −20 db.

Functions obtained for *Sphaerodactylus monensis*, presented in Fig. 14-93, likewise exhibit only limited sensitivity. The best of the three curves shows a rather uniform level around −22 to −25 db over a range of 1500-5000 Hz.

This sphaerodactyline group is consistent in the general form of the sensitivity curve, which exhibits the best response in the high frequencies, chiefly around 3000-5000 Hz. The degree of sensitivity varies considerably within the group, reaching very good or excellent levels in certain species but only moderate levels in others.

These observations on the sensitivity of gecko ears present meaningful relations to subfamily lines. Two of the subfamilies show a fair degree of homogeneity. The eublepharines constitute a uniform assemblage, with

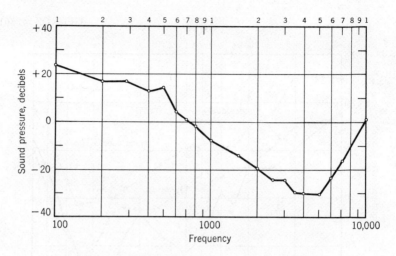

Fig. 14-90. A sensitivity function for *Gonatodes vittatus*.

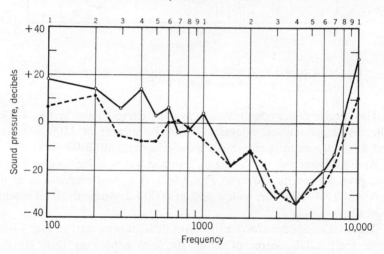

Fig. 14-91. Sensitivity functions for two specimens of *Gonatodes* sp.

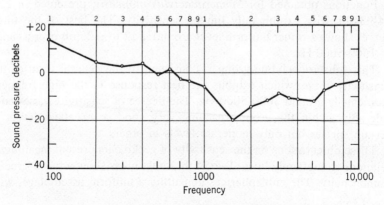

Fig. 14-92. A sensitivity function for a specimen of
Sphaerodactylus f. fantasticus.

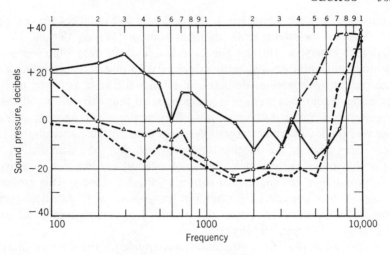

Fig. 14-93. Sensitivity functions for three specimens of
Sphaerodactylus monensis.

very high sensitivity in a narrow band within the medium low frequencies. The sphaerodactylines form a close group with a good to fair degree of sensitivity in the high frequencies. Apart from the *Oedura* genus, the diplodactylines also show great uniformity, with good hearing in the high frequencies. The *Oedura* species present a distinctly different form of sensitivity function, with a broad response extending from the low tones to the medium high range, with a tendency toward bimodality.

The gekkonines, on the other hand, appear as a disparate assemblage. They fall into three subdivisions, two with best sensitivity in the low and middle frequencies and one with best sensitivity for the high tones. The first two groups tend to grade into one another, and are separated only with difficulty in terms of the broadening of the response in the direction of the high frequencies. The third group contains species whose hearing is clearly of the high-frequency type.

The high-frequency ears shown by the sphaerodactylines, most of the diplodactylines, and Group C among the gekkonines present considerable similarities. The size of the sensory structures is no doubt a significant factor, and appears to be related to body size.

SENSITIVITY AND HAIR-CELL POPULATIONS

A persistent problem is the relation between the ear's sensitivity and the size of its complement of hair cells. It seems logical that an increased number of sensory cells would provide improved sensitivity. An early approach to this problem in lizards was made through a correlation computed between sensitivity levels and hair-cell numbers for a group of 13 species belonging to six different families (Wever, Vernon, Crowley, and Peterson, 1965). The correlation coefficients obtained were positive but unimpressive, with values of 0.33 and 0.24 accordingly as the comparisons were made between individual ears or between species averages. From these results it was concluded that hair-cell populations themselves are poor indexes of sensitivity among lizard species, at least across families in which large differences exist in the patterning of the hair cells and in the mechanisms provided for their stimulation. It appeared more likely that a significant relationship between sensitivity and population size might exist within a particular family. This problem therefore was considered further for the geckos, with statistical treatment of a group of 28 species that included representatives of all four subfamilies (Wever, 1974b).

In this effort the first computations were made for the total population of hair cells in relation to peak sensitivity (the best point along the frequency scale), and a correlation of 0.31 was obtained—still a slight relationship, just as had been found in the earlier study. The analysis was then continued with the use of additional measures of sensitivity and with a consideration of the segmental character of the gecko papilla.

Four sensitivity measures were employed: in addition to peak sensitivity, use was made of the mean sensitivity for low tones (100-900 Hz), the mean sensitivity for high tones (1000-10,000 Hz), and the mean sensitivity for all tones. These measures were then correlated with five populations: the hair cells in the dorsal segment, those in the ventral segment with tectorial membrane connections, those in the ventral segment served by sallets, the total in the ventral segment, and the total for the whole cochlea. It was soon clear that no good purpose was served by subdividing the ventral segment, and thereafter this segment was treated as a whole. The data are given in Table 14-II (pp. 542-543) and results of the computations in Table 14-III.

These results show a strong positive relationship between the number of hair cells in the dorsal segment and peak sensitivity, low-tone sensitivity, and the mean sensitivity for all tones. All three of these correlations are statistically significant to the 0.01 level ($p = 0.01$ or less, showing that the probability is no more than 1 in 100 that the obtained correlation is due to chance). The relation to high-tone sensitivity is only slight, however.

The generally similar results obtained for peak sensitivity and low-tone sensitivity are partly to be accounted for by the fact that, in the group studied, a majority of species had their maximums in the low frequencies, and these low-tone maximums were often of considerable magnitude.

TABLE 14-III
SENSITIVITY AND HAIR-CELL POPULATIONS
IN GEKKONIDAE
(Correlations for 28 species)

| Cochlear Region | Sensitivity | | | |
	Peak	Low Tones	High Tones	All Tones
Dorsal segment	0.74[b]	0.77[b]	0.40[a]	0.73[b]
Ventral segment	0.18[a]	0.38[a]	−0.09	0.23
Whole cochlea	0.31	0.38[a]	−0.004	0.35

[a] significant to the 0.05 level
[b] significant to the 0.01 level

For the ventral segment the correlations with sensitivity are generally small, and vanishingly so for the high tones. These correlations for the cochlea as a whole are small also; they largely reflect the relations for the ventral segment because in all gecko species the hair cells in this segment greatly outnumber the others.

Next to be considered was the matter of ciliary orientation for the hair cells. As already described, this orientation varies in a systematic manner in the dorsal segment, though bidirectionality always holds for the ventral segment.

When correlations between sensitivity and hair-cell numbers were computed for 15 species, including 4 eublepharines and 11 gekkonines with unidirectional orientation in the dorsal segment, the results of Table 14-IV were obtained. Most of these correlations are higher than their counterparts in the preceding table, and a particularly close relationship is found between low-tone sensitivity and the numbers of hair cells in

TABLE 14-IV
SENSITIVITY AND HAIR-CELL POPULATIONS
IN GEKKONIDAE
(Correlations for 15 species with unidirectional ciliary orientation
in the dorsal segment)

| Cochlear Region | Sensitivity | | | |
	Peak	Low Tones	High Tones	All Tones
Dorsal segment	0.81[b]	0.87[b]	0.52[a]	0.84[b]
Ventral segment	0.23[a]	0.47	0.08	0.39
Whole cochlea	0.38	0.60[a]	0.19	0.52[a]

[a] significant to the 0.05 level
[b] significant to the 0.01 level

the dorsal segment. Even for the high tones the correlation for this segment becomes significant, though much below that for the other measures that include the low tones.

When the above group of 15 species was enlarged by adding the two gekkonines *Thecadactylus rapicaudus* and *Ptyodactylus hasselquistii guttatus* with partially unidirectional orientation in the dorsal segment, all the correlations were slightly reduced, and when two other gekkonines, *Phelsuma madagascariensis* and *Teratolepis fasciata*, with completely bidirectional orientation in the dorsal segment were included also the correlations were reduced further.

When this computation was carried out on the diplodactylines and sphaerodactylines (five species of diplodactylines and six of sphaerodactylines) with completely bidirectional ciliary orientation in the dorsal segment, the results of Table 14-V were obtained, showing no signifi-

TABLE 14-V

SENSITIVITY AND HAIR-CELL POPULATIONS
IN GEKKONIDAE
(Correlations for 11 species with bidirectional ciliary orientation
in the dorsal segment)

Cochlear Region	Sensitivity			
	Peak	Low Tones	High Tones	All Tones
Dorsal segment	0.11	0.14	0.23	0.21
Ventral segment	0.09	−0.07	0.25	0.11
Whole cochlea	0.09	−0.06	0.26	0.12

cant relationships between sensitivity and hair-cell numbers. Not only are the correlation coefficients small, but the p values indicated complete unreliability of the measures. In this computation the *Oedura* species were excluded, for in these the ciliary orientation though bidirectional is strongly biased toward laterality. If these species are added to the group the correlations are significantly increased (see Wever, 1974b, Table 6).

This analysis has revealed certain systematic relationships between auditory sensitivity and the distribution and orientation of the hair cells in gekkonids. When unidirectional orientation of the ciliary tufts is present in the dorsal segment, as found in eublepharines and most gekkonines, a high correlation exists between the numbers of hair cells in this segment and sensitivity, especially the sensitivity to low tones. Even a moderate degree of unidirectionality, as found in three of the gekkonines and in *Oedura* species among the diplodactylines, has a significant effect on this relationship. A positive correlation of smaller degree shown in these species between sensitivity and the numbers of hair cells in the ventral segment is probably a secondary effect, due to the close relationship between hair-cell numbers in the two segments. This correlation between hair-cell numbers in dorsal and ventral segments was found to be 0.71 for the entire group of 40 species examined, and 0.81 for the group of 20 species with completely unidirectional ciliary orientation in the dorsal segment, with both of these correlations showing extremely high reliability coefficients ($p < 0.0001$).

These results suggest that the elongated basilar membrane of geckos with its relatively large population of hair cells provides the conditions for spatial summation in the action of the hair cells, and thereby gives an improvement in sensitivity. This process of summation is especially advantageous in those species in which the ciliary orientation is unidirectional and the stimulation of the cells is more precisely synchronized. The enhancement of sensitivity is particularly marked for the low tones, no doubt because these tones spread more broadly in their actions on the basilar membrane.

SPECIFICITY IN THE GECKO EAR

The conditions just indicated that make for improved sensitivity through spatial summation have contrary effects upon the specificity of action that is required for tonal discrimination in terms of place along the basilar membrane. For such discrimination it is necessary that the cochlear response to a given tone be localized, at least to the extent that a particular area responds more vigorously to the imposed vibrations than all other areas. The precision with which the tone can be distinguished from others then will depend upon the degree of differential response over the area, and on the fidelity of representation of this differential action in the operation of the auditory nervous system.

It is evident from the dimensional evidence presented above that in the geckos a considerable degree of structural differentiation is present. A special study was made of this differentiation in a sample group of 11 species, with the results presented in Table 14-VI. Shown here for four different cochlear features—basilar membrane width, fundus width, fundus thickness, and the mass of the sensory structures—are the ratios between the dimensions at the maximum and a near minimum (usually at a point 50 μ from the dorsal end). Also given in the table for these specimens are the length of the papilla and the total hair-cell population. The species are listed in the order of the ratios of basilar membrane width, and it will be noted that fundus width and the mass of the sensory structures follow this same order with few exceptions, but fundus thickness fails to do so.

TABLE 14-VI

COCHLEAR DIFFERENTIATION

IN GEKKONIDAE

(Factors of dimensional variation for selected species)

Species	Basilar Membrane Width	Fundus Width	Fundus Thickness	Mass of Sensory Structures	Length of Papilla	Number of Hair Cells
Sphaerodactylus lineolatus	1.9	2.7	2.0	3.3	600	301
Gonatodes albogularis fuscus	2.3	3.2	3.3	3.6	660	514
Oedura ocellata	2.4	3.6	3.4	4.8	1390	1089
Phelsuma madagascariensis	2.6	4.6	2.2	4.9	1440	1464
Diplodactylus vittatus	2.8	3.5	2.6	5.6	960	666
Cyrtodactylus kotschyi orientalis	3.1	3.8	1.6	5.3	635	405
Lucasium damaeum	3.1	4.7	2.5	3.5	970	583
Gehyra variegata	3.7	5.2	1.3	4.6	800	626
Coleonyx variegatus	4.9	7.6	1.8	7.0	820	467
Eublepharis macularius	6.7	10.2	3.4	7.4	1300	814
Gekko gecko	8.9	7.9	1.8	6.0	1730	1625

As will be seen, the two sphaerodactylines show the least differentiation, with basilar membrane ratios around 2 and other factors likewise
varying but little. In *Oedura ocellata* and *Phelsuma madagascariensis*,
the width of the basilar membrane varies only slightly more, fundus
width varies around fourfold, and the mass of the sensory structures
varies nearly fivefold. In a third group containing *Diplodactylus vittatus,
Cyrtodactylus kotschyi orientalis*, and *Lucasium damaeum*, the ratio for
basilar membrane width is around 3, that for fundus width varies from
3.5 to 4.7, and mass varies a little more. The species *Gehyra variegata*
shows a little greater variation than the above. The remaining species,
including *Coleonyx variegatus, Eublepharis macularius*, and *Gekko
gecko*, are almost out of the class of the others, with strikingly larger
ratios. For these species the variation in the width of the basilar membrane extends from 4.9 in *Coleonyx variegatus* to 8.9 in *Gekko gecko*.
Indeed this variation in basilar membrane characteristics is as great in
these species as it is in advanced mammalian ears; in the human ear
this variation is about sixfold. The variation in fundus width is correspondingly great, and that for mass of the sensory structures is large
also.

In contrast to the other factors, fundus thickness varies but little. A
progressive increase is seen in *Diplodactylus vittatus* and to a lesser extent in *Lucasium damaeum*. In *Eublepharis macularius* and to some degree in *Sphaerodactylus lineolatus* this structure increases to a maximum

thickness in the middle of the cochlea and then decreases. In the remaining species this structure quickly increases in thickness near the dorsal end and thereafter remains nearly constant. In general the fundus appears to serve mainly to maintain the stiff plate characteristics of the vibrating structure, and favors the operation of this structure as a rigid beam rather than as a flexible membrane.

This consideration of the mechanical characteristics of the responsive structures of the gecko inner ear supports the assumption that, in many species, this ear is capable of frequency discrimination in terms of local action along the cochlea.

Further evidence for this view was presented for lizards in general, based on observations of the forms of the cochlear potential functions (Wever, 1965a). These potentials show variations in maximum values as a function of frequency, with the low tones displaying the larger maximums. These functions also show a greater rate of departure from linearity at the upper ends of the intensity curves for the higher frequencies. Further, these functions often show secondary maximums in which the curve first rises to a peak and bends downward, then as the stimulus is increased further it flattens out, and finally rises to a second maximum.

These complexities in the intensity functions have been more extensively examined in the ears of mammals, and have been interpreted as reflecting the differential frequency characteristics of these ears (Wever and Lawrence, 1954). In the geckos as well as in other lizards these forms of the intensity functions are prominent, and this evidence supports that based on cochlear structure in indicating the presence of frequency differentiation.

As already pointed out, the effectiveness of a pattern of differential action along the cochlea must depend upon a degree of specificity of cochlear innervation, and upon a transmission of this specificity to higher centers of the auditory nervous system. Clearly, the distinctness of peripheral stimulation and the separateness of the nerve supply are facilitated by the elongation of the cochlea, with an increase in the numbers of hair cells and their nerve fibers. It is significant that the auditory papilla of *Gekko gecko* is one of the most differentiated of all those examined, and is also one of the longest and best provided with hair cells.

The general conclusion that may be drawn from the evidence is that a considerable degree of tonal discrimination in terms of place along the cochlea is highly probable for the eublepharines and the majority of gekkonines, a somewhat lower degree obtains for other gekkonines and the diplodactylines, and the lowest degree holds for the sphaerodactylines.

Cochlear Patterns

With the evidence at hand, a formulation may now be made concerning the patterns of action of tones along the gecko cochlea. It seems likely that for the majority of species the dorsal segment is the principal locus of action of tones in the region of the maximum. For the eublepharines and most of the gekkonines, therefore, the dorsal segment appears to be best adapted to tones in the range of 500-1000 Hz or a little below. A high degree of sensitivity in this range is attained in many species by elongation of the dorsal segment, with relatively little differentiation along it, and by the provision of unidirectional orientation for the ciliary tufts of its hair cells.

The more ventral portion of the cochlea, in which the basilar membrane becomes wider and the mass of the sensory structures becomes greater, would appear to extend the response to still lower tones over a span of one or two octaves, and in some species slightly more.

The action of tones of frequencies above the region of the maximum does not seem to be assisted by the presence of specifically tuned structures, but appears to be mediated by the forcing of response in regions best suited to lower frequencies. Because the best correlations are found between high-tone sensitivity and the numbers of hair cells in the dorsal segment in eublepharines and gekkonines, it seems that high-tone sensitivity in these species is favored by the relatively long and narrow segments found in many of these species.

The assumption that the high tones operate largely by forced response is consistent with the observation that the sensitivity falls off with great rapidity as the frequency rises. This rapid decline is most strikingly shown for the eublepharines and the gekkonines of Group A, but is still present in the gekkonines of Group B and in *Oedura* species, although for these it appears in a higher region of the frequency scale.

The remaining species show forms of sensitivity functions that favor the high tones, but the degree of sensitivity in general is considerably less. In the gekkonines of Group C the dorsal segment is relatively short, and seems best suited to the medium high frequencies. The same appears to be true for the diplodactylines (except *Oedura*) and all the sphaerodactylines. In these (again except for *Oedura*) the dorsal segment is short and the hair cells lack the advantage of unidirectionality. The *Oedura* species are exceptional in that the dorsal segment contains a preponderance of hair cells with lateral ciliary orientation and the ventral segment is strongly bimodal in form.

The above discussion of the capabilities of the gekkonid ear has emphasized sensitivity, but valuable though it is this is not the only essential character of an ear's performance. Other attributes are pitch and intensity discrimination and the analysis of complex tones.

In the absence of discrimination an animal would respond equally well to any acoustic stimulus, and no sound could have meaning or usefulness to life and its preservation. Intensity discrimination provides the possibility of responding to simple patterns and rhythms, but an animal with no further differentiation would be severely handicapped. Pitch discrimination adds enormously to the richness and variety of sounds, and no doubt is provided in some degree in even the most primitive ears. At the most elementary level it is mediated by the repetitive response of the auditory nerve fibers, which are able to maintain a degree of synchronism with the sound waves as long as these waves are low in frequency. There is a limit to this simple discriminatory function, however, imposed by the refractory phase of the nerve fibers. This function in ears provided with simple nerve fibers in limited numbers can operate only for tonal frequencies or noise repetitions of rates up to a few hundreds per second.

The discriminatory function can be extended into the higher frequencies in two ways. One of these is by use of the volley principle, involving a number of nerve fibers responding to different waves in a tonal series in a rotational pattern. This manner of response requires multiple innervation of the cochlear elements and a degree of temporal precision in the firing of the nerve fibers. There is strong evidence for the operation of this principle in the ears of higher vertebrates (Wever, 1949), but its role in reptilian ears is less certain.

A second mode of pitch discrimination involves the place principle, according to which different sensory elements, and consequently their nerve fibers, respond to different frequencies along the audible range. This is in some sense a simple form of discrimination, for it does not depend on temporal precision of neural action as a frequency type of discrimination does, but only on the presence of mechanical variations within the cochlea and on a distinctness of innervation of its different regions. The evidence presented above indicates that the cochlear differentiation required for this form of discrimination is well developed in geckos in comparison with other reptiles, though varying among members of this group as shown.

The analysis of complex sounds is an auditory function closely related to pitch discrimination and commonly considered to be based upon the same or closely similar cochlear and neural processes. A high degree of cochlear differentiation favors this analytic ability. The extent of its development in a given animal depends upon the specificity of the peripheral structures in their patterned responses to sounds and on the maintenance of this specificity along higher neural channels, and also on the analytical capabilities in the higher centers of the nervous system.

The presence of a long, extended basilar membrane in geckos favors this analytic function. Also assisting it is the presence, in certain cochlear regions at least, of a bidirectional orientation of the ciliary tufts of the hair cells. This is true because for a complex sound wave there are numerous reversals of pressure within the period of the fundamental component, and responses at both compressional and rarefactional phases of the wave provide a more effective sampling of the wave form than a single response per period would do.

A question arises as to the advantages of a duplex arrangement in the ventral segment of the cochlea, in which one group of hair cells is served by a tectorial membrane and another by sallets. One consequence of this arrangement is an extension of the dynamic range. The tectorially connected hair cells are the more sensitive, as evidenced by the observation of overstimulation injury in these cells at levels that have no discernible effect on the sallet-connected cells. It follows that the tectorially connected hair cells respond alone at the lowest intensities, and are joined by the sallet-connected component when the intensity reaches sufficiently high levels. This sallet assemblage then continues to operate at the upper levels after the more sensitive elements have reached their limits.

The dynamic range of an ear is also extended when there is frequency differentiation, for when selectivity is present a certain favorable group of hair cells first responds to a particular tone when it is faint, and then as the stimulus intensity is increased other sensory elements join in, until ultimately, at the highest levels, almost the entire cochlea is involved in the response. The mechanics of action of a differentiated cochlea therefore are doubly advantageous, in providing an extended level of pitch discrimination and range, and in the range and delicacy of loudness discrimination as well. The evidence strongly suggests that the geckos have entered into this development of an advanced type of ear.

15. FAMILY PYGOPODIDAE:

THE FLAP-FOOTED LIZARDS

The flap-footed lizards are a small family restricted to Australia and New Guinea, and containing 13 species in seven genera (Bustard, 1970). All are nocturnal and most are of burrowing habit, often living in rock crevices. The forelegs are entirely absent, but traces of hindlegs are usually present in the form of tiny flaps. These animals are reported to be especially difficult to collect and to maintain in captivity, and consequently the family as a whole is poorly known. The best descriptions of the group are those of Shute and Bellairs (1953), Underwood (1957a), and Bustard.

The relations between this family and others have often been in doubt. Boulenger (1885) regarded these lizards as related to the geckos, but Cope (1900) and later Camp (1923) located them among the Diploglossa along with anguids, zonurids (cordylids), and xenosaurids.

Shute and Bellairs then assembled a good deal of evidence in favor of the early view of Boulenger, showing close relationships with the geckos. More recent writers, including McDowell and Bogert (1954), Underwood, and Bustard are in accord with this position.

The similarities between pygopodids and gekkonids are not obvious to casual observation, for the elongate, practically legless body form of the pygopodids makes them almost snakelike in appearance, in contrast with the compact body and sturdy legs of the geckos. More critical study, however, reveals many points of resemblance, especially in features of vertebrae, skull, cloacal structure, and especially in the eye and ear.

Shute and Bellairs placed special emphasis upon correspondences of ear structure. They pointed out the presence in both groups of an elongated basilar membrane and auditory papilla, with a tapered form from dorsal to ventral ends. They stressed particularly the presence in the pygopodids of an elevation of the limbus to form a curving lip that overhangs the auditory papilla and supports the tectorial membrane, a characteristic well known in the geckos, but present, they believed, in no other group but the pygopodids. In fact, however, a third lizard family, the Teiidae, possesses this form of limbic elevation also, but Shute and Bellairs, though they had examined a specimen of *Tupinambis* and had seen what they described as a small and poorly developed limbus of this type, were disposed to ignore it.

Underwood, in his account of the pygopodid ear, added new observations on middle ear structure. In most species there is a direct connection between columella and extracolumella much as in the geckos. He pointed especially to the course of the internal carotid artery and its branches. The stapedial artery arises anterior to the columella, a point of origin found in many geckos but not in other lizard families. From this and other evidence Underwood agreed that the systematic position of the Pygopodidae is adjacent to the Gekkonidae.

Miller (1966a), in his comprehensive survey of the form of the lizard cochlear duct, included three species of pygopodids and noted especially the resemblances between limbus and auditory papilla in these and the gecko species. He pictured the auditory papilla of *Lialis burtonis* as elongate and slightly sinuous, with its greatest expansion toward the anteroventral end.

Baird (1970a) in his description of the ear of Pygopodidae included most of the features already mentioned. He observed that the round window does not lie in a fossa but directly faces the tympanic cavity. Also in most species the saccule and cochlear duct are nearly equal in size, although in *Aprasia* the saccule is the smaller of the two.

These observations have revealed the general form of the ear in the Pygopodidae and some of the relations to other lizards, but many details are lacking. The ear of *Lialis burtonis* (Gray, 1834) was studied further, with measurements of the ear's performance in terms of cochlear potentials (Wever, 1974c). The observations were made on a single specimen, a juvenile of 1.4 grams weight, a snout-vent length of 7.8 cm and total length of 15 cm, and a mid-body diameter of 5 mm.

ANATOMICAL OBSERVATIONS

OUTER AND MIDDLE EAR

The external ear opening is oval, measuring 0.2×0.4 mm in the specimen examined, and leads inward a little obliquely to an expanded tympanic cavity. The tympanic membrane forming the outer boundary of this cavity measured about 1.2×1.4 mm. Its location is indicated in the cross-sectional view of Fig. 15-1.

Meatal Closure Muscle.   A closure muscle for the external ear opening is present in this species and appears in Fig. 15-1 as cut across in two places. It is seen as a moderately compact bundle of fibers in the mass of skin anterior to the meatus and a scattered band deep in the wall of the meatus posterior to the opening.

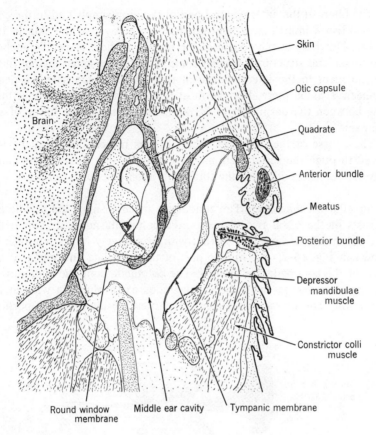

Fig. 15-1. A section through the right ear region of a specimen of
Lialis burtonis. Scale 30×. From Wever, 1974c.

This muscle forms a closed loop around the meatus as was represented in Fig. 6-6. The relations to the meatus are shown in Fig. 15-1,
with the posterior bundle close to the edge of the opening and the anterior bundle farther away. The posterior bundle turns to run close beneath the ventral border of the opening and joins the other bundle anteriorly.

The change from a compact bundle of oval cross section running
along the anterior side of the meatus to a broad band along its posterior
side occurs dorsal to the meatal opening: the fibers running posteriorly
along the dorsal side of the meatus fan out on the dorsoposterior wall.

Over most of the course of the muscle, the number of fibers is nearly
constant between 55 and 60, but toward the ventral end of the posterior
part of the muscle this number increases to about 70 with the appearance of many fine fibers.

The fibers at the anteroventral end of the loop simply end in the connective tissue immediately beneath the dermal layer of the skin in this region. Nowhere in the course of the muscle is there a specific anchorage to skeletal structures, though some of the more anteromedial fibers run adjacent to the lateral flange of the quadrate alongside the dense connective tissue there. For the most part this muscle runs about midway between the dermal layer of the skin and the dense lining of the tympanic cavity.

The course of the individual fibers of this muscle could not be followed through the serial sections. It is possible that single fibers could traverse the entire loop, as the total path is only about 2 mm long.

Columellar Mechanism. The conductive mechanism of the middle ear consists of the usual two elements, the columella and extracolumella. The columella is short and sturdy, with a relatively large footplate, as shown in Fig. 15-2. The middle part of the footplate and much of the shaft of the columella are ossified. The columella continues laterally as cartilage and merges with the cartilaginous extracolumella without interruption. The extracolumella has a particularly heavy body portion,

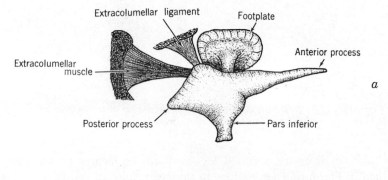

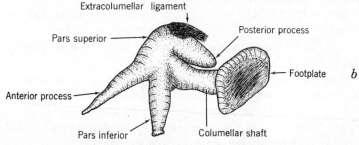

FIG. 15-2. Two drawings of the columellar system in *Lialis burtonis*, based on a wax reconstruction. *a*, dorsolateral view; *b*, anteromedial view. Scale 35×. From Wever, 1974c.

which sends off four processes. A rounded elevation, shown in both parts of Fig. 15-2, is the pars superior, to which two accessory structures are attached. These are a conical extracolumellar ligament that arises from the opisthotic portion of the cochlear capsule and inserts on the dorsal and medial surface of the pars superior, and an extracolumellar muscle that arises from the ceratohyal cartilage at its attachment to the paroccipital process and extends to an insertion on the posterior end of the pars superior. These relations are further indicated in Fig. 15-3, based on a photomicrograph.

A blunt pars inferior extends anteroventrally to the middle of the tympanic membrane. The anterior process is long and tapering and runs forward along the upper portion of this membrane. The posterior process, which is short and blunt, extends to the membrane's posterior edge. The extracolumellar muscle is well developed and contains about a hundred fibers. The surface area of the tympanic membrane is estimated as 1.4

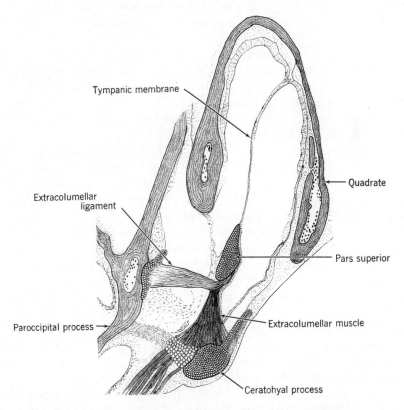

FIG. 15-3. A section of a specimen of *Lialis burtonis* showing the extracolumellar muscle and related structures.

sq mm and that of the stapedial footplate as 0.083 sq mm, giving an areal ratio of about 17.

INNER EAR

The inner ear structures follow a pattern that corresponds closely to that already described for the gekkonids. The general arrangement has been indicated in Fig. 15-1, which gives a cross section through the ventral part of the cochlea, cutting across the round window. A section through the middle of the cochlea is shown in Fig. 15-4, and passes through the stapedial footplate. This section is ventral to the body portion of the extracolumella and cuts through its pars inferior, the posterior process, and the edge of its shaft.

A section through the dorsal part of the cochlea, represented in Fig. 15-5, shows the form of the cochlear limbus with its well-developed lip. The tectorial membrane extends from the edge of the lip and runs across the cochlear duct to the auditory papilla to make the connections to the hair cells. A thickening of this membrane close to the limbic lip, known as the spindle body, is found elsewhere only in the geckos.

Auditory Papilla.   As in many other lizards, and as invariably seen in the geckos, the basilar membrane is an elongated strip, narrow at its dorsal end and increasing in width toward the ventral end. This membrane, viewed in cross section in Fig. 15-5, is thick in its middle portion and greatly thinned out at its edges. This form is further indicated in Fig. 15-6, along with other structural details.

The papilla consists of an outer capsule formed by processes extending from certain of the supporting cells whose bases lie along the surface of the fundus, with reinforcement by slender pillars from other supporting cells more centrally located. The hair cells are sustained at their upper ends between these pillar processes.

In the dorsal region of the cochlea, as seen in Fig. 15-6, there are only a few rows of hair cells (three in the region shown), and all their ciliary tufts are attached to the slightly thickened border of the tectorial membrane.

Toward the ventral end of the cochlea the number of rows of hair cells increases. At a point about a quarter of the distance along the cochlea a new type of tectorial structure, the sallet, makes its appearance and thereafter forms the connections to the ciliary tufts of the more lateral hair cells, while the tectorial membrane continues to serve the more medial cells. This condition is represented in Fig. 15-7.

Also in this ventral region there is a change in the manner of tectorial connection to the ciliary tufts: the tectorial membrane no longer makes these connections directly, but forms a series of finger processes that

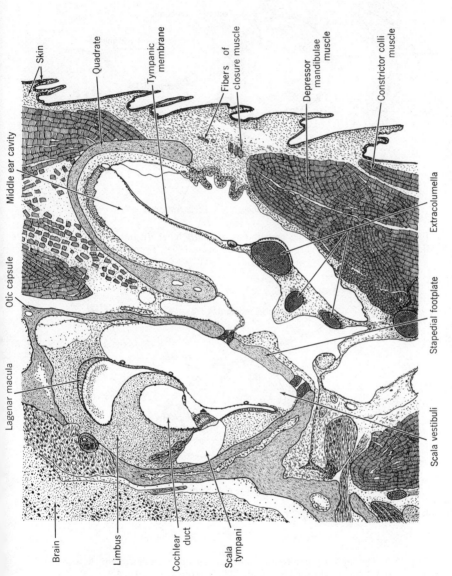

Skin

Quadrate

Tympanic membrane

Fibers of closure muscle

Depressor mandibulae muscle

Constrictor colli muscle

Middle ear cavity

Otic capsule

Lagenar macula

Brain

Limbus

Cochlear duct

Scala tympani

Scala vestibuli

Stapedial footplate

Extracolumella

FIG. 15-4. The right ear of *Lialis burtonis* seen in frontal section near the middle of the cochlea. Scale 75×.

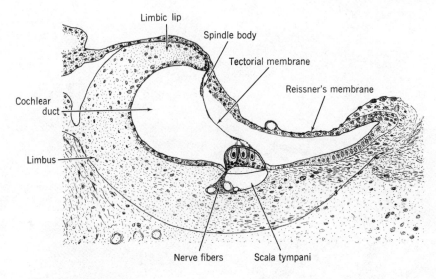

Fig. 15-5. The cochlear duct and limbus of *Lialis burtonis*, in a frontal section through the dorsal part of the cochlea. Scale 175×. From Wever, 1974c.

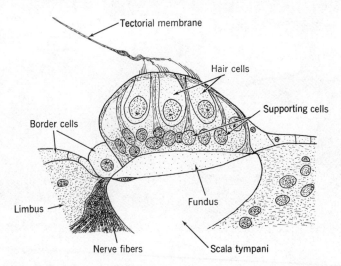

Fig. 15-6. The auditory papilla of *Lialis burtonis* in a section from the dorsal part of the cochlea. Scale 700×. From Wever, 1974c.

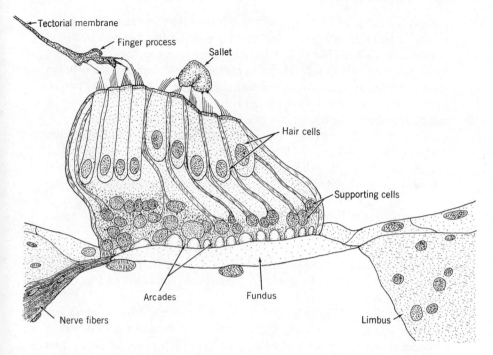

Fig. 15-7. The auditory papilla of *Lialis burtonis*, in a section from the ventral part of the cochlea. Scale 600×. From Wever, 1974c.

extend over the medial hair cells and make the connections. There is one such finger process to serve the more medial cells of each transverse row (including four cells at the level shown in Fig. 15-7), and correspondingly there is a separate sallet for the more laterally disposed cells of each of these rows (five cells as shown in Fig. 15-7).

A further feature seen in the papilla close to its ventral end is a peculiar form of connection to the surface of the fundus. In this region the supporting cells form a series of arcades whose bases are in contact with the fundus. This form of attachment appears to be a relatively weak one, for in a few sections of one ear these structures were torn away from the fundus—something not seen in the dorsal part of this cochlea or elsewhere in other lizard species.

Ciliary Orientation Patterns.   The ciliary tufts of the hair cells have patterns of orientation that vary systematically along the cochlea. These tufts take two positions, with the kinocilium located on either the medial or the lateral side of the array as seen in a transverse section. In the dorsal region of the cochlea (Figs. 15-5 and 15-6) where the sallets are absent, the most medial ciliary tuft in a transverse row has its kinocilium on the lateral side of the tuft, and the most lateral tuft has its kinocilium disposed medially. Hair cells lying between these two, where there are three in a row, will usually alternate between these two positions. When there are four in a row, the usual pattern is for the most medial two to have laterally disposed kinocilia and the most lateral two to have the contrary orientation, but exceptions to this rule are common.

When the sallets appear farther ventrally, there are two orientation patterns (Fig. 15-7), with a medial group of hair cells with tectorial connections having a bidirectional pattern like that just described, and a lateral group of cells with sallet connections having another such pattern.

What has just been described is the familiar pattern of the gecko cochlea.

Cochlear Dimensions. Measurements of the cochlear structures were made on the ear of *Lialis burtonis*, with results indicated in Figs. 15-8 and 15-9.

Part *a* of Fig. 15-8 shows the varying width of the basilar membrane along its length of 1.16 mm. This width increases at the dorsal end at a moderate rate, reaches a maximum at the 300 μ point of 78 μ, then after a slight narrowing it increases further to a broad maximum that approaches 100 μ, after which it narrows slightly and suddenly ends. The width of the fundus follows a similar course, though the secondary maximum around the 300 μ point is less noticeable. After the maximum is reached in the middle of the ventral region, the width decreases gradually and then suddenly at the ventral end. The thickness of the fundus varies somewhat irregularly over the dorsal half of the cochlea, but is nearly uniform over the ventral half.

The area of the papilla, shown in *d* of Fig. 15-9, increases progressively along the cochlea to a maximum near the ventral end. The cellular portion of this structure has much the same form.

The rows of hair cells are shown in *e* of Fig. 15-9 separately for tectorially and sallet-connected cells, and also in total. The rows with tectorial connections increase a little irregularly from 2 to 4 or 5 in the dorsal segment, then decline to 2 and 3 along the first half of the ventral segment, and rise to 4 over most of the latter part of this segment. The rows of sallet-connected hair cells about equal the others in the initial part of the ventral segment, but thereafter exceed these except for occasional points of equality, until at the very end they disappear just before the others do.

Hair-cell density is represented in *f* of Fig. 15-9, without separation into tectorially and sallet-connected cells, and shows a progressive in-

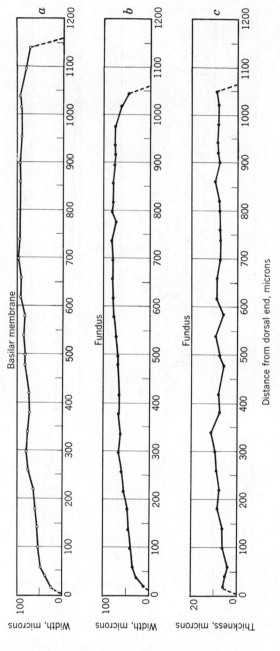

FIG. 15-8. Cochlear dimensions in *Lialis burtonis*. *a*, width of the basilar membrane; *b*, width of the fundus; *c*, thickness of the fundus.

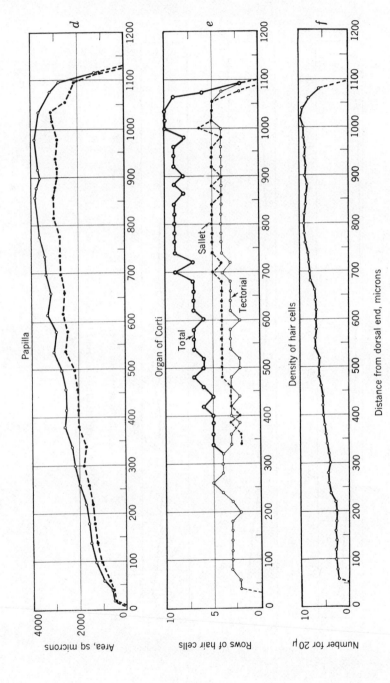

Fig. 15-9. Cochlear dimensions in *Lialis burtonis*. *d*, size of the auditory papilla; *e*, row structure of the hair cells; *f*, density of hair cells.

crease to a maximum in the ventral region, and then continuation with little change almost to the ventral end.

The differentiation seen in these cochlear structures is only of moderate degree. The variation in the width of the basilar membrane from near the dorsal end to the maximum is only about twofold. The variation in the width of the fundus is a little greater, about threefold. The thickness of the fundus varies irregularly, with maximum changes of about threefold, but the form of this variation hardly contributes to the graduation of the cochlea. The size of the auditory papilla varies progressively along the cochlea over a range of about fivefold. Altogether these features show a fair degree of differentiation, but less than was observed in most of the geckos.

A pictorial representation of the form of the basilar membrane is shown in Fig. 15-10. Here the same conventions are used as for the geckos to indicate the cochlear segmentation. The dorsal segment covers about a fifth of the total, with a transitional zone separating it from the long ventral segment. The pattern corresponds to that seen in the diplodactylines and sphaerodactylines (and a few divergent gekkonines), in which the orientation is bidirectional throughout the dorsal segment and doubly so in the ventral segment.

Lialis burtonis

FIG. 15-10. The cochlear segmentation pattern in *Lialis burtonis*.

AUDITORY SENSITIVITY

The results of cochlear potential measurements on the specimen of *Lialis burtonis* are presented in Fig. 15-11. The best sensitivity extends over a broad region from 600 to 4000 Hz, within which there are two peaks, one at 700 Hz and another around 2000-2500 Hz. For lower and especially for higher tones, the sensitivity declines rapidly.

The level of sensitivity shown is poor, and the question arises whether this low level is characteristic of the species. Additional specimens of *Lialis burtonis* should be investigated in this manner, along with other members of the family.

The form of the sensitivity function, with the trend toward better response in the higher tones, is similar to that found in the diplodactylines and sphaerodactylines, as well as in one group of gekkonines, as described in the preceding chapter.

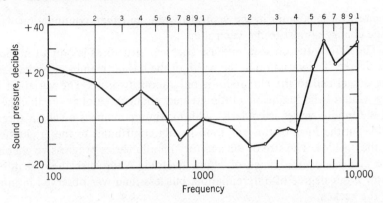

FIG. 15-11. A sensitivity function for a specimen of *Lialis burtonis*.
From Wever, 1974c.

RELATIONS TO THE GECKOS

The points of correspondence between auditory structure and function in the single pygopodid examined and in the gecko group may now be summarized.

1. A meatal closure muscle is present in *Lialis*, but it is of the purse-string type, resembling that found in the diplodactylines, yet somewhat simpler in that the muscle is nearly uniform in its loop around the meatus.

2. An extracolumellar muscle is present in much the same form as in all the geckos, but is found in no other lizards.

3. The cochlear structures show many points of resemblance. A limbic lip is present as in geckos and teiids. The segmentation of the cochlea, with a dorsal segment served by the tectorial membrane and a ventral segment divided longitudinally into two parts, with tectorial connections to the hair cells medially and sallet connections laterally, follows the typical gecko pattern. The ciliary orientation pattern corresponds to that of two main groups of gekkonids, as has been noted.

4. The tectorial membrane shows a thickening close to its point of detachment from the limbic lip, known as the spindle body and present in all the geckos. This structure has not been seen in any other lizards.

5. The species *Hoplodactylus pacificus*, considered to be one of the most primitive of the gecko family, was especially studied for purposes of comparison with *Lialis*. The common features noted above were all found, but one significant difference was noted. *Lialis burtonis* lacks the general thickening of the free portion of the tectorial membrane found in *Hoplodactylus* and all the other diplodactylines.

6. The size of the hair-cell population in *Lialis burtonis* is within the range found in geckos. These lizards show wide variations, with some species having as few as 350 hair cells, certain ones, such as *Gekko gecko* and *Gekko smithi*, going as high as 1600 to 2100, and the great majority having populations in the region of 500 to 800. Six ears of the species *Hoplodactylus pacificus* gave counts from 629 to 756, with a mean of 695, with which the specimen of *Lialis burtonis*, having 665 in one ear and 719 in the other, is in good agreement.

Cochlear Differentiation.   In comparison with other lizard species for which the dimensions of the cochlear structures have been studied, the differentiation in *Lialis burtonis* must be regarded as intermediate in character. It is definitely greater than that found in iguanids and anguids, but less than generally seen in the gekkonids.

If the general form of the auditory papilla, the size of the hair-cell population, and the degree of differentiation along the cochlea may be taken as measures of the level of development of the ear, then we must consider *Lialis burtonis* as advanced over the majority of lizards, but hardly reaching the stage generally attained among the gekkonids. The many similarities between this ear and that of the geckos are in support of the often expressed view that these are gecko relatives.

16. FAMILY GERRHOSAURIDAE:
THE PLATED LIZARDS

The gerrhosaurids are large-bodied lizards inhabiting two neighboring regions, the southern part of Africa and the island of Madagascar. Six genera and 25 species are about equally divided between these two localities. These lizards are ground dwellers, who prefer rough, rocky hills that provide an abundance of hiding places. They feed mainly on insects.

The common name of plated lizards applies more appropriately to the genus *Gerrhosaurus* in which the body is encased in rectangular scales with dense bony centers. The regular geometrical arrangement of this armor is shown for the species *Gerrhosaurus major major* in Fig. 16-1.

FIG. 16-1. A specimen of *Gerrhosaurus m. major*. Drawing by Anne Cox.

In early classifications the Gerrhosauridae and Cordylidae (then called Zonuridae) were well separated. Cope (1900) bracketed the gerrhosaurids with the teiids, lacertids, and skinks in a division of Lepto-glossa (Scincomorpha), whereas he placed the zonurids in the Diplo-glossa along with the anguids, xenosaurids, and pygopodids. McDowell and Bogert (1954) presented convincing evidence for the close relation-ship of gerrhosaurids and cordylids, and regarded both as skink-like forms. Romer (1956) agreed with this position and went so far as to include them in a common family, the Cordylidae, which he then divided into the subfamilies Gerrhosaurinae and Cordylinae. Others have recognized separate families of Gerrhosauridae and Cordylidae within the general group of Scincomorpha, and this arrangement will be followed here, as it seems best suited for a treatment of the ear.

Six species belonging to two genera were available for study. These are *Gerrhosaurus f. flavigularis* (Wiegmann, 1868)—6 specimens; *Gerrhosaurus m. major* (Duméril, 1851)—3 specimens; *Gerrhosaurus n. nigrolineatus* (Hallowell, 1857)—2 specimens; *Gerrhosaurus v. validus* (A. Smith, 1849)—6 specimens; *Zonosaurus madagascariensis* (Gray, 1829)—1 specimen; and *Zonosaurus ornatus* (Gray, 1829)—2 specimens.

ANATOMICAL OBSERVATIONS

OUTER AND MIDDLE EAR

As shown in Fig. 16-1 for *Gerrhosaurus m. major*, the external ear opening is a broad oval slit that exposes only the anterior portion of the tympanic membrane.

The middle ear apparatus is of the iguanid type, with no special features apart from a rather heavy construction.

INNER EAR

A lateral view of the inner ear region of *Gerrhosaurus v. validus*, based upon a series of sagittal sections, is presented in Fig. 16-2, and shows the usual inverted pyramidal form of the labyrinthine enclosure, which contains the cochlear duct ventrally, with the auditory papilla running along its posterior wall. The otic capsule has thick, substantial walls, and the cochlear duct is relatively large.

The fundus of the auditory papilla, shown here in long section, has a tapered form, increasing toward its ventral end, whereas the sensory structures that it bears exhibit a double swelling, one near the dorsal end that has a spindle shape and another ventrally forming a broad oval. This ventral portion is surmounted by a peculiarly formed culmen, consisting of a dense cluster of fingerlike processes and a small round mass of spongiose material on its anterior side. A connection to the tectorial membrane, which is a tapered ribbon, is made at the dorsal end of the fingerlike cluster.

Also seen in this picture are the lagenar macula extending along the anterior wall of the cochlear duct and the broad saccular macula above.

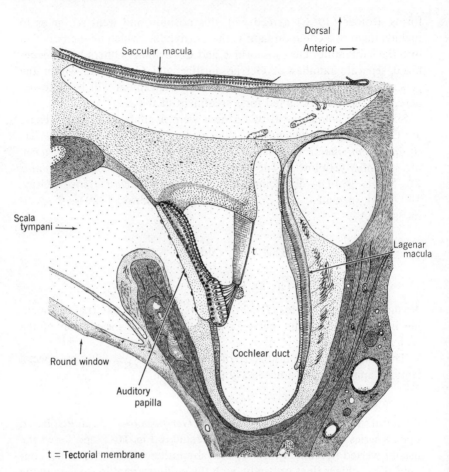

Fig. 16-2. A lateral view of the inner ear region of *Gerrhosaurus v. validus*. Scale 75×.

A more comprehensive picture of the gerrhosaurid auditory papilla is presented in Fig. 16-3, which is a reconstruction based on sections of the right ear of a specimen of *Gerrhosaurus n. nigrolineatus*. The view is from dorsal, lateral, and posterior directions, and shows the auditory papilla over its whole extent, lying over the basilar membrane and partially obscuring it, and bearing along most of its length a row of closely adjoining sallets. A tectorial membrane is present along most of the cochlea, even at the dorsal end, but here it is only a narrow band on the limbic surface. Ventrally this membrane widens progressively, extending farther and farther laterally toward the papilla and lifting away from the limbic surface on its outer edge, until finally, in the far ventral region, it connects with the culmen lying over the papilla.

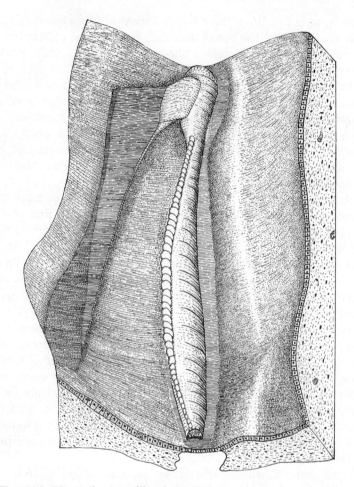

Fig. 16-3. The auditory papilla of *Gerrhosaurus n. nigrolineatus*, re-constructed from serial sections. Dorsal is below, lateral to the right. Scale 85×. From Wever, 1970a.

Along the dorsal and middle portions of the cochlea, the ciliary tufts of the hair cells are connected to the line of sallets. At the ventral end, where the tapering papilla undergoes its broad expansion, the sallets come to an end and the ciliary tufts of these ventral cells are attached to the undersurface of the culmen. We thus distinguish two segments of the papilla, a long dorsal segment with sallet connections to the hair cells and a short ventral segment whose hair cells are served by a teth-ered culmen. Between these two is a short area within which the cilia of the hair cells stand free, without any restraining connections.

As Fig. 16-3 shows, the sallets form a continuous series. Seen in cross section, as in Fig. 16-4, these bodies appear as low mounds lying close over the hair cells. The tips of the ciliary tufts of these cells attach either to the undersurface or, more often, to the pointed ends of skirting processes sent off around the sallet's lower border.

In cross sections such as Fig. 16-4 the sallet appears as a solid mass, but careful examination at high power reveals a reticular structure as shown by sallets in general (Chapter 4). Toward the dorsal end of the cochlea the sallets are relatively isolated, though there is always a commingling of their longitudinal fibers. Toward the ventral end of the cochlea this interweaving of fibers increases, and adjacent sallets become closely joined.

A cross section of the cochlear capsule and its contents, taken at right angles to the views of Figs. 16-2 and 16-3, is shown in Fig. 16-5. The level of this section is near the ventral end of the cochlea where the tectorial membrane makes its connection to the papilla. It will be noted that the medial limbus here remains flat, showing no bulge or lip on its lateral surface as found in most lizards. The epithelial layer is moderately thickened, and the tectorial membrane arises at a grazing angle from this surface.

The form of the culmen varies with species. Usually it clearly shows two portions, a lower basal part composed of relatively dense material that lies close over the hair cells and sends out a number of thorn-like prongs to make connection with the ciliary tufts, and an upper portion made up of delicate sponge-like material of highly variable form. In some species this spongiose part makes the connection to the tectorial membrane, but in others, as seen in Fig. 16-2, it is displaced to one side and appears to have no major role, and the connection is made directly to the base portion.

Other forms of culmen structure are represented in Fig. 16-6. Above in this series is shown the relatively simple structure of *Gerrhosaurus f. flavigularis*. The culmen is small, with the basal part comblike and extended by a long tapered spongiose portion that connects with the tectorial membrane.

In the likewise simple culmen of *Gerrhosaurus m. major*, several long fingers constitute the base portion and extend to the hair-cell surface. From the common fusion of the upper ends of these fingers a slender spongiose portion connects to the thick tectorial membrane. In the culmen of *Gerrhosaurus n. nigrolineatus*, the basal part has relatively short fingers and the spongiose part is relatively large. The culmen of *Zonosaurus madagascariensis* has a compact basal portion with very short projections to which the ciliary tufts attach, and a prominent bulging spongiose portion from one end of which a tapered process connects to the tectorial membrane. The culmen structure in *Zonosaurus ornatus*

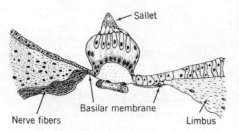

FIG. 16-4. The auditory papilla of *Gerrhosaurus m. major*, seen in cross section near the middle of the cochlea.

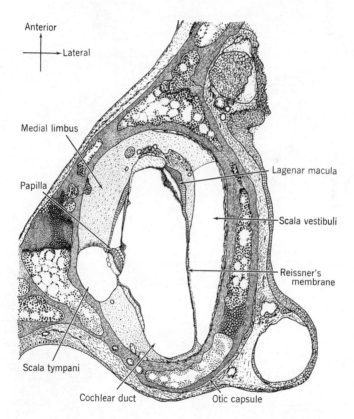

FIG. 16-5. Frontal section of the cochlear capsule of *Gerrhosaurus v. validus* near the ventral end of the cochlea.

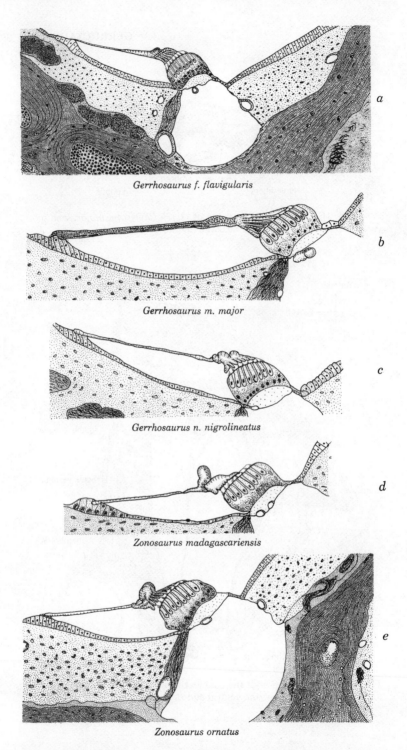

Gerrhosaurus f. flavigularis

Gerrhosaurus m. major

Gerrhosaurus n. nigrolineatus

Zonosaurus madagascariensis

Zonosaurus ornatus

FIG. 16-6. The auditory papilla in five species of Gerrhosauridae, seen in frontal section in the ventral segment and showing various types of culmens. Lateral is above and anterior to the left. Scale 250×.

is similar, but the basal projections are longer and the spongiose portion somewhat larger.

Cochlear Dimensions.   Dimensional measurements on a specimen of *Gerrhosaurus n. nigrolineatus* are given in Figs. 16-7 and 16-8. The first of these figures represents in part *a* the varying width of the basilar membrane over its considerable length of 1.26 mm. The width increases at the dorsal end, at first rapidly and then slowly and a little irregularly, until it reaches a maximum of 178 μ over a region about halfway along the cochlea. Thereafter the width decreases rapidly to a minimum of 77 μ, which marks off the dorsal segment of the cochlea from the short ventral segment. The variation in width over the dorsal segment amounts to 2.3-fold. Beyond the minimum the width increases to a second maximum of 166 μ, and then falls away as the papilla ends.

The width of the fundus, shown in *b* of Fig. 16-7, takes a closely similar course, but attains its maximum somewhat sooner. It reaches a value of 99 μ at a point 400 μ from the dorsal end, then declines to a uniformly low value in the region of 920 to 1040 μ, with the minimum of 19 μ at the 1000 μ point. The variation in this dorsal region is 5.2-fold. The ventral expansion then appears, with a maximum attained of 95 μ, after which the fundus quickly disappears.

The thickness of the fundus, shown in *c* of this figure, rises first rapidly and then slowly to a flat maximum of 28 μ in the region of 520 to 720 μ. Thereafter there is a considerable decrease until this structure almost disappears—the thickness at the 980 μ point falls to 1.7 μ. The variation in the dorsal segment amounts to 16.5-fold. Then the thickness increases rapidly to a new maximum of 30 μ, which is about the same as the first one, and soon falls away.

In Fig. 16-8 are shown the variations in area of the papillar structure including the fundus (solid line) and of the sensory part only (broken line). The whole structure increases rapidly at first and then more slowly to a maximum of 7600 sq μ in the middle of the dorsal segment. The decline beyond the maximum is rapid, until a minimum of 1750 sq μ is reached, a variation of nearly fivefold. Thereafter the ventral expansion appears, with a sharp rise to a maximum of 7750 sq μ, and finally a rapid fall as the papilla ends. The cellular portion follows a similar course.

The distribution of the hair cells, shown in part *e* of this figure, follows a course much like the width of the basilar membrane. The number of longitudinal rows of hair cells begins with 4, rises abruptly to 10, then varies irregularly between 11 and 7 and stabilizes at 6 in the middle of the cochlea. Thereafter the number falls in steps to a level of 3. Finally in the ventral segment the number rises to 8, falls to 7 for a short distance, and then the papilla ends.

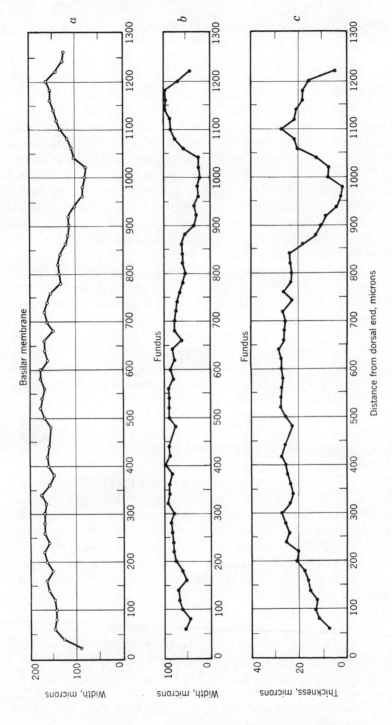

FIG. 16-7. Cochlear dimensions in a specimen of *Gerrhosaurus n. nigrolineatus*, showing the basilar membrane and fundus. From Wever, 1970a.

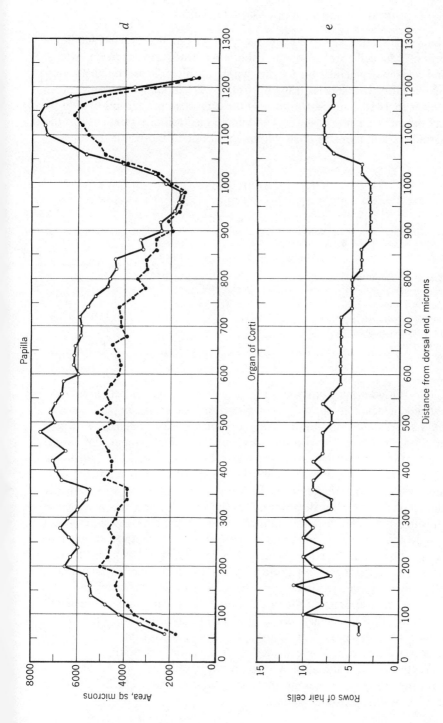

FIG. 16-8. Size of the auditory papilla and hair-cell distribution in *Gerrhosaurus n. nigrolineatus*. From Wever, 1970a.

Number of Hair Cells. For the six species examined the average numbers of hair cells are shown in Table 16-I, listed separately for dorsal and ventral segments and for the whole cochlea. These numbers vary significantly with species, ranging from 529.5 in *Gerrhosaurus f. flavigularis* to 1418.2 in *Gerrhosaurus m. major*. The proportion of cells in the ventral segment, however, exhibits no clear species relationship; the variations are about as great within species as in the whole group.

TABLE 16-I

HAIR-CELL POPULATIONS

IN GERRHOSAURIDAE

Species	Number of Ears	Dorsal Segment	Ventral Segment	Total	Percentage Ventral
Gerrhosaurus f. flavigularis	6	471.5	58.0	529.5	11
Gerrhosaurus m. major	4	1260.5	157.7	1418.2	10
Gerrhosaurus n. nigrolineatus	4	770.0	86.7	856.7	11
Gerrhosaurus v. validus	6	841.2	95.8	937.0	10
Zonosaurus madagascariensis	2	665.5	107.5	773.0	14
Zonosaurus ornatus	2	534.5	74.5	609.0	12

Ciliary Orientation Patterns. The patterns of ciliary orientation were studied in detail in specimens of *Gerrhosaurus m. major* and *Gerrhosaurus v. validus*, and the remaining species were examined more generally to confirm the picture thus obtained. Throughout the dorsal segment of the cochlea, where the hair cells are connected to sallets, the ciliary tufts show a bidirectional orientation. In many sections the cells are divided at the middle of the transverse row, with those on the medial side showing laterally disposed kinocilia and those on the lateral side showing medially disposed kinocilia. Sometimes the division is not equal, and there is a moderate preponderance of laterally facing hair cells. Thus, when four transverse rows are present, the cells in the row next to the outside may alternate between lateral and medial orientation, while those in the remaining rows maintain their regular orientations.

In the ventral segment the orientation is essentially unidirectional: all or nearly all of the ciliary tufts in this region have the kinocilium on the lateral side. Exceptions are most often found for the most lateral cell in a transverse row; this cell may show a medial orientation.

Within the transitional zone, for the few rows of cells in which no covering structure is present, the orientation appears to be bidirectional in a somewhat irregular pattern.

COCHLEAR POTENTIAL STUDIES

Sensitivity curves for two specimens of *Gerrhosaurus f. flavigularis* are shown in Fig. 16-9. Both animals give best responses in the middle frequencies, with one attaining a sensitivity around −60 db in two regions, at 500-600 and at 1500 Hz. The other animal is best in the same region of frequency, with peaks of sensitivity of −52 and −47 db for the two bands indicated. The decline of sensitivity in the high tones is rapid, around 40 db per octave, and in the low frequencies this decline is 30 db per octave in one of the ears and of about the same order, though irregular, in the other.

Results for two specimens of *Gerrhosaurus m. major* are given in Fig. 16-10. These two curves have much the same form, but one runs noticeably below the other except in the uppermost range. The level of sensitivity attained is very great for tones between 300 and 1500 Hz, and in the better of these ears runs around −70 to −74 db. In these ears the tests were extended downward to 50 Hz, and the curves are found to hold their courses well in these low frequencies. The roll-off rates are around 30 db per octave at the low end of the range and approximately 40 db per octave at the high end.

Observations on two specimens of *Gerrhosaurus n. nigrolineatus* are represented in Fig. 16-11. One of these animals showed excellent performance in the region of 300 to 1000 Hz, attaining −80 db at 700-800 Hz. The other animal seems poor by comparison, yet its record of −48 to −55 db in this same region is an excellent one.

Curves for two specimens of *Gerrhosaurus v. validus* are shown in Fig. 16-12. The two functions are in good agreement in indicating excellent sensitivity in the range from 400 to 1500 Hz, with one ear reaching −58 db at 1000 Hz. The roll-off rates are 30 db per octave at the low end of the range and 40 db per octave at the high end.

The two *Zonosaurus* specimens are shown together in Fig. 16-13, with the solid line representing *madagascariensis* and the broken line representing *ornatus*. There is a fair amount of agreement in the forms of these curves and in the maximum sensitivity attained, though the *madagascariensis* species is superior over most of the range. The sensitivity declines very rapidly at the low end of the range. A response around −60 db between 400 and 1000 Hz as shown by both these species represents excellent hearing.

In general, the gerrhosaurids exhibit sensitivity of a high order in a range that varies somewhat with species, but in all lies in the medium low to the medium high frequencies.

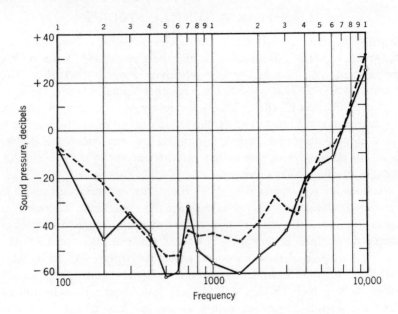

FIG. 16-9. Sensitivity funtions for two specimens of
Gerrhosaurus f. flavigularis.

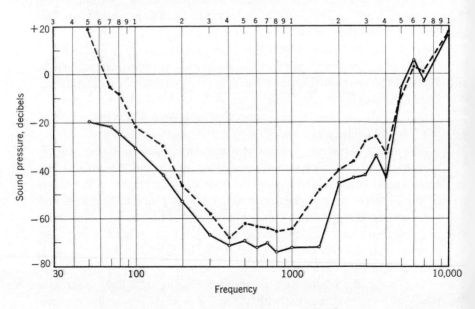

FIG. 16-10. Sensitivity functions for two specimens of *Gerrhosaurus m. major.*

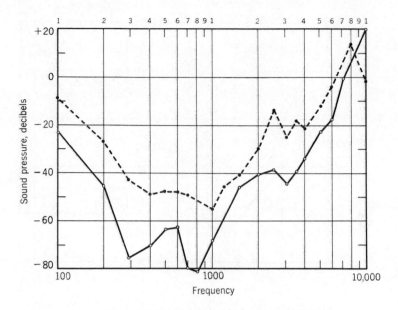

FIG. 16-11. Sensitivity functions for two specimens of
Gerrhosaurus n. nigrolineatus.

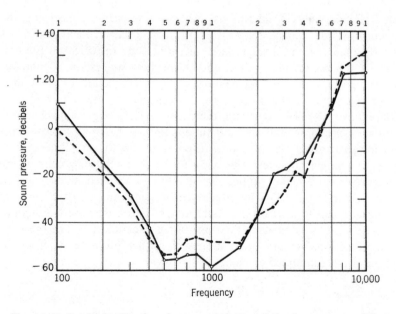

FIG. 16-12. Sensitivity functions for two specimens of *Gerrhosaurus v. validus.*

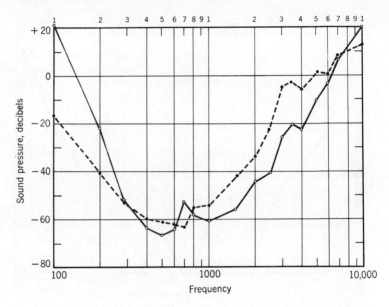

Fɪɢ. 16-13. Sensitivity functions for a specimen of *Zonosaurus mada-gascariensis* (solid line) and one of *Zonosaurus ornatus* (broken line).

Some Functional Considerations. A question arises concerning the function of the ventral segment of the gerrhosaurid cochlea with its broad expansion and a single large tectorial body to which all the hair cells of the region make connection. The best suggestion is that this ventral portion of the cochlea serves for the detection of faint low-frequency sounds. This hypothesis is supported by the following considerations: (1) The cochlear dimensions, including relatively large widths of basilar membrane and fundus and considerable mass of the moving structures, should favor a response to low tones. (2) The presence of a single large tectorial body to which all the hair cells of the ventral segment make connection should aid the sensitivity through spatial summation. (3) The attachment of the culmen to the tectorial membrane should increase the restraining effect on the cilia and tend to hold the various elements to the same phase of action. (4) The presence of unidirectional ciliary orientation will further promote synchronous action of the stimulated elements.

17. FAMILY XANTUSIIDAE:
THE NIGHT LIZARDS

The xantusiids, commonly known as night lizards, make up a small group of about 15 species found in two well-separated regions of North America: in the West Indies and in limited and often narrowly localized areas in the southwestern United States, Mexico, and Central America. This family has attracted special attention because of many structural peculiarities that make uncertain the relations within the group and the position with respect to other lizard families.

In his early system of lizard classification, Cope (1900) regarded the xantusiids as allied to the lacertids, and placed the two together in a suborder Leptoglossa, along with teiids, skinks and gerrhosaurids. Camp, somewhat later (1923), indicated the complexity of xantusiid characteristics by relating them both to the gekkonids and to the scincomorphs. Most other writers have emphasized one or the other of these two relationships; thus Romer (1956), Miller (1966a), and Etheridge (1967) placed the xantusiids with the scincomorphs, whereas McDowell and Bogert (1954), Savage (1963), and St. Girons (1967) located them adjacent to the geckos. Bezy (1972), however, in a series of studies of chromosome patterns found the greatest similarities with the teiids.

Relationships within the family have also presented some problems. Bezy recognized three genera: *Xantusia, Lepidophyma,* and *Cricosaura.* Others have accepted two additional genera: *Klauberina,* which Savage split away from *Xantusia* as *Klauberina riversiana;* and *Gaigea,* proposed by Smith (1939) but by others included with *Lepidophyma.* In the present treatment of the ear the three-genera system of Bezy seems the most suitable.

Xantusia, according to this system, contains three species—*henshawi, riversiana,* and *vigilis*—though a number of subspecies are recognized: two for *henshawi* and seven for *vigilis. Lepidophyma* contains 11 species, with subspecies in two of these. *Cricosaura* is monotypic, with the single species *typica.*

Bezy expressed the opinion that the *Lepidophyma* species, which inhabit tropical regions, are the most primitive ones, and the *Xantusia* species, found in relatively temperate regions, are specialized in many respects. *Cricosaura typica* is isolated in a small area of Cuba and has never been studied in any detail, but Savage regarded it as distinctive and placed it in a subfamily of its own.

For the present study, the available specimens included three species of *Xantusia* and three of *Lepidophyma*, as follows: *Xantusia henshawi* (Stejneger, 1893)—5 specimens; *Xantusia riversiana* (Cope, 1883)—6 specimens; *Xantusia vigilis* (Baird, 1858)—5 specimens; *Lepidophyma flavimaculatum* (A. Duméril, 1851)—2 specimens; *Lepidophyma smithi* (Boucourt, 1876)—2 specimens; and *Lepidophyma sylvaticum* (Taylor, 1939)—1 specimen. The identification of *Lepidophyma flavimaculatum* and *Lepidophyma sylvaticum* is considered tentative.

ANATOMICAL OBSERVATIONS

OUTER AND MIDDLE EAR

An external ear opening is present as a vertical oval as shown for *Xantusia henshawi* in Fig. 17-1. The tympanic membrane lies about a millimeter below the surface, leaving only an extremely shallow meatal cavity.

FIG. 17-1. A specimen of *Xantusia henshawi*.
Drawing by Anne Cox.

The middle ear mechanism of xantusiids takes two forms. In *Xantusia* species it is of the iguanid type, with an internal process connecting the extracolumella to the rim of the quadrate. In *Lepidophyma* species the internal process is lacking and only a fold of mucous membrane provides lateral support for the columella; this mechanism therefore must be regarded as of the scincid type as defined above (Chapter 6).

Xantusia Type of Middle Ear. The middle ear was studied in detail in *Xantusia henshawi* and is represented in Figs. 17-2 and 17-3, the first of which is a sketch of a dissected specimen and the second a reconstruction from serial sections. The sketch shows the form of the structure as seen through a wide opening in the floor of the mouth and pharynx, and represents the long slender bony columella running from its foot-plate in the oval window to a junction with the cartilaginous extra-columella. This junction is a firm union without discontinuity. The heavy shaft of the extracolumella then gives off the internal process, which is a narrow, fan-shaped cartilage that extends to the edge of the quadrate and has a loose anchorage there. The extracolumella continues as a slender rod that curves anterolaterally and gives off four processes, which Versluys called collectively the insertion piece and which makes the connection to the tympanic membrane. Three of these processes are shown in the figure; the fourth is the anterior process, an extremely thin rod of cartilage running mainly in a dorsal direction between the layers of the tympanic membrane and hidden from view in this figure.

The pars inferior is seen running along the inner surface of the tym-panic membrane to its midpoint. The pars superior, which is integral with it, extends anteromedially and serves as the insertion for the extra-columellar ligament. This ligament runs to the paroccipital process, and partially encloses the intercalary cartilage, which is attached to that process. The posterior process appears as a small rounded knob.

The relations between these structures are displayed in further detail in Fig. 17-3. Of particular interest is the course of the internal process, which swings around the end of the quadrate and expands as a terminal ball resting on the medial surface of the bone.

Lepidophyma Type of Middle Ear.   The middle ear structures were examined in two species of *Lepidophyma*. The general form is much like that of *Xantusia* species except for the absence of the internal proc-ess. As shown for *Lepidophyma flavimaculatum* in Fig. 17-4, a stiff fold

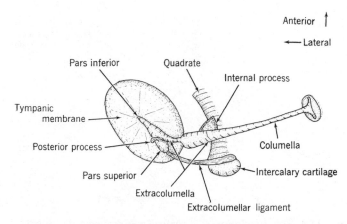

Fig. 17-2. The middle ear mechanism of *Xantusia henshawi*, from a dissected specimen.

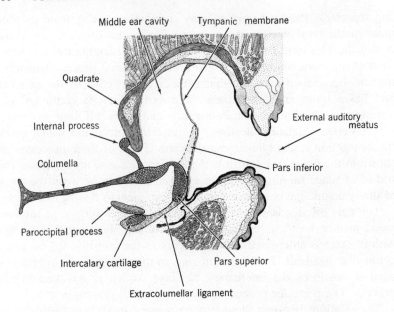

FIG. 17-3. The middle ear of *Xantusia henshawi*, reconstructed from several successive serial sections. The right ear is represented as viewed from behind at the general level of the columella. The pars inferior, which appears more ventrally, is indicated in outline. Scale 25×.

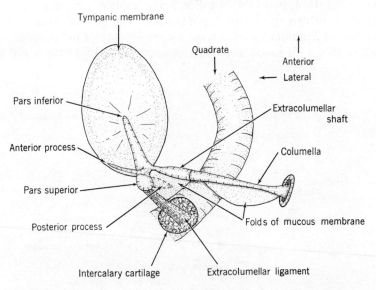

FIG. 17-4. The middle ear of *Lepidophyma flavimaculatum*, from a dissected specimen. The right side is shown in a ventral and postero-lateral view.

of mucous membrane extends from the quadrate along the extracolumella, and another fold lies along the inner portion of the columella, affording considerable lateral support. The anterior process is relatively large at its base and tapers to a thin rod along the posteroventral edge of the tympanic membrane. A posterior process was not seen in the dissected specimen, but was located in serial sections of this species and is indicated in the figure by broken lines.

The middle ear of *Lepidophyma smithi* also was studied both by dissection and in serial sections. A drawing is shown in Fig. 17-5, based primarily on sections through the dorsal portion of this structure, but with the pars inferior, which is more ventrally located, added in outline.

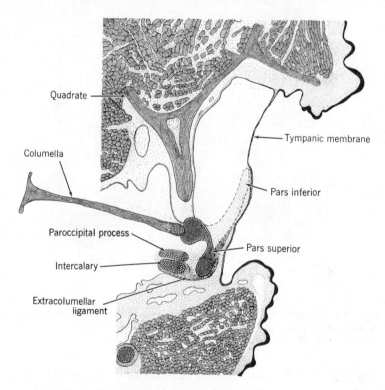

FIG. 17-5. The middle ear region of *Lepidophyma smithi*, from serial sections. The pars inferior is added in outline. Scale 25×.

INNER EAR

The general form of the inner ear in Xantusiidae is indicated in Fig. 17-6, which shows the dorsal end of the cochlea of *Xantusia riversiana* in cross section. The cochlear limbus is well developed, but its anterior portion lacks any marked elevation. Reissner's membrane at this level

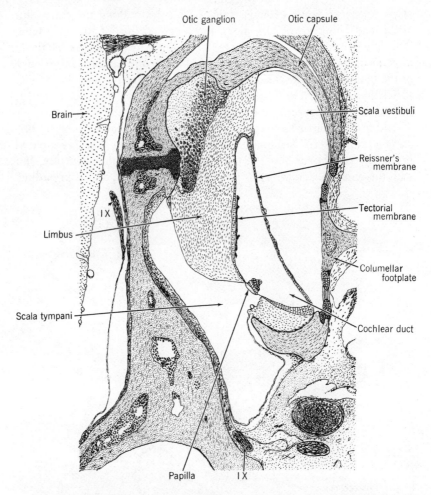

Fig. 17-6. The inner ear region of *Xantusia riversiana*, in a frontal section through the dorsal segment. The glossopharyngeal nerve, cut across at two places, is indicated by IX. Scale 50×.

is only moderately thickened, but it increases to a thickness of 10-15 μ toward the middle of the cochlea, and still more ventrally it reaches 30-36 μ, and finally, close to the ventral end it becomes 40-60 μ in thickness. Such thickening of Reissner's membrane is most prominent in the *Xantusia* species, but also occurs in *Lepidophyma* species.

When the means of ciliary restraint is considered, it is found that the division into two structural types as seen in the middle ear holds for the inner ear as well. The *Xantusia* species present a form of cochlear structure much like that described in Chapter 16 for the Gerrhosauridae: a line of sallets runs over the hair cells along the main part of the cochlea, and then at the ventral end a tethered culmen appears to make the ciliary connections for all the hair cells there. In the *Lepidophyma* species the arrangement is like that seen in the skinks and cordylids, in which the hair cells of the dorsal segment of the cochlea are served by sallets and those of the ventral segment are surmounted by a culmen that stands free, without any tectorial membrane connection. These two types of inner ear structure will now be described in detail.

Xantusia Type of Inner Ear. The cross section through the dorsal region of the cochlea in *Xantusia riversiana* as presented in Fig. 17-6 shows the relations of the structures at this level, with the auditory papilla surmounted by a sallet. The papilla, shown in further detail in Fig. 17-7, has an asymmetrical form, with its anteromedial side nearly twice as high as the posterolateral one. The hair cells are obliquely positioned over the fundus, and their outer ends form a concave surface in the hollow of which the sallet is contained. More ventrally the form of the papilla changes, becoming more nearly symmetrical, and the surface concavity disappears. This form is represented in Fig. 17-8.

The sallets are of variable form, but in general appear to represent an incomplete fusion of a number of elements of elongate type, each with one end directed away from the papilla and the other end, usually sharply pointed, forming a connection with one of the ciliary tufts of the hair cells. Sometimes these elements are of tear-drop form, with the rounded ends fused into a general mass, and at other times these elements are more spindle shaped. One of the more regular forms of sallet structure is shown in Fig. 17-9. Often it is possible to see the connection between the pointed tip of the sallet element and the longest cilium of the ciliary tuft, which is no doubt the kinocilium.

A little ventral to the level shown in Fig. 17-9 the sallets increase considerably in size, and thereafter they maintain this larger size with little change over most of the dorsal segment. The fragmented appearance of the sallets continues. It is likely that as the number of longitudinal rows of hair cells increases there is a corresponding increase in the parts of which the sallets are composed, though a verification by actual counting was hardly possible.

Cochlear Segmentation. — Of special significance are the changes of limbus and tectorial membrane that are associated with the segmentation of the cochlea: with the termination of the sallets and the appearance of the culmen at the ventral end. Near the dorsal end of the cochlea the tectorial tissue appears as a small mass of undifferentiated material, and then a little more ventrally this material forms a thin membrane over the middle portion of the medial limbus, where there is a moderate

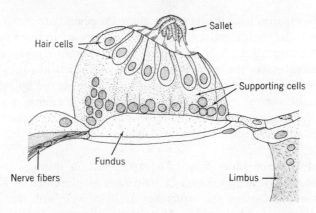

Fig. 17-7. Cross-sectional view of the auditory papilla of *Xantusia riversiana*, at a level close to the dorsal end of the cochlea. Scale 500×.

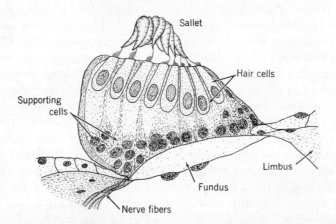

Fig. 17-8. Cross-sectional view of the auditory papilla of *Xantusia riversiana* near the middle of the cochlea. Scale 500×.

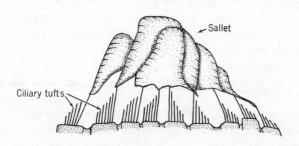

Fig. 17-9. The sallet structure and ciliary connections in the dorsal region of the cochlea in *Xantusia riversiana*. Scale 1300×.

elevation and a more marked thickening of the epithelial layer. This membrane continues ventrally, showing occasional local thickenings, until a little beyond the middle of the cochlea the most posterior of these thickenings sends forth a thin reticular membrane that leaves the limbic surface and extends in the direction of the auditory papilla. This membrane widens as it approaches the papilla, and then a thickened bar at its distal edge expands as a mass of frothy material. This material constitutes the upper portion of the culmen, and connects to a base of dense, finely reticular material whose lower portion forms skirting projections that make connections with the ciliary tufts of the hair cells.

These cochlear structures are shown for a specimen of *Xantusia henshawi* in Fig. 17-10, based upon a reconstruction model. Here the view is anterodorsal and slightly lateral, and the auditory papilla is rep-

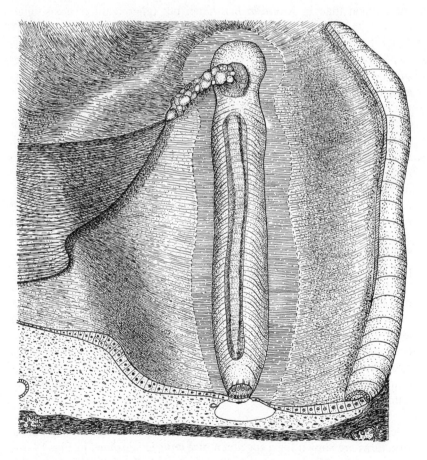

FIG. 17-10. A reconstruction of the inner ear of *Xantusia henshawi*. Scale 165×.

resented as lying on the basilar membrane (indicated by transverse hatching) with its dorsal end below and its bulbous ventral end above. A broad line of sallets runs along the greater part of the papilla, with the hair cells hidden beneath it. After a gap in the ventral region (which is rather wide in this species), the culmen appears with its rounded base over the papillar surface and its spongiose upper portion curving away in a dorsomedial direction to an attachment to the pointed end of the tectorial membrane. This membrane is shown arising from the medial part of the limbus far dorsally and widening as it extends ventrally, and at the same time lifting away from the limbus along its lateral edge.

Other *Xantusia* species show the same general relations, with only minor variations in detail. A picture of the otic capsule and its contents in the region of the ventral segment is shown for *Xantusia riversiana* in Fig. 17-11. The tectorial membrane is attached at a lower position on the culmen than in *Xantusia henshawi*, but other features are similar.

Inner Ear in *Lepidophyma*. In many respects the inner ear of the *Lepidophyma* species resembles that just described. The papillar structure is similar, and over the main part of the cochlea the chain of sallets over the hair cells presents the same picture. A cross section from near the middle of the cochlea is shown in Fig. 17-12.

The medial limbus in these species forms a moderate elevation of broad rounded shape; lying upon it in the dorsal region of the cochlea is a tectorial membrane of almost the same form and extent as in *Xantusia* species, although sometimes with fewer local thickenings. In the middle region of the cochlea the membrane may cover as much as two-thirds of the epithelial layer over the medial limbus. Then as the ventral region is approached this membrane dwindles and disappears, except at times for a small mass located on the posterior declivity of the mound. Still more ventrally, in the transitional zone just before the culmen appears, all traces of tectorial material are lost.

In this transitional region the auditory papilla diminishes in size, loses its sallets and then its hair cells, and only a small papillar framework made up of supporting cells remains on the surface of the fundus. Then a little farther ventrally the papilla reappears, grows much larger than before, and is surmounted by a culmen of upright, regular form. This structure with its tall, shako-like top piece covering a dense base portion is shown in Fig. 17-13 for a specimen of *Lepidophyma flavimaculatum*.

Cochlear Dimensions. Dimensional measurements of the cochlear structures are shown for a specimen of *Xantusia henshawi* in Fig. 17-14. Part *a* of this figure represents the varying width of the basilar mem-

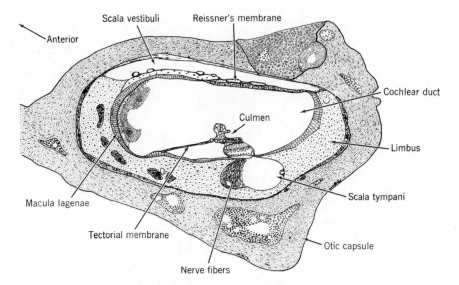

FIG. 17-11. The inner ear of *Xantusia riversiana* at the level of the ventral segment. The drawing represents a composite of five successive sections. Scale 65×.

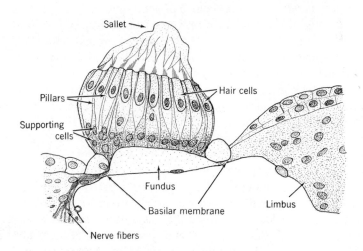

FIG. 17-12. The auditory papilla seen in cross section through the middle of the dorsal segment in *Lepidophyma flavimaculatum*. Scale 400×.

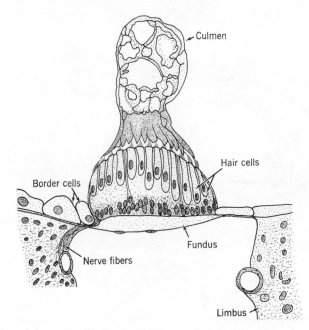

FIG. 17-13. The auditory papilla of *Lepidophyma flavimaculatum* as seen in the ventral segment of the cochlea. Scale 300×.

brane over its relatively short course of about 650 μ. The width increases rapidly at the dorsal end to a maximum of 150 μ, and then declines through the middle region, reaching a plateau around 350-500 μ from the dorsal end. Then comes a moderate decline to a minimum in the transitional zone where the papilla is much reduced, and finally a rise to a sharp peak in the short ventral segment. The width of the fundus, shown in *b* of this figure, takes much the same course but with less variation in the ventral region.

The thickness of the fundus increases to a maximum in the dorsal region and thereafter declines progressively to the transitional zone, beyond which a moderate peak appears.

The size of the auditory papilla, shown in part *d* of the figure, increases rapidly from the dorsal end to an irregular maximum in the middle of the cochlea, after which there is a nearly symmetrical decline to a minimum, followed by a particularly sharp peak at the ventral end. The cellular part of the papilla, represented by the broken curve, follows a closely similar course.

The varying rows of hair cells along the cochlea are indicated in *e* of this figure. The number rises rapidly from 3 at the dorsal end to 7, varies between 6 and 8 over most of the remainder of the cochlea, and in the ventral region falls to 4 and then peaks at 8, after which the papilla quickly ends.

Similar measurements were made on a specimen of *Lepidophyma smithi*, as shown in Figs. 17-15 and 17-16.

The upper curve in part *a* of Fig. 17-15 shows the width of the basilar membrane over its length of about 1000 μ. This width increases at the dorsal end to a rather irregular but approximately flat maximum that extends over about a third of the dorsal segment and then declines rather steadily to a low value in the transitional zone. Thereafter the width rises rapidly to a maximum a little greater than the preceding one, and then declines sharply as the papilla ends. The width of the fundus follows the same course.

The thickness of the fundus, shown in *b* of this figure, rises rapidly at the dorsal end and then gradually to a broad maximum; thereafter it declines slowly over most of the dorsal segment, and then falls abruptly at the end of this segment. Beyond the minimum in the transitional zone there is a rapid rise to a flat peak in the ventral segment.

The area of the papilla increases rapidly and then more slowly, with some irregularity, to a maximum near the middle of the cochlea, declines somewhat and rises again, and finally falls abruptly to a minimum in the transitional zone. Then in the ventral segment there is a rapid rise to a high peak, and an almost equally rapid fall as the papilla ends.

The distribution of the hair cells along the cochlea is represented in Fig. 17-16. Part *d* shows the rows of hair cells, which begin at 6 and 7, vary up to 10 at one point in the dorsal region, and then show a general decline over the remainder of this segment, reaching zero in the transitional zone. In the ventral segment the number increases rapidly to a peak of 12 rows, and then falls away.

Hair-cell density follows much the same course as the row arrangement, showing a somewhat disturbed maximum near the beginning of the dorsal segment and a general decline thereafter to a sharp fall at the end of this segment. A nearly symmetrical peak appears in the ventral segment.

For both the *Xantusia* and *Lepidophyma* species, the segmentation of the cochlea is clearly evident, especially in the variations of papillar size and in the hair-cell distribution.

For all species studied, the differentiation of the cochlea is only moderate, apart from the demarcation of dorsal and ventral segments.

For *Xantusia henshawi* the variation in width of basilar membrane and fundus over the dorsal segment is about twofold. The variation of fundus thickness is of about the same order. Papillar area varies more widely, but consists essentially of the end tapering of a simple oval structure.

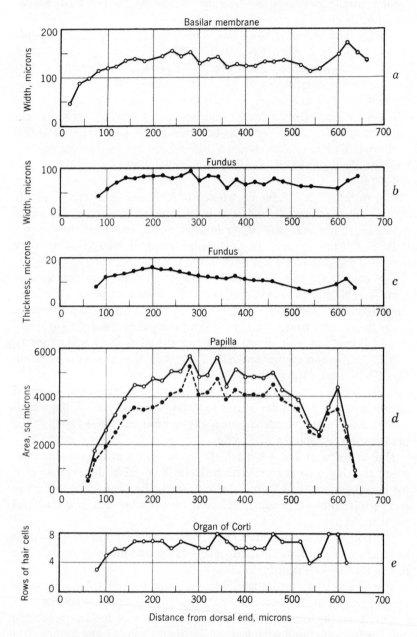

FIG. 17-14. Cochlear dimensions in a specimen of *Xantusia henshawi*.

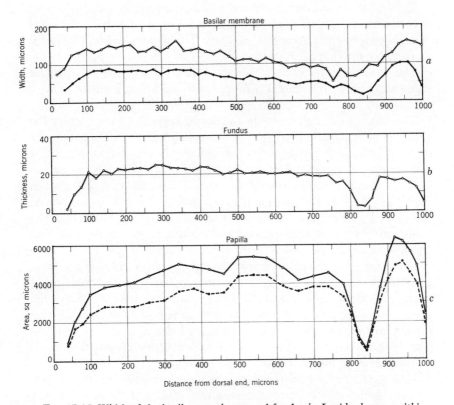

FIG. 17-15. Width of the basilar membrane and fundus in *Lepidophyma smithi*.

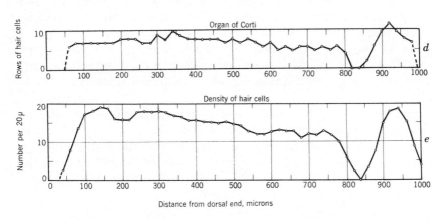

FIG. 17-16. Hair-cell distribution in *Lepidophyma smithi*.

For *Lepidophyma smithi* the width of basilar membrane and fundus over the dorsal segment is less than twofold, and fundus thickness changes only slightly except at the ends. The papillar area varies more markedly, by perhaps 2.5-fold.

The principal differentiation of these ears is between the two cochlear segments, with their different arrangements for hair-cell stimulation.

Ciliary Orientation Patterns.   In the dorsal segment of the cochlea the ciliary pattern is bidirectional, with those hair cells on the medial side of the papilla having ciliary tufts facing laterally (the longer stereocilia and the kinocilium on the lateral side), and with those cells on the lateral side having the tufts facing medially. This condition is shown in Figs. 17-7 to 17-9. The division between the two orientations is approximately in the middle of the papilla; the number of tufts facing one way or the other varies somewhat from one transverse row to the next, but on the average the two orientations approach equality.

In the far ventral regions, where the culmen is present, this situation changes and the orientation becomes strongly biased in the direction of laterality. Indeed, as a rule all the ciliary tufts face laterally except the one on the extreme lateral end of its row, which faces medially. Occasionally this most lateral tuft also is oriented laterally. Somewhat infrequently the two most lateral tufts have a medial orientation. Within the transitional zone the orientation is bidirectional.

Thus the hair cells over the greater part of the cochlea form two populations, and probably respond to contrary phases of the sound waves acting on the ear, whereas the cells in the region of the ventral expansion and served by the culmen mainly constitute a single population and respond at the same phase of the stimulus. This condition of synchronous action should make for high sensitivtiy within some particular region of frequency, determined by the dimensions of this ventral part of the cochlea. The dimensional data presented above suggest that this favored region for the ventral segment is in the low frequencies.

Numbers of Hair Cells.   The results of the counting of hair cells in those specimens prepared histologically are shown in Table 17-I. The hair-cell numbers are shown separately for the dorsal and ventral segments, and also in total. Averages are given for the different species, even though the number of ears is small. The last column indicates the proportion of cells in the ventral segment.

The size of the hair-cell population is smallest in *Xantusia vigilis*, averaging 309.2 for the four ears examined, and of these 19% are in the ventral segment. In *Xantusia henshawi* the total population is a little

TABLE 17-I

HAIR-CELL POPULATIONS IN XANTUSIIDAE

Species	Number of Ears	Dorsal Segment	Ventral Segment	Total	Percent-age Ventral
Xantusia henshawi	6	289.3	44.3	333.7	13
Xantusia riversiana	6	491.0	101.8	592.8	17
Xantusia vigilis	4	251.0	58.2	309.2	19
Lepidophyma flavimaculatum	2	781.0	108.5	889.5	12
Lepidophyma smithi	2	593.5	85.5	679.0	13
Lepidophyma sylvaticum	2	514.0	71.5	585.5	12
Mean of Xantusia species	16	355.4	69.4	424.8	16
Mean of Lepidophyma species	6	629.5	88.5	718.0	12

larger, averaging 333.7, but fewer of these are in the ventral segment. *Xantusia riversiana* has a significantly larger population, averaging 592.8, and also a good proportion, about 17%, in the ventral segment. The 16 xantusiid ears taken together give an average total of 424.8 hair cells, with 16% of these in the ventral segment.

In the *Lepidophyma* species the hair-cell population is generally larger than in any of the *Xantusia* species, though there is some overlapping with *Xantusia riversiana*, and the average of the three species (including six ears) is 718.0, of which 12% are in the ventral segment.

AUDITORY SENSITIVITY

Measurements of sensitivity in terms of the cochlear potentials were made for all species. Results for two specimens of *Xantusia henshawi* are presented in Fig. 17-17. These two ears agree in showing a high degree of sensitivity in the middle frequencies, from 500 to 3000 Hz, where the best points reach −40 to −50 db. The rate of decline in sensitivity is about 25 db per octave for the low tones and about 35 db per octave for the high tones.

Results for *Xantusia vigilis* are shown in Fig. 17-18. These curves exhibit somewhat greater sensitivity than the foregoing, with the best regions reaching −55 to −65 db. Most favored are the medium low frequencies, between 300 and 1000 Hz.

Sensitivity curves for three specimens of *Xantusia riversiana* are shown in Fig. 17-19. These functions agree in the upper frequencies, but differ considerably in the lower frequencies. The best of these ears attains the very high level of −80 db between 400 and 500 Hz, whereas the poorest ear only reaches −33 db at 1000 Hz. The roll-off rate for

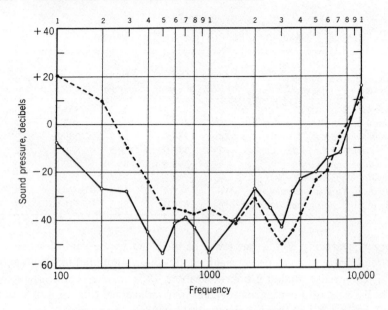

FIG. 17-17. Sensitivity functions in two specimens of *Xantusia henshawi*.

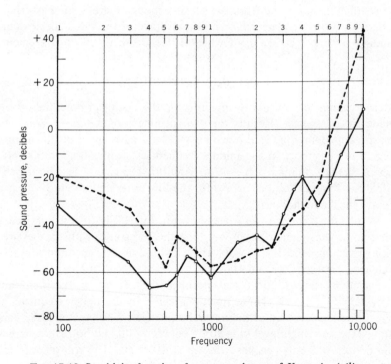

FIG. 17-18. Sensitivity functions for two specimens of *Xantusia vigilis*.

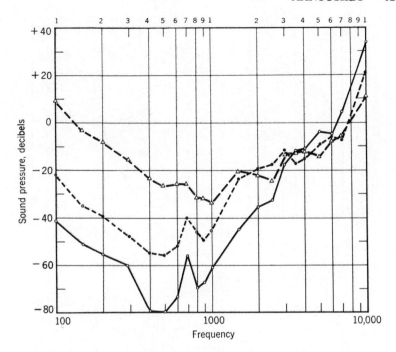

Fig. 17-19. Sensitivity functions for three specimens of *Xantusia riversiana*.

the low tones is moderate, around 10-20 db per octave, and for the high tones is about 30 db per octave at first and then rises to 50 db per octave at the end.

Sensitivity functions for two specimens of *Lepidophyma smithi* are shown in Fig. 17-20. These curves agree in the high frequencies, but differ greatly in the lower range. One specimen attains a sensitivity of −47 db in the region of 1500-2000 Hz and falls off rapidly for lower and higher tones, whereas the other shows unusually high sensitivity over the range 300-3000 Hz, with a peak of −57 db at 1500 Hz.

A function for *Lepidophyma sylvaticum* is shown in Fig. 17-21. This ear attains a maximum sensitivity of −56 db at 700 Hz, and is still excellent in response between 300 and 1500 Hz. The roll-off rate in the high frequencies is about 25 db per octave, and in the low frequencies is 50 db per octave.

In general these xantusiid species exhibit excellent sensitivity within the medium low-frequency range; three species—*Xantusia henshawi, Xantusia vigilis*, and *Lepidophyma smithi*—maintain their sensitivity well into the medium high frequencies as well.

No clear relationship was found between the numbers of hair cells in dorsal or ventral segments and peak sensitivity in these ears.

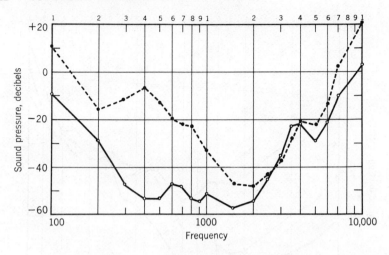

Fig. 17-20. Sensitivity functions for two specimens of *Lepidophyma smithi*.

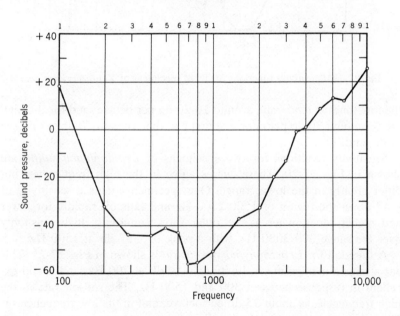

Fig. 17-21. A sensitivity function for a specimen of *Lepidophyma sylvaticum*.

18. FAMILY SCINCIDAE:
THE SKINKS

The skinks form one of the largest lizard families, if not the very largest. Their classification presents many problems, and the number of designated species varies among authorities, with as many as 600, divided among about 60 genera, recognized by some. These animals are widely distributed, and especially numerous in tropical parts of the Eastern Hemisphere, but relatively few are found in North America and Europe. About 17 species occur in the United States and 7 species in Europe.

Most skinks are ground dwellers and occupy areas such as the forest floor with an abundance of surface litter in which to hide. A small number have adapted to desert sand areas, and a few others are semiaquatic, living alongside streams. A good many have adopted burrowing habits, often accompanied by a reduction and occasionally a loss of the legs, together with other changes that make them serpentine in form and behavior. Skinks are shy and retiring, and for this reason are seen more rarely and less often collected for study than their actual numbers would seem to justify.

Available for the present study were 57 specimens belonging to 17 species and representing 8 genera. This is a small sample for so large and widespread a group, but its size reflects the difficulties of collecting these secretive animals. The following specimens were examined by means of cochlear potentials, and with a few exceptions were further investigated anatomically: *Acontias plumbeus* (Bianconi, 1849)—2 specimens; *Chalcides c. striatus* (Cuvier, 1829)—2 specimens; *Chalcides mionecton* (Boettger, 1874)—2 specimens; *Chalcides ocellatus* (Forskal, 1775)—2 specimens; *Eumeces gilberti* (Van Denburgh, 1896)—4 specimens; *Eumeces multivirgatus* (Hallowell, 1857)—2 specimens; *Eumeces obsoletus* (Baird and Girard, 1852)—4 specimens; *Eumeces skiltonianus* (Baird and Girard, 1852)—4 specimens; *Leiolopisma virens* (Peters, 1881)—2 specimens; *Mabuya carinata* (Schneider, 1861) —2 specimens; *Mabuya* cf. *homalocephala* (Wiegmann, 1828)—3 specimens; *Mabuya macularia* (Blyth, 1853)—2 specimens; *Mabuya quinquetaeniata margaritifer* (Peters, 1854)—2 specimens; *Mabuya striata* (Peters, 1844)—6 specimens; *Riopa sundevalli* (A. Smith, 1849)—3 specimens; *Scincella laterale* (Say, 1823)—5 specimens; and *Trachy-dosaurus r. rugosus* (Gray, 1825)—10 specimens.

The *Leiolopisma* specimens presented the unusual feature of green blood plasma, and were probably of the subspecies *anolis* from the Solomon Islands (see Greer and Raizes, 1969). The *Trachydosaurus* specimens represent one of the giant skink species of Australia.

ANATOMICAL OBSERVATIONS

OUTER AND MIDDLE EAR

In most skinks the ear opening is small and often slit-like, with the tympanic membrane deep and well protected. This opening is shown for one of the *Mabuya* species in Fig. 18-1.

FIG. 18-1. A skink, *Mabuya* sp.
Drawing by Anne Cox.

In *Acontias plumbeus*, a burrowing form, the body is elongate, with no trace of legs, and there is no meatal opening or tympanic membrane. The rather limited sensitivity of its ear has been shown in Fig. 6-54.

In all the other skinks studied, the middle ear mechanism is well developed. This mechanism is distinctive, characterized by absence of the internal process of the extracolumella and absence of the extracolumellar muscle. The form of this mechanism was treated in detail in Chapter 6.

INNER EAR

In the skinks the cochlear duct is elongated and the auditory papilla is correspondingly extended. The limbus is only moderately developed, and its medial portion does not present any special elevation as found in geckos, pygopodids, and teiids. As may be seen in Fig. 18-2, which shows a section across the cochlear duct near its dorsal end in *Chalcides c. striatus*, the limbus has only a slight convexity medial to the papilla, accompanied by a moderate thickening of the epithelial layer. Over the middle part of this thickened epithelium lies the vestigial tectorial membrane, at this level showing thickened areas near the edges.

The auditory papilla, presented in further detail in Fig. 18-3, has a thick fundus and is surmounted by a sallet of relatively compact structure. The papilla is suspended between the two edges of the limbus, with thin portions of the basilar membrane along both sides. This position differs from that found in most lizard ears, in which the inner border of the papilla lies close to the edge of the medial limbus and is free to respond to sound stimulation chiefly at its outer edge, producing a rocking motion about the inner attachment. In skinks the suspension of the papilla favors more of an up-and-down movement in response to sound pressures.

Auditory Papilla.   The form of the sensory structures varies along the cochlea, and we may recognize dorsal and ventral segments, with an intermediate zone between.

The dorsal segment is the most extensive, including about three-fourths of the length of the cochlea. It is distinguished by the presence of sallets as shown in cross section in Figs. 18-2 and 18-3, which lie along this segment over the hair cells in a closely connected series, much like the beads on a string. In the skinks these elements are more tightly joined than in most other species containing these structures, such as the geckos. In *Trachydosaurus r. rugosus* the interconnections of adjoining sallets are particularly close, and a careful examination, best made in longitudinal sections, is necessary to reveal the segmental character of the series.

The sallets most commonly consist of a number of tear-drop or wedge shapes in a more or less fused mass, much as described earlier for xantusiids. This structure is indicated for the species *Chalcides c. striatus* in Fig. 18-3.

The interval between dorsal and ventral segments varies greatly among species, and often is extremely short. It is marked by the disappearance of the chain of sallets and the presence of hair cells whose

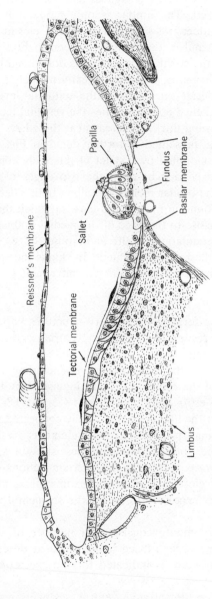

Fɪɢ. 18-2. The inner ear of *Chalcides c. striatus* as seen near the dorsal end of the cochlea. Scale 215×. From Wever, 1970b.

ciliary tufts extend freely without any restraining mechanism. This zone usually is further indicated by a slight constriction of the basilar membrane and fundus.

At the far end of the cochlea there is usually an expansion of the basilar membrane and other structures, which marks the appearance of the ventral segment. The outstanding feature is the presence of the culmen, which can be regarded as a single sallet of great size that lies over the hair cells of this region and makes connection with all the ciliary tufts. This structure and its relations within the otic capsule are shown in Fig. 18-4 for a specimen of *Mabuya striata*. As described earlier in the discussion of the xantusiids and gerrhosaurids, and represented in detail in Fig. 18-5, the culmen consists of two distinct portions, a dense base that lies close over the ends of the hair cells and a light frothy upper portion that extends freely into the cochlear duct. Spike-like projections on the lower surface of the base portion make most of the connections to the ciliary tufts of the hair cells.

As the papillar structures are followed from dorsal to ventral ends, certain other variations are commonly found, though with differences among species. At the dorsal end, as shown in Fig. 18-2, the papilla has an asymmetrical form, with its medial portion relatively tall and raised by the thicker edge of the wedge-shaped fundus. In this region the hair cells have a radiate arrangement, with their ciliated ends converging in a deep trough at the papillar surface, overlaid by a relatively small sallet of nodular form.

A little farther ventrally this radiate arrangement of the hair cells grows less and disappears, the cavity decreases, and the upper surface becomes nearly flat. Thus the papilla as a whole becomes more symmetrical. This form has been shown in Fig. 18-3 for *Chalcides c. striatus* and is displayed for *Mabuya striata* in Fig. 18-6.

The sallets at this level show a regular shape, that of a rounded peak usually with flared edges. The tectorial membrane remains as a thin sheet over the medial limbus, and is non-functional.

Toward the end of the dorsal segment the papilla in many species maintains its form with little change apart from a moderate reduction in size, and then almost abruptly, in a very short transition zone, it goes over into the ventral segment. In other species the transition zone is longer and a number of changes appear, usually beginning within the ventral portion of the dorsal segment. These changes were studied in detail in specimens of *Chalcides* and *Mabuya*, and will now be described.

Near the end of the dorsal segment there is a sudden broadening of the papilla, evidently as an extension of its medial side, and in this extension there are supporting cells and their pillar processes, but few or no hair cells. The hair cells, except for an occasional one or two, re-

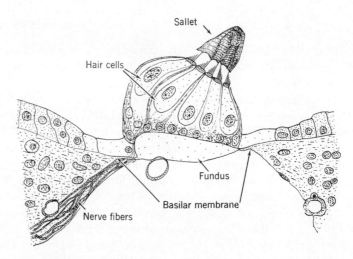

FIG. 18-3. The auditory papilla of *Chalcides c. striatus* at a level near the middle of the dorsal segment. From Wever, 1970b.

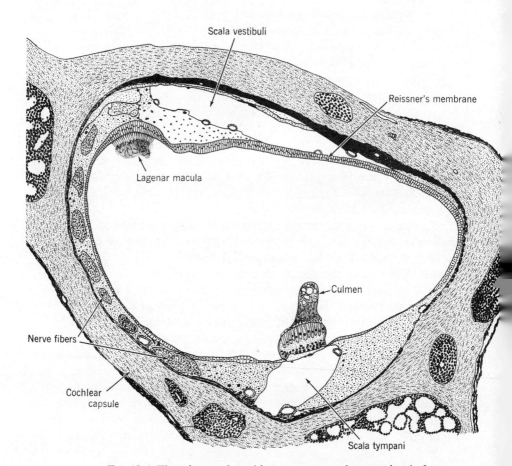

FIG. 18-4. The otic capsule and its contents near the ventral end of the cochlea in *Mabuya striata*. Scale 100×. From Wever, 1971c.

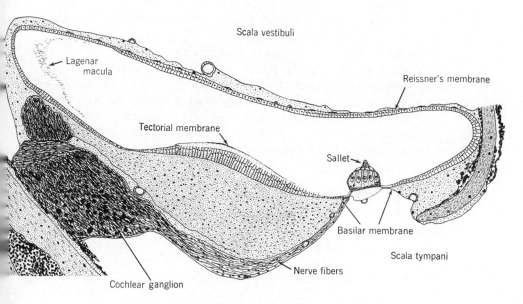

FIG. 18-5. The auditory papilla and culmen in *Chalcides c. striatus*.
From Wever, 1970b.

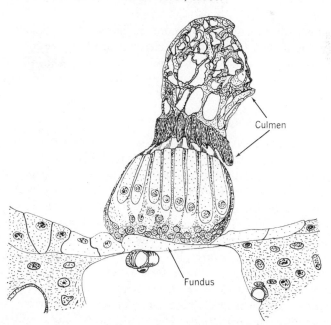

FIG. 18-6. A frontal section through the middle of the cochlear duct in
Mabuya striata. Scale 100×. From Wever, 1971c.

main crowded together on the lateral side of the papilla, or even seem displaced toward this side, with the sallet above them on this side as Fig. 18-7 shows. Often the sallet develops a considerable tilt in a lateral direction as seen in this figure.

With these alterations of papillar structure there is also a change in the internal makeup of the sallets. Instead of the usually compact form, these bodies appear as a somewhat orderly tangle of strands and ribbons. These elements have a generally vertical orientation, with their lower ends mostly connected to the ciliary tufts and their upper ends expanded somewhat and curled around in arcuate fashion. These elements are fused at various places, most often near their upper ends so that the structure maintains a particular form. This form has a certain degree of consistency along the papilla, and is held despite accidental disruptive forces to which the tissues are sometimes exposed. Sallets that have been torn away from the papilla will maintain their characteristic form.

In some species, as in *Eumeces obsoletus* described earlier (Wever, 1967b), the sallets in the ventral part of the dorsal segment may be even more loosely structured than those just indicated, and consist of an array of nearly separate vertical strands, as shown in Fig. 18-8. The elements in this complex appear to be attached below to a ciliary tuft and extend somewhat irregularly upward with tendril-like twistings of their upper ends.

The culmen appears first on the more medial side of the papilla, sometimes showing its first traces at the same level at which the more lateral hair cells with their free-standing cilia are just ending. Thus this ventral restraining mechanism seems in a sense to be added on to the dorsal sallet system.

COCHLEAR DIMENSIONS

Measurements of the cochlear structures were made in one specimen each of four species of skinks, representing as many genera.

Chalcides c. striatus. — Results for a specimen of *Chalcides c. striatus* are presented in Fig. 18-9. Part *a* of this figure shows the width of the basilar membrane along its extent of 930 μ. This width rises to 95 μ at the extreme dorsal end of the cochlea, with a little wavering maintains a magnitude between 80 and 98 μ over most of its course, then declines to a minimum of 74 μ in a region 700-740 μ from the dorsal end. Finally this width increases to a ventral maximum of 110 μ. The degree of uniformity of this structure over its whole middle course is remarkable.

The fundus, shown in *b* of this figure, follows a similar course, but with somewhat greater variations. The width varies between 60 and 76 μ over its dorsal half, falls to about 42 μ toward the ventral region, and

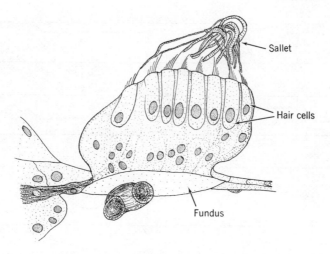

FIG. 18-7. The auditory papilla in *Mabuya quinquetaeniata margaritifer* near the ventral end of the dorsal segment. Scale 500×.

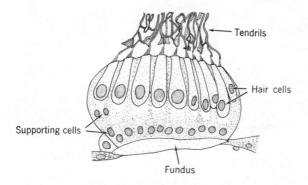

FIG. 18-8. The auditory papilla of *Chalcides mionecton* near the ventral end of the dorsal segment.

finally rises to a new maximum at the ventral expansion. The differentiation of this thickened portion of the basilar membrane, though a little greater than that of the entire membrane, is still very limited.

The thickness of the fundus, shown in *c* of the figure, rises rapidly to a maximum of 22 µ in the early dorsal region, and thereafter declines rapidly until it levels off and even rises slightly near the ventral end. This structure is comparatively well differentiated, varying about fourfold over its course.

The size of the papilla is represented in *d* of this figure. The area of the whole structure (solid line) increases to a maximum at a point about

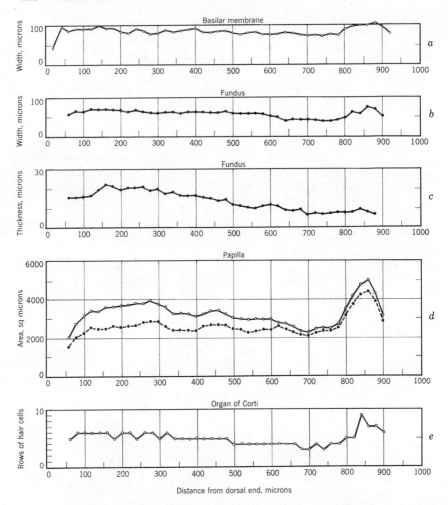

Fig. 18-9. Cochlear dimensions in *Chalcides c. striatus.*

one-fourth of the way along the cochlea, then declines in a wavering manner to a minimum at the 700 μ point, after which it displays a prominent peak in the ventral area. The cellular part of the structure (broken line) follows the same course, but with smaller variations. The ratio between the minimum and the dorsal maximum is about 1.7, and between minimum and ventral maximum is 2.3.

The number of longitudinal rows of hair cells varies along the cochlea in a manner indicated in *e* of this figure. This number rises quickly to 6 at the dorsal end, varies between 6 and 5 until the middle of the cochlea is reached, then drops to 4 for a distance and even to 3 in the region before the ventral expansion. In this expansion the number rises to a peak of 9, and then subsides as the papilla ends.

Altogether this ear presents a picture of considerable uniformity, except for its ventral expansion.

Eumeces skiltonianus. — Dimensional observations on a specimen of *Eumeces skiltonianus* gave results shown in Fig. 18-10. Though this papilla is somewhat shorter than the one just examined, it presents a closely similar form. The width of the basilar membrane increases rapidly to a maximum of 122 μ near the dorsal end, then narrows gradually to a broad minimum of about 95 μ in the region between 460 and 720 μ from the dorsal end. Thereafter is a moderate rise to a maximum of 121 μ, after which the membrane ends.

The fundus is nearly uniform in width at first, varying only from 74 to 60 μ until the midpoint is passed, after which it declines to a minimum of 29 μ at the 680 μ point. The width then increases to a maximum of 71.5 μ and falls rapidly. The thickness of the fundus follows a somewhat similar course, with a value of 12-15 μ in the dorsal region and falling to 4.5 μ at the 720 μ point. Thereafter this thickness increases to a maximum of 10.5 μ and then falls away at the end.

The area of the papilla rises to a maximum in the dorsal region, declines at first slowly and then rapidly, with many irregularities, to a minimum at the 680 μ point, and finally the elevation characterizing the ventral region appears. In this region a sharp peak is reached, and then the decline is abrupt. The areal ratio between the dorsal maximum of 3460 sq μ and the minimum of 2085 sq μ is 5/3, and between the ventral peak of 4040 sq μ and the minimum is nearly twofold.

The number of rows of hair cells increases quickly at the dorsal end of the cochlea to 7, stays around 6, with a few variations, until the middle of the cochlea is passed, and then falls to 4 where it remains until the ventral region is reached. In the ventral region this number rises to 9 and promptly falls as the papilla ends.

Mabuya striata. — This species has a relatively long papilla, which in the specimen studied for Figs. 18-11 and 18-12 measured 1260 μ. At the dorsal end the width of the basilar membrane rapidly rises to a maximum of 161 μ and then promptly falls, at first rapidly and then more slowly, to a minimum of about 100 μ in the early part of the dorsal segment. There is a gradual increase to a second broad maximum near the middle of the cochlea, and thereafter a general decline to a value of 100 μ once more around the 900-1000 μ region. From here the course is upward to a maximum of 163 μ, followed by an almost precipitate decline to the end. The ventral expansion occupies a portion 240 μ long, or about a fifth of the length of the papilla. The variation in width from the maximum near the dorsal end to the minimum points around 300 and 900 μ is only 1.6-fold.

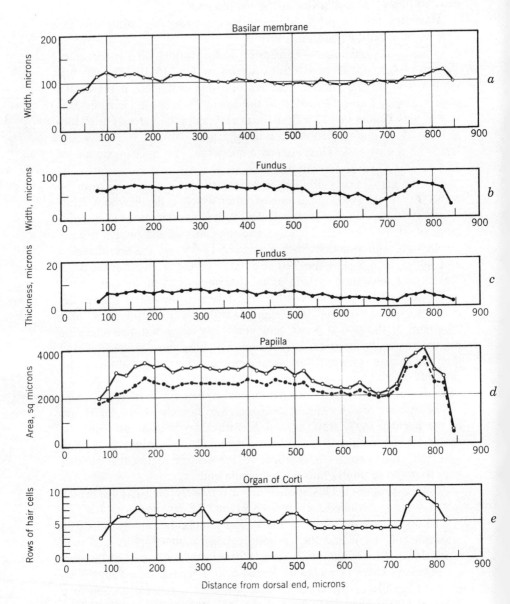

FIG. 18-10. Cochlear dimensions in *Eumeces skiltonianus*. From Wever, 1970b.

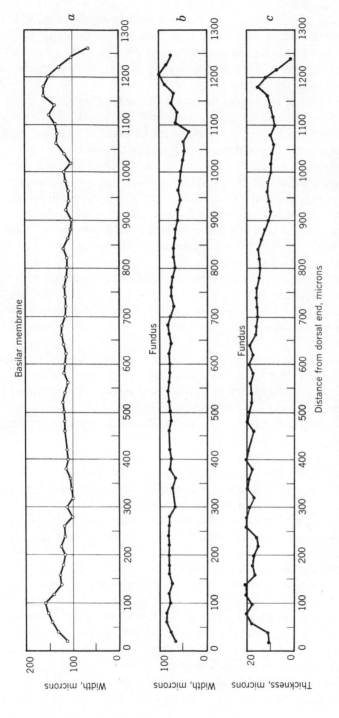

FIG. 18-11. Cochlear dimensions in *Mabuya striata*; basilar membrane and fundus. From Wever, 1970b.

The width of the fundus, shown in *b* of this figure, is remarkably constant around 80 μ over more than half the length of the cochlea, and then shows a decline to a minimum of 38 μ near the beginning of the ventral expansion. Then it rises sharply to a maximum, and as rapidly falls away at the end.

The thickness of the fundus has much the same course as its width: it varies between 18 and 21 μ for about half the length of the papilla, then declines progressively to a minimum of about 8 μ; at the ventral end it rises sharply to a maximum and quickly terminates.

The upper part of Fig. 18-12 represents the cross-sectional area of the auditory papilla, with the solid line showing the whole structure and the broken line showing the cellular portion only. The difference between these two curves represents the area of the fundus. These curves show some irregularities along their courses, but in general are similar. The structure rises rapidly to a maximum at the dorsal end, falls away in undulating fashion to a minimum in a region about 300 μ from the dorsal end, rises irregularly through the middle of the cochlea, then falls to another minimum at the end of the dorsal segment. In the ventral region the whole structure increases rapidly from a minimum of 4600 sq μ to a maximum of 9700 sq μ, and then falls precipitously as the papilla ends.

The lower graph of this figure shows the number of longitudinal rows of hair cells and their variations along the cochlea. This number begins at 3, rises rapidly to 8, then falls to 6 where it remains for a considerable distance, well beyond the middle of the cochlea. After some wavering, the number falls to 5, dips at one point to 4, in the ventral region climbs to 11, and then quickly falls away.

On the whole, this papilla exhibits considerable uniformity in its dorsal half, after only moderate enlargement near the dorsal end. The basilar membrane is remarkably uniform over its whole middle course. The width and thickness of the fundus are uniform over the dorsal half of the cochlea, with declines over the remainder and the usual expansion near the ventral end. Altogether this ear seems only moderately differentiated.

Scincella laterale. — This relatively short papilla (**Fig. 18-13**) presents the same general picture seen in the other skinks. The resemblance to *Eumeces skiltonianus* is particularly close, although the magnitudes of the structures are somewhat different. The width of the basilar membrane increases to a broad maximum of 70 μ in the dorsal region, then declines fairly regularly to a low level in the region 500-650 μ from the dorsal end, after which it rises to a moderate peak.

The width of the fundus takes much the same course as that of the whole basilar membrane. The thickness of the fundus is noticeably

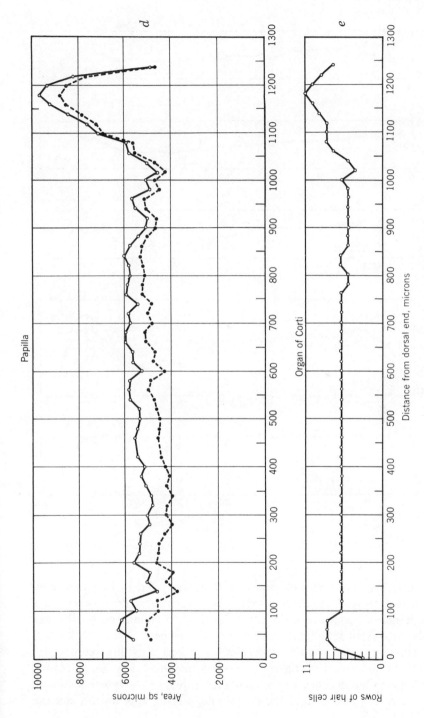

Fig. 18-12. Papillar area and rows of hair cells in *Mabuya striata*. From Wever, 1970b.

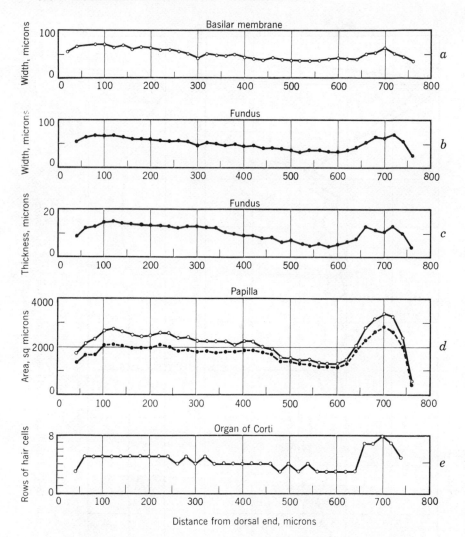

Fig. 18-13. Cochlear dimensions in *Scincella laterale*. From Wever, 1970b.

greater than that seen in *Eumeces skiltonianus*; it rises to a maximum of about 18 μ in the early dorsal region, then falls progressively to a minimum of about 5 μ for a nearly fourfold variation, and finally rises substantially in the middle of the ventral segment.

The size of the papilla is represented in part *d* of this figure. It increases rapidly to a maximum of 2740 sq μ in the early dorsal region, and thereafter declines slowly and then rapidly to a broad minimum in the region 600 μ from the dorsal end. Finally appears the increase to a maximum that marks the ventral region. The cellular portion of the structure follows much the same course.

The rows of hair cells begin at 3, increase abruptly to 5 and remain at this level over the first quarter of the cochlea, then subside with a few waverings to 3 in the intermediate region. At the ventral end there is a rapid rise to 10, after which the papilla ends.

Trachydosaurus r. rugosus. — The papilla in this species is by far the longest in the group, attaining a length of nearly 2 mm, and among all the lizards examined is exceeded only by that of *Gekko smithi* and *Varanus bengalensis nebulosus*. In other respects, however, this ear shows no special features. As indicated in *a* of Fig. 18-14, the width of the basilar membrane rises rapidly to a maximum around 175 μ in the dorsal region, and then declines, with several undulations, all the way to the end of the dorsal segment, where a minimum is reached

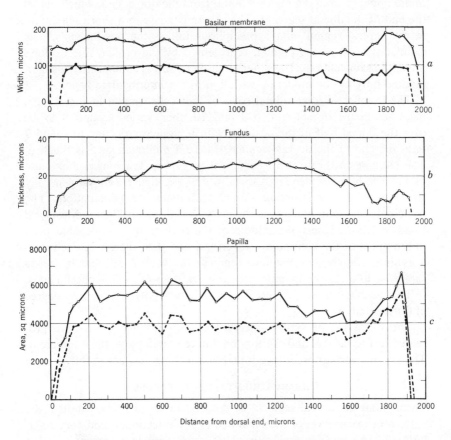

FIG. 18-14. Cochlear dimensions in *Trachydosaurus r. rugosus*. Note the compression of the abscissa scale.

around 125 μ. Then appears the rapid rise to a maximum in the ventral region, which reaches a somewhat higher level than in the dorsal region and maintains it through most of this short segment. The width of the fundus follows almost exactly this same course. The differentiation of width is only slight.

The thickness of the fundus, shown in *b* of this figure, has a very different form. It increases at first rapidly and then more slowly to a broad maximum over the middle of the cochlea, then declines rapidly through the remaining region, with only a moderate upswing within the ventral segment. This structure is well differentiated along the cochlea, varying about threefold.

The upper curve in *c* of Fig. 18-14 represents the area of the papilla, which rises with great rapidity at the dorsal end and reaches an irregular maximum around 6300 sq μ, after which it gradually decreases. A minimum appears around a region 1600 μ from the dorsal end, which marks the end of the dorsal segment. Then there is a rise to a sharp maximum in the ventral region, and a precipitous fall as the papilla ends. The cellular portion of the papilla, represented by the broken line, follows much the same course. The differentiation of size is comparatively slight, marked only by the moderate variation in the intermediate zone.

The number of longitudinal rows of hair cells, shown in Fig. 18-15, increases at the dorsal end to 9 and 10, then varies around 8 to 10 over the dorsal half of the cochlea, with a moderate decline to 6 and 7 until the end of the dorsal segment, where a minimum of 5 is reached. The increase in the ventral segment then occurs, reaching a maximum of 11. The hair-cell density, shown in *e* of this figure, follows generally the form of the row structure, though with many local fluctuations. The principal maximum appears in the early dorsal region, then a downward trend is followed near the middle of the cochlea by a noticeable peak in the midst of the continuing waverings, and then there is a decline to a minimum in the intermediate zone. In the ventral segment is seen a rapid rise to a maximum and finally a precipitous fall as the papilla ends.

This cochlea, though of great length, shows only a moderate degree of differentiation.

The forms of the basilar membranes of six skink species are presented in Fig. 18-16, and the limited differentiation in width is apparent. The species variations in length and width, however, are striking.

CILIARY ORIENTATION PATTERNS

Throughout the dorsal segment of the papilla, in all skinks studied, the pattern of orientation of the ciliary tufts is bidirectional, and very regularly so. The most lateral cells of an array seen in transverse section have medially directed kinocilia, and the most medial cells have laterally directed ones, with the dividing line usually in the middle of the array.

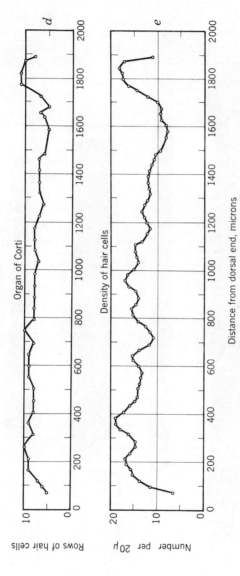

Fig. 18-15. Hair-cell distribution in *Trachydosaurus r. rugosus.*

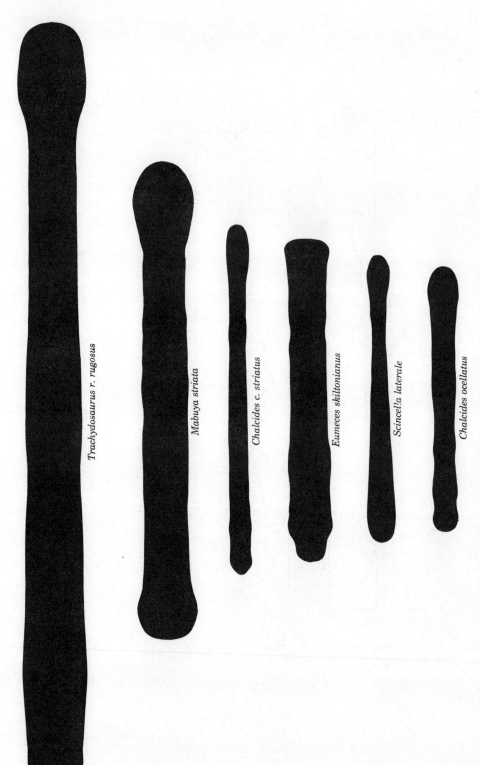

Trachydosaurus r. rugosus

Mabuya striata

Chalcides c. striatus

Eumeces skiltonianus

Scincella laterale

Chalcides ocellatus

Fig. 18-16. Forms of the basilar membrane in six species of skinks. Scale 250×.

From Wever 1970b.

In the ventral segment, where the ciliary contacts are made by the culmen, this pattern is bidirectional also, but is irregular. Rows were occasionally found with complete lateral or with dominantly medial orientation, but the usual condition is one in which reversals of direction occur two to four times along a line of 8-10 cells, ordinarily with the terminal cells of the row having the usual contrary directions, but not always.

The ciliary pattern, though irregular, is by no means random; the reversals along the row of cells are not nearly as numerous as chance would produce. Often the distribution is best described as in one direction with two or three tufts reversed, or as a patchy arrangement in which three or four adjacent cells have their tufts facing in the same direction, and adjoining groups show the reverse direction.

Miller (1974) examined the auditory papilla in four species of skinks by use of the scanning electron microscope, and described the orientation pattern in much the same way as indicated here. He noted that the tips of the kinocilia in *Mabuya* are expanded to a spheroidal form in the region beneath the culmen, but are of a different form along the dorsal segment.

Numbers of Hair Cells

The skink cochleas examined in this study differ greatly in the size of the hair-cell population, the number varying with species from less than 400 in *Eumeces multivirgatus* and *Scincella laterale* to as many as 1400 in specimens of *Trachydosaurus r. rugosus*. The average numbers are given for all species in Table 18-I with particulars for the three segments of the papilla. Also shown are the ranges for the different species groups.

SENSITIVITY MEASUREMENTS

The observations on auditory sensitivity in terms of cochlear potentials will be presented in a series of 17 figures. Results for two specimens of *Chalcides c. striatus* are presented in Fig. 18-17. These curves are in reasonably good agreement in the low frequencies and for tones above 6000 Hz, but differ in the region from 1500 to 6000 Hz. For the solid-line curve, the best sensitivity is in the low frequencies, from 400 to 600 Hz, and a broad secondary maximum representing almost the same degree of sensitivity lies between 1500 and 3500 Hz with its best point at 2000 Hz. The best sensitivity reached is rather poor.

Sensitivity curves for two specimens of *Chalcides mionecton* are presented in Fig. 18-18. These curves are in good agreement in indicating best performance in the medium high frequencies, between 1000 and 3500 Hz, with one of the specimens reaching levels around −46 to

TABLE 18-I

HAIR-CELL POPULATIONS IN SCINCIDAE

Species	Number of Ears	Dorsal Segment Range	Mean	Intermediate Zone Range	Mean	Ventral Segment Range	Mean	Whole Cochlea Range	Mean
Chalcides c. striatus	4	336-410	380	6-22	14	62-83	72	425-494	466
Chalcides mionecton	4	317-391	364	0-16	10	68-105	89	422-493	463
Chalcides ocellatus	4	406-560	476	9-12	11	114-153	129	529-706	616
Eumeces gilberti	5	338-457	399	0-7	4	39-59	49	383-516	451
Eumeces multivirgatus	4	284-332	309	0-10	6	36-45	39	327-381	355
Eumeces obsoletus	4	386-468	421	5-13	8	60-90	77	459-587	506
Eumeces skiltonianus	5	347-410	383	0-15	7	31-67	56	410-481	446
Leiolopisma virens	4	456-562	505	9-13	12	24-45	32	495-606	549
Mabuya homalocephala	4	427-578	515	4-16	11	54-98	76	507-675	602
Mabuya macularia	2	901-916	908	14-15	14	128-151	139	1058-1067	1062
Mabuya quinquetaeniata	2	716-724	720	16-19	17	85-88	86	817-831	824
Mabuya striata	9	572-902	814	0-12	6	85-185	146	707-1094	966
Riopa sundevalli	4	372-417	397	0-3	2	85-118	100	460-528	499
Scincella laterale	10	267-328	301	5-31	18	38-69	55	333-408	374
Trachydosaurus r. rugosus	8	876-1256	1105	0-9	5	74-192	153	1101-1447	1265
Average of 15 species			533.1		9.7		86.5		629.6

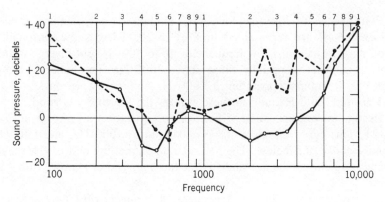

FIG. 18-17. Sensitivity functions for two specimens of *Chalcides c. striatus.*
From Wever, 1970b.

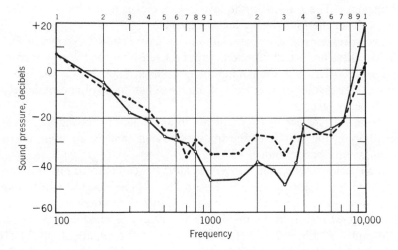

FIG. 18-18. Sensitivity functions for two specimens of *Chalcides mionecton.*

−48 db. The roll-off in the low frequencies is at a uniform rate of 15 db per octave, but reaches a very high rate of about 60 db per octave in the uppermost frequencies.

A specimen of *Chalcides ocellatus,* whose sensitivity curve is presented in Fig. 18-19, shows a broad range of high sensitivity in the span between 300 and 2000 Hz, with the best point at 700 Hz. The sensitivity declines rapidly in the low frequencies at a rate of 50 db per octave, and somewhat irregularly in the high frequencies at a rate averaging a little over 20 db per octave. This species is clearly superior to others of this genus for which sensitivity observations were made.

Four species of *Eumeces* have been studied. Two specimens of *Eumeces gilberti* are represented in Fig. 18-20. These show reasonably good agreement in the low frequencies up to 800 Hz, but diverge in the high frequencies, especially those above 3000 Hz. The better of these ears has a primary maximum at 1000 Hz and secondary ones at 600 and around 3500 Hz, and in these regions the sensitivity is good.

Results for one specimen of *Eumeces multivirgatus* are shown in Fig. 18-21. This curve attains a sharp maximum at 1000 Hz, and falls off rapidly above and below. The roll-off rate at the low end of the range is about 15 db per octave and that at the high end is about 20 db per octave.

Sensitivity curves for two specimens of *Eumeces obsoletus* are presented in Fig. 18-22. These curves are in good agreement except at the maximum and at the high-frequency end. The better curve attains a maximum of −47 db around 600-700 Hz, which is an excellent level of response. The roll-off in the low frequencies is at 30 db per octave, and is about the same in the high frequencies.

Two specimens of *Eumeces skiltonianus* gave results as shown in Fig. 18-23. These are broad functions with best sensitivity in the region from 500 to 2500 Hz, and a maximum of −44 db at 1500 Hz for the better animal. The decline at both ends of the range has a rate of approximately 20 db per octave.

These four species of *Eumeces* are in agreement in having their best sensitivity in the middle of the frequency range, and in attaining maximums a little better than −40 db.

The species *Leiolopisma virens* has already been pointed out as a rare type from the Solomon Islands. Figure 18-24 presents results from one of the two specimens studied. The curve shows only moderate sensitivity with the best region in the middle of the scale, and maximum points of −24 and −23 db at 700 and 1500 Hz. Roll-off rates are about 20 db per octave at the low end of the range and somewhat higher, around 30 db per octave, at the high end.

Five species of *Mabuya* were studied, with sensitivity data obtained on four of them. The species tentatively identified as *Mabuya homolocephala* gave results represented in Fig. 18-25. These two curves show rather sharp maximum regions around 700 Hz, and a secondary maximum at 1500 Hz for one of these ears and at 2000 Hz for the other. The sensitivity for the better ear reaches −48 db, and is at −43 db for the other ear, representing very good acuity. The roll-off rate is about 30 db per octave at both ends of the frequency scale.

Two specimens of *Mabuya macularia* gave the curves of Fig. 18-26. The better of these ears shows a sharp maximum of −76 db at 700 Hz, a rapid fall for the tones immediately above, and then for still higher

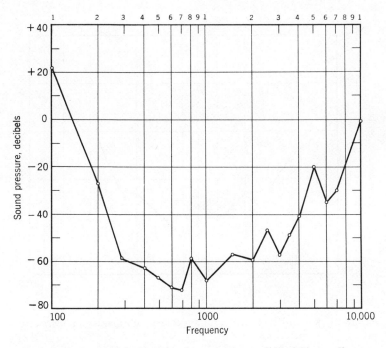

FIG. 18-19. A sensitivity function for a specimen of *Chalcides ocellatus*.
From Wever, 1970b.

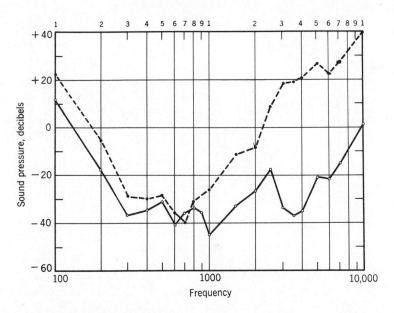

FIG. 18-20. Sensitivity functions for two specimens of *Eumeces gilberti*.

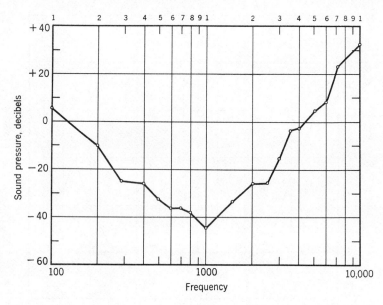

Fig. 18-21. A sensitivity function for a specimen of *Eumeces multivirgatus*.

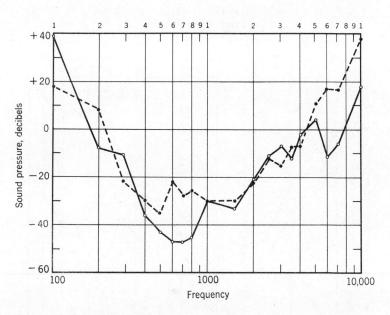

Fig. 18-22. Sensitivity functions for two specimens of *Eumeces obsoletus*.

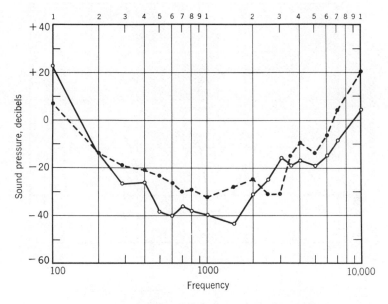

FIG. 18-23. Sensitivity functions for two specimens of
Eumeces skiltonianus. From Wever, 1970b.

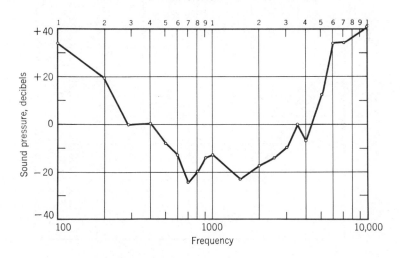

FIG. 18-24. A sensitivity function for a specimen of *Leiolopisma virens*.

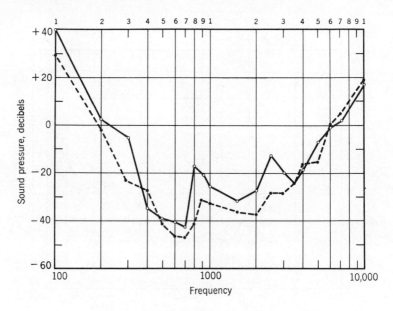

FIG. 18-25. Sensitivity functions for two specimens of
Mabuya cf. *homalocephala*.

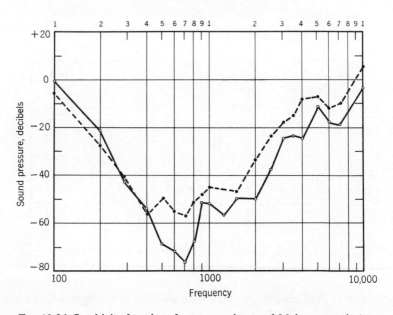

FIG. 18-26. Sensitivity functions for two specimens of *Mabuya macularia*.

tones a return of sensitivity until a further decline appears above 2000 Hz. The other ear follows this one closely except in the maximum region, where the sensitivity only reached −57 db—though this is still a very high level. The roll-off rate is 30 db per octave at the low-frequency end of the range, but is considerably less in the high-frequency region—around 15 db per octave on the average.

A specimen of *Mabuya quinquetaeniata margaritifer* gave the function of Fig. 18-27. This curve is like the two preceding ones in exhibiting a sharp maximum in the medium low frequency region and a secondary maximum in the medium high range. This primary maximum, however, is at a somewhat lower frequency (500 Hz), and the secondary one is at 2000 Hz. The degree of sensitivity shown at the primary maximum is −53 db, which represents excellent performance. The roll-off in the low frequencies averages about 30 db per octave, but in the high frequencies the curve is irregular, averaging about 15 db per octave above 2000 Hz and then increasing to 50 db at the upper end.

The above three species show similar forms of sensitivity functions, with a strong maximum in the medium low tones and a secondary maximum in the region 1½ to 2 octaves above.

Results for three specimens of *Mabuya striata* are shown in Fig. 18-28, and their curves are in good agreement. A broad region of excellent sensitivity is shown between 500 and 800 Hz. The highest level of sensitivity attained is −56 db. Above 2000 Hz the sensitivity declines rapidly, with an interruption around 5000-7000 Hz, where each curve passes through a small secondary maximum and then resumes the upward swing at the high end of the range. The roll-off rate in the low frequencies varies around 20-30 db per octave. This rate at the upper end is variable also, running between 30 and 50 db per octave in the region of 2000-5000 Hz and then returning to 30-40 db per octave after the secondary maximum is passed.

All four *Mabuya* species exhibit two maximum regions, though in *Mabuya striata* these two are much more separated along the scale than in the others. The maximum sensitivity attained by these ears in the low and middle range is outstanding.

Sensitivity curves for two specimens of *Riopa sundevalli* are represented in Fig. 18-29. The two curves have much the same form, with maximums in the medium high frequencies, though for one of them the sensitive region is somewhat broader. For the better curve the maximum point reaches −33 db at 2000 Hz, and for the other it reaches −27 db at 2500 Hz. The roll-off rate is irregular, but approximates 15 db per octave at the low end of the range, and is markedly higher, between 40 and 50 db per octave, at the upper end. This species shows only a moderate degree of sensitivity.

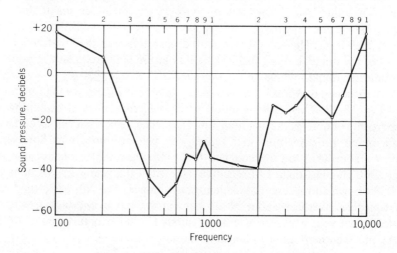

FIG. 18-27. A sensitivity function for a specimen of
Mabuya quinquetaeniata margaritifer.

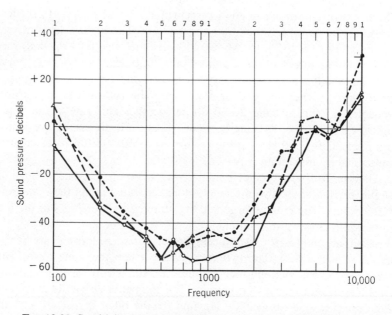

FIG. 18-28. Sensitivity functions for three specimens of *Mabuya striata*.
From Wever, 1970b.

The small skinks of the species *Scincella laterale* gave results that are divergent and difficult to explain. Two types of function were obtained, represented by the curves of Fig. 18-30. One type, shown in the solid-line curve, has its best sensitivity in the high frequencies, with a maximum around 3000 to 4000 Hz. The other type, represented by the broken curve, has its best sensitivity in the low frequencies, reaching a sharp maximum at 700 Hz. In the low frequencies, up to 1500 Hz, this ear shows greater sensitivity than the other, reaching −50 db at 700 Hz, but from 3000 Hz upward the relation is reversed and the other ear is the more sensitive. For both these ears the measurements were made twice, and for each the two sets of observations were in close agreement. Of five animals studied, two gave functions like those shown by the solid line and the three others gave functions like the dashed line. This confusing result needs to be checked further on larger numbers of specimens to determine whether the present observations represent accidental variations of some kind, or whether two subpopulations actually exist within this species. Nothing was found in the histological study of these ears to assist in resolving this question.

Results for a specimen of *Trachydosaurus r. rugosus* are shown in Fig. 18-31. This function displays a sensitive region between 300 and 1000 Hz with a maximum at 500 Hz, where a level of −56 db is reached. A secondary maximum appears in the high frequencies between 3000 and 4000 Hz, with a peak of −34 db at 3500 Hz. The roll-off rate at the low end of the range is 30 db per octave. Above the secondary maximum in the upper frequencies, this roll-off becomes very rapid and reaches 60 db per octave, though it is not sustained beyond 7000 Hz.

Results similar to those just described were obtained in other specimens of *Trachydosaurus*. In all there were two maximum regions as shown here, and in other respects the curves followed the same form. An average function for ten specimens is presented in Fig. 18-32, which also indicates the dispersion in terms of the standard deviation. The form of the average curve is much like that shown for the single animal in the foregoing figure, except that in the averaging process the peaks are rounded off. This rounding is most marked in the region of 200 to 600 Hz, where the primary peaks appear in varying locations on the frequency scale, as clearly indicated by the large values of the standard deviation for these tones. The variability is particularly small at 800 and 1000 Hz, between the primary and secondary peak regions.

A representative of the burrowing skinks, *Acontias plumbeus*, which lacks a tympanic membrane and external ear opening, was described in Chapter 6 and its sensitivity curve was presented in Fig. 6-54. The sensitivity is poor, as expected from the reduced form of sound-conduc-

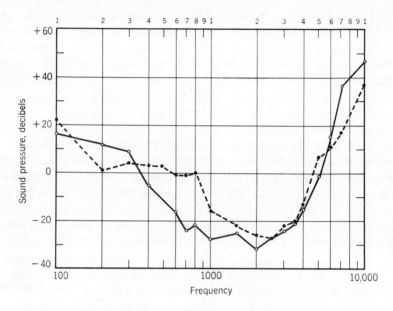

FIG. 18-29. Sensitivity functions for two specimens of *Riopa sundevalli*.

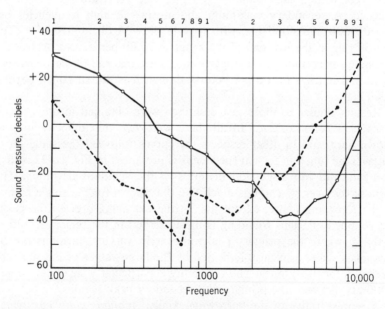

FIG. 18-30. Sensitivity functions for two specimens of
Scincella laterale. From Wever, 1970b.

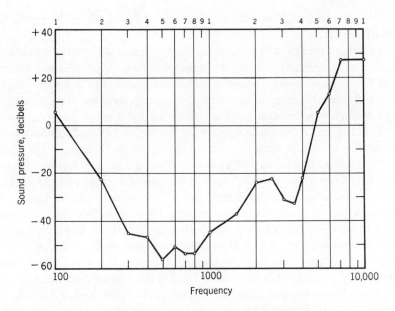

FIG. 18-31. A sensitivity function for a specimen of
Trachydosaurus r. rugosus.

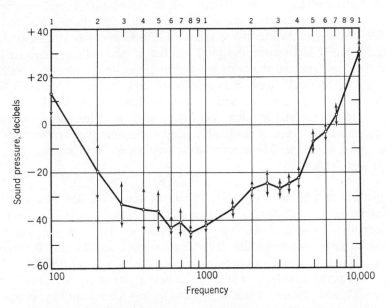

FIG. 18-32. Average results on 10 specimens of *Trachydosaurus r. rugosus.*
The double arrows show the size of the standard deviation.

tive mechanism, and shows a maximum of only -8 db in the region of 400-1000 Hz.

The skink species within the sample investigated show large variations in degree of sensitivity, from *Acontias plumbeus, Chalcides c. striatus,* and *Leiolopisma virens,* which must be rated poor, to *Chalcides ocellatus* and *Mabuya macularia,* which are clearly superior. The *Eumeces* and *Mabuya* species examined are all in the excellent to superior class.

These skinks show variations also in the region of frequency in which sensitivity is best. A majority, including two of the *Eumeces* and three of the *Mabuya* species as well as *Trachydosaurus,* are best in the medium low frequencies, *Riopa sundevalli* and *Chalcides mionecton* favor the high frequencies, and the others fall in the middle of the range. A marked degree of duplexity, represented by a secondary region in the high tones in addition to a primary one in the low or middle frequencies, is seen in more than half of these ears, and an interruption in the course of the curve in the high tones, suggestive of this sort of duplexity, appears in several other functions. In general, and despite the few exceptions noted, these skinks can be characterized as having hearing of better than ordinary quality. In their avoidance of predators and other perils, the auditory sense no doubt plays a significant role.

Hair Cells and Body Size.   There is a strong positive correlation between body size and the size of the hair-cell population. The correlation coefficient obtained for a group of 42 specimens belonging to 14 species was 0.71 between the total number of hair cells as counted in the right cochlea and the body weight, and had nearly the same value (0.70) for hair-cell numbers in relation to body length measured from snout to vent. The close agreement between these two correlations is of course to be expected because of the close relation of length and weight in animals maintained in reasonably good condition; this correlation for the group was 0.90. These correlation coefficients were statistically of high significance.

Hair Cells and Sensitivity.   Determinations were made also of the relations between sensitivity as determined by cochlear potential measurements and the size of the hair-cell population. These determinations were made in two ways, for individual animals and for the different species groups. In each procedure the sensitivity was regarded in two ways, as peak sensitivity, which is the best point along the sensitivity curve taken regardless of frequency, and as low-tone sensitivity, considered as the point just above the region of rapid roll-off at the low-frequency end of the range.

TABLE 18-II

SENSITIVITY AND HAIR-CELL POPULATIONS
IN SCINCIDAE
Correlation coefficients

Individual Specimens (N = 38)	Dorsal Segment	Ventral Segment	Total
Peak sensitivity	0.45 (p = 0.005)	0.51 (p = 0.001)	0.57 (p = 0.0002)
Low-tone sensitivity	0.42 (p = 0.009)	0.45 (p = 0.005)	0.53 (p = 0.0006)
Species groups (N = 15)			
Peak sensitivity	—	0.63 (p = 0.01)	0.54 (p = 0.04)
Low-tone sensitivity	—	0.59 (p = 0.02)	0.57 (p = 0.03)

The calculations for a group of 38 animals treated individually gave the correlation coefficients shown in the upper part of Table 18-II, which range between 0.42 and 0.57. Other determinations in which the animals were first combined in groups according to species gave closely similar results, with correlation coefficients between 0.54 and 0.63. All these coefficients had a high degree of statistical significance as indicated by the p-values.

For both procedures the results for dorsal and ventral segments and for the cochlea as a whole hardly differ from one another, and all indicate a positive relationship of moderate degree.

It appears that a large number of hair cells is an advantage to sensitivity within a limited frequency range in these ears in which the structural differentiation is severely limited. This advantage evidently is derived by a summation of sensory and perhaps also of neural elements.

A sallet and culmen system of hair-cell stimulation is clearly an effective way of providing the necessary restraint of the ciliary tufts of the hair cells in the stimulation process, and these elements at the same time through their attachments contribute to the linkage of these cells. It seems doubtful, however, that this system would permit any great amount of frequency discrimination. The skink ear, with its low degree of structural differentiation, seems primarily adapted to the reception of faint sounds with little regard to frequency specificity, except perhaps for a difference between dorsal and ventral segments of the cochlea.

19. FAMILY CORDYLIDAE:

THE GIRDLE-TAILED LIZARDS

The cordylids constitute a moderately large family of about 50 species and subspecies restricted to Africa, and mainly in its southern part. Though a few species extend as far north as Ethiopia, the great majority are in South Africa, especially in Cape Province and the Transvaal (FitzSimons, 1943; Loveridge, 1944). There are four genera: *Cordylus*, *Pseudocordylus, Platysaurus*, and *Chamaesaura*. Members of the first three genera are strong-bodied, with stout legs, sturdy tails provided with rings of projecting spikes, and a large ear opening. In *Chamaesaura* the legs are short or rudimentary and even absent in front, and the body form is elongated and snake-like. A drawing of one of these species, *Pseudocordylus subviridis*, is shown in Fig. 19-1.

FIG. 19-1. A specimen of *Pseudocordylus subviridis*.
Drawing by Anne Cox.

The present study included 12 species and subspecies, with eight forms of *Cordylus*, one of *Pseudocordylus*, and three of *Platysaurus*. These species and the number of specimens examined were as follows:
Cordylus cataphractus (Boie, 1828)—2 specimens; *Cordylus giganteus* (A. Smith, 1844)—4 specimens; *Cordylus jonesi* (Boulenger, 1891)—2 specimens; *Cordylus polyzonus* (A. Smith, 1838)—5 specimens; *Cordylus tropidosternum* (Cope, 1869)—5 specimens; *Cordylus*

vittifer (Reichenow, 1887)—7 specimens; *Cordylus warreni breyeri* (van Dam, 1921)—4 specimens; *Cordylus warreni depressus* (Fitz-Simons, 1930)—7 specimens; *Pseudocordylus subviridis transvaalensis* (FitzSimons, 1943)—8 specimens; *Platysaurus guttatus minor* (Fitz-Simons, 1930)—3 specimens; *Platysaurus guttatus rhodesianus* (Fitz-Simons, 1943)—2 specimens; and *Platysaurus minor* (FitzSimons, 1930)—6 specimens. All the animals were used for cochlear potential measurements, and most were prepared histologically for the study of details of ear structure.

ANATOMICAL OBSERVATIONS

OUTER AND MIDDLE EAR

In all the cordylid species studied, the external ear opening is relatively large, of oval form, and leads to a tympanic membrane that is deeply situated and thus well protected.

The middle ear is of the iguanid type without special features.

INNER EAR

In cordylids as in skinks the basilar papilla is elongated, slightly curved, and doubly tapered. It attains a maximum size in the dorsal region, tapers down to a small diameter toward the ventral end, and then suddenly expands to a large club-like termination.

Also as in skinks the tectorial tissue that in many other lizards arises from the medial limbus as a tectorial membrane to extend over the auditory papilla is here only vestigial in character. It runs along the limbic surface in this region as a variable and largely unstructured mass with no relations to the auditory hair cells.

This tectorial tissue more anteriorly, in the lagenar portion of the cochlear duct, appears more nearly normal. It forms a membrane over the epithelial surface of the anterior part of the limbus and appears to take its usual part in the formation of the lagenar macula.

The modification of the cochlear portion of the tectorial tissue is more extreme in the cordylids than was observed in the skinks, though functionally the result is the same: the ciliary tufts of the cochlear hair cells lack the direct mechanical restraint commonly provided by a tectorial membrane, and alternative mechanisms are employed.

In the cordylids, as in the skinks, the cochlea contains a long series of sallets. These are small nodules of tectorial tissue forming a closely articulated band through the dorsal and middle portions of the cochlea, to which the tips of the ciliary tufts of the hair cells are attached. In the ventral region of the cochlea this string of sallets comes to a end, and soon thereafter a single large body, the culmen, makes its appearance, covering all the hair cells in this region.

It is convenient, therefore, to distinguish three portions of the auditory papilla: a dorsal segment covered by the line of sallets, a ventral segment in which the culmen appears, and between these two a short intermediate zone. In the intermediate zone the sallets are absent, and the auditory papilla is beginning to increase in size in anticipation of the great enlargement at the ventral end. Though the cordylid cochlea lacks a tectorial membrane, the sallets and culmen provide a form of restraint for the cilia of the hair cells through the operation of inertia and inertia-like forces.

Auditory Papilla.   The position of the auditory papilla within the cochlear duct, and relations to adjoining structures, are shown in Fig. 19-2 for a specimen of *Cordylus warreni depressus*. This drawing is based upon several successive sections that were oriented to pass through the head nearly in a sagittal plane, cutting through the auditory papilla longitudinally. The cochlea and lagena occupy the ventral part of the otic cavity, with the saccule making up the larger anterodorsal part. The utricular macula is just out of the picture in an anterior direction. Only a small portion of the scala vestibuli is shown here; this chamber extends widely in a lateral direction, where its wall is pierced by the oval window containing the footplate of the columella.

The cochlear duct lies between the scala vestibuli and scala tympani. A large round window niche is separated from the scala tympani by the round window membrane as shown.

The auditory papilla is seen as slightly curved, with its dorsal end above and to the left and its ventral end, surmounted by the prominent culmen, below and to the right. Miller (1966a) represented a similar curvature for the species *Cordylus jonesi*.

The papilla is suspended at its two ends between portions of limbic tissue, but it must be borne in mind that other limbic tissue holds the edges of the basilar membrane, which extend toward and away from the plane of the drawing. In fact, the limbic plate that supports the auditory papilla has only a narrow slit-like opening that separates it into lateral and medial portions. The lagenar macula extends over the surface of other limbic tissue covering the anteroventral wall of the otic capsule.

The large limbic mass above and to the right of the papilla as seen in Fig. 19-2, usually referred to as the medial limbus, shows on its surface a dark stripe that represents the rudiments of the tectorial membrane.

Another view of the auditory papilla and its immediate surroundings is given in Fig. 19-3, in a drawing based upon a reconstruction of the

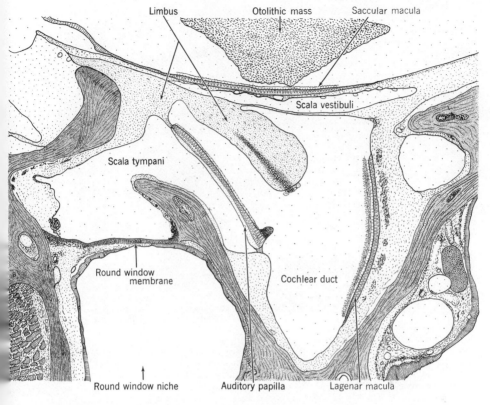

FIG. 19-2. Lateral view of the inner ear of *Cordylus warreni depressus*.
Dorsal is above, anterior to the right. Scale 40×.

right ear of a specimen of *Cordylus vittifer*. As pictured, the papilla has
its dorsal end on the left, increases to a maximum diameter in this dorsal region, then grows progressively smaller, and reaches a minimum as
the ventral region is approached. At the ventral end is a great expansion
to a club-like form.

A line of sallets runs along the greater part of the papilla, ending in
the region of the ventral expansion. A few root-like processes extend the
line of sallets a little farther into the intermediate zone. At the ventral
end the culmen covers the whole body of hair cells there. The hair cells
are not shown in this drawing except for a single row at the dorsal end.
The dark band along the top of the picture represents tectorial tissue
on the surface of the medial limbus.

A distinctive feature is the middle location of the auditory papilla
over the basilar membrane (represented by cross hatching). In most
lizards the papilla is asymmetrically placed, lying close along the edge

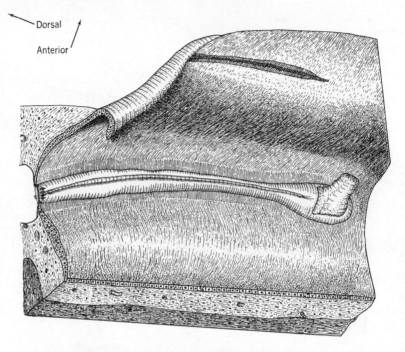

FIG. 19-3. Reconstruction drawing of the auditory papilla of *Cordylus vittifer*. The right ear is shown from lateral and slightly anterior directions, and rotated about 65° counterclockwise. Scale 85×. From Wever, 1970a.

of the medial limbus, and a wide, flexible band of membrane is found along the lateral border of this structure. In the cordylids, as was found for the skinks, the papilla is flexibly sustained on both sides.

This manner of suspension of the basilar membrane, along with other features of the auditory papilla, is more clearly shown in Fig. 19-4, which represents a transverse section across the mid-dorsal region of the cochlea of *Cordylus vittifer*. Here the auditory papilla is nearly round in cross section, and its sallet has a low bell-like form. Between the rows of hair cells are prominent pillars that arise from the supporting cells below.

The epithelial cells are low over both parts of the limbus, and especially over the medial part close to the papilla. The small border cells are shown, one on the medial side of the papilla and two on the lateral side, covering the thin edges of the basilar membrane.

A view of the auditory papilla in the intermediate zone is given in Fig. 19-5. In this zone the line of sallets is ended, but for a short distance their restraining effects are extended by thin root-like processes

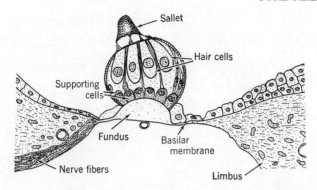

FIG. 19-4. The auditory papilla of *Cordylus vittifer* in a frontal section, taken from the mid-dorsal region. Scale 340×. From Wever, 1970a.

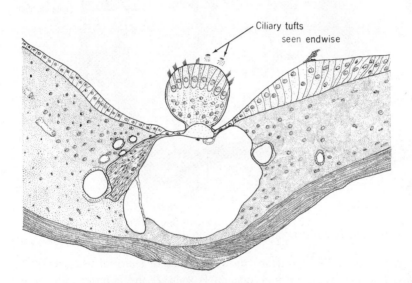

FIG. 19-5. The auditory papilla of *Cordylus vittifer* in a frontal section through the intermediate zone. Scale 200×.

(shown in Fig. 19-3) to which a few of the hair cells make their ciliary connections. In the section shown, two of the ciliary tufts were strongly bent over in a dorsal direction to connect to the root processes.

Figure 19-6 gives a cross-sectional view of the cochlea of a specimen of *Cordylus cataphractus* showing the auditory papilla in the region of its ventral expansion. The relations to other structures of the otic capsule are indicated also. Note the thick portions of the limbus firmly based on the bony walls of the otic capsule. In this region the epithelial

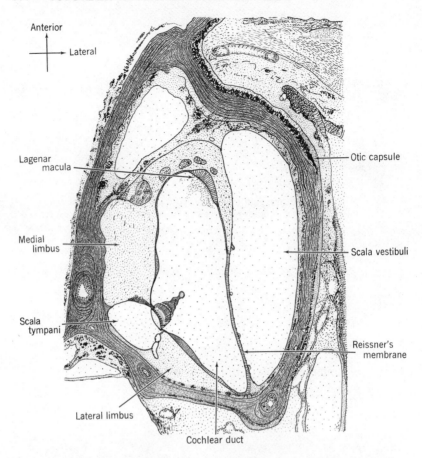

FIG. 19-6. The otic capsule and its contents in *Cordylus cataphractus*, in a frontal section through the ventral region of the cochlea.

layer over the limbic surface shows no trace of tectorial tisue. The ventral portion of the lagenar macula is seen at the anterior side of the cochlear duct.

A more detailed view of the auditory papilla in the ventral region is presented in Fig. 19-7 for a specimen of *Cordylus vittifer*. The papilla is large, with 10 longitudinal rows of hair cells. The tectorial structure is made up of two portions as noted earlier: a base portion of considerable density and a cap portion of light, sponge-like construction with numerous vacuities both large and small. Projections on the base extend to the ciliary tufts of the hair cells at this ventral end of the cochlea.

Tectorial Tissues.   In a few specimens the tectorial tissues were examined in considerable detail. In a specimen of *Cordylus vittifer*, as shown

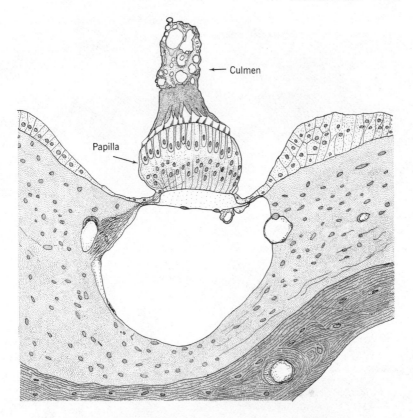

FIG. 19-7. The culmen of *Cordylus vittifer*, seen in a frontal section
at its greatest height. Scale 250×.

in *a* of Fig. 19-8, a tectorial mass was first seen within the medial sulcus
near the dorsal end of the cochlea, evidently representing the lagenar
portion of this tissue. This mass lies over the surface of the epithelial
covering of the limbus at this anterior end of the cochlear duct. More ventrally, as seen in *b* of this figure, the mass becomes reduced in size and
moves posteriorly, seemingly following the region of most pronounced
thickening of the epithelial layer. Still farther ventrally this mass continues to diminish as shown in *c* of this figure, and in the region of the ventral expansion it disappears altogether. An exceedingly thin membrane
evidently remains along the epithelial surface over the anterior portion of
the medial limbus, though only occasionally can it be seen with certainty.
In favorable preparations, in a region toward the ventral end of the cochlear duct, this membrane may be seen to lift off the limbic surface and
extend anterolaterally to the end of the cochlear duct where it expands
as the tectorial reticulum of the lagenar macula.

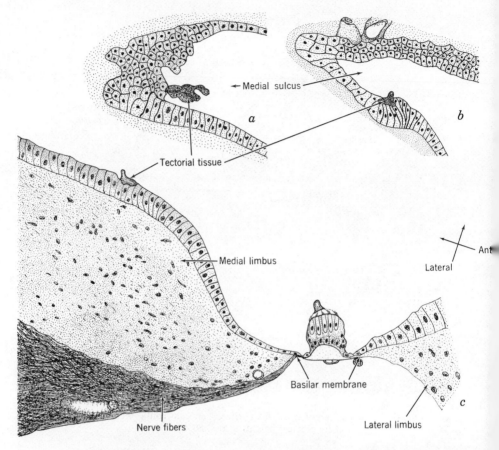

FIG. 19-8. Three views of the tectorial tissues along the cochlea. *a* and *b*, this tissue at the medial sulcus of the cochlear duct; *c*, this tissue farther ventrally, shown in relation to the limbus and auditory papilla.

Most cordylid species show forms of tectorial tissue much like those just described, but a few exhibit some variations. In *Cordylus cataphractus* small amounts of tectorial tissue are present at the dorsal end of the cochlea in two places. One mass occurs in the usual location at the extreme anterior end (no doubt representing the lagenar portion), and another appears near the middle of the cochlea on the moderately thickened epithelial covering of the limbus there (probably representing the cochlear portion). More ventrally these two masses of tissue enlarge, approach one another, and finally join. Part *a* of Fig. 19-9 shows these two masses at a point just dorsal to their fusion.

A little ventral to the junction just referred to, this specimen showed an unusual feature: in a narrow region the tectorial tissue formed a

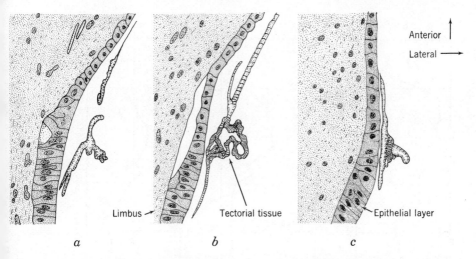

FIG. 19-9. Tectorial tissues in a specimen of *Cordylus cataphractus*, shown at three successive levels from dorsal to ventral. Scale 500×.

ribbon of curiously convoluted shape, as represented in *b* of this figure. It appears that in this instance the tissue manifests its capabilities of membrane production, though in an abortive form. Elsewhere this tissue showed essentially a lack of structure as indicated in part *c*.

When viewed at low and medium magnifications the line of sallets appears as a continuous band. At high powers, however, a segmental character is revealed, represented by variations in the density and courses of fibers within this structure. These iterative formations are the sallets, and their number seems to correspond closely to the transverse rows of hair cells. This sallet structure was considered in detail in Chapter 4.

In a specimen of *Cordylus warreni depressus* sectioned longitudinally, the spacing of the sallets was measured at several points along the papilla, and varied from 7 to 19 μ, with the larger spacings in the dorsal region.

DIMENSIONAL MEASUREMENTS

Observations on the dimensions of the cochlear structures were made for two *Cordylus* species and for one each of the other two genera studied.

Cordylus vittifer. — Results for a specimen of *Cordylus vittifer* are shown in Figs. 19-10 and 19-11. Part *a* of Fig. 19-10 represents the varying widths of the basilar membrane along its extent of 1120 μ. This membrane increases rapidly in width at the dorsal end to reach a maxi-

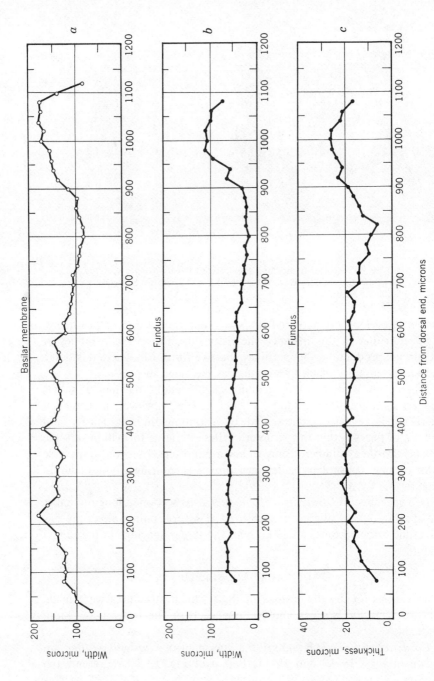

Fig. 19-10. Cochlear dimensions in a specimen of *Cordylus vittifer*; basilar membrane and fundus. From Wever, 1970a.

Distance from dorsal end, microns

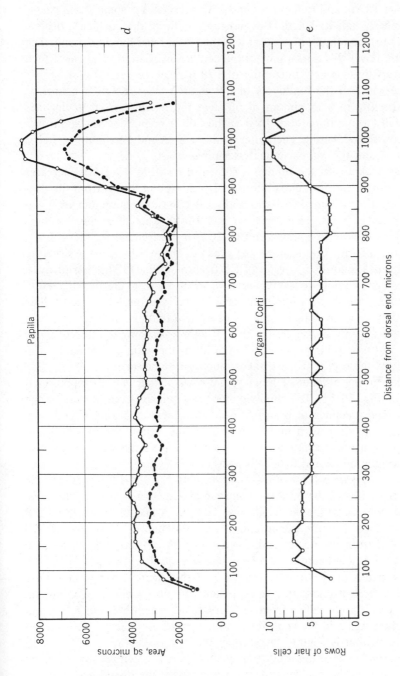

Fig. 19-11. Size of auditory papilla and hair-cell distribution in *Cordylus vittifer*. From Wever, 1970a.

mum of 183 μ, and remains around 150 μ, except for some sharp peaks, over nearly half its length. Beyond the middle of the cochlea a decline occurs to a minimum of 83 μ at a point 800 μ from the dorsal end. The variation over this dorsal part of the cochlea is 2.2-fold. The curve then rises rapidly to a new maximum of 178 μ near the ventral end.

The width of the fundus (b of this figure) takes a similar course. It rises rapidly to a maximum of 66 μ at the dorsal end, then declines gradually to a minimum of 15 μ at the 800 μ point. The variation over this course is 4.4-fold. Thereafter a rise to a second maximum of 111 μ appears, and the papilla quickly terminates.

The thickness of the fundus is shown in c of Fig. 19-10. It increases at the dorsal end to a maximum of 21 μ at a point 280 μ from the dorsal end, after which there is a gradual decline to a minimum of 5.4 μ at the 830 μ point. This variation is 3.9-fold. Thereafter the thickness increases rapidly to a maximum of 26 μ in the ventral region.

Part d of Fig. 19-11 shows the areas of the whole sensory structure (solid line), and of the organ of Corti (broken line). The first of these curves rises rapidly to a maximum of 4200 sq μ in the dorsal region, and then declines to a minimum of 2240 sq μ at the 820 μ point. The variation over the dorsal region is nearly twofold. Thereafter a striking increase appears in the size of this structure, reaching a value of 8740 sq μ, and then falling with equal rapidity as the papilla terminates. The curve for the organ of Corti takes much the same form.

The varying number of longitudinal rows of hair cells is represented in e of Fig. 19-11. The number increases quickly from 3 to 7 at the dorsal end, declines to 6 and 5, oscillates between 5 and 4, and then continues to drop to a minimum of 3 in the region between 800 and 880 μ from the dorsal end. In the ventral region this number rises to a maximum of 10 and then falls away.

Cordylus warreni breyeri. — Curve a of Fig. 19-12 represents the varying widths of the basilar membrane in a specimen of *Cordylus warreni breyeri*, in which the length was 1220 μ. A maximum width of 188 μ is attained near the dorsal end and a minimum of 74 μ around 870 to 990 μ from the dorsal end, for a range of 2.5-fold. The curve then rises to a new maximum of 174 μ in the ventral region.

The width of the fundus (curve b of this figure), varies in similar fashion, from a maximum of 96 μ near the dorsal end to a minimum of 23 μ at a point 990 μ from the dorsal end, which is about a fourfold variation. Again in the ventral region the curve rises sharply to a second maximum that is slightly larger than the first one.

The thickness of the fundus (c of this figure), is remarkably constant over much of the cochlea. After a rapid rise at the dorsal end it varies around 20 to 24 μ until the halfway point is reached, after which it de-

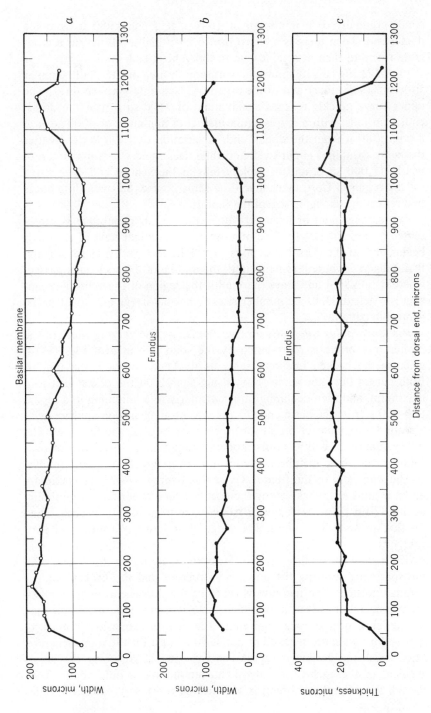

Fig. 19-12. Cochlear dimensions in *Cordylus warreni breyeri*; basilar membrane and fundus. From Wever, 1970a.

clines gradually to a minimum of 16.6 μ at a point 950 μ from the dorsal end. Then there is a sharp rise to 29 μ and a fall first at a moderate rate and then at a rapid one to the ventral end.

Part *d* of Fig. 19-13 shows the variation in area of the whole papillar structure (solid line) and of the organ of Corti only (broken line). The upper curve quickly reaches a maximum of 7100 sq μ near the dorsal end, then falls, with a few irregularities, to a minimum of 4000 sq μ at a point 780 μ from the dorsal end, a variation of 1.8-fold. Thereafter the curve exhibits a slight swell and then rises rapidly to a maximum of nearly 11,000 sq μ, and as rapidly falls at the ventral end. The curve for the organ of Corti takes much the same course, but with differences that reflect the changing size of the fundus.

The arrangement of hair cells along the basilar membrane is represented in *e* of Fig. 19-13. The number of longitudinal rows of hair cells begins at 5 at the dorsal end, rises to 8, falls to 6 and rises to 8 again, then falls to 5 where it remains until the middle of the cochlea is passed, where it falls to 4 and stays there until the region of the ventral expansion is reached. Here the number rises somewhat irregularly to 11 at the ventral maximum.

Pseudocordylus subviridis transvaalensis. — The varying width of the basilar membrane is represented by the solid line in *a* of Fig. 19-14, whereas the broken line represents its thickened portion, the fundus. Both curves have the same form, rising sharply at the dorsal end to a maximum, and then declining rather regularly to a minimum in a region 800-900 μ from the basal end. Then the ventral expansion appears, with a second maximum of broad form nearly as large as the first one. The differentiation of width is only moderate, amounting to only about two-fold for the whole membrane and a little over threefold for the fundus.

The thickness of the fundus (*b* of this figure), rises to a maximum about a third of the way along the cochlea, then declines to a minimum at the 870 μ point, after which it increases to a maximum larger than the first at the 1000 μ point. The range of variation is a little over two-fold.

The size of the auditory papilla is represented in *c* of this figure, with the solid line showing the complete structure and the broken line its cellular portion. The increase is rapid at the dorsal end, with a maximum reached at 200 μ, after which there is a moderate drop and then a long, slow decline to a minimum. The ventral expansion is marked, with a flat maximum reached of about 8000 sq μ for the total structure, after which the decline is rapid. The variation of area over the dorsal segment of the cochlea is only slight, amounting to only about 50%, though the ventral maximum is about three times as great as the minimum.

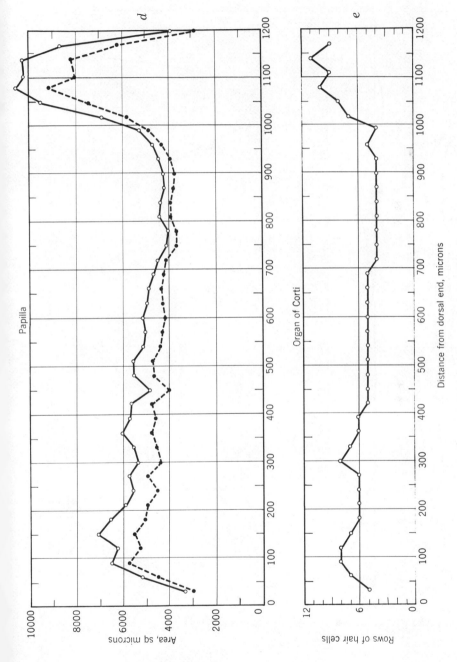

FIG. 19-13. Size of auditory papilla and hair-cell distribution in *Cordylus warreni breyeri*. From Wever, 1970a.

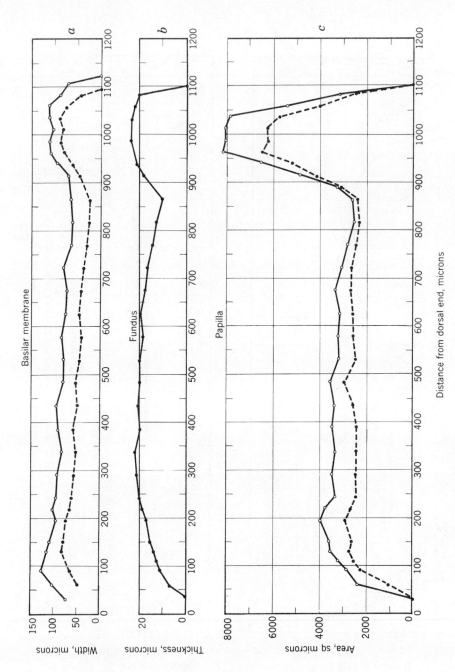

FIG. 19-14. Cochlear dimensions in *Pseudocordylus subviridis transvaalensis*.

The distribution of hair cells along the papilla is represented in *d* of Fig. 19-15. The number of longitudinal rows of hair cells rises quickly at the dorsal end to 7, remains around 6 to about the middle of the cochlea, then falls in two steps to 4 at the minimum region. Thereafter the rise is rapid to a peak of 10, after which the papilla ends.

The density of hair cells (*e* of this figure) follows much the same course, with the dorsal maximum reaching 14 cells per 20 μ, then falling slowly at first and then more rapidly to a minimum of 5, after which a second maximum of about 14 is attained.

Platysaurus minor. — As Fig. 19-16 shows, the cochlear structures of *Platysaurus minor* undergo the same kinds of variations just seen for other species, but on a much smaller scale. The basilar membrane in the specimen illustrated is only 580 μ long, and other dimensions are reduced also, though not to the same degree. As part *a* of this figure indicates, the basilar membrane quickly attains a maximum width of 130 μ in the dorsal region, then declines progressively to a minimum of 53 μ at a point about two-thirds of the distance along the cochlea, and finally rises to a new maximum of 105 μ in the ventral region.

The fundus as shown in *b* of this figure begins with a width of 64 μ and steadily narrows to a minimum of 12 μ at a point 400 μ from the dorsal end, and then rises to a maximum of 67 μ near the ventral end.

The thickness of the fundus, shown in *c* of this figure, attains a broad maximum over the dorsal region around 14 μ, and falls to a minimum of 4 μ at the 400 μ point. This thickness then rises to a second maximum of about 17 μ in the ventral region.

The areas of the auditory papilla and of its main component, the organ of Corti, are shown in *d* of this figure. For the whole structure the maximum near the dorsal end is 3000 sq μ, beyond which there is a progressive decline to a minimum of 1100 sq μ at the 380 μ point, and then a rapid rise to a peak of 4700 sq μ followed by an equally steep decline. The organ of Corti takes a closely similar course.

The number of longitudinal rows of hair cells (*e* of Fig. 19-16), begins at 6 and declines in successive steps to 3 at the end of the dorsal segment, rises once more to a maximum of 6, and falls again as the papilla terminates.

All the cordylids examined present much the same picture, with a long dorsal segment in which the structures quickly reach a maximum size and then progressively decline to a minimum, followed by a short ventral segment in which these structures rapidly increase to a second maximum (often a very large one) and as rapidly fall away as the papilla ends. The long dorsal segment is distinguished by the presence of a line of closely interconnected sallets, and the ventral segment contains a single tectorial mass, the culmen, to which all the hair cells make

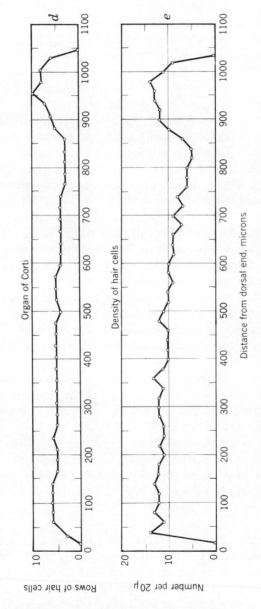

Fig. 19-15. Hair-cell distribution in *Pseudocordylus subviridis transvaalensis*.

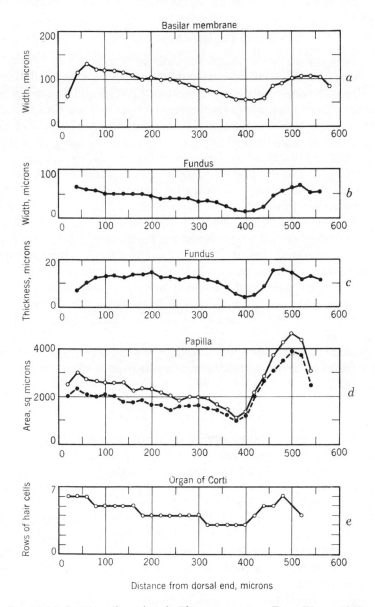

Fig. 19-16. Cochlear dimensions in *Platysaurus minor*. From Wever, 1970a.

ciliary connections. The differentiation shown in the dorsal segment is moderate in amount, though the ventral segment often is strikingly expanded.

CILIARY ORIENTATION PATTERNS

The ciliary orientation is bidirectional throughout the dorsal segment, with the more medial cells having laterally directed kinocilia and the more lateral ones having medially directed kinocilia. The middle cells in the transverse rows often alternate their directions, with perhaps a slight preponderance of lateral positions. In the ventral segment, beneath the culmen, the orientation is almost entirely lateral, though occasionally the most lateral cell in a transverse row will show medial orientation.

In the short intermediate zone the ciliary orientation is probably bidirectional, but this feature is difficult to determine with certainty. The difficulty comes from the fact that these ciliary tufts are usually bent over in making their connections to the thread-like extensions of the sallets, and thus are cut across at oblique angles (Fig. 19-5).

Though determinations of ciliary orientation did not present any serious difficulties in these species (apart from the cells of the intermediate zone as just mentioned), the question whether the kinocilia have terminal bulbs could not be answered with certainty. In a few instances an enlargement at the tip seemed to be present, but this feature was not generally evident.

A special discriminatory function that has been suggested for the ventral segment in Gerrhosauridae and Scincidae seems likely for the cordylids also.

AUDITORY SENSITIVITY

Sensitivity curves for two specimens of *Cordylus cataphractus* are given in Fig. 19-17. Very good response is shown in the middle range from 400 to 2000 Hz, with the maximum around 1000 Hz where for one animal it reaches −46 db. The roll-off is rapid, around 35-40 db per octave, at both low and high ends of the range.

Results for two specimens of *Cordylus giganteus* are presented in Fig. 19-18, and show excellent sensitivity with a maximum near −60 db around 1000 Hz. The roll-off in the high frequencies is at a rate of 40 db per octave, and is greater in the low frequencies where it exceeds 60 db per octave.

Observations on two specimens of *Cordylus jonesi* are presented in Fig. 19-19. These two curves agree well in showing best sensitivity in the region of 400 to 3000 Hz, with a peak of −57 db at 1500 Hz. The roll-off rates in the upper frequencies are between 40 and 60 db per octave and for the lower frequencies are around 40 db per octave.

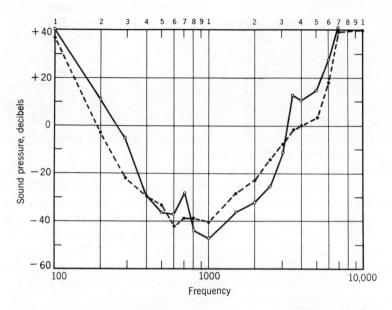

FIG. 19-17. Sensitivity functions for two specimens of *Cordylus cataphractus*.

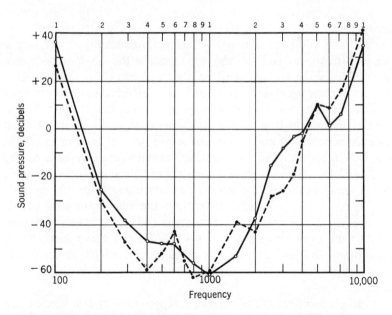

FIG. 19-18. Sensitivity functions for two specimens of *Cordylus giganteus*.

Curves for two specimens of *Cordylus polyzonus* are shown in Fig. 19-20. As in the foregoing species the sensitivity is greatest in the middle frequencies, and shows a maximum between −51 and −57 db. Roll-off rates are about 40 db per octave at the high-frequency end of the range and approach 60 db per octave at the low-frequency end.

Sensitivity curves for *Cordylus tropidosternum* are shown in Fig. 19-21. The maximum region, as before, is in the lower middle frequencies, centered at 800-1000 Hz for one animal and peaking lower, at 500 Hz, for the other. The degree of sensitivity is less than in the preceding species, but is still very good, around −40 db.

Results for two specimens of *Cordylus vittifer* are presented in Fig. 19-22, and show a broad region of best sensitivity between 400 and 2500 Hz, with a peak for one of these ears at 1500 Hz, where a maximum of −53 db is attained. The high-frequency roll-off is rapid, around 60 db per octave, and at the low-frequency end is about 40 db per octave. Results for two additional specimens of this same species, from a batch of animals obtained at a later time, are shown in Fig. 19-23. The curves are similar, but the sensitivity shown in the midfrequency region is somewhat greater, reaching −60 db.

Sensitivity curves for two specimens of *Cordylus warreni breyeri* are shown in Fig. 19-24. These curves are displaced downward on the frequency scale compared with the preceding ones, with the best region extending from 300 to 1500 Hz. The better of these ears reaches −60 db at 500 and 1000 Hz. The roll-off rates are about 40 db per octave at the high-frequency end and about the same at the low-frequency end.

Results for three specimens of another variety of this same species, *Cordylus warreni depressus*, are given in Fig. 19-25. These curves show a greater breadth in the sensitive region, which extends from 290 to 2500 Hz, and for the best ear the maximum is around −60 db. Roll-off rates vary between 40 and 60 db per octave at both ends of the scale.

Fig. 19-26 shows results for another cordylid genus, represented by *Pseudocordylus subviridis transvaalensis*. For this animal the sensitivity curve is broad in the middle frequencies, extending from 400 to 2000 in one of the ears shown and to 3000 Hz in the other, with some irregularities between. The degree of sensitivity in this region is mostly around −40 db with a maximum point of −49 db for one curve and −53 db for the other. Roll-off rates are irregular, around 30-40 db per octave in the high frequencies and close to 60 db per octave in the low frequencies.

A third cordylid genus *Platysaurus* is represented by two species, one with two subspecies. The first of these is *Platysaurus guttatus minor*, a curve for which is shown in Fig. 19-27. This curve shows a marked displacement toward the high frequencies in comparison with all the pre-

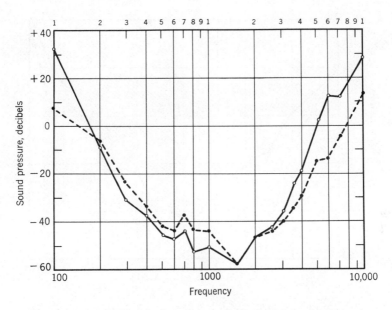

FIG. 19-19. Sensitivity functions for two specimens of *Cordylus jonesi*.

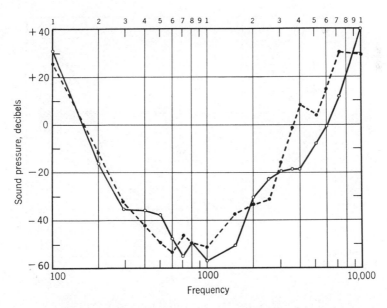

FIG. 19-20. Sensitivity functions for two specimens of *Cordylus polyzonus*.

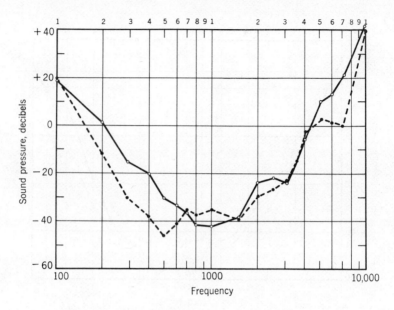

FIG. 19-21. Sensitivity functions for two specimens of *Cordylus tropidosternum.*

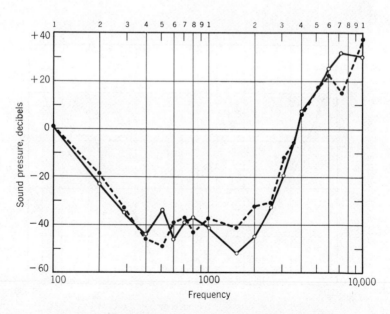

FIG. 19-22. Sensitivity functions for two specimens of
Cordylus vittifer. From Wever, 1970a.

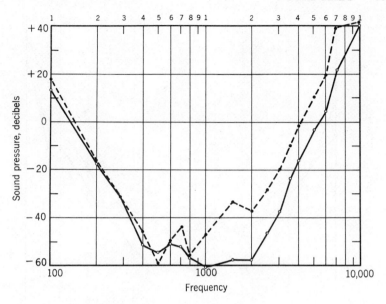

FIG. 19-23. Sensitivity functions for two additional specimens of
Cordylus vittifer.

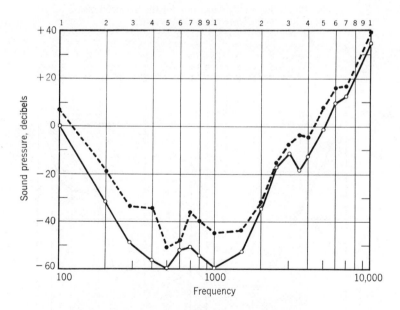

FIG. 19-24. Sensitivity functions for two specimens of
Cordylus warreni breyeri. From Wever, 1970a.

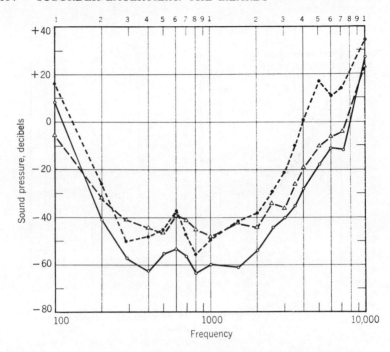

FIG. 19-25. Sensitivity functions for three specimens of
Cordylus warreni depressus.

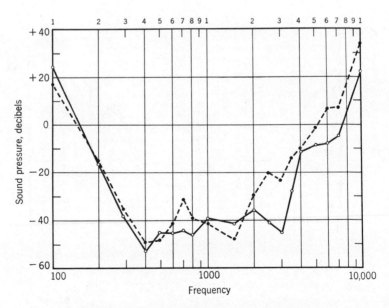

FIG. 19-26. Sensitivity functions for two specimens of
Pseudocordylus subviridis transvaalensis.

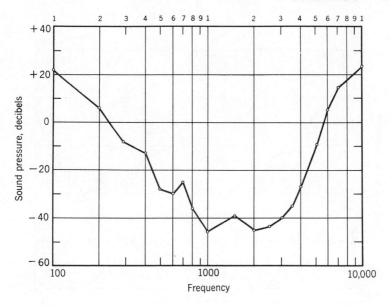

FIG. 19-27. A sensitivity function for a specimen of
Platysaurus guttatus minor.

ceding ones. The sensitive region is from 800 to 3500 Hz, with peaks
at 1000 and 2000 Hz, where the sensitivity reaches −45 db. Roll-off
rates are about 60 db per octave at the upper end of the range and 30
db per octave at the lower end.

The subspecies *Platysaurus guttatus rhodesianus* is represented in Fig.
19-28. The curves are similar in form to the foregoing, with best response in the same region of frequency and peaks for the two ears at
1000 and 2000 Hz. Roll-off rates are about 40 db per octave at the
upper end and 30 db per octave at the lower end.

The final curves of this cordylid series are for *Platysaurus minor,*
shown in Fig. 19-29. The sensitivity shown is only fair, with the maximum around −25 db for one specimen and −10 db for the other. The
roll-off rates are relatively low and somewhat irregular at both ends of
the range.

In summary, the *Cordylus* and *Pseudocordylus* species examined show
good to excellent sensitivity in the mid-frequency region, without any
very marked variations among species. The *Platysaurus* group shows an
upward displacement of the region of best frequencies, with good though
not outstanding levels of sensitivity in two subspecies and a definitely
inferior level in a third species. The departure of this third species,
Platysaurus minor, from all the others ought to be checked by further
tests when additional specimens are available.

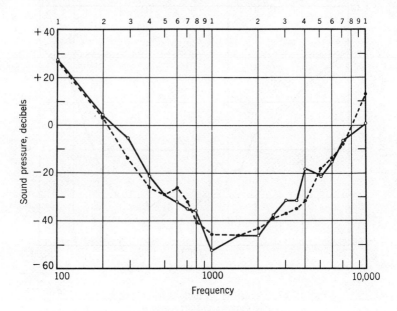

FIG. 19-28. Sensitivity functions for two specimens of
Platysaurus guttatus rhodesianus.

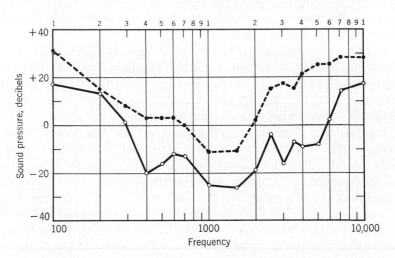

FIG. 19-29. Sensitivity functions for two specimens of *Platysaurus minor.*
From Wever, 1970a.

HAIR-CELL POPULATIONS

The numbers of hair cells contained in the auditory papilla were determined for many of the cordylid specimens, and are reported as species averages in Table 19-I, along with sensitivity data. The hair-cell numbers are shown separately for the two main segments of the papilla and for this structure as a whole. The cells in the intermediate zone are included in the dorsal segment.

TABLE 19-I

SENSITIVITY AND HAIR-CELL POPULATIONS
IN CORDYLIDAE

Species	Number of Specimens	Sensitivity		Number of Hair cells		
		Peak	Low Tone	Dorsal Segment	Ventral Segment	Total
Cordylus cataphractus	2	−44.5	−39.0	512	70	582
Cordylus giganteus	2	−63.0	−61.5	588	105	693
Cordylus jonesi	2	−55.5	−44.0	332	89	421
Cordylus polyzonus	4	−57.2	−53.0	448	98	546
Cordylus tropidosternum	3	−38.7	−38.3	509	71	580
Cordylus vittifer	3	−46.7	−40.0	499	96	595
Cordylus warreni breyeri	4	−48.5	−48.0	487	110	597
Cordylus warreni depressus	2	−57.0	−57.0	550	99	649
Platysaurus guttatus minor	3	−29.7	−29.7	241	29	270
Platysaurus guttatus rhodesianus	2	−49.0	−38.0	389	58	447
Platysaurus minor	2	−18.5	− 8.5	256	36	292
Pseudocordylus subviridis transvaalensis	6	−45.7	−39.5	428	66	494

The largest hair-cell population is found in *Cordylus giganteus*, with *Cordylus warreni depressus* close behind. Then come the other *Cordylus* species, except *jonesi*. The *Pseudocordylus* species is next in order, and finally come the three *Platysaurus* species along with *Cordylus jonesi*.

This order, beginning with the large *Cordylus giganteus* and ending with the small *Platysaurus* species, at once suggests a relation between hair-cell population and bodily development. This relation was explored by calculating the correlation coefficients between total numbers of hair cells and both body weight and the size as measured by length from snout to vent. The results for the 35 specimens available, representing 12 species and subspecies, are shown in Table 19-II. A positive relation of moderate magnitude is found for both measures of bodily development. As expected, body weight and size are themselves strongly

TABLE 19-II

CORRELATIONS WITH BODILY DEVELOPMENT
IN CORDYLIDAE

Total hair cells, related to:	r	p
Body weight	0.49	0.004
Length, S-V	0.53	0.002
Body weight vs. Length, S-V	0.89	0.0001
N = 35		

correlated. All these coefficients are highly significant statistically as indicated by the small values of p.

Relations to Sensitivity. The relations between size of the hair-cell population and auditory sensitivity as measured in terms of the cochlear potentials were explored for this cordylid group. The calculations were made in two general ways, for the individual 35 specimens for which both sensitivity measurements and hair-cell counts had been made, and for the 12 species and subspecies represented. For both types of treatment separate determinations were made for the numbers of hair cells in the dorsal and ventral segments, and for the total of these two. (Again the hair cells in the intermediate zone were included in the dorsal segment.) Two measures of sensitivity were employed, one representing the peak sensitivity, which is the greatest sensitivity attained anywhere along the frequency range, and the other the low-tone sensitivity, defined as the best reading at the low end of the range just above the region of rapid roll-off of the ear's response function. For most of these lizards, in which the best sensitivity is in the low frequencies, these two measures are similar.

The results of these calculations are shown in Table 19-III. As will be seen, there is a moderate positive relationship between sensitivity and the size of the hair-cell population. The statistical values are highly reliable as shown by the p-values. No significant differences are indicated for the different cochlear segments, or between the two measures of sensitivity.

The treatment by species averages gives somewhat higher correlations by minimizing the effects of individual variation in some degree. There may be a suggestion in these results that the relation of sensitivity to the number of hair cells in the ventral segment is a little closer than that found for the dorsal segment. This difference, though slight, deserves attention because it appears with good statistical reliability, despite the small size of the ventral segment. This segment contains only

TABLE 19-III

SENSITIVITY AND HAIR-CELL POPULATIONS
IN CORDYLIDAE

A. Individual Treatment; N = 35				
Number of Hair Cells	Peak Sensitivity		Low-tone Sensitivity	
	r	p	r	p
Dorsal segment	0.52	0.0015	0.53	0.0012
Ventral segment	0.51	0.002	0.54	0.0007
Whole cochlea	0.55	0.0006	0.56	0.0004

B. Species Average Method; N = 12				
Number of Hair Cells	Peak Sensitivity		Low-tone Sensitivity	
	r	p	r	p
Dorsal segment	0.68	0.015	0.67	0.018
Ventral segment	0.77	0.003	0.76	0.004
Whole cochlea	0.71	0.10	0.70	0.010

about 16% of the total population, and thus its contributions to the measured potentials might be expected to be largely swamped by the others. Further study will be needed to establish this point.

Meanwhile, it is suggested that the ventral segment with its large, overtowering culmen serves primarily for the reception of low tones. The considerable width of the basilar membrane in this region, and the mass of the papilla together with that of the culmen, are features well in accord with this view. This group of hair cells ought to be so well intercoupled by the common attachments to the culmen as to respond much as a unit, thereby attaining a high degree of sensitivity for the tones within its range. The essentially unidirectional pattern of ciliary orientation for this group of hair cells, permitting them to operate in phase, should contribute further to the sensitivity of this part of the cochlea.

The dorsal segment of the cochlea exhibits only a moderate degree of differentiation, so that relatively little separation of responses along the cochlea for different tonal frequencies can be expected. The cordylid ear, therefore, appears to be well adapted to the effective detection of sounds over a wide range in the middle frequencies, without a great deal of frequency resolution.

PART III. THE REMAINING REPTILES

SNAKES, AMPHISBAENIANS, *SPHENODON*, TURTLES, AND CROCODILIANS

20. SUBORDER SERPENTES:

THE SNAKES

The snakes share many features with the lizards, and it is generally agreed that they have been derived from common ancestors. For a long time it was believed that this origin was from the platynotan group of lizards, a group that in Camp's (1923) classification included the Varanidae (the monitor lizards) along with three other lizard families now extinct.

Doubt was cast on this platynotan hypothesis for the origin of snakes by Bellairs (1949) and in more certain terms by Bellairs and Underwood (1951) and Bellairs (1972), who called attention to many points of difference, especially in skull structure, between snakes and modern varanids, and suggested that if snakes were derived from this lizard group it was from early types now unknown. Others who have discussed the problem in recent years are likewise doubtful of anything more than a remote or collateral relationship (McDowell and Bogert, 1954). The particular origin of snakes thus seems to be uncertain.

From present evidence we perceive the snakes as an assemblage of forms of ancient origin, undoubtedly saurian in character, which exhibit many features in common with the lizards and show a consistent absence of others that in some sense may be regarded as simplifications or degenerations. Indeed, this last condition of an absence of features that are characteristically present in lizards is so common that the definition of snakes as formulated by many authors is essentially a series of negative statements. Thus a list of eight characters given by Dowling (1959) to define the suborder Serpentes included only a single positive one, a reference to the form of the articulation of the vertebrae—and this feature may be found also in a few of the lizards, though not throughout a family. Otherwise he characterized the snakes as lacking legs and pectoral girdles, without movable eyelids, without an external ear opening, without an upper temporal arch in the skull, lacking a number of bones of the skull (notably the epipterygoid, jugal, and squamosal), and lacking the ability to regenerate the tail when it is lost. This definition of the suborder is further embarrassed by the consideration that all these reductions of structure are encountered in some degree also in certain of the lizards, though never do the reductions appear in the consistent pattern presented by forms identified as snakes.

Underwood (1967) made a particular effort to delineate positive features belonging exclusively to the snakes. He listed 16 such features, and though many of these might be considered of a minor character, the assemblage in its totality provides evidence that this reptilian group is distinctive and has remained so for a long period of time.

The classification of snakes has long been the subject of active discussion, especially in regard to the major subdivisions and the relationships within them. About 400 genera are recognized, and these include perhaps as many as 3000 species, so that from the standpoint of diversity and specialization the snakes must be regarded as a highly successful group of reptiles.

Some of the most notable specializations of snakes have to do with the development of a wide gape and independent control of the two halves of the lower jaw, which make possible the swallowing of relatively large living prey (Gans and Oshima, 1952; Gans, 1961, 1974).

Dowling's classification of the Serpentes, presented in 1959 and revised in 1967 and 1975, may be regarded as one of the more conservative arrangements, and includes 11 families in three divisions as indicated in Table 20-I. One of these families, the Colubridae, contains the vast majority of species, about 2500. Most of the common non-poisonous snakes are in this family. A few others in this family are moderately poisonous (more particularly to their prey), with enlarged teeth near venom ducts at the rear of the mouth. The highly venomous snakes, closely similar to the colubrids except for the presence of anterior fangs and well-developed poison sacs, are located in the families Elapidae and Viperidae.

Snakes are adapted to a variety of habitats. Most are terrestrial, living in grassy, bushy, forest, or desert areas, and often utilizing root or rock crevices or the burrows of other animals as hiding places. A fair number are aquatic, and a great many are burrowers. They occur all over the world, most abundantly in tropical areas; in lesser numbers they extend even into cold regions, to the borders of the arctic circle. An unusual species that is thoroughly adapted to an aquatic existence is the elephant trunk snake of Java, represented in Fig. 20-1.

THE EAR AND HEARING IN SNAKES

An obvious characteristic of snakes is the lack of any external ear opening or other superficial indication of an auditory mechanism. This condition, and the absence of obvious responses to sounds, has led to the widespread opinion that snakes lack a sense of hearing, or at any rate

TABLE 20-I

CLASSIFICATION OF SNAKES

(After Dowling, 1975)

Typhlopoidea (blindsnakes)
    Family Uropeltidae (shieldtailed snakes)
    Family Leptotyphlopidae (slender blindsnakes)
    Family Anomalepididae (primitive blindsnakes)
    Family Typhlopidae (typical blindsnakes)
Booidea (primitive snakes)
    Family Boidae (pythons and boas; 4 subfamilies)
    Family Aniilidae (pipesnakes)
    Family Tropidophiidae (neotropical woodsnakes)
    Family Bolyeriidae (?) (Mauritius snakes)
Colubroidea (advanced snakes)
    Family Colubridae (typical harmless snakes)
      subfamily Xenodontinae (neotropical snakes; 5 tribes)
      subfamily Lycodontinae (Old World swampsnakes; 11 tribes)
      subfamily Colubrinae (terrestrial and arboreal snakes; 6 tribes)
      subfamily Natricinae (watersnakes and allies; 3 tribes)
    Family Viperidae (vipers and adders; 4 subfamilies)
    Family Elapidae (cobras and allies; 5 subfamilies)

are seriously deficient in auditory ability. Partly for this reason there has been relatively little study of the snake's auditory mechanism and few attempts to investigate their hearing.

The only serious attempt to measure hearing in snakes by a behavioral method is that of Manning (1923), who worked with the diamondback rattlesnake and a few other vipers. He placed the snake in a box provided with a viewing hole and produced loud tones over a range from 43 to 2752 Hz in an effort to elicit startle reactions or other indications of hearing. In most instances this attempt failed, but occasionally the diamondback rattlesnakes produced their characteristic rattle sound in response to aerial tones of 43 Hz, though not to higher tones. A timber rattlesnake responded to 43 and 86 Hz. More frequent reactions were obtained in these two rattlers when vibratory stimuli were applied to the frame of the box. Manning concluded that hearing was defective in these animals, especially for aerial sounds. It is clear that under the conditions of this experiment all the responses might have been tactual rather than auditory. It must be borne in mind, however, that the failure of an animal to make any overt response to a sound is no certain evidence of his inability to hear it.

No further attempt to study hearing in snakes by behavioral techniques has been reported, though mention may be made of an experiment by Crawford and Holmes (1966) that can be considered an approach to this problem. In this experiment, specimens of the rat snake,

FIG. 20-1. The elephant trunk snake, *Acrochordus javanicus*.
Drawing by Anne Cox.

Elaphe obsoleta, were placed in a two-compartment box and exposed in one of these to a light and also to a buzzer that was attached to the floor and produced both sound and vibrations. Successful performance in the situation consisted of leaving the stimulating compartment and passing to the neighboring one that was unlighted and somewhat isolated from the acoustic and vibratory stimuli. The results gave evidence of learning, for the time of escape from the starting compartment was reduced as the trials continued. The authors regarded the mechanical vibrations as the aversive stimulus in this situation, but the degree to which sounds were also effective is not clear.

Three electrophysiological studies of hearing in snakes have been carried out. The first of these, by Adrian (1938), was an attempt to record eighth nerve potentials in the grass snake in response to loud sounds, but the results were negative.

Wever and Vernon (1960) studied five species of common colubrids by means of cochlear potentials. The active electrode was a steel needle, insulated except at the tip, and inserted in a fine hole drilled through the dorsal wall of the otic capsule so as to enter the perilymph space. Responses were recorded to aerial and bone-conducted sounds over a range of 20 to 3000 Hz. Best sensitivity was found in the low-frequency range from 80 to 700 Hz, varying somewhat with species. Within this range a good level of sensitivity was indicated, especially in the pine snake, *Pituophis m. melanoleucus*. Results of this investigation indicated that the snake ear is directly responsive to aerial sounds, and showed further that the quadrate bone plays an essential role in their reception. Evidence was also obtained for differential action of aerial sound on the two ears: when the sound was directed from one side of the head the response was greater in the ear on that side. This evidence suggests that snakes should be able to localize the source of a sound. Further, these experiments were designed to provide a test of the long-standing hypothesis that snakes hear aerial sounds primarily or exclusively by their conduction to the head through the substrate on which the head rests. In these tests the head was held in contact with a surface, yet the nearer ear was found to be much more strongly stimulated than the averted one, which would not be the case if the transmission was indirect, by way of the substrate. These experiments established the hearing of snakes on much the same basis as in other animals, and demonstrated a high degree of sensitivity to aerial sounds within a limited range in the low frequencies. Some of the details of these experiments will be presented later.

Some time afterward, a number of tests on snake hearing were carried out by Hartline and Campbell (1968, 1969) and Hartline (1969) by recording responses from the midbrain. These experiments were made mainly on crotalids, but included boids and colubrids as well. The results on all species were generally similar and, like the observations of Wever and Vernon, indicated hearing within a range limited to the low frequencies, and with a strong peaking of sensitivity in a narrow region within this range. These experimenters also reported responses to sounds in a second receptor system designated as the somatic system and probably mediated by sense organs of the skin. This somatic system is responsive to both aerial and mechanically applied vibrations, and in this experiment appeared to cover a wider frequency range than the ear. This system was found to be less responsive than the ear in the ear's best region by 30 db or so, and it failed to exhibit any marked frequency selectivity.

A sensitivity function obtained by Hartline for the rattlesnake *Crotalus viridis* is reproduced in Fig. 20-53 below for comparison with the cochlear potential observations on this species in the present study.

The species available for the present investigation and the number of specimens of each are listed by families:

UROPELTIDAE

Rhinophis drummondhayi (Wall, 1921)—1 specimen.

TYPHLOPIDAE

Typhlops pusillus (Barbour, 1914)—2 specimens; *Typhlops schlegeli mucrosa* (Peters, 1854)—2 specimens.

BOIDAE

Corallus annulatus (Cope, 1876)—1 specimen; *Corallus enydris* (Linnaeus, 1758)—5 specimens; *Epicrates cenchria* (Linnaeus, 1758)—1 specimen; *Eryx conicus* (Schneider, 1801)—2 specimens; *Eryx johni* (Russell, 1801)—1 specimen.

COLUBRIDAE

Acrochordus javanicus (Hornstedt, 1787)—2 specimens; *Elaphe guttata* (Linnaeus, 1766)—1 specimen; *Elaphe o. obsoleta* (Say, 1823)—2 specimens; *Farancia abacura* (Holbrook, 1836)—1 specimen; *Lampropeltis getulus holbrooki* (Stejneger, 1902)—1 specimen; *Natrix sipedon pictiventris* (Cope, 1895)—2 specimens; *Natrix s. sipedon* (Linnaeus 1758)—5 specimens; *Oxybelis aeneus* (Wagler, 1824)—2 specimens; *Pituophis melanoleucus* (Daudin, 1803)—13 specimens; *Thamnophis s. sirtalis* (Linnaeus, 1758)—5 specimens; *Thamnophis s. sauritus* (Linnaeus, 1758)—1 specimen.

VIPERIDAE

Crotalus viridis (Rafinesque, 1818)—2 specimens.

ELAPIDAE

Pelamis platurus (Linnaeus, 1766)—2 specimens.

Five specimens of *Pituophis melanoleucus* were obtained from New Jersey and are of the subspecies *melanoleucus*; one was from Florida and is considered to be *mugitus*; and seven were from the Pacific coast and probably are *annectens*.

The 19 species investigated, belonging to 15 genera, obviously constitute a limited sample of this large and widely distributed group of reptiles. Yet the present results show general agreement, especially among the boids and colubrids, in the form of the auditory structures and their principles of operation. Greater divergences appear in the three species of typhlopoids examined, and further study of this group and of additional representatives of the elapids and viperids is clearly desirable.

SOUND RECEPTION AND TRANSMISSION

The sound receptive and conduction system of snakes presents some unusual features, studied most thoroughly in the colubrids *Natrix s. sipedon, Lampropeltis getulus holbrooki*, and *Pituophis melanoleucus*.

A view of the right side of the head of a specimen of the common water snake, *Natrix sipedon* (*a* of Fig. 20-2), shows the lack of an outer ear and ear opening. A middle ear mechanism lies beneath the

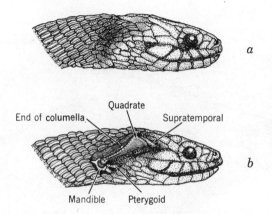

FIG. 20-2. *a*, the right side of the head of the common water snake, *Natrix s. sipedon*; *b*, this specimen dissected to expose the quadrate and its attachments. About natural size.

skin and muscle layers at the side of the head, partially exposed in *b* of this figure. It consists of the quadrate bone and a columella auris that extends from the inner surface of the quadrate to the oval window of the cochlea.

The quadrate, of roughly triangular form, is largely free from the skull and is loosely held in an oblique position by ligamentary connections at its two ends. Its anterodorsal end has a capsular connection with the lower edge of the supratemporal, which itself (in this species) is flexibly attached over the parietal region of the skull. The posteroventral end of the quadrate is attached to the articular portion of the mandible and receives medial bracing by the pterygoid flange.

Further details of this conductive mechanism are shown in Fig. 20-3, in a view from a more anterior and ventrolateral position. The columella has a long, tapered shaft, which anteriorly expands to form a particularly large footplate, and posteriorly changes into a cylindrical cartilaginous portion. The posterior part makes a connection with an articulatory process of the quadrate.

The footplate enters deep into the otic capsule, and ordinarily is concealed by connective tissue that continues the edges of the opening to form a conical sheath around this forward part of the columellar shaft. In Fig. 20-4 this membranous covering has been removed to reveal a portion of the deep-lying footplate.

A more comprehensive picture of the columella and its relations to adjoining structures is provided by Fig. 20-5, which shows a cross section through the ear region of the king snake, *Lampropeltis getulus holbrooki*. The section passes through the entire columella, showing the

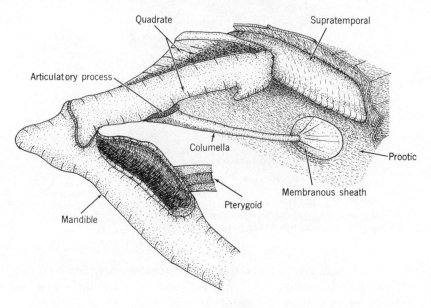

Fig. 20-3. Details of the columella and its attachments in *Natrix s. sipedon*. The view is more anterior and ventrolateral than in the preceding figure. Scale 5×.

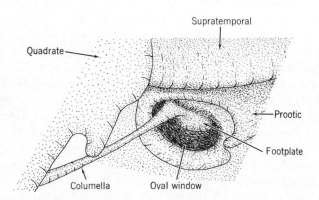

Fig. 20-4. The same specimen after removal of the membranous sheath over the columellar footplate. Scale 10×.

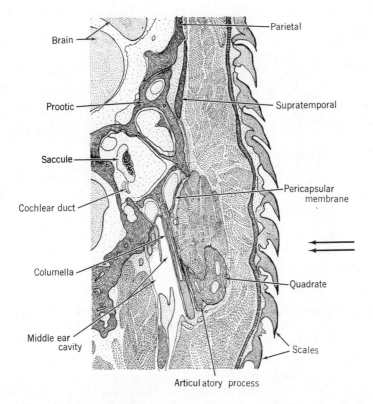

Brain

Prootic

Saccule

Cochlear duct

Columella

Middle ear
cavity

Parietal

Supratemporal

Pericapsular
membrane

Quadrate

Scales

Articulatory process

Fig. 20-5. A frontal section through the ear region of
Lampropeltis getulus holbrooki. Scale 12.5×.

deep location of its footplate in the otic capsule and the connection of
its cartilaginous end with the articulatory process of the quadrate.

Only the posteroventral end of the quadrate is visible here; the main
portion of this bone, which extends to the supratemporal, lies at a more
dorsal level. Extending medially is the bony articulatory process of the
quadrate with its inner surface flattened and covered with a thin layer
of cartilage and connective tissue. Adjacent is the articulatory end of the
columella, which is largely cartilaginous, and the two joint surfaces are
held in close contact by a connective tissue capsule. Within the limits
imposed by this capsular enclosure the two parts slide freely on one another. A more detailed picture of this columellar articulation (with the
connective tissue capsule removed) is given in Fig. 20-6.

Articulatory Process of the Quadrate. In all the colubrids examined
the articulation between columella and quadrate is a simple junction between the two cartilaginous surfaces, with contact maintained by the

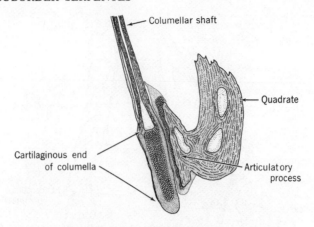

FIG. 20-6. The articulation between quadrate and columella in
Lampropeltis getulus holbrooki. Scale 25×.

connective tissue sheath as just indicated for *Lampropeltis*. This form
was seen also in *Crotalus viridis*. For the boids included in the study,
however, this articulation takes a different form. The columellar and
quadrate surfaces are never in contact, but are separated by one or more
cartilaginous disks, with the whole within a common connective tissue
sheath. This condition is represented in Fig. 20-7 for a specimen of *Eryx
johni*, in which a large, flat cartilaginous plate, along with a small wedge
of cartilage at the anterior end, are inserted between columellar and
quadrate elements. Dorsal to the region shown, the structure is more
complex still, and the intervening material consists of five or six thin
slips of cartilage. Similar relations were found in *Eryx conicus* and
Epicrates cenchria.

In two *Corallus* species the arrangement is even more involved: the
elements instead of forming a simple lateral stack are strung out obliquely to produce a short column. In *Corallus enydris*, as shown in
Fig. 20-8, there are two intervening cartilages within the heavy connective tissue sheath. The osseous part of the columella runs forth from this
complex body at an angle of about 45° as indicated.

This articulation is much the same in *Corallus annulatus*, as shown
in Fig. 20-9.

The identity of the "intervening cartilage" in snakes is uncertain. It
has been regarded as the intercalary (de Beer, 1937; Kamal and Hammouda, 1965), but McDowell (1967) considered it to be the internal
process of the columella together with a part of the extracolumella.

In the boids these articulations present two or more surfaces of sliding
contact, all within a common capsule, and evidently provide effective

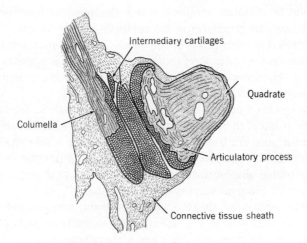

FIG. 20-7. The quadrate-columella articulation in the boa, *Eryx johni*. Scale 100×.

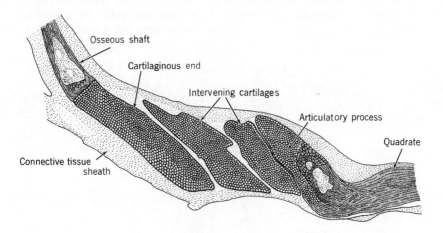

FIG. 20-8. The quadrate-columella articulation in *Corallus enydris*. Scale 37.5×.

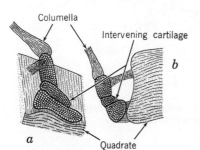

FIG. 20-9. The quadrate-columella articulation in *Corallus annulatus. a*, posterior view; *b*, lateral view.

transmission of quadrate motions with a minimum of longitudinal restraint. The simpler structures in the colubrids appear to serve the same function, probably with equal effectiveness. As the boids are accepted as a separate and in some ways more primitive stock, it appears that the colubrids in their development have lost the intervening elements without seriously reducing the mobility of the joint.

In these species a careful removal of the capsule allows the two parts to separate at once; there is no indication of direct attachments. Therefore these surfaces are free to slide on one another within the confines of the capsule. This longitudinal mobility no doubt has the purpose of preventing undue displacements of the columella that might pull the footplate out of the oval window of the cochlea.

The small burrowing snake *Typhlops pusillus* shows some striking variations. As indicated in Fig. 20-10, the columella consists of a relatively large spoon-shaped footplate and a short head process. The footplate lies entirely inside the otic capsule and nearly fills its cavity. Its posterior end is flattened, and articulates with a tablelike enlargement of the posterior floor of the capsule. The slender headpiece extends a short distance through an opening in the capsular wall (the oval window) and connects through a heavy ligament with the end of the quad-

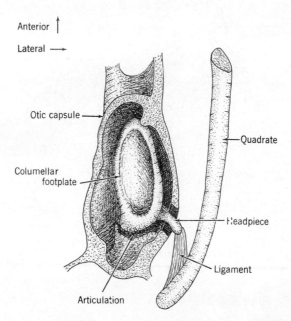

FIG. 20-10. The ear of *Typhlops pusillus*, reconstructed from serial sections. Scale 100×.

rate, which is a long, slender bone whose anterior end articulates with the mandible.

In List's (1966) treatment of the Typhlopidae there are indications of a firm connection between columella and quadrate, and the form of this connection in different species needs to be investigated further.

The sound-conductive mechanism in the uropeltid burrowing snake *Rhinophis drummondhayi*, shown in Fig. 20-11, is similar to that just described in *Typhlops pusillus* in that the columellar footplate is contained in the otic capsule, but differs in the form of connection to the quadrate. The small headpiece of the columella emerges through the oval window and connects with the quadrate by way of two small cartilages, the larger of which lies in a depression near the end of the quadrate. These cartilages along with the osseous end of the columella are

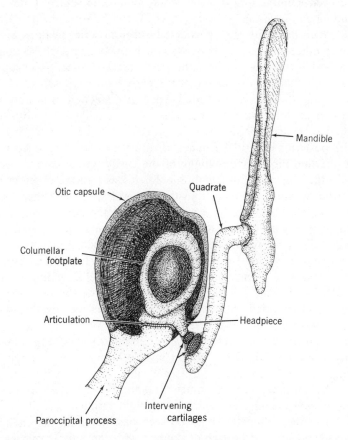

FIG. 20-11. The ear of *Rhinophis drummondhayi*. Drawn from a model based on serial sections. Scale 37×.

enclosed in a sheath of connective tissue. This linkage between columella and quadrate resembles that in the boas.

MIDDLE EAR CAVITY

As was shown in Fig. 20-5, immediately below the quadrate and extending forward along the columella is a greatly flattened middle ear cavity.

For nearly two centuries, since the first description of the ear of snakes by Geoffroy in 1778, it has been supposed that these reptiles lack a tympanic space as well as an external ear opening and tympanic membrane, and that the conductive structures are simply wedged between the muscle layers at the side of the head.

Yet by careful technique an air cavity in the ear of snakes can be demonstrated. This cavity was first discovered by dissection in living (anesthetized) specimens of *Pituophis melanoleucus annectens* and has since been identified by this dissection method in several other species including *Pituophis m. melanoleucus, Natrix s. sipedon,* and *Corallus annulatus.* The cavity may of course be seen in serial sections as in Fig. 20-5 in other snake species dealt with here, but in such sections it is not possible to determine whether the observed space is occupied by air or body fluids; most observers evidently have supposed (as I did for a time) that this was a fluid-filled space. A dissection procedure is necessary to explore the true character of the cavity. The most suitable method of revealing this cavity is by removing an area of skin dorsal to the posterior end of the quadrate and then dissecting deeper to expose the thin membranous lining of the cavity. Great care is necessary to avoid tearing this thin membrane, for if it is torn the observer will be uncertain whether the air within the exposed space was there to begin with or entered through the tear. When the membrane is kept intact and freed of overlying tissues, its transparency allows the air space within to be readily perceived. The membrane when opened reveals the cavity more completely, and further probing shows the forward extent of this narrow space all along the course of the columella, ending at the otic capsule. Fluid is not normally found within the cavity, but the space is narrow and crowded with blood vessels and nerves, together with the long shaft of the columella.

In a specimen of *Natrix sipedon* a particularly good exposure of the cavity was obtained. The skin and muscle layers were removed in two regions, posterior to the quadrate and also anterior to it, with care to leave the lining membranes intact. Looking through the exposed membrane of one opening and introducing light through the other, the silhouette of the columella could be seen passing along the cavity. This element appeared to be free of contact with the muscle masses of the region, though the separation was small.

This cavity is referred to here as an air space on the assumption that it is filled with air from the outside as in other animals with gaseous middle ear spaces, but this point remains unproved. A Eustachian tube or its equivalent has not been identified.

The presence of a middle ear air space is of great significance for the functioning of the snake's ear, for even a narrow gas layer beneath the quadrate and along the course of the columella provides mobility for these structures and relieves them of the large frictional forces that embedment in tissue and fluids would entail.

This evidence also lends weight to the argument that the snake's ear was derived from one in early ancestral forms having a more conventional reptilian structure.

PROCESSES OF SOUND RECEPTION

Two aspects of the operation of the sound receptive and conductive mechanism of snakes will now be considered: the initial operation of the surface tissues in response to sound pressures, and the nature of the action by which vibrations are conveyed through the columella to the fluids of the cochlea.

Initial Actions of Sounds.   It is obvious that the primary sound receptive surface is the skin, together with the underlying muscle layers. Secondarily it can be shown that the sound pressures exerted on these superficial tissues are communicated to the large surface of the quadrate bone, and lateral movements of this bone are transmitted to the columella to produce displacements of its footplate in the oval window.

The evidence on this matter was obtained in two series of experiments in which cochlear potentials were observed before and after a variety of surgical modifications of ear structure. One series was carried out by Wever and Vernon (1960) on five species of common colubrids: *Pituophis m. melanoleucus, Thamnophis s. sauritus, Thamnophis s. sirtalis, Natrix s. sipedon,* and *Natrix sipedon pictiventris.* The second series was carried out by Wever and Strother (1974) on *Pituophis melanoleucus annectens.* These observations showed no significant variations from one species to another, and the results can be summarized for all.

Role of Skin and Muscle Layers. — The removal of skin and portions of the muscle layers over the quadrate until its surface was practically bare had little or no effect upon the reception of sounds. After removal of these tissues the sound waves could act more directly on the quadrate surface, but their effectiveness appeared to be unaltered. Thus the mass of the superficial tissues removed in this procedure is unimportant in relation to the whole mass of tissue involved in the response to sounds— a mass evidently provided mainly by the quadrate.

The Quadrate Connections. — Cutting the ligamentary attachments of the quadrate to the mandible and to the supratemporal also had no appreciable effect. It may be concluded that these attachments are sufficiently loose to permit the small amplitudes of motion made by the quadrate in its responses to sounds.

Removal of the Quadrate. — Removal of the quadrate, with the columella left in place, had a marked effect upon sound reception: the responses declined by amounts varying between 10 and 18 db for the low tones, and by amounts that could not be determined for the high tones because responses for these became too small to measure. The important role of the quadrate in sound reception is thereby established.

Form of Quadrate Motion. — An experiment was carried out (Wever and Strother, 1974) to ascertain the mode of movement of the quadrate in response to sounds. There are several possibilities. One that seemed reasonable was that, when this bone was acted on by acoustic forces, it might execute a rocking motion about one of its attachments, say the one at its posterolateral end. To test the possibilities, the surface of this bone was driven with a mechanical vibrator with its tip applied normal to the surface at numerous points along the length of the bone, and at each application the electrical input to the vibrator was adjusted to obtain a standard amount of cochlear potential (0.1 μv) from the electrode in the otic capsule. This required input was found not to vary with the position of driving.

Thus the quadrate is not pivoted at either end, or otherwise restrained in its vibratory motions. Its articulations evidently are sufficiently loose to permit the small amplitudes of motion produced by sounds. The quadrate acts as a floating plate, free to vibrate as a whole in response to acoustic pressures. Accordingly, no lever action is involved in the transmission of the movements of this plate to the end of the columella; this ossicle evidently is carried along in simple fashion with the vibrations of the quadrate above it.

EFFECTS OF OPENING THE MOUTH

As is well known, the snake's skull is greatly modified so as to permit a wide gape for the swallowing of large prey. The quadrate plays a major role in this gaping process, as along with the pterygoid it forms the articulation for the lower jaw. The mouth opening process has received particular attention in front-fanged forms such as the rattlesnakes in which the further function is served of setting the fangs in striking position (Klauber, 1939). The question arises whether the movements of the quadrate during mouth opening impose stresses on the columella or otherwise affect its functioning in the transmission of sounds to the inner ear.

Observations were made on an anesthetized specimen of *Pituophis melanoleucus annectens* by recording cochlear potentials in response to sounds with the mouth closed and also with it widely opened. A suitably thick piece of heavy rubber tubing was used to hold the mouth open to produce a gape of about 20 mm, which in the small specimen used (head diameter 13 mm and jaw length 35 mm) was close to the maximum extent without undue forcing. Drawings based on photographs of the bony structure in a dissected specimen with mouth open and closed are shown in Fig. 20-12, parts *a* and *b*.

Tests were made at 5 different frequencies from 100 to 3000 Hz. The results showed large variations for every manipulation of the jaws, but no systematic difference between open and closed positions. Though small displacements of the posteroventral end of the quadrate were observed during mouth movements, and slight torsions of the footplate could be seen, the sound transmission characteristics of the system appeared to be unaltered.

This stability of the sound transmission system in the face of potentially great disruptive forces is no doubt assured by the peculiar form of connection between quadrate and columella as described above, in which the end of the columella makes firm lateral contact with the quadrate surface within a connective tissue enclosure that permits free sliding between these elements in a longitudinal direction.

THE SUBSTRATE HYPOTHESIS

A common view, already referred to, is that hearing in snakes occurs generally, or even exclusively, as a reception of vibrations of the substrate. It is said that vibrations transmitted to the body or head when in contact with the ground pass to the quadrate and then are conducted through the columella to the cochlea.

There is no question that vibrations communicated to the head, when of sufficient magnitude, are effective in stimulating the ear. Such stimulation occurs in all ears, including our own.

In the experiments of Wever and Vernon already referred to, tests were made with a mechanical vibrator applied in a number of ways. This stimulation was effective when the vibrator point was placed on any part of the skull, such as the dorsal surface and the posterior end of the lower jaw. As might be expected, the most favorable location was the quadrate. When the vibrator was moved just off this bone the responses fell away markedly, and they declined still further when more remote locations over soft tissues were used. Further observations made by Wever and Strother (1974) on specimens of *Pituophis melanoleucus annectens* have verified and extended these earlier ones. Three forms of vibratory stimulation were compared: the vibrating point was applied in

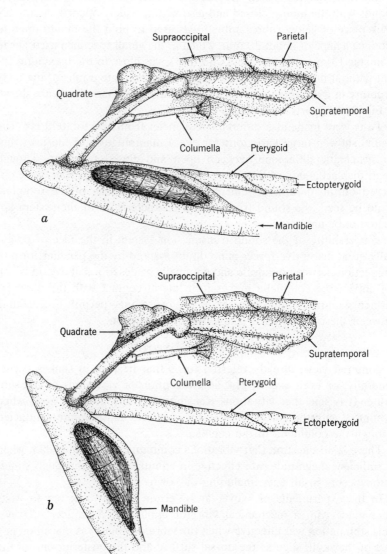

Fig. 20-12. Effects of mouth opening on the quadrate in
Pituophis melanoleucus annectens. a, mouth closed; *b*, mouth open.

turn (1) to the dorsal surface of the skull at the midline of the parietal bone, (2) to the end of the retroarticular process of the mandible, and (3) to the middle of the quadrate perpendicular to its surface. The results are shown in Fig. 20-13. The quadrate position is generally the most effective, though in the region of best sensitivity the retroarticular location is equally favorable, and at two frequencies is even better. The parietal position for the most part shows poorer sensitivity than the others by 10 db or more, except in the region of 500-900 Hz where there is no significant variation among these three modes of stimulation.

From the evidence it is clear that the substrate hypothesis can be accepted only within strict limits. A snake with the head in firm contact with the ground, or one within a burrow with the head pressed against floor or wall might be expected to receive earthborne sounds such as hoof beats, digging, and the like. But the critical question is to what extent are the more important events in the environment represented in ground vibrations? Sounds originating in the air and at appreciable distances could not be received in this manner. Aerial waves on striking the ground are largely reflected, to an extent determined both by the degree of packing of the soil and the angle of incidence; but for the most favorable conditions, with soft soil and a position directly overhead, the attenuation can be estimated as of the order of 60 db or more. It is obvious that the snake's ear would be nearly useless if it were limited to the perception of substrate vibrations.

Some observations relevant to this question were made in the experiments of Wever and Vernon. A snake was placed upon a table, with the underside of the head in close contact with its wood surface, and aerial sounds were presented through a tube whose open end was located 5 cm away from the side of the head, first on the same side as the ear in which cochlear potentials were being recorded, and then on the opposite side. If the sounds were first transmitted to the table and then these vibrations were communicated to the head, it would make no difference how the head was oriented. On the other hand, if the sounds were acting directly on the tissues of the head—specifically on the quadrate—the location facing the source should be the more favorable. This was indeed the case, as Fig. 20-14 shows. Direct orientation provides greater sensitivity by amounts varying with frequency from 4 to 17 db and averaging 12.6 db over the range indicated. At 2000 Hz the averted position no longer gave measurable potentials. Clearly the sounds are being received directly from the air, and the averted position shows a loss because the quadrate is then in the sound shadow cast by the snake's head. Under these conditions any contribution by substrate vibrations is entirely negligible. The inference from these observations is that in a natural situation, with the snake resting its head on the ground, aerial sounds will be

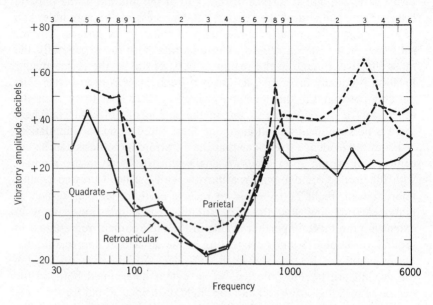

FIG. 20-13. Sensitivity functions in *Pituophis melanoleucus annectens* for three methods of vibratory stimulation.

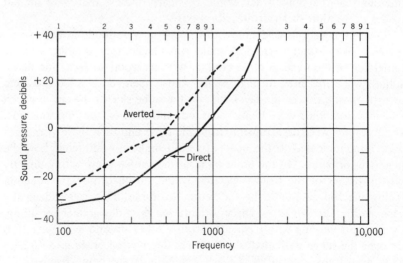

FIG. 20-14. Sensitivity to aerial sounds in *Pituophis m. melanoleucus* with the sounds presented on the same side as the ear under test (direct) and on the opposite side (averted). From Wever and Vernon, 1957.

heard through their direct action, by way of skin and quadrate, and substrate conduction is less by a considerable amount, probably as much as 60 db.

IMPEDANCE MATCHING PROBLEM

In most ears the response to aerial sounds is facilitated by the provision of one or more forms of mechanical transformer action in the middle ear. In mammalian ears, in which this problem has been most thoroughly studied, the initial receptor surface is the tympanic membrane, whose low elastic resistance and slight mass make a good match with the acoustic properties of the aerial medium so that this surface takes up the vibratory motions of the air particles with little loss. The transformer action through which the acoustic energy at the tympanic membrane is effectively transmitted to the fluids of the cochlea is then achieved in at least two ways: by a concentration upon the relatively small area of the stapedial footplate of the sound pressures applied over the large area of the tympanic membrane, and by a lever action of the ossicular chain through which the amplitude of motion at the tip of the malleus near the middle of the tympanic membrane is reduced, and the force correspondingly increased, in the motion of the footplate. In cats, in which precise measurements of these actions have been made (Wever and Lawrence, 1954), the hydraulic increase of force from tympanic membrane to footplate amounts to about 18-fold (the effective ratio of the two areas), and the ossicular lever action amounts to about 2.5-fold, giving a total transformation of about 45 times. This transformation gives an improvement of sensitivtiy of the order of 28 db as ascertained by cochlear potential measurements made before and after surgical disruption of the middle ear mechanism. The principle involved is one of acoustic impedance matching: the transfer of vibratory energy from one medium to another depends upon the relation between their acoustic impedances. When these impedances are different the sound energy is partially reflected from the boundary between the two media, and the amount of this loss in transmission is proportional to the square root of the impedance ratio. This loss between two media can be avoided by the use of a mechanical transformer that makes the two impedances effectively equal.

In lizards, as already noted, the middle ear performs this transformer action in an effective manner; measurements made in *Gekko gecko*, for example, showed an improvement of sensitivity by this means of the order of 30 db over the whole acoustic range, and by as much as 40 db through the frequency region in which this ear is most acute (Werner and Wever, 1972).

The same acoustic problem is present in the snake ear, though some of the conditions are different. Instead of a light and flexible receptive membrane, the snake ear presents a surface of scaly skin beneath which lie a layer of muscle and the quadrate bone: an arrangement hardly ideal for facilitating the entrance of aerial sounds. This surface must present an acoustic impedance considerably greater than that of the aerial medium, and for all tones an inital loss of sound energy by reflection must be accepted.

However, this receptive surface in snakes is not completely rigid, and for a limited range of tones in the low frequencies the surface is rendered acoustically "soft" because at these frequencies the mass of the quadrate and associated tissues together with the stiffness with which this mass is maintained provide the conditions for resonance. The stiffness is determined by the mode of suspension of the quadrate and its connections to adjoining tissues, and also by the presence of the air cushion of the middle ear space below. Accordingly, within a narrow band of frequencies we can consider the quadrate plate with its mass of skin and muscle tissues as vibrating with considerable effectiveness in response to incident sounds.

In this situation there is still a need for transformer action in the middle ear to match the impedance of the quadrate system with the greater impedance of the inner ear. The amount of this transformation will of course be less than in conventional types of ears, but the same types of mechanism can be expected to serve this function.

Areal Ratios. — The quadrate presents a considerable surface for the action of sounds, and its area is great relative to that of the footplate of the columella. This relation varies with species and with individual size; in a specimen of *Pituophis m. melanoleucus* the quadrate surface was determined as 43.8 sq mm and the area of the footplate was 3.8 sq mm, giving a ratio of 11.8.

Lever Action. — It is conceivable that the columella could have a pistonlike form of motion, vibrating longitudinally in and out of its window, or it could execute a rocking motion with one edge of the footplate as the axis, or it could present some combination of these two motions. From a consideration of the anatomical relations the rocking form of action seems the most likely, and this form is the common one in other types of ear.

Evidence on this question was sought in a specimen of *Pituophis melanoleucus annectens* by driving the columella with a mechanical vibrator. After the distal end of the columella was freed from its attachment to the quadrate, this end was driven in two ways, with the driving point at the end in line with the columellar shaft (axial driving) and with this point applied to the side of the shaft and thus at right angles to the axis (lateral driving). The results are represented in Fig. 20-15, plotted to show the vibratory amplitude, in decibels relative to a zero level of 1 mμ, required to produce a standard output of 0.1 μv.

As shown, the axial mode of driving is generally the more effective. In the most sensitive region of the ear, at 300-400 Hz, this axial mode is better by an amount of 25 db.

Two inferences can be drawn from this evidence. In the first place it is clear that the columella does not simply rock back and forth on an axis near the middle of its footplate, for then the action would merely set up eddies of fluid flow over the inner surface of the footplate, and no fluid displacement through the cochlea would occur. The most probable mode of motion is a rocking on one edge, as the anatomical relations suggest. This motion involves a lever action of the reducing type as found in mammalian and other ears. The length of the columellar shaft and the width of the footplate indicates that this reduction is about 2:1. Such a reduction, with the attendant increase in force, would contribute to the desired impedance transformation.

When the columella is forced into longitudinal motion, as it is in this experiment under the conditions of axial driving, the sensitivity is obviously greater than for lateral driving, because the volume displacement for axial driving is much greater; there is no lever reduction in

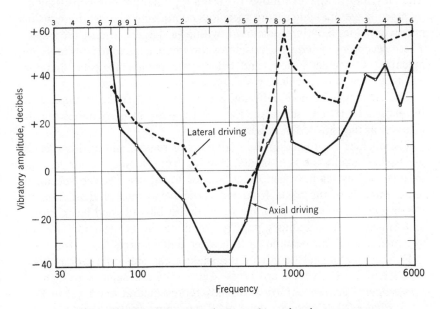

FIG. 20-15. Sensitivity functions in *Pituophis melanoleucus annectens* for two modes of vibratory stimulation of the columellar shaft.

this instance. This does not signify, however, that the axial mode of motion is the normal one or even the most effective under natural conditions.

There is reason to believe, from the evidence at hand—chiefly the high levels of sensitivity attained—that the snake's ear provides a suitable solution for the impedance matching problem. No measurements are available for the impedance of the quadrate system and the reptilian inner ear, but by analogy with mammalian ears it seems likely that the inner ear impedance is many times greater than that of the quadrate system, and a large transformer ratio is needed for matching these two. An areal ratio between quadrate and footplate of about 12 and a lever ratio in columellar action of 2 gives a combined ratio of 24, providing an impedance transformation of 24^2 or 576. This amount of transformation seems of the proper order of magnitude for its purpose.

It is of interest to point out that the relatively large size of the columellar footplate in snakes may have functional significance in this relation. A smaller footplate, with the quadrate surface maintained at its usual dimensions, might well produce an excessively large areal ratio—and too large a transformer ratio would reduce the transfer of acoustic energy to the same extent as one that is too small.

INNER EAR

The labyrinth of snakes follows the usual reptilian pattern, consisting of an assemblage of eight sensory receptors comprising the membranous or endolymphatic labyrinth enclosed within a largely osseous capsule. These receptors include the cristae of the three semicircular canals, the maculae of saccule, utricle, and lagena, the papilla neglecta, and the auditory papilla or cochlea.

The right membranous labyrinth of the colubrid snake *Pituophis m. melanoleucus* is represented in Fig. 20-16, with an inset indicating its location at the side of the head.

The cochlea is situated in the midposterior region, and in this species the basilar membrane is elongate, running from dorsoposterior to ventroanterior, just as has been described for lizard ears. In the specimen represented here the angle from the vertical was measured as 23°. In this species the basilar membrane faces almost laterally and has a rather uniform taper. The auditory papilla lies along the anterodorsal side of this membrane, and occupies a little over half its area.

Further anatomical details of the snake's ear are indicated in Figs. 20-17 and 20-18, which are cross sections of the same specimen of *Lampropeltis getulus holbrooki* already represented at one level in Fig. 20-5. Figure 20-17 shows a level 0.28 mm more ventral than Fig. 20-5, and the scale is enlarged to show the cochlear region in detail.

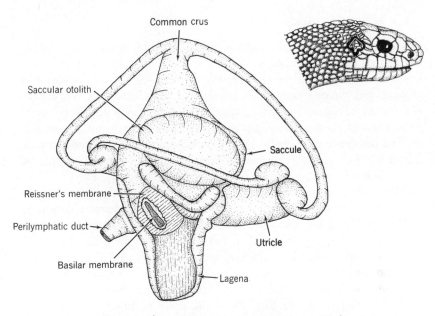

FIG. 20-16. The labyrinth of *Pituophis m. melanoleucus.* The inset shows
the location in the head. Scale for main drawing 12×.

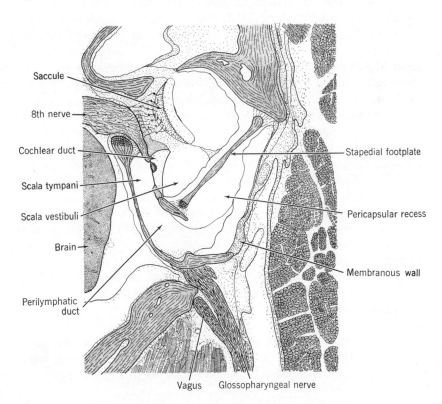

FIG. 20-17. A frontal section through the ear region in
Lampropeltis getulus holbrooki. Scale 30×.

The footplate of the columella is seen in position in the oval window, but the section has passed below the columellar shaft. This footplate has a deep position in the otic capsule: it lies close to the cochlea and saccule, and leaves a large space posterolateral to its surface. This space is enclosed partly by the bony capsule and partly by a relatively heavy membrane of connective tissue alluded to earlier as forming a sheath around the columellar shaft. The enclosed space was designated by de Burlet (1934) as the pericapsular sinus, and was regarded by him as an extension of the perilymphatic system of the labyrinth.

At the level represented in Fig. 20-17 the pericapsular space is separated from the cranial cavity by a bony wall. More ventrally, however, as Fig. 20-18 shows, the wall between brain cavity and pericapsular recess is penetrated, and the perilymphatic duct enters the brain cavity and expands to form the perilymphatic sac. This sac contains a loose network of connective tissue trabeculae, which forms a permeable barrier between perilymphatic duct and the subarachnoid space of the brain.

At the level shown in Fig. 20-18 the scala vestibuli communicates with the scala tympani by way of the helicotrema. The cochlear duct is enlarged, but the cochlear endorgan itself fails to extend this far ventrally. The section has entered the lagenar portion of the cochlear duct, and the nerve bundles shown in this section on the scala tympani side of the cochlear duct are running to the lagenar macula farther ventrally.

Fluid Mobilization Process.   The snake ear lacks a round window for mobilization of the cochlear fluids and employs a reentrant fluid circuit for this purpose. The relations already shown in Fig. 20-17 are represented schematically in Fig. 20-19, with arrows indicating the paths of vibratory fluid flow. As the footplate moves inward, a displacement of fluid passes along a course through scala vestibuli and cochlear duct to the scala tympani, involving the basilar membrane in this path; the return flow follows along the pericapsular recess to the outer face of the footplate.

An alternative fluid pathway is present from scala vestibuli to scala tympani by way of the helicotrema (see Fig. 20-18). It constitutes a bypass to transmission through the basilar membrane and reduces sensitivity in some degree. This bypass is a characteristic of all ears, and perhaps has the functions of preventing the local accumulation of fluid and of protecting the cochlear structures against stimulation of excessive magnitude.

Auditory Papilla.   In snakes, as in other reptiles, the cochlear duct contains two sensory endorgans, the auditory papilla and the lagenar macula. Instead of being more or less adjacent as in lizards, these two

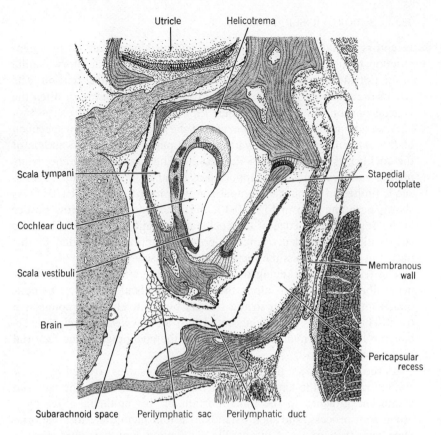

Fig. 20-18. A section through the same specimen at a more ventral level.

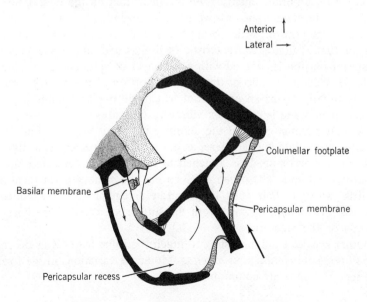

Fig. 20-19. The reentrant fluid circuit in snakes.

receptors in snakes are well separated, and usually marked off by a constriction of the middle part of the duct, and with the auditory papilla in the dorsal portion and the lagenar macula in the ventral portion. The size of these two divisions varies greatly with species, but most often the lagenar portion is the larger (Miller, 1968).

A section across the dorsal portion of the cochlear duct of a specimen of *Pituophis melanoleucus mugitus*, which passes through the middle of the cochlea, is represented in Fig. 20-20. The basilar membrane, bearing the auditory papilla, is supported as usual by limbic tissue. This tissue is limited in amount in comparison with lizards (Baird, 1970a). As shown here and also in Fig. 20-17, the limbus forms a narrow border at the posterolateral edge of the cochlear duct and extends as a broader strip with an expanded triangular portion anterior and medial to this duct, but never exhibits the massive character seen in lizards.

The triangular thickening on the medial side transmits the cochlear nerve fibers, and extends along the side of the cochlear duct to be continued anteriorly adjacent to the saccule by relatively loose connective tisue. The limbus is never elaborated into an elevated mound or overhanging lip that in many lizards serves as the anchorage of the tectorial membrane.

As the figure shows, the limbus forms a deep sulcus covered by a low epithelium along the anterior border of the cochlear duct. The tectorial membrane attaches to the inner portion of this epithelial layer by a broad root process, and then extends as a thin membrane to a tectorial plate over the surface of the papilla. This plate continues over most of the extent of the papilla except for its very ends. A somewhat thickened Reissner's membrane encloses the cochlear duct on the lateral side.

The basilar membrane has two portions, usually of about equal width. The anterior portion, which bears the papilla, is particularly thin, lacking the fundus that is characteristic of lizards and amphisbaenians. The posterior portion is covered with a low layer of epithelial cells. The papilla lies close to the lip of the anterior portion of the limbus, and evidently this lip forms an axis about which the papilla rotates when the basilar membrane is driven by vibratory pressures.

A more detailed view of the auditory papilla is given in Fig. 20-21. The tectorial plate appears here, as in most other species, as a ribbon of dense reticular tissue. It is attached to the tectorial membrane near its anteromedial end. The organ of Corti has the same general form as in lizards, with the hair cells in moderately regular rows, sustained by processes extending upward from the supporting cells whose bases are mainly on the membrane.

Other species examined showed much the same features as described for *Pituophis melanoleucus mugitus*. Moderate variations in the form of the tectorial plate are common.

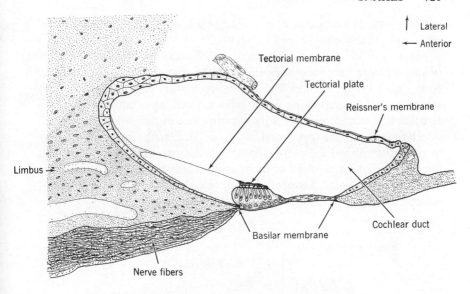

↑ Lateral
← Anterior

Tectorial membrane

Tectorial plate

Reissner's membrane

Limbus →

Cochlear duct

Basilar membrane

Nerve fibers

FIG. 20-20. The cochlear duct in *Pituophis melanoleucus mugitus,*
seen in cross section. Scale 120×.

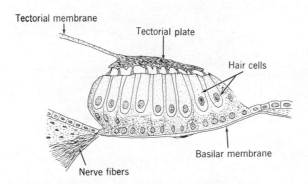

Tectorial membrane

Tectorial plate

Hair cells

Basilar membrane

Nerve fibers

FIG. 20-21. The auditory papilla in *Pituophis melanoleucus mugitus.*
Scale 375×.

In most species this plate is a medium-heavy reticulum with pits on the underside, and the walls of these pits extend as narrow spinous processes to which the tips of the ciliary tufts of the hair cells usually make their connections. Commonly, at both ends of the papilla the tectorial plate is reduced in width and thickness, and either breaks up into two or three small nodules or forms a single oval or rounded body. In *Oxybelis aeneus* the plate is relatively thin and tenuous. In *Farancia abacura* it takes a brushlike form, consisting of a central cone of delicate mesh from which several narrow processes extend over the papilla to connect with its hair cells. A sketch representing this structure near the middle of the papilla is shown in Fig. 20-22.

Miller (1968), after an extensive investigation of the general form of the cochlear duct in snakes, in which he covered about 200 genera, commented on the remarkable constancy of this structure throughout the group. In his survey he was unable to find any anatomical characteristics that would relate in a meaningful way to taxonomic groupings. A relation to habitat was found, however, in that the burrowing forms in general exhibited the more elongate basilar papillae. The longest and largest papillae were found in members of three groups, the Loxocemidae, Uropeltidae, and Xenopeltidae, all of which are burrowers and considered to be among the more primitive snakes.

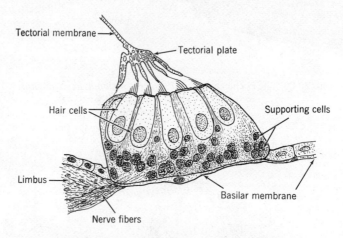

FIG. 20-22. The auditory papilla in *Farancia abacura*.
Scale 500×.

COCHLEAR DIMENSIONS

Although the general form of the auditory structures is highly regular, the dimensions vary widely. Detailed dimensional determinations were made in eight species covering the range of types. These were *Rhinophis drummondhayi* representing the Uropeltidae; *Epicrates cenchria* and *Eryx johni* representing the Boidae; *Pituophis melanoleucus mugitus, Farancia abacura*, and *Oxybelis aeneus* among the Colubridae; *Acrochordus javanicus* as representative of the Acrochordini (a tribe of Natricinae); and *Crotalus viridis*, the one member of the Viperidae included in the study.

Rhinophis drummondhayi. — Dimensional measurements on a specimen of *Rhinophis drummondhayi* gave the results of Figs. 20-23 and 20-24. In *a* of Fig. 20-23 are shown the varying width of the basilar

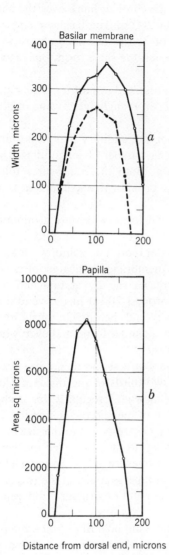

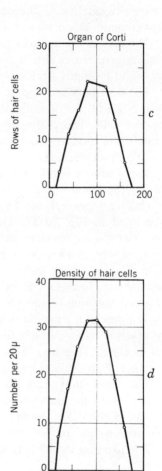

Distance from dorsal end, microns

FIG. 20-23. Cochlear dimensions in the uropeltid snake *Rhinophis drummondhayi*; basilar membrane and papilla.

Distance from dorsal end, microns

FIG. 20-24. Cochlear dimensions in *Rhinophis drummondhayi*; hair-cell distributions.

membrane (solid line) and of the papilla (broken line) over the short length of this membrane, which is about 200 μ. The basilar membrane increases rapidly to a maximum width of 355 μ, and then almost symmetrically falls away. The papilla has much the same form, but reaches its maximum at a more dorsal position.

The area of the auditory papilla is represented in part b of this figure. This area increases almost abruptly to a surprisingly large value of 8200 sq μ and promptly declines.

The distribution of the hair cells along the auditory papilla is shown in Fig. 20-24. Part c indicates the varying number of longitudinal rows of hair cells, and d the varying density. Both these measures follow closely the form of the papilla, with a prominent maximum near the middle of the cochlea.

This auditory papilla has a narrow oval form, with only a minimum of differentiation.

Epicrates cenchria. — Dimensional data on the rainbow boa are presented in Fig. 20-25. The basilar membrane, indicated at a, has considerable asymmetry; its width increases rapidly at first and then more slowly, and reaches a maximum around 200 μ in the ventral region. Thereafter this width declines slowly and then precipitously to the ventral end. The width of the papilla has a similar form, and everywhere is about half that of the basilar membrane.

The area of the papilla increases rapidly at first and then more slowly to reach a maximum at a point about two-thirds of the distance along the cochlea. This area is maintained somewhat irregularly for a short distance, then declines with great rapidity to the end of the papilla.

The distribution of the hair cells roughly corresponds to the size of the papilla, with the number of rows running around 6-8 over the whole middle portion. The density of these cells likewise is fairly regular along the mid-region, apart from a few dips, but tapers rapidly at the two ends.

The papilla exhibits only a moderate amount of structural differentiation.

Eryx johni. — Dimensional data on the sand boa, *Eryx johni*, are presented in Fig. 20-26. In part a of this figure the upper curve shows the width of the basilar membrane as determined at various distances from the basal end, and the lower curve shows the width of the papilla. The basilar membrane has an oval form, 2½ times as long as it is wide, and the papilla, occupying the more anterior region of the membrane, is about four times as long as it is wide.

The size of the papilla, measured as the varying area seen in cross section, varies regularly as indicated in part b of this figure.

The pattern of distribution of the hair cells is shown in parts c and d of this figure. The number of longitudinal rows increases from 4 at the

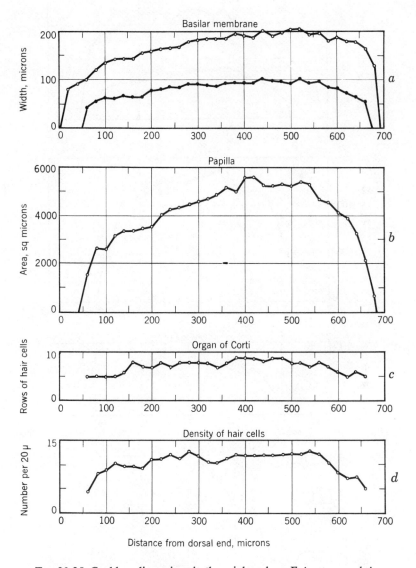

Fig. 20-25. Cochlear dimensions in the rainbow boa, *Epicrates cenchria*.

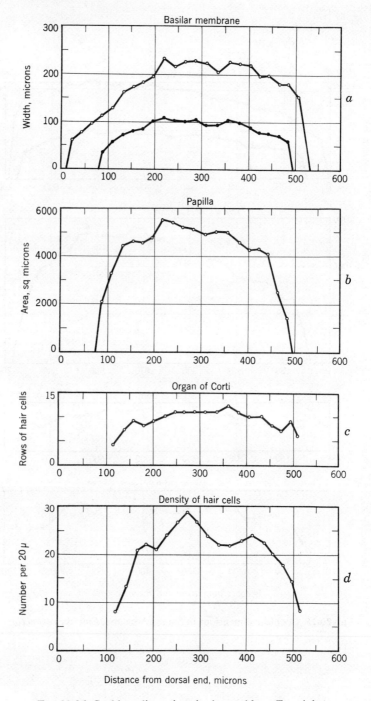

FIG. 20-26. Cochlear dimensions in the sand boa, *Eryx johni.*

dorsal end of the papilla to 11 or 12 in the middle region, and then falls toward the ventral end. The density of these cells, represented as the number appearing in a 20 μ section, rises rapidly to a maximum of 28 near the middle of the cochlea, falls and then recovers, and falls again over the ventral portion of the papilla.

This ear, like the preceding, presents only a moderate degree of differentiation, mainly shown in the variation of width of the basilar membrane over its dorsal half.

Next to be considered are three species regarded as more advanced and belonging to the Colubridae.

Pituophis melanoleucus mugitus. — Fig. 20-27 shows the Florida pine snake, *Pituophis melanoleucus mugitus.* The basilar membrane is among the longest of the species studied and reaches about 850 μ. The width increases progressively to a maximum around 200 μ in the ventral region, just as in the preceding species. Thereafter the decline is rapid, until the membrane ends. The papilla takes much the same course, attaining about half the width of the membrane.

The area of the papilla increases rapidly at first and then more slowly to a maximum in the ventral region, and then declines rapidly.

The number of longitudinal rows of hair cells increases gradually from 3 to 7 or 8 somewhat beyond the middle of the cochlea, maintains this level fairly well for a short way, and then declines. The density of these cells takes much the same course except for a moderate trough near the middle of the cochlea.

This papilla, though noticeably more elongated than the preceding ones, shows the same moderate degree of differentiation.

Farancia abacura. — The mud snake, a burrower, is represented in Fig. 20-28. The form of the basilar membrane and papilla is much like that shown by the boid *Epicrates cenchria*, except that the length is somewhat less. The width of the basilar membrane increases sharply, then more slowly, to a maximum of a little over 200 μ, maintains this level over a considerable distance, and then falls abruptly. The width of the papilla is similar but is less than half that of the membrane; the maximum is reached near the middle, and then the decline is irregular and moderate for a way, and at the end becomes abrupt.

The area of the papilla increases rapidly to a maximum a little beyond the middle and then falls rapidly, though with some fluctuations.

The number of rows of hair cells increases by successive steps from 3 to 6 near the middle of the papilla, reaches a maximum of 7, and then declines over the ventral region. Density of the hair cells follows a different course, more nearly resembling the areal pattern of the papilla, so that a progressive rise to a maximum near the middle of the papilla is seen, and thereafter there is a regular decline that becomes abrupt at the ventral end.

The cochlear differentiation is slightly greater than in the preceding species, largely due to the varying size of the auditory papilla.

Oxybelis aeneus. — The vine snake, an arboreal colubrid, presents a strikingly different picture from the foregoing. As shown in Fig. 20-29, the cochlea is extremely short. The basilar membrane rises rapidly to a maximum width of 130 μ, then as rapidly falls away, in a total length of about 160 μ. The papilla shows a nearly flat maximum around 60 μ, and its total length is about 140 μ.

The area of the papilla increases rapidly to a sharp maximum well beyond its middle, then almost precipitously falls to zero.

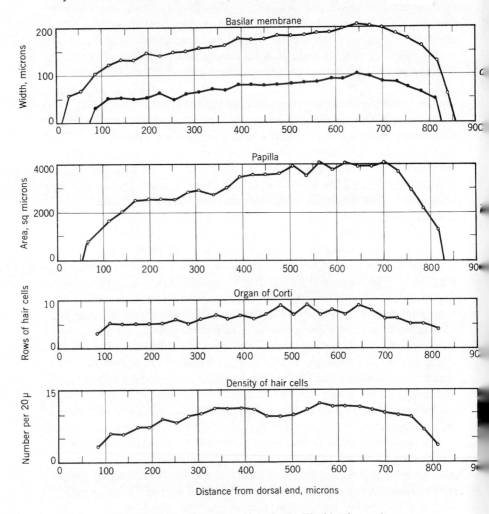

FIG. 20-27. Cochlear dimensions in the Florida pine snake, *Pituophis melanoleucus mugitus.*

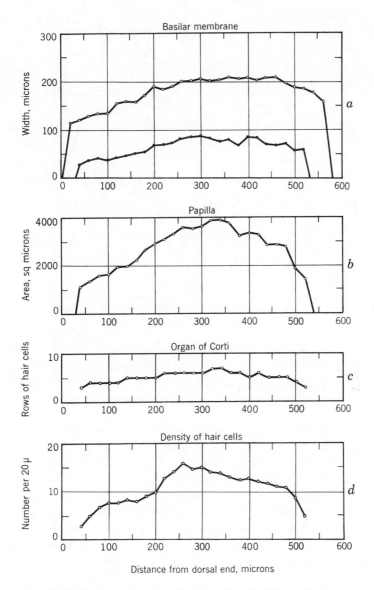

Fɪɢ. 20-28. Cochlear dimensions in the mud snake, *Farancia abacura.*

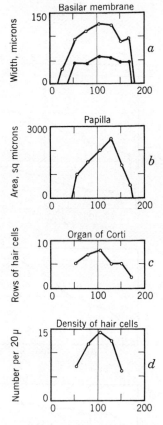

FIG. 20-29. Cochlear dimensions in the vine snake, *Oxybelis aeneus*.

Distance from dorsal end, microns

The number of rows of hair cells rises from 5 to a maximum of 8, then falls away in two steps. The density of the hair cells increases rapidly to 14 per 20 μ section at the middle of the papilla, and shows a nearly symmetrical decline to the ventral end.

This cochlea presents a very low level of spatial differentiation.

Acrochordus javanicus. — The elephant trunk snake, pictured in Fig. 20-1, presents many curious features and its systematic relationships remain uncertain. Hoffstetter (1965) assigned it to a separate family Acrochordidae parallel to the colubrids, whereas Underwood (1967) placed it alongside the boids in his infraorder Henophidia (primitive snakes), largely on account of its form of arterial system. In Dowling's classification (1975) this species is given a less prominent position as a tribe in the family Natricinae among the Colubridae.

The two specimens examined were in poor condition, and the results of the cochlear potential tests had uncertain validity. The anatomical study revealed some unusual characters, as shown in Figs. 20-30 to 20-33.

The width of the basilar membrane (solid curve of Fig. 20-30), increases rapidly from the basal end and reaches a large maximum of about 630 μ in the ventral region, after which it declines rapidly. The papilla at the extreme dorsal end rests entirely on the limbus, and then more ventrally it extends onto the membrane, occupying at first more than two-thirds of its width. This fraction grows less toward the ventral end of the cochlea, because the papilla widens less rapidly than the membrane does.

The size of the auditory papilla is represented in Fig. 20-31. Two curves are shown, one (solid line) indicating the area of the papilla proper, and another (broken line) including a series of blood vessels that over the major part of the cochlea run along the tympanic side of the basilar membrane, becoming particularly numerous around the maximum region of the papilla. These vessels probably represent a branch-

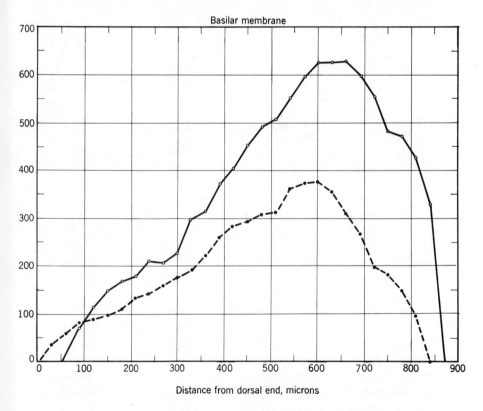

FIG. 20-30. Dimensions of the basilar membrane in the elephant trunk snake, *Acrochordus javanicus*.

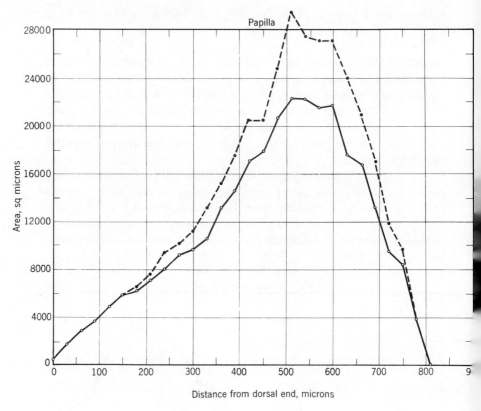

FIG. 20-31. The size of the auditory papilla in *Acrochordus javanicus*.
A number of blood vessels are included in the area indicated by the broken curve.

ing of the single vessel seen at the two ends of this cochlea; as many as
eight of these branches may be found in the middle region.

The distribution of hair cells in this cochlea is represented in Fig. 20-
32. Above in this figure are shown the longitudinal rows of hair cells,
which increase progressively to a maximum of 63 in the ventral region.
The density of the hair cells follows a corresponding form, and reaches
the large value of 107 cells per 30 μ section.

This cochlea is the largest among all the snakes examined.

Crotalus viridis. — The prairie rattlesnake, in the family Viperidae,
has only a slightly elongated cochlea. As shown in Fig. 20-33, the basilar membrane increases in width rapidly to a flat maximum around
220 μ, then in the ventral region declines in nearly symmetrical fashion.
The auditory papilla, with a length of 340 μ, has a flatter form and attains a maximum of about 100 μ.

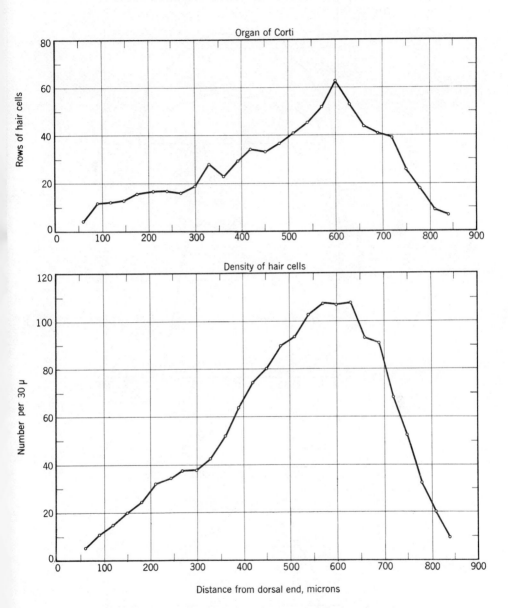

Fig. 20-32. Hair-cell distribution in *Acrochordus javanicus*.

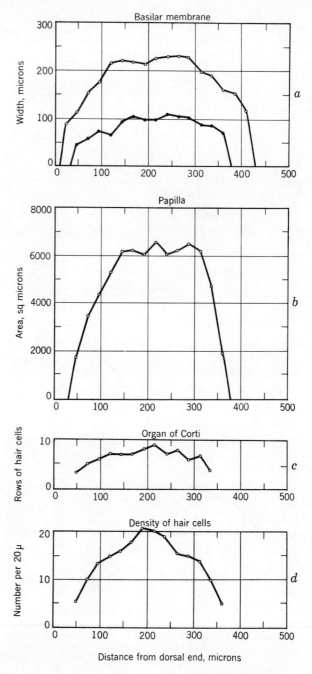

FIG. 20-33. Cochlear dimensions in the prairie rattlesnake, *Crotalus viridis*.

This papilla is the largest after that of *Acrochordus*. Its area increases rapidly to an irregular, broad maximum around 6300 sq μ, then falls almost precipitously.

The number of rows of hair cells increases progressively from 3 to 9, then falls irregularly to the end of the papilla. The density of these cells shows a sharp and nearly symmetrical maximum near the middle of the papilla.

A comparison of the forms and sizes of the basilar membranes in the eight species studied in detail is presented in Fig. 20-34. Here outlines are drawn with the assumption of bilateral symmetry. The variety of forms among these species is striking. In the four longest among the elongate types of basilar membrane a moderate taper may be seen, but in the others the form is more compact: nearly circular in *Oxybelis*, oval in *Crotalus* and *Rhinophis*, and pear-shaped in *Acrochordus*. The form in *Rhinophis* is most unusual in that the width of the membrane exceeds its length.

The Problem of Cochlear Differentiation

For the most part, these snake species exhibit only slight to moderate degrees of differentiation, though there are large variations in the degree of elongation of the cochlea. These differences are apparent in Fig. 20-34. The compact forms in *Rhinophis drummondhayi*, *Oxybelis aeneus*, and *Crotalus viridis* are in sharp contrast to the more extended structures of the others. *Acrochordus javanicus* has an intermediate form, with considerable length and a large amount of variation over this extent. The extended forms show variations mainly at their ends, so that the elongation appears to give only limited benefit to differentiation. Among these, *Pituophis melanoleucus mugitus* shows the greatest taper, though it fails by much to approach that of *Acrochordus*. Yet apart from the sharp dorsal tip the variation in width in *Pituophis* is less than twofold; it is of the order of fourfold in *Acrochordus*.

As is generally true, the elongation of the auditory papilla in these snakes has two benefits apart from presenting the possibility of variations in tuning. It increases the number of hair cells, thereby aiding the sensitivity through spatial summation, and it provides at least the opportunity of specific nerve connections to the various elements and consequently the most effective transmission to higher centers of whatever differential actions are present.

With the exception of *Acrochordus*, however, the spatial distribution of response along the cochlea can only be slight, and for the most part all tones to which these ears are sensitive must spread broadly. The presence of a tectorial plate over nearly the whole auditory papilla interconnects the entire array of hair cells, and tends to unify their action.

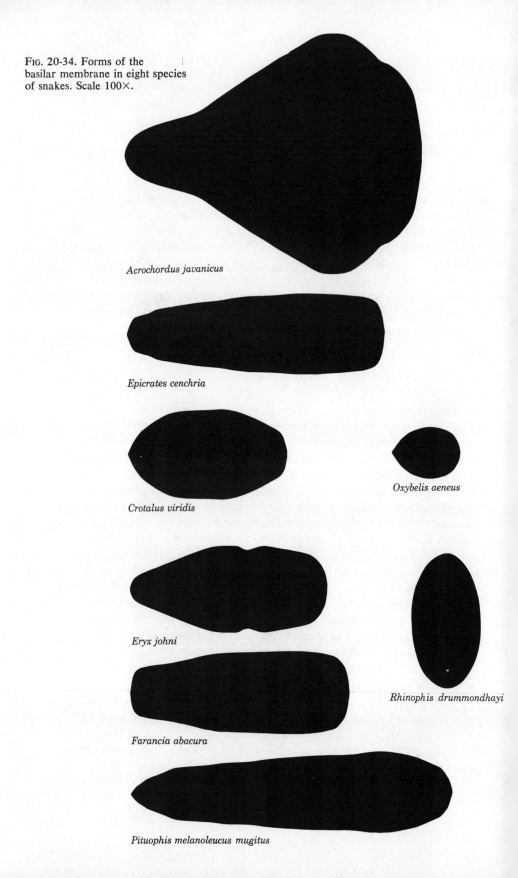

FIG. 20-34. Forms of the basilar membrane in eight species of snakes. Scale 100×.

Acrochordus javanicus

Epicrates cenchria

Crotalus viridis

Oxybelis aeneus

Eryx johni

Rhinophis drummondhayi

Farancia abacura

Pituophis melanoleucus mugitus

Numbers of Hair Cells.   The number of hair cells was determined in all the specimens for which serial sections were obtained, with the results shown in Table 20-II. The size of these populations shows no clear relation to families. The numbers vary greatly, but most are in the range between 169 and 432—all except *Typhlops pusillus, Oxybelis aeneus,* and *Acrochordus javanicus.* The counts in the two *Typhlops* specimens were difficult to make because of poor preservation, and the numbers given should be taken only as indicating an order of magnitude. *Oxybelis aeneus* has a particularly small auditory papilla, and showed the smallest hair-cell population of all the snakes studied. The considerable difference between the two subspecies of *Pituophis melanoleucus* is significant. These two forms are reported to differ also in general habit and behavior. The number of hair cells in *Acrochordus javanicus* is impressive.

TABLE 20-II

HAIR-CELL POPULATIONS IN SNAKES

Species	Number of Ears	Mean number of Hair Cells
Rhinophis drummondhayi	2	162.0
Typhlops pusillus	2	158.0
Corallus enydris	6	209.5
Epicrates cenchria	2	363.5
Eryx conicus	2	275.0
Eryx johni	2	413.0
Elaphe guttata	2	384.0
Elaphe o. obsoleta	4	186.0
Farancia abacura	2	247.5
Lampropeltis getulus holbrooki	2	242.0
Oxybelis aeneus	2	57.5
Pituophis melanoleucus annectens	2	226.5
Pituophis melanoleucus mugitus	2	357.0
Thamnophis s. sirtalis	1	193.0
Acrochordus javanicus	1	1476 + 35[a]
Crotalus viridis	4	202.2

[a] limbic hair cells

Ciliary Orientation Patterns.   An effort was made to determine the orientation patterns of the ciliary tufts of the hair cells along the auditory papillae of the specimens studied. This determination in snakes is difficult because the cilia are slender and their resolution is close to the limits of the light microscope. Usually only a few of the tufts appearing in a given section could be seen clearly enough to show the orientation with complete confidence. The ones that did so nearly always had the longest cilia on the posterior side. The infrequent exceptions, in which the longest cilia were on the anterior side of the tuft, were always situated at the extreme posterior end of the hair-cell row.

This orientation is consistent with that commonly found in lizards for hair cells beneath a tectorial plate. As reported above, this orientation is nearly always lateral. (In snakes the cochlea is rotated in such a manner as to make posterior correspond to lateral in lizards.)

Previous reports on ciliary orientation in the ears of snakes first indicated that this orientation was entirely unidirectional (Mulroy, 1968a; Baird, 1970a), but later Mulroy (1968b) found in two colubrid species that a few hair cells at the end of the papilla had a contrary orientation.

AUDITORY SENSITIVITY

Cochlear potential sensitivity, determined with a needle electrode inserted into the perilymph space of the otic capsule, is shown for a number of species in the following figures, arranged by families.

UROPELTIDAE

Rhinophis drummondhayi. — A sensitivity curve for the small burrowing snake *Rhinophis drummondhayi* is shown in Fig. 20-35. Best sensitivity is at 100 Hz, and the curve falls for higher tones and for the two lower tones used in the tests. The decline of sensitivity becomes particularly rapid above 1500 Hz. This ear must be regarded as having poor sensitivity.

TYPHLOPIDAE

Typhlops schlegeli mucrosa. — A sensitivity function for a large typhlopid species from Swaziland in southeast Africa is shown in Fig. 20-36. The best sensitivity is in the low frequencies up to 1000 Hz, with a sharp maximum at 200 Hz, where a level of −6 db is reached. In the higher frequencies the decline is at a rate of 20 db per octave. The performance must be rated as poor.

BOIDAE

Corallus enydris. — Results for two specimens of the tree boa are given in Fig. 20-37. These curves show a definite maximum at 300 Hz, with rapid decline on either side. In the high frequencies this decline seems to reach a limit, and thereafter the curve varies irregularly. For one of these ears the slope beyond 600 Hz has the extraordinary value of nearly 90 db per octave. The better of these two animals shows very good sensitivity, amounting to −36 db (for the standard response of 0.1 μv) for the frequency 300 Hz. The sensitivity is fair over a range from 100 to 400 Hz.

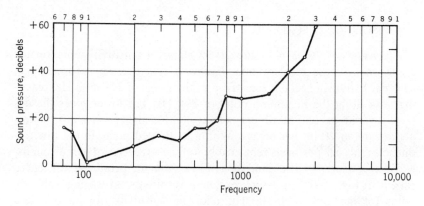

FIG. 20-35. Sensitivity to aerial sounds in *Rhinophis drummondhayi.*

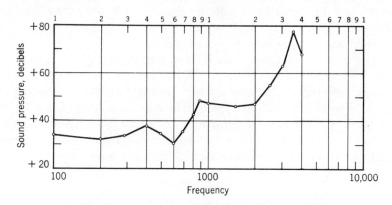

FIG. 20-36. Sensitivity in a specimen of the blindsnake
Typhlops schlegeli mucrosa.

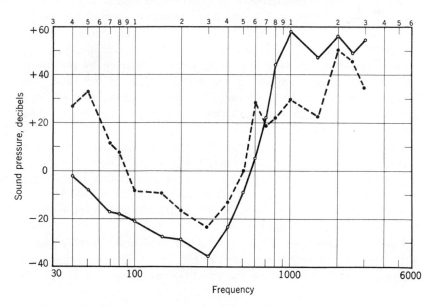

FIG. 20-37. Sensitivity in two specimens of the tree boa, *Corallus enydris.*

Epicrates cenchria. — Figure 20-38 shows a sensitivity function for a specimen of the rainbow boa, *Epicrates cenchria*, an arboreal type. This ear exhibits excellent sensitivity in the range of 150-500 Hz, reaching −44 db at the maximum point of 290 Hz. For lower tones the decline is rapid, around 60 db per octave, and for higher tones it is moderate in rate, at 20 db per octave. Within its limited range in the low frequencies this ear performs remarkably well as a receiver of aerial sounds.

Eryx conicus. — The sand boa, *Eryx conicus*, produced the curve shown in Fig. 20-39, in which excellent sensitivity, averaging −46 db, holds for the tones 200-500 Hz, and fair sensitivity obtains over the range 80-700 Hz. The lower tones show a decline of about 20 db per octave, and the higher tones a decline around 30 db per octave, increasing rapidly as the frequency rises. Like the foregoing, this species has excellent sensitivity in the low-tone range.

Eryx johni. — In this species of sand boa, as Fig. 20-40 shows, the sensitivity reaches a maximum at 290 Hz, and the curve is nearly symmetrical about this point over a range from 100 to 700 Hz. For higher tones the sensitivity declines rapidly. The maximum reached is −24 db, representing only moderate sensitivity. Additional specimens of this species ought to be studied to ascertain whether the present results are typical, in view of the good performance of the other boas.

COLUBRIDAE

Elaphe guttata. — The corn snake, *Elaphe guttata*, one of the colubrids, gave the sensitivity function shown in Fig. 20-41. The performance is very good over a range of 100-500 Hz, averaging about −40 db and reaching −45 db at 200 Hz. The sensitivity declines rapidly for tones outside this range, with a rate around 20 db per octave at the low end of the tested range and one of 50 db per octave at the high end.

Observations were made also on this same specimen by the use of vibratory stimulation applied to the dorsal surface of the skull, with results shown in Fig. 20-42. Despite the difference in the units of measurement (amplitude of motion as compared with sound pressure), these curves are closely similar in form.

Elaphe o. obsoleta. — Sensitivity curves for two specimens of the rat snake, *Elaphe o. obsoleta*, are shown in Fig. 20-43. These animals display only moderate sensitivity, with the best region around 150-600 Hz, and one of the curves reaches a maximum of −27 db at 400 Hz. The roll-off in the low frequencies is 30 db per octave, and in the high frequencies is about 25 db per octave.

Farancia abacura. — A specimen of the mud snake, a burrower in wet areas, gave the sensitivity curve of Fig. 20-44. The best region is in the range 150-500 Hz, where the sensitivity averages −44 db and the

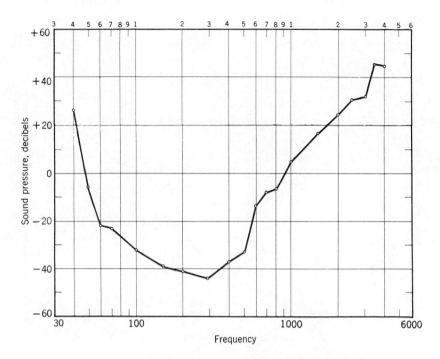

FIG. 20-38. A sensitivity function in the rainbow boa, *Epicrates cenchria*.

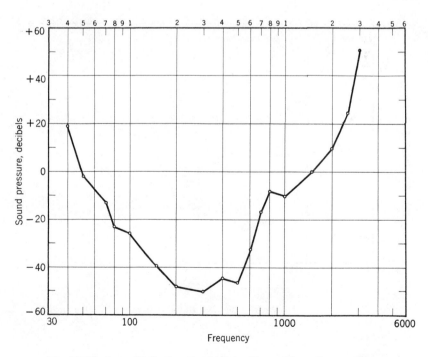

FIG. 20-39. A sensitivity function in Russell's sand boa, *Eryx conicus*.

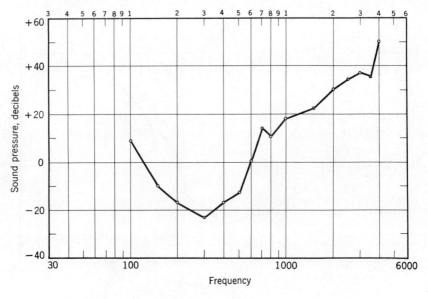

Fig. 20-40. A sensitivity function in John's sand boa, *Eryx johni*.

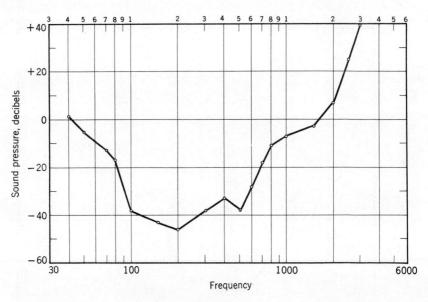

Fig. 20-41. An aerial sensitivity function in the corn snake, *Elaphe guttata*.

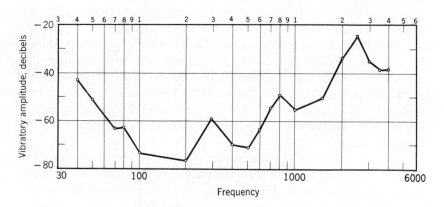

FIG. 20-42. A vibratory sensitivity function in *Elaphe guttata*.

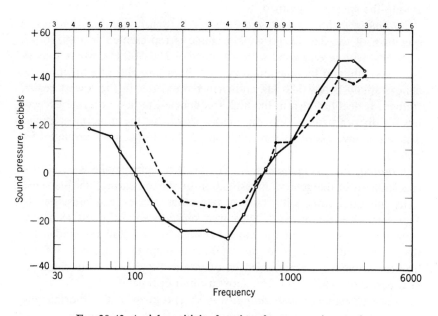

FIG. 20-43. Aerial sensitivity functions for two specimens of
the rat snake, *Elaphe o. obsoleta*.

maximum is −49 db at 400 Hz. The sensitivity is good from 80 to 700 Hz. For lower tones the decline is 30 db per octave and for higher tones it is about 50 db per octave. This ear is highly effective within the low-tone range.

Lampropeltis getulus holbrooki. — The speckled king snake produced the results shown in Fig. 20-45. The sensitivity is excellent over the range 150-600 Hz, with a maximum of −52 db at 200 Hz, and is still good for the wider range of 100-800 Hz. The decline in the low frequencies is at a rate of about 30 db per octave and in the high frequencies at a rate of about 40 db per octave. This ear also is highly effective within its range.

Oxybelis aeneus. — The vine snake, *Oxybelis aeneus*, is an extremely slender arboreal form. Also it is somewhat venomous, with venom ducts at the rear of the mouth. Its sensitivity curve, as shown in Fig. 20-46, has a very flat form with the best response at 400 Hz, where it amounts to −5 db. The sensitivity varies only moderately over the range up to 700 Hz, though it becomes noticeably worse for higher tones.

This ear is poorest in sensitivity, apart from the typhlopoid snakes, among the species examined.

Pituophis melanoleucus annectens. — A sensitivity function for a specimen of the San Diego gopher snake is represented in Fig. 20-47. The sensitivity is very good within a range of 150-500 Hz, where it averages −32 db, and the maximum is −37 db at 290 Hz. The function declines rapidly below 150 Hz and then flattens out in the lowest region studied. It declines also in the high frequencies, at first at a rate exceeding 60 db per octave, and then more slowly in the uppermost range. This ear shows excellent sensitivity in the low frequencies around 150-500 Hz.

Pituophis m. melanoleucus. — The northern pine snake was the first species used in the general investigation of snake hearing, and its sensitivity function, along with those of two other species, was presented by Wever and Vernon in 1960. The curve, reproduced in Fig. 20-48, shows very good sensitivity for the tones between 80 and 200 Hz, and a fair degree of acuity for all tones up to 600 Hz except for a notch at 400 Hz. The low-frequency response is well maintained to 30 Hz, while the higher tones, those above 700 Hz, show a rapid decline.

Pituophis melanoleucus mugitus. — A specimen of the Florida pine snake gave the sensitivity function of Fig. 20-49. This ear shows good sensitivity over a range of 200-600 Hz, and a maximum of −36 db at 500 Hz. The curve at its lower end declines at a rate of about 15 db per octave, and at its upper end at a rate of about 20 db per octave. Like the other pine snakes, this species exhibits good sensitivity for low tones.

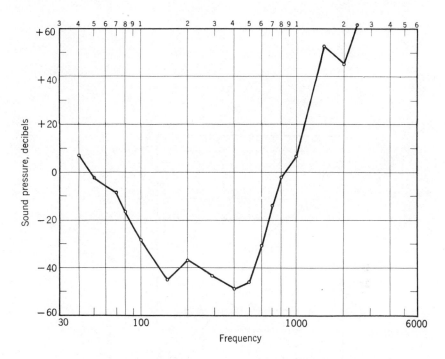

FIG. 20-44. A sensitivity function for the mud snake, *Farancia abacura.*

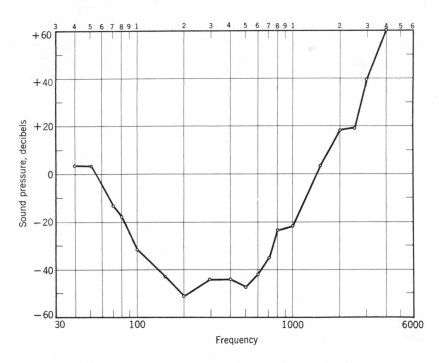

FIG. 20-45. A sensitivity function for the speckled king snake,
Lampropeltis getulus holbrooki.

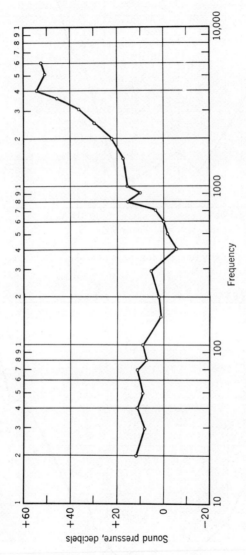

FIG. 20-46. A sensitivity function for the vine snake, *Oxybelis aeneus.*

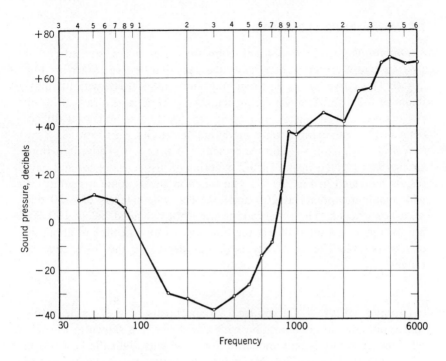

FIG. 20-47. A sensitivity function for the San Diego gopher snake, *Pituophis melanoleucus annectens.*

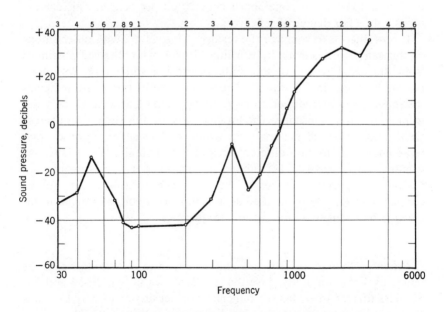

FIG. 20-48. A sensitivity function for the Northern pine snake, *Pituophis m. melanoleucus.* Data from Wever and Vernon, 1960.

Thamnophis s. sirtalis. — The common garter snake, studied earlier (Wever and Vernon, 1960), gave the sensitivity curve shown in Fig. 20-50. This curve has its best region around 200-500 Hz, with the maximum at −24 db. At lower frequencies the decline is at a rate of 20 db per octave, and at higher frequencies the decline is about 40 db per octave. This ear exhibits only moderate sensitivity.

Natrix s. sipedon. — The water snake, *Natrix s. sipedon*, is a third species investigated in the earlier experiment just referred to, and gave results indicated in Fig. 20-51. The function shows some irregularities, with maximum points at 200 and 400 Hz, where levels of −30 and −26 are reached. The observations were not extended below 100 Hz, but for tones beyond 1000 Hz the function shows a decline of about 20 db per octave. The sensitivity is of fair degree in the low-frequency range.

VIPERIDAE

Crotalus viridis. — Curves for two specimens of the prairie rattlesnake, obtained in collaboration with Carl Gans, are shown in Fig. 20-52. The sensitivity is only moderate for these ears, with the best region lying between 100 and 300 Hz, where the better ear attains a level of −19 db. The sensitivity declines very rapidly for lower tones, with a cut-off rate between 40 and 60 db per octave, and though irregular for the higher tones the decline approaches 20 db per octave. This ear exhibits only a fair degree of sensitivity.

It is of interest to compare the results of Fig. 20-52 with observations on this same species made by Hartline (1969), who obtained potentials from the tectal region of the midbrain during stimulation with tone bursts, as shown in Fig. 20-53. The solid curve of this figure indicates the sound pressure required to produce responses that were visually detectable on an oscilloscope. The broken curve on this same graph shows the sound pressures necessary to produce detectable responses that represented activity in a different area of the midbrain. These latter responses are considered as somatic in character and presumably are mediated by receptor organs in the skin. These results testify to the considerable superiority of the ear in sound reception, and agree in indicating best sensitivity for the low frequencies.

ELAPIDAE

Pelamis platurus. — This sea snake, the only representative of its family available for study, gave the results shown in Fig. 20-54 and 20-55. The first of these figures indicates the sensitivity to aerial sounds over a range from 30 to 5000 Hz. Sensitivity is best at the low end of the range, and decreases continuously, rapidly at first until 100 Hz is

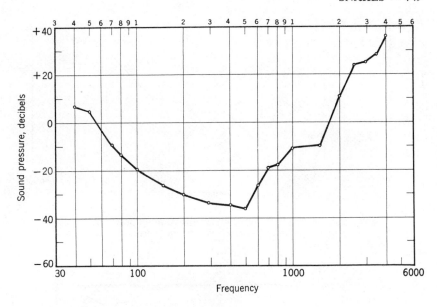

FIG. 20-49. A sensitivity function for the Florida pine snake,
Pituophis melanoleucus mugitus.

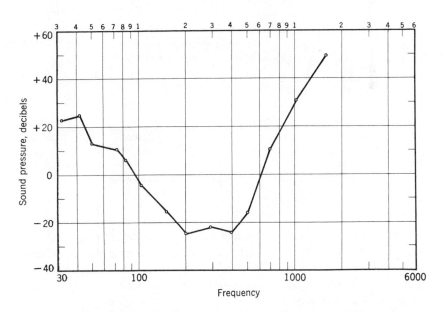

FIG. 20-50. A sensitivity function for the common garter snake,
Thamnophis s. sirtalis. Data from Wever and Vernon, 1960.

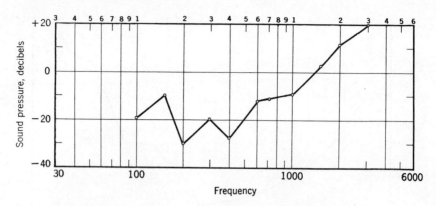

Fig. 20-51. A sensitivity function for the common water snake,
Natrix s. sipedon. Data from Wever and Vernon, 1960.

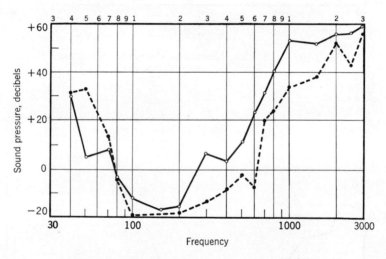

Fig. 20-52. Sensitivity functions for two specimens of the
prairie rattlesnake, *Crotalus viridis.*

reached—where it levels off through the low and middle tone range—
and then rapidly again in the higher frequencies. Even at 30 Hz, where
the sensitivity is best, it must be rated as only moderately good, and the
question must be raised whether these ears are well adapted to the
reception of aerial sounds.

This animal was tested also by the use of vibratory stimuli applied
to the surface of the head over the quadrate, with results shown in Fig.
20-55. This function also indicates poor sensitivity.

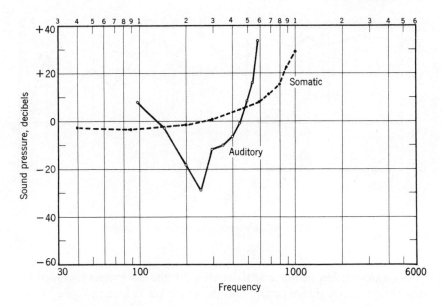

FIG. 20-53. Neural responses of *Crotalus viridis* to sounds as mediated by auditory and somatic receptors. Data from Hartline, 1969.

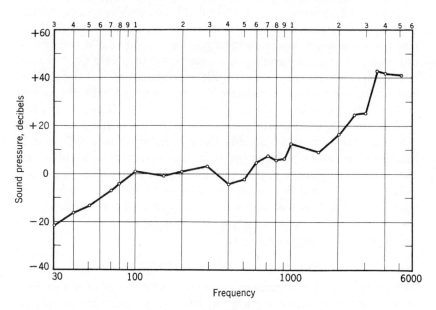

FIG. 20-54. A sensitivity function for the sea snake, *Pelamis platurus,* for aerial sounds.

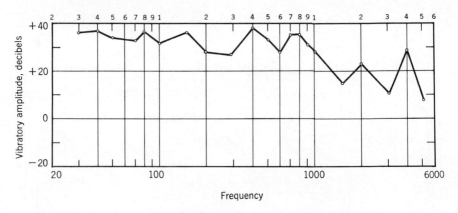

Fig. 20-55. Vibratory sensitivity in *Pelamis platurus*.

An examination of the sensitivity data for the 18 species included in the present study shows clearly that the snake ear is a band-limited low-frequency receptor. Its performance is largely restricted to a relatively narrow range centered usually around 200-400 Hz, though in certain species the best point is shifted as low as 90 Hz, or as high as 500 Hz.

The level of sensitivity attained within this band varies greatly. Among the species examined, the performance is poorest in the uropeltid *Rhinophis drummondhayi*, where the best point is +2 db at 300 Hz. Only a little better is *Oxybelis aeneus*, with a maximum of −5 db (at 400 Hz), and *Typhlops schlegeli mucrosa* with a maximum of −6 db (at 200 Hz). Then come *Crotalus viridis* with a maximum of −19 db (at 100 Hz), and *Pelamis platurus* with best sensitivity of −23 db (at 30 Hz).

The remaining species are included in a sensitivity range of −24 to −52 db, and average −38.3 db. There is no significant difference in maximum sensitivity between the boids and colubrids included in this study: the four boids average −38.5 db and the eight colubrids average −38.2 db. Of course the sample is too small to permit any safe generalization for these groups as a whole, but neither the averages nor the distributions within these two groups give any suggestion of a systematic difference in the ear's capabilities.

SENSITIVITY AND HAIR-CELL POPULATIONS

The relation between maximum sensitivity and the number of hair cells in the auditory papilla is a matter of interest. A correlation was determined for the 14 species for which sufficient data were available, and a coefficient of 0.61 was obtained. This correlation was statistically significant ($p = 0.02$), but it must be pointed out that its high value was mainly determined by the colubrids in the group, and was especially contributed to by *Oxybelis aeneus* with its very poor sensitivity and small hair-cell population. A much larger number of species will be needed to explore this relationship in a fully satisfactory manner.

AN EVALUATION OF HEARING IN SNAKES

In view of the peculiar structural characteristics of the snake's ear, it is surprising that the sensitivity within the favored portion of the acoustic range is as high as it is found to be. Within its restricted range in the low frequencies, the performance of this ear in most species is about on a par with that observed in the majority of lizards.

This unexpected performance is clearly the result of a number of special adaptations that have in large measure compensated for the loss of the external ear opening and tympanic membrane in this group of reptiles. These adaptations include both peripheral and cochlear processes.

The Peripheral Processes. The special features responsible for the effective reception and transmission of aerial sounds have already been given close attention and here need only be summarized.

The skin and muscle mass at the side of the head, along with the quadrate beneath, constitute the receptive surface for aerial sounds. The muscle layer over the quadrate is relatively thin, and it has been shown that this layer together with the skin may be removed without any appreciable alteration in the ear's performance. The quadrate itself is the significant element in the initial reception of sound.

The freeing of the quadrate from the skull and its elastic suspension over the side of the head have provided an acoustic receptor plate whose mobility is assured by the presence beneath of a shallow middle ear cavity.

The natural frequency of this receptive structure will be determined by the mass and elasticity of the tissues. The experiments described above appear to eliminate the mass of skin and muscle layers in this consideration, and to leave the quadrate as the critical mass factor. These experiments indicate further that the ligamentary attachments of the quadrate are insignificant as elastic factors, and what remains seems to be the resiliency of the tissues underlying the quadrate, and perhaps primarily the compressibility of the thin air cushion beneath it. Also some contribution to the elasticity, though no doubt a minor one, can be expected from the columella and the attachment of its footplate in the oval window, together with the reaction of the cochlear contents beyond. Thus a second significant role, in addition to that of ossicular mobility, can be found for the middle ear cavity that the snake's ear has retained.

The natural frequency that these conditions determine is indicated as varying somewhat with species, but usually lies between 100 and 700 Hz, and most often is at 300 Hz.

The quadrate operates in response to sounds in a simple fashion, in the manner of a floating plate, and transmits its vibrations through a special form of articulation to the outer end of the columella. This articulation provides effective transmission of lateral movements but permits slippage along the axis of the columella, thereby reducing displacements that might be disruptive to the attachments of the footplate in the oval window. This arrangement serves to maintain the integrity of the system in the presence of accidental forces of various kinds, and under the great stresses produced when the snake strikes at prey with wide open mouth and then distends the head in swallowing.

The motion transferred to the columella evidently produces a rocking action of the footplate, probably about the posteromedial edge of the oval window. Because the length of the columellar shaft is usually about twice the width of the footplate, the columella acts as a reducing lever, decreasing the vibratory amplitude to about half and doubling the force exerted on the cochlear fluids.

The large quadrate surface relative to the area of the columellar footplate provides a further force amplification, which from measurements in a specimen of *Pituophis m. melanoleucus* amounts to about 12 times.

Inner Ear Processes.   The volume displacements of the columellar footplate are transmitted to the cochlear fluids, and a vibratory fluid flow is set up around a circuit that passes through the basilar membrane, involving this membrane and the auditory papilla borne upon it.

The area of the basilar membrane is small relative to that of the columellar footplate. Accordingly, the footplate motion is greatly magnified in the action of the basilar membrane and hence in the displacement of the hair cells. From measurements on a specimen of *Pituophis m. melanoleucus*, in which the footplate area was 3.83 sq mm and the total basilar membrane area was 0.16 sq mm, this displacement amplification is estimated as at least 23.3 times. No doubt the actual amplification is greater than this, because the basilar membrane can not be expected to yield equally over its entire surface. It should be deflected most along its middle portion, thus producing a rocking motion of the papilla at its anterior attachment to the limbus (see Fig. 20-20). The most posterior hair cells therefore should be the ones most strongly stimulated. Further, if there is a degree of differentiation along the cochlea, as seems likely in many species, the action should be more vigorous in certain places.

This differential action, even in the most elongate cochleas, is not expected to be more than moderate. The dimensional variations never approach the degree found in many lizards. Moreover, the presence of the tectorial plate over almost the whole papilla, interconnecting the ciliary tufts of nearly all the hair cells, tends to unify the action. This unifying influence aids the sensitivity by bringing into concerted response a large population of hair cells. The mainly unidirectional orientation of the ciliary tufts continues this concerted action at the next stage of neural excitation, so that the great majority of nerve fibers supplying the hair cells may be expected to fire in close synchrony.

It seems likely that the special adaptations of the snake's ear make it mainly suited for the function of sound detection within this ear's limited frequency range. Other more complex functions, such as frequency discrimination and tonal analysis, probably are not served in any large degree through a spatial differentiation of the action along the cochlea. Indeed there is hardly any need for such differentiation within the limited low-tone range that this ear displays. Frequency following in the cochlear nerve discharge, even in these ectothermal animals, ought to be entirely adequate to represent tonal quality at this initial level of the auditory nervous system. The limits of discrimination, therefore, will be determined by the degree of development of higher levels of the nervous system. Unfortunately, nothing is known about the discriminative capabilities of these ears, and further consideration of the problems here raised must await the development of this kind of understanding.

21. SUBORDER AMPHISBAENIA:

THE AMPHISBAENIANS

The amphisbaenians constitute a small group of little-known, wormlike reptiles. They are true burrowers, constructing their own tunnels underground, often to considerable depths, and only rarely coming to the surface. Indeed, some species probably never leave their burrows of their own accord, and the ones that do so come forth at night or in dim light. Accordingly these animals are seldom seen, and are known to few persons apart from the specialists. They lack a common name in English, though the Germans call them "Doppelschleichen"—two-way crawlers —referring, as the latinized name does also to their ability to move freely forward and backward in their narrow tunnels. As an example, the large South American species *Amphisbaena alba* is represented in Fig. 21-1.

Fig. 21-1. The large South American species *Amphisbaena alba*. Drawing by Anne Cox.

The relations of these reptiles to other groups have always been uncertain. Linnaeus (1758) classified them among the snakes, and most others during the next century followed this lead. Later authors have been more inclined to group them with the lizards, usually as a special family or superfamily (Camp, 1923; Romer, 1956).

These reptiles assuredly belong among the Squamata, for they show many features in common with both lizards and snakes, and yet there are numerous points of difference. The best current opinion is that they merit independent status as a suborder or order coordinate with the Sauria and Serpentes (Gans, 1969, 1974). Certainly their sound recep-

tive system, which is unique in form and surprisingly effective in operation, provides evidence consonant with this distinctive treatment.

The amphisbaenians are active predators, feeding on worms, arthropods, crustaceans, and the like—probably any small creatures that they
can find and capture in their subterranean haunts. They are limbless except for three species of the genus *Bipes*, which have short but well-
developed forelegs. They vary considerably in size; body length runs
from 8 to 60 cm, and the diameter is from 3 to 25 mm. The skin of the
trunk is divided into a series of rings, each formed of a number of
roughly rectangular plates, and the skin is loose enough to permit movements of rectilinear crawling, one of their most important means of
locomotion. The head is greatly modified, with close union of the bones
of the skull and special elaboration of the snout to a conical or shovel-
like form to facilitate their excavation and packing of soil in tunnel
building.

The suborder Amphisbaenia contains 23 genera divided among 130
to 140 species, together with several subspecies. Gans (1974) recognized the four families Amphisbaenidae, Bipedidae, Rhineuridae, and
Trogonophidae.

Much the largest of these families is Amphisbaenidae, with representatives in both hemispheres and heavy concentrations in South America and southern Africa. The Bipedidae include a single genus with
three species occurring in Mexico. The Rhineuridae are only a remnant,
containing the single species *Rhineura floridana*, limited to north-cen-
ral Florida. The Trogonophidae include four genera with only six species, and occupy eastern and northwestern fringes of the African con-
tinent.

These groups are distinguished by such features as tooth structure,
the shape of the skull, and movement patterns used in digging. For the
Trogonophidae the teeth are acrodont, with their bases attached to the
crest of the jaw bones, which lack a dental groove. For the others the
teeth are pleurodont, with the bases attached along the sides of a shallow
groove.

The Bipedidae as just mentioned have well-developed forelegs, provided with formidable claws. No other amphisbaenians show even a
trace of limbs.

The Rhineuridae present several distinctive features: a vestigial
though still marginally functional eye, a snout of spatulate, horizon-
tally oriented form, a nuchal fold, and traces of pectoral shields (Gans,
1967).

The present study includes 14 species, with nine of these belonging to
the Amphisbaenidae, three to the Trogonophidae, and one each to the
other two families, though no living specimens of *Rhineura* were ob-

tained and this species could not be tested by means of cochlear potentials.

The species and the number of specimens examined are the following:

AMPHISBAENIDAE

Amphisbaena alba (Linnaeus, 1758)—3 specimens; *Amphisbaena darwini trachura* (Cope, 1885)—1 specimen; *Amphisbaena gonavensis leberi* (Thomas, 1965)—5 specimens; *Amphisbaena manni* (Barbour, 1914)—7 specimens; *Blanus cinereus* (Vandelli, 1797)—3 specimens; *Chirindia langi* (FitzSimons, 1939)—2 specimens; *Cynisca leucura* (Duméril and Bibron, 1839)—3 specimens; *Monopeltis c. capensis* (A. Smith, 1848)—2 specimens; *Zygaspis violacea* (W.C.H. Peters, 1854) —3 specimens.

BIPEDIDAE

Bipes biporus (Cope, 1894)—8 specimens.

RHINEURIDAE

Rhineura floridana (Baird, 1858)—3 specimens (preserved).

TROGONOPHIDAE

Agamodon anguliceps (Peters, 1882)—2 specimens; *Diplometopon zarudnyi* (Nikolski, 1907)—3 specimens; *Trogonophis wiegmanni* (Kaup, 1830)—3 specimens.

Nearly all these specimens were obtained by Carl Gans and studied in collaborative experiments (Gans and Wever, 1972, 1974, 1975; Wever and Gans, 1972, 1973).

Early observations on the ear of amphisbaenians were made by Versluys (1898), who extended some work of Stannius (1880), Peters (1880, 1882), and Smalian (1885). These investigators agreed that in this animal a tympanic membrane and tympanic cavity are lacking. They identified the middle ear structures: the columella or stapes in the oval window and the forward-extending extracolumella with its attachment to the skin of the face. But because of the absence of a tympanic membrane Versluys concluded that these structures had no significance for hearing. Thereafter for a long time the amphisbaenian ear received little further attention, until Gans (1960) described the columella and extracolumella in further detail, with careful consideration of their relations to the skull, and suggested—contrary to early opinion—that the middle ear structures have the function of conveying sound vibrations to the cochlea. Further he developed the hypothesis that in the amphisbaenians the ear constitutes the primary sense through which this animal detects and pursues its prey. For a creature spending most or all of its life in a labyrinth of tunnels underground, the other senses are almost excluded from significant function in the food finding process, but sounds produced by a worm or insect breaking through a tunnel wall should readily be detected and at least roughly localized. Experimental evidence on this question will be given later, after consideration of further details of ear structure in the various species.

MIDDLE EAR

What may be regarded as the standard form of sound conductive mechanism in these animals, though certain species show significant departures from it, will now be described, with *Amphisbaena manni* as the main example.

A side view of the skull of *Amphisbaena manni*, with the middle ear mechanism left in place, is presented in Fig. 21-2. The oval window lies far back on the skull, below the quadrate, and faces laterally and ventrally. Figure 21-3 gives a detailed picture of the columellar footplate in its window and the articulation of its head portion with the extracolumella. The oval window and footplate are relatively large, covering the whole anterolateral surface of the otic capsule.

The shaft of the columella extends ventrolaterally and slightly anteriorly, but hardly projects beyond the surrounding rim of the window. These relations are further indicated in Fig. 21-4; in addition the skin is represented with extensive thickening of its dermal layer anteriorly and the penetration of this layer by the extracolumella. The thickening occurs in the region covered by the second infralabial skin shield, an enlarged shield along the lower jaw as shown in Fig. 21-5.

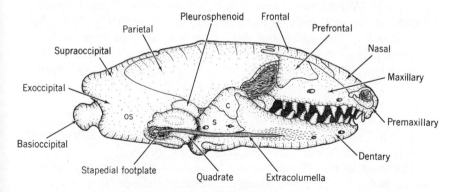

FIG. 21-2. The skull of *Amphisbaena manni*, with the middle ear mechanism in situ. *c*, coronoid; *s*, supra-angular; *os*, the sphenoid region, including the opisthotic.

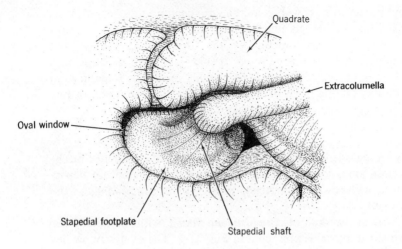

FIG. 21-3. The oval window region of *Amphisbaena manni*. Scale 32×.
From Wever and Gans, 1973.

Anterior to its articulation with the columella, the extracolumella runs
through dense connective and muscle tissues, passing first along a little
furrow in the dorsolateral surface of the quadrate. More anteriorly this
rod of cartilage emerges freely into the subcutaneous space. Its anterior
end continues as cartilage through the deep portion of the dermal thickening, then turns into a thin thread of connective tissue and terminates.

The picture just presented for *Amphisbaena manni* is repeated in
much the same form in seven other species examined, including *Amphisbaena alba, Amphisbaena darwini trachura, Chirindia langi, Cynisca
leucura, Monopeltis c. capensis, Zygaspis violacea*, and *Trogonophis
wiegmanni* (Wever and Gans, 1973). The columellar footplate varies
greatly in size and shape in these species, as shown in Fig. 21-6. The
three species of *Amphisbaena* show close similarity of form, with an
anterior portion that is curved on its medial surface to make a close
articulation with the prootic wall. This similarity extends in lesser degree
to *Trogonophis* and *Monopeltis*. The remaining footplates vary considerably in form and are the smallest encountered.

There are differences also in the directions in which the footplates
face. For *Chirindia langi, Zygaspis violacea*, and *Cynisca leucura*, the
direction is much more anterior than for the others.

The footplate is held in its window by an annular ligament, represented
in two places in Fig. 21-4. Along the anteromedial portion of the footplate this ligament is thin, forming a tight connection, and along its posterolateral portion the gap is wider and the ligamentary connection ap-

Anterior ↑

Lateral →

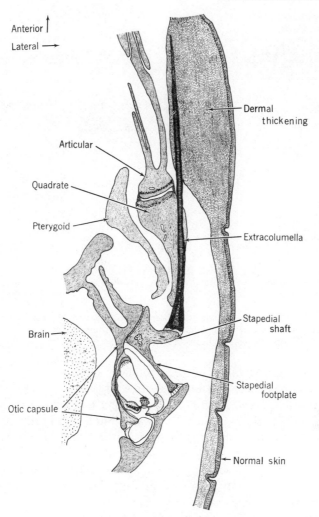

Dermal
thickening

Articular

Quadrate

Pterygoid

Extracolumella

Brain

Stapedial
shaft

Stapedial
footplate

Otic capsule

Normal skin

FIG. 21-4. The ear of *Amphisbaena manni* in a frontal section.
Scale 25×. From Wever and Gans, 1973.

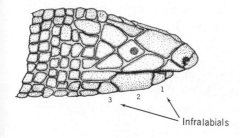

3   2   1

Infralabials

FIG. 21-5. The head of a specimen of
Amphisbaena manni, showing the
scalation. The eye is seen as a dark
area, covered by a scale.

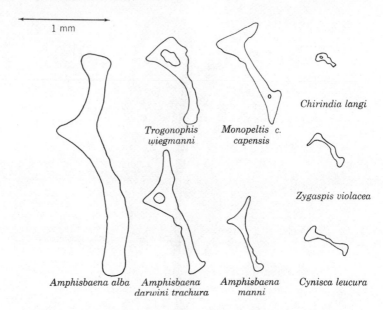

FIG. 21-6. The columellar footplate in eight species of amphisbaenians. Drawn from frontal sections through their midportions. Scale 25×. From Wever and Gans, 1973.

parently looser. Mobility of the footplate appears to be somewhat limited anteromedially and relatively free elsewhere.

In *Cynisca leucura, Zygaspis violacea*, and *Chirindia langi*, the columellar shaft is relatively long. In *Cynisca* the length nearly equals the width of the footplate, in *Zygaspis* it is a little longer, and in *Chirindia* this shaft is about twice as long as the footplate (Fig. 21-7). In these three species there is no joint cleft between columella and extracolumella; the osseous shaft goes over into cartilage without any physical discontinuity. In *Trogonophis wiegmanni* the shaft is short, equal to about half the width of the footplate. It runs forward obliquely and turns into the extracolumella without interruption.

All the species of *Amphisbaena* show short shafts similar to that represented for *Amphisbaena manni*. The shaft runs laterally and ventrally, and ends in a flat surface to which the extracolumella is attached.

Monopeltis c. capensis lacks a shaft. As Fig. 21-8 shows, the anterolateral surface of the footplate bears a large rounded boss that serves for the articulation of the extracolumella. The joint surfaces are well defined, as illustrated. This observation is in disagreement with Kritzinger's statement (1946), based on the examination of an animal of indeterminate size, that in this species the columella and extracolumella are fused.

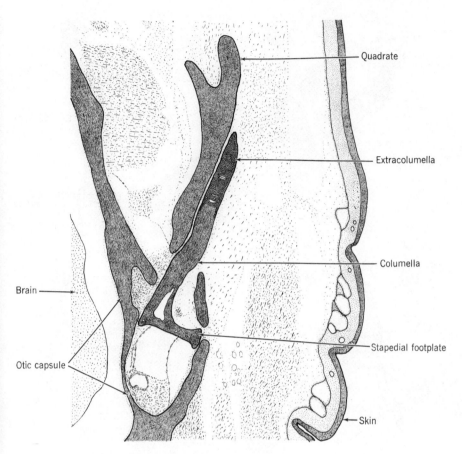

Fig. 21-7. The otic capsule and columellar system of *Chirindia langi*, in frontal section. Scale 75X.

In the three species of *Amphisbaena* studied and in *Monopeltis c. capensis* the extracolumella arises at a somewhat lateral position and runs almost directly in an anterior direction along the external surface of the quadrate. In *Chirindia langi, Zygaspis violacea*, and *Cyniscus leucura* the course of the extracolumella is at first anterolateral and then becomes more anterior beyond the quadrate.

In *Cynisca leucura* and *Amphisbaena alba* the extracolumella extends farther forward, but terminates well behind the most posterior tooth. In *Chirindia langi* this process almost reaches the last tooth, and in *Amphisbaena manni* and *Amphisbaena darwini trachura* it extends past this tooth. In *Monopeltis c. capensis* it goes still farther, almost to the level of the next to last tooth.

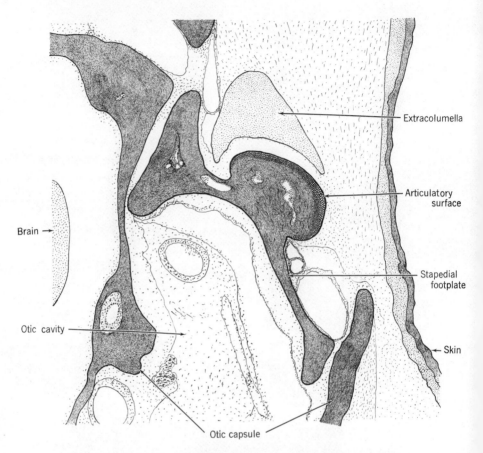

FIG. 21-8. The ear region of *Monopeltis c. capensis*, in frontal section. Scale 75×. From Wever and Gans, 1973.

In most species studied the extracolumella enters the dermal pad in the region of the face and runs anteriorly just below its inner surface, as shown in Figs. 21-4 and 21-9. In *Zygaspis violacea* and *Amphisbaena darwini trachura* this cartilage runs a little more laterally, well below the inner surface of the dermal pad, and in *Amphisbaena alba* it is almost in the middle of the pad as shown in Fig. 21-10.

In all these animals the anterior end of the extracolumella makes a firm connection with the thickened inner layer of the skin. Therefore this element is capable of being set in vibration by sound pressures exerted on the skin surface in the region of the face, and conveys these vibrations to the cochlea as further described below.

Middle Ear in *Bipes biporus*.   The species *Bipes biporus*, best known
along with the other two species of this genus for the possession of forelegs, is also distinctive in the form of its sound-receptive mechanism.
This mechanism lacks the extracolumella and consists only of a columellar element that extends directly to the skin in the region of the
neck constriction. This region is indicated in Fig. 21-11, where a lateral
view is given of the anterior portion of one of these animals.

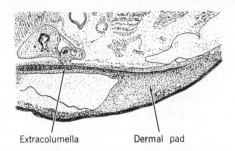

FIG. 21-9. The facial region of *Cynisca leucura*,
showing the dermal pad and course of the
extracolumella along it. From Wever and Gans, 1973.

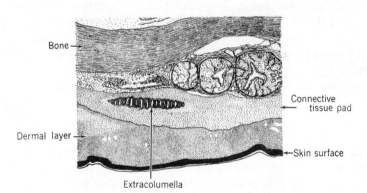

FIG. 21-10. The facial region in *Amphisbaena alba*.
Scale 20×. From Wever and Gans, 1973.

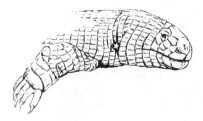

FIG. 21-11. The anterior portion of a
specimen of *Bipes biporus*, with the
location of the columella indicated by
a star. Drawing by Anne Cox.

The oval window in this species lies in the posterolateral wall of the otic capsule, rather than in the anterolateral wall as in most other amphisbaenians. As Fig. 21-12 shows, this location is ventral to the quadrate, and the head of the columellar shaft extends laterally beyond the surface of this bone. These relations are shown in further detail in the cross-sectional view of Fig. 21-13. As this figure indicates, the footplate lies in the large oval window and sends its shaft posterolaterally and a little ventrally toward the skin surface. The inner portion of the columella is osseous, and gives way without physical discontinuity to an outer cartilaginous head portion. The end of this head process flares to form a terminal disk with a concave surface. The cartilage of the disk shows a moderate amount of calcification.

Held within the shallow concavity of the disk is a pad of connective tissue, and a thick ligamentary band runs anteriorly over this pad to attach to the capsule of the pterygo-quadrate joint and the retroarticular process of the mandible, as shown in Fig. 21-14. This ligamentary band is an extension of two flat muscles as shown. One of these muscles was traced along the posterolateral region of the skull to the level of the posterior semicircular canal, where it has its final attachment on the prootic, though many of its fibers were found to attach to the side of the skull along its course. The second muscle extends posteriorly and somewhat dorsally as indicated, but its complete course was difficult to follow. In the dissection its fibers were lost in the complex musculature of the shoulder region; this muscle probably arises in the superficial fascia of this region.

The terminal disk with its ligamentous pad lies directly beneath the infolded skin of the neck constriction, as Fig. 21-14 shows. The areas of skin anterior and posterior to this constriction are attached to deep tissues by means of thick ligamentary sheets that follow the pattern of the annular rings, but by careful dissection it was established that such attachments are absent for the vertical strip of skin within the lateral part of the neck constriction. This area of skin is relatively free, and when exposed to sound pressures it undergoes vibratory motions that are transmitted to the terminal disk below and then through the columella to the cochlea.

Middle Ear in *Blanus cinereus*.   The sound-conductive mechanism of *Blanus cinereus* bears considerable resemblance to that of *Bipes biporus*. The oval window is located far back on the skull, in a recess formed by the overhanging quadrate and the articular portion of the mandible, and the nearly round columellar footplate faces ventrally, laterally, and a little posteriorly, as shown in Fig. 21-15. The short, blunt shaft of the columella quickly goes over into cartilage, and the end of this cartilag-

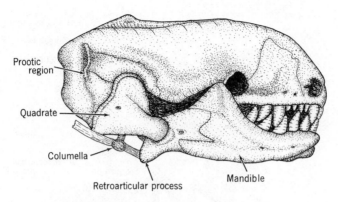

FIG. 21-12. The skull of *Bipes biporus*, with the columella in situ. Scale 10×. From Wever and Gans, 1972.

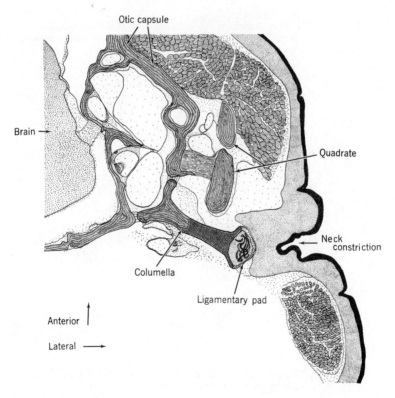

FIG. 21-13. Transverse section of the ear region of *Bipes biporus*.

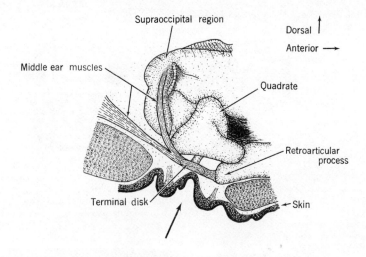

FIG. 21-14. Terminal disk of the columella of *Bipes biporus* and its connections. The large arrow suggests the action of sound waves. From Wever and Gans, 1972.

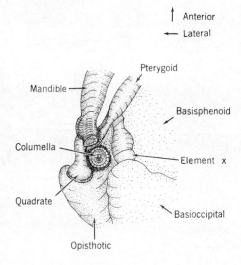

FIG. 21-15. The ear region of *Blanus cinereus* in a ventral view; right side. Scale 8×. From Gans and Wever, 1975.

inous portion flares at the end and shows a concave outer surface, just as described for *Bipes*. Then, in distinction from *Bipes*, a further cartilaginous element is present in the form of a roughly rectangular plate, which extends a short distance posteriorly but mainly runs anterolaterally. These relations are shown in the cross section of Fig. 21-16, and the form of the whole columellar structure is represented in Fig. 21-17.

As seen in Fig. 21-16, the main portion of the terminal plate passes anterior to the depressor mandibulae muscle into a dense mass of connective tissue that adjoins the moderately thickened dermal layer in this region.

The footplate of the columella and inner portion of its shaft are osseous. The cartilaginous portion of the shaft shows light calcification in its flared outer end, as does the cartilage of the terminal plate. The terminal plate is attached to the end of the shaft in one region, but only lightly, for it comes away easily.

The sound-receiving surface in *Blanus* is located far back on the head, though not as far posterior as in *Bipes*. Sound pressures exerted on the lateral skin about the level of the first body annulus will be transmitted through dermal and subdermal tissues to the terminal plate and thereby to the columella, finally reaching the fluids of the cochlea.

Middle Ear in *Rhineura floridana*.   In its basic features the middle ear in this species resembles that found in most amphisbaenians, but the

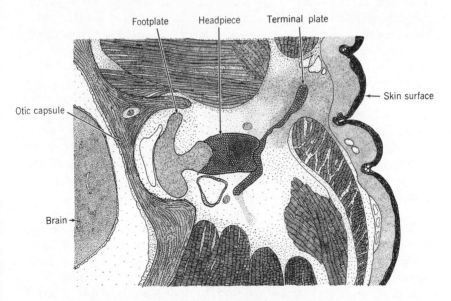

FIG. 21-16. Frontal section of the middle ear region of *Blanus cinereus*. Scale 50×. From Gans and Wever, 1975.

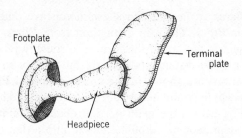

FIG. 21-17. The columella and its junction with
the terminal plate in *Blanus cinereus*.
From Gans and Wever, 1975.

connections made to facial and lip regions are of distinctive form. Figure 21-18 shows the columellar system in a lateral view, and a detailed sketch of the isolated mechanism is given in Fig. 21-19. A relatively large columellar footplate faces laterally and slightly anteriorly, and sends off a slender shaft that first runs laterally and a little dorsally and then swings in a nearly anterior direction, expanding greatly as it does so to a blunt spatulate form. The headpiece articulates by means of dense fibrous material with a short extracolumella composed of lightly calcified cartilage. This extracolumella forms two kinds of connections with the skin. One of these connections is like that found in most other amphisbaenians, in which the extracolumella extends to a specially thickened dermal layer in the facial region. In this species, however, the dermal thickening is only moderate and the extracolumella penetrates it only a short ditance, as seen in Fig. 21-20. The second form of connection, already represented in Fig. 21-18, is highly elaborate and clearly more effective: the extracolumella is continued by a dense mass of ligamentary fibers that runs forward along the face and divides into two branches. One branch extends along the lower jaw, and the other, which is somewhat larger, swings dorsally and then anteriorly along the upper jaw. These fibrous extensions are firmly attached to spongy tissue running along upper and lower lips, and also are closely bound to the overlying skin. It is easily demonstrated that gross movements of the facial skin of the area or of the fibrous masses themselves are communicated to the columellar footplate.

Middle Ear of the Trogonophidae.   In *Trogonophis wiegmanni* the otic capsule is located far back in the skull, and the large oval window faces ventrolaterally. The columellar footplate sends its relatively short shaft anterolaterally and a little ventrally. This footplate and the initial portion of the shaft are osseous, and then the shaft expands to form a broad cartilaginous portion, which connects to a cartilaginous extracolumella.

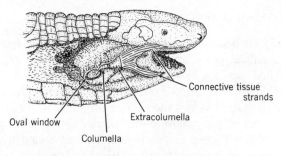

Fig. 21-18. The sound-receptive mechanism in
Rhineura floridana.

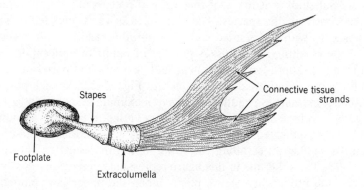

Fig. 21-19. Detail of the columellar system in *Rhineura floridana*. Scale 12×.

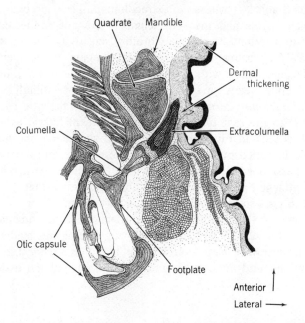

Fig. 21-20. A frontal section through the ear region of *Rhineura floridana*, showing the extracolumellar connection to the skin. Scale 25×.

There is no joint between these elements, but a partial cleft in a region of hyalin cartilage provides a clear demarcation between regions of lightly calcified cartilage in headpiece and extracolumella.

The extracolumella tapers down rapidly, continues the anterolateral direction at first, then swings more anteriorly around the quadrate, and ends at the level of the quadrato-mandibular articulation, as shown in Fig. 21-21. It is still deep in this region, well separated from the skin, but is continued by a broad band of dense connective tissue that runs obliquely into the dermal layer of the third infralabial region of the face. A few fibers of this connective tissue band extend also to the upper jaw.

A special study was made of the regional pattern of sound reception in a specimen of this species. The skin surface was stimulated at various positions with the blunted end of a vibrating needle, with observation of the resulting potentials. A number of points of application were used along the supra- and infralabials and also in a line posterior to the mouth region, as indicated in the sketch of Fig. 21-22. At each position the contact pressure was adjusted to give maximum response for a tone of 400 Hz, and the stimulus amplitude was varied to produce an output of 0.2 μv. As indicated, a line of 11 positions was used along the upper lip and a similar series along the lower lip, with results as shown in Fig. 21-23. The solid line in this figure represents points along the lower jaw, and the broken line shows points along the upper jaw, plotted in terms of distance from the tip of the jaw. There is little difference among these positions as long as they are well forward, ahead of the region (positions 8 and 20) where the extracolumella lies deep below the skin and muscle layers and the skin segments are not connected with it. From positions 9 and 21 on, the vibratory sensitivity falls off sharply to a level 12-15 db below its value in the face region. It is evident that the columellar mechanism with its regional connections to the skin provides a very effective means for the detection of vibratory stimuli.

In *Diplometopon zarudnyi* as in *Trogonophis* the otic capsules are located far to the rear of the head, in prominent lateral extensions of the skull as seen in Fig. 21-24. The sound-conductive mechanism has been well described by Gans (1960). There is an enormous columella with its head portion articulating with a platelike extracolumella, to which is attached a thick bundle of connective tissue that divides anteriorly, with portions extending to the facial regions.

The forms of the columella and its attached extracolumella are represented in Fig. 21-25 and 21-26, showing lateral and medial aspects. These drawings were made from a model based on a reconstruction from serial sections, and indicate the massive character of the ossicular elements.

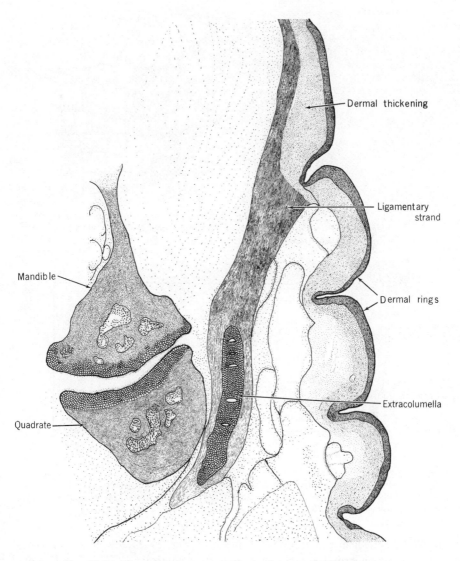

FIG. 21-21. The columellar connection to the dermal thickening in *Trogonophis wiegmanni*. From Wever and Gans, 1973.

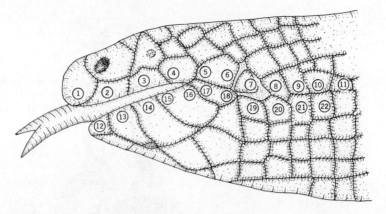

FIG. 21-22. The left side of the head of a specimen of *Trogonophis wiegmanni* showing the scalation and locations used for vibratory stimulation. From Gans and Wever, 1972.

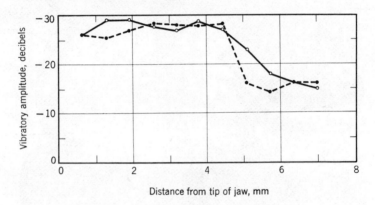

FIG. 21-23. Sensitivity in a specimen of *Trogonophis wiegmanni* stimulated at points along the upper jaw (broken line) and along the lower jaw (solid line). From Gans and Wever, 1972.

As shown in the sectional drawing of Fig. 21-27, the footplate goes over into a broad conical shaft that runs anterolaterally and ventrally. The end of this shaft is cartilaginous and is extended ventrally as a small rod of cartilage whose interior is partly calcified, and whose anterior surface is flattened and articulates with the shaft of a spatulate extracolumella that extends forward into the thickened dermal layer of the skin over the lateral surface of the head. The posterior third of this extracolumella remains cartilaginous, and its outer surface is heavily calcified. The anterior portion retains a core of simple, clear cartilage while the outer layer is osseous and becomes solid near the end. This

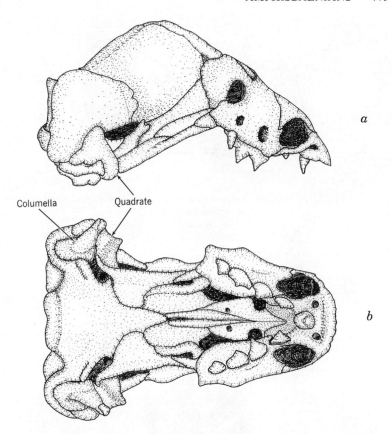

Fig. 21-24. The skull of *Diplometopon zarudnyi* in (*a*) lateral and (*b*) ventral views. After Gans, 1960; from drawings by S. B. Mc-Dowell.

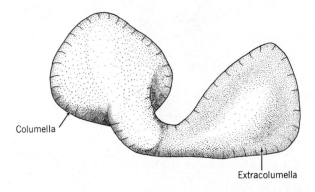

Fig. 21-25. The columellar mechanism in *Diplometopon zarudnyi* in a lateral view. Scale 25×. From Gans and Wever, 1975.

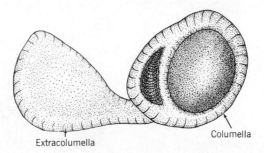

FIG. 21-26. The columellar mechanism of *Diplometopon zarudnyi* in a medial view. Scale 25×.

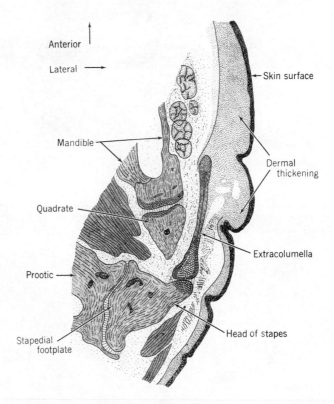

FIG. 21-27. The columellar mechanism of *Diplometopon zarudnyi* in a frontal section. Anterior is above, lateral to the right.

osseous end is embedded in the deep part of a greatly thickened dermal layer in the region of the second and third infralabial segments. This element completely traverses the third infralabial and enters the second infralabial a short distance, reaching a point well ahead of the quadrato-mandibular articulation. Because of the considerable thickening of the skin, the mechanical coupling of this element appears to extend in marked degree through the entire region of the second infralabial and into the first as well. There is also a connection between the connective tissue band and areas of the upper lip region (Gans, 1960).

The species *Agamodon anguliceps* was studied by Peters (1882), who described and pictured the skull. Two specimens were later examined by Gans (1960), who described the ear structures in considerable detail. The otic capsules are located in lateral bulges at the extreme posterior end of the skull, as shown in Fig. 21-28. The form is closely similar to that of *Diplometopon*, including the ventrolateral location of the oval window and the deep oblique position of the columella. The head process of the columella extends ventrolaterally and also anteriorly, and connects to the extracolumella (removed in these drawings) by a definite articulation.

The complete columellar structure is shown in Fig. 21-29, sketched from a model based on a reconstruction from serial sections. The extracolumella extends forward and expands as a thin triangular plate, which in these juvenile specimens (at least) consists entirely of cartilage. Posteriorly this plate lies deep beneath the skin and a thin muscle layer, and as it runs forward it comes closer to the surface and, near its anterior end, passes ahead of the muscle layer and enters the deepest part of the dermal thickening. It enters the most posterior part of the second infralabial segment, but penetrates only a short distance, as shown in Fig. 21-30.

The great expansion of this anterior blade of the extracolumella in the deep skin layer provides an effective coupling with the lateral and ventrolateral surface of the head.

Origin of the Extracolumella in Amphisbaenia.   Considerable attention has been given to the manner of development of the extracolumella in the amphisbaenians, and the relations between this element and the one of the same name in lizards. An early suggestion by Fürbringer (1919, 1922) is that the amphisbaenian extracolumella is derived from the epihyal portion of the hyoid apparatus, but Camp (1923) argued against this view.

Observations made in the course of our experiments are pertinent to this question. In most amphisbaenians an extracolumella is present, and the hyal horn of the hyoid apparatus is much reduced, represented only

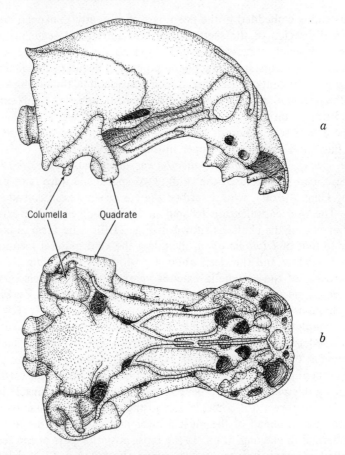

Columella   Quadrate

FIG. 21-28. The skull of *Agamodon anguliceps* viewed (*a*) laterally and (*b*) ventrally. After Gans, 1960; from drawings by P. Adams.

by short remnants as pictured in *a* of Fig. 21-31 (Wever and Gans, 1972). However, in two species among those examined, *Bipes biporus* and *Blanus cinereus*, as well as in the other species of these genera (Gans, pers. comm.), a different condition is found: there is no extracolumella and the hyal horn of the hyoid system is well developed. In *Bipes* a continuous dorsal horn was followed to the occipital region, where it terminates in the vicinity of the otic capsule as shown in *b* of Fig. 21-31 (Gans, 1974). In *Blanus* likewise an epihyal horn extends to the otic region (Gans and Wever, 1975). Therefore, as pointed out in the publications referred to, it seems reasonable to suppose that in the majority of amphisbaenians the terminal portion of the hyal arch has been converted into an extracolumella, but in *Bipes* and *Blanus*,

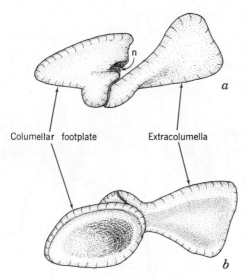

Columellar footplate   Extracolumella

FIG. 21-29. The right columella of *Agamodon anguliceps* in (*a*) lateral and (*b*) medial views. The notch *n* forms an articulation with the wall of the otic capsule. Scale 50×.

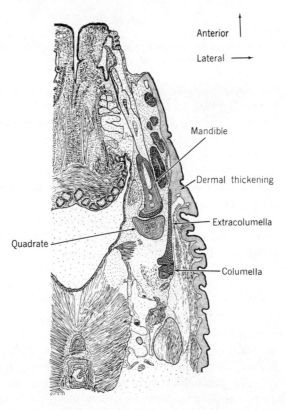

Anterior

Lateral ⟶

Mandible

Dermal thickening

Extracolumella

Quadrate

Columella

FIG. 21-30. A frontal section through the head of *Agamodon anguliceps* showing the anterior course of the extracolumella. Scale 20×.

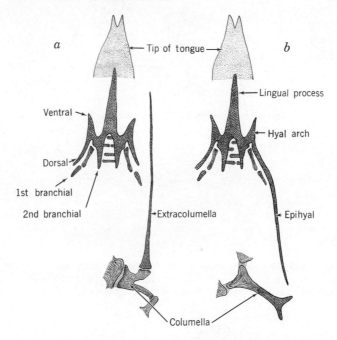

FIG. 21-31. Forms of the hyoid apparatus in most amphisbaenids
(*a*) and in *Bipes* (*b*). After Gans, 1974.

which are generally considered to be more primitive forms, this development has not taken place.

This hyal origin of the extracolumella in most amphisbaenians places these animals in contrast with the lizards, in which the hyoid system is believed not to be homologous to the extracolumella.

The origin of the terminal plate in *Blanus cinereus* presents a further problem. The completeness of the epihyal horn in this species makes it seem improbable that this plate has a hyoid origin; it is more likely that it is a neomorph derived from local fascia. It also seems probable that the further development of such materials has led to the capture of the epihyal horn and has brought about the condition seen in the other amphisbaenian species.

FUNCTION OF THE MIDDLE EAR IN AMPHISBAENIANS

The unusual form of the middle ear mechanism in the amphisbaenians, and its considerable variation among the different families, calls for a special consideration of the performance of this structure in sound reception. A number of experiments were carried out by the use of cochlear potentials, with measurements of sensitivity before and after surgical modifications of the sound-conductive system. For these measurements the sounds were applied through a tube located over the facial region, with the end of the tube close to the surface but not touching, and the narrow gap sealed with cotton fibers soaked in petroleum jelly. This precaution was taken to avoid any possibility of direct mechanical conduction from tube to animal.

Results obtained in a specimen of *Amphisbaena alba* are presented in Fig. 21-32. Here the curve indicates at several frequencies the loss in sound transmission, expressed in decibels, that occurred after the extracolumella was interrupted at a point about a third of the distance from anterior to posterior ends, thereby reducing the extracolumella to two-thirds its normal length. As the graph shows, the transmission of sounds was greatly reduced over most of the frequency range, with the greatest effects in the region of 1500-2000 Hz, where the loss amounted to 53 db. Removal of the portion of the extracolumella that extends to the skin area at the side of the head has greatly reduced the transmission to the ear.

A second experiment on this same species was carried out in two stages. After the normal tests the extracolumella was first cut at an anterior position and the changes noted; then a second cut was made more posteriorly, with further measurements of sensitivity. The two curves of Fig. 21-33 indicate the results: a moderate loss of sensitivity after the first procedure and a further loss after the second. It is clear that the effectiveness of transmission to the ear of sounds applied in the facial region depends on the length of extracolumella available, or perhaps more exactly it depends on the degree of separation of the intact portion of columella from the site of application of the stimulus.

Observations were made also on a specimen of *Amphisbaena darwini trachura*, as shown in Fig. 21-34. In this experiment the extracolumella was severed just anterior to its connection to the columella. The result was a loss of sensitivity of about 40 db on the average for tones up to 1000 Hz and progressively smaller losses at the higher frequencies.

Similar observations were made on a specimen of *Amphisbaena manni*, as shown in Fig. 21-35. In this experiment the extracolumella was severed at a posterior position, just ahead of its junction with the columella. The loss was considerable over the entire frequency range, and especially for tones below 2000 Hz.

Results obtained in a specimen of *Monopeltis c. capensis* are represented in Fig. 21-36. When almost the entire extracolumella was separated as in the preceding experiment, a significant loss was observed, though a smaller one than in the other species. Perhaps the broad head-piece of the columella in this species (see Fig. 21-8) makes more extensive contact with the skin and fascial layers at the side of the head, and provides more direct sound transmission than in other amphisbaenians.

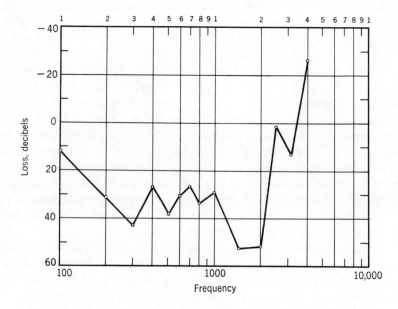

Fig. 21-32. Effects on sensitivity of sectioning the extracolumella in a specimen of *Amphisbaena alba*. Loss is shown in decibels relative to normal (zero line).

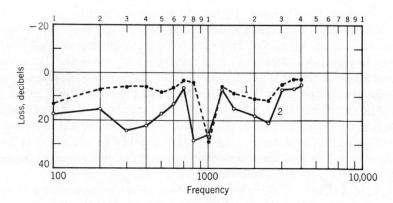

Fig. 21-33. Effects of sectioning the extracolumella in *Amphisbaena alba* in two steps, at an anterior position (curve 1) and then more posteriorly (curve 2).

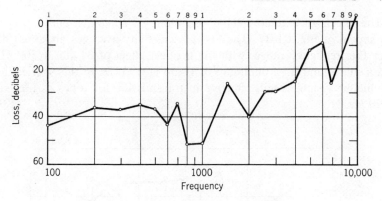

FIG. 21-34. Effects of sectioning the extracolumella in
a specimen of *Amphisbaena darwini trachura*.

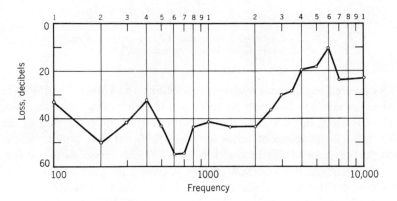

FIG. 21-35. Effects of sectioning the extracolumella in
a specimen of *Amphisbaena manni*.

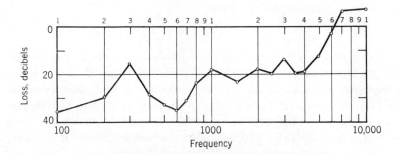

FIG. 21-36. Effects of sectioning the extracolumella in
a specimen of *Monopeltis c. capensis*.

The effects of extracolumellar sectioning in *Diplometopon zarudnyi* are shown in Fig. 21-37. The loss of sensitivity averages about 20 db over the frequency range, with the greatest effect of 31 db at 500 Hz and a sharp jump to a minimum of 10 db at 600 Hz. In this species the columellar system is still effective in transmission for tones as high as 7000 Hz.

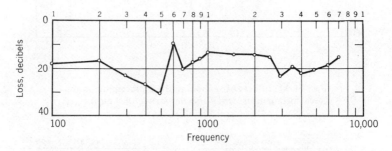

FIG. 21-37. Effects of sectioning the extracolumella in a specimen of *Diplometopon zarudnyi*.

As pointed out in the original reports (Gans and Wever, 1972; Wever and Gans, 1973), these observations show clearly that the extracolumella functions in the conduction of vibrations set up in the skin of the face through the action of aerial sounds. The facial skin, and especially that portion beneath the second infralabial shield, serves as the sound-receptive surface. This portion of skin is greatly thickened, and is flexibly connected with the surrounding skin at the boundaries between shields. Further, this portion of the skin is underlaid by loose connective tissue that provides the freedom of motion required for its receptive action. The amount of freedom necessary for this purpose is only slight, and the resistance to such motion limits the sensitivity, yet permits a good level of auditory performance.

INNER EAR

In general structure the inner ear of amphisbaenians presents great uniformity. The basic pattern is seen in Fig. 21-38, which shows a section through the otic capsule of *Amphisbaena manni*. The auditory organ occupies the posterior portion of the cochlear duct, and the lagenar macula appears in the anterior portion. This section shows the usual ear structures: the scala vestibuli above, separated from the cochlear duct by Reissner's membrane, and the scala tympani below, separated by the basilar membrane. The columellar footplate is particularly large; its annular ligament shows a broad but close connection with the cap-

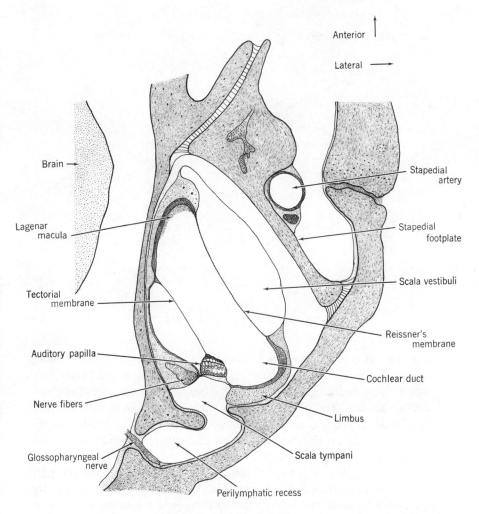

Anterior

Lateral →

Brain →

Stapedial
artery

Lagenar
macula

Stapedial
footplate

Scala vestibuli

Tectorial
membrane

Reissner's
membrane

Auditory papilla

Cochlear duct

Nerve fibers

Limbus

Glossopharyngeal
nerve

Scala tympani

Perilymphatic recess

FIG. 21-38. The otic capsule of *Amphisbaena manni* in frontal section.
Anterior is above, lateral to the right. Scale 75×. From Wever and Gans, 1973.

sular wall anteromedially and a narrow but looser connection posterolaterally.

The basilar membrane is comparatively short and broadly oval.
Among the species examined it is largest in *Amphisbaena alba*. As represented in Fig. 21-39, this species exhibits the usual condition in which
the papilla, resting on the thickened portion of the membrane (the fundus), is displaced toward the medial edge, leaving a band of thin
membrane on the lateral side. The thickness of the fundus varies with
species, but never attains the great thickness found in many of the liz-

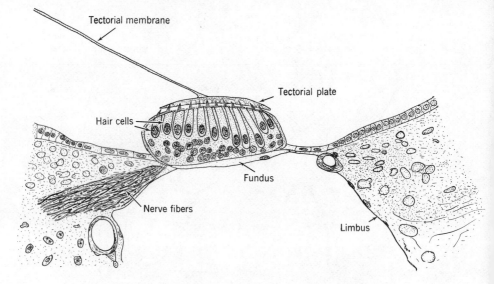

FIG. 21-39. The auditory papilla of *Amphisbaena alba*.
From Gans and Wever, 1972.

ards. The form of the papilla follows the standard pattern for reptiles: accessory cells provide a supporting framework that suspends the hair cells at their neck regions, with their ciliary tufts extending above the papillar surface. The number of rows of hair cells varies with position along the cochlea; it is greatest about the middle, and the number varies with species. In *Amphisbaena alba*, as illustrated, the number of rows may reach 12, but most species have fewer.

A tectorial plate is always present; it extends over the greater part of the papilla, leaving only the two ends uncovered. This plate usually has a nodular structure, with local thickenings that correspond at least roughly with the number of longitudinal rows of hair cells. The lower surface typically contains a number of cup-like excavations or little caverns into which the ends of the ciliary tufts extend, and the tips of these tufts are firmly attached to the roofs or side walls of the caverns.

A tectorial membrane is invariably present, but it is narrow in comparison with the papilla, and is attached to only the middle portion of the tectorial plate. This membrane runs from an anchorage on the medial limbic wall, as seen in Fig. 21-38. In many species this wall is elevated to form a prominent mound in this region, but in others, like *Amphisbaena manni*, it shows only a slight protuberance.

The number of hair cells contained in the auditory papilla varies considerably with species. Table 21-I gives results of population counts in

TABLE 21-I
Hair-cell Populations in Amphisbaenia

Species	Number of Ears	Mean number of Hair Cells
Amphisbaena alba	5	153.6
Amphisbaena darwini trachura	2	102.0
Amphisbaena gonavensis leberi	6	102.0
Amphisbaena manni	10	103.2
Blanus cinereus	4	50.2
Chirindia langi	4	38.2
Cynisca leucura	4	50.0
Monopeltis c. capensis	2	37.5
Zygaspis violacea	5	55.8
Bipes biporus	7	54.1
Rhineura floridana	2	78.5
Agamodon anguliceps	2	73.0
Diplometopon zarudnyi	6	77.5
Trogonophis wiegmanni	4	97.0

a number of specimens belonging to 14 species, and the means extend from 38.2 in *Chirindia langi* to 153.6 in *Amphisbaena alba*.

At the extremes of this range there seems to be a correspondence between hair-cell numbers and body size: *Chirindia langi* is the smallest species studied and *Amphisbaena alba* the largest. A plot of population size against body weight for 19 animals belonging to 9 species is presented in Fig. 21-40, and bears out this relationship. The correlation coefficient computed for this group of animals gave a value $r = 0.65$. However, as is evident from the plot, this correlation is determined mainly by the end values, and when the two species named are omitted from the distribution the correlation coefficient falls to 0.30, a small and not very reliable figure. Evidently for the main group of animals the numbers of hair cells bear little relation to body size.

Differentiation of the Cochlea

In many lizard species, and especially in the geckos, the cochlea is greatly elongated and is differentiated along its extent in such features as the width of the basilar membrane and the mass of the sensory structures borne upon it. Systematic variations occur also in the number of rows of hair cells and the density of distribution of these cells. There is good reason to believe that species with a broad range of physical differentiation along the cochlea possess the capability of pitch and loudness discrimination to a degree beyond that of others in which this differentiation is small. Suggestions regarding the extent of discriminative capacity in amphisbaenians can be derived from similar dimensional measure-

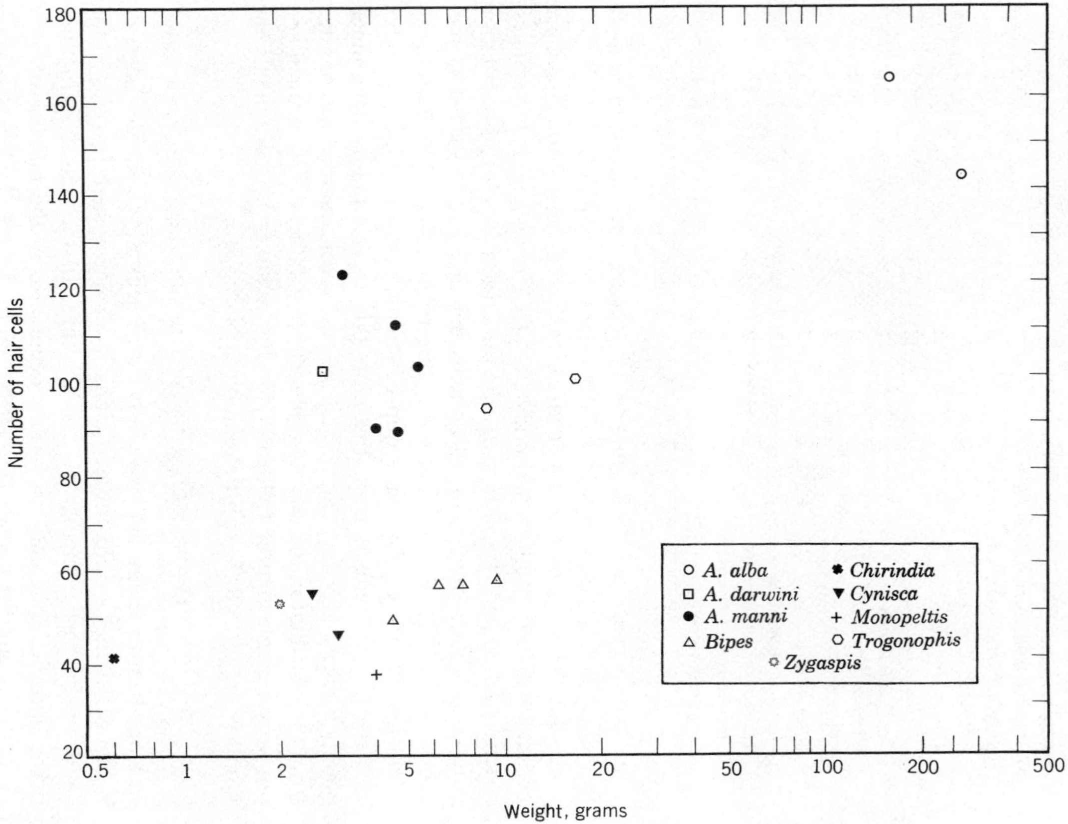

Fig. 21-40. Relations between body weight and hair-cell populations for

Weight, grams

Number of hair cells

180
160
140
120
100
80
60
40
20

0.5 1 2 5 10 20 50 100 200 500

○ A. alba   ✳ Chirindia
□ A. darwini   ▼ Cynisca
● A. manni   + Monopeltis
△ Bipes   ○ Trogonophis
                 ✿ Zygaspis

ments. Such measurements were carried out in a few representative species, and the results are presented in the following seven figures.

Amphisbaena manni. — Dimensional data for a specimen of *Amphisbaena manni* are shown in Fig. 21-41. The basilar membrane is about 170 μ in length, and varies in width in a manner indicated in the upper curve in part *a* of this figure, reaching a maximum width of 150 μ. This membrane has the form of a slightly flattened circle: the membrane is nearly as wide as it is long. The thickened portion of the membrane, the fundus, is about three-fifths the size of the whole structure and has much the same form. The thickness of the fundus, which largely determines the stiffness of the moving structures, is shown in part *b* of this figure. This thickness is of moderate degree and varies but little along the papilla.

The cross-sectional area of the papilla as it varies along the cochlea is indicated in part *c* of the figure. Here the solid curve represents the whole structure, including the cellular portion along with the fundus on which it rests, whereas the broken curve represents the cellular portion only. These curves differ but little because the fundus is comparatively thin. The curves rise rapidly to a maximum in the middle of the cochlea, and then rapidly fall away. The maximum area attained is large for an amphisbaenian ear.

Along the cochlea the hair cells of the organ of Corti appear in longitudinal rows that vary in number from 5 at the two ends to 12 in the midregion (part *d*). The density of these cells varies along the cochlea in much the same manner, reaching a maximum near the middle of about 27 cells per 20 μ section.

Amphisbaena alba. — The largest species studied was *Amphisbaena alba*, and its cochlear dimensions are shown in Fig. 21-42. This basilar membrane is longer than the preceding, though not as wide, and forms a flattened oval with a length a little over twice the width (upper curve of part *a*). The fundus has nearly the same form and a little over two-thirds the width. The thickness of the fundus (part *b*), as for the preceding species, is nearly uniform over its extent.

The area of the papilla (part *c*) increases rapidly at the dorsal end, forms a broad maximum with a depression near the middle, and then as rapidly declines ventrally. The maximum area attained is considerably less than found in *Amphisbaena manni*. The cellular portion as usual follows the same form.

The rows of hair cells, shown in part *d*, increase rapidly at the dorsal end and reach a maximum of 10, subside somewhat in the middle of the cochlea and then increase to 10 again, and finally decline to 7 as the papilla ends. The density of the hair cells rises rapidly to about 17, and maintains a high value over most of the remainder of the cochlea until

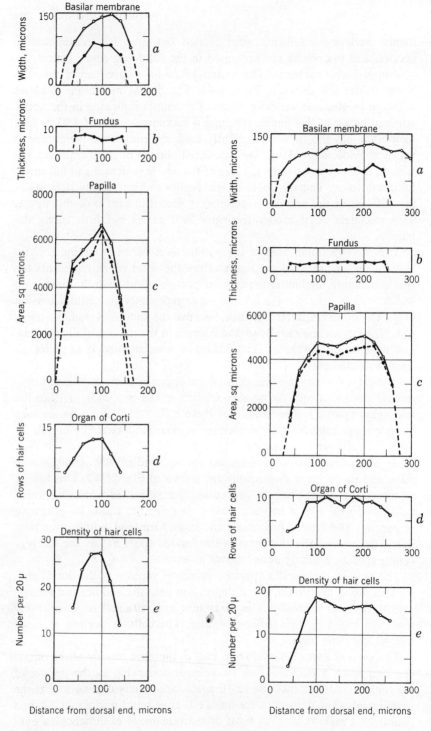

FIG. 21-41. Cochlear dimensions in a specimen of *Amphisbaena manni.*

FIG. 21-42. Cochlear dimensions in a specimen of *Amphisbaena alba.*

it falls somewhat near the ventral end. The maximum is well below that of the preceding species.

In this species the cochlea is longer than in any other examined, but the degree of differentiation is not outstanding.

Chirindia langi. — The species *Chirindia langi* was selected for special attention because it was the smallest among those studied. As Fig. 21-43 shows, its cochlear structures likewise are particularly small. The basilar membrane hardly attains a length of 100 μ, and the width reaches a maximum of only 75 μ. The thickened fundus portion measures 50 μ in the middle of the cochlea. The thickness of this structure is variable and attains values greater than in the preceding species. The area of the papilla is small, with a maximum value of 3700 sq μ in the middle of the cochlea.

The number of rows of hair cells only varies between 5 and 6. The density of these cells, however, increases sharply toward the middle of the cochlea because of a close packing of these cells in the midregion.

Here is a small and relatively simple cochlea.

Blanus cinereus. — In *Blanus cinereus*, shown in Fig. 21-44, the basilar membrane is about 200 μ long and is decidedly asymmetrical in form, increasing in width from the dorsal end to a maximum of 100 μ in the ventral region. The fundus follows this same form. The thickness of the fundus is greater than in the species previously described, and shows a double hump in the middle region.

The papilla is relatively small and only slightly asymmetrical in form.

The number of rows of hair cells is 3 at the two ends of the cochlea, rising to 7 in the middle. The hair-cell density increases with the number of rows.

This is a relatively simple cochlea, lacking any special features apart from the asymmetry of the basilar membrane, which is reflected in only slight degree in the papilla itself.

Bipes biporus. — As shown in Fig. 21-45, the asymmetry of the basilar membrane seen in the preceding species is also present in *Bipes biporus*. This membrane has a length of nearly 200 μ, and in its ventral region reaches a width of 150 μ. The fundus is narrow and nearly symmetrical. Its thickness is uniform over the midregion, tapering sharply at the ends.

The papilla is nearly symmetrical, and reaches a narrow peak near the middle.

The number of rows of hair cells increases from 2 to a maximum of 8 in the middle and a little beyond, and then declines. The density of the hair cells rises to a sharp peak near the middle of the cochlea and declines rapidly thereafter.

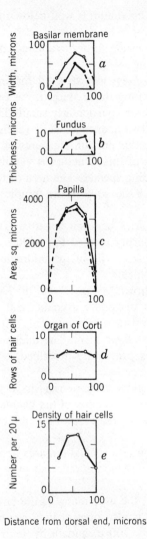

FIG. 21-43. Cochlear dimensions in a specimen of *Chirindia langi*.

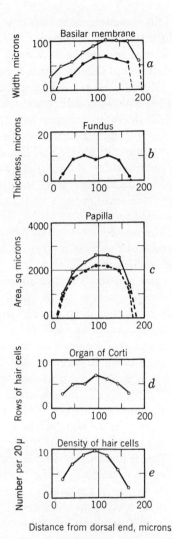

FIG. 21-44. Cochlear dimensions in a specimen of *Blanus cinereus*.

This cochlea also is relatively simple. The asymmetry of the basilar membrane is not reflected to any large degree in the cellular structures.

Diplometopon zarudnyi. — Measurements on a specimen of the acrodont species *Diplometopon zarudnyi*, shown in Fig. 21-46, reveal a curiously irregular form of basilar membrane, with the fundus following the same course. The thickness of the fundus, however, is nearly uniform over the midregion, and tapers sharply at the ends.

The papilla is of moderate size, with a sharp peak near the middle. It shows only a trace of the asymmetry in the form of the basilar membrane.

The number of rows of hair cells increases to a maximum of 10 just beyond the middle of the papilla, and then declines toward the ventral end. The density of the hair cells has nearly the same form, though the maximum appears a little more dorsally.

Rhineura floridana. — Dimensional measurements on a specimen of the sole existing representative of the Rhineuridae are shown in Fig. 21-47. The basilar membrane presents the same asymmetry of form found in four of the preceding species in which these measurements were carried out, with the maximum width appearing in the ventral region. The fundus only partially reflects this form. Its thickness varies only slightly over the middle portion of the cochlea.

The papilla reaches a large size a little beyond its midregion, a size exceeded only in *Amphisbaena manni* among the species examined.

The rows of hair cells vary from 6 to 9, and the hair-cell density rises to a nearly symmetrical peak in the middle of the cochlea.

All these amphisbaenian cochleas exhibit great simplicity of structure, as readily seen in Fig. 21-48, in which outlines are drawn for the basilar membranes of all seven species. For the most part the basilar membrane has the form of a flattened circle or pointed oval, and only in *Amphisbaena alba* does it present a moderate elongation. In *Blanus cinereus, Bipes biporus,* and *Rhineura floridana,* a taper is seen in the form of the basilar membrane, but (as reference to Figs. 21-41 to 21-47 will show) this form is only slightly reflected in the form of the papilla. Even in *Amphisbaena alba,* in which the broad extent of the basilar membrane is found also in the papilla, the middle portion is relatively uniform and the variation is limited to the end regions. These cochleas must be regarded as differentiated only in slight degree. Pitch discrimination and tonal analysis can be aided only marginally by the spatial distribution of response along the cochlea, and must be handled largely by frequency principles—through a neural representation of the frequency characteristics of the stimulating sounds.

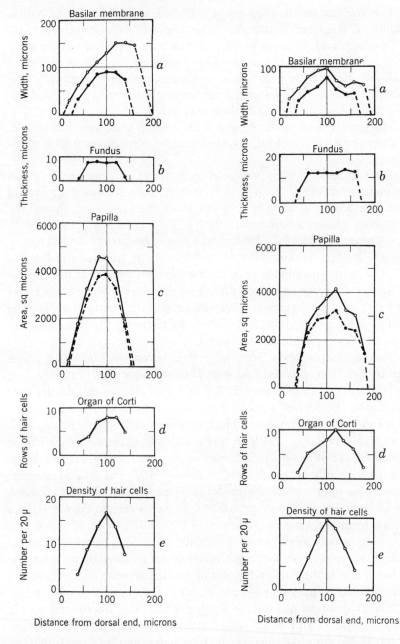

FIG. 21-45. Cochlear dimensions in a specimen of *Bipes biporus*.

FIG. 21-46. Cochlear dimensions in a specimen of *Diplometopon zarudnyi*.

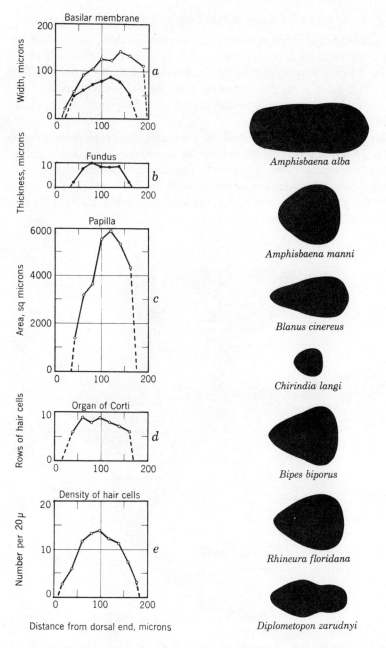

FIG. 21-47. Cochlear dimensions in a specimen of *Rhineura floridana*.

FIG. 21-48. Forms of the basilar membrane in seven species of amphisbaenians.

MODE OF STIMULATION OF THE INNER EAR

The inner ear of amphisbaenians lacks a round window, and stimulation by sounds involves a continuous fluid circuit, as in snakes and turtles and in a few divergent species of lizards. In amphisbaenians this fluid circuit is distinctive in that it passes along the brain cavity over a large portion of its course, a condition found elsewhere only in the lizard *Anniella pulchra*.

This fluid circuit is represented for a specimen of *Amphisbaena manni* in Fig. 21-49, and it takes essentially the same form in all other species. The scala tympani opens posteriorly into a small pocket, the perilymphatic recess, which in turn connects with the brain cavity in a region where the glossopharyngeal nerve runs posterolaterally through the cranial wall on its way to the facial region.

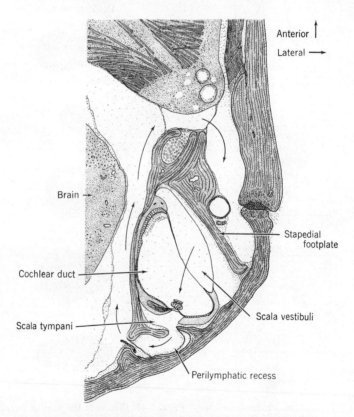

FIG. 21-49. Paths of vibratory fluid flow in the amphisbaenian ear. The section represents a specimen of *Amphisbaena manni*.

From this posterior part of the brain cavity there is a free passage anteriorly along the medulla, then laterally at a place anterior to the prootic; finally the path turns posteriorly to the anterolateral face of the columellar footplate, where the circuit becomes complete. An inward thrust of the columella results in a continuous fluid displacement all along this route: the small quantity of fluid pushed ahead of the inner face of the footplate is restored to its outer face.

Only thin membranes lie across this fluid path. These include Reissner's membrane close to the footplate, the basilar membrane between cochlear duct and scala tympani, a portion of the arachnoid membrane across the opening between perilymphatic recess and brain cavity, and two further connective tissue membranes running anteriorly from the prootic at the upper end of this pathway, as indicated in Fig. 21-49. These thin, flexible membranes are transparent to pressure pulsations, though they would prevent a continuous flow of fluid.

Except for the basilar membrane, these membranes can have only a slight effect upon the vibratory fluid flow. The basilar membrane, however, presents a degree of stiffness that often varies regionally as indicated in a consideration of cochlear dimensions. Its effects on sound transmission therefore can be expected to vary in the different species as a function of frequency.

Other characteristics of the acoustic system must be borne in mind, and especially the properties of the reentrant fluid circuit, as already discussed in Chapter 4. The mass loading that in most ears is provided mainly by the sensory structures lying on the basilar membrane is greatly augmented in reptiles like the amphisbaenians in which a reentrant fluid circuit is present. The fluid in this circuit must be displaced back and forth with the frequency of the sound. Also in amphisbaenians the circuit is more complex than in most other reptiles, in that it passes through the brain cavity and its route is poorly defined. Numerous loose tissues are encountered in this path. These add their masses to the total, but even more important is the friction involved, both at the surfaces of these tissues and in their deformations by the vibratory forces.

Thus a considerable mass loading is imposed upon the action of the cochlear structures, and large frictional forces are introduced as well. These components, together with the stiffness of the moving structures, determine the reactive conditions under which the amphisbaenian ear operates. The tissue mass, together with the quantity of fluid in the reentrant circuit, appears to be the primary condition, and is responsible for the predominantly low-tone sensitivity of these ears. Evidence for this type of sensitivity among amphisbaenians will now be presented.

AUDITORY SENSITIVITY

Sensitivity was determined for the available species in terms of cochlear potentials. The results will be given by families, except no observations could be made on the Rhineuridae.

AMPHISBAENIDAE

Amphisbaena alba. — Sensitivity curves for two specimens of the large South American species *Amphisbaena alba* are shown in Fig. 21-50. This species has the best sensitivity in the low frequencies from 200 to 600 Hz, where levels up to −27 db are reached. For higher tones the response falls off progressively, and at an increasing rate, except for an inversion between 5000 and 7000 Hz.

Amphisbaena darwini trachura. — A single specimen of this species was available, and measurements were made on both ears. For the tests on the left ear the sound tube was sealed over the face at an anterior position, centered over the lower jaw. The resulting curve is shown by the broken line in Fig. 21-51. The best sensitivity is at 1000 Hz, and decreases for lower and especially for higher tones. Tests were then made on the right ear, with the sound cannula located over the second infralabial shield. The sensitivity, represented by the solid line in this figure, is appreciably greater, reaching −20 db, but with the maximum in the same region of frequency.

Amphisbaena gonavensis leberi. — Five specimens of the moderately large species *Amphisbaena gonavensis leberi* from Hispaniola were studied. Sensitivity functions for two of these are shown in Fig. 21-52. The sensitivity is moderately good for the low tones, up to 800-1000 Hz, and falls off rapidly thereafter.

Amphisbaena manni. — Several specimens of the small Hispaniola species *Amphisbaena manni* were examined. Results for two of these are presented in Fig. 21-53. Best sensitivity is in the middle frequencies from 500 to 1500 Hz, with one curve showing a maximum of −21 db at 1000 Hz.

Chirindia langi. — Observations were made on this African species at two different body temperatures as indicated in Fig. 21-54. At the lower temperature of 24.4° C the function reaches its lowest point indicating the greatest sensitivity at 500 Hz, whereas at the higher temperature of 29.4° this point is in the higher frequencies, around 3000 Hz. There is also a shift in the forms of the curves: at the higher temperature the low end is raised and the high end is lowered. Such changes are commonly found in other reptiles, and have been studied most extensively in lizards (Werner, 1972, 1976). Variations of sensitivity with temperature do

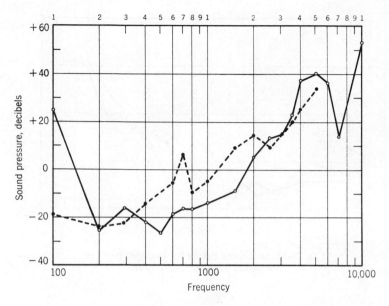

FIG. 21-50. Auditory sensitivity in two specimens of *Amphisbaena alba*.

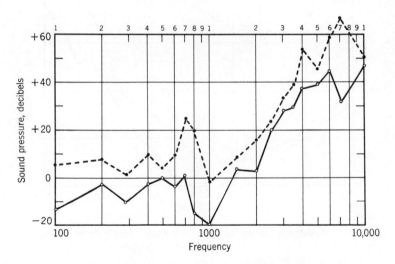

FIG. 21-51. Sensitivity curves for right and left ears of a
specimen of *Amphisbaena darwini trachura*.

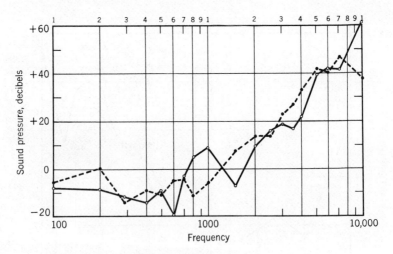

Fig. 21-52. Sensitivity curves for two specimens of
Amphisbaena gonavensis leberi.

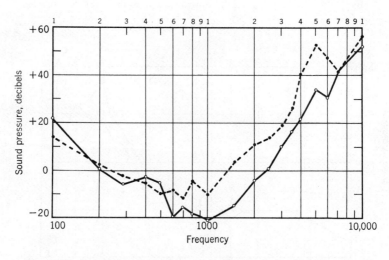

Fig. 21-53. Sensitivity curves for two specimens of *Amphisbaena manni.*

not always occur, however, as will be shown presently in observations on *Diplometopon zarudnyi*.

Cynisca leucura. — A sensitivity curve is given for a specimen of another African species, *Cynisca leucura*, in Fig. 21-55. The best region of response is in the low tones, with peaks at 300 and 500 Hz. Above this region the sensitivity falls off rapidly to 800 Hz, then levels until 2500 Hz is reached, after which there is a further decline. The best sensitivity reaches −29 db.

Monopeltis c. capensis. — Results for two specimens of another African species, *Monopeltis c. capensis*, are presented in Fig. 21-56. These curves agree in showing best sensitivity in the low frequencies. In both specimens the peaks are at 200 Hz and again at 500 or 600 Hz, after which there is a gradual decline in sensitivity at first, and finally an abrupt loss. This ear shows a considerable range of moderately good sensitivity.

Zygaspis violacea. — A sensitivity curve for a specimen of *Zygaspis violacea* is presented in Fig. 21-57. The best response is in the low-frequency range, with peaks at 300 and 500 Hz. For lower tones the decline in sensitivity is rapid, and for higher tones it is more gradual but continuous into the upper range. The maximum sensitivity attained is −22 db.

All the amphisbaenids show close similarities of hearing, with a moderate degree of sensitivity, usually reaching about −20 db in the low-tone region around 300 to 700 Hz. This region may contain two peaks of sensitivity. In *Amphisbaena manni* the best region is a little higher, around 700 to 1000 Hz, and in *Chirindia langi*, when the temperature is elevated above the usual level, the sensitive region shifts upward to 3000 Hz.

BIPEDIDAE

Bipes biporus was the only species studied in this small family. A number of tests were made on this species before it was discovered that the sound-receptive mechanism is of unusual form, consisting, as described above, of a columella alone extending to the skin surface. Because in the early experiments the sound tube was applied to the face as in the other amphisbaenians, the results seemed to indicate relatively poor sensitivity. When the true form of this receptive system became known, the experiments were repeated, with the sound applied more appropriately to the neck region, and then this species was found to be one of the most sensitive among amphisbaenians. Curves obtained by both procedures are given in Fig. 21-58. The posterior position for the sound tube gives a broad curve that shows best sensitivity in the regions of 400-500 and

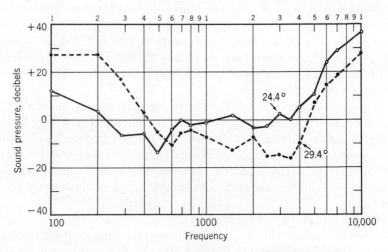

FIG. 21-54. Sensitivity functions for a specimen of *Chirindia langi*, obtained at two different body temperatures. From Gans and Wever, 1972.

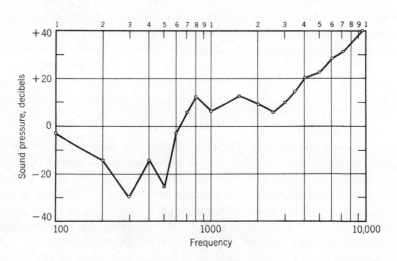

FIG. 21-55. Sensitivity in a specimen of *Cynisca leucura*. From Gans and Wever, 1972.

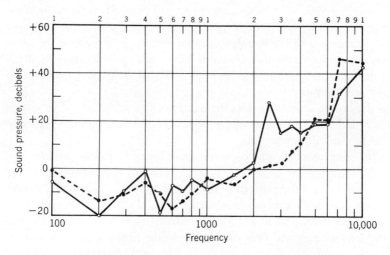

FIG. 21-56. Sensitivity functions in two specimens of
Monopeltis c. capensis. From Gans and Wever, 1972.

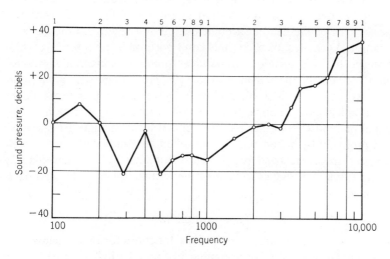

FIG. 21-57. Sensitivity in a specimen of *Zygaspis violacea*.
From Gans and Wever, 1972.

1000 Hz, where levels around −35 db are reached. This is a very creditable degree of sensitivity for a reptilian ear.

Blanus cinereus. — This species, found in a curiously isolated location in North Africa on the Iberian peninsula, is considered to be the most primitive member of the Amphisbaenidae, but as shown above the anatomy of its ear seems similar to that of *Bipes,* and in some sense its structures are intermediate in form between those of *Bipes* and members of the Amphisbaenidae.

As during the sensitivity tests the nature of the sound-receptive system was incompletely known, a number of procedures were used in presentation of the stimulating sounds.

In one specimen the sound cannula was located in three positions, *a,* over the lower lip far anteriorly, *b,* over both lips near the angle of the mouth, and *c,* over the posterior edge of the lower lip. Sensitivity curves for these three locations are presented in Fig. 21-59. The curves show some irregularities, but in general the sensitivity is best in the low and middle frequencies, and falls off rapidly in the high range. It is difficult to choose among the three locations of the sound tube, and it appears, in light of the anatomical studies made later, that all were too far anterior for optimum sensitivity.

Further tests were carried out on a second specimen, with results shown in Fig. 21-60. In this animal the sound cannula was given two positions, *a,* over the middle portion of the lower lip and *b,* over the posterior part of the head at a point behind and ventral to the angle of the mouth. As the results show, this latter position is the most favorable. The curve for this location shows clear points of maximum response around −20 db for tones from 800 to 1500 Hz, though with a variation at 900 Hz, and a considerable decline for lower and higher tones. From the anatomical observations, this position is in the general area of the terminal plate in this species. A more dorsal location than the one used would probably be better, but it seems likely that the whole lateral area in this region of the head is sufficiently stiffened by the thickening of skin and subdermal tissues to involve the terminal plate embedded deep in this layer.

TROGONOPHIDAE

Trogonophis wiegmanni. — Sensitivity functions for two specimens of *Trogonophis wiegmanni* are presented in Fig. 21-61. The sensitivity is best around 300-500 Hz, where the level approaches −36 db, and maintains an acceptable level for the whole range from 100 to 1000 Hz. For higher and lower tones the curve ascends uniformly, indicating progressively poorer sensitivity. The range extends well into the low frequencies, and measurements were made down to 40 Hz.

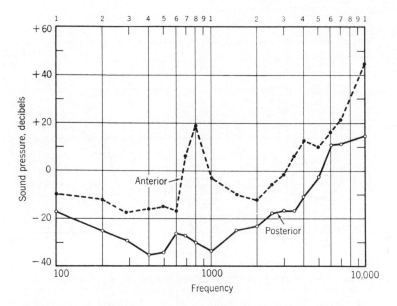

FIG. 21-58. Sensitivity in a specimen of *Bipes biporus* measured with the sound applied at two positions, anteriorly in the facial region and posteriorly in the region of the neck constriction. From Wever and Gans, 1972.

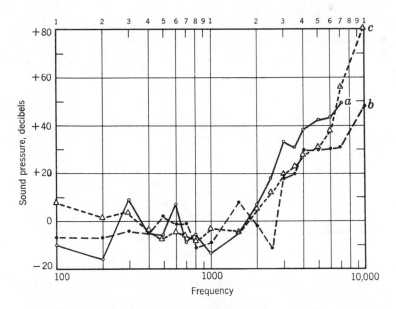

FIG. 21-59. Sensitivity curves obtained in a specimen of *Blanus cinereus* with sound stimuli applied at three positions: *a*, anteriorly over the lower lip; *b*, over both lips farther back; and *c*, posteriorly over the lower lip. From Wever and Gans, 1972.

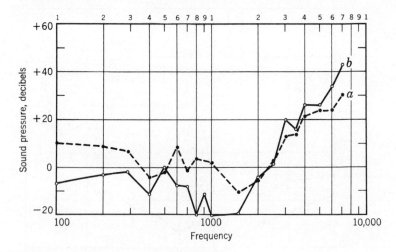

FIG. 21-60. Sensitivity curves obtained in a second specimen of *Blanus cinereus* with the sound tube in two positions: *a*, over the middle of the lower lip and *b*, far posterior beyond the angle of the mouth. From Gans and Wever, 1975.

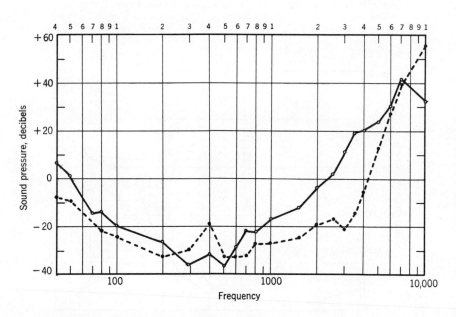

FIG. 21-61. Sensitivity curves for two specimens of *Trogonophis wiegmanni*.

Diplometopon zarudnyi. — For this species the sound tube was located over the side of the face, centered about the region of the second infralabial shield. Sensitivity curves for two specimens are shown in Fig. 21-62. These two curves have much the same form, though they differ in level of sensitivity. The best responses are in the midfrequency region, with the peak at 900 Hz in one and at 1500 Hz in the other.

Additional observations were made on two other specimens, and in these a special study was made of temperature effects (Gans and Wever, 1974). The results obtained in one of these animals, in which four different body temperatures were used, are presented in Figs. 21-63 and 21-64. As will be seen, these curves are closely similar, but with a few variations that seem unrelated to the temperature difference. Sensitivity is best in a region from 700 to 1500 Hz, where a level around −20 to −30 db is attained. The other animal, in which three different body temperatures were employed, gave the results shown in Fig. 21-65. Again the functions show considerable variations, but with little relation to the temperature.

The results shown in Fig. 21-65 for a temperature of 35° C are complicated by a brief exposure to a much higher temperature. After the observations had been completed with the body temperature at 28° and 37°, an attempt was made to use a still higher temperature of 43° C. Soon after the body temperature had been elevated to this point, however, the animal became hyperactive and the observations had to be interrupted. The level was then dropped to 35° C and observations resumed, giving the function shown. For the most part this curve runs above the other two, suggesting that some injury to the ear may have resulted from the exposure to the high temperature.

These results on *Diplometopon* differ from the ones already described for *Chirindia langi*, in which a temperature change of 5° C produced a significant alteration in the ear's responses. Evidently there are large interspecific differences in temperature effects, which need to be explored further.

DIRECTIONAL SENSITIVITY

Whether amphisbaenians are able to perceive the direction of a sound has deep significance for their reactions to prey and predators. A study of directionality was carried out in a specimen of *Monopeltis c. capensis* by locating the sound tube to the left side of the head 5 cm from the surface and recording responses from two different electrodes, one in the saccule on the left side and another in the saccule on the right. The results are shown in Fig. 21-66, and show a consistent advantage (except at 100 and 7000 Hz) for the ipsilateral electrode position. Clearly there is stronger stimulation of the left ear for a sound coming from this side.

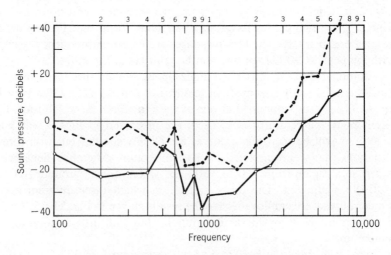

FIG. 21-62. Sensitivity curves for two specimens of
Diplometopon zarudnyi. From Gans and Wever, 1975.

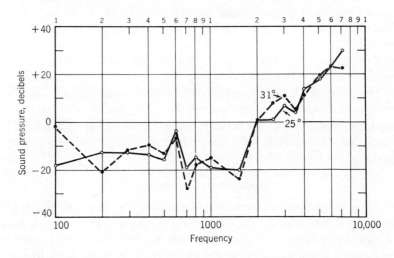

FIG. 21-63. Sensitivity measurements in a specimen of *Diplometopon zarudnyi*
at two body temperatures, 25° and 31° C.

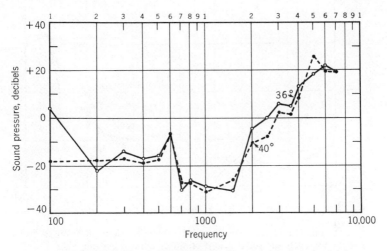

FIG. 21-64. Sensitivity measurements in the same specimen as in the preceding figure, at body temperatures of 36° and 40° C.

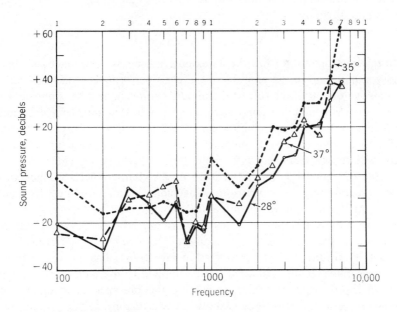

FIG. 21-65. Sensitivity measurements in an additional specimen of *Diplometopon zarudnyi* at three body temperatures.

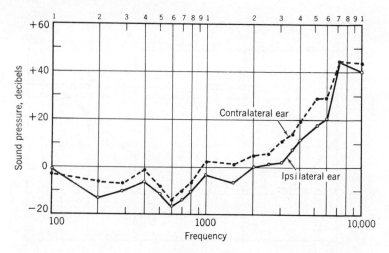

FIG. 21-66. Directional sensitivity in amphisbaenians. Responses are shown for each ear for a sound stimulus applied on one side. From Gans and Wever, 1972.

The amphisbaenian ear thus should have the facility for the perception of sound direction.

VIBRATORY SENSITIVITY

In the investigation of lizard ears extensive use has been made of vibratory stimulation in the analysis of the ear's performance, and especially in an evaluation of the role of the middle ear mechanism in sound reception. It was concluded from such experiments that mechanical stimuli applied to the skull are able to reach the inner ear with only slight relation to the presence and character of the middle ear structures. Sensitivity functions obtained with such stimuli therefore provide, to a fair degree of approximation, a measure of the characteristics of the inner ear. Accordingly, in a few amphisbaenians a series of measurements were made with vibratory stimulation.

A sensitivity function in terms of vibratory stimulation was obtained for one of the specimens of *Trogonophis wiegmanni* in addition to the usual aerial function, as represented in Fig. 21-67. The vibratory stimulus was delivered by a blunted needle applied to the facial surface. The curve is somewhat irregular, indicating the presence of complex resonances, and best sensitivity is shown in two regions, at 500 and around 2000 Hz, with very rapid falling off of sensitivity for the higher tones.

These results, taken together with those for aerial sensitivity (Fig. 21-61), suggest that middle and inner ear factors operate together in determining the region of best aerial sensitivity around 500 Hz. The considerable sensitivity between 1500 and 2500 Hz seen in the vibratory curve is not obviously reflected in the aerial function, unless the flattening in this region of frequency can be considered to show this relationship.

Further vibratory studies were carried out in a specimen of *Bipes biporus*. Figure 21-68 shows results obtained with two different sites of application of the stimulus, one with a blunted needle on the side

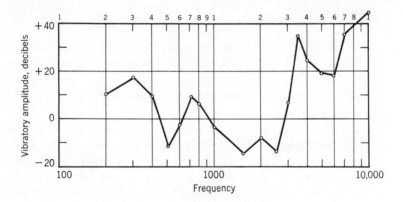

FIG. 21-67. Sensitivity to vibratory stimulation in a specimen of *Trogonophis wiegmanni*. The stimulus magnitudes are expressed in decibels relative to a root-mean-square amplitude of 1 mμ. An aerial curve for the same ear is shown by the broken line in Fig. 21-61.

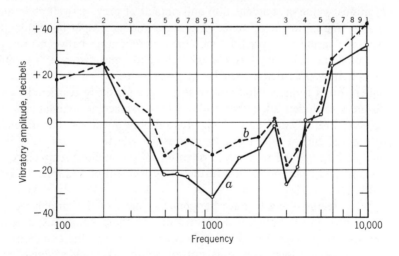

FIG. 21-68. Vibratory sensitivity in a specimen of *Bipes biporus* for two sites of application of the stimulus, *a*, to the face near the columellar disk, and *b*, to the anterior portion of the lower jaw. From Gans and Wever, 1972.

of the head close to the lateral end of the columella and another with this needle anteriorly on the lower jaw. The most posterior position is clearly the better for the middle range of frequencies where sensitivity is best and also at 3000-3500 Hz, where a secondary maximum appears, but it is clear that a mechanical stimulus applied anywhere is readily taken up by the tissues, no doubt mainly by the skull, and reaches the inner ear with a high degree of effectiveness.

HEARING IN THE AMPHISBAENIANS

It is clear from the evidence presented that the amphisbaenian ear operates at an effective level, and though its performance is not the equal of that observed in many other animals, it does remarkably well in view of the unusual forms of receptive and conductive mechanisms present in this group.

The evolutionary development of the amphisbaenian ear presents a provocative problem. It seems likely that in the history of these animals the tympanic membrane was lost, perhaps as a result of the adoption of burrowing habits, as is thought to have occurred also in several lines of lizards. The inner ear was retained, but at first its function was severely limited by the reduction in the reception of aerial sounds, though such sounds could still enter via skin and tissue conduction, along with sounds transmitted through the ground or any other vibrating substances with which the animal might come in contact. Aerial hearing may be of utmost value in the detection and pursuit of prey, as well as in alerting the animal to the presence of enemies; under proper conditions this type of hearing may provide cues for differentiating between prey and predator and also may disclose their direction and distance. The information thus provided would be of great value to an animal living in underground tunnels where other senses are seriously restricted: where vision is excluded altogether and smell is reduced to emanations from objects close at hand and which provide only uncertain and unreliable directional cues. Under these conditions the selective advantage of hearing can be expected to insure a retention of the auditory capability that remains and to bring about a further improvement of this form of sensitivity.

What seems likely to have happened in most of the amphisbaenians is a diversion of the epihyal process of the hyoid apparatus to a new purpose, to provide a connection between the end of the columella and the skin of the face. Also this portion of the skin, by a considerable thickening and by development of a degree of isolation from the surrounding skin, acquired the characteristics of a sound-receptive surface and thereby became an acceptable substitute for a tympanic membrane. The long rod-like extracolumella of the derived Amphisbaenidae has been shown experimentally, as described above, to assist very materially in sound reception. It increases the effectiveness of action of aerial sounds by amounts of the order of 20-40 db, which in terms of the amplitude of displacement produced in the receptor cells is 10 to 100-fold. A displacement of the sound-receptive site forward to the face is no doubt of particular usefulness to an animal living in a labyrinth of tunnels, where occurrences just ahead are of prime significance.

In *Bipes*, usually considered to be a relatively primitive amphisbaenian, an extracolumella has not been differentiated from the hyoid. The columella runs out in a lateral and somewhat ventral direction to make contact with the overlying skin of the neck. Yet this ear also is a highly effective receiver of sounds, as the cochlear potential observations have shown.

In the Trogonophidae an extracolumella is present, and its evolutionary origin probably was common to that of the Amphisbaenidae, but this process does not extend as far forward along the face, and has been subject to some special modifications. In *Trogonophis wiegmanni*, as described, a broad band of connective tissue fibers continues the anterolateral course of the extracolumella and makes a connection to the infralabial region of the face, with a few of its fibers going to the supralabial area. In *Diplometopon zarudnyi* the extracolumella is expanded to a broad trapezoidal plate, much of whose cartilaginous tissue is calcified, and whose anterior portion is encased in bone. From this sturdy plate a thick band of connective tissue fibers runs forward and then divides to attach to the dermal layers of upper and lower lips. In *Agamodon anguliceps* the structure is much the same as in *Diplometopon* (Gans, 1960).

In *Rhineura floridana* the extracolumella is extremely short, and the connective tissue band is greatly extended to make connections with upper and lower labial regions.

In *Blanus cinereus* the arrangement is similar to that of *Bipes* except that a terminal plate is present to form a connection to the skin in a region anterior to that in *Bipes* but well behind the facial area employed in the remaining species.

The evidence shows that all forms of sound-receptive systems occurring among the amphisbaenians are effective for their purpose, and this evidence does not discriminate very well among them in respect to sensitivity and frequency range. The system in *Trogonophis*, in which the anterior extension to the skin consists of fibrous connective tissue, performs quite as well as that of *Amphisbaena manni*, for example, in which the extracolumella itself continues into the skin layer. The transmission in *Bipes biporus* is highly effective also, though here the columella carries out the task alone.

There is perhaps a functional difference between the species *Bipes biporus* and *Blanus cinereus* and the other amphisbaenians, in that in these two the sound-receptive site is far back on the head, and in all remaining species it is on the face, where it seems most useful for detecting events immediately ahead.

Stimulation Process. In view of the physical differences in the sound conduction systems among the different groups of amphisbaenians, and the somewhat different ways in which these systems operate in the reception and transmission of sound, this lack of distinction in performance is surprising. It may well be that further investigation of these systems and better identification of the role of mechanical factors in the determination of sensitivity will indeed reveal some systematic variations.

The anatomical relations give some indications of the probable operation of these conductive mechanisms. In all amphisbaenids (except *Blanus cinereus*) in which the extracolumella joins the end of the columella to the anterior skin thickening of the infralabial region, the process may be conceived of as follows. Alternating sound pressures acting on the skin surface in the facial region move the facial pad in and out, carrying with it the anterior end of the extracolumella embedded in the pad. These movements will be transmitted posteriorly along the length of this cartilaginous rod, probably in a complex manner because the rod is flexible and is under restraints from tissue sheathings along it, and also because near its posterior end it is partially confined in a groove along the quadrate. No doubt there will be an appreciable anteroposterior component of this motion, and this component will be mainly responsible for the effective action.

Such motion exerted at the end of the columellar shaft will bring about the movement of the footplate that is essential for mobilization of the cochlear fluid and a stimulation of the hair cells.

Most likely there is not a simple, piston-like displacement of the footplate, but a rocking motion in which the anteromedial portion of the footplate is largely restrained and its anchorage at this end serves as a pivot, while the posterolateral portion swings along the wall where it is loosely attached. The form of the footplate suggests this type of action, as it is greatly thickened along its anteromedial edge, and is maintained within a correspondingly thick wall of the prootic by relatively short ligamentary fibers. On its posteromedial side, on the other hand, the footplate is thin and is held by longer ligamentary fibers.

Nearly all ears are constructed in this fashion, with the stapes attached so as to rotate on one edge. The biological reason for this arrangement probably derives from a matter of safety in the presence of excessive forces, such as extremely loud sounds and the direct displacements that might be produced by a blow. A footplate structured like a simple piston would have to be loose in its window, and an excessive force would push it into the cochlea and tear away its ligaments as well as damaging the inner ear structures. A strong ligamentary attachment in one region absorbs a part of the disruptive energy and reduces the likelihood of serious damage.

VALUES OF THE AUDITORY SENSE

There is good evidence that these animals make positive use of auditory information. As already pointed out, Gans suggested that these animals must make extensive use of auditory cues in their discovery and capture of prey and in their avoidance of enemies. At first the evidence for this was largely circumstantial: for a burrower the other senses seem of little value until a prey object is actually encountered, and smell and touch come into play. Later observations of the movements of these animals in response to sounds made in different parts of their environment have confirmed these expectations. Gans (1960, 1969) reported the approach of amphisbaenians to the vicinity of insects moving at the surface, and avoidance movements in response to slight digging in the soil above their burrows. Further investigations of the behavior of these animals in the presence of acoustic stimuli should be greatly rewarding.

22. ORDER RHYNCHOCEPHALIA:

SPHENODON PUNCTATUS

The tuatara, *Sphenodon punctatus*, found on a number of small rocky islands off the coast of New Zealand, is the sole living representative of an ancient group of reptiles that flourished in the Triassic and Jurassic periods, some 190-155 million years ago. This species has attracted much attention since its discovery early in the nineteenth century. At first it was regarded as one of the "stem" reptiles, ancestral to all the other living forms, and numerous studies were made, especially of the skeleton, for comparisons with both the living reptiles and others known as fossils. The results of these studies have not borne out the original assumption that *Sphenodon* is in the direct ancestral line of other living reptiles, yet this species is still regarded as primitive in many respects. It remains of great interest as a representative of one of the lines of development taken by descendents of the ancient eosuchians.

This species was once moderately abundant in New Zealand, but has been much reduced by encroachments on its environment and the introduction of destructive mammals. For a time it faced extinction, until rigid governmental protection was provided that promises for it at least a precarious existence.

Though many general anatomical and physiological characters of this species have received attention, the ear and its function have been examined only to a limited extent, and most of this consideration has been given to the middle ear.

Even casual inspection reveals the absence of an external ear opening, and the earlier studies soon led to the conclusion that the middle ear structures also are of unusual form, either primitive as was first suggested or degenerate as later came to be the general view.

Though a number of earlier anatomists, notably Huxley (1869), Peters (1875), and Gadow (1899) discussed the middle ear of *Sphenodon*, the first thoroughgoing accounts were presented by Versluys around the turn of the century. Versluys began his observations in 1898 as a part of his broad treatment of outer and middle ear structures in lizards and then resumed them later, in 1904, in a consideration of the embryological development of these structures.

Wyeth continued the embryological study in *Sphenodon* two decades later (1924), and more general examinations of structure, largely from a comparative point of view, were made by de Beer (1937) and Reinbach (1950). Baird more recently (1970) included a brief description of these structures in his general account of the anatomy of the reptilian ear.

The inner ear of *Sphenodon* has had only passing consideration. Baird (1960) described in some detail the forms and spatial relations of the parts of the labyrinth, with particular concern with the perilymphatic channels. He pointed to a number of variations in the sizes and locations of the labyrinthine structures in relation to these in other reptiles. Thus the saccule is located mainly ventral to the utricle rather than lateral to it, the basilar and lagenar parts of the cochlear duct are about equal in size and separated by a wall constriction, and a limbic bulge is absent.

In a further discussion of these spatial relations Baird concluded, in agreement with Hamilton (1963) and Miller (1966a), that the *Sphenodon* inner ear is the least specialized among existing reptiles. No observations were made concerning details of cochlear structure, and this conclusion rests only on the general spatial relations.

Present observations were made initially on a single specimen of *Sphenodon punctatus* (Gray, 1842) that had been provided by the New Zealand Government for experimental studies by R. G. Northcutt, and that he generously made available for studies on the ear carried out in collaboration with Carl Gans. These studies included the determination of auditory sensitivity in terms of cochlear potentials in the living animal, and thereafter a histological examination of ear structures. There were certain strictures on this histological phase of the study, however, because experiments already in progress on the brain tissues by the Northcutt group required the use of simple formol fixation and removal of the brain in advance of the processing of the ear. In this removal a considerable portion of the right labyrinth was lost or damaged, but the entire left cochlea remained intact along with the peripheral structures on this side of the head. After the removal of the brain, a secondary fixation was carried out by immersion in Maximow's solution, with daily changes over a period of six days, which resulted in a fully acceptable preservation of the tissues. The only deficiency of note, apart from the damage mentioned, consists of a moderate retraction of the tectorial membrane from shrinkage caused by the initial formol fixation. The observations on the middle ear structures have been extended by the use of two additional preserved specimens that became available. A preliminary report on these studies was made by Gans and Wever (1976).

PERIPHERAL STRUCTURES

The absence of an external ear opening in *Sphenodon* has already been noted. Also, contrary to normal expectations, the auditory area is not in the usual temporal region at the side of the head, but is high in the lateral occipital area as indicated in Fig. 22-1, which presents a lateral view of the head. This location, just medial to the posterior projection formed by the junction of squamosal and quadrate, is more precisely indicated in the skull drawing of Fig. 22-2.

The columellar apparatus consists of the usual two elements, the osseous columella and cartilaginous extracolumella, together with a connection to the hyoid. The columella lies in a sulcus formed by the paroc-

FIG. 22-1. The head of *Sphenodon punctatus*, seen from the right side. A star indicates the auditory area. Drawing by Anne Cox.

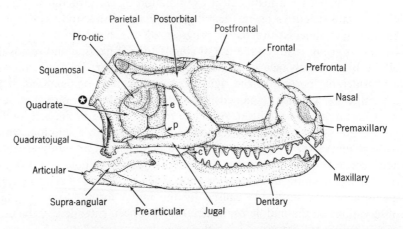

FIG. 22-2. The skull of *Sphenodon punctatus*. The star is at the junction of squamosal and quadrate, which marks the lateral border of the ear region. *e*, epipterygoid; *p*, pterygoid; *c*, coronoid. Scale 1.25×.

cipital process as shown in Fig. 22-3. Its inner end expands in conical fashion to form a particularly large, round footplate, and its shaft extends in a ventrolateral direction, continued by the shaft of the extracolumella to a head portion lying beneath the skin.

This head portion or plate of the extracolumella is of complex form as shown in further detail in Fig. 22-4. This is a nearly posterior view, and shows the outer end of the columella going over into the extracolumella, which expands laterally into three processes and a central portion designated as the headpiece. Previous writers have attempted to homologize these processes with the extracolumellar processes of lizards, but with only limited plausibility; here they are named simply by reference to their terminal attachments. A small process extends dorsally and attaches to the tip of the squamosal as it inserts between the end of the paroccipital process and the quadrate; this is designated as the squamosal process. A second process runs laterally at first and then curls around dorsally and attaches to the quadrate in a region just lateral to the first process; this is indicated as the quadrate process. The third process is at the ventral end of this mass of cartilage, and articulates with the end of the epihyal; this is the hyoid process.

Between the squamosal and quadrate processes is an opening known as Huxley's foramen. According to earlier accounts this opening may

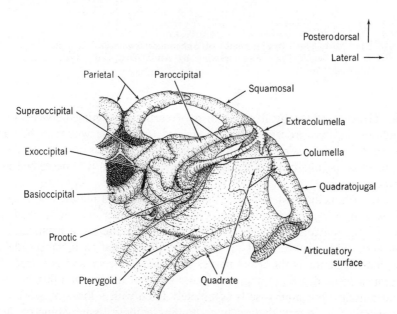

FIG. 22-3. The columellar apparatus of *Sphenodon punctatus*, seen in a posteroventral view of a portion of the skull. Scale 3×.

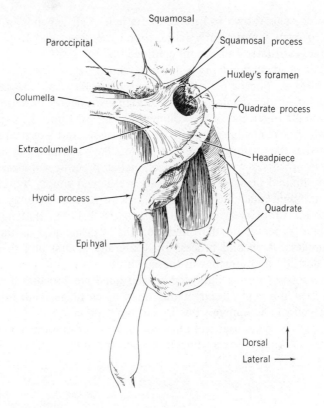

FIG. 22-4. The extracolumella of *Sphenodon punctatus* in a nearly posterior view. Scale 6×. Drawing by Anne Cox. From Gans and Wever, 1976.

be further outlined by a small strip of cartilage running between the ends of the two processes, but no sign of this bridge was seen in the present specimen. From former accounts also, the size of this opening appears to be highly variable, and in some specimens it is described as a mere pinhole.

The columellar system extends across an elongated middle ear cavity as shown in Fig. 22-5. This cavity is enclosed superiorly by a vaulted bony roof and elsewhere by soft tissues, mainly the large masticatory muscles. This cavity is not an open, air-filled space as in most other reptiles. It contains the usual nerves and blood vessels, and is intruded upon in some degree by the sheathing tissues of the surrounding bones and muscles, but otherwise is completely filled with a loose network of connective tissue strands largely containing adipose tissue. Some of the cells in the meshes of this tissue appear to be filled with single fat glob-

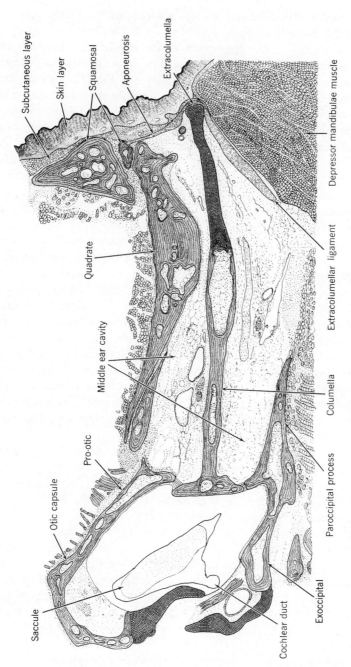

Subcutaneous layer

Skin layer

Squamosal

Aponeurosis

Extracolumella

Depressor mandibulae muscle

Quadrate

Extracolumellar ligament

Middle ear cavity

Columella

Pro-otic

Paroccipital process

Otic capsule

Saccule

Cochlear duct

Exoccipital

Fig. 22-5. A frontal section through the middle ear and otic capsule of *Sphenodon*. Anterior is above, lateral to the right. Scale 10×. From Gans and Wever, 1976.

ules, and others have a central nucleus surrounded by a large number of smaller fat droplets much as described for brown fat.

Versluys (1898) in his description of the middle ear apparatus of *Sphenodon* indicated the presence of a joint, together with a definite cleft, between columella and extracolumella. As Fig. 22-5 shows for the sectioned specimen, the shaft of the columella goes over into that of the extracolumella without interruption, and this condition was verified by examination of the uninterrupted series of sections through this region. Also in two other specimens studied by dissection there was no sign of a joint between these elements. It seems likely that in the specimen examined by Versluys these two elements had been broken apart. This could easily happen from even slight mishandling because a large marrow cavity near the end of the columella leaves the walls at the junction extremely thin.

As the figure shows, the external end of the extracolumella is in immediate contact with a layer of ligamentary tissue known as the aponeurosis. This layer is variable over its extent, in some places appearing as a flat sheet and at others showing irregular bends and thickenings. It extends anteriorly to the outer edges of the squamosal and quadrate bones, where it attaches to their periosteal sheaths. Its lowermost portion runs closely around the edge of the extracolumellar plate and then continues as the extracolumellar ligament. This ligament runs obliquely in a posteromedial direction along the contour of the depressor mandibulae muscle and mainly attaches to its connective tissue sheath, though dorsally it is anchored to the surfaces of the paroccipital and squamosal bones. Versluys homologized this ligament with the extracolumellar ligament of lizards.

In earlier discussions of the middle ear of *Sphenodon* considerable attention was given to the nature and origin of the aponeurosis. Huxley regarded it as a modified tympanic membrane, or more specifically as the persisting middle layer of that membrane. Versluys (1898) presented convincing arguments in favor of this view.

The relations between the extracolumella and the external surface, already indicated in Fig. 22-5, are represented in further detail in Fig. 22-6. This is a lateral view of the structures, showing the head of the extracolumella with its squamosal and quadrate processes extending dorsally, to the right, and the hyoid portion on the left. The body of the extracolumella with its superimposed tissues has been lifted away from the quadrate on the left side of the drawing to expose the bony base more clearly.

Of particular interest are the superficial tissues through which sound pressures must act to involve the columellar mechanism. Over the aponeurosis, shown here only in part, is the mass of the depressor man-

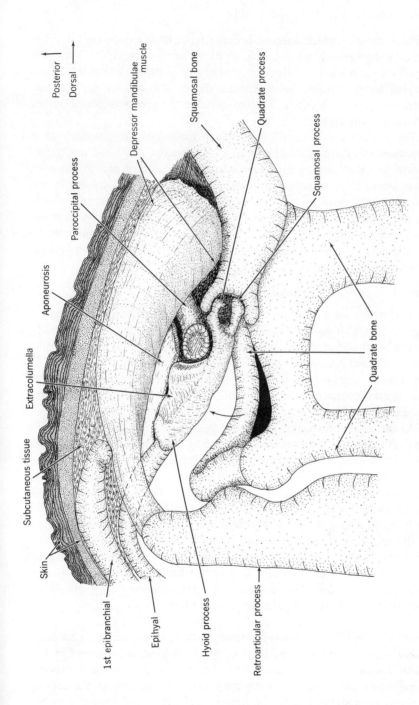

Posterior ←

Dorsal →

Depressor mandibulae muscle

Squamosal bone

Quadrate process

Paroccipital process

Squamosal process

Aponeurosis

Extracolumella

Subcutaneous tissue

Quadrate bone

Skin

1st epibranchial

Epihyal

Hyoid process

Retroarticular process

FIG. 22-6. The middle ear region of *Sphenodon*, seen in a lateral view. The medial part of the extracolumella has been lifted away from its normal seat within the medial and lateral portions of the quadrate as indicated by the curved arrow. Scale 3.7×.

dibulae muscle, which extends broadly from its insertion on the retro-articular process of the mandible to its anchorage largely on the quadrate. Over this muscle is the skin, consisting of dense dermal layers and varying areas of loose subcutaneous tissue. The end of the first epibranchial process is shown inserted between the skin and muscle layers, adjacent to the epihyal.

The middle ear evidently operates in much the same fashion as in a number of lizards without tympanic membranes, described in Chapter 6, with the skin serving as the sound receptive surface and transmitting its movements in response to vibratory pressures to the columella and thereby to the fluids of the inner ear. The aponeurosis perhaps adds in some degree to the stiffness of the dermal surface and aids in transmitting the effects to the extracolumella. Also the depressor mandibulae muscle may enter into this transmission, much as has been seen in some species of chameleons (Chapter 9).

INNER EAR STRUCTURES

The relatively large otic capsule of *Sphenodon* has already been seen in Fig. 22-5, with a good sized saccular cavity and extensive perilymphatic spaces. At the level shown, the cochlear duct is small, but more ventrally it expands and becomes separate from the saccule as shown in Fig. 22-7. The basilar membrane is suspended over the opening in the limbic plate, and lighter connective tissue extends anteriorly to anchor the limbic plate to the wall of the otic capsule.

At the level shown, the scala tympani is continued posteriorly as the perilymphatic sac, with an area enclosed by arachnoid tissue interposed, and with a wide communication with the brain cavity.

No round window is present, and pressure relief for the vibratory movements of the columellar footplate is provided by a reentrant fluid circuit as in snakes, amphisbaenians, and turtles. A somewhat hindered pathway to complete this circuit is found along the course of the glossopharyngeal nerve. The pathway for some distance contains only loose connective tissue and membranes around the nerve at the two ends, and this material can easily yield to permit vibratory movements as suggested in Fig. 22-8. The condition of this pathway in *Sphenodon* is comparable with the one seen in chameleons.

Auditory Papilla. A cross-sectional view of the auditory papilla of *Sphenodon*, taken from the middle of the cochlea, is presented in Fig. 22-9. This structure is located over the anterior portion of the basilar membrane close to the lip of the limbus, and at the ends of the cochlea it encroaches somewhat on the limbus much as found regularly in turtles. Only a thin fundus is present in the midregion, as the figure shows,

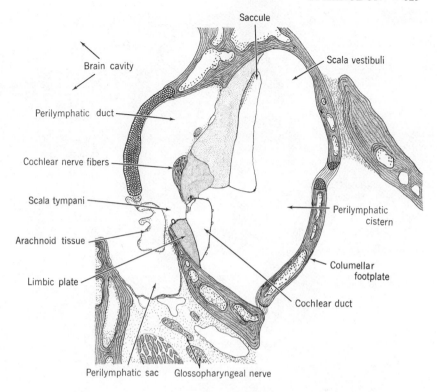

FIG. 22-7. A frontal section through the otic capsule of *Sphenodon*, passing through the auditory papilla near its midregion. Anterior is above, lateral to the right. Scale 25×.

and toward the two ends of the cochlea this portion of the basilar membrane is still thinner.

A tectorial membrane arises in a substantial root process deep in the limbic sulcus, and runs as a rapidly thinning sheet to a heavy tectorial plate. This plate contains numerous deep pits on its undersurface—the side facing the hair cells—leaving long square-edged prongs extending close to the papillar surface. Normally, as seen in other reptiles with corresponding structures, the tufts of the hair cells enter these recesses and connect with their side walls, but in this example, on account of the shrinkage of the tectorial membrane already referred to, the plate was pulled away and its connections were broken. However, the ciliary tufts remained intact in practically all sections as shown here.

The tectorial membrane extends along almost the entire papilla; in this specimen it was absent only in a single section at the dorsal end. It terminates in a tectorial plate at its first appearance dorsally, and

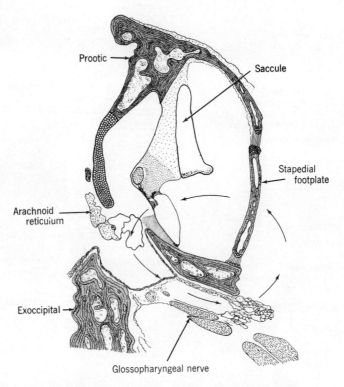

FIG. 22-8. The reentrant fluid circuit in *Sphenodon*.

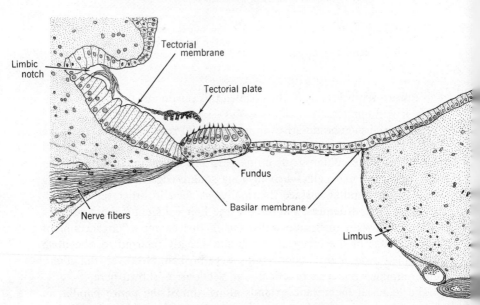

FIG. 22-9. The cochlear region in *Sphenodon*, seen near the middle. The tectorial membrane is contracted, and the tectorial plate is pulled away from its normal position over the auditory papilla. Scale 250×.

this plate continues almost to the ventral end of the cochlea. The plate grows larger as the number of longitudinal rows of hair cells increases, and then diminishes as these rows are reduced to three or four toward the ventral end. In this specimen, at a point about 90 μ from the ventral end, the plate disappears and the tectorial membrane terminates first as a frayed strand of three or four thin fibers, and then shortly thereafter is reduced to a single fiber. With the shrinkage present in this preparation it was not possible to determine whether at the extreme end of the papilla this terminal strand reached the hair cells.

The ciliary orientation of the hair cells was consistently the same throughout the cochlea: with the longest cilia of the tuft (and presumably including the kinocilium) always on the posterior side, which corresponds to the lateral orientation nearly always found in lizards beneath a tectorial plate. A single exception was noted: one of the three hair cells in the papilla at the extreme ventral end had a contrary orientation.

A further distinctive feature, found regularly in turtles as described in the next chapter, is the extension of the auditory papilla beyond the two ends of the basilar membrane so as to rest entirely on the surface of the limbus. In the *Sphenodon* specimen studied, this extension was for a distance of 30 μ at the dorsal end and twice this distance at the ventral end. Thus of the total population of hair cells in this papilla, which amounted to 225, there were 15 on the limbic surface.

Cochlear Dimensions. Dimensional measurements were made on the left ear of this specimen, with results as shown in Fig. 22-10. The width of the basilar membrane, represented by the upper curve in part *a* of this figure, increases rapidly from the dorsal end to a broad maximum of about 350 μ, and then a little more rapidly declines ventrally. This membrane therefore is lozenge shaped, as shown in Fig. 22-11. The width of the papilla, shown in the broken curve in Fig. 22-10, varies markedly from this form. It rises to a maximum near the dorsal end, subsides somewhat, rises once more to a second broader maximum in the middle region, declines toward the ventral end, and then passes through a further maximum before ending abruptly. The area of the papilla, shown in part *b* of this figure, shows the same three maximum regions, with the middle one much exaggerated.

The number of longitudinal rows of hair cells along the cochlea increases irregularly from 4 at the dorsal end to a maximum of 8, and then declines over the ventral third of the cochlea. The hair-cell density more clearly reflects the size of the papilla, with a maximum near the middle.

This ear appears to be only slightly differentiated in a structural sense.

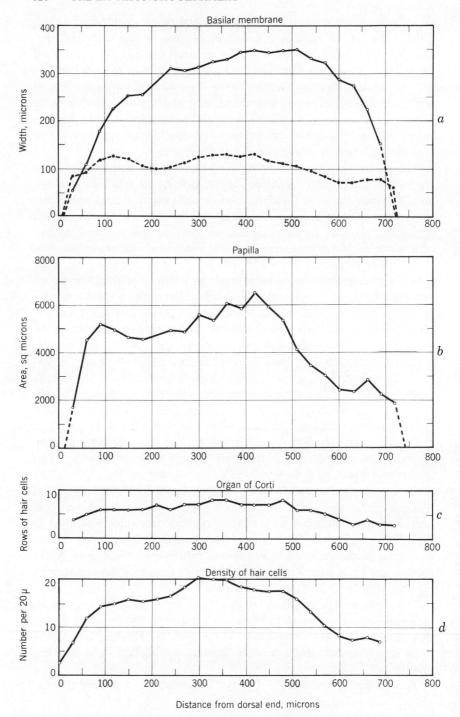

FIG. 22-10. Cochlear dimensions in *Sphenodon*.

FIG. 22-11. Form of the basilar membrane in *Sphenodon*.
Dorsal is to the left. Scale 100×.

AUDITORY SENSITIVITY

The sensitivity of this ear was tested by means of cochlear potentials in response to both aerial and vibratory stimuli, recorded with an electrode inserted into the perilymph space. Aerial sounds were applied through a tube sealed over the skin surface in three regions: at the front part of the throat, at the rear of the throat, and at the neck region. The results are shown in the three curves of Fig. 22-12. The three functions

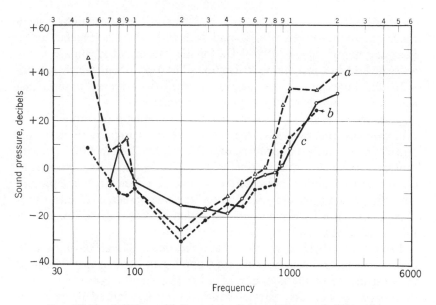

FIG. 22-12. Aerial sensitivity curves in a specimen of *Sphenodon punctatus*, for three places of application of the sound tube: *a*, to the front of the throat; *b*, to the rear of the throat; and *c*, to the neck region. From Gans and Wever, 1976.

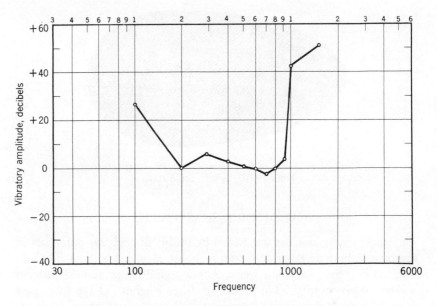

FIG. 22-13. A vibratory sensitivity function for *Sphenodon*.
From Gans and Wever, 1976.

show some irregularities, but agree in indicating best sensitivity over a
range of 100 to 800 Hz, with a rapid decline for lower and higher tones.
Peak sensitivity appears at 200 Hz for the two throat locations, and is
an octave higher for the neck location. The best point, which reaches
−30 db, represents a good degree of sensitivity—indeed a surprising
level for an animal lacking a tympanic membrane.

Tests with vibratory sensitivity, with the vibrating prod applied to the
skin over the quadrate, gave the results shown in Fig. 22-13. Here the
sensitivity is nearly uniform around 0 db from 200 to 900 Hz, with
rapid declines below and especially above. This type of function is considered to be mainly representative of inner ear factors. A comparison
with the preceding figure suggests that the middle ear mechanism provides a special reinforcement of the action of tones at the low end of
this range, around 200 Hz.

From the evidence it appears that *Sphenodon*, though having a conductive mechanism that has suffered from degenerative changes, perhaps
in consequence of the adoption of burrowing habits, possesses a well-
developed cochlea. Its tectorial structures are specialized in much the
same manner as in turtles, snakes, and many others, and, though the
hair-cell population does not attain great size, it is not as severely restricted as in some forms of lizards and amphisbaenians. This ear should

be capable of reporting sounds of moderate intensity levels within a range of about three octaves in the low frequencies. Comparatively little frequency differentiation in terms of specific action along the cochlea can be expected for this ear, but it should serve usefully in the reception of alerting and warning sounds, including the harsh croaking cries that these animals themselves produce.

23. ORDER TESTUDINES:

THE TURTLES

The turtles are an ancient group of reptiles, known from fossils dating from the Triassic period about 200 million years ago. Like other reptiles they are believed to have arisen sometime earlier from the cotylosaurs, but took an independent line of development that they have maintained to the present day. Their possession of body armor—the carapace above and plastron below—no doubt has contributed to their conservatism by shielding them from vicissitudes of the environment and protecting them from predators.

Existing turtles belong to two suborders, the Cryptodira and Pleurodira, with the first of these containing the great majority of species. These two groups are most readily distinguished by their different ways of withdrawing the head; in the Cryptodira the head is pulled backward by a sinuous contortion of the neck, and in the Pleurodira the head and neck are bent sideways. A specimen of the Eastern box turtle, *Terrapene carolina carolina*, one of the Cryptodira, is illustrated in Fig. 23-1, and one of the marsh terrapin, *Pelomedusa subrufa*, belonging to the Pleurodira, is shown in Fig. 23-2.

FIG. 23-1. The Eastern box turtle, *Terrapene c. carolina*. Drawing by Anne Cox.

The Cryptodira, containing 10 families and including about 52 genera and 167 species (Wermuth and Mertens, 1961), are of nearly worldwide distribution, absent only on the continents of Australia and Antarctica, whereas the Pleurodira, containing two families and including about 13 genera and 45 species, are present in only limited regions of South America, Madagascar, Australia, and New Guinea.

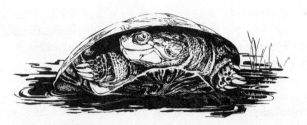

Fig. 23-2. The marsh terrapin, *Pelomedusa subrufa*, seen from in front with the head tucked away in the forepart of the shell. Drawing by Anne Cox.

Turtles are found in a variety of habitats. Many of them live in marshy areas or around freshwater ponds and streams, others are largely land dwellers, and a few are marine.

Until recently it was commonly believed that turtles were deaf, or at most possessed only rudimentary hearing. Some have suggested that turtles cannot hear aerial sounds but are only capable of perceiving ground vibrations. These views are probably accounted for by the inconspicuous form of their eardrums, the absence of any dramatic responses to sounds, and the relative rarity of vocalizations (Gans and Maderson, 1973).

There is now ample evidence not only that turtles are able to hear but that, within their particular range in the low frequencies, many species possess a high degree of sensitivity. The growth of this evidence has been gradual and replete with difficulties, and includes both behavioral and electrophysiological lines of experimentation.

Behavioral Studies

An early experiment is that of Andrews (1915), who trained turtles of the genus *Chrysemys* to approach for food at the sound of a bell but to refrain from doing so at the sound of a whistle. However, Kuroda, in two experiments in 1923 and 1925, failed to confirm these observations in the species designated as *Clemmys japonica* (now *Mauremys japonica*), though in one of his attempts he used the same method as Andrews.

Berger (1924) presented a variety of sounds, including tones, whistles, and noises, to 20 turtles belonging to eight different species, and looked for significant reactions such as head movements, eye movements, or arousal from sleep. The results were essentially negative.

Poliakov in 1930 used a conditioning procedure in which turtles of the common European species *Emys orbicularis* were caused to withdraw the head in response to a variety of stimuli, including bell sounds, noises, and tones. The method consisted of presenting a sound, and then eliciting the head withdrawal response by striking the head with a little hammer attached through a lever to the shell and operated by pulling a cord. Positive results were reported for a number of tones, including 165, 410, and 512 Hz produced by organ pipes and tuning forks, and (somewhat incredibly) a tone of 12,000 Hz produced with a Galton whistle.

These conditioning experiments were repeated some years later by Chernomordikov (1958) and Karimova (1958), both of whom reported essentially negative results. In Chernomordikov's experiments three turtle species were used: *Testudo graeca, Clemmys caspica*, and *Emys orbicularis*, and the conditioning procedure was varied in different experimental series. A direct conditioning to sounds was not obtained, but a conditioned response to a visual stimulus was readily produced. It was noted that the visual response when conditioned could not be inhibited by the presentation of sounds.

Karimova, who used specimens of *Testudo horsfieldi* and *Emys orbicularis*, made an effort to condition leg withdrawal, but without success. Both Karimova and Chernomordikov failed to obtain feeding responses in which the turtle was required, in response to an acoustic signal, to approach an apparatus and pull on a mouthpiece to obtain a morsel of food. When auditory stimuli were combined with gross mechanical vibrations, the conditioning was successful. However, after the conditioning to this combination of stimuli was well established, a presentation of the auditory signal alone was usually ineffective. These authors did not deny the presence of hearing in the turtles examined, but considered the auditory sense to be poorly developed in relation to the behavior required.

A thoroughgoing study of the problem of hearing in turtles by a conditioning method was carried out by Patterson (1966). Like Poliakov he used head withdrawal as the conditioned response, but with shock as the conditioning stimulus. The animals were of the species *Chrysemys scripta*, and included the subspecies *Chrysemys s. scripta, Chrysemys scripta elegans*, and intergrades.

The animals were induced to keep the head extended by the application of steady traction through a cord attached to metal pins inserted into the bone of the lower jaw. The initial training consisted of a number of trials in which a combination of low-frequency auditory and vibratory stimuli was produced with a telephone receiver applied to the anterior region of the carapace. The experimental arrangement was such that the animal could avoid the shock if it reacted to the tone with a slight retraction of the head. After stable conditioning to the sound-vibration complex was obtained, the vibratory stimulus was discontinued and a tone from a loudspeaker was substituted. The response then appeared to the tone alone. After further training to stabilize the tonal response, the intensity was varied systematically for the determination of thresholds.

That these responses were in fact mediated by the ear was proved by sectioning the columella on both sides in certain of the specimens. After this operation the response to tones fell sharply, in amounts up to 60 db. The responses to vibratory stimuli, however, were hardly altered by this sectioning procedure. The quantitative results of these experiments will be presented later, with a comparison with observations on similar species by the cochlear potential method.

ELECTROPHYSIOLOGICAL EXPERIMENTS

The first attempt in turtles to record electrophysiological activity in response to sound stimuli was made by Foà and Peroni in 1930. With an electrode on the eighth nerve of the giant sea tortoise *Caretta caretta*, they stimulated with intense organ-pipe tones beween 16.5 and 132 Hz, but the resulting potentials seemed unrelated to the stimulus frequency. Soon afterward, Wever and Bray (1931) obtained synchronized responses from the eighth nerve of the painted turtle, *Chrysemys picta*, with clearest results for tones below 500 Hz. Similarly, Adrian, Craik, and Sturdy (1938) reported responses from the eighth nerve of a box turtle named as *Cistudo* (perhaps *Emys orbicularis*) and a land turtle *Testudo graeca* on stimulation with tones up to 400 Hz, and in certain specimens for tones up to 800 Hz.

A further series of experiments was carried out by Wever and Vernon in 1956, in which the electrical potentials of the inner ear were used to measure auditory sensitivity in three common turtle species, *Chrysemys p. picta, Chrysemys scripta*, and *Clemmys insculpta*. Details of these sensitivity measurements will be presented in Figs. 23-51, 23-53, and 23-55; at this point it is sufficient to note that the results consistently showed a high degree of sensitivity in the low frequencies between 100 and 700 Hz, and a very rapid decline of response for tones above this range—a form of sensitivity that further experiments (including the behavioral observations of Patterson already referred to) have shown to be a general characteristic of the turtle ear.

VOCALIZATION IN TURTLES

Many have reported the production of vocal sounds by turtles, but these observations are widely scattered in the naturalistic literature. Campbell (1967 *b*) pointed out the need for further observations and recordings of turtle sounds, with attention to the relation between hearing and the frequency range of vocalizations. Gans and Maderson (1973) reviewed what is known about sound production in turtles and other reptiles.

Vocalizing in these and most other reptiles, except many of the geckos, is hardly an outstanding feature of their behavior, but several turtle species, especially among the land turtles, are reported to produce roars, grunts, or moans. This vocalization appears to be most frequent in males engaged in courting or copulative activity. The common descriptions of the sounds suggest that low frequencies are predominant in their compositon, though some species are described as producing whistles or hissing sounds.

A general exploration of auditory structure and levels of sensitivity has been carried out in a number of turtle species. The species examined include representatives of all but four of the existing families. The following list indicates the species studied, grouped by families, and the number of specimens of each.

SUBORDER CRYPTODIRA

Emydidae: *Chrysemys p. picta* (Schneider, 1873)—12 specimens; *Chrysemys rubiventris* (Le Conte, 1830)—1 specimen; *Chrysemys scripta* (Schoepff, 1792)—10 specimens; *Chrysemys scripta elegans* (Wied, 1838)—4 specimens; *Chrysemys scripta troostii* (Holbrook, 1838)—2 specimens; *Clemmys insculpta* (Le Conte, 1830)—3 specimens; *Clemmys guttata* (Schneider, 1792)—1 specimen; *Deirochelys reticularia* (Latreille, 1802)—2 specimens; *Geoemyda pulcherrima manni* (Dunn, 1930)—1 specimen; *Terrapene carolina carolina* (Linnaeus, 1758)—4 specimens; *Terrapene carolina triunguis* (Agassiz, 1857)—2 specimens; *Terrapene ornata* (Agassiz, 1857)—1 specimen.

Testudinidae: *Geochelone carbonaria* (Spix, 1824)—1 specimen; *Testudo horsfieldi* (Gray, 1844)—3 specimens; *Kinixys belliana belliana* (Gray, 1831)—1 specimen.

Chelydridae: *Chelydra serpentina* (Linnaeus, 1758)—3 specimens; *Macroclemmys temminckii* (Troost, 1835)—4 specimens.

Kinosternidae: *Kinosternon scorpioides* (Linnaeus, 1766)—3 specimens; *Kinosternon s. subrubrum* (Lacepede, 1788)—3 specimens; *Staurotypus triporcatus* (Wiegmann, 1828)—1 specimen.

Trionychidae: *Trionyx cartilagineus* (Boddaert, 1770)—2 specimens.

Cheloniidae: *Chelonia mydas* (Linnaeus, 1758)—4 specimens.

SUBORDER PLEURODIRA

Chelidae: *Platemys platycephala* (Schneider, 1792)—2 specimens.

Pelomedusidae: *Pelomedusa subrufa* (Lacepede, 1789)—3 specimens; *Pelusios sinuatus* (A. Smith, 1838)—2 specimens; *Podocnemis expansa* (Schweigger, 1812)—2 specimens; *Podocnemis unifilis* (Troschel, 1848)—2 specimens. Nearly all these specimens were first tested for auditory capability in terms of cochlear potentials and then were prepared histologically for a study of ear structure.

ANATOMY OF THE EAR

The Peripheral Structures

In turtles the sound-receptive and conductive mechanism is well developed. There is a superficial tympanic membrane, a tympanic cavity filled with air, and an ossicular mechanism of two elements, the columella and extracolumella. These structures will be described for the common species *Chrysemys scripta*, with occasional reference to variations in other species.

The Sound-receptive Surface. The tympanic membrane lies at the side of the head, well behind the eye and about the level of the corner of the mouth. It is inconspicuous because its outer layer is simply the skin of the face, whose surface markings in this species continue the striped pattern of the surrounding area. The skin is somewhat thinned in the middle of the tympanic area, and on careful scrutiny this area can usually be recognized as flattened and slightly depressed, or even delicately outlined. Careful palpation reveals the area as relatively soft and yielding.

In Fig. 23-3 this membrane is intentionally emphasized. Note that the rotation of the head makes it appear in a low position.

The inner layer of the tympanic membrane is formed by a plate of cartilage, which is the main part of the extracolumella. Between this plate and the skin is a layer of connective tissue that is increasingly thickened posteriorly and attaches to the bone in this region as the posterior ligament. The drawing of Fig. 23-4 shows the three tympanic layers and their local variations.

In the sea turtle *Chelonia mydas* the middle layer of the tympanic membrane is particularly thick and contains a large amount of fatty tissue. This material serves to couple the thick, horny surface layer to the relatively deep-lying extracolumellar knob (Fig. 23-5).

The mode of action of the tympanic membrane was studied in the common box turtle *Terrapene c. carolina*. This membrane was set in motion by applying to its surface the point of a needle driven by a mechanical vibrator. The cochlear potentials were then observed as the needle was given various locations over the surface of the membrane.

In one series of experiments the vibrating needle was first applied at a point in front of the anterior edge of the extracolumellar disk, and

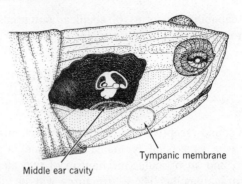

FIG. 23-3. The head of *Chrysemys scripta elegans*, seen in a dorsolateral view, with a deep dissection to expose the labyrinth. The tympanic membrane and a portion of the middle ear are indicated also. About natural size.

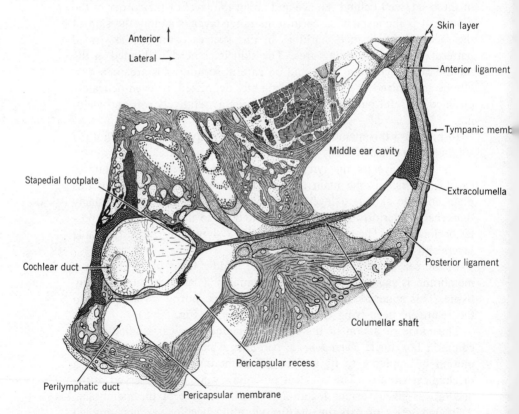

FIG. 23-4. A frontal section through the ear region of *Chrysemys scripta*, at a level that cuts across the entire columella. The location is far ventral with respect to the inner ear, and shows only the lagenar end of the cochlear duct. Scale 7.5×.

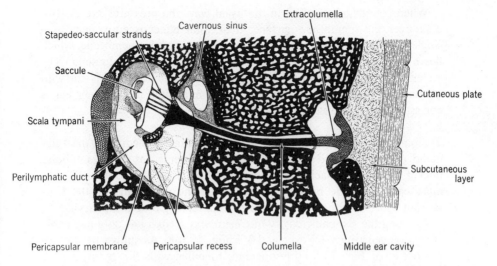

FIG. 23-5. Semischematic representation of the auditory mechanism in *Chelonia mydas*. The right ear is shown in a dorsal view. Scale 2×.

then was moved posteriorly along a line by steps until it was well behind the posterior edge of this disk. At each driving position along this line the alternating current sent through the vibrator was adjusted to produce a constant cochlear potential of 0.3 μv. The results were then expressed in decibels relative to the current required at a point on the anterior edge. These observations are represented in Fig. 23-6, and the figure includes a sketch of the extracolumella on the same scale as the graph to locate the driving positions. It is seen that the actuating current, and therefore the driving amplitude, varies progressively along the anteroposterior diameter of the plate, decreasing to −40.7 db at the extreme posterior edge.

These results signify that the tympanic membrane operates as on a hinge at its posterior edge. This form of motion is to be expected from the fact that the extracolumellar plate is held lightly by fine fibers around most of its circumference, but has a heavy ligamentous attachment at its posterior edge. As Fig. 23-4 has shown, this posterior ligament is particularly heavy, whereas the anterior ligament is thin and attaches to the extracolumellar plate through a section of loose connective tissue.

As the driving position is moved from anterior to posterior edges of the membrane the lever action continually changes. Anterior to the columella the amplitude of movement transmitted to the columella is reduced, immediately over the columella it is unchanged, and at more posterior positions it is increased.

When sounds are applied in the usual way, the pressures are distributed over the membrane surface; their total effect is equivalent to a single pressure applied at the geometrical center of the membrane, which lies a little anterior to the point of attachment of the columella. The system then acts as a reducing lever of moderate value, in this ear with an amplitude ratio of 4:3. The force communicated to the inner ear is correspondingly increased.

Tympanic Cavity.   An air cavity lies immediately medial to the tympanic membrane, and its upper portion continues as a large tubular pocket running far back in the squamosal bone. In the floor of the primary cavity, a little behind the middle of the membrane, is the Eustachian tube, which connects with the lateral angle of the pharynx.

Opening the tympanic cavity in *Chrysemys scripta* has only negligible effects on sensitivity (Wever and Vernon, 1956). As shown by the upper curve of Fig. 23-7, this opening, together with the removal of a part of the lateral wall, had no effect upon the response to low tones and produced only slight variations for high tones. It appears that this cavity does not produce any significant resonance within the turtle's range, as indeed is to be expected from the small dimensions. (A calculation made by treating this cavity as a closed tube indicated a fundamental resonance frequency of about 6,000 Hz.)

Conductive Mechanism.   The ossicular chain contains two elements, as already mentioned. As Fig. 23-4 shows, the extracolumellar plate has a short process on its inner surface that makes a firm junction with the outer end of the osseous columellar shaft. This shaft is long and thin, and takes a curved path through the middle ear cavity and then runs medially to penetrate a portion of the quadrate. It then traverses the pericapsular recess and ends at the oval window of the cochlea. Within the pericapsular recess, this medial end of the columella expands rapidly to produce a large funnel-shaped stapedial footplate.

At the point of leaving the tympanic cavity and entering its canal in the quadrate, the columella is deeply embedded in connective tissue, and this tissue continues as a sheath along the canal. The thin bony rod is therefore guided along its curved path, with freedom of longitudinal motion but with lateral displacement largely prevented.

The columella on emerging from its canal in the quadrate runs close by a cavity known as the cavernous sinus, which contains the internal carotid artery, internal jugular vein, and facial nerve.

The absence of any contact between the tympanic cavity and the otic capsule, which is a peculiarity of turtles, evidently arises through the extensive development of the paroccipital process and quadrate. The

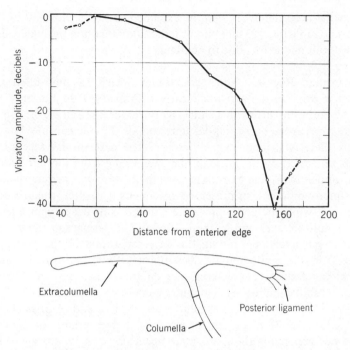

FIG. 23-6. Mode of action of the tympanic membrane in *Terrapene c. carolina*. Amplitude of driving of the membrane surface is represented on the ordinate in decibels relative to the least effective position at the anterior edge. Distances from this position are shown on the abscissa in thousandths of an inch. The cross-sectional view of the extracolumellar disk below is on the same scale as the abscissa. From Wever and Vernon, 1956c.

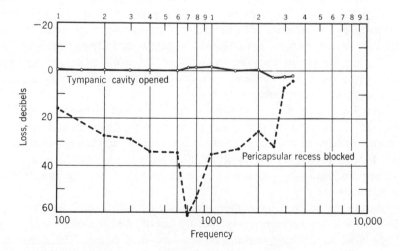

FIG. 23-7. Effects on sensitivity of opening the tympanic cavity (solid line) and blocking the pericapsular recess (broken line) in a specimen of *Chrysemys scripta*. From Wever and Vernon, 1956b.

columella accordingly is greatly extended, and its close-fitting canal makes possible the transmission of longitudinal motion through a slender rod without undue loss due to bending.

Pericapsular Recess.   The pericapsular recess, a fluid-filled cavity shown in Fig. 23-4, was first clearly identified by de Burlet (1934). Functionally, in turtles this fluid cavity serves the same purpose as in snakes, and as the name implies it surrounds the otic capsule, but structurally there is an important difference. The pericapsular recess of turtles is separated by the pericapsular membrane from the perilymphatic duct, rather than being continuous with it. Also the pericapsular fluid is distinctly different; in preserved specimens it is often coagulated into a firm gel, while the perilymph spaces in the same ear remain filled with a clear fluid (Wever and Vernon, 1956). The pericapsular fluid usually stains distinctively, permitting the easy separation of the pericapsular recess from the perilymphatic duct. Baird (1960) performed a number of staining tests to characterize the difference between this fluid and either cerebrospinal or perilymphatic fluids.

In most species the pericapsular recess is clear and open as shown in Figs. 23-4 and 23-8, except for occasional thin strands of connective tissue accompanying the glossopharyngeal nerve in its passage through the cavity. In some, however, there is a considerable amount of this tissue, which forms webs and partial partitions that divide the cavity into numerous connected spaces. This invasion of the cavity is moderate in *Terrapene c. carolina*, but becomes extreme in *Chelonia mydas*.

One specimen of *Terrapene c. carolina* had not only intracapsular tissue following the glossopharyngeal nerve but two strong webs arising from the posterior edge of the oval window and extending to the lateral wall of the recess. These and other minor vanes of tissue separated the cavity into ten or more compartments. These compartments were outlined in varying degree; some seemed to be almost fully enclosed, others only partially so. In another specimen of *Terrapene* the intracapsular tissue was much less abundant. *Pelusios sinuatus* had a small amount of very fine reticular tissue.

In *Chelonia mydas*, as shown in Fig. 23-5 and in further detail in Fig. 23-9, the intracapsular tissue is everywhere abundant and forms a sponge-like mass. A few relatively free channels run through this mass, but for the most part the cavity is filled with loose reticular tissue. In all regions the spaces appear to be interconnected, so that the fluid is continuous.

As this tissue must have about the same density as pericapsular fluid and will be almost completely lacking in stiffness, it can present little impedance to vibratory motion through the fluid circuit. There may be some moderate frictional resistance imposed by the close attachments to the footplate, especially in *Chelonia mydas*.

The perilymphatic duct is enlarged at the level seen in Fig. 23-8 (considerably dorsal to that shown in Fig. 23-4) and then passes through an opening in the posterior wall of the otic capsule to join the scala tympani. The opening into the capsule was referred to by Hasse (1871) as the round window, for he considered it to be homologous to this window in other animals. As the opening in turtles is not covered by a membrane, it is best designated simply as the perilymphatic foramen.

This foramen completes the fluid circuit extending from the inner surface of the stapedial footplate by way of the pericapsular recess to the

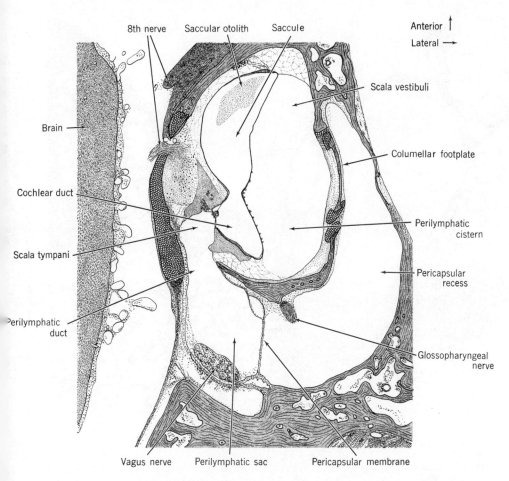

FIG. 23-8. A frontal section through the ear region of *Chrysemys scripta* at a level well dorsal to that of Fig. 23-4. Scale 15×.

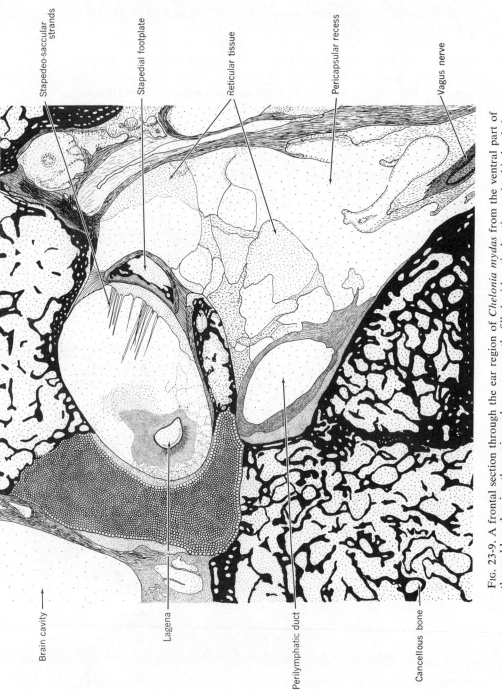

Brain cavity ⟶

Stapedeo-saccular strands

Stapedial footplate

Reticular tissue

Pericapsular recess

Vagus nerve

Lagena

Perilymphatic duct

Cancellous bone

FIG. 23-9. A frontal section through the ear region of *Chelonia mydas* from the ventral part of the cochlea, showing the pericapsular recess partly filled with reticular tissue. At this level the

outer surface of the footplate, with only Reissner's membrane, the basilar membrane, and the pericapsular membrane across the path. In turtles, as already seen in amphisbaenians, snakes, *Sphenodon*, and a few lizards, the mobilization of the cochlear fluids is achieved by a reentrant fluid circuit. This circuit is represented by the arrows in Fig. 4-3 above.

In most of the turtles studied, the form of the pericapsular recess and its relations to adjoining structures are similar to those represented in Figs. 23-4 and 23-8. A variation from this form was described by de Burlet (1934a) in *Emys orbicularis*. In this species the pericapsular recess does not end at its junction with the perilymphatic duct, but continues around this duct and into the brain cavity, where it runs anteriorly alongside the cranial wall, passes around the endolymphatic duct, and ends at the level of the utricle.

A condition similar to the one represented by de Burlet in *Emys* was found in only one species among those studied here, and in a simpler form. In *Terrapene c. carolina*, as shown in Fig. 23-10, the pericapsular recess joins the perilymphatic sac through the pericapsular membrane much as usual, and here becomes greatly reduced. It does not terminate, however, but sends a narrow canal into the brain cavity, which first runs anteromedially between the bony cranial wall and the perilymphatic duct and sac and then continues anteriorly along this wall to expand somewhat and end just below the endolymphatic sac. This cranial extension of the pericapsular recess does not surround the endolymphatic duct or sac, as de Burlet described for *Emys orbicularis*. In *Terrapene ornata* this cranial extension of the pericapsular recess is present only as a remnant. At its medial end this recess makes contact with the perilymphatic sac and sends off a narrow duct that runs a short distance anterior to the sac and ends blindly. This duct is largely filled with loose connective tissue, though a somewhat interrupted channel can be traced along it in which are small amounts of material staining in the manner characteristic of pericapsular fluid.

The only other variation in the pericapsular recess among the species examined is a degree of foreshortening, seen in some of the Pleurodira. In *Pelosios sinuatus* and *Pelomedusa subrufa* this recess is less than half its usual length, terminating close to the posterior border of the columellar footplate as seen in Fig. 23-11. In *Podocnemis expansa* and *Pelomedusa unifilis* it is a little longer, turning medially beyond the footplate and ending soon thereafter. In all instances the perilymphatic duct is extended laterally to meet this recess, and the two are separated by the pericapsular membrane as usual.

These variations have no obvious significance for sound transmission; in all these species a continuous fluid circuit is present for the mobilization of the cochlear structures.

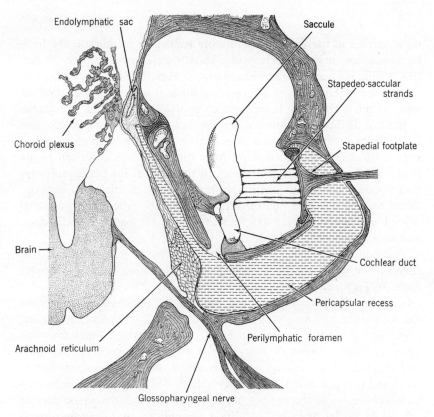

FIG. 23-10. Frontal section through the ear region of *Terrapene c. carolina*, showing the extension of the pericapsular recess into the cranial cavity.

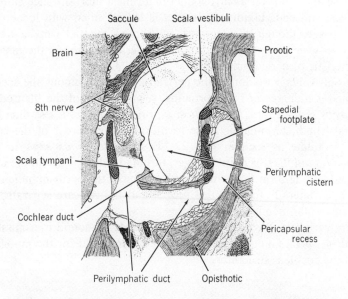

FIG. 23-11. The pericapsular recess in *Pelusios sinuatus*. Scale 12×.

Processes of Sound Reception

How well the receptive and conductive structures of the turtle's ear serve for the transmission of acoustic vibrations to the cochlear hair cells is indicated by the results of sensitivity measurements, the details of which will be presented for various species in a later section. Certain special features of these receptive processes are revealed in a series of experiments carried out earlier (Wever and Vernon, 1956) on the two species *Chrysemys p. picta* and *Chrysemys scripta*.

Action of the Receptive System. The seeming rigidity of the turtle's auditory receptive structures, especially in the larger species, has led some writers to question their efficacy for sound transmission. Thus de Burlet (1934b) expressed doubts whether the tympanic membrane of the green turtle *Chelonia mydas*, which is several millimeters thick in large specimens, has any capability of vibratory motion.

For the *Chrysemys* species this question was answered experimentally. After a normal curve of cochlear potential sensitivity was obtained, the columella was interrupted by cutting out a small piece near its junction with the extracolumella, and the sensitivity measurements were repeated. Results for one of the specimens are given in Fig. 23-12. The loss of continuity of the ossicular chain has produced a reduction in sensitivity varying in amount with frequency from 35 to 74 db and averaging 50 db for the range covered. Average results for a series of eight ears, expressed as a loss in sensitivity resulting from cutting the columella, are presented in Fig. 23-13. A similar experiment was carried out by Patterson using a behavioral method of assessing the changes in sensitivity, as noted above. The two sets of observations are in good agreement.

Reentrant Fluid Circuit. The effectiveness of the fluid circuit in the mobilization of the cochlear fluids was studied in a number of *Chrysemys* specimens. For this investigation it was necessary to open the tympanic cavity and to expose and open the pericapsular recess. Removal of the lateral wall of the tympanic cavity gave access to the medial portion of the pericapsular recess. As already indicated by the upper curve in Fig. 23-7, this opening has only slight effects upon sensitivity. The location of the pericapsular recess was determined by its relation to the cavernous sinus, which in the species studied can be seen as a dark shadow through the bony wall. The wall of the pericapsular recess was penetrated above the sinus, using great care to avoid the sinus itself and its blood vessels.

When a small opening is made in the pericapsular recess its fluid does not immediately escape. The effects on the ear's sensitivity are only slight, as shown by the solid curve of Fig. 23-14. For most tones there

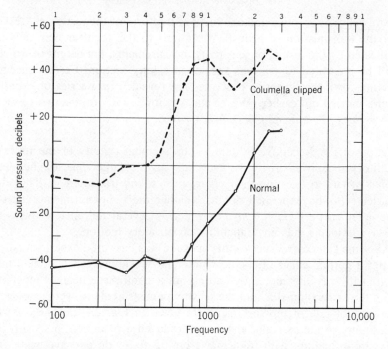

FIG. 23-12. Sensitivity curves in a specimen of *Chrysemys scripta* be-fore and after interrupting the ossicular chain. From Wever and Vernon, 1956b.

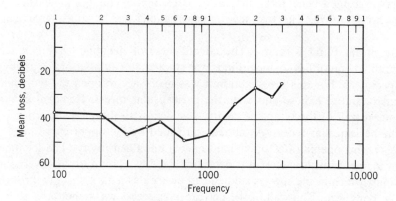

FIG. 23-13. Loss of sensitivity caused by interrupting the ossicular chain, averaged for eight ears of the species *Chrysemys scripta*. From Wever and Vernon, 1956b.

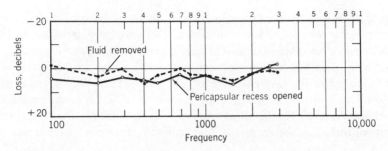

Fig. 23-14. Effects on sensitivity of opening the pericapsular recess (solid line) and removing its fluid (broken line), observed in *Chrysemys scripta*. From Wever and Vernon, 1956b.

were losses up to 6 db, and the highest tones showed slight gains; the average was a loss of 3 db.

The fluid contained in this recess was then removed by inserting small pledgets of absorbent cotton. This removal was practically complete up to the pericapsular membrane. Again the effects on sensitivity were small, as shown by the broken curve of Fig. 23-14. This curve, it should be noted, indicates the additional effects of fluid removal beyond those already produced by opening the recess. The total effect of opening the recess and removing its fluid is obtained by adding the two curves of this figure; this effect averaged 5 db.

A limitation on the fluid displacements produced along the reentrant circuit by vibrations of the stapedial footplate is imposed by impedances along the entire path, including the frictional forces arising at the walls of the enclosure and also in the fluids themselves by reason of their viscosity. Removal of the pericapsular fluid ought to reduce this frictional resistance. That no improvement of sensitivity follows this procedure shows that the impedance imposed by this portion of the fluid circuit is negligible in relation to that of the remainder of the system.

The removal of the pericapsular fluid converts these turtle ears to the more familiar type seen in the great majority of lizards and in all the birds and mammals, in which the vibratory fluid column ends at the flexible membrane of the round window that is bounded externally by air. After drainage of pericapsular fluid the pericapsular membrane plays the role of the round window membrane as a flexible boundary. An improvement in sensitivity, especially in the high frequencies, might well be expected as a result, but this effect was not found. Therefore it seems that other conditions, perhaps within the cochlea itself, enter into the fashioning of the turtle ear as a low-frequency receptor.

The essential role of fluid mobilization in the turtle's ear was demonstrated by an experiment in which the pericapsular fluid was removed up to its bounding membrane, and then this medial end of the pericapsular recess was filled with bone wax. The wax plug extended to the pericapsular membrane, and was applied with care to make good contact with its surface without any injury to it. The effect of this procedure is indicated in the broken curve of Fig. 23-7. The effect is frequency dependent and reached its greatest magnitude of 60 db in the region of 700 Hz. At 3500 Hz the effect was only slight; evidently the conductive system, including the reentrant fluid pathway, does not function well at these high frequencies.

These results suggest that in the low frequencies, up to about 700 Hz, the fluid circuit has resonant properties that enhance the sensitivity, but at the higher frequencies the mass of the moving structures progressively limits the vibratory action.

IMPEDANCE MATCHING IN THE TURTLE EAR

As is well known, the efficiency of the ear in its reception of sounds depends upon the provision in the sound-receptive system of means of matching the impedance of the aerial medium with that of the inner ear structures. More complex ears employ a combination of mechanical transformer actions, including a lever action in the ossicular chain and a hydraulic action arising from the difference in areas between tympanic surface and stapedial footplate. In the cat, for example, the transformations in the middle ear produce a total ratio computed as 60.7, which multiplies the acoustic resistance of the air by 60.7^2 or 3,644, thus suitably matching the impedance of the inner ear fluids (Wever and Lawrence, 1954).*

Turtles seem to lack an ossicular lever as such, but a similar mechanical effect may be produced when the columella is eccentrically attached to the stiff cartilaginous disk forming the inner layer of the tympanic membrane, as shown for *Terrapene c. carolina* in Fig. 23-6. The effect of such a lever appears to be limited, however. Its ratio was determined as 1.33 in *Terrapene c. carolina* as noted above, and in a specimen of *Chrysemys scripta* was measured as 1.37, only a little larger.

More effective in the turtle middle ear is the hydraulic ratio as defined above. In a specimen of *Chrysemys scripta* the area of the extracolumellar disk was measured as 103.5 sq mm and that of the stapedial footplate as 12.1 sq mm, giving a ratio of 8.5. This ratio when combined with the mechanical effect of 1.37 gives a total transformation of

* Tonndorf and Khanna (1972) are of the opinion, on the basis of their holographic studies of the middle ear mechanism, that a lever action in the tympanic membrane itself, of a form suggested early by Helmholtz, adds further to the impedance matching in ears with flexible tympanic membranes. This factor can not operate in turtles with their stiffened membranes, however.

11.7. Such an amount is small in comparison with that obtaining for the cat, and seems clearly inadequate for best reception of aerial sounds.

If the turtle ear were best adapted for the reception of sounds in water it should require no transformer action in the receptive system: water and cochlear fluid are already well matched, and any impedance increase provided by the middle ear would be disadvantageous. Thus it is suggested that the turtle ear represents a compromise between aerial and aquatic reception of sounds, and the transformer ratio of this ear lies between the best values for the two media. It seems likely therefore that this ear suffers a loss due to the need to compromise, but is still serviceable in each medium.

Evidence on this question was sought in a comparison of hearing in air and in water, and the results will be presented in the section on sensitivity below.

INNER EAR

Figure 23-3 gave a view of the turtle's head from the right side, with a deep dissection to show the location of the membranous labyrinth on this side. The prootic and opisthotic bones have been dissected away along with parts of other bones closely adjoining these, which form the labryinthine cavity.

As in other vertebrates, the osseous labyrinthine cavity has a form corresponding to that of the membranous structure within it, especially in the dorsal region. The cavity is larger than its contained structure, and the intervening space is filled with connective tissue of two types, a loose trabecular network and limbic tissue. In turtles the trabecular tissue is especially abundant, whereas limbic tissue is comparatively sparse. As usual, the trabecular network serves to suspend the membranous structures within the walls of the enclosure, forming an effective shock mounting for their delicate parts, and the rather isolated masses of limbic tissue provide more rigid support where it is needed. Notably the limbic tissue serves as a reinforcement of portions of the membranous walls and especially for the sensory areas. The largest mass forms a plate that contains an opening covered by the basilar membrane on which the auditory papilla is supported.

A peculiar feature of the turtle's ear is the presence of a special form of trabecular tissue, the stapedeo-saccular strands, which consist of parallel-running fibers between the stapedial footplate and the wall of the saccule (see Figs. 23-5 and 23-9). These fibers are particularly heavy and strong and are highly elastic. In most species the fibers are largely independent, not forming a network as is usual for this type of tissue elsewhere in the labyrinth. Only occasionally do the fibers show branching or cross connections, except close to their end attachments. Their function is unknown, though Wever and Vernon (1956), who first described them, suggested two possibilities: the transmission of vibratory motion to the saccular macula or the representation of water pressure by simple displacements.

Cochlear Duct. As Fig. 23-8 shows, the cochlear duct lies in the otic capsule just posterior to the saccule, and is separated from it by a constriction of the enclosing wall. The basilar membrane lies in the medial wall, supported by the limbic plate, seen in cross section at this level as two triangular areas. In *Chrysemys scripta* the basilar membrane is a slightly curved oval about 3½ times as long as it is wide, and its long axis stands nearly vertical. In one specimen in which a careful dissection was made, this surface was found to be inclined anteriorly about 15° from the vertical.

Auditory Papilla. The turtle's ear differs from that of all other reptiles in having a marked extension of the auditory papilla beyond the basilar membrane, usually at both dorsal and ventral ends. Accordingly, a good many of the auditory hair cells are not on the basilar membrane, but rest on the solid surface of the limbus (Wever, 1971c). This limbic extension of the hair cells is minimal in the species *Kinixys belliana*.

The changing picture of the turtle's papilla will be followed from dorsal to ventral ends in a specimen of *Chrysemys scripta elegans*, which may be taken as representative of the majority of species.

Sensory epithelium is recognizable in a region of the cochlea well dorsal to the appearance of the limbic hiatus and its covering by the basilar membrane. In this region the first departure from the ordinary lining epithelium of the cochlear duct is a noticeable thickening, in which two types of cells occur instead of the usual simple cuboid or low columnar cells. These cells are readily recognizable as hair cells and supporting cells, and they show the usual characteristics of such cells. The first are tall columnar cells bearing ciliary tufts, with elongate nuclei located toward their bottom ends, and standing well above the limbic surface. The supporting cells rest on the limbus, have their round or ovoid nuclei in a low position, and often send up thin columnar processes between the hair cells. These columns expand somewhat near the surface to support the hair cells at their upper ends.

The limbic hair cells are surmounted by a tectorial membrane, and are innervated by small branches of the cochlear nerve. For the most part these hair cells are smaller in diameter than the ones occurring on the basilar membrane, and their separation is less, so that a great many appear in a relatively small area.

Still more ventrally this patch of ciliated cells is larger and covers a wider area of the limbus, often forming as many as 30 rows. This condition is represented in Fig. 23-15. A simple basilar membrane is present, with the scala tympani below, but the hair cells appear only at its anterior border, where they rest on the limbus and do not appear to be mobile. A relatively heavy tectorial plate makes connections with the cilia of these hair cells.

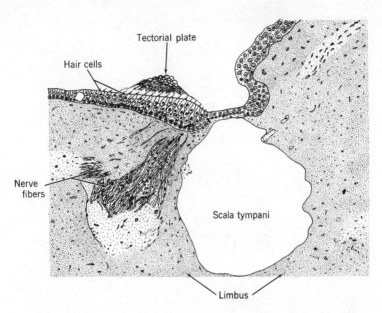

FIG. 23-15. The ear region of *Chrysemys scripta elegans* near the dorsal end of the cochlea with hair cells lying entirely on the limbus. Scale 125×.

A little farther ventrally, as shown in Fig. 23-16, the situation is much the same except that there are now two groups of hair cells, both covered by the expanded tectorial plate. The posterior group contains noticeably larger hair cells, and these cells are served by nerve fibers of a distinct bundle running along the posterior edge of the limbus. The innervation of the anterior group of cells is not evident in this figure, but comes from fibers present in sections just anterior to this one, which send branches in a posterior direction.

Still more ventrally, as shown in Fig. 23-17, the anterior group of hair cells is no longer present. The hair cells in the posterior position are encapsulated and lie on the anterior part of the basilar membrane. The picture presented here is typical of the dorsal region of the cochlea.

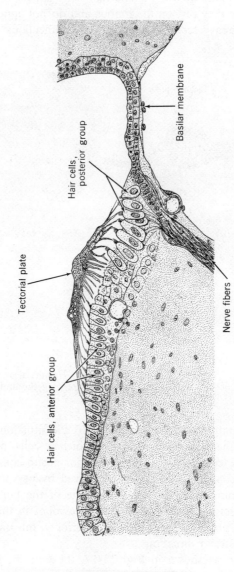

FIG. 23-16. The ear region of *Chrysemys scripta elegans* at a level ventral to that of the preceding figure, where the hair cells form two groups on the limbus. Scale 250×.

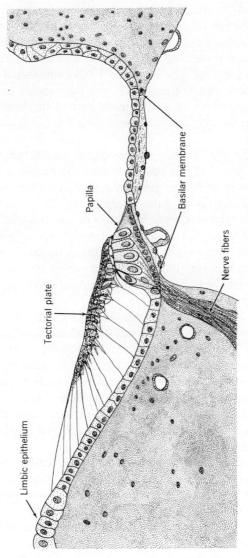

Fig. 23-17. The ear region of *Chrysemys scripta elegans* at a level still farther ventral than in the preceding figure, where a definite papilla is present. Scale 250×.

The tectorial structure is dense at its middle region, forms a cap with a number of little pits on its underside over the hair cells, and is anchored by a great many fine fibers running obliquely posteroanteriorly and somewhat medially to the limbic epithelium.

About a third of the distance along the basilar membrane the structure changes somewhat, taking the form shown in Fig. 23-18, and this altered form continues throughout the ventral region. The tectorial membrane here has a root process on the limbus, and elsewhere has a reticular form. It tapers progressively, and attaches to a substantial plate with pits on the underside that correspond to the numbers of rows of hair cells. The anchoring fibers along the main part of the membrane run posteromedially to the limbic epithelium, with a few going to the end of the papilla itself.

Everywhere from the first appearance of the papilla on the basilar membrane to the ventral region now described, the papilla is located on the anterior portion of the basilar membrane, with its anteromedial border attached along the sharp edge of the inner limbus. The posterior portion of the basilar membrane is covered by a layer of low epithelial cells, and beneath it (toward the scala tympani) is a loosely fibrous layer containing scattered nuclei, thickened in its midregion. A fundus is present under the papilla, but it is very thin, usually measuring no more than 4 μ in thickness. It is relatively flat, not regularly swollen in the middle as is typical of this structure in lizards.

In this species, and in all others studied except *Testudo horsfieldi* and *Kinixys belliana*, the papilla also extends beyond the ventral end of the basilar membrane. As the basilar membrane narrows at this end, the papilla continues and even grows wider. At first it extends a moderate distance onto the inner limbus, and then more ventrally it spreads onto the outer limbus to a progressively increasing extent. One stage of this spread is shown in Fig. 23-19, where only the middle rows of hair cells lie over the basilar membrane and can be considered as having some measure of mobility.

A further stage is shown in Fig. 23-20. Here the basilar membrane is absent, and all the hair cells lie on the limbus. A tectorial membrane is present and makes connections with at least the more anterior hair cells. Also the nerve supply to these cells is maintained.

Still more ventrally this patch of ciliated cells is larger and covers a wide area of the limbus, often forming as many as 30 rows. Thereafter these limbic cells rapidly disappear.

The relative number of limbic cells varies greatly with species, as will be shown presently.

The general form of the papilla in other species is much the same as just described for *Chrysemys scripta elegans*. The dimensions vary, but

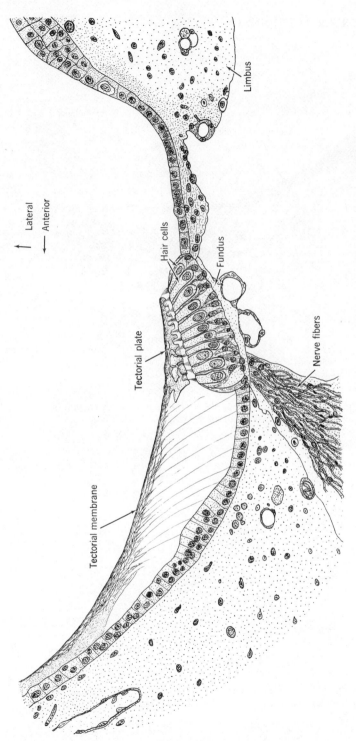

Lateral

Anterior

Limbus

Hair cells

Fundus

Tectorial plate

Nerve fibers

Tectorial membrane

Fɪɢ. 23-18. The auditory papilla of *Chrysemys scripta elegans* at a level about a third of the way along the cochlea. Scale 300×.

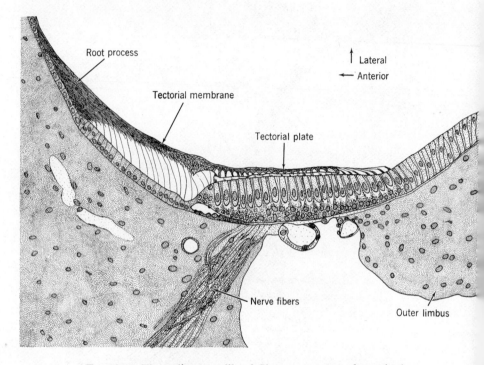

Fig. 23-19. The auditory papilla of *Chrysemys scripta elegans* in the far ventral region of the cochlea. Scale 250×.

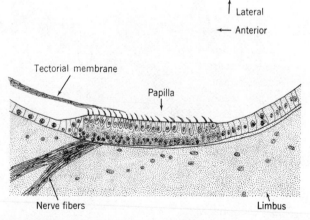

Fig. 23-20. The auditory papilla of *Chrysemys scripta elegans* close to the ventral end where the basilar membrane is absent. Scale 250×.

in all forms examined the papilla rests on the anterior half of the basilar membrane, with the tectorial structure extending from its attachment on the inner limbus, and showing its characteristic expansion to form a terminal plate over the papilla.

The relations between the tectorial structure and the papilla vary somewhat along the cochlea, with minor differences among species. In general, this structure at the dorsal end of the cochlea is only slightly differentiated. It consists of a ribbon of reticular tissue that grows progressively thicker toward its outer end, and here changes abruptly from a reticulum to a more orderly structure that in the sections presents a number of finger processes with thin webs of material between. This form represents a deeply pitted tissue that has been cut lengthwise of the pits. In some instances it has been possible to observe that the ciliary tufts of the hair cells approach the openings of these pits and attach to the edge of a fingerlike projection forming the lateral border.

A little farther ventrally along the cochlea, the outer border of the tectorial ribbon becomes more highly differentiated, and a definite plate may be recognized. At the same time the middle of this ribbon usually thins out, often to only about a micron in thickness, while the innermost edge sends out many fine filaments that form a sort of root process on the surface of the limbus. The anchoring fibers that extend along the medial surface of the ribbon to the limbic epithelium are usually abundant at the dorsal end, but often diminish in numbers toward the middle of the cochlea, and sometimes nearly disappear.

Still farther ventrally the tectorial plate continues to enlarge as the papilla grows in size and the number of longitudinal rows of hair cells increases, though usually this plate remains narrower than the papilla and fails to extend over its outermost edge. The hair cells of the most posterior two or three rows then make their connections to the outer edge of the tectorial plate by means of fine fibers or elongated kinocilia. Fig. 23-21 shows a section of a specimen of *Trionyx cartilagineus* through the ventral region of the cochlea, in which the tectorial plate is unusually extensive and covers all 11 rows of hair cells. This figure shows the deep pits present in the midportion of the tectorial plate and the progressively more shallow pits posteriorly. Also to be noted is the thin form of the tectorial membrane in its middle portion and the root processes on the limbus at the anterior end. Usually, if the anchoring fibers have nearly disappeared around the middle of the cochlea, they tend to reappear toward the ventral end.

Fig. 23-22 is a sectional view of the cochlea of *Podocnemis expansa*, taken near the dorsal end of the basilar membrane. Here a number of hair cells occur on the limbus, and the tectorial membrane extends over both these and the ones contained in the papilla. The outer end of the tectorial structure constitutes a sort of claw, and the inner end breaks up into fine fibers that connect with the hair tufts of the limbic hair cells. Two distinct bundles of cochlear nerve fibers serve the two groups of hair cells.

CILIARY ORIENTATION PATTERNS

Brief reports on orientation of the cilia borne by the auditory hair cells of turtles have been made by Mulroy (1968b) and Baird (1969), from observations with the electron microscope. They indicated that all the ciliary tufts are oriented in one direction, but gave no further details.

Ciliary orientation in the turtle ears of the present study has been examined with the light microscope, though the resolution of this instrument is hardly sufficient for a complete picture. In turtles the cilia of the hair cells are particularly fine, and in most regions the tectorial plate lies so close over the hair cells that it is difficult to obtain a full view of the tufts. In some regions, however, these tufts appeared clearly enough for a judgment of the orientation to be made in terms of the varying length of the stereocilia; the kinocilium itself was not usually identifiable. In nearly all instances in which a firm judgment could be made, the longest stereocilia were on the posterolateral side of the tuft, from which it may be inferred that the kinocilium was on that side also.

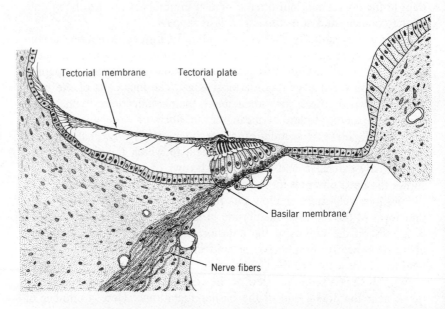

Fig. 23-21. A cross section of the cochlea of *Trionyx cartilagineus*, taken near the ventral end. Scale 200×.

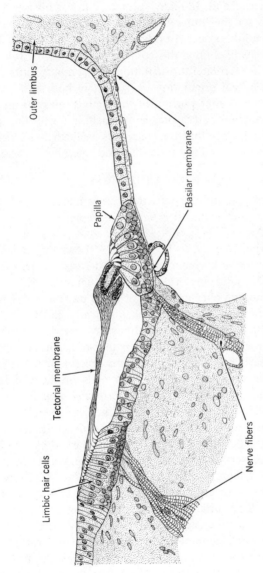

Outer limbus

Papilla

Basilar membrane

Tectorial membrane

Limbic hair cells

Nerve fibers

FIG. 23-22. A cross section of the cochlea of *Podocnemis expansa* near the dorsal end of the basilar membrane. Scale 250×.

Sometimes, however, the opposite orientation was found. In some instances this condition was observed for nearly all the hair cells in a transverse row, except perhaps for the first or second cell in the row at its extreme anterior end. In many specimens it was observed that the ciliary orientation for the limbic hair cells was contrary to that for the hair cells of the papilla proper. In other specimens this observation was difficult, but it seems likely that this contrary relation is the rule.

A posterior or posterolateral orientation of the ciliary tufts in turtles corresponds to a lateral orientation in lizards; in both cases the kinocilium lies toward the outer portion of the limbus, away from the side on which the nerve fibers enter. The arrangement in turtles thus parallels the observation in lizards that the hair cells beneath a tectorial plate nearly always have their ciliary tufts disposed laterally.

COCHLEAR DIMENSIONS

The dimensions of the cochlear structures were measured in 11 of the 25 species studied, and included at least one representative of each of the eight families.

Chrysemys scripta elegans. — Figure 23-23 shows the dimensions of the basilar membrane in a specimen of the red-eared turtle, *Chrysemys scripta elegans.* The solid line in part *a* of the figure shows the varying width of this membrane over its extent of 1050 μ. This width increases progressively in the dorsal region, reaches a nearly flat maximum over the range 300-730 μ, and then falls almost symmetrically toward the ventral end. The form of the membrane is that of a flattened oval.

As described earlier, many hair cells are present in the turtle's ear in an extension of the papilla at the two ends of the cochlea. This extension at the dorsal end in *Chrysemys scripta elegans* has been indicated in Figs. 23-15 and 23-16 and that at the ventral end in Figs. 23-19 and 23-20. The transition from limbic hair cells to the regular papillar hair cells at the dorsal end of the cochlea is marked by two features, a change in the forms of the hair cells and the appearance of a papillar capsule. The first of these features is indicated in Fig. 23-16, where the hair cells of the posterior group are taller and thicker and usually show more prominent ciliary tufts, and the second, appearing in Fig. 23-17, is the presence of a definite capsule enclosing the hair cells. The papilla proper is regarded as beginning with the encapsulated form as represented here. The broken curve in *a* of Fig. 23-23 indicates this structure as appearing at a point 90 μ from the end of the basilar membrane, where it has a width of 100 μ. This width increases somewhat in a ventral direction, becomes nearly constant around 130 μ through the middle of the cochlea, and then rises irregularly toward the ventral end.

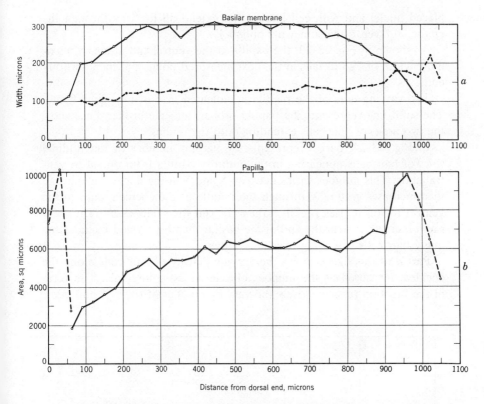

FIG. 23-23. Cochlear dimensions in a specimen of *Chrysemys scripta elegans.* Part *a* shows the varying width of the basilar membrane (solid line) and of the auditory papilla (broken line). Part *b* shows the areas of the papilla proper (solid line) and of its extensions onto the limbus (broken lines).

Near the ventral end of the cochlea, as seen in Fig. 23-19, the papilla lies with its midportion over the narrowing basilar membrane and its edges extending along the limbus in both anterior and posterior directions. The anterior edge is well differentiated from the low-lying epithelial layer that runs along the limbic surface here, but the posterior edge is less distinct, merging with the epithelium that only gradually becomes lower as it runs posteriorly along this portion of the limbic plate. Still farther ventrally, as Fig. 23-20 shows, the papilla comes to rest entirely on the limbus. The demarcation between papilla and epithelium at the ventral end of the cochlea is shown as the ending of the ciliated cells, together with the change from supporting cells with columnal processes running between the hair cells to simple cells with relatively highly placed nuclei and with cell bodies that extend all the way through the epithelial layer, often somewhat obliquely.

As shown in Fig. 23-23, the papilla at the ventral end attains a maximum width of 223 μ, falls at the next measured point, and then abruptly ends.

The varying area of the papilla is indicated in *b* of this same figure. The solid line represents the papilla proper, and the broken lines show the two areas of hair cells extending onto the limbus.

In Fig. 23-24, part *c* represents the varying number of longitudinal rows of hair cells appearing in the auditory papilla from dorsal to ventral ends. This number varies between 5 and 7 in the dorsal region, and then increases gradually through the middle of the cochlea, until in the ventral region it rises rapidly to a peak and then falls away. The most rapid increase is near the end of the basilar membrane, and includes the many hair cells spreading over the limbus.

Part *d* of this figure shows the varying density of hair cells along the cochlea, measured as the number of cells at 20 μ intervals. The form of the function reflects closely the varying number of rows of hair cells.

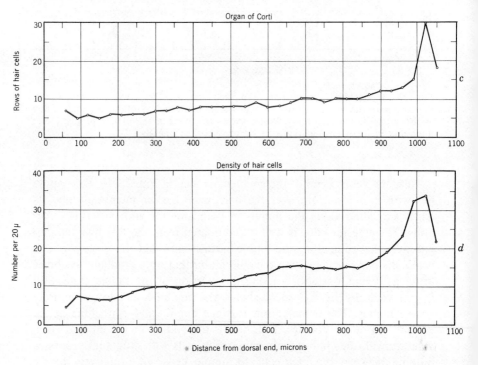

FIG. 23-24. Cochlear dimensions in *Chrysemys scripta elegans*; hair-cell distributions.

The vibratory differentiation of the cochlea mainly involves the varying width of the basilar membrane and the increasing size of the papilla from dorsal to ventral ends. The papilla does not follow the form of the basilar membrane, but increases progressively from dorsal to ventral ends, though only to a moderate extent. Its area changes in like manner, and increases somewhat more markedly at the two ends of the cochlea. In general it appears that the degree of differentiation of this cochlea is only moderate, and a spatial distribution of response for different frequencies of tones can hardly be expected to play a significant role.

Deirochelys reticularia. — The width of the basilar membrane in the chicken turtle is represented by the solid line in part *a* of Fig. 23-25, and shows this width as increasing rapidly in the dorsal region to a maximum of 350 μ at a point 240 μ from the dorsal end. Thereafter the width decreases, and this membrane ends at 780 μ. The auditory papilla has a somewhat different form; it increases rapidly to a width around 140 μ and remains nearly constant, with only a slight decrease until the middle of the cochlea is well past, then falls rapidly to a minimum, after which it rises steeply. This structure has its greatest width at the ventral end, when the basilar membrane has ended and the hair cells lie entirely on the limbus.

The area of the papilla is represented in part *b* of this figure. This area has a peculiar form; there is a rapid though irregular increase to a nearly level value in the dorsal region, a rise to a maximum near the middle of the cochlea, followed by a rapid fall and then, as the end of the basilar membrane is reached, a very steep rise. As the broken curves show, the most rapid changes appear in the limbic hair cells at both ends of the cochlea.

The varying rows of hair cells in this specimen are represented in part *c* of Fig. 23-26. There is a rapid increase in the number of longitudinal rows at the dorsal end of the cochlea, a moderate variation around 11 to 14 through the middle region, then a slight drop followed by an abrupt rise at the ventral end as the limbic cells appear. The function for hair-cell density, shown in *d* of this figure, follows the same pattern.

This ear, like that of the preceding species, shows only moderate differentiation in terms of dimensional characters.

Testudo horsfieldi. — Dimensional data on this species, representative of the family Testudinidae, are presented in Fig. 23-27.

The width of the basilar membrane, shown by the solid line in part *a*, follows a nearly semicircular function, rising progressively to a maximum in the middle of the cochlea and similarly falling away in the ventral region. The width of the papilla, shown by the broken line, follows the same course but with less curvature.

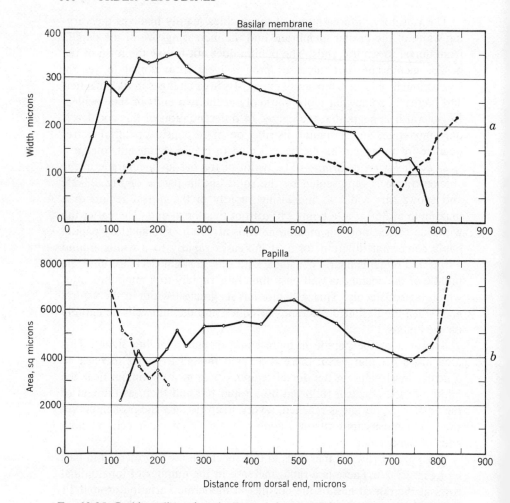

FIG. 23-25. Cochlear dimensions in *Deirochelys reticularia*. In part *a* the solid line
shows the width of the basilar membrane, and the broken line the width of the
papilla. In part *b* the solid line represents the area of the papilla, and the broken
lines indicate extensions of the hair cells onto the limbus.

The size of the papilla, shown in *b* of this figure, after an initial drop
at the dorsal end largely representing the limbic cells in this region, takes
a course much like that shown by the width functions, passing through
a flattened maximum in the middle region.

The number of rows of hair cells, shown in part *c*, rises steadily from
the dorsal end, fluctuates around 8-10 through the middle region, and
falls rapidly at the ventral end. The density of the hair cells shows a
similar function.

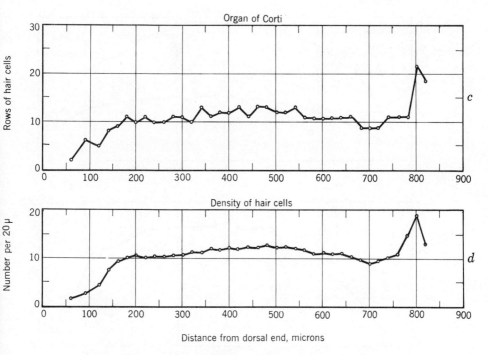

Fɪɢ. 23-26. Cochlear dimensions in *Deirochelys reticularia*;
hair-cell distributions.

This ear is notable in the presence of a comparatively small number
of limbic hair cells at the dorsal end of the cochlea, and a complete
absence of these cells at the ventral end.

This ear shows somewhat more differentiation of papillar structure
than in the two foregoing species. The variation in the width of the
papilla from the middle to the two ends is around fourfold, and the
variation of papillar area also is of about the same order.

Chelydra serpentina. — The results of dimensional measurements on
the common snapping turtle, *Chelydra serpentina*, are presented in Figs.
23-28 to 23-30.

The solid line of Fig. 23-28 shows the varying width of the basilar
membrane. This function rises rapidly in the dorsal region, after some
fluctuations attains a maximum near the middle of the cochlea, and then
declines. The variation over this course is considerable. The width of
the papilla, however, shows a progressive increase in the dorsal region,
a flattening around the middle, and then a further moderate rise toward
the ventral end.

The size of the papilla (Fig. 23-29) exhibits more striking changes.
After a sharp drop at the dorsal end, the function rises, with some ir-

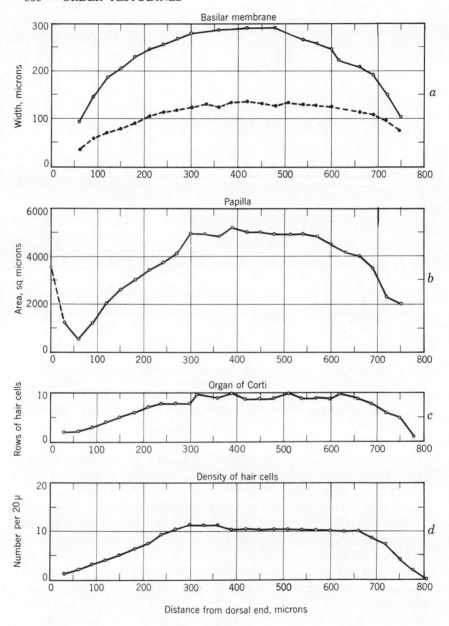

FIG. 23-27. Cochlear dimensions in *Testudo horsfieldi*.

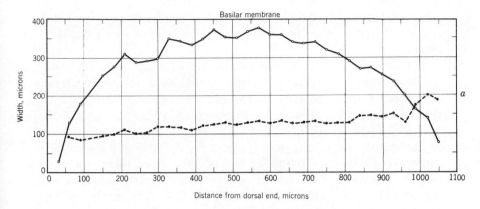

Fig. 23-28. Cochlear dimensions in *Chelydra serpentina*;
width of basilar membrane and papilla.

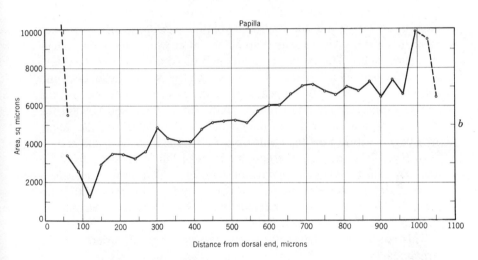

Fig. 23-29. Cochlear dimensions in *Chelydra serpentina*; the solid line shows
the area of the papilla proper, and the broken lines show the masses of
limbic hair cells at the two ends of the cochlea.

regularities, to attain a maximum in the ventral region, then maintains
this general level despite a few fluctuations, and finally rises abruptly near
the end. The limbic hair cells form a large area at the dorsal end, and
this area falls away precipitously as the papilla proper begins. At the
ventral end these limbic cells form a great mass also, but quickly diminish and disappear.

The rows of hair cells, shown in *c* of Fig. 23-30, present a rather
regular increase until the middle of the cochlea is reached, then main-

tain a level around 8-10 in the ventral region, and finally show a marked rise at the ventral end. The hair-cell density presents much the same picture.

This ear exhibits only a moderate degree of dimensional variation.

Macroclemmys temminckii. — Dimensional measurements made on a moderately large specimen of the alligator snapping turtle, *Macroclemmys temminckii,* gave the results presented in Figs. 23-31 to 23-33.

In Fig. 23-31 the solid curve indicates the varying width of the basilar membrane, which increases rapidly at the dorsal end of the cochlea and reaches a large maximum of 640 μ, after which there is a rapid decline. The width of the papilla, shown by the broken line, increases slightly but with considerable irregularity at first, rises to a maximum somewhat beyond the middle of the cochlea, then declines and rises to a second maximum in the ventral region, after which it falls away.

The area of the auditory papilla, shown in Fig. 23-32, increases rapidly, except for one interruption, from the dorsal end to a high maximum of 14,250 sq μ in the ventral region, falls abruptly, then rises once more to a slightly higher level than before, after which the limbic extension supervenes. The largest area attained of 15,000 sq μ represents a very great size indeed. The limbic areas are large at the dorsal end, and continue well beyond the appearance of the basilar membrane and the papilla proper, though this area is rapidly diminishing and beyond the 220 μ point falls below that of the papilla. At the ventral end the limbic area begins with a very high value but falls precipitously to the end.

In Fig. 23-33, part *c* represents the rows of hair cells and part *d* the hair-cell density. These two functions are of similar form, and reflect the bimodal character of the papillar width and areal curves of the two preceding figures. Both the number of rows and the hair-cell density reach high values in the ventral region.

This is a large cochlea with a considerable range of variation in its structures.

Kinosternon scorpioides. — The solid curve in part *a* of Fig. 23-34 represents the varying width of the basilar membrane, and takes a nearly semicircular course, with the maximum near the middle of the cochlea. The broken curve in this figure shows the width of the papilla, which increases only moderately in the dorsal region, reaches a maximum beyond the middle of the cochlea, varies somewhat, then subsides a little in the ventral region, and finally rises rapidly at the ventral end.

The solid curve of part *b* of this figure exhibits many variations, but its general course is upward from the basal end to a maximum somewhat beyond the middle of the cochlea, then it falls toward the ventral end. The broken curve at the dorsal end represents a large mass of hair cells on the limbus, and these cells continue well beyond the point of

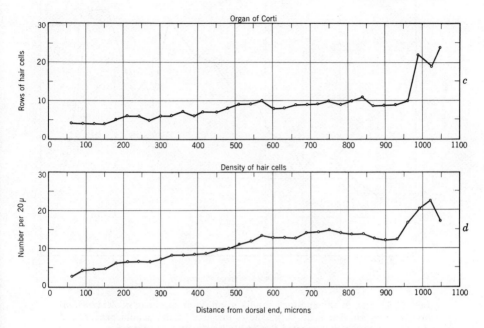

FIG. 23-30. Cochlear dimensions in *Chelydra serpentina*; hair-cell distributions.

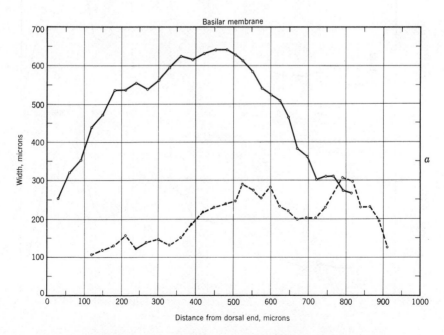

FIG. 23-31. Cochlear dimensions in an adult specimen of *Macroclemmys temminckii*; width of basilar membrane and papilla.

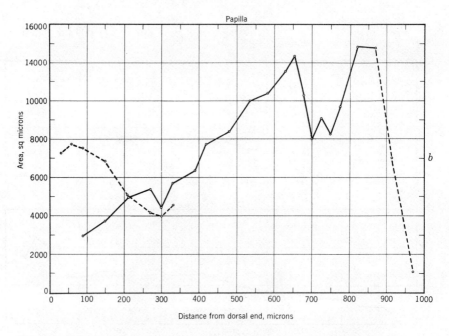

FIG. 23-32. Size of the auditory papilla in an adult specimen of *Macroclemmys temminckii*. The solid curve shows the area of the papilla proper, and the broken curves the areas of the limbic hair cells.

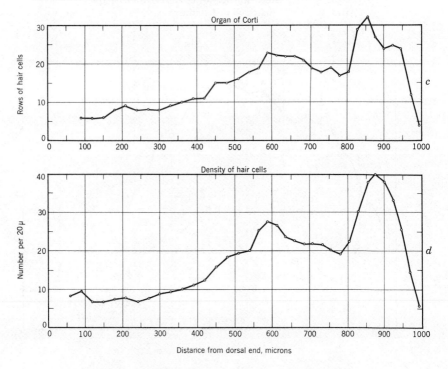

FIG. 23-33. Cochlear dimensions in an adult specimen of *Macroclemmys temminckii*; hair-cell distributions.

appearance of the papilla proper. At the ventral end the limbic cells extend only a short distance beyond the basilar membrane.

The rows of hair cells, represented in part c of Fig. 23-35, increase gradually in numbers from the dorsal end to about the middle of the cochlea, fluctuate about a level around 9, then fall and show a sudden rise at the ventral end. The density of these cells, indicated in part d, follows a similar course.

This cochlea exhibits a low degree of differentiation.

Trionyx cartilagineus. — The width of the basilar membrane, shown by the solid line of part a in Fig. 23-36, rises rapidly at first and then more slowly, and attains a maximum near the middle of the cochlea, followed by a gradual decrease to the ventral end. The width of the papilla, represented by the broken line, increases rapidly near the dorsal end, then varies but slightly to reach a flat maximum in the midregion, and remains nearly flat, with a barely noticeable decrease, until in the ventral region it undergoes a moderate rise.

The papillar area, represented by the solid curve in b of this figure, presents a sharp dip at the dorsal end, then a varying rise and two maximums on either side of the middle of the cochlea. In the ventral region a general level is maintained despite variations, and near the end there is a noticeable increase followed by a fall. A moderately large area of limbic hair cells is found at the dorsal end, which quickly subsides, and a smaller area appears at the ventral end.

In Fig. 23-37 the curve in part c represents the rows of hair cells, and the curve in part d shows the hair-cell density. These two functions are similar, and both exhibit the depression seen at the dorsal end in the papillar area function, a rather gradual rise along the main course of the cochlea, and a rapid rise at the ventral end. This cochlea is somewhat longer than that of most turtles, with the papilla length exceeding 1100 μ.

Chelonia mydas. — Dimensional data for the green turtle, the only marine form examined, are presented in Fig. 23-38. Because this cochlea attains a length of 1750 μ, almost twice that of other turtles studied, it was necessary to condense the scales of the graphs.

The solid line in part a represents the width of the basilar membrane, which increases from the dorsal end to a value around 500 μ in the middle of the cochlea. Considerable variations appear here, but the general trend is a decline toward the ventral end. The auditory papilla exhibits only slight variations apart from irregularities, with a barely significant increase in width toward the ventral end, and then a sharp rise at this end.

The solid line of part b shows the area of the papilla, which despite irregularities displays an almost flat region near the dorsal end, then a

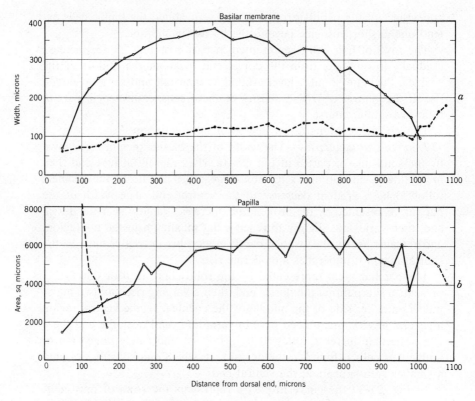

FIG. 23-34. Dimensional data for *Kinosternon scorpioides*. Part *a*, width of basilar membrane (solid line) and papilla (broken line); part *b*, area of the papilla (solid line) and area of limbic hair cells (broken lines).

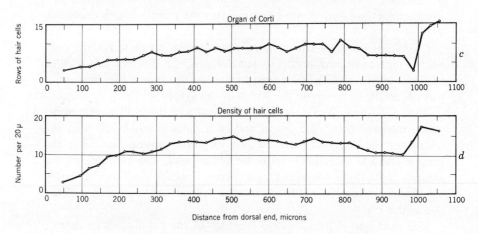

FIG. 23-35. Dimensional data for *Kinosternon scorpioides*; hair-cell distributions.

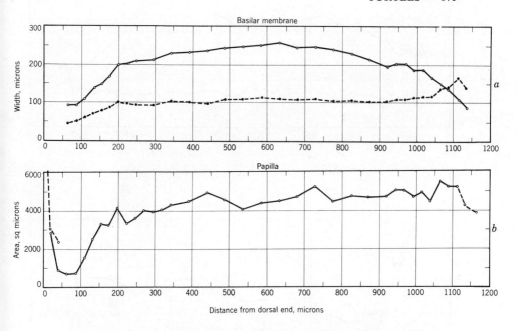

FIG. 23-36. Cochlear dimensions for *Trionyx cartilagineus*. Part *a*, width of the basilar membrane (solid line) and of the papilla (broken line); part *b*, area of the papilla (solid line) and of the limbic hair cells (broken line).

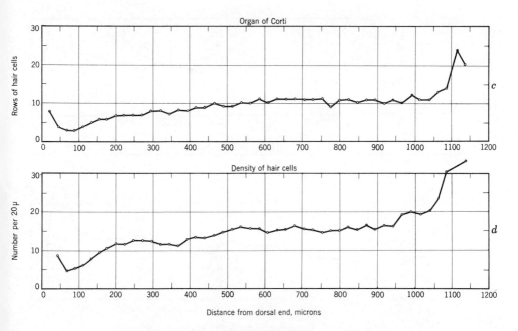

FIG. 23-37. Cochlear dimensions in *Trionyx cartilagineus*; hair-cell distributions.

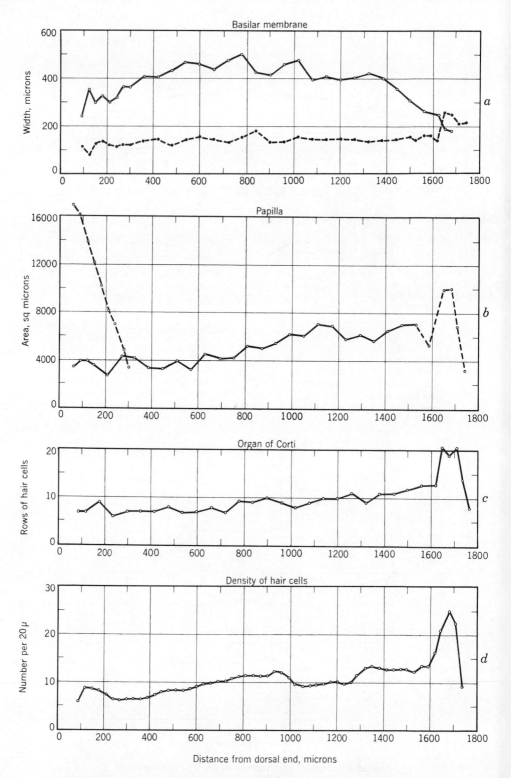

FIG. 23-38. Cochlear dimensions in *Chelonia mydas*. Part *a*, width of basilar membrane (solid line) and papilla (broken line); part *b*, area of the papilla (solid line) and of the limbic hair cells (broken line); part *c*, rows of hair cells; and part *d*, hair-cell density. The abscissa scales are condensed, and also the ordinate scales for parts *a* and *b*.

rise through the middle region, and an uncertain leveling off ventrally. The limbic hair cells at the dorsal end form an enormous area, which continues well beyond the appearance of the basilar membrane and papilla, though rapidly subsiding. At the ventral end of the cochlea there is a considerable mass of limbic hair cells also.

Part c of this figure shows the varying rows of hair cells. Around 7 rows appear in the dorsal region, there are 9 and 10 in the middle, and their number increases moderately and then sharply to a maximum of 22 in the ventral region, after which there is a rapid fall. This maximum of 22 rows is larger than that found in other species, but seems moderate in relation to the large cochlear dimensions. The hair-cell density, shown in part d, follows a course somewhat similar to that of the row pattern. This function over its main extent shows a density comparable with that of other turtle species, and even at the maximum of 25 cells per 20 μ is within the average range.

Platemys platycephala. — We turn now to the Pleurodira, the side-necked turtles, and consider first the only representative available of the family Chelidae. In *Platemys platycephala*, as shown by the solid line in a of Fig. 23-39, the width of the basilar membrane increases rapidly at the dorsal end, then more slowly to a rather flat maximum over the main part of the cochlea, and then in the ventral region declines rapidly. The width of the papilla (broken line) varies only moderately; in the dorsal region it rises steadily, reaches a maximum around 130 μ, then declines slowly to the ventral end.

The area of the papilla, as shown by the solid line of part b, is large even at the beginning at the dorsal end, increases considerably to a maximum near the middle of the cochlea and then falls away somewhat irregularly. At the dorsal end the limbic hair cells form a large area that rapidly diminishes and ends as the papilla begins, and at the ventral end these limbic cells continue for only a short distance beyond the basilar membrane.

The rows of hair cells, as part c of this figure shows, are particularly numerous over the limbus at the dorsal end, but rapidly diminish as the papilla begins. There are fewer of these limbic cells at the ventral end of the cochlea.

Along the papilla the number of rows of hair cells increases gradually through the dorsal region, reaches a maximum of 10 in the middle of the cochlea, then declines somewhat until the ventral end is reached. At this end there is an abrupt increase as the limbic hair cells appear.

The density of the hair cells, shown in d, increases rapidly at the dorsal end, then a slight decline is followed by a steady increase over most of the cochlea. A flat maximum is reached in the ventral region, and finally there is a moderate increase at the ventral end due to the appearance of the limbic hair cells.

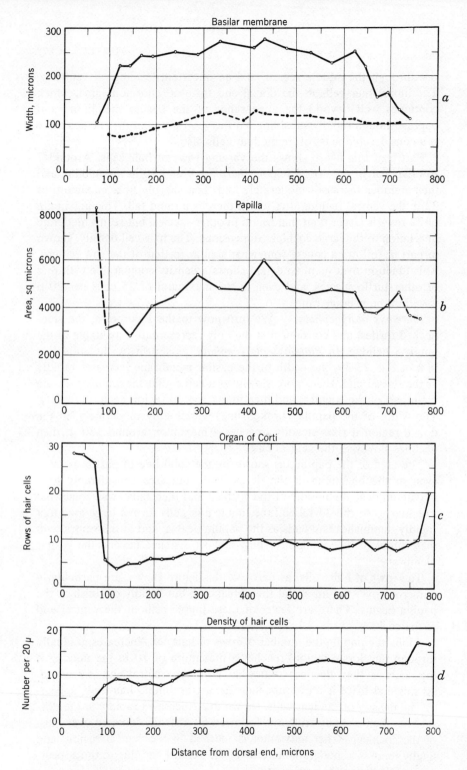

FIG. 23-39. Cochlear dimensions in *Platemys platycephala*. Part *a*, width of basilar membrane (solid line) and papilla (broken line); part *b*, area of the papilla (solid line), and of the limbic hair cells (broken lines); part *c*, rows of hair cells; and part *d*, hair-cell density.

This ear shows only moderate differentiation. A striking feature is the large mass of limbic cells at the dorsal end.

Pelomedusa subrufa. — In this representative of the family Pelomedusidae, the width of the basilar membrane is shown by the solid curve of part *a*, Fig. 23-40. This width increases rapidly at the dorsal end and attains a rather large maximum near the middle of this short cochlea, after which it swiftly falls away. The width of the papilla increases progressively to a maximum in the middle region, then falls, until near the end of the basilar membrane it shows a marked increase.

The area of the papilla, shown by the solid curve of part *b*, increases rapidly to a prominent maximum near the middle of the cochlea, then declines rapidly. The limbic area at the dorsal end is very large, but that at the ventral end only continues the regular level for a moderate distance along the limbus.

The rows of hair cells (part *c*) show a rapid increase at the dorsal end, attain a flat maximum in the middle region, then fall to a trough in the ventral area, from which there is a rise to about the former maximum. The hair cell density, as seen in part *d* of this figure, passes through changes much like the row pattern except at the extreme ventral end.

This is a particularly short cochlea, with a considerable degree of differentiation in comparison with other turtle species.

Podocnemis expansa. — The varying width of the basilar membrane in *Podocnemis expansa* is represented by the solid line in part *a* of Fig. 23-41. This width increases rapidly at the dorsal end and then more slowly with many irregularities, and reaches a large maximum somewhat beyond the middle of the cochlea. In the ventral region this width drops sharply and then recovers in part at the ventral end. The width of the papilla increases progressively over the greater part of the cochlea until a maximum is reached in the ventral region, after which there are large variations leading to a sharp rise well beyond the end of the basilar membrane.

The area of the auditory papilla, shown by the solid curve in part *b* of this figure, increases rapidly from the dorsal end to a sharp peak somewhat beyond the middle of the cochlea, then suffers a rapid drop as the papilla proper ends. At the dorsal end the mass of hair cells on the limbus is very large, and these limbic cells continue well beyond the appearance of the papilla. At the ventral end the limbic hair cells begin as the papilla proper terminates, rise to a sharp peak, and then decline.

The rows of hair cells, represented in *c* of Fig. 23-42, show a fairly steady increase over the greater part of the cochlea and reach a maximum in the ventral region, followed by a moderate decline. The density of the hair cells, shown in *d* of this figure, follows a similar course, but in exaggerated degree.

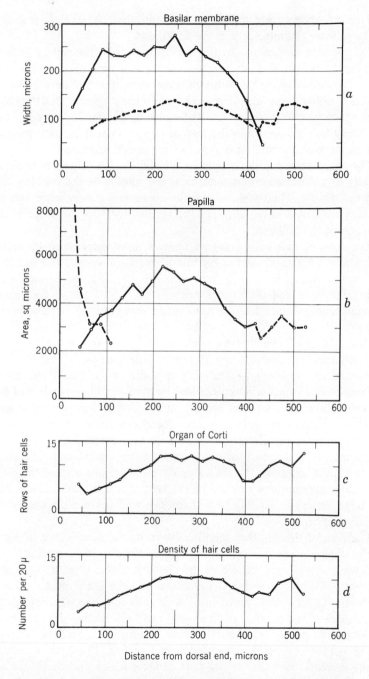

FIG. 23-40. Cochlear dimensions in *Pelomedusa subrufa*. Part *a*, width of the basilar membrane (solid line) and of the papilla (broken line); part *b*, area of the papilla (solid line) and of the limbic hair cells (broken lines); part *c*, rows of hair cells; and part *d*, hair-cell density.

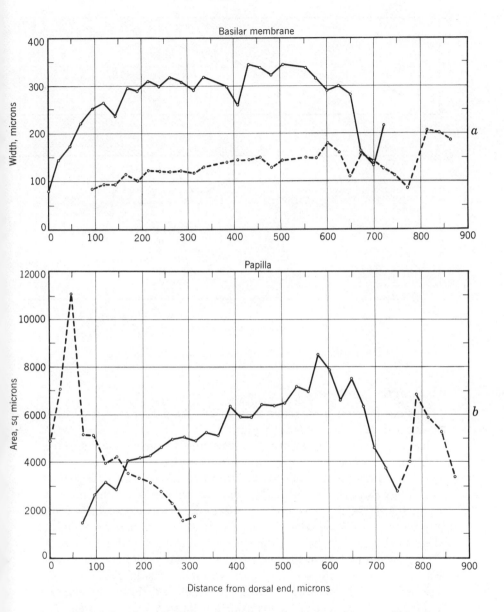

FIG. 23-41. Cochlear dimensions in *Podocnemis expansa*. Part *a*, width of the basilar membrane (solid line) and of the papilla (broken line); part *b*, area of the papilla (solid line) and of the limbic hair cells (broken line).

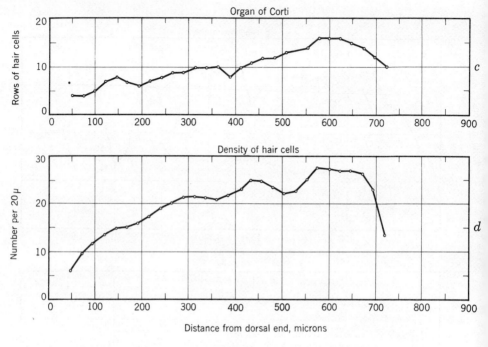

FIG. 23-42. Cochlear dimensions in *Podocnemis expansa*;
hair-cell distributions.

This ear presents a moderate amount of differentiation over a cochlea of a little less than average length.

The eleven turtle species just described provide a fair sampling of the cochlear dimensional characteristics of this reptile group. It is notable that these characteristics fail to exhibit any high degree of consistency. In *Trionyx cartilagineus*, where the basilar membrane has the smallest maximum width, this membrane is very long, exceeded in this group only by that of *Chelonia mydas*. Again, in *Macroclemmys temminckii*, where the basilar membrane has the greatest maximum width, the length of this membrane falls at the median of the group. There are some correlations, however: for the most part the dimensions in *Trionyx cartilagineus* and in *Pelomedusa subrufa* are small, and those in *Chelonia mydas* and *Macroclemmys temminckii* are large. *Chelonia mydas* has the longest basilar membrane (1690 μ), and *Macroclemmys temminckii* the widest (635 μ).

In general the variation in these dimensional characters is only moderate for so wide a range of species; for most of the features shown, the ratio between smallest and largest values approaches fourfold only for the length of the basilar membrane, and for all other features is between two and threefold. This evidence therefore supports the common view that the turtles form a homogeneous group.

Effects of Age and Size.   In one species, *Macroclemmys temminckii*, in which large and small specimens were available, it was possible to obtain some preliminary information on the effects of age and size on the dimensions of the cochlear structures. Small specimens (carapace width of 3.3 cm, body weight of 14.5 grams) were purchased from dealers as "hatchlings," and for convenience will be referred to as such, though their age is unknown and probably amounted to a few weeks. The large specimen whose cochlear dimensions have already been given (see Figs. 23-31 to 23-33) had a carapace width of 20 cm and weighed 2200 grams.

The width of the basilar membrane in the small specimen is shown by the solid curve in part *a* of Fig. 23-43. This curve follows a semicircular course somewhat like that of the adult (Fig. 23-31), but the size is smaller and the form more nearly symmetrical. The width increases, with minor irregularities, to a maximum of 370 μ near the middle of the cochlea, and then falls off a little more slowly toward the ventral end. The maximum in this specimen is about 58% of the adult value.

The length of the basilar membrane in the hatchling is also less: 580 μ as compared with 820 μ in the adult, or about 71% of the adult length.

The width of the papilla, shown by the broken curve in part *a* of this figure, presents a sharp peak at the dorsal end, and then takes a rising course containing two further peaks, along with some minor variations. The two peaks can perhaps be related to the prominent elevations in the adult curve (Fig. 23-31), though the variations are much smaller.

The area of the papilla, shown in *b* of this figure, presents a strikingly different picture from the adult, especially in the limbic extensions of the hair cells. At the dorsal end the limbic area increases abruptly to a very large value of 15,000 sq μ, then falls almost as rapidly. As in the adult, this area continues well beyond the point of appearance of the basilar membrane and of the papilla proper. The area of the papilla then increases, though not as rapidly as in the adult, and reaches a maximum a little beyond the middle of the cochlea. This area subsides somewhat irregularly, and finally gives way to the limbic area at the ventral end.

In Fig. 23-44, part *c* shows the rows of hair cells and part *d* the hair-cell density. These two characters follow similar functions, but are very different from the ones shown for the adult (Fig. 23-33). These functions rise rapidly at the dorsal end, reach a rather flat maximum in the middle and ventral regions, and then fall sharply. There is no sign of

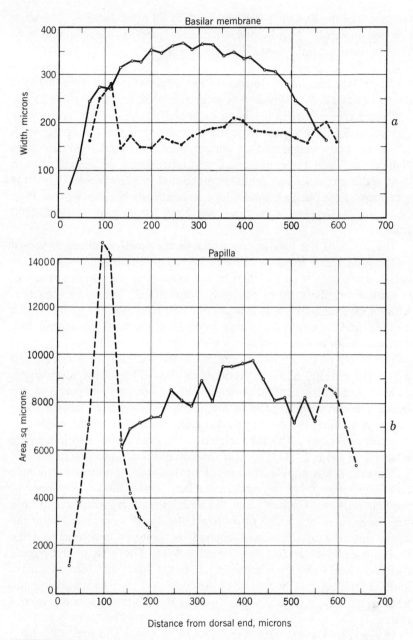

FIG. 23-43. Dimensional data for a hatchling of the species *Macroclemmys temminckii*. Part *a*, width of the basilar membrane (solid line) and papilla (broken line); part *b*, area of the papilla (solid line) and of the limbic hair cells (broken lines).

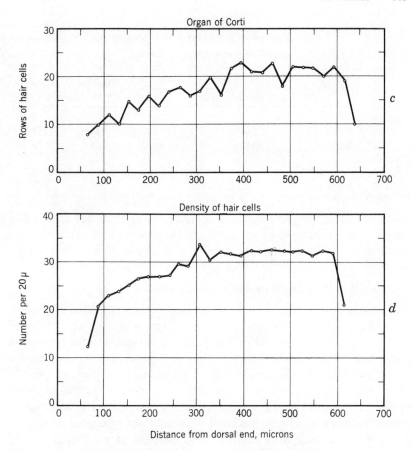

FIG. 23-44. Dimensional data for a hatchling of the species
Macroclemmys temminckii; hair-cell distributions.

the gradual rise at the dorsal end or of the bimodal form in the ventral
region seen in the adult.

Of particular interest are the relations between the number and
distribution of the hair cells in hatchling and adult. A count was made
of these cells with the results shown in Table 23-I.

It is evident that the numbers of hair cells contained in the papilla
proper do not vary significantly between hatchling and adult. There
seems to be a difference in the limbic hair cells, however. In the adult
these are more numerous on the limbus at the dorsal end of the cochlea
—more than twice as many—but fewer at the ventral end. With only
one specimen of each age it is of course unsafe to generalize, but these
variations seem large enough to be significant. It is suggested that the

TABLE 23-I

RELATION OF AGE TO HAIR-CELL NUMBERS
IN *Macroclemmys temminckii*

		Limbic Hair Cells			Papillar Hair Cells	Total Hair Cells
		Dorsal	Ventral	Total		
Adult:						
	Left ear	364	45	409	748	1157
	Right ear	350	62	412	766	1178
Hatchling:						
	Left ear	169	99	268	734	1002
	Right ear	142	85	227	762	989

development of the papilla proper occurs early and reaches a relatively steady state by the time of hatching, but the extension of hair cells over the limbus is a more variable process and may continue for a longer time.

HAIR-CELL POPULATIONS

The numbers of hair cells were determined both over the basilar membrane and in the two limbic regions for all specimens for which a satisfactory histological preparation had been carried out, and the results are presented in Table 23-II. This table shows the numbers of hair cells for dorsal and ventral limbic areas and for these two combined, and finally this combined sum is added to the number of cells on the basilar membrane for a grand total. A column is shown also for the percentage of basilar membrane cells in this total. The data are presented as species averages, even though the number of specimens of each kind is very limited.

It is apparent that the numbers of hair cells in the turtle cochlea reach high levels, comparing favorably with these numbers in other reptiles and even in the advanced lizards, such as the geckos. If we include the limbic hair cells, the numbers exceed a thousand in many instances, and in *Chelonia mydas* are not far below the largest lizard population so far found of about 2500 in *Gekko smithi*. Even if the limbic cells are omitted and only those on the basilar membrane are considered, the numbers remain in the hundreds, from the lowest count of about 250 in a specimen of *Pelomedusa subrufa* to numbers exceeding 800 in *Trionyx cartilagineus* and *Chelonia mydas*, and averaging 538 in the species examined.

This large number of hair cells, occurring in a cochlea that appears to present only a moderate degree of structural differentiation, ought to provide a considerable amount of spatial summation in the action of the cochlea and auditory nerve in response to sounds. If the sound re-

TABLE 23-II

HAIR-CELL POPULATIONS IN TURTLES

Species	Number of Ears	On Dorsal Limbus	On Ventral Limbus	Total Limbus	On Basilar Membrane	Total	Percentage on Basilar Membrane
Chrysemys picta picta	2	237.0	113.0	350.0	401.0	751.0	53
Chrysemys rubiventris	2	223.5	62.5	286.0	487.0	773.0	63
Chrysemys scripta elegans	2	263.0	139.0	402.0	635.0	1037.0	61
Clemmys guttata	2	197.5	9.0	206.5	493.5	700.0	70
Deirochelys reticularia	4	287.0	85.7	372.7	599.7	972.5	62
Geoemyda pulcherrima	2	143.5	13.0	156.5	547.5	704.0	78
Terrapene c. carolina	6	77.3	5.3	82.7	461.7	544.3	85
Terrapene ornata	2	104.0	59.0	163.0	305.0	468.0	65
Geochelone carbonaria	2	311.0	23.0	334.0	701.0	1035.0	68
Testudo horsfieldi	4	44.2	0.2	44.5	494.5	539.0	92
Kinixys belliana	2	6.0	0	6.0	771.0	777.0	99
Chelydra serpentina	2	73.0	25.5	98.5	548.5	647.0	85
Macroclemmys temminckii							
Adult	2	357.0	53.5	410.5	757.0	1167.5	65
Hatchling	2	155.5	92.0	247.5	748.0	995.5	75
Kinosternon scorpioides	4	292.5	90.5	383.0	540.2	923.2	59
Staurotypus triporcatus	2	472.0	118.5	590.5	569.0	1159.5	49
Trionyx cartilagineus	2	162.0	50.5	212.5	819.5	1032.0	79
Chelonia mydas	1	633.0	89.0	722.0	852.0	1574.0	54
Platemys platycephala	4	157.5	53.0	210.5	417.5	628.0	66
Pelomedusa subrufa	4	107.0	81.5	188.5	283.5	472.0	60
Pelusios sinuatus	4	234.0	71.7	305.7	296.7	602.5	49
Podocnemis expansa	2	400.0	254.0	654.0	732.5	1386.5	54
Podocnemis unifilis	4	153.0	176.7	329.7	543.4	873.2	62

ceptive and transmitting mechanism is even reasonably effective we should expect a good level of sensitivity in these animals.

AUDITORY SENSITIVITY

As in the other reptiles, the sensitivity of the turtle ear was measured in terms of the electrical potentials of the cochlea. The recording was carried out by a procedure developed earlier (Wever and Vernon, 1956) in which the active electrode is a steel needle insulated except at the tip, inserted into the superior part of the otic cavity through a minute hole drilled at the suture between opisthotic and supraoccipital bones (see Fig. 23-45) so as to make contact with the perilymphatic fluid.

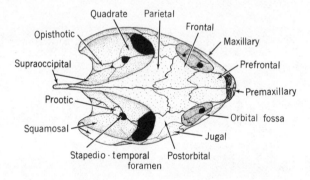

Fig. 23-45. Dorsal view of the skull of *Clemmys insculpta*.

In the earlier experiments a number of electrode positions were tried, including insertion into each of the semicircular canal spaces and the utricle, and in a few instances a placement on the pericapsular membrane. The canal and utricle positions gave closely similar results. The pericapsular location showed improved potentials by an amount of about 5 db, but this small gain hardly justifies the great difficulty of the surgical approach and the modification of the ear's dynamics by a reduction of the pericapsular sinus. Accordingly, the simple needle insertion described was employed in the present experiments.

As in the other reptiles, the body temperature was maintained at 24-25° C, except in a few special instances.

COCHLEAR POTENTIALS

The cochlear potentials in the turtle have the same general form as in other reptiles. As shown in Fig. 23-46 for *Chrysemys p. picta*, these potentials increase nearly linearly as a function of sound pressure at low levels, then at higher levels they begin to bend and approach a maxi-

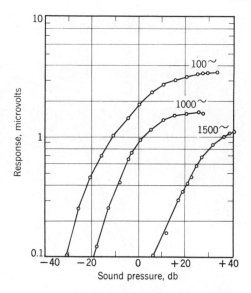

FIG. 23-46. Intensity functions for the cochlear potentials in a specimen of *Chrysemys p. picta* for three tones. From Wever and Vernon, 1956a.

mum. A curve for a specimen of *Terrapene c. carolina*, shown in Fig. 23-47, shows a linear course over a range of 40 db, and then a rapid bending at the upper end.

At high levels of stimulation these functions bend increasingly and pass through a maximum, and the maximum response obtained is a function of frequency. As Fig. 23-48 indicates, the maximum values decline rapidly as the stimulus frequency is increased.

Cochlear Distortion.   The departure from linearity seen in the intensity functions of Figs. 23-46 and 23-47 represents one aspect of cochlear distortion, and arises in part through a conversion of some of the response energy into frequencies that are harmonics of the fundamental tone. This harmonic distortion can be measured with a wave analyzer by tuning to the various components in succession, first recording the output at the fundamental frequency and then at its multiples.

This analysis was carried out in a specimen of *Chrysemys p. picta* during stimulation with tones of 100, 200, 290, 400, 600, 800, 1000, 2000, and 3000 Hz. For all these tones the second harmonic f_2 (with a frequency of 2f i.e., twice the frequency of the fundamental) and the third harmonic f_3 (frequency = 3f) were nearly always present in measurable amounts when the stimulus was raised to a level sufficient to pro-

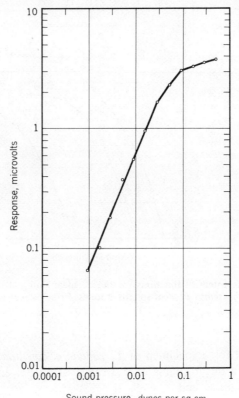

Sound pressure, dynes per sq cm

FIG. 23-47. Intensity function for a specimen of *Terrapene c. carolina* stimulated with a tone of 200 Hz at sound pressures over a range of 55 db.

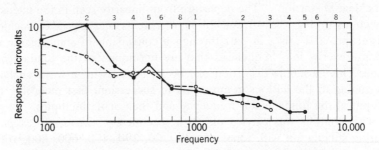

FIG. 23-48. Maximum responses observed in two turtle ears of the species *Chrysemys picta picta*. From Wever and Vernon, 1956a.

duce a fundamental output of 1 μv or higher. When still greater stimulation levels were used, the next two harmonics f_4 and f_5 were commonly found also for tones within the frequency range in which the ear was relatively sensitive. However, it was found unsafe to drive the turtle's ear very far into the region of overloading, especially with the higher frequencies to which the ear is insensitive. Such driving caused severe injury to the cochlea as shown by a reduction in its potentials.

In the animal whose results have just been referred to, the stimulation was applied with caution to avoid such cochlear injury, and at the end of the experiment some of the earlier tests were repeated to make certain that the responses remained stable and unimpaired.

The responses observed at this level of stimulation are represented in Fig. 23-49. The function is nearly linear in its middle region, but shows the initial stage of bending above 1.6 μv. The lower end of the curve also departs slightly from linearity because of the presence of noise. The harmonics produced at the highest level of stimulation are shown at four points for f_2-f_5 at the right of the figure. The amounts of potential at these harmonic frequencies are well below the fundamental response, as indicated. The total of these harmonics was calculated as the root-mean-square (square root of the mean of the squares of the four potentials) and found to be 12% of the fundamental, or 18.4 db below the fundamental.

At the level of routine measurements around 0.1 μv used in all the sensitivity determinations, the distortion is much less, and indeed was not detectable by the recording instruments used.

Temperature Effects. — The performance of the turtle ear as a function of temperature has been little studied. Two experiments may be mentioned, one a short series of measurements on specimens of *Clemmys insculpta* and *Chrysemys scripta* made by Wever and Vernon in 1956 (unpublished), and another by Patterson, Evering, and McNall (1968) on *Chrysemys scripta elegans*.

The earlier observations were made by immersing the turtle's body in water, with the top of the head exposed, and recording the head temperature with a thermocouple probe inserted deep in the superior temporal fossa below the external adductor muscle of the mandible and adjacent to the bony surface (see Fig. 23-45). The temperature was changed slowly by heating or cooling the water bath, and at frequent intervals the magnitude of the cochlear potentials was recorded for a tone of 400 Hz presented at a constant intensity.

For the experiment represented in Fig. 23-50 the series began at the adaptation temperature of 23.3° C (solid line with filled circles), and the animal was gradually cooled to about 8° C, then was warmed progressively (broken line with open circles) until 30° C was reached, and finally cooled to 19.3° C (dashed line with triangles), so that three sweeps were made through the middle temperature region.

The function is complex, and shows three general regions of temperature in which maximum responses occur, and for these regions the points of maximum vary for the different sweeps.

In the middle region from 18 to 25° C, the maximums appear at 19.0, 20.8, and 22.4° C, and all are displaced toward the temperature range to which the animal has most recently been adapted.

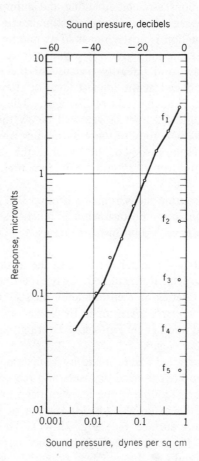

FIG. 23-49. An intensity function for a specimen of *Chrysemys picta picta* on stimulation with a 200 Hz tone, with measurements at the fundamental frequency (f_1) over a wide range, and at four harmonic frequencies (f_2 to f_5) at the strongest intensity used.

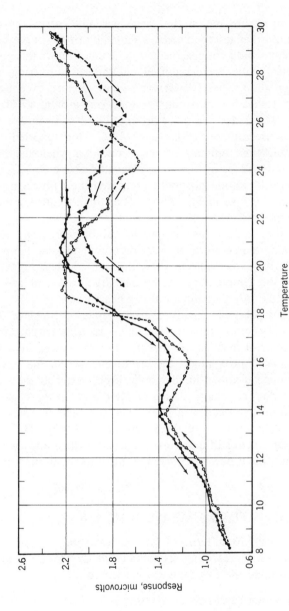

Fig. 23-50. Temperature effects on cochlear potentials in a specimen of *Clemmys insculpta*. The abscissa shows the temperature in the region of the skull, and the ordinate indicates potentials produced by a 200 Hz tone of constant intensity. The course of the temperature changes is indicated by the arrows.

In the low-temperature region, the maximums are seen at 13.8 and 14.1° C, and the descending and ascending sweeps are in close agreement.

In the high-temperature region the range was not carried beyond 30° C, but this and other tests indicated a flattening of the curve here, so that the maximum would not be expected to lie much above this point.

Generally similar results were reported by Patterson, Evering, and McNall (1968) on the species *Chrysemys scripta elegans*, except that these investigators found two maximum regions, one around 17-20° C (with a mean of 19° C) and the other around 29-32° (mean 30° C), with a minimum between these regions near 24° C. They reported difficulty in maintaining their animals in good condition when the body temperature was reduced below 14° C or elevated above 35° C, and the low-temperature limitation no doubt prevented their observing the low maximum region shown in Fig. 23-50. Their high-frequency maximum was well defined.

These investigators studied the relations to rate of heart beat, which also varies with body temperature. By varying body and head temperature independently they were able to show that the cochlear potentials are a function of the temperature of the ear structures, and are not directly dependent upon body temperature or heart rate.

The physiological processes involved in a change of cochlear temperature are obviously complex; Patterson and his associates argued from the bimodal character of their curves that two processes were involved, and suggested that one peak might be due to changes in the oxygen supply and resulting variations in metabolic processes, and the other peak might reflect physical changes such as alterations in the elasticity of membranes and in the viscosity of fluids in the cochlea, or variations in blood pressure. The presence of three maximums, as Fig. 23-50 has shown, makes the problem still more involved, and further study will be necessary to identify the processes concerned. Most likely all these processes are cochlear.

SENSITIVITY MEASUREMENTS

Measurements of sensitivity in terms of the cochlear potentials were carried out in 14 species representing six families of Cryptodira, and on five species representing the two families of Pleurodira.

Suborder Cryptodira; Family Emydidae

Chrysemys p. picta. — Cochlear potential curves for three specimens of the Eastern painted turtle, *Chrysemys p. picta*, are presented in Fig. 23-51. A fourth curve on another animal, shown in Fig. 23-52, repre-

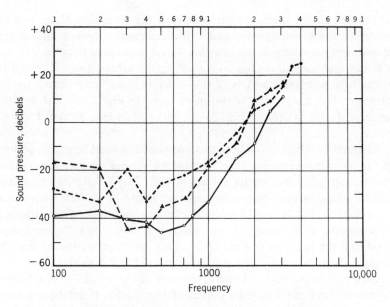

FIG. 23-51. Sensitivity curves for three turtles of the species *Chrysemys picta picta*. From Wever and Vernon, 1956a.

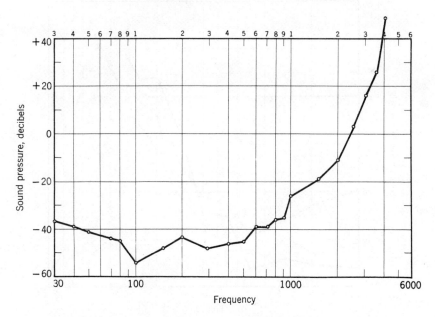

FIG. 23-52. Sensitivity function for a specimen of *Chrysemys picta picta* in which the measurements were extended far into the low frequencies.

sents more recent measurements in which the frequency range was extended at the low-frequency end. All these results show a high degree of auditory sensitivity in the low-tone range, especially between 100 and 900 Hz, and a rapid decline in sensitivity for higher tones. The roll-off at the high-frequency end is of the order of 25 db per octave for the curves of Fig. 23-51, and is still more rapid in Fig. 23-52. At the low-frequency end, as seen in Fig. 23-52, this rate is low, around 6 db per octave.

Chrysemys scripta. — Three specimens of the pond terrapin, perhaps including subspecies or intergrades of the *scripta* group, were studied earlier and gave the results shown in Fig. 23-53. Like the preceding species, these turtles show very good sensitivity in the low frequencies, up to 600 or 700 Hz, and rapidly declining response to higher tones.

A specimen belonging to the subspecies *Chrysemys scripta troostii* produced the function of Fig. 23-54. This curve resembles the ones already shown, with a definite maximum around 300-500 Hz.

Clemmys insculpta. — Three specimens of the wood turtle, *Clemmys insculpta*, gave the sensitivity curves of Fig. 23-55. Here the sensitivity

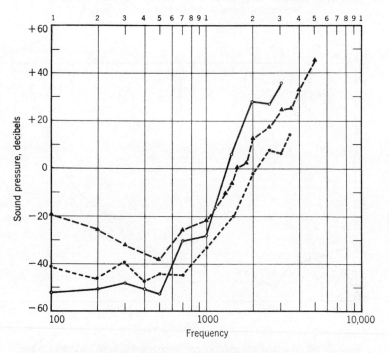

FIG. 23-53. Sensitivity functions for three specimens of *Chrysemys scripta* (not identified as to subspecies). From Wever and Vernon, 1956a.

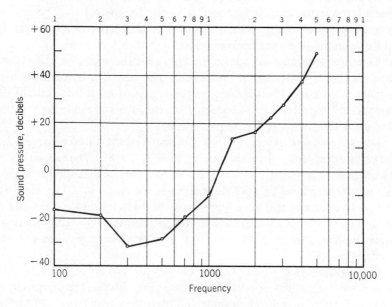

FIG. 23-54. A sensitivity curve for a specimen of
Chrysemys scripta troostii.

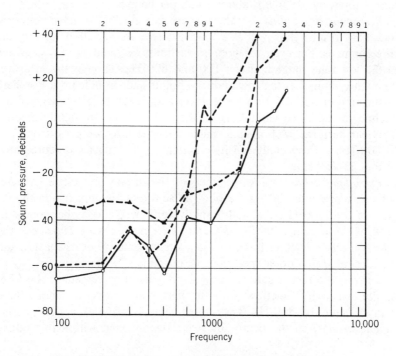

FIG. 23-55. Sensitivity curves for three turtles of the species
Clemmys insculpta. From Wever and Vernon, 1956a.

is excellent in two individuals and only fair in a third over the low-frequency range up to 700 Hz, and falls rapidly thereafter. The loss of sensitivity for the high tones, though somewhat irregular, seems to be of the order of 40 db per octave.

Clemmys guttata. — A sensitivity curve for the spotted turtle, *Clemmys guttata*, is presented in Fig. 23-56. The response is excellent over a range from 20 to 1500 Hz, and in the region of 80-150 Hz attains the remarkably high level of −66 db. This ear is exceptional both in the degree of sensitivity and in the wide range.

Deirochelys reticularia. — Two specimens of the chicken turtle, *Deirochelys reticularia*, gave the curves of Fig. 23-57. The sensitivity is excellent for the range 100-600 Hz, and in a part of that span it reaches the astonishing level of −67 db, and over the range of 50 to 1000 Hz the performance is still very high. From 600 Hz upward the decline of sensitivity is fairly rapid, around 30 db per octave, but creditable hearing is indicated up to 2000-3000 Hz. The agreement between the two animals is close.

Geoemyda pulcherrima. — A specimen of *Geoemyda pulcherrima manni* gave sensitivity results as shown in Fig. 23-58. The sensitivity is excellent, attaining a level around −60 db over a span from 150 to 500 Hz, and is still very good for the wider range between 70 and 1500 Hz. For lower tones the decline is at the rate of 20 db per octave, and for higher tones the decline is about 40 db per octave.

Terrapene c. carolina. — The Eastern box turtle, *Terrapene c. carolina*, gave results shown in Fig. 23-59. There are variations among the three animals, but even the poorest ear must be rated as very sensitive in the low-tone range between 100 and 600 Hz. Though the measurements are limited at the low end, the points shown indicate a loss here at the rate of about 20 db per octave, and in the high frequencies the decline for all three animals is of about this order or slightly greater. This ear performs well over a range of about three octaves, from 100 to 800 Hz. An especially sensitive animal of this species is represented in Fig. 23-78 below.

Terrapene ornata. — A specimen of the ornate box turtle gave results as represented in Fig. 23-60. The sensitivity is only fair in the region of 300-500 Hz, and falls rapidly for lower and higher tones. In view of the excellent performance of other members of *Terrapene* the question arises whether this specimen is representative of its species, and it is desirable that other specimens be examined.

The emydid turtles studied show good general agreement in the form of the sensitivity function, with the best region covering about three octaves from 100-800 Hz. They agree also (with the exception of *Terrapene ornata*) in the degree of sensitivity attained within this range,

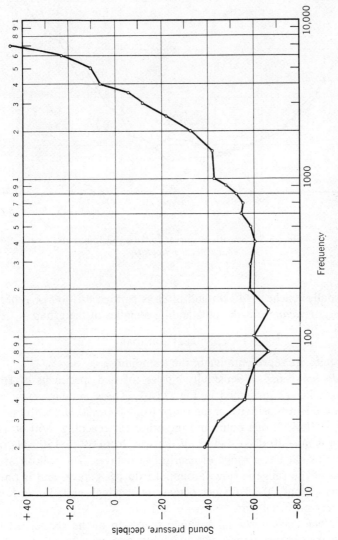

Fig. 23-56. A sensitivity curve for a specimen of *Clemmys guttata*.

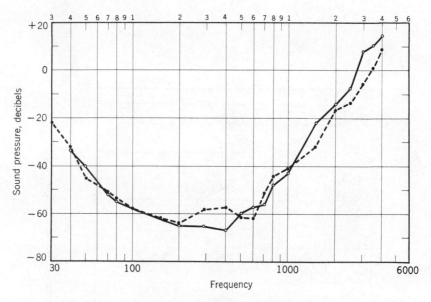

Fɪɢ. 23-57. Sensitivity curves for two specimens of
Deirochelys reticularia.

which usually reaches −40 db and often is as high as −60 or −65 db.
Deirochelys reticularia is the outstanding member of the group.

Fᴀᴍɪʟʏ Tᴇsᴛᴜᴅɪɴɪᴅᴀᴇ

Three species are representative of the testudinids.

Testudo horsfieldi. — Sensitivity curves for two specimens of Hors-
field's terrapin are presented in Fig. 23-61. These animals show very
good to excellent sensitivity in the range from 100-800 Hz, with one ear
reaching −50 db at one point and the other ear reaching −60 db. The
sensitivity is at −20 db or better all the way from 50 to 1500 Hz, and
evidently this ear has a range of nearly five octaves. The roll-off at the
lower end of the range is slow, around 15 db per octave, and is only a
little faster, about 20-25 db per octave, at the high end. For a turtle,
this is a very proficient ear.

Geochelone carbonaria. — A single specimen of the red-legged tortoise, *Geochelone carbonaria*, was available, and gave the results shown
in Fig. 23-62. The sensitivity is only fair, with the best region between
80 and 400 Hz, where the level is mostly around −12 db and reaches
−20 db at one point. The roll-off has a rate of about 20 db per octave
above 400 Hz. These results, if typical of the species, indicate hearing
of mediocre order.

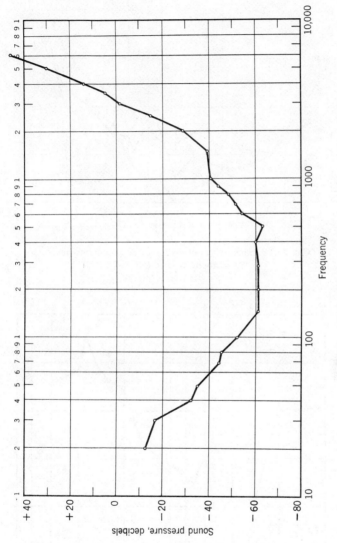

Fig. 23-58. A sensitivity curve for a specimen of
Geoemyda pulcherrima manni.

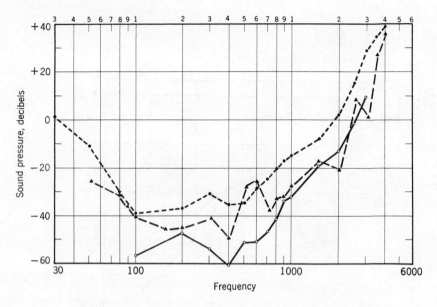

FIG. 23-59. Sensitivity curves for three specimens of
Terrapene c. carolina.

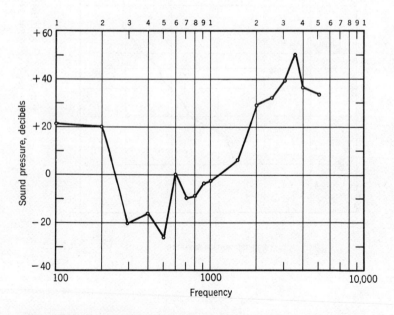

FIG. 23-60. A sensitivity function for a specimen of *Terrapene ornata.*

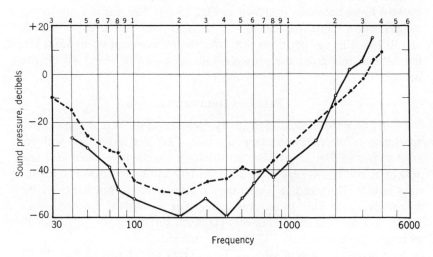

FIG. 23-61. Sensitivity curves for two specimens of *Testudo horsfieldi*.

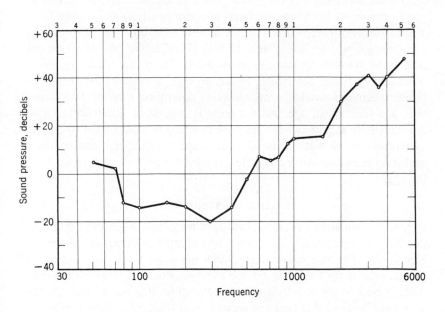

FIG. 23-62. A sensitivity curve for a specimen of *Geochelone carbonaria*.

Kinixys b. belliana. — A sensitivity function for a specimen of Bell's Eastern hinged tortoise is shown in Fig. 23-63. This ear is good to excellent over a range from 30 to 600 Hz, and reaches a level of −58 db at 150 Hz. This ear performs well over the entire range up to 1500 Hz, but beyond this point the response falls at a rate of about 40 db per octave.

FAMILY CHELYDRIDAE

Chelydra serpentina. — Sensitivity measurements were made on two specimens of the common snapping turtle, *Chelydra serpentina*, with results shown in Fig. 23-64. The sensitivity is excellent between 200 and 1000 Hz, and is still very good down to 100 and up to 1500 Hz. Beginning at 600 Hz there is a moderately rapid decline of sensitivity as the frequency rises. This is an outstanding ear within the low-frequency range, and the peaks at 400 or 500 Hz reach surprising levels of −68 db in one of these ears and −72 db in the other.

Macroclemmys temminckii. — Two curves are shown in Fig. 23-65 for the alligator snapping turtle, *Macroclemmys temminckii*, one (broken line) for an adult specimen weighing 2200 grams and another (solid line) for a hatchling of 14.5 grams. The two functions are in reasonably good agreement. Both indicate fairly effective hearing over a range from 100 to 700 Hz, where the average is around −25 db. The roll-off in the high frequencies, beginning around 600 Hz, is at the rate of 30 db per octave. In the high frequencies, from 1000 Hz on, the hatchling shows considerably better sensitivity, by an average amount of 20 db. Additional specimens would be necessary to determine whether this difference is a function of age, but the point is suggestive. Elsewhere, within the sensitivity regions where auditory function is most useful, there are variations but no notable trend, and it appears that this ear attains a useful level of operation early, perhaps by the time of emergence from the egg.

FAMILY KINSTERNIDAE

Kinosternon scorpioides. — Two specimens of the scorpion mud turtle, *Kinosternon scorpioides*, were investigated with results shown in Fig. 23-66. The two curves have much the same form, but one shows greater sensitivity by 12 db on the average. For both curves the sensitivity is best in the low-frequency range, up to 800 Hz. In one ear the sensitivity level in this range is around −40 db, and in the other ear is more variable between −43 and −54 db. The roll-off in the upper frequencies is between 20 and 30 db per octave. This ear must be rated as very good to excellent in its range.

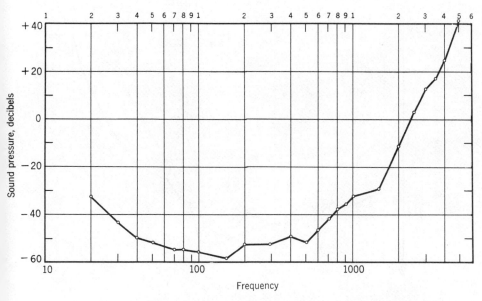

Fɪɢ. 23-63. A sensitivity curve for a specimen of *Kinixys belliana belliana*.

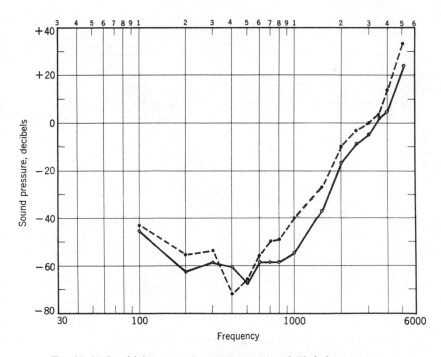

Fɪɢ. 23-64. Sensitivity curves for two specimens of *Chelydra serpentina*.

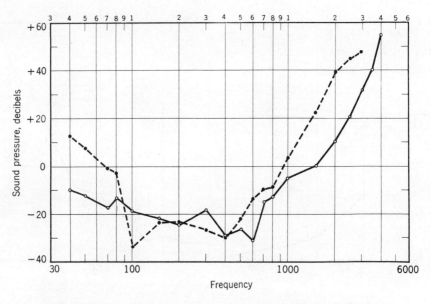

FIG. 23-65. Sensitivity curves for two specimens of *Macroclemmys temminckii*, one a hatchling (solid line) and the other an adult (broken line).

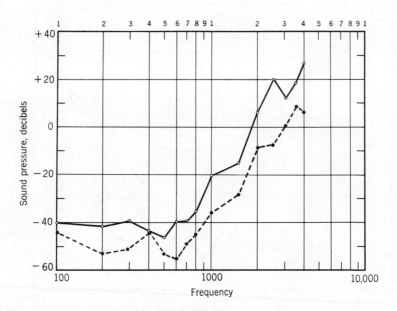

FIG. 23-66. Aerial sensitivity curves for two specimens of *Kinosternon scorpioides.*

The more sensitive of these two ears was studied also with vibratory stimuli, produced with a mechanical vibrator whose plunger ended in a blunt point that was applied to the middle of the tympanic membrane. The resulting curve, given in Fig. 23-67, indicates the vibratory amplitude, expressed in decibels relative to a zero level of 1 millimicron (rms), required to produce the standard response of 0.1 μv for the frequencies indicated. In terms of this stimulus the greatest sensitivity is at 500 Hz, with good performance from 400-1000 Hz. As in the aerial curve, the sensitivity declines rapidly in the high-frequency range beginning at 600 Hz, and then flattens beyond 2000 Hz. Because a mechanical vibrator forces the movement of the conductive mechanism almost independently of the impedance of this mechanism, the results principally reflect the characteristics of the inner ear. Thus it appears that the cochlear impedance, along with other cochlear factors such as hair-cell distribution, provide the most favorable conditions for response around 500 Hz, and at higher frequencies these conditions grow less favorable at a rapid rate.

Kinosternon s. subrubrum. — An aerial response function for the common mud turtle, *Kinosternon s. subrubrum*, is shown in Fig. 23-68. The sensitivity is excellent over a broad range at the low-frequency end of the spectrum, averaging −45 db for tones between 100 and 500 Hz and reaching a peak of −48 db at 400 Hz. Beyond 700 Hz the response falls progressively at a rate of about 30 db per octave.

Staurotypus triporcatus. — A specimen of the giant musk turtle, *Staurotypus triporcatus*, gave the sensitivity function of Fig. 23-69, in which a fair level of sensitivity is found for the range 80 to 800 Hz, with a good peak at 500 Hz. The sensitivity is maintained reasonably well up to 1500 Hz, but thereafter falls rapidly, at a rate averaging 40 db per octave. The performance of this ear is only moderately good, though its range is rather wide.

FAMILY TRIONYCHIDAE

Trionyx cartilagineus. — Sensitivity curves for two specimens of *Trionyx cartilagineus*, the only representatives of their family obtained for this study, are presented in Fig. 23-70. The two curves are in good agreement, and indicate excellent sensitivity, especially between 200 and 800 Hz, where one of these ears reaches a level of −57 db. The sensitivity is very good for the entire span of 3½ octaves from 80 to 1000 Hz, but the sensitivity is declining progressively at a rate of 20 db per octave below the maximum region, and at a higher rate of 30 db per octave above this region. This is an excellent ear, though clearly selective in its tuning.

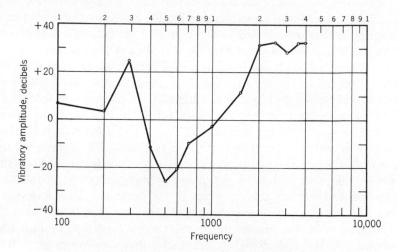

FIG. 23-67. A vibratory sensitivity curve for a specimen of *Kinosternon scorpioides* (the same animal whose aerial curve is shown by the broken line in the preceding figure). The vibratory amplitude is expressed in decibels relative to 1 millimicron.

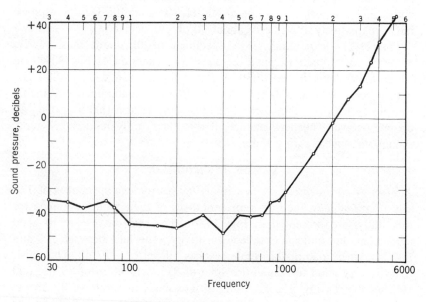

FIG. 23-68. An aerial sensitivity curve for a specimen of *Kinosternon s. subrubrum.*

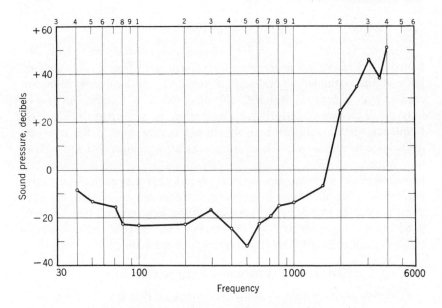

FIG. 23-69. A sensitivity curve for a specimen of
Staurotypus triporcatus.

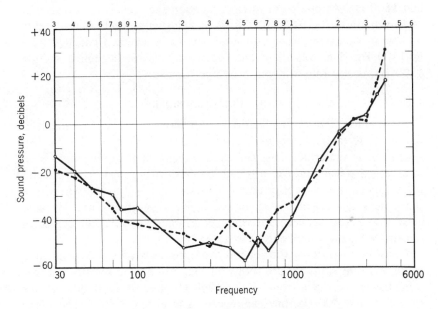

FIG. 23-70. Sensitivity curves for two specimens of
Trionyx cartilagineus.

FAMILY CHELONIIDAE

Chelonia mydas. — Results for two specimens of the green turtle, obtained from Hawaiian fishermen, are presented in Fig. 23-71. One of these animals attained a good level of sensitivity in the middle low-tone region, with a peak at 400 Hz. The other curve is similar, but shows only a fair degree of sensitivity. The decline of sensitivity in the high frequencies has a rate of about 40 db per octave, and in the low frequencies is a little slower. Three specimens were studied also with vibratory stimulation (Fig. 23-72). The lateral surface of the head was explored with the vibrating tip, and the place found that gave the greatest potentials. This place was a particular scale in the middle of the temporal region. The three curves show much the same general trend, which is an improvement up to 300-500 Hz and then a rapid decline. These results reflect primarily the characteristics of the cochlea.

SUBORDER PLEURODIRA; FAMILY CHELIDAE

Platemys platycephala. — Two specimens of *Platemys platycephala*, as shown in Fig. 23-73, gave sensitivity curves of generally similar form, though with many irregularities. The degree of sensitivity is very high in the range from 60 to 600 Hz, and is better than −40 db between 50 and 1500 Hz. At one point in each function the sensitivity reaches the astonishing level of −67 db. This is an unusually excellent ear and is notable in the extension of high sensitivity as far up the frequency scale as 1500 Hz. The roll-off in the upper frequencies, beginning around 1500 Hz, is at a rate of about 50 db per octave.

FAMILY PELOMEDUSIDAE

Pelomedusa subrufa. — Sensitivity functions for two specimens of the marsh terrapin, *Pelomedusa subrufa*, are shown in Fig. 23-74. Despite some irregularities these curves are in agreement in indicating good to excellent sensitivity over a broad low-frequency range. The better of these ears runs at a level of about −40 db from 100 to 900 Hz with a sharp peak that reaches −55 db at 300 Hz. The other curve averages close to −35 db between 150 and 700 Hz and shows a fair degree of sensitivity all the way from 40 to 1500 Hz.

Pelusios sinuatus. — Curves for two specimens of the African mud turtle, *Pelusios sinuatus*, are shown in Fig. 23-75. The sensitivity is excellent over the range from 100 to 1000 Hz, and in the middle of this range, around 400 Hz, one ear reaches a level of −57 db. The sensitivity declines below 100 Hz, and more markedly, at a rate around 40 db per octave, in the high frequencies.

Podocnemis expansa. — Results for two specimens of the Arrau river turtle, *Podocnemis expansa*, are shown in Fig. 23-76. These two differ somewhat in sensitivity, but agree in showing excellent response over the whole low-tone range up to 1500 Hz. In one the sensitivity is relatively uniform around −40 db from 40 to 1000 Hz, peaks to −47 db at 1500 Hz, and then declines. In the other animal the sensitivity runs between −50 db and −61 db between 100 and 1500 Hz, and then rapidly declines for the higher tones. This species is remarkable for the broad range of sensitivity.

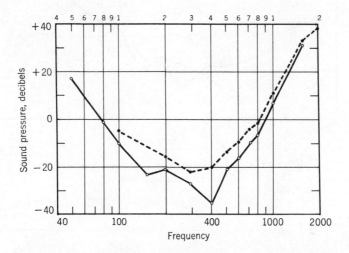

FIG. 23-71. Sensitivity curves for two specimens of *Chelonia mydas*. From Ridgway *et al.*, 1969.

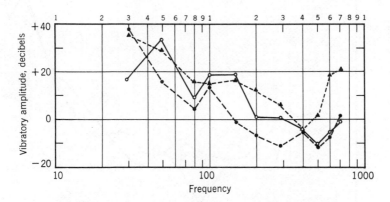

FIG. 23-72. Vibratory sensitivity curves for three specimens of *Chelonia mydas*. From Ridgway *et al.*, 1969.

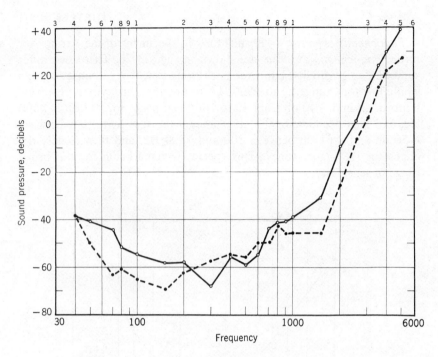

FIG. 23-73. Aerial sensitivity curves for two specimens of
Platemys platycephala.

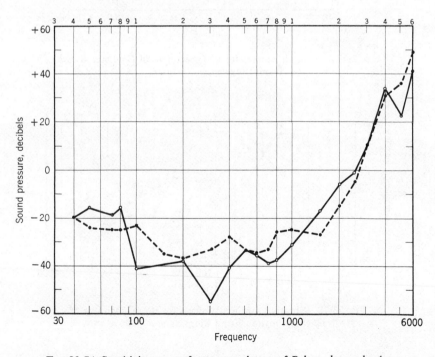

FIG. 23-74. Sensitivity curves for two specimens of *Pelomedusa subrufa.*

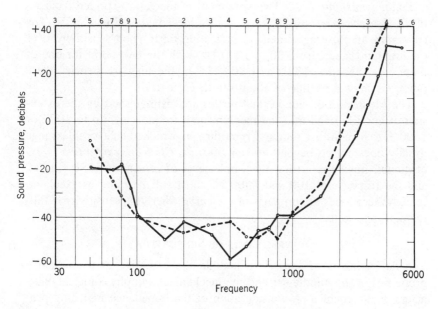

FIG. 23-75. Sensitivity curves for two specimens of *Pelusios sinuatus*.

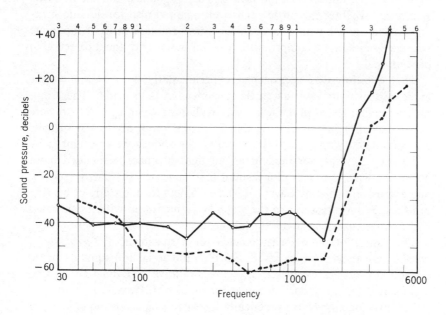

FIG. 23-76. Sensitivity curves for two specimens of *Podocnemis expansa*.

Podocnemis unifilis. — Two specimens of the yellow-spotted Amazon turtle, *Podocnemis unifilis*, gave results presented in Fig. 23-77. The curves are in close agreement in indicating fairly good sensitivity over the low-tone range, from 30 to 1000 Hz, with the best point for one of these ears reaching −32 db at 200 Hz. In the high frequencies these functions show a decline of about 40 db per octave.

The Pleurodira or side-necked turtles are distinguished by the extension of good hearing into the medium high frequencies, up to 1000 or 1500 Hz, in contrast to the Cryptodira in which the favorable range usually does not go beyond 700 or 800 Hz. *Chelydra serpentina* is exceptional, however, in that the whole level of sensitivity is very great, and the response at 100 and 1500 Hz must still be rated as excellent, though these tones are on parts of the curve where the sensitivity is falling at a rapid rate.

Determinants of Sensitivity

The determinants of sensitivity include the receptive and conductive properties of the middle ear and the mechanical and physiological characters of the cochlea. A consideration of the impedance matching provided by the middle ear mechanism has led to doubts whether this matching even approaches the ideal for aerial sounds, and the question was raised whether the turtle ear is well adapted also for aquatic hearing, so that its ear with a mechanical transformation ratio of moderate size may represent a compromise between aerial and aquatic reception of sounds.

This question was approached experimentally by determining the auditory sensitivity in a few turtle specimens in both media. These animals were first tested in the usual way with aerial sounds, and then were submerged in water, with only the tip of the snout above the surface, and stimulated with underwater sounds. The sounds were produced by a dynamic loudspeaker held in an air-tight chamber suspended close above the water surface, in a fashion first suggested by Harris (1967), and extensively used by Fay (1969) and others for the study of hearing in fishes. Sound pressures in the water were measured with a calibrated hydrophone placed near the turtle's head.

The results of these measurements for a specimen of *Terrapene c. carolina* are shown in Fig. 23-78. For both the aerial and aquatic curves, the sound pressure, expressed in decibels relative to 1 dyne per sq cm, is indicated for a standard cochlear potential of 0.1 μv.

The two functions have practically the same shape. The aquatic curve is somewhat disturbed by variations that no doubt represent reflections in the water tank causing differences in pressure between ear and hydrophone locations, but the course of the curve is sufficiently clear. In this ear the sensitivity is greater for aerial sounds by an amount of about

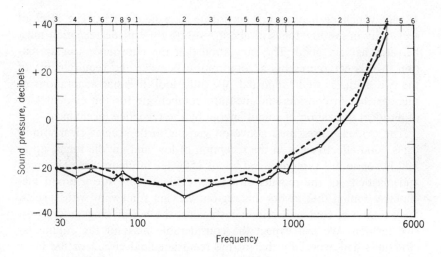

FIG. 23-77. Sensitivity curves for two specimens of *Podocnemis unifilis*.

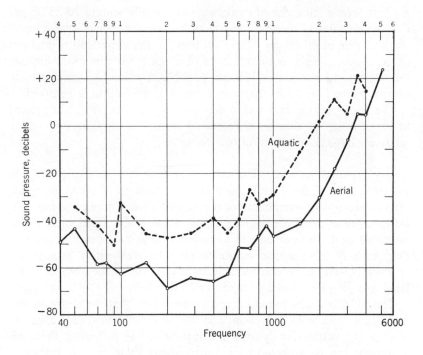

FIG. 23-78. Sensitivity functions for a specimen of *Terrapene c. carolina* determined by aerial stimulation (solid line) and by underwater stimulation (broken line).

18 db, which perhaps is not surprising in a land turtle. The level of sensitivity in the low tones of about −40 db for aquatic reception must be rated as very good. The suggestion that the impedance transformation in this ear represents a compromise between the requirements of the two media is well supported. We must look to other adaptations of this ear that in considerable measure compensate for the evident lack of an ideal utilization of sound energy by the receptive mechanism.

The evidence presented above on various turtle species clearly indicates good effectiveness in the reception of low and middle tones, up to 800 Hz or so in Cryptodira and about 1500 Hz in Pleurodira. The mass and elasticity of the moving structures evidently provide a broad mechanical tuning that favors this region, varying somewhat with species. Such tuning of course discriminates against the higher tones in progressive fashion. We can expect the considerable mass of the columellar structures and especially that of the reentrant fluid circuit of the inner ear to have much to do with this rapid roll-off in the upper frequencies. Nearly all the sensitivity curves exhibit two distinct rates of decline in the high frequencies, one of moderate rate, usually around 20-25 db per octave immediately above the region of best response, and another, usually at a rate of 40 db per octave or more, at the upper end. This form of the function suggests the presence of at least two conditions limiting the response to high tones. It is not possible at present to identify these factors, though no doubt the mass factor just referred to is involved.

The good performance within the low-frequency range can be related to two further conditions, a degree of inner ear amplification and a relatively large complement of hair cells.

Inner-ear Amplification.   The columellar footplate is large relative to the total area of the basilar membrane, so that the fluid volume displaced through the movements of the footplate when applied to the basilar membrane produces displacements of increased magnitude. Measurements made in three ears of *Chrysemys scripta* indicate the order of magnitude of this cochlear amplification. The areal ratios, represented in Table 23-III, show considerable individual differences. When translated into gain these values vary from 25 to 32 db.

This calculation is based on two assumptions, that the entire basilar membrane undergoes deflection, and that the mechanical impedance of driving and responsive systems are matched. The second of these assumptions can be accepted as at least roughly valid, but the first must be taken with reservation.

The edges of the basilar membrane are attached to the limbus, and the restraints of these attachments are transmitted to the adjacent areas to a degree that falls off with distance, so that the border areas of the membrane are relatively fixed and only the interior areas are free. This

TABLE 23-III

COCHLEAR AMPLIFICATION IN *Chrysemys scripta*

Animal No.	Footplate Area (sq mm)	Basilar Membrane Area (sq mm)	Ratio	Gain (db)
998 R	4.16	0.228	18.3	25.2
998 L	4.14	0.201	20.6	26.3
999 L	4.85	0.111	41.9	32.4

edge restraint will of course be most marked where the membrane is narrow and at the two ends. Also, if there is any significant degree of frequency selectivity along the turtle cochlea only certain areas of the basilar membrane will be readily responsive to a particular tone. These conditions make the effective area of the basilar membrane less than the anatomical area, and increase the amplifier ratio. Accordingly, the gains indicated in Table 23-III must be regarded as minimum values; the actual amounts may be somewhat greater.

On the basis of the dimensional characteristics already described, the turtle cochlea does not give the impression of great selectivity, and accordingly the amount of gain through cochlear amplification may be only moderately greater than these calculations indicate.

This inner-ear amplification must be contemplated in the face of another condition that is in some degree counteracting: the amount of spatial summation in the cochlear and neural activity. If hair cells over a broad area of the cochlea are involved in the response, this summation will be extensive, whereas a degree of specificity would limit this summation.

The relatively large number of hair cells in the turtle cochlea, as shown in Table 23-II, and especially those cells lying on the basilar membrane, provide the basis for this spatial summation both in the cochlear response and in neural action at cochlear and higher levels.

Sensitivity and Numbers of Hair Cells.   The above considerations lead to the expectation that auditory sensitivity should relate to the number of cochlear hair cells. This point was investigated in a group of 16 turtle species by the correlation method. Correlation coefficients were calculated, using two measures of sensitivity and three determinations of hair-cell numbers.

The two sensitivity measures were (1) the most sensitive point along the cochlear potential function (peak sensitivity), and (2) the mean of all points within 5 db of this maximum (mean best sensitivity). The three measures of hair-cell populations were (1) the number on the basilar membrane, (2) the number on the limbus, and (3) the total of these two groups.

The results of these computations are surprising: all the correlations are negative. As seen in Table 23-IV, both the peak and mean sensitivity measures bear a reverse relationship to the numbers of hair cells on the basilar membrane, though these coefficients are small and with poor reliability. For the hair cells on the limbus the relations are more meaningful: for these the coefficients approach significance, and show a tendency for large numbers of limbic hair cells to appear in ears with the lower degrees of sensitivity.

TABLE 23-IV
SENSITIVITY AND HAIR-CELL POPULATIONS
FOR 16 TURTLE SPECIES
(Correlation Coefficients)

	Peak Sensitivity	Mean Best Sensitivity
Hair cells on the basilar membrane	$r = -0.24$	$r = -0.26$
	$p = 0.37$	$p = 0.33$
Limbic hair cells	$r = -0.44$	$r = -0.58$
	$p = 0.09$	$p = 0.02$
Total hair cells	$r = -0.47$	$r = -0.51$
	$p = 0.06$	$p = 0.04$

No very satisfactory explanation can be offered for these relationships. It appears, as has already been concluded for lizards, that a comparatively small number of hair cells can determine the level of sensitivity, and the addition of further hair cells may not improve this function, though it may have other results such as extending the dynamic range of the ear.

AN EVALUATION OF THE TURTLE EAR

Sensitivity and Intensity Range. In turtles the ear is well developed, and in most species the sensitivity is good to excellent within the low-frequency range. The maximum sensitivity in Cryptodira is usually in the region of 100 to 700 Hz, and in Pleurodira extends perhaps an octave higher. The basilar membrane has the form of an oval, usually two or three times as long as it is wide, and the papilla lies close to the limbus on its inner side. When the basilar membrane moves in response to oscillating sound pressures the papilla evidently swings along its limbic edge, with the outer half of the membrane undergoing the principal flexure. The bodies of the hair cells are carried along in the papillar motions, but the tips of their ciliary tufts, because of their attachments to the tectorial plate, are largely restrained. Consequently there is relative motion between cilia and cell body, and the cell is stimulated.

The function of the limbic hair cells presents a further problem. As has been shown, these cells in most species occur in considerable numbers at both ends of the cochlea, and often nearly equal the numbers of hair cells on the basilar membrane. They are less numerous, however, in *Chelydra serpentina* and *Trionyx cartilagineus*, and are relatively few in *Terrapene c. carolina, Testudo horsfieldi*, and *Kinixys belliana*. In these last three species the limbic hair cells are nearly absent at the ventral end of the cochlea.

That these limbic cells have a significant function is indicated by their considerable numbers in the great majority of species, by their connections to the tectorial membrane, and by their rich innervation.

The presence of hair cells off the basilar membrane has been observed also in other reptile species, though only in small numbers: at the dorsal end of the cochlea in varanid lizards and at both dorsal and ventral ends in the snake *Acrochordus javanicus* and in *Sphenodon*.

Limbic hair cells have not been found in crocodilians, but they exist in large numbers in birds (Wever, 1971c). In the owl *Bufo virginianus* these cells number about 3500 in a total of 12,000, and thus constitute more than a fourth of the hair-cell population.

The mode of stimulation of these limbic hair cells was discussed in Chapter 4; it is suggested that their involvement is indirect, by a transmission from basilar membrane first to the tectorial membrane through its ciliary connections to the papillar hair cells, and then along this membrane to the ciliary tufts of the limbic hair cells.

It seems obvious that a displacement of the ciliary tuft of a cell or simply of its kinocilium, with the cell body remaining at rest, will be quite as stimulating as the more familiar movement of the cell body with the cilia remaining relatively immobile. Indeed, this "reversed" form of hair-cell stimulation is the rule in amphibians, in which all the hair cells rest on the limbic base, and the ciliary tufts of these cells are deflected through their attachments to tectorial tissues suspended in the fluid closely adjacent to their surface (Wever, 1973c).

Under the conditions just described for the limbic hair cells of turtles it seems clear that the motions transmitted through the tectorial membrane will be small relative to those involving the hair cells on the basilar membrane. The basilar membrane cells, therefore, will provide the principal representation of the stimulating sounds when the intensity is low or moderate, and the limbic cells will add their contribution to the response when the intensity reaches high levels. This involvement of the limbic cells should begin to be significant in the range of intensity at which the basilar membrane cells, and the auditory dendrites that serve them, are reaching their maximum output. The effect of the limbic cells, therefore, is an extension of the dynamic range of the ear.

This extension of intensity range seems particularly important in these ears because of the limited differentiation along the basilar membrane. As pointed out earlier (Wever, 1965a), the high degree of differentiation present in the more advanced ears provides a large amount of range extension because for low and moderate tones the principal action is in one region while other parts of the cochlea are relatively little involved. And then, when the intensity is raised and the hair cells of the peak region begin to reach their limits, the outlying areas come in more strongly.

RELATIONS TO BEHAVIORAL FUNCTION

It is fortunate that reliable behavioral observations are available on one turtle species, so that comparisons can be made with cochlear potential measurements. Patterson's results on four animals by the conditioning procedure described earlier are shown in Fig. 23-79. These curves agree closely in indicating a high degree of sensitivity to tones over the range of 200-700 Hz, where the average threshold is about −30 db. For lower tones the sensitivity declines progressively at a rate around 20 db per octave, and for higher tones it falls off almost precipitously, approaching 80-100 db per octave.

Patterson (1966) noted the close correspondence between his re-

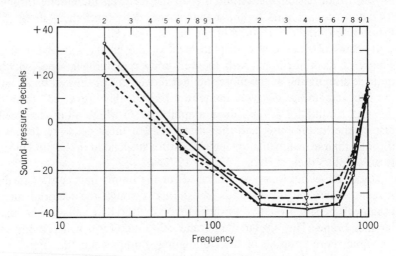

Fig. 23-79. Auditory sensitivity in four specimens of *Chrysemys scripta* as determined by Patterson by a conditioned response method. The ordinate indicates the sound pressure, in decibels relative to 1 dyne per sq cm, required at various frequencies to produce a threshold response. Redrawn from Patterson, 1966.

sults and the electrophysiological observations that had been made earlier on the same species (*Chrysemys scripta*) by Wever and Vernon, and presented a figure containing the two functions. Gulick and Zwick (1966) made the same comparison, and added further electrophysiological data on four specimens.

This comparison was made again in Fig. 2-6, where the solid curve presents more recent results on a specimen of *Chrysemys scripta elegans* in which the cochlear potential measurements were extended downward to 20 Hz, and the broken curve represents the mean thresholds for Patterson's four animals. It must be borne in mind that the cochlear potentials do not have a threshold, and the level of response of 0.1 μv is quite arbitrary. What is to be observed is the degree of correspondence in the forms of the two functions. This correspondence is close: for both types of function the best sensitivity is in the same region of frequency, and the sensitivity falls off rapidly for tones above and below this region.

It is further of interest that the effects of surgically eliminating the turtle's conductive mechanism are of about the same magnitude as observed by cochlear potential and behavioral methods. The results obtained on clipping both columellas in one of Patterson's specimens are shown in Fig. 23-80, and agree well with those shown earlier in

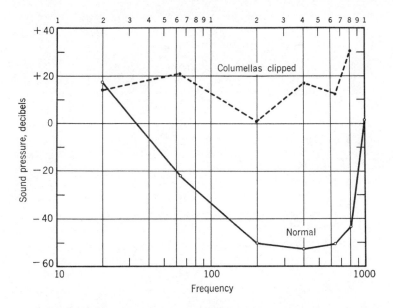

FIG. 23-80. Effects of columella sectioning on behavioral responses to sounds in *Chrysemys scripta*. Thresholds before and after this procedure are represented as in the preceding figure. Redrawn from Patterson, 1966.

Figs. 23-12 and 23-13 using cochlear potential observations on this same species.

This correspondence between behavioral and electrophysiological determinations of auditory sensitivity was noted in Chapter 2, and indicated as providing a substantial basis for the use of cochlear potential observations in assessments of the functional characteristics of reptilian ears.

24. ORDER CROCODILIA:

THE CROCODILES

The crocodiles now living are a small remnant of a group of reptiles that arose toward the close of the Triassic period and underwent a great expansion during the following period, the Cretaceous, about 100 million years ago. From fossil records about 18 families are recognized (Wermuth and Mertens, 1961), of which only three, the Crocodylidae, Alligatoridae, and Gavialidae, have living representatives.

The Crocodylidae include three genera that comprise about 13 species, found in tropical and subtropical areas around the world. The Alligatoridae have four genera and seven species, mainly tropical but extending into the warmer regions of the temperate zone. The Gavialidae are represented by a single genus and species, restricted to India and Burma. All the crocodilians are seriously endangered, and the gavials are close to extinction.

The crocodilian ear was examined and described with varying degrees of accuracy by a number of early anatomists. The first thoroughgoing account was that of Hasse in 1873, who studied two specimens of *Crocodylus niloticus*. Retzius in 1884, from observations on *Alligator mississippiensis*, provided the most complete description, with some corrections of Hasse's work and many additional details. No extensive treatment of the ear in this group of reptiles has appeared in the present century, though there have been numerous studies of particular features.

Studies of the hearing abilities of crocodilians have been few. Observations of behavior have afforded general evidence of sound perception. Loud roaring and bellowing sounds are produced by males, especially in the breeding season, and other individuals appear to react to them. Berger in 1924 examined a single specimen of the blunt crocodile, *Osteolaemus tetraspis*, and noted changes of breathing rate in response to intense noises. Beach in 1944 reported hissing or roaring in specimens of *Alligator mississippiensis* when stimulated with low-frequency tones and noises.

More specific evidence of auditory sensitivity in these animals has been obtained by electrophysiological methods. Adrian, Craik, and Sturdy in 1938 recorded both cochlear and auditory nerve potentials in an alligator in response to various tones. Wever and Vernon in 1957 used the cochlear potential method in extensive measurements on *Caiman crocodilus*, with results to be mentioned later.

The present study includes measurements of auditory sensitivity in terms of the cochlear potentials, followed by an examination of ear structure by dissection and in serial sections. For this purpose a limited number of crocodilian species were available, as follows.

ALLIGATORIDAE

Alligator mississippiensis (Daudin, 1803)—3 specimens; *Caiman crocodilus* (Linnaeus, 1758)—14 specimens.

CROCODYLIDAE

Crocodylus acutus (Cuvier, 1807)—3 specimens.

All the above were young animals, mostly of 40-100 cm total length. In addition to the specimens mentioned, a number of *Caiman crocodilus* were used in gross dissection for general orientation and the study of middle ear structures.

ANATOMY OF THE CROCODILIAN EAR

EXTERNAL EAR

In crocodiles the tympanic membrane is located in the floor of a shallow chamber on the dorsolateral surface of the head, immediately behind the eye. This chamber is referred to as the external auditory meatus or perhaps better as the meatal cavity, for it bears little resemblance to the tubal opening commonly found in other reptiles. The chamber is covered by a pair of flaps of modified skin, or earlids, whose surface resembles the surrounding skin area, and only close examination reveals the presence of an ear opening. These earlids are evident in Fig. 24-1.

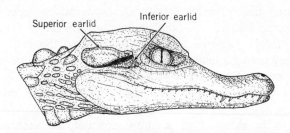

FIG. 24-1. The head of a juvenile specimen of *Caiman crocodilus*. Scale ½×.

The superior earlid is much the larger of the two, and extends along the whole meatal cavity, broad posteriorly and narrowing toward the anterior end. It is attached along its upper margin to the lateral edge of the squamosal bone by a stiff ligamentous hinge. Its lower edge, except for the anterior portion, lies close over the border of the facial skin, and a thin fleshy strip extends to form a seal along this border. The narrow anterior portion, which constitutes about a fourth of the length, lies over the inferior earlid, and under most conditions a gap is left between these two for the entrance of sound.

The inferior earlid is triangular in form, wider anteriorly and tapering to a point posteriorly, so that it fills out the reduced portion of the upper earlid. The slit ordinarily left between the two earlids in the small specimens examined varied from 1-2 mm in width and 8-10 mm in length. The usual resting position for an undisturbed animal, with the slit open, is represented in Fig. 24-1.

Movements of the superior earlids are rarely seen, but have been reported in animals that are greatly excited or were stimulated, as described by Shute and Bellairs (1955), by the injection of water into the meatal cavity. On this injection the lids were flickered up and down repeatedly for a few seconds, evidently as a means of clearing away the water.

Movements of the inferior earlid are common. A shifting of the upper edge of this lid in a dorsoposterior direction to meet the superior earlid is readily elicited by touching the skin in this region or by submerging the head. When the head sinks below the water surface the earlid closes, but a little later, if the animal remains submerged, it may open again, as noted by Garrick and Saiff (1974). When the animal rises to the surface this closing and immediate reopening of the auricular slit is repeated, and the reaction evidently protects the ear against the entrance of water during passage through the air-water interface. When the animal is submerged a bubble of air ordinarily will be trapped in the cavity, and the tympanic membrane remains free of water loading.

Garrick and Saiff also observed that there is a close linkage between eye blinking and earlid closure; the movements of the inferior earlid appear to be synchronized with eyelid motion.

The earlid musculature of crocodiles was first seriously investigated by Killian (1890), though apparently he was unaware of the presence of the inferior earlid. Shute and Bellairs (1955) treated this subject in considerable detail.

The Superior Earlids and Their Action. For the superior earlid Shute and Bellairs identified two muscles, a levator auriculae superior and a

corresponding depressor muscle. These muscles are attached by ligamentous strands to a core of dense connective tissue that runs through
the main portion of the lid and gives it great stiffness. Shute and Bellairs
called this core the auricular plate, and compared it with the tarsal
plate of mammalian eyelids. It is composed of dense connective tissue
with large collagenous fibers forming a feltwork with much crossing and
interweaving, but with a majority of the fibers having a common orientation, most generally running from the upper margin of the lid obliquely downward and forward. This plate is attached at its posterior end
to the postorbital bone, and forward along its whole length to the
squamosal.

Shute and Bellairs described the actions of the two muscles of the
superior earlid in a manner illustrated in Fig. 24-2. They regarded the
more ventral fibers of the levator muscle as mainly responsible for
rotating the lid upward and away from the skin surface overlaying the
quadrate, and indicated the depressor muscle as producing a contrary
rotation.

In the present study these muscles were examined in specimens of
Caiman crocodilus both by dissection methods and in serial sections.
The main body of the levator is revealed by raising the posterior por-
tion of the earlid with forceps as indicated in Fig. 24-3. This muscle

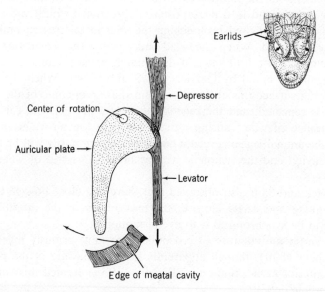

FIG. 24-2. Schematic representation of the action of the superior earlid muscles (right side). Simplified from Shute and Bellairs, 1955,
Fig. 4 b. The inset drawing, a front view of the head, gives the orientation.

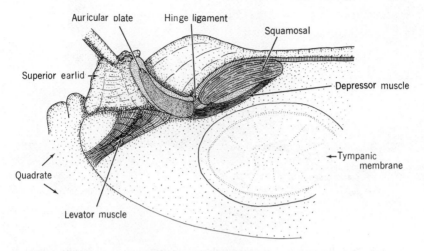

FIG. 24-3. The superior earlid muscles in *Caiman crocodilus*, seen on the right side in an anterolateral view after removal of the anterior three-fourths of the earlid and raising the remaining portion. Scale 5×.

lies beneath and behind the posterior fourth of this lid, and shows two closely connected bellies that run to an anchorage in dense fascia on the surface of the quadrate. This anchorage is far posterior, well beyond the main part of the muscle.

The muscle is followed farther in Fig. 24-4; it runs from its quadrate anchorage medially beneath the broad inner edge of the auricular plate and attaches as shown to both the squamosal bone above and the dorsal surface of this plate. The auricular plate attachment is made by heavy ligamentous tissue that runs along the surface of the plate and penetrates deeply into its substance. The action appears to be twofold. Straightening of the muscle crowds the inner portion of the earlid away in an upward and somewhat lateral direction, and this dorsolateral force, because of the hinge restraint, is converted into an upward rotation. At the same time the more superficial fibers exert tension on ligamentous strands running in the body of the plate, which stiffens the outer portion of the plate and causes it to rotate upward with the main body. This action was easily demonstrated in dissected specimens by pulling on the end of the muscle near its quadrate anchorage.

The depressor for the superior earlid, already seen in Fig. 24-3, is shown in further detail in Fig. 24-5. The auricular plate is attached to the squamosal by a heavy ligament that serves as a hinge along a fold of skin at the upper border of the earlid. The depressor muscle inserts along the lower surface of the plate, and its main bundle of fibers runs

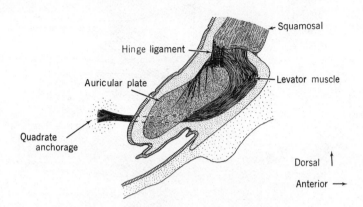

FIG. 24-4. Reconstruction of the right auricular plate with its levator muscle, based on serial sections of *Caiman crocodilus*.

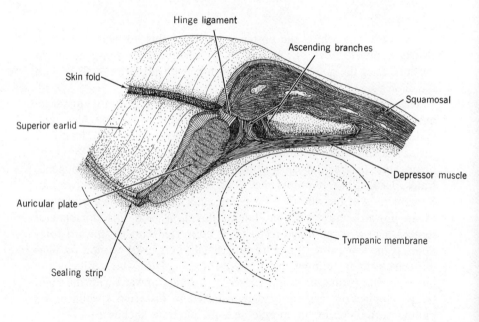

FIG. 24-5. Details of the depressor muscle of the superior earlid in *Caiman crocodilus*, from the same viewpoint as in Fig. 24-3, but with the earlid depressed.

medially to attach to the fascia along the medial undersurface of the squamosal. Other muscle fibers run from the medial surface of the auricular plate to the adjacent part of the squamosal just medial to the hinge region, and in the specimen studied most thoroughly there were two groups of these fibers.

It is clear that the superior earlid is pulled in a medial direction and held firmly against the surface of the side of the head by the action of the principal bundle of fibers. When the head is submerged the water pressure will assist the sealing action to prevent a flooding of the middle ear cavity.

A third superior earlid muscle is represented in Fig. 24-6, indicated as the lining muscle. This muscle arises from the anterior surface of the paroccipital process just lateral to the extracolumellar muscle and runs

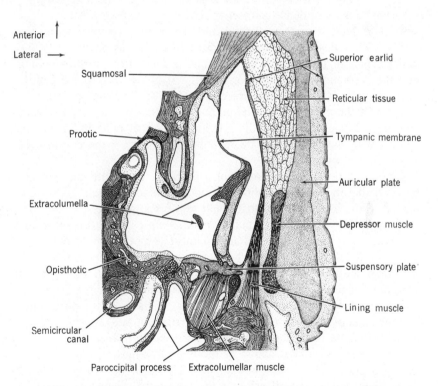

FIG. 24-6. A frontal section through the right middle ear region of *Crocodylus acutus*. The superior earlid is shown along most of its length, with its auricular plate and thick reticular lining. A muscle extends from the paroccipital process to the posterior portion of this lining. An extracolumellar muscle runs from a deep recess to a suspensory plate to which the posterior edges of the tympanic membrane and extracolumellar structure are attached. Scale 10×.

forward to an insertion along the inner lining of the superior earlid. Because this muscle is attached only to loose reticular tissue it can hardly be concerned with gross movements of the lid. A suggestion is that this muscle produces jerky movements of the lid lining that clear away fluid or foreign particles.

The Inferior Earlids. Shute and Bellairs also investigated the action of the inferior earlid. Working mainly on *Alligator mississippiensis*, they identified a ribbon-like muscle (which they called the depressor auriculae inferior) arising from the infra-orbital septum and running transversely along the floor of the orbit. Over most of its course this muscle bears a close relation to the levator bulbi, and more anteriorly it is related also to the depressor palpebrae inferior muscle.

The depressor auriculae inferior muscle inserts on the inner edge of a Y-shaped mass of dense connective tissue that Shute and Bellairs considered to have a certain identity and which they referred to as the ypsilon. They described a dorsal arm of this body as anchored to the undersurface of the postorbital bone and a dorsolateral arm extending into the substance of the inferior earlid. As they envisaged the action, a contraction of the muscle exerts a lateral tension that produces a rotation of the ypsilon about its point of anchorage, with a downward and inward displacement of the inferior earlid that pulls its lip out of contact with the superior earlid, opening up a slit for the entrance of sound into the meatal cavity.

Closure of this slit poses more of a problem, but Shute and Bellairs believed that they had identified a strand of smooth muscle above and medial to the inferior earlid that could serve this purpose.

Present observations made mainly on *Caiman crocodilus* confirm the observations of Shute and Bellairs in most respects, but show a few points of difference. As Fig. 24-7 indicates, the condensation of connective tissue described as the ypsilon can be made out in a general way, but this tissue merges into the lateral skin and inferior earlid tissue. Also in *Caiman* the connection of this mass of tissue to the postorbital is only tenuous, consisting of a few thin strands, and a swivel action about a definite anchorage here does not seem possible. Accordingly, the main effect of a contraction of the depressor muscle appears to be simply a pulling downward and slightly inward of the whole body of dense tissue, including all arms of the connective tissue condensation that Shute and Bellairs called the ypsilon together with the main mass of the inferior earlid and much of the lateral skin. The long ascending strands of tissue (the dorsal arm of the ypsilon) connecting to the postorbital will be stretched in the process.

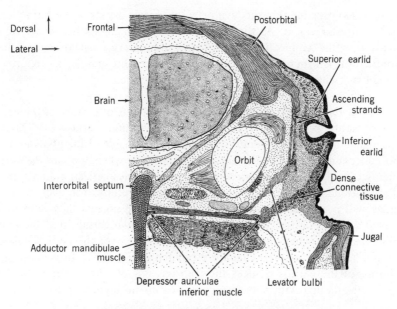

Dorsal

Lateral →

Frontal —

Postorbital

Superior earlid

Brain →

Ascending strands

Inferior earlid

Orbit

Interorbital septum —

Dense connective tissue

Adductor mandibulae muscle

Jugal

Depressor auriculae inferior muscle

Levator bulbi

FIG. 24-7. A transverse section through the right ear region of
Caiman crocodilus, seen from behind. Scale 5×.

This displacement downward of the inferior earlid does not involve the superior lid to any significant extent both because of the discontinuity of tissue at the notch between these two structures and because the superior lid is closely connected to the postorbital: posteriorly the postorbital extends all the way to the notch and effectively immobilizes a large part of this superior earlid.

The hypothesis offered by Shute and Bellairs for elevating the inferior earlid and closing the meatal slit cannot be supported, at least for *Caiman*. No trace could be found of the bundle of smooth muscle fibers that they described. Moreover, no other active mechanism for closing this slit could be identified. A simple and effective means is present, however: it is the restoration of the position of the tissues through their own elasticity, occurring when the depressor muscle is relaxed. This tissue has the same general character as the auricular plate of the superior earlid; it consists of a dense feltwork of collaginous fibers, and differs from that plate only in that here the fibers appear even more tightly twisted and interwoven. The strands that extend dorsally to the postorbital bone, and are put in stretch when the depressor muscle contracts, will no doubt contribute materially to the restoration of the resting state of the structure when this muscle relaxes.

It is also possible that the levator bulbi muscle assists in the retraction of the inferior earlid. At the level shown in Fig. 24-7, this muscle extends into the connective tissue below the inferior earlid, and if its contraction is synchronized with that of the earlid muscle, as reported by Garrick and Saiff, these two will be synergistic.

Effects of Earlid Closure on Hearing.   The effects of closure of the earlids on sound reception were studied by Wever and Vernon (1957) in specimens of *Caiman crocodilus* by measuring cochlear potentials under normal conditions, with the slit open, and then when the earlids were tightly closed. The closed condition is usual in anesthetized animals, but it was found necessary to check this point carefully, for even a tiny crack will permit nearly normal hearing. For the normal measurements a bent wire was inserted between the lids to make certain that the gap remained open when the sound tube was sealed over the area. Typical results of these measurements are presented in Fig. 24-8. In general the closure of the earlids reduces the transmission of aerial sounds by 10 to 12 db, though with some irregularities in the high frequencies.

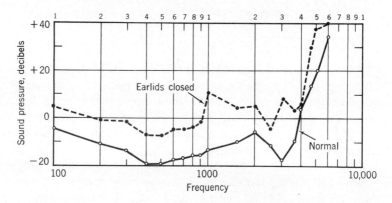

FIG. 24-8. Effects of earlid closure on the aerial reception of sounds in *Caiman crocodilus*. After Wever and Vernon, 1957.

MIDDLE EAR

On removal of the earlids the tympanic membrane is seen, as in Fig. 24-9, as a broad oval, facing dorsolaterally. Visible through the semi-transparent surface, and producing a prominent bulge at its center, is the end of the cartilaginous extracolumella.

The extracolumella has the form of a flattened pyramid, with a shaft extending from its apex. As seen from within, as in Fig. 24-10, it can best be described as consisting of a thin triangular vane whose base is in continuous contact with the tympanic membrane from its middle to

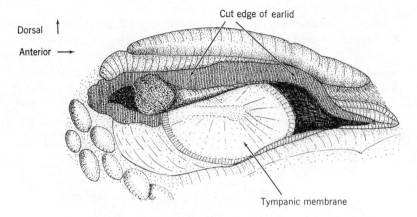

Dorsal

Anterior →

Cut edge of earlid

Tympanic membrane

FIG. 24-9. The tympanic membrane of *Caiman crocodilus*, exposed by removal of the earlids. Scale 3.5×.

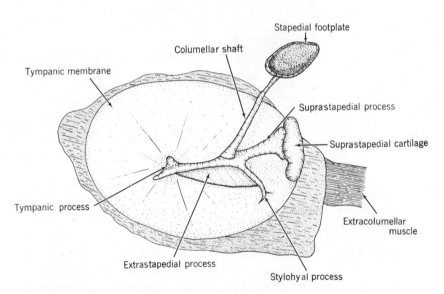

Stapedial footplate

Columellar shaft

Tympanic membrane

Suprastapedial process

Suprastapedial cartilage

Tympanic process

Extracolumellar muscle

Extrastapedial process

Stylohyal process

FIG. 24-10. The tympanic membrane and columella of *Caiman crocodilus* removed from the skull and viewed from a ventrolateral direction. The names of the parts largely follow Parker, 1883. Scale 5×.

the posterior edge, and a ribbon-like strut that runs from the dorsoposterior edge to meet the apex of the triangular portion. The columellar shaft extends inward from this junction.

After the columella is cut near its junction with the extracolumella and the tympanic membrane is removed, a view into the cochlea is obtained as in Fig. 24-11. A natural opening in the quadrate affords a view into the middle ear space, and both oval and round windows may be seen in *a* of this figure. The stapedial footplate is partially obscured in this view, but with some manipulation of the specimen it appears as shown in part *b*. The round window membrane is readily recognized in the living animal because of the bright red color produced by a network of blood vessels on its surface.

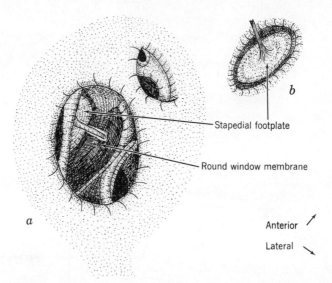

FIG. 24-11. At *a*, the cochlear region of *Caiman crocodilus* after removal of the tympanic membrane; and at *b*, a detail of the stapedial footplate. Scale 5×.

The performance of the tympanic membrane and columellar system in sound transmission was evaluated in the experiments of Wever and Vernon (1957) on *Caiman crocodilus* by observing the effects on the cochlear potentials of cutting the columella. In these experiments access to the middle ear cavity was obtained by drilling a hole in the parietal bone over a transverse passage that runs from one middle ear space to the other above the cranial capsule. This opening, when enlarged a little laterally, provides a view of the round window, on whose surface a recording electrode was placed, and also reveals the middle portion of the columella. After normal tests of sensitivity were carried out (usually with the parietal opening plugged with bone wax), the columella was severed and the tests repeated. Usually a small piece of the columella was removed to prevent the reestablishment of contact between the severed ends.

The results of this experiment are presented in Fig. 24-12. Loss of the peripheral receptive mechanism produces a decrease in sensitivity of 47 to 60 db for most tones, rising to 70 db at 1000 Hz, and then declining somewhat in the upper frequencies. After the columella clipping it was no longer possible to obtain a measurement at 6000 Hz under the conditions of this test. The mean loss over the range up to 5000 Hz was 52 db. It is evident that the middle ear mechanism of the crocodile is highly effective in the reception of sounds.

Interaural Transmission.   The interconnection of the middle ear cavities through the transverse passage described above is a peculiar feature in crocodilians. The effect on sound transmission between the two ears was measured by Wever and Vernon (1957) by recording from the round window of one ear and presenting sounds first through a tube sealed over this ear and then similarly presenting them to the other ear. The results are shown in Fig. 24-13. It is clear that sounds pass readily across the head and affect the two ears almost equally. The differences shown are small and vary in direction along the frequency scale; the average difference was 0.2 db. These variations probably represent complex resonances in the air passages. This interconnection of the two ears is of interest in relation to the matter of sound localization. It ap-

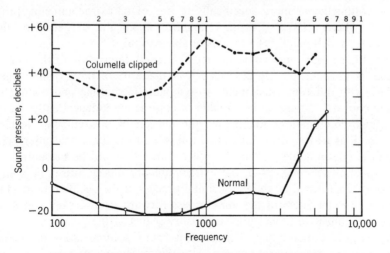

FIG. 24-12. Effects on sensitivity of cutting the columella in *Caiman crocodilus*. From Wever and Vernon. 1957.

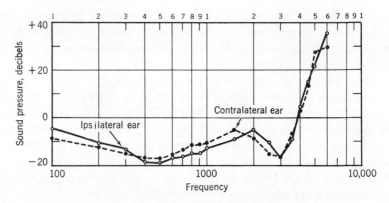

Fig. 24-13. Interaural transmission of sounds in *Caiman crocodilus*.
Data from Wever and Vernon, 1957.

pears that binaural intensity, which is one of the important cues for
sound direction in most ears, will be of doubtful service in crocodilians.
Other cues, such as time or phase difference between the sounds received in the two ears, may of course be utilized.

Extracolumellar Muscle.   The ear of crocodiles contains an intrinsic
muscle, usually designated as the stapedius, but which functionally appears to be a tensor tympani. Our knowledge about this muscle has
been in a state of confusion ever since its discovery, which is credited
to Cuvier in 1824. Early writers, including Peters (1868), Huxley
(1869), and Cuvier himself, considered the muscle to be attached to
the columella, or more specifically to one of its extracolumellar processes. Hasse (1873b) seemed doubtful about the existence of this
muscle, and Retzius (1884) was unable to find it.

Killian (1890) presented the most detailed description of the muscle
and its attachments. He described it as made up of three distinct bundles, with different attachments to tympanic membrane and columella,
and considered these to be capable of contracting independently and in
various combinations to regulate the tension of the tympanic mechanism. This treatment is difficult to follow and the functional interpretation is questionable. A much simpler conception will be presented.

As shown in Fig. 24-6, the extracolumellar muscle arises in a cup-
like recess formed by the paroccipital process of the occipital bone. The
short parallel fibers of the muscle pass forward to insert on a mass of
dense connective tissue (the suspensory plate) at the posterior edge of
the tympanic membrane. This plate is firmly secured along its medial
edge to the paroccipital process, with reinforcement by a projection of
the opisthotic, whereas the lateral edge is relatively free. A contraction

of the muscle therefore will produce a rotation of the suspensory plate about its medial edge, exerting tension on the tympanic membrane together with the extracolumellar process attached near by.

This tympanic muscle in its contractions can be expected to reduce the transmission of sounds through the ear, as has been demonstrated for middle ear muscles in other animals. The complex form of this muscle, with several groups of fibers running in somewhat different directions, which led Killian to postulate a variety of actions, is a suitable construction for a muscle whose function is to immobilize the middle ear mechanism rather than to produce any considerable displacement of its parts. Unfortunately no specific experiments have been carried out on this action in crocodilians.

Inner Ear

The location of the crocodilian labyrinth in the head is indicated in Fig. 24-14. As shown here and in further detail in Fig. 24-15 (taken from Retzius), this structure follows the usual reptilian pattern. The basilar papilla occupies the main portion of the cochlear duct, and the lagenar endorgan appears at the extreme ventral end. This endorgan closely resembles the ones found in the utricle and saccule, and no doubt serves an equilibrial function as these others do.

A distinctive feature, seen in all the crocodiles, is a sharp bend near the middle of the cochlear duct, which divides the auditory papilla into an upper portion that runs nearly straight alongside the round window, and a lower portion that extends ventrally below the "knee."

The form and course of the basilar membrane were studied in detail in a specimen of *Caiman crocodilus* in which the whole cochlea was carefully dissected free, and the basilar membrane was exposed throughout its course by the removal of Reissner's membrane. The form in relation to the head is represented in Fig. 24-16. The basilar membrane

Fig. 24-14. Location of the labyrinth in the head of *Caiman crocodilus*.

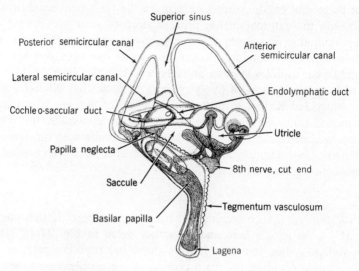

FIG. 24-15. The labyrinth of *Alligator mississippiensis.*
Redrawn from Retzius, 1884. Scale 5×.

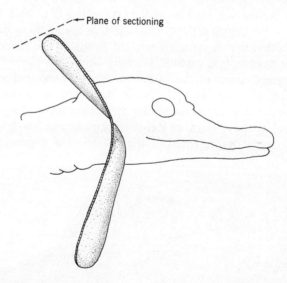

FIG. 24-16. Form of the basilar membrane in *Caiman crocodilus.*
The membrane is enlarged 15×, and shown on an outline of the head
of about natural size.

begins near the posteromedial edge of the oval window and runs forward beside this window along a line about 25° from the long axis of the head and at the same time extending laterally about 28°. Then it passes around the knee in a smooth curve and continues ventrally, finally showing a posterior inclination of about 15°. Near its upper end the membrane surface is about 20° from the horizontal, with the lateral edge raised, and then it turns rapidly around the bend until in the ventral region it is facing almost anteriorly.

In spite of its curiously twisted form, the crocodilian cochlea presents no serious problem for sectioning. The block of tissue was positioned with the nose end rotated downward 25°, so that the cutting for the posterodorsal region was nearly transverse to the basilar membrane. As the sectioning continued through the block, the angle relative to the basilar membrane became somewhat oblique beyond the knee region, increasing to about 28° from the transverse near the ventral end. This degree of obliquity does not distort the appearance of the structures unduly, and measurements of the width of the basilar membrane remain within acceptable limits even at the ventral end. It is estimated that with this manner of sectioning the apparent width varies from the true value by about 2% for the upper half of the membrane, and for the remainder the error rises to a maximum value of about 12%. This discrepancy is not much greater than may be expected from errors of measurement and individual differences, and no effort has been made to correct for it in the reported data.

The relations between the cochlear duct and neighboring structures are indicated in Fig. 24-17. The middle ear cavity is large and of complex form; here it is shown on the anterolateral, lateral, and posterior sides of the otic capsule, and it also extends far dorsally. Indeed, the ramifications of this air space represent an approach to the extensive pneumatization of the inner ear region of birds.

The tympanic space is connected with the pharynx by a Eustachian tube system of unusual form. This system was carefully studied by Owen as long ago as 1850, but even now the reasons for its complexity, if there are any, are hardly understood. Colbert (1946) presented a clear description of this structure.

Below the deep portion of the middle ear cavity on each side is a funnel-shaped space that Owen called the rhomboidal sinus, and between this sinus and a final opening into the pharynx, which is common to the two sides, there are three slender air passages. One of these is the lateral Eustachian tube, which runs directly from the sinus to join the one on the other side near the pharyngeal end. The two others leave the sinus by small openings, run medially, and each joins its corresponding tube from the opposite side to form anterior and posterior median

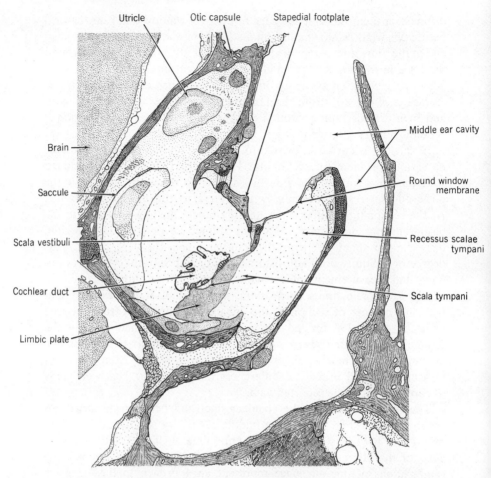

Fig. 24-17. A frontal section through the dorsal part of the otic capsule of a specimen of *Caiman crocodilus*, showing middle ear and labyrinthine structures. The middle ear cavity extends halfway around the otic capsule. Scale 12×.

tubes. These two then fuse to form a common median tube, which is joined by the two lateral tubes, and this final passage, now outside the skull and with only membranous walls, opens into the pharynx.

Owen hesitatingly tried to explain this complex Eustachian tube system of crocodiles as permitting hearing by an animal lying nearly submerged, with only eyes and external nares exposed, by the transmission of sounds through this tube system. This idea has not been taken seriously, and the purpose that Owen suggested, like the primary one of pressure equalization between the two sides of the tympanic membrane, would be better served by a single larger tube. This complex system has been traced back through all known fossil forms of the crocodilians as far as the Protosuchia (Langston, 1973), and its persistence to the present without essential change is indeed remarkable.

The columella enters a deep recess in the otic capsule on the way to its seat in the oval window, where its footplate is held by an annular ligament, parts of which are indicated in Fig. 24-17. In the section shown, which is taken from the laterodorsal end of the cochlea, Reissner's membrane exhibits a number of folds and convolutions. This form of the membrane, more familiar in the cochlea of birds, is often referred to as the tegmentum vasculosum because of the presence of many capillaries at the bottom of the folds.

Limbic tissue is abundant in the cochlear region, forming a dense plate with an opening over which the basilar membrane is suspended. As seen here this opening is only a shallow trough, but farther ventrally it deepens and breaks through as the scala tympani. This scala is particularly large, and its anterolateral extension may properly be designated as the recessus scalae tympani. It is bounded at its anterior end by the round window membrane.

Further details of cochlear structure are shown in Fig. 24-18. Here the scala vestibuli is above and the scala tympani below, with the cochlear duct between, walled off by Reissner's membrane and various tissues along the basilar membrane.

Reissner's membrane at the level shown lacks the folds seen more dorsally, but its medial portion is greatly thickened. This membrane in its main part consists of two cell layers, a relatively thick one facing the cochlear duct and an extremely thin one on the scala vestibuli side. Capillaries appear in varying numbers along its extent, always on the vestibular side.

The basilar membrane is a fibrous structure, anchored at its edges to the lips of the medial and lateral parts of the limbic plate. Beneath this membrane are two prominent thickenings of relatively loose tissue, indicated as the tympanic lamella and most likely corresponding in function to the fundus of lizards and some other reptiles. Sometimes over a narrow region there are three of these thickened areas. A large blood vessel is always found near the medial end of the tympanic lamella, which Retzius regarded as corresponding to the spiral vein of the mammalian cochlea.

Within the cochlear duct and supported on the basilar membrane are a number of distinct cellular structures. At the medial end, over the surface of the limbus, is a ribbon of tall cylindrical cells, the limbic epithelium. Next to these cells are similar cells, but much lower, that are known as the inner sulcus cells (Retzius), and these grade into

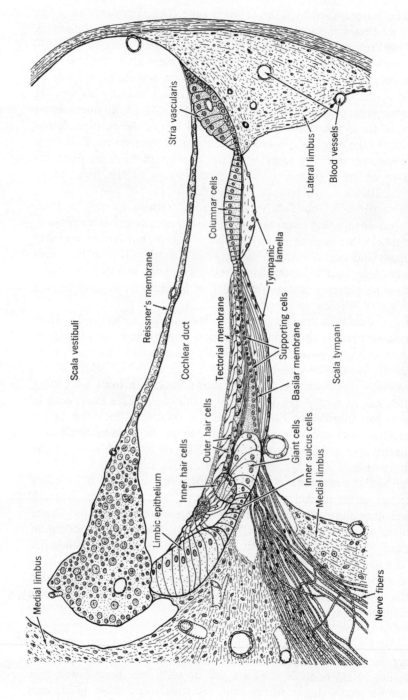

Fig. 24-18. A frontal section through the middle of the cochlear duct of *Caiman crocodilus*. Scale 175×. From Wever, 1971c.

other cells, also generally similar in form but distinguished by their great size; they are here designated as giant cells.

On the lateral edge of the basilar membrane is a mound of cells of columnar form, usually of graded height as shown. Outside of these, and resting on the vestibular face of the lateral limbus, is the stria vascularis. It consists of two layers, low cuboid or sometimes flatter cells adjacent to the limbus, and tall cylindrical cells above these. One or more small blood vessels may be seen, situated between these two layers. In one of the specimens studied a stain was used that differentiated between these two cell layers, coloring the tall ones blue and those of the other layer pink; and this same distinction was found for the two layers of Reissner's membrane but in the reverse order: the cells of the thin upper layer of Reissner's membrane were stained pink and those of the thick lower layer were stained blue. Thus it is suggested that the stria vascularis represents a continuation and elaboration of Reissner's membrane, forming a portion that is folded back on the limbus.

Of central interest are the remaining structures on the basilar membrane, which constitute the organ of Corti. There are two distinct groups of hair cells, as Retzius observed earlier. A compact group of inner hair cells appears at the medial end, lying close above the inner sulcus cells. These are tall cylindrical cells, inclined medially, and bearing on the free end a prominent ciliary tuft.

Over the main part of the basilar membrane, up to the region of the columnar cells, are several rows of outer hair cells. These cells have a distinctive form: they are low and tear-shaped; Retzius compared them with little boats that "swim" on the sensory surface. A relatively large nucleus is contained in the more swollen portion of the cell, and close to the pointed lateral end is a ciliary tuft that usually curls around in a medial direction.

In the specimens examined, the transition from inner to outer hair cells was abrupt; no intermediates between the two types were observed.

Strands of the tectorial membrane make attachments to the ciliary tufts of both inner and outer hair cells. This membrane arises from the limbic epithelium as a dense reticular structure, and continues across the hair cell rows as a laminated sheet. The number of laminae decreases progressively as the structure extends laterally, and fine strands continue to be given off to the hair cells along the way, until the structure ends as a single thin net.

The outer hair cells are supported by processes arising from a series of cells whose bodies rest on the basilar membrane. These cells send off slender columns that run obliquely to the hair cells, where they expand to form a broad surface, or pedestal, as shown in Fig. 24-19 and in more detail in Fig. 24-20.

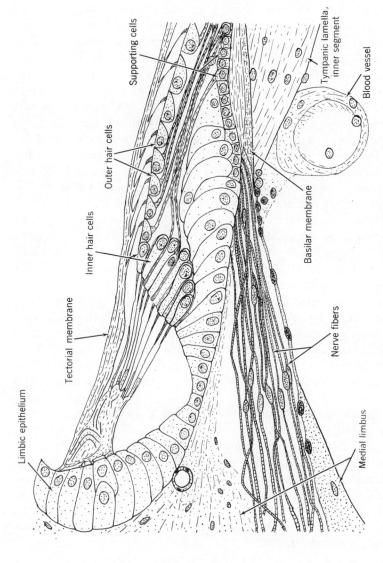

Supporting cells

Outer hair cells

Inner hair cells

Tectorial membrane

Limbic epithelium

Medial limbus

Nerve fibers

Basilar membrane

Tympanic lamella, inner segment

Blood vessel

FIG. 24-19. Enlarged view of the medial portion of the preceding section. Scale 500×

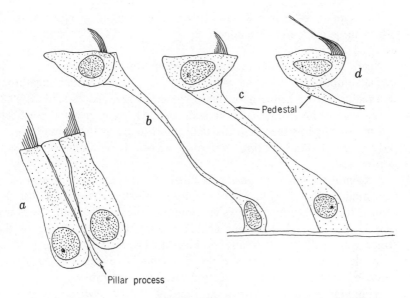

F<small>IG</small>. 24-20. Hair cells in *Alligator mississippiensis*. *a*, inner hair cells, *b*, *c*, *d*, outer hair cells. Supporting cells are shown for *b* and *c*, and the pedestal portion of such a cell for *d*. A tectorial strand is attached to the cilary tuft in *d*. Lateral is to the right.

The manner of support of the inner hair cells could not be determined with certainty by light microscope observations. Long, slender processes from the supporting cells reach these hair cells, and at times these processes could be seen extending between them (as in *a* of Fig. 24-20) and expanding around their neck portions in the manner familiar for the supporting processes of lizards and other reptiles. In some instances, however, it appeared that small pedestal types of supports were present also, at least for some of these cells. The point needs further study with the electron microscope.

Retzius, who saw the supporting cells and their extended processes going to the outer hair cells, was of the opinion that at their terminal ends they send off branch processes that pass between these hair cells to the exposed surface, but he was unable to obtain convincing evidence for this condition. Present observations with the light microscope reinforce Retzius's position by clearly indicating the presence of material separating the bodies of these cells. According to Baird (1974), who examined crocodile specimens with the electron microscope, the intervening material consists of microvilli arising from the supporting processes. Thus the early impressions of Retzius are substantiated.

The support of hair cells on a pedestal is unusual in reptiles; in all others these cells are suspended only at their necks by expanded processes of the supporting cells.

Nerve fibers from the cochlear ganglion run through spaces in the medial limbus and penetrate the medial edge of the basilar membrane in a region known as the habenula perforata, where they lose their medullary sheaths and continue as naked fibers. Often many of these fibers can be followed further with the light microscope as they pass between the sulcus cells and continue obliquely across the columns of the supporting cells.

A frontal section through the cochlea of *Alligator mississippiensis* is presented in Fig. 24-21. The structure is much the same as just described for *Caiman crocodilus*. The internal sulcus cells at this level are less developed, and there are no giant cells. The medial part of the tympanic lamella is especially prominent. The nerve fibers extending from the habenula perforata to the hair cells were well stained in this specimen, and a few of these are indicated in the figure.

A similar section is shown in Fig. 24-22 for *Crocodylus acutus*. Some of the internal sulcus cells are moderately large, though they fail to reach the size observed in *Caiman crocodilus*.

COCHLEAR DIMENSIONS

Caiman crocodilus. — In Fig. 24-23 the solid curve shows results of measurements of the varying width of the basilar membrane along the cochlea of a specimen of *Caiman crocodilus*. This width increases rapidly at the dorsal end to a maximum of about 500 μ, then falls somewhat and follows a slowly increasing course until a second maximum of about 600 μ is reached near the ventral end, whereupon the cochlea ends abruptly. The range of variation from the 400 μ point to the ventral end is only moderate. The total length indicated for this ear is nearly 4 mm.

The number of rows of outer hair cells is shown in Fig. 24-24 by the solid curve. This number rises rapidly to a maximum around 24 in the dorsal region, declines somewhat, and then undergoes a slow rise, with moderate irregularities, over most of the cochlea, and finally there is a rapid rise to a maximum of 34 near the ventral end. The fall thereafter is rapid. The inner hair cells do not appear at the extreme dorsal end, but beginning with the 280 μ point the number in the transverse rows rises rapidly to a maximum of 12. Thereafter there is a progressive decline to a broad minimum around 7 or 8, and then a moderate rise to a maximum of 17 rows at the ventral end. The inner hair cells extend a little farther ventrally than the outer hair cells do.

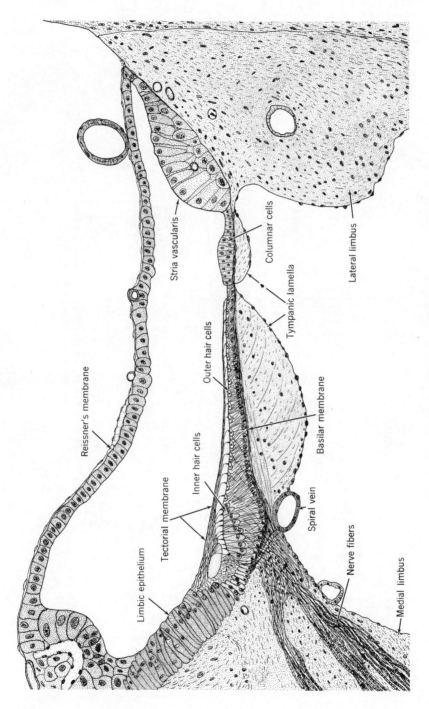

Fig. 24-21. A frontal section through the cochlear duct of *Alligator mississippiensis*. Scale 175×.

Stria vascularis

Columnar cells

Lateral limbus

Tympanic lamella

Reissner's membrane

Outer hair cells

Basilar membrane

Tectorial membrane

Inner hair cells

Limbic epithelium

Spiral vein

Nerve fibers

Medial limbus

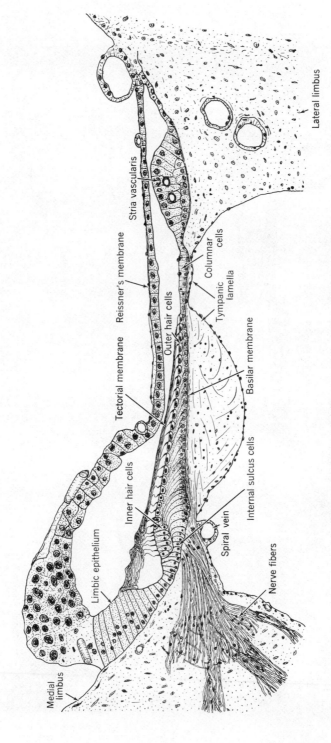

Fig. 24-22. A frontal section through the cochlear duct of *Crocodylus acutus*. Scale 175×.

Lateral limbus

Stria vascularis

Columnar cells

Tympanic lamella

Reissner's membrane

Basilar membrane

Tectorial membrane

Outer hair cells

Limbic epithelium

Inner hair cells

Internal sulcus cells

Spiral vein

Nerve fibers

Medial limbus

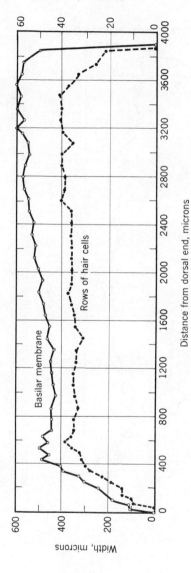

Fig. 24-23. The width of the basilar membrane (solid line) and the total number of rows of hair cells (broken line) along the cochlea of a specimen of *Caiman crocodilus*. Here, and in similar graphs to follow, the ordinate scale is more expanded than the abscissa.

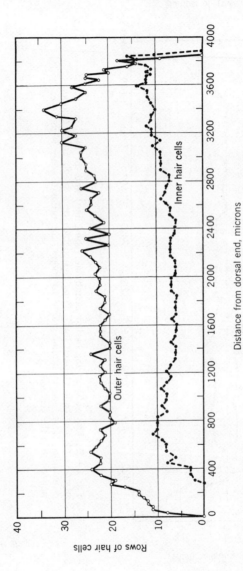

Fig. 24-24. The number of rows of outer hair cells (solid line) and inner hair cells (broken line) in the same specimen of *Caiman crocodilus* shown in the preceding figure.

The complete array of hair cells along the cochlea (outer plus inner rows) has been shown by the broken curve of Fig. 24-23, and it is evident that this distribution follows closely the form of the basilar membrane.

The density of the hair cells, represented as the number per 24 μ section along the cochlea, is shown in Fig. 24-25. The three curves indicate this function separately for inner and outer hair cells and for the total of the two types. For all three categories there are two prominent maximums, one dorsal and the other far toward the ventral end, with the outer hair cells reaching their peak sooner than the inner hair cells in each region. These density variations are much greater than the variations in numbers of transverse rows, and represent increased crowding within the rows.

Alligator mississippiensis. — The variations in width of the basilar membrane along the cochlea of *Alligator mississippiensis* are represented in Fig. 24-26. The form of this function is like that shown for *Caiman* except that the rise at the dorsal end is much less rapid, and the maximum attained in the ventral region is somewhat broader.

The rows of hair cells, plotted in Fig. 24-27, present a different picture. The bimodal character, shown clearly in *Caiman*, is here rather inconspicuous. The rows of outer hair cells increase progressively in the dorsal region to a maximum at the 1200 μ point, decline slightly and then increase slowly to a maximum in the midventral region, after which there is a rapid fall. Again the inner hair cells are absent at the dorsal end of the cochlea and only begin at the 460 μ point. From here the number increases to a maximum near the 1500 μ point, falls away slightly and levels off over the middle of the cochlea, and then increases slowly to a maximum in the far ventral region.

The variations in hair-cell density along the cochlea for this species are represented in Fig. 24-28. As in *Caiman*, these variations exhibit a bimodal form, though less pronounced. The outer hair cells reach a maximum density in the dorsal region, fall away somewhat and then rise to a lower maximum in the ventral region, after which there is a progressive decline. The inner hair cells reach a maximum in the dorsal region that is only a little farther along the cochlea than for the outer hair cells, fall to a minimum in the middle of the cochlea, then rise progressively to a rather high maximum far toward the ventral end. The function for total hair cells resembles the outer hair-cell curve in the dorsal region, and presents an irregularly flat maximum over the ventral half of the cochlea.

Crocodylus acutus. — The width of the basilar membrane in *Crocodylus acutus*, as represented in Fig. 24-29, shows a relatively uniform taper in comparison with the preceding species. The curve increases

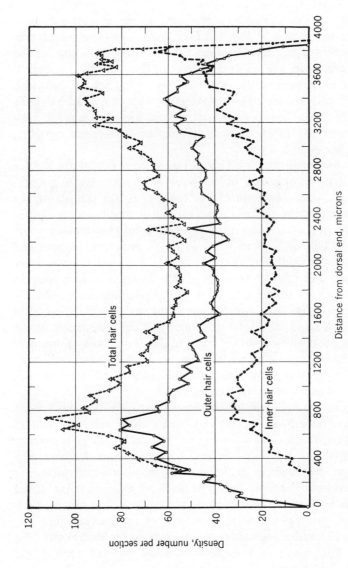

FIG. 24-25. The varying density of hair cells along the cochlea of *Caiman crocodilus*, shown for outer and inner hair cells and for their sum. The density is expressed as the number of cells in a section 24 μ thick.

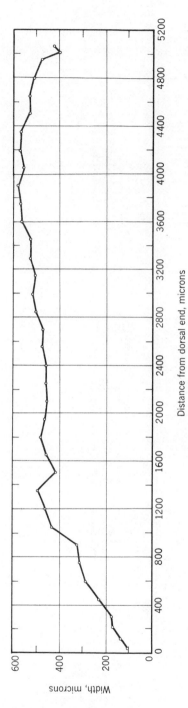

Fig. 24-26. The width of the basilar membrane in a specimen of *Alligator mississippiensis.*

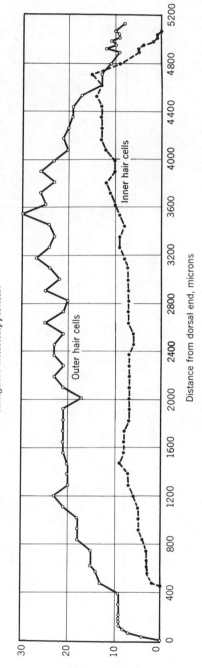

Fig. 24-27. The number of rows of hair cells in a specimen of *Alligator mississippiensis*, shown for outer and inner hair cells.

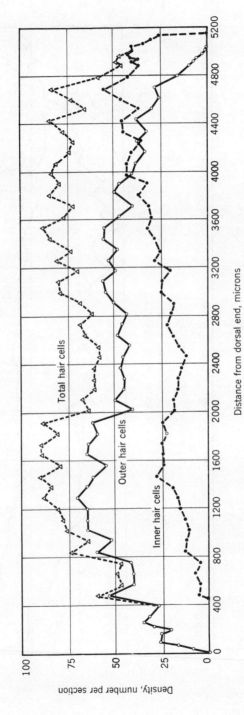

FIG. 24-28. Hair-cell density in *Alligator mississippiensis*, shown for outer, inner, and total hair cells. Density is represented as the number of cells per 30 μ section.

regularly from the dorsal end, except for a little variation in the ventral region, to a maxiumum almost at the extreme end.

The rows of hair cells, shown in Fig. 24-30, present a similar picture, though with greater irregularities. The rows of outer hair cells rise rapidly at the dorsal end, then a little more slowly to a maximum in the ventral region, after which the decline is abrupt. Unlike the other two species, the inner hair cells appear almost at the very dorsal end. These rows of cells increase slowly to a flat maximum, are maintained with only a slight decrease over the middle part of the cochlea, and in the ventral region rise slowly and then rapidly to a sharp maximum at the ventral end, after the outer hair cells have disappeared.

The density of these cells (Fig. 24-31) follows much the same pattern. For the outer hair cells this function increases to a maximum in the dorsal region, falls moderately and holds reasonably well to a point a little beyond the middle of the cochlea, then increases to a new level over much of the ventral region, after which there is a precipitous decline. The inner hair-cell density grows slowly and then almost imperceptibly to a level around the middle of the cochlea, after which there is a rapid increase and finally a very abrupt one near the end of the cochlea. The combined function closely follows that for the outer hair cells over most of the cochlea, and then reflects the rapid increase in inner hair cells toward the ventral end.

Cochlear Differentiation.   These dimensional observations on the three species of crocodilians indicate a considerable degree of cochlear differentiation. The forms of the basilar membrane for the three specimens are compared in Fig. 24-32. The basilar membrane is graduated in width along its course by an amount of five- or sixfold, mainly appearing at the dorsal end, which is the order of magnitude of the variation found in most mammalian ears. The arrangement of the rows of hair cells follows much the same pattern in these species, with some interesting variations among the inner and outer hair cells, including a marked resurgence of inner hair cells toward the ventral end. The distribution of hair cells over the cochlea is striking, with two distinct areas of high density in *Caiman*, and marked differences in the outer and inner hair-cell patterns in the other two species. All the dimensional variations along the cochlea are consistent with the view that the crocodilian ear possesses a substantial degree of frequency discrimination in terms of the spatial localization of response. The fact that the differentiation occurs over a relatively short cochlea may raise the question whether the spatial distribution of tonal action is as effective in producing a specificity of neural response in these animals as it is in the more elongated cochleas of mammals. The answer to this question depends in large part

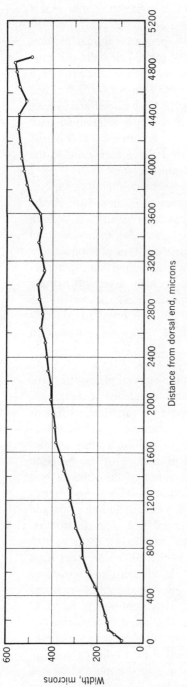

Fig. 24-29. The width of the basilar membrane in a specimen of *Crocodylus acutus.*

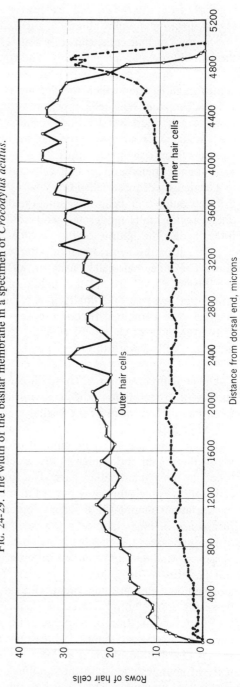

Fig. 24-30. The rows of outer and inner hair cells in a specimen of *Crocodylus acutus.*

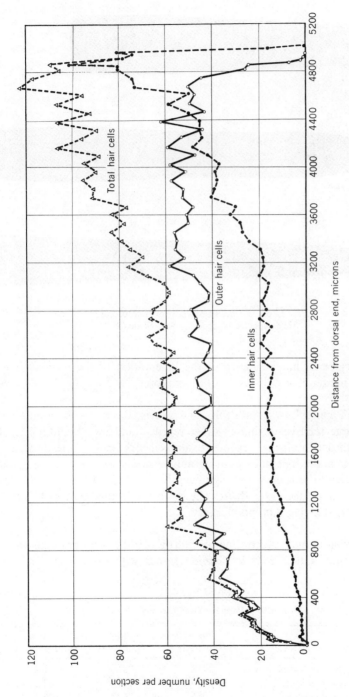

FIG. 24-31. Hair-cell density in *Crocodylus acutus*, expressed as the number of cells per 24 μ section.

Caiman crocodilus

Alligator mississippiensis

Crocodylus acutus

Fig. 24-32. Forms of the basilar membrane in
three species of crocodilians. Scale 20×.

on the degree of independence of the innervation of the cochlear elements, a matter that still needs to be explored.

Hair-cell Populations.   As Table 24-I shows, the hair-cell population is very large relative to that of other reptiles, and the magnitude is of the same order in all three species examined. For each of the three, the number of outer hair cells considerably exceeds the number of inner hair cells, and the ratio varies significantly: 2.27 for *Alligator*, 2.05 for *Caiman*, and 1.7 for *Crocodylus*. The totals are in this same order, with *Crocodylus* showing a preponderance.

Ciliary Orientation.   According to Mulroy (1968), who examined both inner and outer hair cells in *Caiman fuscus* with the electron micro-

TABLE 24-I
HAIR-CELL POPULATIONS IN CROCODILIA

Species	Inner Hair Cells	Outer Hair Cells	Total Hair Cells
Alligator mississippiensis	3312	7560	10,872
Caiman crocodilus	3800	7775	11,575
Crocodylus acutus	5066	8613	13,678

scope, the orientation of the ciliary tufts is unidirectional, with the kinocilium always located away from the neural limbus. These observations were confirmed by Baird (1970). The present observations on all three species are in agreement.

AUDITORY SENSITIVITY

Sensitivity measurements in terms of cochlear potentials were carried out on all the specimens available except a few of the caimans used for special anatomical study. The procedure followed that of Wever and Vernon (1957), with some refinements of recording equipment that permitted observations at lower response levels.

An active electrode in the form of a silver bead was placed on the round window membrane, which was exposed by drilling a hole in the parietal bone over the transverse passage between the two middle ear cavities. This hole was located close to the ear to be stimulated (usually the left) but with care to avoid injury to tissues attached to the tympanic membrane near by. A suitably placed opening provides a clear view of the round window for electrode placement. Figure 24-33 shows the location of the hole made for the left ear of a specimen of *Caiman crocodilus*. The round window is viewed with an operating microscope containing a light directed along the optical axis, and the direction is indicated by a short arrow in the drawing. In the specimen prepared for this illustration the tympanic membrane on the right side was broadly exposed by removing the superior earlid and a portion of the squamosal ridge along which it is attached.

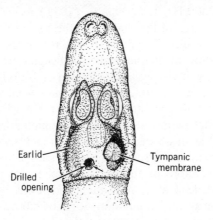

Earlid

Tympanic membrane

Drilled opening

FIG. 24-33. The approach to the left round window for electrode placement in crocodilians. A wide opening on the right side shows the tympanic membrane. From Wever, 1971b.

In crocodilians the electrical potentials of the cochlea as recorded at the round window vary nearly linearly at lower levels as a function of sound pressure, as in other animals. Some sample curves are presented in Fig. 24-34, taken from the early study by Wever and Vernon.

Caiman crocodilus. A sensitivity function for a specimen of *Caiman crocodilus* is shown in Fig. 24-35. Best sensitivity appears in the region of 300 to 2000 Hz, and there is a rapid decline for lower and especially for higher tones. The testing was not extended beyond 5000 Hz because of a danger of injury to the ear.

In two other animals, as shown in Fig. 24-36, the sensitivity was greater over the region of 200 to 2000 Hz, and was maintained sufficiently well to permit an exploration of the upper frequencies. At this high end of the range, the sensitivity is decreasing at a rate of 40 db per octave.

Auditory Nerve Responses. — A few studies have dealt with neural responses in *Caiman crocodilus* in response to sounds. Weisbach and Schwartzkopff (1967) recorded from neurons in the striate area of the forebrain, and obtained responses to tones from 75 to 5000 Hz, with the greatest sensitivity between 1500 and 2000 Hz. G. A. Manley (1970b) worked with this same species and recorded from three areas of the medulla: nucleus angularis, nucleus magnocellularis lateralis, and nucleus magnocellularis medialis. Single neural units in these areas responded to tones over a range of 70 to 2900 Hz, with greatest sensitivity in the band 800-1200 Hz. The units showed marked variations of sensitivity with frequency, with best points at some characteristic frequency for each. Only rarely were units found with characteristic frequencies above 2000 Hz. J. A. Manley (1971) recorded in this species from the auditory area of the midbrain (the torus semicircularis), with generally similar results. The range of the responses was 70-1850 Hz, with most units showing characteristic frequencies around 100-500 Hz, and the greatest sensitivity was in the region of 1000 Hz.

In both the last two studies, with recordings from medulla and midbrain, the distribution of auditory units over the sensitive area presented a three-dimensional pattern depending on stimulus frequency in much the same manner as has been shown in birds and mammals. There is a general type of tonotopic arrangement suggesting a specificity of neural pathways, and therefore of the primary activity of the cochlea. This specificity is evidenced at or near threshold levels of stimulation and rapidly decreases as the stimulation is raised above minimum levels. Thus it appears that in these reptiles, as in higher vertebrates, the cochlea presents a degree of tuning, probably of a mechanical sort, that varies regionally.

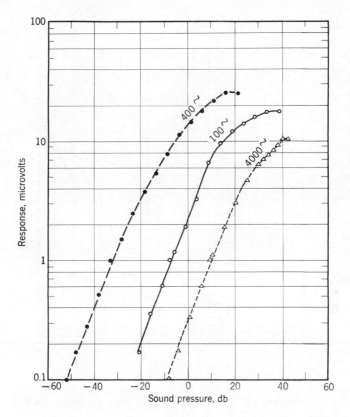

FIG. 24-34. Intensity functions for a specimen of *Caiman crocodilus*, measured for three frequencies. From Wever and Vernon, 1957.

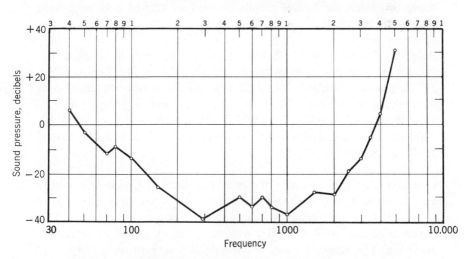

FIG. 24-35. A sensitivity function for a specimen of *Caiman crocodilus*. From Wever, 1971b.

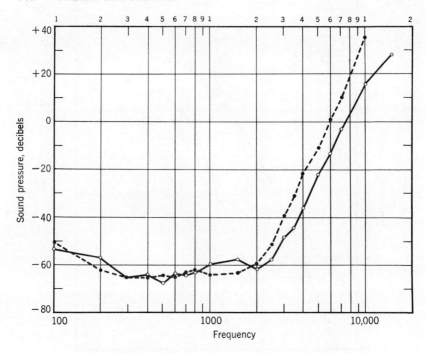

FIG. 24-36. Sensitivity functions for two additional specimens of *Caiman crocodilus* in which the measurements were extended into the high frequencies. From Wever, 1971b.

This evidence is consistent with the dimensional data presented above, in which the basilar membrane and its related structures have been found to vary in a systematic manner.

Alligator mississippiensis. Sensitivity curves in terms of cochlear potentials for two specimens of the American alligator are shown in Fig. 24-37. The two functions are similar in indicating best sensitivity over the range of 100 to 1000 Hz, with one of these ears attaining a level of −66 db. A good level of sensitivity is maintained over a range of 50-3000 Hz. The tones above this range suffer a rapid loss, at a rate around 50 db per octave.

Crocodylus acutus. Results for two specimens of *Crocodylus acutus* are presented in Fig. 24-38. In this species the region of best sensitivity is shifted along the frequency scale in comparison with the others. This region is around 300-3000 Hz, with the peak in one of these ears at 3000 Hz. The sensitivity can be considered excellent, measuring −40 db or better for the range 300-3000 in one ear and 100-6000 in the

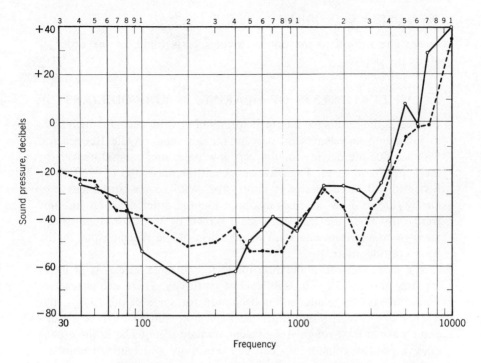

FIG. 24-37. Sensitivity functions for two specimens of *Alligator mississippiensis.* From Wever, 1971b.

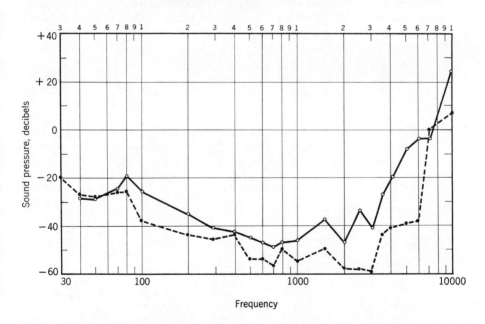

FIG. 24-38. Sensitivity functions for two specimens of *Crocodylus acutus.* From Wever, 1971b.

other. The roll-off in the upper frequencies is rapid, of the order of 40 db per octave.

AN EVALUATION OF HEARING IN CROCODILIANS

In general the results for the three species of crocodilians are much alike in indicating excellent sensitivity in the low and middle frequencies, with a moderate decline for the very low tones and a rapid cut-off for the upper frequencies. The performance at the maximum of −60 db or better for all three species is impressive. With so few specimens it is hardly safe to draw conclusions about familial differences, but the upward shift along the frequency scale shown by the *Crocodylus* specimens in comparison with the two members of the Alligatoridae merits some consideration.

The large hair-cell population found for all three species is of interest also, in relation to the high level of sensitivity. These two characters do not always correlate, as has been seen for some of the lizards, but when sound receptive and other mechanical conditions of the ear's operation are at least roughly equivalent, as they seem to be in the crocodiles, a positive relationship between sensitivity and hair-cell numbers is to be expected.

In general, the crocodilian ear must be rated as an excellent instrument for the perception of sounds in comparison with other reptilian ears, and ranks well also with the ears of many birds and mammals.

PART IV. CONCLUDING OBSERVATIONS

25. THE REPTILIAN EAR: ITS BIOLOGICAL AND EVOLUTIONARY SIGNIFICANCE

The part played by hearing in the life of an animal is often difficult to determine, and this is especially true of lower forms such as the reptiles. In most of the birds and mammals even casual observations reveal the role of the auditory sense in such activities as mating, the care of the young, and the maintenance of social relations, but corresponding indications in the reptiles are uncommon. It seems likely that this lesser role for the auditory sense as seen in reptiles is real, and is not simply a reflection of our lack of insight into reptilian life processes.

Some instances are known, however, in which hearing plays a conspicuous role in reptilian activities. Many of the geckos are strongly vocal, and the males produce sounds to announce their presence in a territory and to repel other males. Some species also produce cries as a part of their defensive behavior, and it is a reasonable assumption that they hear their own vocalizations.

The amphisbaenians, because of their confirmed burrowing habits, appear to rely on hearing to a greater extent than other reptiles. Some species respond to the sounds of insects moving at the surface of the ground or breaking through the walls of their tunnels. They appear to localize such sounds, and sometimes come to the surface to prey upon the invaders (Gans, 1969).

Some species of turtles produce sounds, and there have been occasional reports of chorusing behavior, though the functions of these vocalizations are obscure. For the crocodiles the significance of sounds is more obvious. The males often produce roars and bellows in the mating season, and a smaller animal responds to the signals of a dominant one by submission and flight. Both members of a mated pair of *Crocodilus niloticus* react to the chirping of the young immediately before hatching (when the incubated eggs must be dug out of the nest) and for a few days afterward when parental care is essential for survival (Pooley and Gans, 1976).

Such indications of the specific use of sounds by reptiles are convincing, but are found in comparatively few species. The great majority of these animals appear to be mute, and rarely respond to even loud sounds in any obvious way. It seems likely that the auditory organ of reptiles serves mainly as an alerting sense, informing the animal of some occurrence in its immediate vicinity. A sound stimulus often is responded to initially by quiescence, which no doubt generally has a protective value.

This protective role for the ear is significant even though the tonal range in many species is severely limited, often (as in the snakes and many turtles) being restricted to a narrow band in the low frequencies. Most natural events are signalled by noises whose acoustic energies are distributed widely over the frequency spectrum, so that a band-limited receptor, if reasonably sensitive within its range, can still be effective in the detection of danger.

THE REPTILIAN EAR IN EVOLUTION

With the evidence now available, the development of the vertebrate ear turns out to be a far more complex series of events than has usually been considered. Also the reptilian ear takes a role of special significance in relation to the higher type of auditory receptor shared by man and the mammals (Wever, 1974a). The course of thinking on this problem is best followed historically.

During the latter half of the nineteenth century, when the doctrine of evolution of species was receiving its basic formulation, it was assumed that organs and functional structures of all kinds must have developed along a generally continuous course. The ear was considered as conforming to this principle of continuity, and attempts were made to trace a single line of development from the early invertebrates through the vertebrate series to its complex forms in mammals and man. This tracing turned out to be difficult, and many large gaps were encountered, yet there seems to have been little doubt about the general principle then and thereafter down to the present day.

Two forms of theory of the ear's origin emerged, one known as the otocyst hypothesis and another identified as the lateral line hypothesis. For the most part these explanations existed as alternatives, but some authors attempted to combine them into a common theory.

The Otocyst Hypothesis. In the bodies of many invertebrates, especially among the coelenterates and ctenophores, are little sacs in which a mass of dense particles, usually a form of calcium carbonate, is held in close contact with the long ciliary tufts of a patch of sensory cells. At first these organs were called otocysts and an auditory function was assigned to them. This was done because of their resemblance to the macular organs of vertebrates, which then, along with other receptors within the otic capsule, were considered to be hearing organs. The ears of vertebrates were regarded as representing stages along the course of development from these early otocyst organs. When later it was found that the macular organs serve an equilibratory function, providing information concerning head position and motion, the invertebrate structures were similarly reinterpreted and were renamed as statocyst organs. It was still possible, however, to maintain that the statocyst of the invertebrates was the forerunner of the vertebrate labyrinth, and that the ear arose later in the developmental process out of the non-auditory labyrinth of early vertebrates.

The Lateral Line Hypothesis. Near the end of the nineteenth century a second hypothesis arose in which the vertebrate ear was considered to have its origin in the lateral line apparatus of fishes. The lateral line, a system of canals running deep in the skin along the sides of the body in a continuous line from head to tail and with extensive branching in the head region, contains along the floor of its canals a series of patches of sensory cells surmounted by a narrow cupula, and communicating with the surface through fine pores. This structure attracted much attention, and numerous characters were attributed to it. At different times it was regarded as involved in the senses of touch, taste, and smell, and even as participating in the development of the eye, and among the mechanical stimuli that were thought to affect it were water movements and sound vibrations. Ayers (1892) was mainly responsible for promulgating the theory that this structure is the original form of the vertebrate ear. As he conceived it, the labyrinth and cochlea are a particular part of the lateral line that has sunk below the surface of the head.

The most recent advocacy of this hypothesis is that of van Bergeijk (1966, 1967), who presented three lines of evidence that he considered to support this view of the ear's origin. These lines of evidence included (1) the embryological development of the otic capsule and ear out of the lateral sensory placode, (2) similarities among the hair cells of lateral line organs and ears, and (3) correspondences in the innervation patterns of these sense organs.

The Embryological Argument. The embryological observations of Wilson (1891) and Wilson and Mattocks (1897) in salmon and sea bass are cited in this relation. These authors described the appearance early in embryonic growth of a line of ectodermal cells along the neural keel on each side of the head and body, which constitutes the lateral placode. As this placode develops it divides into anterior and posterior parts, and then the anterior part divides again, and out of these segments three groups of sensory structures are formed. The anterior portion gives rise to a number of sense organs mainly associated with the gills. The middle portion forms a hollow spherical body that sinks below the surface and forms the vestibular labyrinth. The posterior portion ramifies in the skin layer as the lateral line apparatus.

Observations of this kind in the teleost fishes have been repeated many times, and it is well established that labyrinth and lateral line arise from masses of cells that lie close together along the ectodermal placode. This evidence does not indicate, however, that the labyrinthine structures develop from the same cells that produce the lateral line. Indeed, the lateral line portion of the placode separates early, before the masses that produce the head organs and the auditory sac have split apart. It is erroneous to state, as has often been done, that the labyrinth has its origin in the lateral line. The cells of origin of these two organs are distinct from the beginning, as soon as they become identifiable as such; what these two cell masses have in common is their ectodermal character and adjacent location along the lateral placode. These relations do not signify genetic identity or the derivation of one mass from another.

Histological Similarities.   The sensory endorgans distributed along the lateral line system contain patches of ciliated cells, and such cells are found also in the sensory epithelium of labyrinth and ear. This common feature has been pointed to as evidence for an origin of labyrinthine and aural sense organs in the lateral line, but such a conclusion overstates the case. Ciliated cells are of wide distribution. They occur in mucous membranes, on the gills, and in many other epithelial organs. Their presence in lateral line and ear only represents the basic epithelial character of these tissues.

Much more significant than these cell similarities is the presence in the endorgans of the lateral line as well as in labyrinthine and cochlear organs of cupulae and cupula-like bodies, which are in intimate contact with the ciliary tufts of the hair cells and are instrumental in the stimulation of these cells by fluid motion in their vicinity. These bodies consist of a reticulum of fine fibrous material that stains in a distinctive manner and can be identified as tectorial tissue. It is the presence of this tissue, and its close connection with the cilia of the hair cells, that makes all these sensory structures responsive to mechanical stimuli.

The Innervation Argument.   In further support of the idea that the ear had its origin in the lateral line apparatus it has been said that these organs have a common innervation: that the nerve fibers supplying their sensory epithelium have their origin in the same nuclei of the brain. Van Bergeijk cited the work of Herrick (1897, 1899, 1901) as a basis for this assertion. Herrick examined the innervation patterns of a number of fishes, and traced to the tuberculum acusticum and cerebellum the dendrites of eight different kinds of peripheral sense organs including those in maculae, cristae, and lateral line organs, and also others of uncertain function such as the shallow and deep pit organs of siluroid fishes and the Savi vesicles of the torpedo. However, Herrick was not able to determine the central courses of these nerve fibers. He followed them into the brain and often to the same group of ganglia, but there they became so intermingled as to be no longer identifiable. This limit to observation does not signify that these fibers lose their identity or become interchangeable in function. A degree of intermingling of neural elements is common in the nervous system, yet these elements maintain their specificity of function. An inability to separate them is no indication of an identity of function or of a derivation of one group from another.

THE ORIGIN OF THE VERTEBRATE EAR

In this consideration of origins it is necessary to be quite clear about what we shall call an ear and hearing. Numerous attempts have been made to define this organ and its function, but with only limited success. No one has any doubt about what the ear is in man and his close relatives among the mammals; the difficulty arises when consideration is given to lower forms, and especially the invertebrates, in which the organs responding to vibratory stimuli take a variety of unfamiliar forms. Also in a number of animals there are several different types of endorgans, all responsive to mechanical vibrations, such as the receptive surfaces of the skin and deep tissues, the many elements of the lateral line system, and several types of pit organs in the head region of primitive fishes. The question is whether any or all of these shall be regarded as auditory organs, and their responses as belonging to the sense of hearing.

A formal definition of hearing was presented earlier (Wever, 1974a) as follows: "Hearing is the response of an animal to sound vibrations by means of a special organ for which such vibrations are the most effective stimulus." This special organ is the ear.

The critical distinction made in this definition, which separates the ear from all other means of receiving vibratory stimuli, is that this organ is particularly adapted to its purpose, and is more responsive to sounds than to any other form of energy. This condition excludes the skin, which though responsive to mechanical vibrations at relatively slow rates is far more sensitive to steady pressures and pressure gradients. Embraced by this definition are a great variety of organs in insects and spiders that respond specifically to sounds, many extending into the ultrasonic frequencies. Whether the lateral line apparatus is included remains in some doubt because we still have too little understanding of this organ; probably its function is to be regarded as hearing of a special type.

In all statements of otocyst and lateral line theories of the ear's origin it is implied, and sometimes explicitly stated, that the original organ first gave rise to the labyrinth and that the ear was elaborated out of this structure. This second assertion, the derivation of the ear from the nonauditory labyrinth, when separated from the first, can be firmly supported. All ears in fact, including some ten or more different kinds that may be found among the invertebrates, are developed from simpler types of mechanoreceptors. Probably in most instances the ears of insects were derived from kinesthetic organs: the sensillae that constitute the receptive elements in these ears appear to be modified chordotonal organs, which in simpler form are widely distributed over the bodies of insects, connecting across joints and body segments and serving for the sensing of relative movements between parts of the body. An association of a number of chordotonal organs with a flexible membrane at the surface of the body constitutes a serviceable ear. Curiously, there is never any indication that invertebrate ears have been derived from statocyst organs.

Among the vertebrates the ear also is derived from a motion sensing structure, though one of a more complex kind. This derivation likewise was not a single event, with a continuous course of development along the vertebrate series, but occurred separately in three different vertebrate groups, and in these it took place on at least five separate occasions. The evidence that this is the case comes from a study of the types of existing ears and their modes of stimulation by sounds (Wever, 1974a, 1976).

The Ears of Teleost Fishes

Four classes of fishes are recognized: Agnatha (including the extinct ostracoderms and living cyclostomes—the hagfishes and lampreys); Placodermi (known only as fossils); Chondrichthyes (sharks and rays); and Osteichthyes (bony fishes). Though all have labyrinths, ears are present only in the last named, and are best developed in the most advanced of these, the teleosts.

The labyrinth contains a number of sensory endorgans, which structurally are of several kinds and functionally may be divided into two groups accordingly as they serve the equilibrial sense and the sense of hearing.

Two structural types of sense organ are readily recognized, and appear in a remarkably uniform pattern. These are the cristae and maculae, both types serving most generally for the perception of head position and movement. The cristae, first found in simple form in the hagfishes and lampreys, are three in number from the advanced fishes on, with one in each of the semicircular canals located in an enlargement (ampulla) adjacent to the utricle. These organs consist of a mound of sensory cells whose ciliated ends are surmounted by a cupula, a tectorial body with little canals on its underside into which the kinocilia of the hair cells penetrate. The cupula is located in the ampullary space so as to form a flexible barrier across it, and is displaced when acted upon by fluid pressures produced when the head is rotated in the plane of the canal.

The macular organs are three in number also in all forms from the advanced fishes to the primitive mammals, and are located in three regions of the vestibule: in the utricle, saccule, and lagena. These organs contain compact masses of sensory and supporting cells, with the ciliated ends of the sensory cells surmounted by a plate or network of tectorial tissue, and over this tissue, and closely bound to it, is a mass of calcareous particles. This calcareous mass is known as a statolith (more appropriately when, as in the bony fishes, the particles are fused into a single body). Because the density of the calcareous mass exceeds that of the surrounding tissues, there are gravitational and inertial effects when the head takes different positions in space or is subjected to sudden displacements. The shearing forces exerted by the loading body on the ciliary ends of the hair cells produce a relative motion so that the kinocilium of the hair cell is bent relative to the cell body, and the cell is stimulated.

In the advanced mammals this organ is present in the utricle and saccule, but in these a lagena is absent.

An additional sense organ was discovered by Retzius in fishes, and though he at first confused it with the papilla basilaris he finally called it the macula neglecta (because it had escaped notice until then). It is of widespread occurrence among the lower vertebrates and appears more sporadically in the higher vertebrates, including several mammalian species.

This organ is best identified by its location in the labyrinth. It is most often found in the floor of the posterior sinus of the utricle, but frequently has other locations in this general region, including the wall of the utriculosaccular canal.

Though often duplex in the bony fishes, this organ elsewhere is single and has a relatively simple form. It consists of a patch of hair cells sustained in the usual way by supporting cells lying on a bed of limbic tissue and sending up slender columns that expand at their upper ends to embrace the necks of the hair cells. The ciliated ends of the hair cells are overlaid by a mass of tectorial tissue that embeds and firmly holds the tips of the kinocilia. Usually this tectorial layer is a simple reticulum of moderate density, but in the Amphisbaenia and a few lizards there is an overlay of calcareous particles. As the presence of these statolithic coverings characteristic of macular organs is exceptional, the organ is perhaps better called a papilla.

There have been indications, from observations on elasmobranchs, that the papilla neglecta is responsive to vibratory stimuli (Lowenstein and Roberts, 1951; Tester, Kendall, and Milisen, 1972; Fay, Kendall, Popper, and Tester, 1974), but whether it performs as an auditory receptor in any true sense remains uncertain. In most species, and perhaps in all, it does not lie in a path of vibratory fluid flow and can be expected to be insensitive to aerial and aquatic sounds.

Other papillae, as will be noted presently, are found in the labyrinths of amphibians and higher forms that unquestionably are auditory in function.

The ear of fishes is derived from the macular organs. The derivation is a simple process, as it does not require any special alteration of the basic structure, but consists merely of a switching in the nervous system whereby the sensory output of the macular organ is interpreted as acoustic information rather than as cues to equilibrium.

This switching appears to take two forms among different groups of fishes, involving different macular organs. In the majority of fishes it concerns the saccular macula and perhaps the lagenar macula as well (von Frisch, 1938); but in the clupeids, according to the best evidence, it concerns the utricular macula (Wohlfahrt, 1936).

The evidence for this transformation of a macular organ into an ear, and thus the existence of two types of maculae that are functionally distinct, is particularly clear in a group known as the Ostariophysi and including many species of Cyprinidae (carp) and Siluridae (catfish), extensively studied by von Frisch and his associates. In these animals a removal of saccule and lagena on both sides has little or no effect on equilibrium but causes a loss of responses to sounds. On the other hand, the bilateral removal of the part of the utricle that contains the statolith (and removal also of two and sometimes all three of the ampullae) produces serious and persistent disturbances of equilibrium, yet hearing in normal form is readily demonstrated by conditioning tests. No corresponding experiments have been carried out on the clupeids, but in these forms an air tube from the swim bladder runs to the wall of the utricle, and may be expected to enhance the sensitivity of its macular organ to sounds.

Though, as already mentioned, no significant changes in the maculae themselves have been found in these fishes, there are often developments of secondary structures, such as the air tube of clupeids just referred to, that produce an improvement in the acoustic sensitivity of these transformed macular organs. In the Ostariophysi a chain of small bones derived from the ribs, known as Weberian ossicles, extends on each side of the body from the anterior end of the swim bladder to a fluid recess adjacent to the saccule and lagena. That this connection improves the response to sounds is well substantiated: there is a significant loss of sensitivity when this Weberian chain is interrupted. These developments were probably secondary to the reorientation of neural connections that transformed the macular organ into an auditory receptor.

This transformation becomes understandable when we consider that the original macular organ, though best suited to stimulation by postural changes and sudden head movements, is also responsive to mechanical vibrations, at least when these are of large magnitude. Ordinarily this vibratory form of stimulation would be secondary, and even disruptive (as is known to be the case for the human ear when excessive sounds affect the vestibular receptors), and normally would tend to be suppressed. The organ becomes an auditory receptor when this suppression is removed, the vibratory effects are transmitted along sensory channels so as to be integrated into behavior patterns, and the earlier postural responses are inhibited. This adaptation of old mechanisms to new uses is one of the most common ways in which evolution proceeds.

The sound stimulation process in these ears appears to follow the form characteristic of macular organs, which is an inertial reaction arising from the density difference between the general body tissues, including the base structures on which the hair cells are borne, and the otolithic masses to which the cilia of these cells are connected through tectorial strands. When sound is transmitted through the water to the body of the fish (whose tissues are largely transparent to such vibrations), the hair cells are caused to vibrate along with their base structures, but the otolithic mass, with a density nearly twice that of the remaining tissues, takes up the vibration to a lesser extent and with a time lag. Accordingly there is a relative motion between cell body and ciliary tuft that is stimulating.

The presence of accessory structures in many of these fish species evidently augments this acoustic action. An air cavity adjacent to a macula increases the differential between the otolith and its surroundings and adds to its mobility. A chain of ossicles from the ear region to the swim bladder as found in ostariophysans is a particularly effective inertial device serving this same purpose.

THE AMPHIBIAN EAR

The amphibians now living belong to three orders, the Anura (frogs and toads), Caudata (salamanders), and Gymnophiona (apodans or caecilians). The ears of these forms have been studied only to a limited extent, and chiefly in the anurans (Wever, 1973c). In the anurans and many of the salamanders there are two types of inner ear structures, the amphibian and basilar papillae, whereas in the apodans only the amphibian papilla is present (Wever, 1975; Wever and Gans, 1976).

These papillae in amphibians consist of an epithelial structure contained in a recess formed by the limbus. This structure is made up of hair cells and supporting cells, in a general form now familiar, in which the ciliated ends of the hair cells are surmounted by tectorial tissues that extend into a fluid channel that is a path of vibratory fluid flow through the cochlea. Stimulation of the hair cells occurs when the tectorial tissues lying in the fluid path take up the vibratory movements and transmit them to the cilia.

This transfer of vibratory motion by way of tectorial tissues to ciliary tufts is accomplished in two ways. In most papillae the tectorial tissue is a spongy mass containing numerous interconnected canals with thin membranous walls. On the side facing the sensory epithelium, this mass presents an array of little caverns into which the cilia of the hair cells enter and make connections. This tectorial mass is more tenuous in its outer portions and evidently is acoustically well matched to the fluid, so that it readily takes up the vibratory motions (Wever, 1973c).

The second mode of vibratory transfer involves a sensing membrane. In the amphibian papilla of anurans a thin sheet of tectorial material lies partway across the path of vibratory fluid flow, extending from a rigid attachment along one edge to the main mass of hair cells lying in the sensory recess. This membrane is set in motion by the fluid vibration and communicates the motion to the hair cells to which its medial edge attaches. This arrangement appears to add to the sensitivity of a certain favored group of hair cells.

In the basilar papilla of anurans this design has been carried farther, and the whole structure consists of a semilunar membrane across the lumen of the fluid duct, halfway obstructing it, and leading to an array of hair cells resting on the limbus at the inner boundary of this membrane.

Of the two acoustic organs in amphibians the one known as the amphibian papilla appears to be the most fundamental, occurring in all species. The basilar papilla appears in functional form only in the anurans and some of the salamanders; it is absent in the caecilians.

The origin of these receptors in amphibians is uncertain. There is a degree of resemblance to a crista organ, with a large caniculated mass of tectorial tissue lying over the hair cells, and an absence of otolithic particles. Thus it seems possible that these papillae are transformed cristae or crista-like organs. If so, these structures must have been derived from supernumerary cristae, for the amphibians all have the usual three crista organs in their standard places in the ampullae of the semicircular canals. An alternative explanation is that these papillar organs are new developments out of the basic labyrinthine epithelium.

Along with this emergence of inner ear structures, the anuran amphibians also acquired serviceable middle ears, so that in most species the hearing is well developed.

THE REPTILIAN EAR

The auditory papilla of reptiles appears to be a novel structure, with no close relation to the inner ears of fishes and amphibians. In this ear the acoustic pathway through the cochlea traverses an opening in the limbus, with a basilar membrane suspended over the opening so as to be maximally exposed to the fluid motions.

The hair cells are borne on the basilar membrane and are carried along in its movements. A restraint of the ciliary tufts of these cells is then necessary to produce the relative motion required for stimulation. Mechanically this arrangement is just the contrary of the one seen in amphibians: in amphibians the ciliary tuft is deflected while the cell body remains at rest, whereas in the reptiles the cell body is made to move and the ciliary tuft is restrained.

As has already been seen, this ciliary restraint in reptilian ears is achieved in a variety of ways, but usually is produced directly or indirectly by a tectorial membrane whose inner edge is attached to the limbus.

This type of ear has outstanding advantages, many well realized among the reptiles themselves and others shown more obviously in birds and mammals, whose ears clearly have followed the reptilian design.

The advantages include features of sensitivity, frequency range, and discriminative capability, which appear in a degree of development that for mechanical reasons is unattainable in the ears of fishes and amphibians. A larger amount of acoustic energy is transferred to a basilar membrane lying athwart the path of vibratory fluid flow than to masses or strands of tectorial tissue suspended in the fluid. A basilar membrane can be extended into a ribbon with varying mechanical characteristics along its course, so that differential tuning can be introduced for its successive portions. In this manner the acoustic range can be broadened, and at the same time tonal discrimination becomes a possibility when specific nerve connections are made to the different sensory elements. These developments appear to have begun in some of the reptiles, and to have been carried forward in the birds and mammals.

How the reptilian ear arose out of the non-auditory labyrinth is not altogether clear, but most likely it represents a mutation involving the general labyrinthine epithelium. In the course of the present study two instances have come to light in which supernumerary papillae are present. One of these (Wever, 1974a) was found in the left cochlea of a specimen of the cordylid lizard *Platysaurus minor*. As shown in Fig. 25-1, this papilla lies on the medial limbus alongside the functional papilla. It consists of hair cells and supporting cells, both of typical form, contained in a capsule. As seen in further detail in Fig. 25-2, the ciliary tufts are surmounted by a sallet of somewhat irregular form. Only in two respects does this structure depart from normal: it rests on the solid plate of the limbus, and it is not innervated.

Another example of a supernumerary papilla was discovered in the right ear of a specimen of the snake *Corallus enydrus*. A cross section of the cochlear region is shown in Fig. 25-3. The normal papilla bears a tectorial plate, which more dorsally becomes greatly thickened and connects through a tectorial membrane to a broad root process on the limbic sulcus.

The supernumerary papilla is smaller than the other, and rests on the posterior portion of the limbus. It also bears a tectorial plate as shown in further detail in Fig. 25-4, but this plate is not as large or as well formed as the normal one. Thin strands of tectorial material connect it to the epithelial surface posteriorly. The hair cells and supporting cells appear of normal form. The hair cells have tufts containing numerous cilia, graduated in length in the usual way. The supporting cells form well-defined columnar processes that extend between the hair cells, expanding at their upper ends.

It will be noted that this supernumerary papilla lies on the part of the limbus away from the cochlear nerve, whereas the one shown for *Platysaurus minor* is on the same side as the nerve. There is no trace of nerve fibers in either of these papillae.

The appearance of these supernumerary papillae is evidence that the epithelium in these ears is capable of differentiating in a complex manner. It seems evident that if one of these structures arose on a flexible membrane in the path of fluid vibrations, and if it acquired a nerve supply, it would constitute a true auditory receptor. It is easy to con-

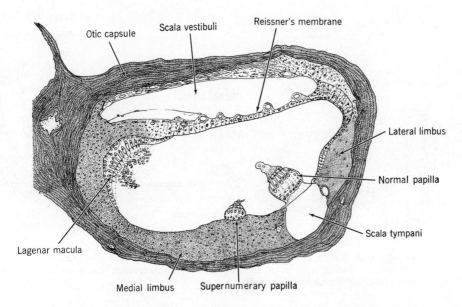

FIG. 25-1. A frontal section through the otic capsule of a specimen of the cordylid lizard *Platysaurus minor*, showing a supernumerary papilla alongside the normal one. (The left ear was reversed so as to appear as a right.) Scale 125×.

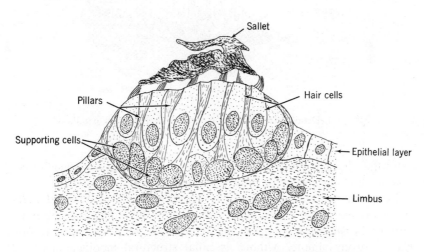

FIG. 25-2. An enlarged view of the supernumerary papilla of the preceding figure. Scale 1250×.

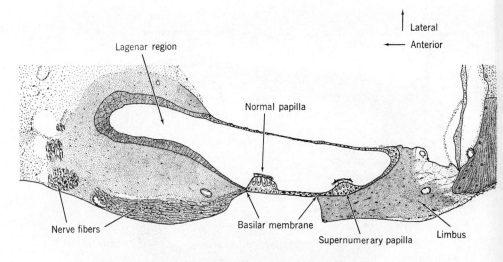

FIG. 25-3. A frontal section through the cochlear region of a specimen of the snake *Corallus enydrus*. A supernumerary papilla rests on the postero-lateral area of the limbus. Scale 100×.

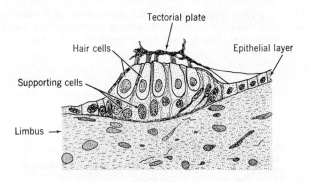

FIG. 25-4. Detailed representation of the supernumerary papilla of *Corallus enydrus*. Scale 500×.

ceive that the reptilian ear could have emerged in such a manner some-time in the long history of the cotylosaurs.

In summary, it appears that the vertebrate cochlea emerged independently out of the non-auditory labyrinth in three different vertebrate groups: in fishes, amphibians, and reptiles. In the fishes its derivation was relatively simple, without essential structural modifications, and consisted primarily of a neural switching to provide a new interpretation of sensory information. It seems to have appeared separately in two different groups of fishes, producing saccular (or saccular-lagenar) and utricular forms of derived ears.

In the amphibians the derivation was either by transformation of an existing structure (perhaps a crista organ) or a new emergence out of the labyrinthine epithelium. The result was a new type of ear, operating on principles very different from the inertial devices of fishes: a sensing of vibratory fluid motions transmitted through tectorial sensors in the fluid to the ciliary tufts of the hair cells, while the bodies of these cells remain at rest.

The reptilian ear is yet another emergence, most likely a new derivation from the labyrinthine epithelium. It resulted in an ear operating in a manner distinctly different from anything that had gone before. Contrary to the condition in amphibians, the hair cells in the reptilian ear are borne on a basilar membrane that lies athwart the path of vibratory fluid flow, so that these cells are moved bodily while their ciliary tufts are restrained.

The ears of birds and mammals followed the pattern of the reptilian ear. This seems to be true despite the fact that avian and mammalian branches of the vertebrate line have arisen separately from early reptiles and have remained apart for many millions of years.

As the treatment of the foregoing chapters has brought out, the reptilian ear while maintaining its basic mechanical pattern has developed in a number of ways, especially in respect to the means of restraint of the ciliary tufts of the hair cells. This diversification is greatest among the lizards, and it seems likely that the early ancestors of this suborder were the ones in which the original differentiation occurred.

We have noted four general forms of ciliary restraint that occur among lizards, often in combination and varying in different papillar regions. These are the simple tectorial membrane with direct attachments to the ciliary tufts, a tectorial membrane that subdivides in dendritic fashion to send strands to the ciliary tufts, a tectorial membrane that terminates in a plate structure that makes the ciliary connections, and inertial bodies—sallets and culmens—that rest over the cilia, sometimes with an attachment through a tectorial membrane to the limbus but more often without such an anchorage.

The remaining reptiles have specialized along two of these lines of ciliary restraint. Most of them—the snakes, amphisbaenians, *Sphenodon*, and turtles—make use of the tectorial membrane and plate system. The crocodiles, however, employ the dendritic form of tectorial membrane, found among lizards only in the chameleons.

The birds have adopted the dendritic system in essentially the same form in which it occurs in crocodilians.

The mammals have utilized a relatively simple tectorial membrane with a moderate amount of thickening that produces a form intermediate in character between the thin membranes of lizards and the heavy tectorial plates of most reptiles. Thus the mammalian structure can be considered either as a simple tectorial membrane that has thickened somewhat and come to rest over all or most of the hair cells, or as a tectorial plate that became reduced and simplified in this same service.

It is interesting that the sallet and culmen mode of restraint through the use of inertial and inertia-like forces, though extensively utilized in four lizard families and combined with other systems in several others, has no representation outside this lizard group. It is an intriguing speculation, however, whether inertial forces may combine with direct mechanical restraint in the operation of the thickened tectorial membranes in mammalian ears.

SOME OPEN HORIZONS

What lies ahead in the investigation of the reptilian ear? The observations that have produced the principal content of this book have necessarily been limited: in the number of species examined, in the size of the sample for each one, and in the experimental procedures carried out.

SPECIES LIMITATIONS

The kinds and numbers of specimens have been limited primarily by availability, a restriction increasingly felt during the course of the study as reptile populations all over the world have steadily declined. The results obtained may be regarded as providing a good general picture of the structures and performances of the ears in the different reptile groups, for as has been seen these characteristics generally follow a common pattern within a family. Yet variations appear, as strikingly seen in the Xantusiidae, which separate into two groups with respect to both middle ear and inner ear anatomy, and in occasional idiosyncrasies as encountered in *Leiolepis belliana* within the Agamidae. Also in this survey there are a number of gaps and doubtful areas that deserve further attention.

Among the lizards, specimens are lacking in five groups often accorded familial or subfamilial status: Anelytropsidae, Dibamidae, Feyliniidae, Lanthanotidae, and Shinisaurinae. The first three of these are regarded as closely allied to the skinks (Romer, 1956), and their ear structures can be expected to follow the scincid form in a general way. Yet variations might well appear in these, and further study is needed. Another family, the Pygopodidae, has had minimal representation by the single species *Lialis burtonis*, and the examination of further species in this group is highly desirable. Among the Chamaelonidae the genus *Chamaeleo* is moderately well covered, but species belonging to other genera of this family still need to be investigated.

The ear structures and their performances should be explored further in several groups of snakes. Of special interest are the burrowing blindsnakes, which include the Typhlopidae, studied somewhat inadequately in two species, and three other typhlopoid families that have not been studied at all. Also attention ought to be given to the relations between the levels of ear development and the ecological niches occupied by various species of boids and colubrids, which would require the addition of a good many species.

An important area, barely touched upon, is that of the aquatic reptiles, such as the water turtles and the sea snakes. There is a need to explore further the possibility, suggested in one of the turtle experiments, that the ears of these reptiles may occupy a compromise position between aerial and aquatic adaptations.

Procedural Developments. For a small number of species the indications about auditory function provided by cochlear potential measurements have been supplemented by the results of other investigators using different procedures, including the recording of neural responses and behavioral determinations of sensitivity. These further measurements have been few, and obviously need to be extended.

Neural Studies. — The existing neural studies of auditory capabilities in reptiles, though severely limited in scope, give a picture that corresponds generally to the one provided by cochlear potential measurements. A closer comparison on the basis of more extensive measurements might disclose some variations reflecting the effects of higher levels of integration in the nervous system, and this problem needs to be pursued further. These experiments are arduous and time-consuming, and there are difficulties in maintaining a specimen in good condition long enough to obtain records on a sufficiently large number of neural elements, but such experiments on a few representative species would be decidedly worthwhile.

Behavioral Test Methods. — The behavioral determination of the ear's sensitivity has achieved a degree of success only in turtles. Other reptiles need to be brought into the picture to evaluate more fully the results obtained by electrophysiological methods.

Earlier experiments designed to produce conditioned responses in reptiles were reviewed by Davidson and Richardson (1970) and by Colnaghi (1970, 1972). Of these experiments only one employed sound as the conditioned stimulus, and this was in combination with light. A number of attempts to condition reptiles to sounds alone have been made at the Auditory Research Laboratories in conjunction with the present study of the reptilian ear.

Exploratory experiments dealing with several species of lizards were carried out by Julaine Kinchla (1971) over a period of two years, largely by the use of operant conditioning techniques. These experiments used such rewards as food, water, and heat, but no very satisfactory results were obtained.

Further experiments on lizards were made by George L. Colnaghi. He tried to train *Anolis carolinensis* specimens to escape from a shock chamber in response to an acoustic signal, and in other experiments sought to produce specific behaviors, such as running and mouth-opening, in response to a sound as a means of avoiding shock.

Colnaghi (1972) also attempted with *Crotaphytus collaris* to use changes in respiratory rhythm as an indication of hearing, by presenting shock as the unconditioned stimulus and sound as a conditioned stimulus both alone and in combination with light flashes. Evidence for conditioning was obtained, but the required number of trials was very large: about three days of testing were needed for a single stimulus, so that a complete threshold determination was estimated as requiring something like a year's time. This work expenditure, along with the hazards of maintaining the animal in good condition over this period of time, makes the method impractical.

Lee W. Klepper (1973) sought to condition the heart rate in the lizard *Crotaphytus insularis*, with a tone as the conditioned stimulus and shock as the unconditioned stimulus. After presenting 20 trials a day for periods up to 38 days, only two out of seven animals showed any indications of conditioning, and these indications were minimal.

None of the above conditioning experiments on lizards resulted in the determination of auditory thresholds.

Three different attempts have been made to obtain conditioned responses to sounds in turtles, with the intent of confirming and extending the observations of Patterson (1966). The latest of these, by James K. Lang and Kathleen Lang, is still in progress at this writing, and gives great promise of success.

This account of failures with lizards and limited successes with turtles emphasizes a need for further understanding of reptile behavior in all its manifestations as a background for the development of conditioning and other tests of hearing ability. No doubt improved behavioral methods for hearing determinations in reptiles will be worked out in time, but much insightful research effort lies ahead.

APPENDIX A

ABBREVIATIONS AND

DEFINITIONS

ASA standard level
: The reference (for power) $= 10^{-16}$ watts per sq cm (ASA refers to the American Standards Association).

cf
: (compare with)—used to designate a specimen for which the species identification is uncertain, but which resembles the species named.

db
: (decibels) The difference between two sound pressures P_1 and P_0, expressed in decibels, is $N_{db} = 20 \log_{10} P_1/P_0$.

Hz
: (hertz) $=$ acoustic frequency in cycles per second.

N
: (number).

mµ
: (millimicrons) $=$ nanometers $= 10^{-9}$ meters.

µ
: (micron)—a unit of length $= 10^{-6}$ meters; also more generally $= 1$ millionth.

µv
: (microvolts) $= 10^{-6}$ volts.

p
: $=$ the probability that an observed relationship is due to chance. In general, measures are regarded as acceptable when $p \gtrless 0.05$ (said to be significant to the 5% level), are significant when $p \gtrless 0.01$, and highly significant when $p \gtrless 0.001$.

P_0
: $=$ a reference level for sound pressure, often chosen arbitrarily. The level used in this book is "zero level," defined as 1 dyne per sq cm. This level is 73.8 db above the ASA standard level.

P_1
: $=$ any observed value of sound pressure (expressed in dynes per sq cm).

r
: $=$ correlation coefficient. It varies from $+1$ (perfect relation) through 0 (no relation) to -1 (perfect inverse relation).

rms
: (root-mean-square) $=$ the square root of the sum of the squares of a series of measures. For a sound wave, the rms amplitude is the square root of the integrated amplitudes through a cycle of the wave; for a sine wave the rms amplitude $=$ the maximum amplitude divided by $\sqrt{2}$.

Throughout this book all measurements of periodic quantities (sound waves, vibratory amplitudes, cochlear potentials) are rms values unless otherwise indicated.

sp   (species)—often added to a genus name and used to designate a specimen for which the species is unknown.

s-v   (snout-to-vent)—commonly used to measure the lengths of reptiles; more reliable than total length for animals that may lose portions of the tail.

APPENDIX B

FORMULAS FOR HISTOLOGICAL

SOLUTIONS

I.   Reptilian Ringers Solution
Stock solution:

Sodium chloride (NaCl)	130 g
Potassium chloride (KCl)	2.8 g
Sodium phosphate, monobasic (Na H$_2$ PO$_4$·H$_2$O)	4.0 g

Dissolve in 1000 ml distilled water, then add Calcium
     chloride (CaCl$_2$)   2.4 g
Filter; store in refrigerator.
For use, add 1 part of stock solution to 19 parts distilled water.

II.   Maximow Fixative Solution
Solution A:

Mercuric bichloride (HgCl$_2$)	25 g
Potassium dichromate (K$_2$Cr$_2$O$_7$)	25 g
Sodium sulfate (Na$_2$SO$_4$·10 H$_2$O)	10 g
Distilled water	1000 ml

Solution B:
     Formaldehyde, full strength (saturated solution of HCOH in
     water)
Mix 9 parts of solution A with 1 part of solution B just before
using. Filter if any precipitate is present. This material is highly
toxic; avoid all contact with the skin.

III.   Staining Solutions
1. HAO Stain
Solution A:

Commercial Delafield's hematoxylin (or a mixture made up from formulas available in histological texts)	100 ml
80% ethyl alcohol	400 ml
glacial acetic acid	4 ml

Solution B:

Azocarmine	1.5 g
60% alcohol	250 ml

Solution C:

Orange G	5.0 g

Dissolve in distilled water 50 ml

Then add 95% alcohol 200 ml

Procedure:

(1) Stain sections in bulk with Solution A for 2 hours, remove without rinsing, barely cover with 70% alcohol and keep overnight.

(2) Differentiate in weak acid solution (4-8 drops HCl in 100 cc of 60% alcohol).

(3) Rinse in lithium carbonate solution (10 ml saturated solution of Li_2CO_3 in 50% alcohol) to neutralize acid.

(4) Rinse in 50% alcohol.

(5) Stain in Solution B for 30-60 sec, or until celloidin is lightly colored.

(6) Place in 5% solution of phosphotungstic acid in 50% alcohol for 5-10 min, or until sections are light pink.

(7) Rinse in 50% alcohol; pass through 60% to 70% alcohol.

(8) Stain heavily in Solution C (½ to 1 min).

(9) Rinse in 80% alcohol.

(10) Pass through 90% to 95% alcohol.

(11) Drain, blot, place in clearing oil mixture (3 parts cedarwood oil to 1 part bergamot oil).

(12) Mount, 1 section to a slide, using damar in xylene.

2. Held's phosphomolybdic acid stain

Hematoxylin 1 g

Dissolve in 100 cc of 70% ethyl alcohol.

Add 15 g phosphomolybdic acid.

Let stand 1 month.

Then add NaOH (solid grains) to bring pH to 7.6.

3. Pollak's trichrome stain (see McClung's Handbook, 1950, p. 251).

4. Masson's trichrome stain (*op. cit.*, p. 249).

Note: all alcohol referred to is ethyl alcohol (C_2H_5OH).

REFERENCES

Adrian, E. D., 1938. The effect of sound on the ear in reptiles, *J. Physiol.*, *92*, 9P-11P.

Adrian, E. D., Craik, K.J.W., and Sturdy, R. S., 1938. The electrical response of the auditory mechanism in cold-blooded vertebrates, *Proc. Royal Soc. London*, Ser. B., *125*, 435-455.

Alexander, I. E., and Githler, F. J., 1951. Histological examination of cochlear structure following exposure to jet engine noise, *J. Comp. Physiol. Psychol.*, *44*, 513-524.

———, 1952. Chronic effects of jet engine noise on the structure and function of the cochlear apparatus, *J. Comp. Physiol. Psychol.*, *45*, 381-391.

Andrews, O., 1915. The ability of turtles to discriminate between sounds, *Bull. Wisconsin Nat. Hist. Soc.*, *13*, 189-195.

Avery, D. F., and Tanner, W. W., 1971. Evolution of the iguanine lizards (Sauria, Iguanidae) as determined by osteological and myological characters, *Brigham Young Univ. Sci. Bull.*, *12*, 1-79.

Axtell, R. W., 1972. Hybridization between western collared lizards with a proposed taxonomic arrangement, *Copeia*, 707-727.

Ayers, H., 1892. Vertebrate cephalogenesis, *J. Morphol.*, *6*, 1-360.

Baird, I. L., 1960. A survey of the periotic labyrinth in some representative recent reptiles, *Kansas Univ. Sci. Bull.*, *41*, 891-981.

———, 1964. Some features of the aural anatomy of the turtle, *Trionyx spiniferus*, *Amer. Zoologist*, 4, 396. (Abstr.)

———, 1967. Some histological and cytological features of the basilar papilla in the lizard, *Anolis carolinensis*, *Anat. Rec.*, *157*, 208-209. (Abstr.)

———, 1969. Some findings of comparative fine structural studies of the basilar papilla in certain reptiles, *Anat. Rec.*, *163*, 149. (Abstr.)

———, 1970a. The anatomy of the reptilian ear, in C. Gans and T. Parsons, eds., *Biology of the reptilia*, New York, Academic Press, 2, 193-275.

———, 1970b. A preliminary report on light and electron microscopic studies of a crocodilian basilar papilla, *Anat. Rec.*, *166*, 274. (Abstr.)

———, 1974. Anatomical features of the inner ear in submammalian vertebrates, in W. D. Keidel and W. D. Neff, eds., *Handbook of sensory physiology*, V-1, Auditory system, Berlin and New York, Springer-Verlag, 159-212.

Baird, I. L., and Marovitz, W. F., 1971. Some findings of scanning and transmission electron microscopy of the basilar papilla of the lizard *Iguana iguana*, *Anat. Rec.*, *169*, 270. (Abstr.)

Baird, I. L., and Winborn, W. B., 1966. A preliminary report on some aspects of the fine structure of the cochlear duct in certain reptiles, *Anat. Rec.*, *154*, 449 (Abstr.)

Beach, F. A., 1944. Responses of captive alligators to auditory stimulation, *Amer. Naturalist*, *78*, 481-505.

Beck, A., 1890. Die Bestimmung der Localisation der Gehirn und Rücken-marks functionen vermittelst der elektrischen Erscheinungen, *Centralbl. f. Physiol., 4*, 473-476.

Békésy, G. von, 1960. *Experiments in hearing*, New York, McGraw-Hill Book Co.

———, 1967. *Sensory inhibition*, Princeton, N.J., Princeton University Press.

Békésy, G. von, Wever, E. G., Rahm, W. E., Jr., and Rambo, J.H.T., 1961. A new method of perfusion for the fixation of tissues, *Laryngoscope, 71*, 1534-1547.

Bellairs, A. d'A., 1949. The anterior brain-case and interorbital septum of Sauropsida, with a consideration of the origin of snakes, *J. Linn. Soc. Zool., 41*, 482-512.

———, 1972. Comments on the evolution and affinities of snakes, in K. A. Joysey and T. S. Kemp, eds., *Studies in vertebrate evolution*, 157-172, New York, Winchester Press.

Bellairs, A. d'A., and Underwood, G., 1951. The origin of snakes, *Biol. Revs., 26*, 193-237.

Beneden, E. van, 1882. Recherches sur l'oreille moyenne des Crocodiliens et ses communications multiples avec le pharynx, *Arch. de Biol., 3*, 497-560.

Berger, K., 1924. Experimentelle Studien über Schallperzeption bei Rep-tilien, *Zeits. f. vergl. Physiol., 1*, 517-540.

Berman, D. S., and Regal, P. J., 1967. The loss of the ophidian middle ear, *Evolution, 21*, 641-643.

Bezy, R. L., 1972. Karyotypic variation and evolution of the lizards in the family Xantusiidae, *Contrib. in Science, Los Angeles County Nat. Hist. Mus.*, No. 227, 29 pp.

Bogert, C. M., and Martín del Campo, R., 1956. The Gila monster and its allies, *Bull. Amer. Mus. Nat. Hist., 109*, 238 pp.

Boulenger, G. A., 1885. *Catalogue of the lizards in the British Museum (Natural History)*, 3 vols., 2nd ed., 1885, 1887, London.

Brattstrom, B. H., 1965. Body temperature of reptiles, *Amer. Midland Naturalist, 73*, 376-422.

Brock, G. T., 1940. The skull of the chameleon *Lophosaura ventralis* (Gray); some developmental stages, *Proc. Zool. Soc. London*, Ser. B, *110*, 219-241.

Broom, R., 1924. On the classification of the reptiles, *Bull. Amer. Mus. Nat. Hist., 51*, 39-65.

———, 1935. On the structure of the temporal region in lizard skulls, *Ann. Transv. Mus., 18*, 13-22.

Bustard, H. R., 1964. Defensive behavior shown by Australian geckos Ge-nus *Diplodactylus, Herpetologica, 20*, 198-200.

———, 1965. Observations on Australian geckos, *Herpetologica, 21*, 294-302.

———, 1970. *Australian lizards*, Sydney and London, Collins, 162 pp.

Camp, C. L., 1923. Classification of the lizards, *Bull. Amer. Mus. Nat. Hist.*, *48*, 289-481.

Campbell, H. W., 1967a. The effects of temperature on the auditory sensitivity of lizards, Ph.D. Diss., Univ. of Calif. at Los Angeles, 78 pp.

———, 1967b. Stop, look, listen: acoustic behavior of turtles, *Internat. Turtle and Tortoise Soc. J., 1*, 13-14, 44.

———, 1969. The effects of temperature on the auditory sensitivity of lizards, *Physiol. Zool., 42*, 183-210.

Carroll, E. H., 1969. Origin of reptiles, in C. Gans, A. d'A. Bellairs and T. S. Parsons, eds., *Biology of the Reptilia*, London and New York, Academic Press, *1*, 1-44.

Chernomordikov, V. V., 1958. On the physiology of the auditory analyzer in turtles, *Zhurnal Vysshei Nervnoi Deyatel I. P. Pavlova, 8*, 102-108.

Clason, E., 1871. Die Morphologie des Gehörorgans der Eidechsen, *Hasse's Anat. Stud., 1*, 300-376.

Colbert, E. H., 1946. The Eustachian tubes in the Crocodilia, *Copeia*, 12-14.

———, 1955. *Evolution of the vertebrates*, New York, John Wiley and Sons.

———, 1965. *The age of reptiles*, New York, W. W. Norton and Co.

Colnaghi, G. L., 1970. The problem of auditory conditioning in lizards, 54 pp., Research Report, Department of Psychol., Princeton Univ. [Unpubl.]

———, 1972. Stimulus control of respiration in the lizard *Crotaphytus collaris*, 60 pp., Research Report, Auditory Research Laboratories, Princeton Univ. [Unpubl.]

Cope, E. D., 1864. On the characters of the higher groups of Reptilia Squamata—and especially of the Diploglossa, *Proc. Acad. Nat. Sci. Philadelphia, 16*, 224-231.

———, 1892. The osteology of the Lacertilia, *Proc. Amer. Philos. Soc., 30*, 185-221.

———, 1900. The crocodilians, lizards, and snakes of North America, *Ann. Rep. U. S. Nat. Mus. for 1898*, Part II, 151-1294.

Crawford, F. T., and Holmes, C. E., 1966. Escape conditioning in snakes employing vibratory stimulation, *Psychonomic Sci., 4*, 125-126.

Crowley, D. E., 1964. Auditory responses in the alligator lizard, *J. Auditory Res., 4*, 135-143.

Danilewsky, B., 1891. Zur Frage über die elektromotorischen Vorgänge im Gehirn als Ausdruck seines Thätigkeitszustandes, *Centralbl. f. Physiol., 5*, 1-4.

Davidson, R. E., and Richardson, A. M., 1970. Classical conditioning of skeletal and autonomic responses in the lizard (*Crotaphytus collaris*), *Physiol. and Behav., 5*, 589-594.

De Beer, G. R., 1937. *The development of the vertebrate skull*, Oxford, England, Clarendon Press.

De Burlet, H. M., 1934a. Zur vergleichenden Anatomie und Physiologie des perilymphatischen Raumes, *Acta oto-laryngol., 13*, 153-187.

De Burlet, H. M., 1934b. Vergleichende Anatomie des statoakustischen Organs, in L. Bolk, E. Göppert, E. Kallius, and W. Lubosch, eds. *Handbuch der vergleichenden Anatomie der Wirbeltiere*, Sect. 2, Part II, 1293-1432.

Deiters, O., 1862. Über das innere Gehörorgan der Amphibien, *Arch. f. Anat. Physiol.*, 262-275, 277-310.

Dowling, H. G., 1959. Classification of the Serpentes: a critical review, *Copeia*, 38-52.

―――, 1967. Hemipenes and other characters in colubrid classification, in Symposium on colubrid snake systematics, Miami Beach, Florida, June 20, 1966, *Herpetologica, 23*, 137-168.

―――, 1975. A provisional classification of snakes, in H. G. Dowling, ed., *1974 Yearbook of herpetology*, New York, Herpetological Information Search Systems, 167-170.

Earle, A. M., 1961a. The middle ear of *Holbrookia maculata*, the northern earless lizard, *Copeia*, 68-74.

―――, 1961b. An additional note on the ear of *Holbrookia maculata, Copeia*, 355.

―――, 1961c. The middle ear of *Holbrookia* and *Callisaurus, Copeia*, 405-410.

Edmund, A. G., 1969. Dentition, in C. Gans, A. d'A. Bellairs, and T. S. Parsons, eds., *Biology of the Reptilia*, London and New York, Academic Press, *1*, 117-200.

Engelbrecht, D. van Z., 1951. Contributions to the cranial morphology of the chamaeleon *Microsaura pumila* Daudin, *Ann. Univ. Stellenbosch, 27*, Sect. A, 3-31.

Etheridge, R., 1964. The skeletal morphology and systematic relationships of Sceloporine lizards, *Copeia*, 610-631.

―――, 1966. The systematic relationships of West Indian and South American lizards referred to the iguanid genus *Leiocephalus, Copeia*, 79-91.

―――, 1967. Lizard caudal vertebrae, *Copeia*, 699-721.

Evans, L. T., 1936. The development of the cochlea in the gecko, with special reference to the cochlea-lagena ratio, *Anat. Rec., 64*, 187-199.

Fay, R. R., 1969. Behavioral audiogram for the goldfish, *J. Auditory Res., 9*, 112-121.

Fay, R. R., Kendall, J. I., Popper, A. N., and Tester, A. L., 1974. Vibration detection by the macula neglecta of sharks, *Comp. Biochem. Physiol., 47* A, 1235-1240.

Fernandez, C., 1958. Postmortem changes in the vestibular and cochlear receptors (guinea pig), *Arch. of Otolaryngol., 68*, 460-487.

FitzSimons, V. F., 1943. The lizards of South Africa, *Transvaal Mus. Memoir*, No. 1, 267-282, 408-481.

Flock, A., and Wersäll, J., 1962. A study of orientation of the sensory hairs of the receptor cells in the lateral line organ of fish, with special reference to the function of the receptors, *J. Cell. Biol., 15*, 19-27.

Foà, C., and Peroni, A., 1930. Primi tentativi di registrazione delle correnti d'azione del nervo acustico, *Archivi di Fisiol.*, *28*, 237-241; *Valsalva*, 6, 105-109.

Frank, G. H., 1951. Contributions to the cranial morphology of *Rhampholeon platyceps* Günther, *Ann. Univ. Stellenbosch*, *27*, Sect. A., 33-37.

Frank, G. H., and Smit, A. L., 1974. The early ontogeny of the columella auris of *Crocodilus niloticus* and its bearing on problems concerning the upper end of the reptilian hyoid arch, *Zoologica Africana*, 9, 59-88.

Frankenberg, E., 1974. Vocalizations of males of three geographical forms of *Ptyodactylus* from Israel (Reptilia: Sauria: Gekkoninae), *J. Herpetol.*, *8*, 59-70.

Freedman, R., 1947. The area of the auditory sense organ in the cat, Thesis, Princeton Univ. [Unpubl.]

Frisch, K. von, 1938. Über die Bedeutung des Sacculus und der Lagena für den Gehörsinn der Fische, *Zeits. f. vergl. Physiol.* 25, 703-747.

Frisch, K. von, and Stetter, H., 1932. Untersuchengen über den Sitz des Gehörsinnes bei der Elritze, *Zeits. f. vergl. Physiol.*, *17*, 686-801.

Fürbringer, M., 1919. Über das Zungenbein der Reptilien, *Bijdragen Tot de Dierkunde k. zoolog. geneetschap. Amsterdam*, Feest Nummer für Kerbert, 195-212.

―――, 1922. Das Zungenbein der Wirbeltiere insbesondere der Reptilien und Vögel, *Abh. heidelb. Akad. Wiss.*, math.-nat. Kl., Abt. B, *11*, 164 pp.

Gadow, H., 1888. On the modifications of the first and second visceral arches, with especial reference to the homologies of the auditory ossicles, *Phil. Trans. Royal Soc. London*, *179*, 451-485.

―――, 1901. The evolution of the auditory ossicles, *Anat. Anz.*, *19*, 396-411.

Gans, C., 1952. The functional morphology of the egg-eating adaptations in the snake genus *Dasypeltis, Zoologica*, *37*, 209-244.

―――, 1960. Studies on amphisbaenids (Amphisbaenia, Reptilia), *Bull. Amer. Mus. Nat. Hist.*, *119*, 131-204.

―――, 1961. The feeding mechanism of snakes and its possible evolution, *Amer. Zoologist*, *1*, 217-227.

―――, 1967a. A check list of Recent amphisbaenians (Amphisbaenia, Reptilia), *Bull. Amer. Mus. Nat. Hist.*, *135*, 61-105.

―――, 1967b. *Rhineura, Catalogue of American amphibians and reptiles*, *42*, Amer. Soc. Ichthyol Herpetol.

―――, 1969. Amphisbaenians—reptiles specialized for a burrowing existence, *Endeavour*, *28*, 146-151.

―――, 1973a. Uropeltid snakes—survivors in a changing world, *Endeavour*, *32*, 60-65.

―――, 1973b. Sound production in the Salientia: mechanism and evolution of the emitter, *Amer. Zoologist*, *13*, 1179-1194.

―――, 1974. *Biomechanics, an approach to vertebrate biology*, Philadelphia, J. P. Lippincott Co., 261 pp.

Gans, C., and Maderson, P.F.A., 1973. Sound producing mechanisms in Recent reptiles: review and comment, *Amer. Zoologist*, *13*, 1195-1203.

Gans, C., and Oshima, M., 1952. Adaptations for egg eating in the snake *Elaphe climacophora* (Boie), *Amer. Mus. Novitiates*, No. 1571, 1-16.

Gans, C., and Pandit, H., 1965. Notes on a herpetological collection from the Somali Republic; V. The amphisbaenian genus *Agamodon* Peters, *Mus. Royal Afrique Cent.* (134), 71-86.

Gans, C., and Wever, E. G., 1972. The ear and hearing in Amphisbaenia (Reptilia), *J. Exper. Zool., 179*, 17-34.

———, 1974. Temperature effects on hearing in two species of Amphisbaenia (Squamata, Reptilia), *Nature, 250*, 79-80.

———, 1975. The amphisbaenian ear: *Blanus cinereus* and *Diplometopon zarudnyi, Proc. Natl. Acad. Sci. USA, 72*, 1487-1490.

———, 1976. The ear and hearing in *Sphenodon punctatus, Proc. Natl. Acad. Sci. USA, 73*, 4244-4246.

Garrick, L. D., and Saiff, E. I., 1974. Observations on submergence reflexes of *Caiman sclerops, J. Herpetol., 8*, 231-235.

Goldby, F., 1925. The development of the columella auris in the Crocodilia, *J. Anat., 59*, 301-325.

Gray, Peter, 1954. *The microtomist's formulary and guide*, New York, Blakiston Co.

Greer, A. E., and Raizes, G., 1969. Green blood pigment in lizards, *Science, 166*, 392-393.

Grüneberg, H., Hallpike, C. S., and Ledoux, A., 1940. Observations on the structure, development and electrical reactions of the internal ear of the Shaker-1 mouse (*Mus musculus*), *Proc. Royal Soc. London*, Ser. B., *129*, 154-173.

Guinan, J. G., Jr., and Peake, W. T., 1967. Middle-ear characteristics of anesthetized cats, *J. Acoust. Soc. Amer., 41*, 1237-1261.

Gulick, W. L., and Zwick, H., 1966. Auditory sensitivity of the turtle, *Psychol. Record, 16*, 47-53.

Hamilton, D. W., 1960. Observations on the morphology of the inner ear in certain gekkonoid lizards, *Univ. Kansas Sci. Bull., 41*, 983-1024.

———, 1963. Posterior division of the eighth nerve in *Lacerta vivipara, Nature, 200*, 705-706.

———, 1964. The inner ear of lizards, I. Gross structure, *J. Morphol., 115*, 255-271.

Harris, G. G., 1967. As reported by W. A. van Bergeijk, in W. D. Neff, ed., *Contributions to sensory physiology*, New York and London, Academic Press, *2*, 30-31.

Hartline, P. H., 1969. Electrical responses to sound and vibration in the brains of snakes, Ph.D. Thesis, Univ. of Calif. at San Diego, 140 pp.

Hartline, P. H., and Campbell, H. W., 1968. Hearing and vibration detection in snakes, *Proc. Internat. Union. Physiol. Sci., 7*, 183. [Abstr.]

———, 1969. Auditory and vibratory responses in the midbrains of snakes, *Science, 163*, 1221-1223.

Hasse, C., 1871. Das Gehörorgan der Schildkröten, *Hasse's Anat. Stud., 1*, 225-299.

————, 1873a. Die vergleichende Morphologie und Histologie des häutigen Gehörorganes der Wirbelthiere, *Hasse's Anat. Stud., 1*, Suppl., 1-96.

————, 1873b. Das Gehörorgan der Crocodile nebst weiteren vergleichend anatomischen Bemerkungen über das mittlere Ohr der Wirbelthiere und dessen Annexa, *Hasse's Anat. Stud., 1*, 679-750.

————, 1873c. Die Morphologie des Gehörorgans von *Coluber natrix, Hasse's Anat. Stud., 1*, 648-678.

Hepp-Reymond, M-C., 1966. Action potentials of the auditory nerve in the Tokay gecko (*Gekko gecko*) and their relations to cochlear potentials, Ph.D. Thesis, Princeton University.

Hépp-Reymond, M-C., and Palin, J., 1968. Patterns in the cochlear potentials of the Tokay gecko (*Gekko gecko*), *Acta otolaryng., 65*, 270-292.

Herrick, C. J., 1897. The cranial nerve components of teleosts, *Anat. Anz., 13*, 425-431.

————, 1899. The peripheral nervous system of the bony fishes, *Bull. U.S. Fish Comm. for 1898, 18*, 315-320.

————, 1901. The cranial nerves and cutaneous sense organs of the North American siluroid fishes, *J. Comp. Neurol., 11*, 117-249.

Hoffstetter, R., 1964. Observations sur l'osteologie et la classification des Acrochordidae (Serpentes), *Bull. Mus. Natn. Hist. Nat.*, 2nd ser., (1965), *36*, 677-696.

Hoofien, J. H., 1964. Geographic variability of the common chamaeleon in Israel, *Israel J. Zool., 13*, 136-137.

Huxley, T. H., 1869. On the representative of the malleus and the incus of the *Mammalia* in the other Vertebrata, *Proc. Royal Soc. London*, 391-407.

Iordansky, N. H., 1968. Muscles of the external ear in some lizards [in Russian], *Zool. zhurnal (Akad. nauk. SSSR), 47*, 1730-1732.

Jorgensen, C. D., Orton, A. M., and Tanner, W. W., 1963. Voice of the leopard lizard *Crotaphytus wislizeni* Baird and Girard, *Proc. Utah Acad. Sci. Arts Letters, 40*, 115-116.

Kästle, W., 1964. Verhaltenstudien an Taggeckonen der Gattungen *Lygodactylus* und *Phelsuma, Zeits. f. Tierpsychol., 21*, 486-507.

Kamal, A. M., and Hammouda, H. G., 1965. The columella auris of the snake, *Psammophis sibilans, Anat. Anz., 116*, 124-138.

Karimova, M. M., 1958. The conditioned reflex characteristics of the auditory analyser in turtles, *Zhurnal Vysshei Nervnoi Deyatel I. P. Pavlova, 8*, 96-102.

Kennedy, M. C., 1974. Auditory multiple-unit activity in the midbrain of the Tokay gecko (*Gekko gecko* L.), *Brain Behav. Evol., 10*, 257-264.

Khanna, S. M., and Tonndorf, J., 1972. Tympanic membrane vibrations in cats studied by time-averaged holography, *J. Acoust. Soc. Amer., 51*, 1904-1920.

Kiang, N.Y.S., Watanabe, T., Thomas, E. C., and Clark, L. F., 1966. *Discharge patterns of single fibers in the cat's auditory nerve*, Cambridge, Mass., M.I.T. Press, 154 pp.

Killian, G., 1890. Die Ohrmuskeln des Krokodiles, *Jena Zeits. f. Naturwiss.*, *24*, 632-656.

Kinchla, Julaine, 1971. Key pressing as a heat-light escape response by the lizard *Eublepharis macularius*, Research Report, Auditory Research Laboratories. [Unpubl.]

King, W., and Thompson, F. G., 1968. A review of the American lizards of the genus *Xenosaurus* Peters, *Bull. Florida State Mus.*, *12*, 93-123.

Kinsler, L. E., and Frey, A. R., 1962. *Fundamentals of acoustics*, 2nd ed., New York, John Wiley and Sons.

Klauber, L. M., 1972. *Rattlesnakes*, 2nd ed., 2 vols., Berkeley and Los Angeles, Univ. of Calif. Press.

Klepper, L. W., 1973. Auditory heart rate conditioning in the lizard, *Crotaphytus insularis*, Thesis, Princeton Univ. [Unpubl.]

Kluge, A. G., 1967. Higher taxonomic categories of gekkonid lizards and their evolution, *Bull. Amer. Mus. Nat. Hist.*, *135*, 59 pp.

Konishi, M., 1969. Hearing, single-unit analysis, and vocalizations in songbirds, *Science, 166*, 1178-1181.

———, 1970. Comparative neurophysiological studies of hearing and vocalizations in songbirds, *Zeits. f. vergl. Physiol.*, *66*, 257-272.

Kritzinger, C. C., 1946. The cranial anatomy and kinesis of the South African amphisbaenid *Monopeltis capensis* Smith, *South African J. Sci.*, *42*, 175-204.

Kuhn, —., 1882. Über die häutige Labyrinth der Reptilien, *Arch. f. mikr. Anat.*, *20*, 271-361.

Kuile, E. ter, 1900. Die Übertragung der Energie von der Grundmembran auf die Haarzellen, *Pflügers Arch. ges. Physiol.*, *79*, 146-157.

Kuroda, R., 1923. Studies on audition in reptiles, *J. Comp. Psychol.*, *3*, 27-36.

———, 1925. A contribution to the subject of hearing in tortoises, *J. Comp. Psychol.*, *5*, 285-291.

Langston, W., Jr., 1973. The crocodilian skull in historical perspective, in C. Gans and T. S. Parsons, eds., *Biology of the Reptilia*, London and New York, Academic Press, *4*, 263-284.

Leydig, F., 1872. *Die in Deutschland lebenden Arten der Saurier*, Tübingen, 262 pp.

Linné, Charles, 1802. *A general system of nature, 1*, p. 636, Trans. by William Turton from the last ed. of *Systema naturae*.

List, J. C., 1966. Comparative osteology of the snake families Typhlopidae and Leptotyphlopidae, *Illinois Biol. Monog.*, No. 36, 1-112.

Loveridge, A., 1942. Revision of the African lizards of the family Gerrhosauridae, *Bull. Mus. Comp. Zool., Harvard, 89*, 485-543.

———, 1944. Revision of the African lizards of the family Cordylidae, *Bull. Mus. Comp. Zool., Harvard, 95*, 118 pp.

———, 1947. Revision of the African lizards of the family Gekkonidae, *Bull. Mus. Comp. Zool., Harvard, 98*, 468 pp.

———, 1957. Check list of the reptiles and amphibians of East Africa, *Bull. Mus. Comp. Zool., Harvard, 117*, 153-362.

Lowenstein, O., and Roberts, T.D.M., 1951. The localization and analysis of the responses to vibration from the isolated elasmobranch labyrinth, *J. Physiol., 114,* 471-489.

McClung's handbook of microscopical technique, 1950. R. McC. Jones, ed., 3rd ed., New York, Paul B. Hoebler, Inc.

McDowell, S. B., Jr., 1961. On the major arterial canals in the ear region of testudinoid turtles, and the classification of the Testudinoidea, *Bull. Mus. Comp. Zool., Harvard, 125,* 23-39.

————, 1967. The extracolumella and tympanic cavity of the "earless" monitor lizard, *Lanthanotus borneensis, Copeia,* 154-159.

McDowell, S. B., Jr., and Bogert, C. M., 1954. The systematic position of *Lanthanotus* and the affinities of the anguinomorphan lizards, *Bull. Amer. Mus. Nat. Hist. 105,* 142 pp.

McGill, T. E., 1959. Auditory sensitivity and the magnitude of the cochlear potential, *Ann. Otol. Rhinol. Laryngol., 68,* 193-207.

Mahendra, B. C., 1938. Some remarks on the phylogeny of the Ophidia, *Anat. Anz., 86,* 347-356.

Malan, M. E., 1945. Contributions to the comparative anatomy of the nasal capsule and the organ of Jacobson of the Lacertilia, *Ann. Univ. Stellenbosch,* Sect. A, *24,* 69-137.

Manley, G. A., 1970a. Frequency sensitivity of auditory neurons in the caiman cochlear nucleus, *Zeits. f. vergl. Physiol., 66,* 251-256.

————,1970b. Comparative studies of auditory physiology in reptiles, *Zeits. f. vergl. Physiol., 67,* 363-381.

————, 1972a. Frequency response of the ear of the Tokay gecko, *J. Exper. Zool., 181,* 159-168.

————, 1972b. The middle ear of the Tokay gecko, *J. Comp. Physiol., 81,* 239-250.

————, 1972c. Frequency response of the middle ear of geckos, *J. Comp. Physiol., 81,* 251-258.

————, 1974. Activity patterns of neurons in the peripheral auditory system of some reptiles, *Brain, Behav. Evol., 10,* 244-256.

Manley, J. A., 1971. Single unit studies in the midbrain auditory area of *Caiman, Zeits. f. vergl. Physiol., 71,* 255-261.

Manning, F. B., 1923. Hearing in rattlesnakes, *J. Comp. Psychol., 3,* 241-247.

Maximow, A., 1909. Über zweckmassige Methoden für cytologische und histogenetische Untersuchungen am Wirbeltierembryo, mit spezieller Berücksichtigung der Celloidenschnittserien, *Zeits. f. wiss. Mikros. u. mikr. Techn., 26,* 177-190.

Mertens, R., 1963. Liste der rezenten Amphibien und Reptilien: Helodermatidae, Varanidae, Lanthanotidae, *Das Tierreich,* Lief. 79, 26 pp.

————, 1966. Liste der rezenten Amphibien und Reptilien: Chamaeleonidae, *Das Tierreich,* Lief. 83, 37 pp.

————, 1971. Rie Rückbildung des Tympanum bei Reptilien und ihre Beziehung zur Lebensweise, *Senckenb. Biol., 52,* 177-191.

Mertens, R., and Wermuth, H., 1955. Die rezenten Schildkröten, Krokodile und Brückenechsen, *Zool. Jahrb.*, Abt. f. Syst., *83*, 323-440.

Miller, M. R., 1966a. The cochlear duct of lizards, *Proc. Calif. Acad. Sci.*, *33*, 255-359.

———, 1966b. The cochlear duct of lizards and snakes, *Amer. Zoologist, 6*, 421-429.

———, 1966c. The cochlear ducts of *Lanthanotus* and *Anelytropsis* with remarks on the familial relationship between *Anelytropsis* and *Dibamus*, *Occas. Papers Calif. Acad. Sci.*, No. 60, 15 pp.

———, 1968. The cochlear duct of snakes, *Proc. Calif. Acad. Sci., 35*, 425-476.

———, 1973a. A scanning electron microscope study of the papilla basilaris of *Gekko gecko, Zeits. f. Zellforsch., 136*, 307-328.

———, 1973b. Scanning electron microscope studies of some lizard basilar papillae, *Amer. J. Anat., 138*, 301-330.

———, 1974a. Scanning electron microscopy of the lizard papilla basilaris, *Brain, Behav. Evol., 10*, 95-112.

———, 1974b. Scanning electron microscope studies of some skink papillae basilares, *Cell. Tiss. Res., 150*, 125-141.

Miller, M. R., Kasahara, M., and Mulroy, M., 1967. Observations on the structure of the cochlear duct limbus of reptiles, *Proc. Calif. Acad. Sci., 35*, 37-52.

Mittleman, M. B., 1942. A summary of the iguanid genus *Urosaurus, Bull. Mus. Comp. Zool., Harvard, 91*, 103-181.

Mulroy, M. J., 1968a. Ultrastructure of the basilar papilla of reptiles, Ph.D. Diss., Univ. of Calif. at San Francisco.

———, 1968b. Orientation of hair cells in the reptilian auditory papilla, *Anat. Rec., 160*, 397. [Abstr.]

———, 1974. Cochlear anatomy of the alligator lizard, *Brain, Behav. Evol., 10*, 69-87.

Mulroy, M. J., Altmann, D. W., Weiss, T. F., and Peake, W. T., 1974. Intracellular electric responses to sound in a vertebrate cochlea, *Nature, 249*, 482-485.

Norris, K. S., and Lowe, C. H., 1951. A study of the osteology and musculature of *Phrynosoma m'calli* pertinent to its systematic position, *Bull. Chicago Acad. Sci., 9*, 117-125.

Oerlich, T. M., 1956. The anatomy of the head of *Ctenosaura pectinata* (Iguanidae), *Misc. Publ., Mus. Zool.*, Univ. Mich., No. 94, 122 pp.

Osawa, G., 1898. Beiträge zur Anatomie der *Hatteria punctata, Arch. f. mikros. Anat., 51*, 481-691.

Owen, R., 1850. On the communications between the cavity of the tympanum and the palate in the Crocodilia (gavials, alligators and crocodiles), *Phil. Trans. Royal Soc. London*, 140, part 1, 521-527.

Parker, W. K., 1883. On the structure and development of the skull in Crocodilia, *Trans. Zool. Soc. London, 11*, 263-310.

———, 1885. On the structure of the skull in the chameleons, *Trans. Zool. Soc. London, 11*, 77-105.

Parvulescu, A., 1964. Problems of propagation and processing, in W. N. Tavolga, ed., *Marine bio-acoustics*, New York, Macmillan Co., 87-100.

Patterson, W. C., 1965. Hearing in the turtle, Ph.D. Diss., Univ. of Delaware.

———, 1966. Hearing in the turtle, *J. Auditory Res., 6*, 453-464.

Patterson, W. C., Evering, F. C., and McNall, C. L., 1968. The relationship of temperature to the cochlear response in a poikilotherm (*Pseudemys scripta elegans*), *J. Auditory Res., 8*, 439-448.

Patterson, W. C., and Gulick, W. L., 1966. A method for measuring auditory thresholds in the turtle, *J. Auditory Res., 6*, 219-227.

Paulsen, K., and Mares, W., 1971. Zur Morphologie des äusseren Ohres und des Mittelohres von *Gekko gecko, Zeits. f. wiss. Zool., 183*, 97-113.

Peters, W., 1875. Über die Gehörknöchelchen und ihre Verhältniss zu dem ersten Zungenbeinbogen bei *Sphenodon punctatus, Monatsber. Akad. Wiss. Berlin für 1874*, 40-45.

———, 1882. Über eine neue Art und Gattung der Amphisbaenoiden, *Agamodon anguliceps, Sitzungsber. k. preuss. Akad. Wiss.*, Berlin, 579-584.

Poliakov, K., 1930. Zur Physiologie des Riech- und Höranalysators bei der Schildkröte *Emys orbicularis, Russkii fiziol. zhurnal, 13*, 161-177.

Pooley, A. C., and Gans, C., 1976. The Nile crocodile, *Scientific Amer., 234*, 114-124.

Rahm, W. E., Jr., Strother, W. F., and Gulick, W. L., 1958. The stability of the cochlear response through time, *Ann. Otol. Rhinol. Laryngol., 67*, 972-977.

Rand, A. S., 1963. Notes on the *Chamaeleo bitaeniatus* complex, *Bull. Mus. Comp. Zool., Harvard, 130*, 1-29.

Raslear, T. G., 1974. The use of the cochlear microphonic response as an indicant of auditory sensitivity, *Psychol. Bull., 81*, 791-803.

Reinbach, W., 1950. Über den schalleitenden Apparat der Amphibien und Reptilien, *Zeits. f. Anat. Entwickl., 114*, 611-639.

Retzius, G., 1880. Zur Kenntnis des inneren Gehörorgans der Wirbelthiere, *Arch. f. Anat. Physiol., Anat. Abt.*, 235-244.

———, 1884. *Das Gehörorgan der Wirbelthiere*, Vol. 2: *Das Gehörorgan der Reptilien, der Vögel und der Säugethiere*, Stockholm.

Ridgway, S. H., Wever, E. G., McCormick, J. G., Palin, J., and Anderson, J. H., 1969. Hearing in the giant sea turtle, *Chelonia mydas, Proc. Natl. Acad. Sci. USA, 64*, 884-890.

Romer, A. S., 1956. *Osteology of the reptiles*, 772 pp. Chicago, Univ. of Chicago Press.

———, 1933. *Vertebrate paleontology*, 2nd ed. 1945; 3rd ed. 1966, Chicago, Univ. of Chicago Press.

———, 1959. *The vertebrate story*, 4th ed., Chicago and London, Univ. of Chicago Press.

St. Girons, H., 1967. Morphologie comparée de l'hypothese chez les Squamata: Données complémentaires et apport a la phylogénie des reptiles, *Ann. des Sci. Nat. Zool.*, Paris, Ser. 12, *14*, 229-308.

Savage, J. M., 1958. The iguanid lizard genera *Urosaurus* and *Uta* with remarks on related groups, *Zoologica, 43*, 41-54.

———, 1963. Studies on the lizard family Xantusiidae, IV, the genera, *Contrib. in Sci., Los Angeles County Mus.*, No. 71, 38 pp.

Schmidt, K. P., 1950. Modes of evolution discernible in the taxonomy of snakes, *Evolution, 4*, 79-86.

Schmidt, R. S., 1964. Phylogenetic significance of lizard cochlea, *Copeia*, 542-549.

Shute, C.C.D., and Bellairs, A. d'A., 1953. The cochlear apparatus of Geckonidae and Pygopodidae and its bearing on the affinities of these groups of lizards, *Proc. Zool. Soc. London, 123*, 695-709.

———, 1955. The external ear in Crocodilia, *Proc. Zool. Soc. London, 124*, 741-749.

Siebenrock, F., 1893. Das Skelet von *Brookesia superciliaris* Kuhl, *Sitzungsber. Akad. Wiss., Wien*, Abt. I, *102*, 71-118.

Sill, W. D., 1968. The zoogeography of the Crocodilia, *Copeia*, 76-88.

Smalian, C., 1885. Beiträge zur Anatomie der Amphisbaeniden, *Zeits. f. wiss. Zool., 42*, 126-202.

Smith, H. M., 1939. Notes on Mexican reptiles and amphibians, *Zool. Ser., Field Museum Nat. Hist., 24*, 15-35.

Stannius, H., 1956. *Die Wirbelthiere*, in V. Siebold and H. Stannius, *Handbuch der Zootomie*, 2nd ed., p. 163.

Suga, N., and Campbell, H. W., 1967. Frequency sensitivity of single auditory neurons in the gecko *Coleonyx variegatus, Science, 157*, 88-90.

Symposium on colubrid snake systematics, Miami Beach, Florida, June 20, 1966, *Herpetologica*, 1967, *23*, 137-168.

Tester, A. L., Kendall, J. I., and Milisen, W. B., 1972. Morphology of the ear of the shark genus *Carcharhinus*, with particular reference to the macula neglecta, *Pacific Sci., 26*, 264-274.

Toerien, M. J., 1948. The evolution of the auditory conducting apparatus, *Proc. Royal Soc. Med.*, Sect. Otol., *41*, 877-888.

———, 1963. The sound-conducting systems of lizards without tympanic membranes, *Evolution, 17*, 540-547.

Tonndorf, J., and Khanna, S., 1972. Tympanic membrane vibrations in cats, studied by time-averaged holography, *J. Acoust. Soc. Amer., 51*, 1904-1920.

Tumarkin, A., 1949. On the evolution of the auditory conducting apparatus, *J. Laryngol. Otol., 63*, 119-140.

———, 1955. On the evolution of the auditory conducting apparatus, *Evolution, 9*, 221-243.

Underwood, G., 1954. On the classification and evolution of geckos, *Proc. Zool. Soc. London, 124*, 469-492.

———, 1957a. On lizards of the family Pygopodidae: a contribution to the morphology and phylogeny of the Squamata, *J. Morphol., 100*, 207-268.

———, 1957b. *Lanthanotus* and the anguinomorphan lizards: a critical review, *Copeia*, 20-30.

————, 1967. *A contribution to the classification of snakes*, 179 pp., Publ. No. 653, London, Brit. Mus. (Nat. Hist.).

————, 1970. The eye, in C. Gans and T. S. Parsons, eds., *Biology of the Reptilia*, 2, 1-97, New York, Academic Press.

————, 1971. Introduction to reprint ed. of C. L. Camp, *Classification of the lizards*, vii-xvii, Soc. Stud. Amphib. Rept.

Van Bergeijk, W. A., 1966. Evolution of the sense of hearing in vertebrates, *Amer. Zoologist*, 6, 371-377.

————, 1967. The evolution of vertebrate hearing, in W. D. Neff, ed., *Contributions to sensory physiology*, 2, 1-49, Berlin, Heidelberg, New York, Springer-Verlag.

Van Beneden, E., 1882. Recherches sur l'oreille moyenne des Crocodiliens et ses communications multiples avec le pharynx, *Arch. de Biol.*, 3, 497-560.

Vanzolini, P. E., 1951. A systematic arrangement of the family Amphisbaenidae (Sauria), *Herpetologica*, 7, 113-123.

Versluys, J., Jr., 1898. Die mittlere und äussere Ohrsphäre der Lacertilia und Rhynchocephalia, *Zool. Jahrb., Abt. Anat.*, 12, 161-406.

————, 1904. Entwicklung der Columella auris bei den Lacertiliern, *Zool. Jahrb., Abt. Anat.*, 19, 107-188.

Walls, G. L., 1942a. *The vertebrate eye and its adaptive radiation*, 785 pp., Bull. 19, Cranbrook Inst. of Sci.

————, 1942b. The visual cells and their history, in J. Cattell, ed., *Visual mechanisms*, Biological symposia, 7, 203-251.

Weber, E., and Werner, Y. L., 1977. Vocalizations of two snake-lizards (Reptilia: Sauria: Psygopodidae), *Herpetologica*, 33, 353-363.

Weisbach, W., and Schwartzkopff, J., 1967. Nervöse Antworten auf Schallreiz im Grosshirn von Krokodilen, *Die Naturwiss.*, 54, 650.

Weiss, T. F., Mulroy, M. J., and Altmann, D. W., 1974. Intracellular responses to acoustic clicks in the inner ear of the alligator lizard, *J. Acoust. Soc. Amer.*, 55, 606-619.

Wermuth, H., 1965. Liste der rezenten Amphibien und Reptilien: Gekkonidae, Pygopodidae, Xantusiidae, *Das Tierreich*, Lief. 80, 246 pp.

————, 1967. Liste der rezenten Amphibien und Reptilien: Agamidae, *Das Tierreich*, Lief. 86, 127 pp.

————, 1968. Liste der rezenten Amphibien und Reptilien: Cordylidae (Cordylinae und Gerrhosaurinae), *Das Tierreich*, Lief. 87, 30 pp.

————, 1969. Liste der rezenten Amphibien und Reptilien: Anguidae, Anniellidae, Xenosauridae, *Das Tierreich*, Lief. 90, 41 pp.

Wermuth, H., and Mertens, R., 1961. Schildkröten, Krokodile, Brückenechsen, Jena, G. Fischer Verlag, 422 pp.

Werner, F., 1902. Prodromus einer Monographie der Chamaeleonten, *Zool. Jahrb., Abt. f. Syst.*, 15, 295-460.

————, 1911. Liste der rezenten Amphibien und Reptilien: Chamaeleontidae, *Das Tierreich*, Lief. 27, 52 pp.

Werner, Y. L., 1961. The vertebral column of the geckos (Gekkonoidea), with special consideration of the tail, Ph.D. thesis, Hebrew University of Jerusalem, 257 pp. [Unpubl.]

————, 1965. Über die israelischen Geckos der Gattung *Ptyodactylus* und ihre Biologie, *Salamandra, 1*, 15-25.

————, 1968. The function of the inner ear in lizards: temperature effects in Gekkonoidea and Iguanidae, Proc. 11th Ann. Meeting, Society Stud. Amphib. Rept., *J. Herpetol., 2*, 178.

————, 1969. The evolution of vocalization in "higher lower vertebrates"; questions raised by recent studies of hearing in lizards, Report NATO Advanced Study Inst., *Vertebrate evolution: mechanism and process*, Robert College, Istanbul, Turkey, Aug. 1969, p. 33.

————, 1972. Temperature effects on inner-ear sensitivity in six species of iguanid lizards, *J. Herpetol., 6*, 147-177.

————, 1976. Optimal temperatures for inner-ear performance in gekkonoid lizards, *J. Exper. Zool., 195*, 319-351.

Werner, Y. L., and Wever, E. G., 1972. The function of the middle ear in lizards: *Gekko gecko* and *Eublepharis macularius* (Gekkonoidea), *J. Exper. Zool., 179*, 1-16.

Wersäll, J., 1967. Physiological action of the sensory hairs, in S. Iurato, *Submicroscopic structure of the inner ear*, Oxford, Pergamon Press, Ltd. 209-210.

Wersäll, J., Kimura, R., and Lindquist, P. G., 1965. Early postmortem changes in the organ of Corti (guinea pig), *Zeits. f. Zellforsch., 65*, 220-237.

Weston, J. K., 1939. Observations on the comparative anatomy of the VIIIth nerve complex, *Acta oto-laryngol., 27*, 457-497.

Wever, E. G., 1949. *Theory of hearing*, New York, John Wiley & Sons, 484 pp.

————, 1959. The cochlear potentials and their relation to hearing, *Ann. Otol. Rhinol. Laryngol., 68*, 975-989.

————, 1965a. Structure and function of the lizard ear, *J. Auditory Res., 5*, 331-371.

————, 1965b. The degenerative processes in the ear of the Shaker mouse, *Ann. Otol. Rhinol. Laryngol., 74*, 5-21.

————, 1966. Electrical potentials of the cochlea, *Physiol. Revs., 46*, 102-127.

————, 1967a. The tectorial membrane of the lizard ear: types of structure, *J. Morphol., 122*, 307-319.

————, 1967b. The tectorial membrane of the lizard ear: species variations, *J. Morphol., 123*, 355-371.

————, 1967c. Tonal differentiation in the lizard ear, *Laryngoscope, 77*, 1962-1973.

————, 1968a. The lacertid ear: *Eremias argus*, Proc. Natl. Acad. Sci. USA, *61*, 1292-1299.

————, 1968b. The ear of the chameleon: *Chamaeleo senegalensis* and *Chamaeleo quilensis*, *J. Exper. Zool., 168*, 423-436.

————, 1969a. The ear of the chameleon: the round window problem, *J. Exper. Zool., 171,* 1-6.

————, 1969b. The ear of the chameleon: *Chamaeleo höhnelii* and *Chamaeleo jacksoni, J. Exper. Zool., 171,* 305-312.

————, 1969c. Cochlear stimulation and Lempert's mobilization theory, *Arch. Otolaryngol., 90,* 720-725.

————, 1970a. The lizard ear: *Cordylus, Platysaurus,* and *Gerrhosaurus, J. Morphol., 130,* 37-56.

————, 1970b. The lizard ear: Scincidae, *J. Morphol., 132,* 277-292.

————, 1971a. The ear of *Basiliscus basiliscus* (Sauria: Iguanidae); its structure and function, *Copeia,* 139-144.

————, 1971b. Hearing in the Crocodilia, *Proc. Natl. Acad. Sci. USA, 68,* 1498-1500.

————, 1971c. The mechanics of hair-cell stimulation, *Ann. Otol. Rhinol. Laryngol., 80,* 786-804.

————, 1971d. The lizard ear: Anguidae, *J. Auditory Res., 11,* 160-172.

————, 1973a. The function of the middle ear in lizards: *Eumeces* and *Mabuya* (Scincidae), *J. Exper. Zool., 183,* 225-239.

————, 1973b. The function of the middle ear in lizards: divergent types, *J. Exper. Zool., 184,* 97-126.

————, 1973c. The labyrinthine sense organs of the frog, *Proc. Natl. Acad. Sci. USA, 70,* 498-502.

————, 1973d. Closure muscles of the external auditory meatus in Gekkonidae, *J. Herpetol., 7,* 323-329.

————, 1973e. Tectorial reticulum of the labyrinthine endings of vertebrates, *Ann. Otol. Rhinol. Laryngol., 82,* 277-289.

————, 1974a. The evolution of vertebrate hearing, in W. D. Keidel and W. D. Neff, eds., *Handbook of sensory physiology,* Vol. V-1, Auditory system, New York, Springer-Verlag, 423-454.

————, 1974b. The lizard ear: Gekkonidae, *J. Morphol., 143,* 121-165.

————, 1974c. The ear of *Lialis burtonis* (Sauria: Pygopodidae); its structure and function, *Copeia,* 297-305.

————, 1974d. Sound reception, in *Encyclopaedia Britannica,* 15th ed., *17,* 39-51.

————, 1975. The caecilian ear, *J. Exper. Zool., 191,* 63-72.

————, Origin and evolution of the ear of vertebrates, in R. B. Masterton, M. E. Bitterman, C. B. G. Campbell, and N. Hotton, eds., 1976. *Evolution of brain and behavior in vertebrates,* Hillsdale, N.J., Lawrence Erlbaum Assoc., 89-105.

Wever, E. G., and Bray, C. W., 1931. Auditory responses in the reptile, *Acta oto-laryngol., 16,* 154-159.

Wever, E. G., Bray, C. W., and Lawrence, M., 1941. The nature of cochlear activity after death, *Ann. Otol. Rhinol. Laryngol., 50,* 317-329.

Wever, E. G., Crowley, D. C., and Peterson, E. A., 1963. Auditory sensitivity in four species of lizards, *J. Auditory Res., 3,* 151-157.

Wever, E. G., and Gans, C., 1972. The ear and hearing in *Bipes biporus, Proc. Natl. Acad. Sci. USA, 69,* 2714-2716.

Wever, E. G. and Gans, C., 1973. The ear in Amphisbaenia (Reptilia); further anatomical observations, *J. Zool.*, London, *171*, 189-206.

———, 1976. The caecilian ear: further observations, *Proc. Natl. Acad. Sci. USA, 73*, 3744-3746.

Wever, E. G., and Hepp-Reymond, M-C., 1967. Auditory sensitivity in the fan-toed gecko, *Ptyodactylus hasselquistii, Proc. Natl. Acad. Sci. USA, 57*, 681-687.

Wever, E. G., Hepp-Reymond, M-C., and Vernon, J. A., 1966. Vocalization and hearing in the leopard lizard, *Proc. Natl. Acad. Sci. USA, 55*, 98-106.

Wever, E. G., and Lawrence, M., 1949a. The patterns of response in the cochlea, *J. Acoust. Soc. Amer., 21*, 127-134.

———, 1949b. The functions of the round window, *Ann. Otol. Rhinol. Laryngol., 57*, 579-589.

———, 1952. Sound conduction in the cochlea, *Ann. Otol. Rhinol. Laryngol., 61*, 824-834.

———, 1954. *Physiological acoustics*, Princeton, N.J., Princeton Univ. Press, 454 pp.

———, 1955. Patterns of injury produced by overstimulation of the ear, *J. Acoust. Soc. Amer., 27*, 853-858.

Wever, E. G., and Peterson, E. A., 1963. Auditory sensitivity in three iguanid lizards, *J. Auditory Res., 3*, 205-212.

Wever, E. G., Peterson, E. A., Crowley, D. E., and Vernon, J. A., 1964. Further studies of hearing in the gekkonid lizards, *Proc. Natl. Acad. Sci. USA, 51*, 561-567.

Wever, E. G., Rahm, W. E., Jr., and Strother, W. F., 1959. The lower range of the cochlear potentials, *Proc. Natl. Acad. Sci. USA, 45*, 1447-1449.

Wever, E. G., and Strother, W. F., 1974. The middle ear mechanism in *Pituophis melanoleucus annectens* [Unpubl.]

Wever, E. G., and Vernon, J. A., 1956a. The sensitivity of the turtle's ear as shown by its electrical potentials, *Proc. Natl. Acad. Sci. USA, 42*, 213-220.

———, 1956b. Sound transmission in the turtle's ear, *Proc. Natl. Acad. Sci. USA, 42*, 292-299.

———, 1956c. Auditory responses in the common box turtle, *Proc. Natl. Acad. Sci. USA, 42*, 962-965.

———, 1957. Auditory responses in the spectacled caiman, *J. Cell. Comp. Physiol., 50*, 333-339.

———, 1958. Auditory responses in reptiles, *Symposium Proc.*, Office of Naval Research, Pensacola, Florida, March, 1958, pp. 191-196.

———, 1960. The problem of hearing in snakes, *J. Auditory Res., 1*, 77-83.

Wever, E. G., Vernon, J. A., Crowley, D. E., and Peterson, E. A., 1965. Electrical output of lizard ear: relation to hair-cell population, *Science, 150*, 1172-1174.

Wever, E. G., Vernon, J. A., Peterson, E. A., and Crowley, D. E., 1963. Auditory responses in the Tokay gecko, *Proc. Natl. Acad. Sci. USA, 50*, 806-811.

Wever, E. G., and Werner, Y. L., 1970. The function of the middle ear in lizards: *Crotaphytus collaris* (Iguanidae), *J. Exper. Zool., 175*, 327-342.

Wilson, H. V., 1891. The embryology of the sea bass *Serranus atrarius, Bull. U.S. Fish Comm. for 1889, 9*, 209-277.

Wilson, H. V., and Mattocks, J. E., 1897. The lateral sensory anlage in the salmon, *Anat. Anz., 13*, 658-660.

Wittmaack, K., and Laurowitch, Z., 1912. Über artifizelle postmortale und agonale Beeinflussung der histologischen Befunde im membranosen Labyrinth, *Zeits. f. Ohrenheilk., 65*, 157-189.

Wohlfahrt, T. A., 1936. Das Ohrlabyrinth der Sardine (*Clupea pilchardus* Walb.) und seine Beziehung zur Schwimmblase und Seitenlinie, *Zeits. Morphol. Ökol. Tiere, 31*, 371-410.

Wyeth, F. J., 1924. The development of the auditory apparatus in *Sphenodon punctatus, Phil. Trans. Royal Soc. London*, Ser. B, *212*, 259-368.

Zug, G. R., 1971. The distribution and patterns of the major arteries of the iguanids and comments on the intergeneric relationships of iguanids (Reptilia: Lacertilia), *Smithsonian Contrib. Zool.*, No. 83, 23 pp.

INDEX

Library of Congress Cataloging in Publication Data

Wever, Ernest Glen, 1902-
  The reptile ear.

  Includes index.
  1. Reptiles—Physiology.   2. Reptiles—
Anatomy.   3. Ear.   I. Title.
QL669.2.W48   '   598.1'04'1825   78-51204
ISBN 0-691-08196-4